SURFACTANT SCIENCE AND TECHNOLOGY

RETROSPECTS AND PROSPECTS

T0315380

SURFACTANT SCIENCE AND TECHNOLOGY

A Festschrift in honor of
Dr. Kash Mittal

Edited by
Laurence S. Romsted
Rutgers University, New Brunswick, New Jersey, USA

RETROSPECTS AND PROSPECTS

CRC Press
Taylor & Francis Group
Boca Raton London New York

CRC Press is an imprint of the
Taylor & Francis Group, an **informa** business

Cover Image: Cross section of a gemini micelle and probe molecule courtesy of Michel Laguerre and Massimiliano Porrini, Institut Européen de Chimie et Biologie (IECB), Bordeaux, France. A molecular dynamics simulation of a chemical trapping probe, 4-hexadecyl-2,6-dimethylbenzenediazonium ion, nestled in a cross section of micelle composed of decanediyl-α,ω-bis (dodecyldimethylammonium chloride) or 10-2-10 2Cl (aggregation number = 27). Depicted are chloride ions (red), a spaghetti-like core of decyl tails (golden yellow), gemini nitrogens (steel blue), and the chemical probe with carbons (cyan), hydrogens (white), and diazonio group (dark blue). For esthetic purposes, various groups in the gemini micelle cross section are not quite to scale. (Examples of probe applications: X. Gao et al., *Langmuir* **2013**, *29*, 4928; Q. Gu et al., *J. Coll. I. Sci.*, **2013**, *400*, 41; and Y. Zhang et al., *Langmuir*, **2013**, *29*, 534.)

CRC Press
Taylor & Francis Group
6000 Broken Sound Parkway NW, Suite 300
Boca Raton, FL 33487-2742

First issued in paperback 2021

First issued in hardback 2019

© 2014 by Taylor & Francis Group, LLC
CRC Press is an imprint of Taylor & Francis Group, an Informa business

No claim to original U.S. Government works

ISBN 13: 978-1-03-223580-6 (pbk)
ISBN 13: 978-1-4398-8295-5 (hbk)

Library of Congress Cataloging-in-Publication Data

Surfactant science and technology : retrospects and prospects / edited by Laurence Romsted.
    pages cm
    Summary: "Written by the stars of surface and colloid chemistry, this edited work covers developments in the field of association colloid chemistry. The book presents an overview of the direction of the field and gives insight into the forces controlling surfactant and polymer self-assembly. The text addresses numerous research areas, including rheology, surfactant ionic liquids, foams, forces responsible for structural changes of micelles, polymeric surfactants, phase separation, surfactant interactions with solid surfaces, protein and enzymesurfactant interactions, solubilization, mesophases in concentrated solutions, enhanced oil recovery, surfactants for liquid CO2, and biobased surfactants"-- Provided by publisher.
    Includes bibliographical references and index.
    ISBN 978-1-4398-8295-5 (hardback)

    1. Surface chemistry. 2. Surface active agents. 3. Colloids. I. Romsted, Laurence, editor of compilation.
    QD506.S847 2014
    668'.1--dc23

Visit the Taylor & Francis Web site at
http://www.taylorandfrancis.com

and the CRC Press Web site at
http://www.crcpress.com

Dedication

IN HONOR OF KASHMIRI LAL MITTAL

A special symposium was held in honor of the publication of Kashmiri Mittal's 100th edited book as a part of the 18th Surfactants in Solution (SIS) meeting in 2010 in Melbourne, Australia. The title of the symposium, *Surfactant Science and Technology: Retrospects and Prospects*, became the title of this Festschrift. Speakers, and others unable to attend the meeting, were invited to present overviews of their research areas at the meeting and in the chapters of this book recalling the past, summarizing the present, and projecting into the future. The book title reflects the major and continuing contributions of Kashmiri, or Kash as he is known to most. For nearly four decades he has brought together seasoned and new scientists and students of surfactant chemistry at international meetings, in journals, and in books. The totality of his contributions is one of a kind.

Kash was born in the village of Kilrodh, district Rohtak, Haryana, India, in 1945. He obtained a BSc degree from Panjab University in Chandigarh, India, in 1964, and an MSc degree (First Position) in chemistry from the Indian Institute of Technology, New Delhi, in 1966 (thesis title: "Ion Exchange Capacity of Clays"). Shortly thereafter, he migrated to the United States and earned a PhD degree in colloid chemistry in 1970 at the University of Southern California, Los Angeles (thesis title: "Factors Affecting Ultracentrifugal Stability of Oil-in-Water Emulsions," advisor: Professor Robert D. Vold). Kash completed two postdoctoral positions, one at Pennsylvania State University and one at the University of Pennsylvania. He joined the IBM Corporation in 1972 as a postdoctoral fellow and investigated adhesion science until 1994. Since leaving IBM, he has been teaching and consulting worldwide in the broad area of adhesion science.

Kash's professional contributions and awards are listed at the end of this dedication. Kash's edited books and journals have reported scientific progress in colloid and adhesion sciences (see list of published edited books) and chronicle the spawning of future work from past accomplishments. The Surfactants in Solution meetings he has chaired have provided forums for scientists to meet, talk, and sometimes plan new collaborations. Equally important, the meetings enhance the education of scientists in training and provide them an opportunity to display their work. Kash has received numerous awards for his contributions. Some numbers give a feel for the extent of Kash's unique contributions to stimulating scientific research up to 2013: edited books, 112; SIS meetings, 20; journals created, 2; member of editorial boards, 13; and awards, 18. But in reality, the outcomes from this exchange of information and ideas by so many people over many decades cannot be quantified.

I am most familiar with the biennial Surfactants in Solution meetings (see list of SIS meetings), which is, as some of us say, Kash's baby. The first SIS meeting was in 1976 in Albany, New York, and the 20th will be in Coimbra, Portugal, in June 2014. The fundamental and applied research talks presented at these meetings for nearly forty years are a running record of the growth and progress in surfactant chemistry. In 1976, the emphasis of the research talks and the companion books published in 1977 was on the properties of optically transparent micellar solutions, their structures, and internal organization, i.e., hydrocarbon-like cores and polar interfaces, physical properties of the aggregates such as size and shape, adsorption of ions and molecules, and their effects on chemical reactivity. At the meeting in Coimbra in 2014, the proposed topics include many of the same areas, but with the advancement in instrumentation, new understanding is reaching new areas at a greater level of molecular detail. For example, talks on wetting, interfacial tension and rheology, monolayers and films, electrokinetic phenomena, phase behavior, self-assembly, thermodynamics and kinetics, association colloids, liposomes, and emulsions. But also on niosomes, colloidosomes, polymerizable surfactants, biosurfactants, surfactants in pharmacy, energy production, and nanotechnology. Surfactant chemistry is critical to virtually every aspect of the organization of life and of materials science. The SIS meetings have been in the forefront of these advances.

However, SIS meetings do even more because, unlike meetings organized by chemical societies in specific countries, these meetings have been held in many countries (see list of SIS meetings) routinely bringing together people from around the globe. The meetings have also intentionally involved and recognized the contributions of women and young scientists. The poster sessions also provide graduate students from many countries the opportunity to display their work in an international setting, also a unique contribution. Speaking for myself, I was a newly minted PhD just turned postdoc in 1976 when I attended the first, and also my first, SIS meeting. I found that the opportunity to listen and learn from and present to the people whose research I was reading was awesome.

In addition to all these contributions, Kash has published approximately 75 papers in the areas of surface and colloid chemistry, adhesion, polymers, and surface cleaning. He has also served on the advisory boards and committees of numerous international conferences, has given seminars and lectures on many aspects of adhesion science and technology around the globe, and is listed in numerous biographical references including *American Men and Women of Science, International Who's Who of Contemporary Achievement, Men of Achievement, Who's Who in Frontier Science and Technology, Who's Who of Intellectuals, Who's Who in the East (USA),* and *Who's Who in Technology Today.*

Congratulations, Kash, for carving out a unique and productive career that is of extraordinary value to the surfactant and adhesion sciences.

Laurence S. Romsted
Department of Chemistry and Chemical Biology (CCB) Rutgers
The State University of New Jersey
New Brunswick, New Jersey
romsted@rutchem.rutgers.edu
http://chem.rutgers.edu/romsted_laurence_s

PROFESSIONAL CONTRIBUTIONS OF KASHMIRI LAL MITTAL

PUBLISHED BOOKS, HONORS, RECOGNITION AND AWARDS, JOURNAL EDITORIAL
BOARD MEMBERSHIPS, JOURNALS INITIATED, AND SIS MEETINGS

A. Edited Books

	Title	Year
1.	*Adsorption at Interfaces*	1975
2.	*Colloidal Dispersions and Micellar Behavior*	1975
3.	*Micellization, Solubilization, and Microemulsions, Vol. 1*	1977
4.	*Micellization, Solubilization, and Microemulsions, Vol. 2*	1977
5.	*Adhesion Measurement of Thin Films, Thick Films and Bulk Coatings*	1978
6.	*Surface Contamination: Genesis, Detection and Control, Vol. 1*	1979
7.	*Surface Contamination: Genesis, Detection and Control, Vol. 2*	1979
8.	*Solution Chemistry of Surfactants, Vol. 1*	1979
9.	*Solution Chemistry of Surfactants, Vol. 2*	1979
10.	*Solution Behavior of Surfactants: Theoretical and Applied Aspects, Vol. 1*	1982
11.	*Solution Behavior of Surfactants: Theoretical and Applied Aspects, Vol. 2*	1982
12.	*Physicochemical Aspects of Polymer Surfaces, Vol. 1*	1983
13.	*Physicochemical Aspects of Polymer Surfaces, Vol. 2*	1983
14.	*Adhesion Aspects of Polymeric Coatings*	1983
15.	*Surfactants in Solution, Vol. 1*	1984
16.	*Surfactants in Solution, Vol. 2*	1984
17.	*Surfactants in Solution, Vol. 3*	1984
18.	*Adhesive Joints: Formation, Characteristics and Testing*	1984
19.	*Polyimides: Synthesis, Characterization and Applications, Vol. 1*	1984
20.	*Polyimides: Synthesis, Characterization and Applications, Vol. 2*	1984
21.	*Surfactants in Solution, Vol. 4*	1986
22.	*Surfactants in Solution, Vol. 5*	1986
23.	*Surfactants in Solution, Vol. 6*	1986
24.	*Treatise on Clean Surface Technology, Vol. 1*	1987
25.	*Surface and Colloid Science in Computer Technology*	1987
26.	*Particles on Surfaces 1: Detection, Adhesion and Removal*	1988
27.	*Opportunities and Research Needs in Adhesion Science and Technology*	1988
28.	*Particles in Gases and Liquids 1: Detection, Characterization and Control*	1989
29.	*Surfactants in Solution, Vol. 7*	1989
30.	*Surfactants in Solution, Vol. 8*	1989
31.	*Surfactants in Solution, Vol. 9*	1989
32.	*Surfactants in Solution, Vol. 10*	1989
33.	*Particles on Surfaces 2: Detection, Adhesion and Removal*	1989
34.	*Metallized Plastics 1: Fundamental and Applied Aspects*	1989
35.	*Polymers in Information Storage Technology*	1989
36.	*Particles in Gases and Liquids 2: Detection, Characterization and Control*	1990
37.	*Acid-Base Interactions: Relevance to Adhesion Science and Technology*	1991
38.	*Particles on Surfaces 3: Detection, Adhesion and Removal*	1991
39.	*Metallized Plastics 2: Fundamental and Applied Aspects*	1991
40.	*Surfactants in Solution, Vol. 11*	1991

B. General Chair, Biennial Surfactants in Solution (SIS) Meetings

1976	Albany, USA	1996	Jerusalem, Israel
1978	Knoxville, USA	1998	Stockholm, Sweden
1980	Potsdam, USA	2000	Gainesville, USA
1982	Lund, Sweden	2002	Barcelona, Spain
1984	Bordeaux, France	2004	Fortaleza, Brazil
1986	New Delhi, India	2006	Seoul, South Korea
1988	Ottawa, Canada	2008	Berlin, Germany
1990	Gainesville, USA	2010	Melbourne, Australia
1992	Varna, Bulgaria	2012	Edmonton, Canada
1994	Caracas, Venezuela	2014	Coimbra, Portugal

C. Honors, Awards and Special Recognitions in Surfactant and Adhesion Sciences (reverse chronological order):

- Special Issue of Colloids and Surfaces A: Physicochemical and Engineering Aspects (Volume 391, Issues 1–3, 2011) dedicated to him to honor the publication of his 100th edited book.
- Special Symposium on "Surfactant Science and Technology: Retrospects and Prospects" in Melbourne, Australia (November 2010), organized in his honor to commemorate the publication of his 100th edited book. The current volume represents the Festschrift from this event dedicated to him.
- Special Symposium on "Recent Advances in Adhesion Science and Technology" organized in honor of publication of his 100th edited book at the American Chemical Society (ACS) meeting in Boston, August 2010, documented in a Festschrift from this event dedicated to him.
- Special Issue of *Particulate Science and Technology—An International Journal*, Vol. 25, No. 1 (Jan./Feb. 2007), dedicated to him on his 60th birthday.
- *Advances in Colloid and Interface Science* (Vols. 123–126, 2006), dedicated to him in his honor on his 60th birthday.
- Title of Doctor *honoris causa* awarded by the Maria Curie–Sklodowska University, Lublin/Poland (2003).
- Establishment of the biennial *Kash Mittal Award* by the worldwide surface and colloid science community (awarded to peer-assessed scientists active in this field) in recognition of his large contributions to the field of colloid and interface chemistry (2002).
- *Adhesives Age Award* (1997).
- *Adhesives Award* of ASTM International Committee D-14 (1997).
- *John A. Wagnon Technical Achievement Award* of the International Microelectronics and Packaging Society (IMAPS) (1977).
- *Thomas D. Callinan Award* of the Dielectric Science and Technology Division of the Electrochemical Society (1995).
- The 1st International Congress on Adhesion Science & Technology held in his honor on his 50th birthday in Amsterdam, the Netherlands, October 1995 (235 papers from 38 countries were presented).
- Special issue (Vol. 13, Nos. 3 and 4, July–December 1995) of *Particulate Science and Technology—An International Journal* dedicated to him on the occasion of his 50th birthday.
- *Adhesives Age* (September 1995)—An interview with him recognizing his contributions marked by the Amsterdam *1st International Congress on Adhesion Science & Technology*.
- *Robert L. Patrick Fellow* title of the Adhesion Society (1990).
- *Charles B. Dudley Award* of ASTM International (1990).
- "Recognition Plaque for Continued Leadership and Distinguished Professional Service" presented by the international surface and colloid science community comprising prominent scientists from 51 countries at the 6th International Symposium on Surfactants in Solution (SIS), New Delhi (1986).
- Invitation by the International Advisory Panel and Chinese Review Commission (under the auspices of the World Bank) as a Project Specialist to visit Shanghai Jiao Tong University, China (1985).

D. Journal Editor

In 1986, he founded the *Journal of Adhesion Science and Technology* (JAST) and was its Editor-in-Chief until April 2012. In February 2013, he started a new journal, *Reviews of Adhesion and Adhesives* and also a new book series entitled *Adhesion and Adhesives: Fundamental and Applied Aspects*.

E. Member of Editorial Boards of Journals and Encyclopedia:

1. *Adhesives Age*
2. *Advances in Colloid and Interface Science*
3. *Journal of Adhesion*
4. *Journal of Coatings Technology*
5. *Journal of Polymer Materials*
6. *Journal of Surface Science and Technology*
7. *Particulate Science and Technology: An International Journal*
8. *Precision Cleaning*
9. *Progress in Organic Coatings*
10. *Solid State Technology*
11. *Southern Brazilian Journal of Chemistry*
12. *Surface Innovations*
13. *Encyclopedia of Surface and Colloid Science*

Contents

PART I Theory of Self-Assembly and Ion-Specific Effects

PART II Surfactants at Solid–Liquid Interfaces

PART III   Polymeric Surfactants and Polymer/ Surfactant Mixtures

PART IV   Biosurfactants

PART V   Formulation and Application of Surfactant Aggregates

PART VI   *Formulation and Application of Emulsions*

Preface

A celebration was held at the 18th Surfactants in Solution (SIS) meeting in November 2010 in Melbourne in honor of Kashmiri Mittal's 100th edited book.

Those who participated in the symposium are leaders in the fields of surfactant-based, physical, organic, and materials chemistries, and many agreed to contribute a chapter to this book. Some chapters are contributed by others who wanted to participate in the meeting, but were unable to attend. The authors were asked to give an overview of their research areas and to include sections on past, present, and future directions. The authors updated and revised their manuscripts as needed in 2012. The cumulative result in this volume is a broad perspective on the current developments in and future of surfactant science and technology.

The next SIS will be held in Coimbra, Portugal, in June 2014, the 20th biennial meeting over about two score years. During this time, the field of surfactant chemistry has expanded dramatically and has evolved considerably, aided by the development of modern instrumentation and new experimental techniques that permit exploration of surfactant properties in both the bulk and at molecular levels and by simulation.

The physical properties of surface-active agents, commonly known as surfactants, amphiphiles, detergents, or soaps, are governed by covalently joining two opposite chemical properties in one molecule: a water-insoluble hydrophobic tail, typically composed of linear hydrocarbon chemically bonded to polar or ionic headgroup and counterion. These surfactant monomers or unimers self-assemble into a plethora of aggregate structures such as micelles, microemulsions, vesicles, and emulsions depending on solution composition, but they also form surfactant monolayers at the air, liquid, and solid interfaces. A large variety of aggregate mesophases may be formed including aqueous and reverse micelles, flexible rod-like and hexagonal structures, cubic and lamellar phases and bicontinuous regions, and vesicles. The equilibrium sizes and shapes of surfactant aggregates are governed by a delicate balance of forces such as coulombic, dispersion, hydrogen bonding, hydration, and dipole–dipole and dipole–charge, whose strengths are typically ≤ 20 kJ/mol. Biological systems, e.g., protein coiling and stability, and the formation of biological membranes, depend on the same basic forces, and basic research in surfactant assemblies is motivated by the realization that they are "simple" models of the more complex biological ones.

Applied surfactant research is focused on tuning the properties of surfactant assemblies by identifying the optimal combination of surfactants and additives for particular applications. The number and type of applications are almost limitless and include some of the most important industrial, medical, and personal applications: washing and cleaning, microelectronics, viscosity control, speeding reactions, stabilizing drugs, drug delivery, compartmentalization, cosmetics, enhanced oil recovery, and foods.

This book is divided into six parts. An asterisk (*) indicates the corresponding author.

PART I. THEORY OF SELF-ASSEMBLY AND ION-SPECIFIC EFFECTS (TWO CHAPTERS)

R. Nagarajan provides an extensive overview of the development of the theory of micellization, and R. I. Slavchov,* S. I. Karakashev, and I. B. Ivanov introduce a model for interpreting ion-specific effects on aggregate properties including adsorption, micellization, and thin liquid films.

PART II. SURFACTANTS AT SOLID–LIQUID INTERFACES (FIVE CHAPTERS)

The focus is on interactions of surfactant solutions with solid supports. E. Chibowski,* M. Jurak, and L. Holysz used different methods to deposit lipid layers and measured contact angles to understand the hydrophobic/hydrophilic changes in the lipid layer. R. Miller,* V. B. Fainerman, V. Pradines, V. I. Kovalchuk, N. M. Kovalchuk, E. V. Aksenenko, L. Liggieri, F. Ravera, G. Loglio, A. Sharipova, Y. Vysotsky, D. Vollhardt, N. Mucic, R. Wüstneck, J. Krägel, and A. Javadi evaluated equilibrium and dynamic surface tensions and other methods to obtain information on the arrangement of molecules at interfaces. N. Ivanova and V. M. Starov* provide an overview of surfactant effects on dynamic and spreading phenomena. V. I. Kovalchuk, E. K. Zholkovskiy, M. P. Bondarenko, and D. Vollhardt* have studied the mechanism of pattern formation on solid surfaces with high contrast resolution at the nanometer scale. M. Ferrari,* F. Ravera, L. Liggieri, and L. Navarini show that tensiometry provides new information on the contributions of beverage components to *espresso* flavor.

PART III. POLYMERIC SURFACTANTS AND POLYMER/SURFACTANT MIXTURES (THREE CHAPTERS)

M. C. Morán,* D. Costa, M. da Graça Miguel, and B. Lindman used their general understanding of the interactions of DNA and oppositely charged molecules, e.g., cationic surfactants, to create novel DNA-based materials. C. W. Park, H.-M. Yang, S. R. Yoon, and J.-D. Kim* prepared multifunctional poly(amino acid) graft polymers for improved drug delivery and diagnosis and report on the morphological transitions of the aggregates. Th. Tadros discusses the classification of polymeric surfactants and illustrates their adsorption isotherms and the effects on emulsification in terms of a combination of intermolecular interactions and steric stabilization.

PART IV. BIOSURFACTANTS (TWO CHAPTERS)

G. Biresaw describes the preparation, properties, and applications of farm-based biosurfactants synthesized from natural products, and their surface and interfacial properties are discussed. P. Somasundaran,* P. Patra, J. D. Albino, and I. M. Nambi describe the preparation, structures, and properties of "greener" biosurfactants obtained from bacteria.

PART V. FORMULATION AND APPLICATION OF SURFACTANT AGGREGATES (FIVE CHAPTERS)

D. Libster, A. Aserin, and N. Garti* describe the structural features of lyotropic liquid crystals, their application for solubilization and drug and biomacromolecule release, and the use of cell-penetrating peptides to enhance drug penetration of the skin. M. Win, P. Lang, M. Vashishtha,* and D. O. Shah review the properties of microemulsions, their efficacy in drug delivery, and controlled release by tailoring the microemulsion properties. K. Holmberg characterizes the structure/function relationships for hydrotropes, their ability to improve the solubility of poorly soluble organic compounds, and their use in formulations by destabilizing surfactant liquid crystals and changing interfacial curvature. P. Brown, C. Butts, and J. Eastoe* discuss the possibility that ionic liquids can be combined to generate tunable and selective reaction media and provide a focus on new directions. J. Texter provides a detailed overview of the origin, properties, and future developments of stimuli-responsive surfactants including topics such as surfmers and inisurfs, biosurfactants, photochromic and pH-sensitive surfactants, and their diverse applications for dispersion stabilization, electrode coating, and shape memory of polymers.

VI. FORMULATION AND APPLICATION OF EMULSIONS (THREE CHAPTERS)

J.-L. Salager,* A. Forgiarini, and J. Bullón review the development in designing emulsion properties for particular applications for foods, agrochemicals, water treatment, asphalts, and paints. P. M. Mwangi and D. N. Rao* describe the role of surfactants in the oil and gas industries such as enhanced oil recovery, drilling, spill remediation, oil flotation, and emulsion breaking. C. A. Miller discusses the surfactant mechanisms for soil removal, considering the conversion of soils into microemulsions and the possibility of spontaneous emulsification.

This book reflects the cumulative wisdom of a number of major contributors to the broad field of surfactant chemistry. The chapters should be useful to both neophytes (as a general introduction) and veteran researchers (as a commentary on current status). Those engaged in surfactant chemistry in a variety of industries and academic disciplines including surface and colloid science, chemical engineering, cosmetics, pharmaceuticals, biomedical, and nanotechnology will find this book of immense interest.

Laurence S. Romsted
Department of Chemistry and Chemical Biology (CCB) Rutgers
The State University of New Jersey
New Brunswick, New Jersey

Acknowledgments

Numerous people have made this book possible. I appreciate the time, effort, and patience of all the authors for completing and polishing their chapters. The book made it to print because of the unstinting support and critical editing assistance of my wife Jean; periodic computer and software repairs by my son Eric and by John Furnari of CCB at Rutgers; the steady support of Barbara Glunn (Senior Editor) and Amber Donley (Project Coordinator), who helped guide this book to completion; and finally Patrick Hartley who organized the SIS meeting in Melbourne in 2010. I also thank C. A. Bunton for all his wonderful personal support in the early part of my career and Kashmiri Mittal for his continuing friendship over multiple decades.

Editor

Laurence (Larry) S. Romsted was born in Chicago, Illinois, USA, in 1941. He is the grandchild of immigrants, and at the end of World War II, his parents moved to one of the suburbs, where he learned much in school but little about the world. He entered DePauw University, Greencastle, Indiana, in 1959 with a taste for doing chemistry fostered first by an excellent high school science teacher. Second, by an introduction to colloids and food emulsions in a summer job at Kraft Foods Research Labs in his hometown that was amplified by stories about scientists making cool discoveries that helped people and Sputnik. And third, the quixotic hope that a science career meant a minimal amount of writing work.

In 1964, he joined Eugene Cordes' group, at Indiana University, Bloomington, and began graduate research on micellar catalysis as a simple model for enzymatic catalysis—that proved complex, but solvable–eventually. Uncertain about a career in chemistry and totally opposed to the Vietnam War, he joined the Peace Corps and went to the Philippines to reflect on his life, do useful service, and avoid being drafted. He returned to graduate school, but still ambivalent about his future, he dropped out again, married, and taught chemistry part-time in a community college in Ohio. Faced with a forever part-time position and vexed by a research problem he had left unfinished, he again returned to Indiana University, where Cordes said in essence, "Here's a desk, good luck." Eighteen months later, in December 1975, he defended his PhD thesis entitled, "Rate Enhancements in Micellar Systems." To his own (and Cordes') amazement, he had developed a new pseudophase model for ionic micelle effects on the rates of chemical reactions in which the micellar surface is treated as a specific ion exchanger. Pseudophase models remain the primary basis for interpreting reactivity in ionic association colloids. In 1976, Romsted joined C. A. Bunton's group at UC Santa Barbara where he had an extraordinarily fruitful postdoc, developed an appreciation for collaboration from his ornery but caring advisor who became his scientific father, and experienced the deep pleasure of formulating, testing, and carrying ideas to fruition. Doing chemistry, he later realized, had become his art.

In the fall of 1980, he took an assistant professor position at Rutgers, The State University of New Jersey, and, except for the first 3 years, occupied the same office to this day, a personal record for staying in one place. Here he developed the chemical trapping method for estimating changes in interfacial molarities of anionic and neutral nucleophiles with changes in micellar properties. His publication record is modest, 92 to date, but replete with long, data-rich, papers. He has also given about 90 invited talks at international meetings and another 90 or so university seminars. In 1991, he published a singularly important paper in *Accounts of Chemical Research*, "Ion Binding and Reactivity at Charged Aqueous Interfaces," that spread the catalysis model to the world (559 citations and counting). The Account was coauthored with C. A. Bunton (UCSB), F. Nome (UFSC), and F. Quina (USP), each of whom has made major contributions to the current understanding of micellar catalysis.

His current research is in three areas: (a) continued development of the chemical trapping method to better understand the relationships between interfacial compositions and association colloid properties; (b) creating a new method with Carlos Bravo-Díaz (University of Vigo, Spain) for determining the distributions of antioxidants in intact, opaque emulsions from measured rate constants that provide new insight into the polar paradox; and (c) a novel project applying the chemical trapping method to determine protein topologies at biomimetic interfaces. To do one's art and work with bright, engaged students is not such a bad life—most days.

Contributors

Eugene V. Aksenenko
Institute of Colloid Chemistry and Chemistry
of Water
Kiev, Ukraine

John D. Albino
Earth and Environmental Engineering
Columbia University
New York, New York

Abraham Aserin
The Institute of Chemistry
The Hebrew University of Jerusalem
Jerusalem, Israel

Girma Biresaw
Bio-Oils Research Unit
NCAUR-ARS
United States Department of Agriculture
Peoria, Illinois

Mykola P. Bondarenko
Institute of Bio-Colloid Chemistry
Kiev, Ukraine

Paul Brown
Department of Chemical Engineering
Massachusetts Institute of Technology
Cambridge, Massachusetts

Johnny Bullón
Laboratorio de Formulación
Universidad de Los Andes
Mérida, Venezuela

Craig Butts
School of Chemistry
University of Bristol
Bristol, United Kingdom

Emil Chibowski
Department of Physical Chemistry—Interfacial
Phenomena
Faculty of Chemistry
Maria-Curie Sklodowska University
Lublin, Poland

Diana Costa
CICS—Centro de Investigação em Ciências da
Saúde
Universidade da Beira Interior
Covilhã, Portugal

Julian Eastoe
School of Chemistry
University of Bristol
Bristol, United Kingdom

Valentin B. Fainerman
Donetsk Medical University
Donetsk, Ukraine

Michele Ferrari
Istituto per l'Energetica e le Interfasi—CNR
Genoa, Italy

Ana Forgiarini
Laboratorio de Formulación
Universidad de Los Andes
Mérida, Venezuela

Nissim Garti
Casali Institute of Applied Chemistry,
The Institute of Chemistry
The Hebrew University of Jerusalem
Jerusalem, Israel

Krister Holmberg
Chalmers University of Technology
Chemical and Biological Engineering
Göteborg, Sweden

Lucyna Holysz
Department of Physical Chemistry—Interfacial
    Phenomena
Faculty of Chemistry
Maria-Curie Sklodowska University
Lublin, Poland

Ivan B. Ivanov
Laboratory of Chemical Physics and
    Engineering
Sofia University
Sofia, Bulgaria

Natalia Ivanova
Department of Physics
Tyumen State University
Tyumen, Russia

Aliyar Javadi
Max Planck Institute of Colloids and
    Interfaces
Potsdam/Golm, Germany

Malgorzata Jurak
Department of Physical Chemistry—Interfacial
    Phenomena
Faculty of Chemistry
Maria-Curie Sklodowska University
Lublin, Poland

Stoyan I. Karakashev
Department of Physical Chemistry
Sofia University
Sofia, Bulgaria

Jong-Duk Kim
Department of Chemical and Biomolecular
    Engineering
Korea Advanced Institute of Science and
    Technology
Daejeon, Republic of Korea

Nina M. Kovalchuk
Institute of Bio-Colloid Chemistry
Kiev, Ukraine

Volodymyr I. Kovalchuk
Institute of Bio-Colloid Chemistry
Kiev, Ukraine

Jürgen Krägel
Max Planck Institute of Colloids and
    Interfaces
Potsdam/Golm, Germany

Paul Lang
Rehrig Pacific Company
Vernon, California

Dima Libster
Casali Institute of Applied Chemistry,
    The Institute of Chemistry
The Hebrew University of Jerusalem
Edmond J. Safra Campus, Givat Ram
Jerusalem, Israel

Libero Liggieri
Istituto per l'Energetica e le Interfasi—CNR
Genoa, Italy

Björn Lindman
Physical Chemistry 1
University of Lund
Lund, Sweden

Giuseppe Loglio
University of Florence
Sesto Fiorentino
Firenze, Italy

Maria da Graça Miguel
Departamento de Quimica
Universidade de Coimbra
Coimbra, Portugal

Clarence A. Miller
Department of Chemical and Biomolecular
    Engineering
Rice University
Houston, Texas

Reinhard Miller
Max Planck Institute of Colloids and
    Interfaces
Potsdam/Golm, Germany

M. Carmen Morán
Departament de Fisiologia
Facultat de Farmàcia
Universitat de Barcelona
Barcelona, Spain

Nenad Mucic
Max Planck Institute of Colloids and Interfaces
Potsdam/Golm, Germany

Paulina M. Mwangi
Craft & Hawkins Department of Petroleum
    Engineering
Louisiana State University
Baton Rouge, Louisiana

Ramanathan Nagarajan
Molecular Sciences and Engineering Team
Natick Soldier Research, Development and
    Engineering Center (NSRDEC)
Natick, Massachusetts

Indumathi M. Nambi
Department of Civil Engineering
Indian Institute of Technology Madras
Chennai, Tamil Nadu, India

Luciano Navarini
Illycaffe SpA, R&D
Trieste, Italy

Chan Woo Park
Department of Chemical and Biomolecular
    Engineering
Korea Advanced Institute of Science and
    Technology
Daejeon, Republic of Korea

Partha Patra
Earth and Environmental Engineering
Columbia University
New York, New York

Vincent Pradines
Laboratoire de Chimie de Coordination
Toulouse, France

Dandina N. Rao
Craft & Hawkins Department of Petroleum
    Engineering
Louisiana State University
Baton Rouge, Louisiana

Francesca Ravera
Istituto per l'Energetica e le Interfasi—CNR
Genoa, Italy

Jean-Louis Salager
Laboratorio de Formulación
Universidad de Los Andes
Mérida, Venezuela

Dinesh O. Shah
Shah-Schulman Center for Surface Science and
    Nanotechnology
Dharmsinh Desai University
Nadiad, Gujarat, India

and

Department of Chemical Engineering and
    Department of Anesthesiology
University of Florida
Gainesville, Florida

Altynay Sharipova
Kazakh National Technical University
Almaty, Kazakhstan

Radomir I. Slavchov
Department of Physical Chemistry
Sofia University
Sofia, Bulgaria

Ponisseril Somasundaran
Earth and Environmental Engineering
Columbia University
New York, New York

Victor M. Starov
Department of Chemical Engineering
Loughborough University
Loughborough, United Kingdom

Tharwat Tadros
Consultant
Wokingham, Berkshire, United Kingdom

John Texter
Polymers and Coatings
Eastern Michigan University
Ypsilanti, Michigan

Manu Vashishtha
Shah-Schulman Center for Surface Science and
    Nanotechnology
Dharmsinh Desai University
Nadiad, Gujarat, India

Dieter Vollhardt
Max Planck Institute of Colloids and Interfaces
Potsdam/Golm, Germany

Yuri Vysotsky
Donetsk National Technical University
Donetsk, Ukraine

Maung Win
Department of Biomedical Engineering
University of Minnesota
Minneapolis, Minnesota

Rainer Wüstneck
Max Planck Institute of Colloids and
    Interfaces
Potsdam/Golm, Germany

Hee-Man Yang
Korea Atomic Energy Research Institute
Daedeok-daero
Daejeon, Republic of Korea

Se Rim Yoon
Amore-Pacific Corporation
R&D Center
Daejeon, Republic of Korea

Emiliy K. Zholkovskiy
Institute of Bio-Colloid Chemistry
Kiev, Ukraine

Part I

Theory of Self-Assembly and Ion-Specific Effects

1 One Hundred Years of Micelles

Evolution of the Theory of Micellization

Ramanathan Nagarajan

CONTENTS

1.1 INTRODUCTION

Surfactant molecules are composed of a polar head group that likes water and a nonpolar tail group that dislikes water, thus contributing to an intrinsic duality in their molecular characteristics. Despite their mutual antipathy, the head and tail groups of the surfactant cannot leave one another because they are covalently connected. The dilemma faced by these molecules is resolved in nature by the intriguing phenomenon of molecular self-assembly, wherein the amphiphiles self-assemble into three-dimensional structures with distinct and separate regions composed of the nonpolar parts and the polar parts, having minimal contact with one another.

Numerous variations are possible in the types of the head group and tail group of surfactants. For example, the head group can be anionic, cationic, zwitterionic, or nonionic. It can be small and compact in size or it could be an oligomeric chain. The tail group can be a hydrocarbon, fluorocarbon, or a siloxane. It can contain straight chains, branched or ring structures, multiple chains, etc. Surfactant molecules with two head groups (Bola surfactants) are also available and there are dimeric surfactants with two head groups and two tail groups with a covalent linkage connecting those (Gemini surfactants). Furthermore, the head and the tail groups can be polymeric in character, as in the case of block copolymers. This variety in the molecular structure of surfactants allows for extensive variation in their solution and interfacial properties and their practical applications. Therefore, it is not surprising that the drive to discover the link between the molecular structure of the surfactant and its physicochemical behavior has had a long history.

The existence of surfactant molecules in the form of self-assembled aggregates was first suggested by McBain in 1913 [1] based on his studies on how the conductivity of a solution of soap molecules changes with the concentration. Soap solutions exhibit even lower osmotic activity than would be predicted if one assumed that soap existed in solution as simple undissociated molecules. Soap solutions also conduct the electric current far better than would be expected from the observed osmotic effects. Attempting to explain these anomalies, McBain, in 1913, suggested that the fatty soap ions aggregated in solution. Such colloidal aggregations of ions, which were termed micelles, would explain both the low osmotic activity and relatively high conductivity of soap solutions.

Understanding the aggregation properties of surfactant molecules has nearly a hundred years of history. The early work, until about 1950, was significantly focused on experimental methods to identify the size and shape of the aggregates. McBain [2–5], Adam [6], Harkins [7–11], Hartley [12–15], and Philippoff [16,17] are among the pioneers who proposed different structural models for micelles. The x-ray diffraction technique was extensively applied and there were many qualitative structural models proposed for the aggregates.

The first theory, published in 1949 by Debye [18–20], explained why micelles form and why there are finite-shaped micelles. In the period between 1950 and 1956, this work stimulated a number of theoretical studies by Hobbs [21], Ooshika [22], Reich [23], and Halsey [24], who attempted to rectify some of the fundamental theoretical aspects as to how an equilibrium system should be described and also drew attention to the role of surface energy, which was missing in the Debye model. From 1955 to 1965, further advances in theory emerged and were guided by the methods of statistical mechanics. Specifically, the theories of Hoeve and Benson [25] and Poland and Scheraga [26,27] have provided some key concepts that have eventually been integrated into the current models. Also during this time, a light-scattering technique was applied to surfactant solutions and experimental evidence for the formation of rodlike micelles was generated.

The work of Poland and Scheraga explicitly introduced the role of hydrophobic interactions in micelle formation. This concept was quantitatively included in the free energy model proposed by Tanford [28–30] during 1970 to 1974. Tanford's model had the simplicity of the earlier theory of Debye, and was also consistent with a rigorous statistical thermodynamic formulation of the aggregation process occurring in the surfactant solution. The free energy model incorporated all important physical chemical factors and provided an explanation for why micelles form, why they grow, and why they remain finite in terms of clearly identified opposing interactional efforts. Israelachvili et al. [31] used the framework of the Tanford free energy model along with molecular packing considerations inside surfactant aggregates to develop a geometry-based approach to predict the formation of different shapes of aggregates. They were able to explain the formation of spherical and cylindrical micelles and spherical bilayer vesicles, as well as transitions between these structures based on a combination of general thermodynamic principles, Tanford's free energy model, and the geometric constraints imposed by molecular packing considerations. At this stage, one could argue that the general principles of surfactant self-assembly were sufficiently well-established due to the contributions of Tanford and Israelachvili et al., in particular (Figure 1.1).

In 1976, Kash Mittal organized a symposium on *Micellization, Solubilization and Microemulsions* [32–37], which became the precursor to the current biennial symposia *Surfactants in Solution*. These series of symposia have given impetus to practically all of the subsequent research in surfactant science and technology. In addition to stimulating theoretical work, the *Surfactants in Solution* symposia, since 1976, have also contributed to the extensive development of novel applications using surfactants beyond the classic areas of detergency, emulsions, foams, and dispersions. Chief among these is the use of surfactants for material synthesis, especially the use of reverse micelles as nanoreactors for nanoparticle synthesis including metal oxides, metals, and quantum dots and the use of surfactant liquid crystals as templates for the synthesis of mesoporous materials that vastly extended the pore size beyond the small range possible with zeolites.

The period following Tanford's work has seen a number of key theoretical advances. First, theoretical models began to be developed with a view to a priori predict the self-assembly properties from the molecular structure of the surfactant. Second, theoretical models began to be applied to mixtures of surfactants and also to surfactants that had hydrophobicity arising from

(a)   (b)   (c)

FIGURE 1.1 (a) James W. McBain. (Courtesy of Stanford University.) (b) Peter Debye. (Courtesy of Cornell University Archives.) (c) Charles Tanford. (Copyright 2009, The Protein Society.)

fluorocarbons and siloxanes. Third, self-assembly phenomena such as solubilization and micro-emulsions were described using theoretical approaches used for surfactant aggregation so that a unified view of self-assembly in multicomponent systems was made possible. Fourth, some key features of the free energy models, including the calculation of electrostatic interactions in ionic surfactants, and treatment of how hydrophobic chains pack and arrange inside aggregates of different shapes, were more fundamentally treated. Fifth, the theory of self-assembly was quantitatively applied to novel surfactants such as Bola surfactants and Gemini surfactants. Sixth, self-assembly phenomena at gas–liquid and solid–liquid interfaces began to be modeled. Finally, the free energy models of the analytical form were improved and sometimes combined with free energy models developed by molecular dynamic simulations so that the predictive models can be truly a priori.

In this chapter, we focus mainly on the theoretical evolution of free energy models for surfactant self-assembly. The models chosen for discussion are important in the sense that they have affected the evolution of theory and have also influenced our own work in developing predictive models for a range of self-assembly phenomena. Because of the structure of this chapter, a number of scientists who have contributed to experimentally identifying important phenomena or theoretically modeling one or another specific feature of the self-assembly will go unmentioned. Their important contributions and influence over the evolution of theory are not to be ignored and hopefully are visible through the extensive citations of their work in our previous theoretical articles.

1.2 EARLY QUALITATIVE STRUCTURAL MODELS OF MICELLES

The early discussions on aggregate shapes (Figure 1.2) focused on two types proposed by McBain [4,5], a lamellar (disk) aggregate with 50 to 100 molecules and a small oligomeric surfactant cluster with approximately 10 molecules. The lamellar micelle consists of two layers of soap molecules or ion pairs partially dissociated and arranged side by side, with the two hydrocarbon layers inside. Harkins and coworkers [7–10] interpreted their x-ray results as confirming the existence of the lamellar McBain micelle. Furthermore, Hess and coworkers proposed the existence of McBain lamellar micelles that repeat in parallel arrangement, separated by layers of water thus giving an x-ray long spacing equal to twice the length of the molecule plus that of the layer of water [38,39]. This micelle was referred to as the Hess micelle (Figure 1.2).

Harkins considered how molecular shape influences molecular packing at interfaces [6]. He initially favored the lamellar aggregate model and also proposed a cylindrical lamellar form in which

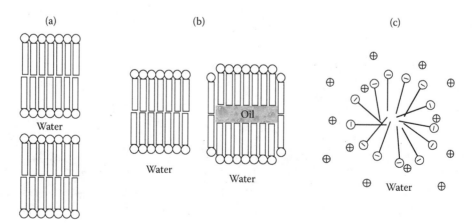

FIGURE 1.2 (a) Hess lamellar micelle made of two McBain lamellar micelles; (b) Harkins cylindrical micelle model showing his idealized cross-section of a soap micelle without solubilized oil and with solubilized oil; (c) Hartley model of a spherical micelle.

the surfactant molecules are oriented parallel to the axis of the cylinder, with the head groups constituting the two end surfaces of the cylinder (the cylinder being merely a small section of the lamellar aggregate) [7–9]. Harkins also suggested that although a cylinder seems to represent most of the properties of the micelle, it is imperfect in that it represents too large an interface at the side between hydrocarbon groups and water. Thus, it does not seem improbable that a model, which lies between a cylinder and a cylindrical type of spheroid, may be found to more accurately represent the energy relationships. This would amount to a distortion of the side of the cylinder with which polar groups cover (to some extent) the nonpolar hydrocarbon chains [10].

Philippoff [16,17] analyzed osmotic activity, specific conductivity, and x-ray data obtained at various electrolyte and surfactant concentrations and for various surfactant tail lengths and identified two regimes with respect to micellar shapes. In the first regime, which corresponds to zero or small amounts of electrolytes, micelles are practically spherical in the sense that they need not be true spheres but can be cubes, short cylinders, prisms, or spheroids. In the other regime, which corresponds to large electrolyte concentrations and longer tail lengths of surfactants, the aggregates are anisometric and large. Philippoff considered that it is improbable for small micelles to reorganize into large micelles with increasing salt and surfactant concentrations and therefore postulated the large structures to be secondary aggregates of the preformed primary micelles.

As opposed to the lamellar structures, Adam [6] arrived at spherical micellar structure based on how molecules pack at interfaces depending on their molecular shapes. Adam suggested that "molecules larger at their polar ends will naturally pack into a curved film having the hydrocarbon side concave and the water-attracting side convex." Hartley [12–15] proposed the existence of larger spherical micelles with approximately 50 surfactant molecules. Hartley suggested that the "aggregates are essentially liquid and since they will tend to present the minimum surface to the water, they will presumably be roughly spherical and of the largest radius consistent with none of the heads being submerged in the paraffin interior." Harkins, who initially supported the lamellar micelle model and also proposed a cylindrical–lamellar structure, eventually considered the possibility of spherical shape for micelles [11]. Corrin demonstrated [40] from an analysis of the x-ray diffraction data from the Harkins laboratory that a lamellar model is not required to explain the x-ray diffraction patterns obtained from solutions of long-chain ionic surfactants. Qualitatively, the diffraction patterns could be satisfactorily interpreted by a model of spherical micelles whose relative position can be represented by a radial distribution function. He argued that although this does not prove the validity of the spherical micelle model, it indicates that x-ray measurements alone do not allow one to decide between the lamellar and spherical models.

During that stage of development in surfactant research, no definitive theory of micelle formation yet existed. The following statement from Philippoff [17] in his 1951 article makes the case. "Having reviewed the field, we can make only the negative statement that there is at present no theory to account for the causes of micelle formation which can interpret the whole of the experimental evidence. Micelle formation is independent of the sign of the charge of the micelle forming ion. Micelles are partially ionized, therefore a theory must account for this. Micelles form with nonionizing detergents in water, and with some ionizing detergents in hydrocarbons (aerosol OT, diethyl hexyl sodium sulfosuccinate), showing that a charge is not essential for the phenomenon. A straight-chain compound is not necessary for micelle formation. Sodium deoxycholate with a single ionizing group on a sterol skeleton forms micelles as well as divalent aerosol OT in hydrocarbons with four branched chains to one ionic group. The commercially important micellar arylalkyl sulfonates have a complicated structure of the hydrocarbon part of the molecule. Likewise, an extended hydrophilic group as in the polyethylene oxide derivatives is not prohibitive even to x-ray structures. Tween 40, polyethylene oxide sorbitan monopalmitate, with even three hydrophilic chains, forms micelles. Mixed micelles are also readily formed by detergents of the same general structure, differing only in chain. The only common principle left is the segregation of the hydrophilic and hydrophobic parts of a molecule." It is in this context, that theories began to be developed to explain one or more features of surfactant aggregates.

1.3 FIRST QUANTITATIVE THEORY OF MICELLE DUE TO DEBYE

The first quantitative, molecular theory of the formation of micelles was proposed by Debye [18–20]. He considered ionic surfactants consisting of a hydrocarbon chain with a charge at one end. To create a micelle with the lamellar form proposed by McBain, Debye proposed accounting for two different kinds of energy. There is a gain in energy because a number of hydrocarbon tails are removed from the surrounding water and brought into contact with each other in the micelle. He considered the molecular forces of importance in this process to be relatively short-range forces of the van der Waals kind. To bring the charged ends of the monomers nearer to each other on both flat surfaces of the lamellar micelle requires energy to overcome the long-range electrical forces. Accordingly, Debye concluded that the interplay between short-range van der Waals forces and long-range electrostatic forces are responsible for the equilibrium structure of the micelle.

1.3.1 MONOMER–MICELLE EQUILIBRIUM AND CRITICAL MICELLE CONCENTRATION

Debye started with the classic mass action equilibrium between simple ions and micelles. Consider the following reversible association equilibrium between fatty soap ions A (monomers) and micelles A_g, where g is the number of fatty ions present in a micelle and K_g is the equilibrium constant for the monomer–micelle association:

$$gA \overset{K_g}{\Leftrightarrow} A_g \tag{1.1}$$

If X_g denotes the concentration of micelles, X_1 the concentration of unaggregated paraffin chains, and X_T the total concentration of fatty ions, with all concentrations expressed in mole fractions, then

$$X_T = X_1 + gX_g \tag{1.2}$$

Denoting the micellization equilibrium constant K_g in terms of a concentration X_C in the form, $K_g = X_C^{1-g}$, the application of the mass action law assuming unit activity coefficients leads to

$$X_g = K_g \left(X_1 \right)^g = X_C \left(\frac{X_1}{X_C} \right)^g, \quad X_T = X_1 + gX_C \left(\frac{X_1}{X_C} \right)^g \tag{1.3}$$

From this mass action law, for large enough g, one can see that when $X_1 < X_C$, the micelle concentration X_g will remain negligibly small and X_T will be practically equal to X_1. To prevent the divergence of the right-hand side of the equation, it is clear that X_1 will become equal to X_C but never exceed it. Therefore, when micelles form, X_1 will be practically equal to X_C. These considerations indicate that if g is large enough, the soap is practically unaggregated up to the concentration X_C and, above that concentration, all excess soap will appear in the form of micelles. On this basis, one can identify X_C as the critical micelle concentration (cmc).

1.3.2 WORK OF FORMATION OF MICELLE

To estimate the energy change on micelle formation, Debye considered the work necessary to create an aggregate from single molecules. One part of this work is electrical and it was estimated as follows. The total charge on the micelle surface will be proportional to $\sigma_e R^2$, where R is the radius of the micelle surface viewed as a circular disk and σ_e is the constant surface charge density. The potential at the rim of the disk will be proportional to $\sigma_e R$. Therefore, for a differential change in the disk radius R, the change in surface charge is $2\pi R dR \sigma_e$ and additional work should be done against the Coulomb forces proportional to $R^2 dR\sigma_e^2$. The total electrical work involved in building up the disk is thus proportional to R^3. Because the surface area R^2 is proportional to the number of molecules g constituting the micelle, the electrical energy to be overcome in creating a micelle of size g is

$$W_e = g^{3/2} w_e \qquad (1.4)$$

where w_e is a fundamental electrical energy (absorbing all other constants independent of the size g). Due to the long-range character of the electrical forces, this work increases faster than the number of molecules in the micelle. The second work is related to the energy gained by bringing g hydrocarbon tails in contact (which involves only short-range molecular forces) with one another. It was represented as

$$W_m = -g w_m \qquad (1.5)$$

with the introduction of another fundamental molecular energy w_m. The negative sign indicates that this energy represents a favorable process of bringing the surfactant's tail from contact with water to contact with other tails.

1.3.3 EQUILIBRIUM MICELLE

The total energy $W = W_e + W_m$ has a minimum for a certain value $g = g_0$ and at this point, the energy W_0 of the micelle is negative. This means that the micelle is more stable than g_0 separate molecules and that work is required to either increase or decrease the equilibrium number g_0. The energy contributions W_e and W_m and the total energy W are shown in Figure 1.3 as functions of micelle aggregation number g. The total energy W is negative with a shallow minimum W_0 at g_0.

From the minimization of the total energy W, one gets the equilibrium aggregation number g_0 and the energy W_0 corresponding to this equilibrium aggregate.

$$g_0 = \left[\frac{2}{3} \frac{w_m}{w_e} \right]^2, \quad W_0 = g_0^{3/2} w_e - g_0 w_m = -\frac{1}{3} g_0 w_m \qquad (1.6)$$

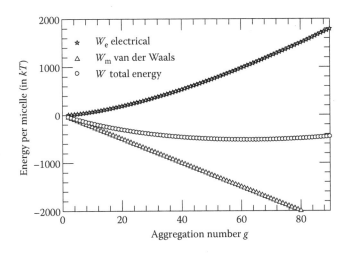

FIGURE 1.3 Debye model energy contributions to micelle formation. The electrostatic energy increases faster than the increase in the van der Waals energy as g increases and thereby contributes to the finite size of micelles.

Equating the energy of formation of the equilibrium micelle to the equilibrium constant for the aggregation in the mass action model, and recognizing that the number of molecules g_0 in the equilibrium aggregate will be much larger than 1, one gets an expression for the cmc.

$$W_0 = -kT \ln K_g = (g_0 - 1)\, kT \ln X_C \approx g_0\, kT \ln X_C$$

$$kT \ln X_C = \frac{W_0}{g_0} = -\frac{1}{3} w_m \tag{1.7}$$

Knowing the fundamental energy constants w_m and w_e, one can determine the equilibrium micelle size g_0 from Equation 1.6 and the critical concentration X_C from Equation 1.7. Alternatively, knowing the experimentally determined micelle aggregation number and the cmc, one can determine the fundamental energy constants w_m and w_e appearing in the Debye model.

Debye evaluated the physical relevance of the characteristic energy parameters in his model by considering that for dodecylamine hydrochloride at 25°C, the experimental cmc is 0.0131 M (converts to mole fraction $X_C = 2.36 \times 10^{-4}$) and the aggregation number is $g_0 = 66$. For these experimental values, the model shows, $W_0/g_0 = 8.36\ kT$, $w_m = 25\ kT$, and $w_e = 2.2\ kT$. He noted that the heat of vaporization per dodecane molecule is $25\ kT$, and that it has a correspondence to the characteristic short-range energy parameter w_m. He calculated w_e for a disk covered with a constant charge density in a medium of dielectric constant ε, to be

$$w_e = \frac{4}{3}\sqrt{\frac{2}{\pi}}\,\frac{e^2}{\varepsilon\sqrt{a}} \tag{1.8}$$

where a is the area per head group and e is the electronic charge. Taking the estimate of Harkins et al. [8] that the surface area occupied by one head group at the aggregate surface is 27 Å^2, and assuming for the effective dielectric constant (the average of water and that of a hydrocarbon), he obtained $w_e = 2.8\ kT$ compared against $2.2\ kT$ obtained from the experimental cmc and micelle size data.

1.3.4 SIZE VARIANCE OF EQUILIBRIUM MICELLE

Debye also made an assessment of the distribution of micelle aggregation numbers by expressing the total energy W of the micelle in the vicinity of $g = g_0$ as a function of $(g - g_0)$. From a Taylor expansion of the energy function W around g_0, he obtained

$$W = \left[W + (g - g_0)\frac{\partial W}{\partial g} + \frac{(g - g_0)^2}{2}\frac{\partial^2 W}{\partial g^2} \right]_{g_0} = W_0 \left[1 - \frac{3}{4}\left(\frac{g - g_0}{g_0} \right)^2 \right] \tag{1.9}$$

This equation is rearranged to obtain the fluctuation in the micelle aggregation number

$$g - g_0 = g_0 \sqrt{\frac{4}{3}\frac{W - W_0}{W_0}} \tag{1.10}$$

Introducing natural thermal fluctuations into this expression, $W - W_0 = kT$ and Equation 1.6 for W_0, Debye showed that the fluctuation in the micelle aggregation number is

$$g - g_0 = \sqrt{4g_0\frac{kT}{w_m}} \tag{1.11}$$

For the experimental values of g_o and w_m determined for the dodecylamine hydrochloride surfactant, $g - g_o$ is of the order of 3.3, indicating that the micelle aggregation number is relatively narrowly dispersed.

Debye's first theoretical model thus introduced the concept of opposing forces (long-range electrical versus short-range van der Waals). It showed the driving force for micelle formation to be the attractive van der Waals interactions whereas the finite size of the micelle is determined by repulsive electrostatic forces. It identified a cmc with the recognition that it remains practically constant. From the minimization of the energy of a single micelle, it showed that the micelles are small. From the Taylor expansion analysis of the total energy, the model suggested that the small micelles are narrowly dispersed in their sizes. Because the model was based on minimizing the energy of a single micelle, it can be viewed as a precursor to what later evolved as the pseudophase model of micelle formation. However, it is different from the pseudophase model because in a pseudophase, the energy of a molecule that constitutes the pseudophase is a minimum whereas in the Debye model, the energy of the micelle (and not of a single soap molecule within it) is a minimum.

1.4 EVOLUTION OF THEORIES ADDRESSING THE SHORTCOMINGS OF THE DEBYE MODEL

The pioneering theory of Debye stimulated other theoretical studies on micellization; some of which attempted to improve on the Debye theory without questioning the basic model whereas others attempted to correct some of the critical omissions of the Debye theory. Debye had already noted that the role of the addition of electrolyte to the surfactant solution needed to be worked out. He argued that because in an electrolyte solution, every charge will be surrounded by an excess of ions of the opposite sign, its electrical action will be screened out for larger distances. Therefore, an added electrolyte will screen the action of the charges on the micelle, reduce the electrical work W_e, and therefore increase the equilibrium size of the micelle. This analysis was quantitatively developed by Hobbs [21], who extended the Debye theory to a solution containing electrolytes by applying the Debye–Hückel approximation to estimate the salt-induced change in the ionic interaction energy. He obtained expressions for the increase in the micelle size and decrease in the cmc as a function of the added salt concentration. However, the validity of the Debye model was not questioned. Hobbs also estimated the electrostatic interactions between the two charged surfaces of the lamellar micelle. In the absence of any salt, he estimated that these interactions may account for 20% of the total electrostatic energy and could thus reduce the equilibrium micelle size and increase the cmc.

Ooshika [22] and Reich [23] questioned some of the fundamental concepts underlying the Debye theory. In building a theory based on McBain's lamellar aggregate, Debye did not include any energy contribution to account for the fact that the curved surface of the aggregate is exposed to water. The need for adding such a surface energy component to the micelle energy was pointed out by Ooshika and Reich. Ooshika showed that it is the surface energy, rather than the van der Waals interactions between tails proposed by Debye, which opposes the repulsive head group interaction energy in determining the size of the micelle. Ooshika and Reich also questioned the validity of minimizing the energy of a single micelle as a criterion for equilibrium rather than minimizing the energy of the entire system, which would also include the solution entropy.

1.4.1 OOSHIKA MODEL

Ooshika [22] pointed out that in the equilibrium state, there must be aggregates of various sizes and shapes, and their distribution is to be determined by statistical mechanics. However, if the size (aggregation number) of a micelle is large, the mixing entropy of the micelles in the solution is negligible when compared against that of monomers, and the cmc may be regarded as the point where a kind of phase change occurs (a view recognized as the pseudophase model for micelles). Therefore, the cmc is not appreciably affected by the distribution of the micelle sizes. The finiteness of micelles at the cmc has been attributed by Debye to the counterbalance of the van der Waals attractive energy of the hydrocarbon tails and the Coulomb-repulsive energy of the polar heads. Ooshika argued that in ordinary phase changes, the phase that has a lower free energy tends to grow infinitely because of the surface energy and not because of the additive van der Waals energy. Hence, it is clear that one must add to the free energy of micelle formation, the surface energy between the micelle and water. The surface energy of the micelle is proportional to the circumferential area $2\pi R H$ that is exposed to water, where H denotes the thickness of the lamellar aggregate. Because the volume $\pi R^2 H$ of the micelle is proportional to the aggregation number g, the circumferential area $2\pi R H$ will be proportional to $g^{1/2}$. Adding this contribution to the energy model of Debye, Ooshika proposed

$$W = W_e + W_m + W_s = g^{3/2} w_e - g w_m + g^{1/2} w_s \tag{1.12}$$

where w_e and w_m are constants determined by particular surfactants as defined earlier, and w_s is a characteristic surface energy constant.

To take into account the system entropy contribution to the total free energy, Ooshika considered the system composed of N_1 soap molecules in the solution, N_2 soap molecules incorporated into the micelles, and N_w water molecules. The free energy of mixing of these components in the solution is written as

$$F_{mix} = -kT \left[N_1 \ln \frac{N_1}{N_w + N_1 + \dfrac{N_2}{g}} + N_w \ln \frac{N_w}{N_w + N_1 + \dfrac{N_2}{g}} + \frac{N_2}{g} \ln \frac{N_2/g}{N_w + N_1 + \dfrac{N_2}{g}} \right]$$

$$F_{mix} \approx -kT \left[N_1 \ln \frac{N_1}{N_w + N_1} + N_w \ln \frac{N_w}{N_w + N_1} \right] \tag{1.13}$$

where k is the Boltzmann constant and T the absolute temperature. The three terms appearing in the first equality in Equation 1.13 represent the entropic contributions from the monomeric surfactant, solvent water, and micelles, respectively. Because the aggregation number g is appreciably larger than 1, the number of micelles N_2/g will be much smaller than N_1 and N_w. Therefore, the first equality in Equation 1.13 can be simplified to obtain the second equality as shown. Effectively, the relatively smaller contribution from the entropy of micelles is neglected. Combining the interaction energy (Equation 1.12) and the free energy of mixing (Equation 1.13), Ooshika obtained (for the free energy of the system) the expression,

$$F = -kT \left[N_1 \ln \frac{N_1}{N_w + N_1} + N_w \ln \frac{N_w}{N_w + N_1} \right] + N_2 \left(\frac{W}{g} \right) \tag{1.14}$$

The cmc, $X_C = N_1/(N_1 + N_w)$ and the equilibrium micelle size (the aggregation number g_0) are determined by minimizing the total free energy with respect to the independent variables N_1 and g, holding the total soap molecules $N_1 + N_2$ and water molecules N_w at constant values, respectively. One obtains,

$$kT \ln X_C = \frac{W_0}{g_0}, \quad g_0 = \frac{w_s}{w_e} \tag{1.15}$$

Introducing the expression for W_o (Equation 1.12 with $g = g_o$) in Equation 1.15, the cmc can be expressed as

$$kT \ln X_C = 2\sqrt{w_e w_s} - w_m, \quad g_o = \frac{w_s}{w_e} \tag{1.16}$$

Clearly, in contrast with the Debye theory, the equilibrium micelle size g_o is obtained from the counterbalance of the surface energy and the electrostatic head group interactions. Correspondingly, the cmc also shows an explicit dependence on the surface energy.

1.4.2 REICH MODEL

The Debye and Ooshika models, in which the electrostatic head group interactions play a central role, are not applicable to nonionic detergents. Reich [23] proposed that a general theory of micelles should be developed, starting from a treatment of nonionic surfactants for which the electrostatic interactions are not relevant. He also criticized the Debye theory for the same reasons as Ooshika, namely, the stable micelle size must be that which results in minimum free energy for the system and not the work of formation of a micelle. He argued that the growth of micelles will involve a decrease in the total number of independent species (singly dispersed surfactant molecules and micelles) in the system, and hence will involve a decrease in total system entropy, which must be taken into account. Furthermore, Reich suggested that the van der Waals energy of hydrocarbon chain per surfactant molecule must increase as the micelle grows, arguing that if it remained constant, as in the Debye model, then growth beyond a dimer would not occur. This last feature is an early recognition by Reich of the need for cooperativity of aggregation, characteristic of micelle formation, and is discussed in detail in Section 1.8.

Reich expressed the free energy of the system described by the equilibrium relation in Equation 1.1 as

$$\ln X_g = \ln K_g + g \ln X_1 = -\frac{\Delta G^o}{kT} + g \ln X_1 = \frac{T\Delta S^o - \Delta E^o}{kT} + g \ln X_1 \tag{1.17}$$

where ΔS^o and ΔE^o are the entropy and enthalpy changes associated with micelle formation. For calculating ΔS^o, he assumed a constant entropy change s per molecule, on micelle formation, which is independent of the size of the micelle. To calculate the energy change ΔE^o for the nonionic detergent, he considered the changes experienced by the aliphatic hydrocarbon tail of the detergent on micelle formation. For an aliphatic hydrocarbon tail in water, the molecule will show a tendency to fold up so that the segments of the chain can escape from water and remain in contact with each other. He assumed that each hydrocarbon tail was a tightly packed sphere, excluding any water and with a surface area A_H. Of this surface area A_H, a portion a_p is covered by the polar group, whereas the remainder, $A_H - a_p$, will be the exposed hydrocarbon surface. The surface energy of this single molecule is denoted by Υ. Micelle formation causes a fraction of the original exposed hydrophobic surface to be protected from water. This protected fraction multiplied by $(-\Upsilon)$ provides the surface energy change on aggregation. For g surfactant molecules in the nonaggregated state, the exposed hydrocarbon surface area is $g(A_H - a_p)$. For g molecules in spherical micelles, the exposed hydrocarbon surface area is $(g^{2/3}A_H - ga_p)$. Therefore, the fraction of hydrocarbon surface exposed to water that is eliminated on aggregation is $(gA_H - g^{2/3}A_H)/(gA_H - ga_p)$. Reich defined aggregates of the size g_{com} at which the surface becomes completely covered with polar groups as "complete micelles," namely,

$$\frac{gA_H - g^{2/3}A_H}{gA_H - ga_p} = \frac{A_H}{A_H - a_p}\left(1 - g^{-1/3}\right) = 1 \quad \text{at} \quad g = g_{com} \tag{1.18}$$

If an aggregate grew larger than the complete micelle size and still remained spherical, the surface would not be able to accommodate all the polar groups. Some polar groups would have to be buried in the interior (this is unlikely energetically). Such large aggregates would presumably be flattened sufficiently to create enough extra surface to accommodate all the polar groups. Thus, as g increases beyond the complete micelle size g_{com}, the aggregates become larger and flatter. The fraction of the hydrocarbon surface eliminated remains at 1.

Accordingly, for values of g less than that of the complete micelle, Equation 1.17 becomes

$$\ln X_g = \frac{Tsg + gY\left[\dfrac{A_H}{A_H - a_P}\right]\left[1 - g^{-1/3}\right]}{kT} + g \ln X_1 \tag{1.19}$$

whereas for values g exceeding that of the complete micelle (fractional surface removed is 1 in this case), it becomes

$$\ln X_g = \frac{Tsg + gY}{kT} + g \ln X_1 \tag{1.20}$$

Reich was able to calculate and plot a micelle size distribution based on the above equation for a nonionic surfactant. To estimate the entropy change s per molecule, he calculated the standard entropy change for condensing a hydrocarbon molecule from vapor state to liquid state, holding the density constant. He recognized that there would be some contribution from the entropy change associated with the polyoxyethylene type head group of the surfactant, although it could not be estimated. He calculated the energy parameter Y from consideration of the macroscopic interfacial free energy for a hydrocarbon.

The calculated size distribution function (Figure 1.4) showed (i) how the concentrations of the micelles of sizes different from the complete micelle fall off sharply; (ii) how the chemical potential of the surfactant, or explicitly, the concentration of the unaggregated surfactant controls the micelle

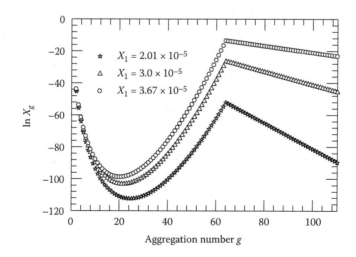

FIGURE 1.4 Size distribution of micelles calculated by the Reich model for a nonionic detergent at different monomer concentrations. Parameters used in the computations are $A_H = 4a_P$, $s = -20\,k$, and $Y = 30\,kT$, all taken from the article by Reich. (From I. Reich, *J. Phys. Chem.*, 60, 257–262, 1956.)

size distribution; (iii) the sharpness of the cmc and the narrow size distribution of the micelles; and (iv) that the size of the complete micelle depends on the ratio a_P/A_H. All these features are qualitatively valid for what we know to be the properties of nonionic detergents. However, because of the definition of the complete micelle, the most probable micelle size always remained that of the complete micelle without being affected by the total surfactant concentration.

Reich also discussed the problem of the McBain micelle. In this case, the micelle is assumed to grow in two dimensions, with the hydrocarbon chains aligned parallel to each other and the polar groups covering the flat faces. Irrespective of the size of such an aggregate, the hydrocarbon–water interface will never be eliminated completely because the detergent molecules at the edge will have their hydrocarbon chains exposed to water. In developing an expression for the energy change in the McBain micelle, which follows similar considerations as the spherical micelle, Reich showed that the formation of the McBain micelle would indeed correspond to a condition of phase separation, a conclusion that we currently understand for lamellar aggregate shapes.

1.5 EMERGENCE OF A THEORY FOR RODLIKE MICELLES

Early discussions on micelle shapes did not include consideration of the rodlike micelles as we recognize them presently. Philippoff [17] analyzed osmotic activity, specific conductivity, and x-ray data obtained at various electrolyte and surfactant concentrations and for various surfactant tail lengths and identified two regimes with respect to micellar shapes. In the first regime, which corresponds to zero or a small amount of electrolyte, micelles are practically spherical in the sense that they need not be true spheres but can be cubes, short cylinders, prisms, or spheroids. In the other regime, which corresponds to large electrolyte concentrations and longer tail lengths of surfactants, the aggregates are anisometric and large. Philippoff considered that it is improbable for small micelles to reorganize into large micelles with increasing salt and surfactant concentrations and therefore postulated the large structures to be secondary aggregates of the preformed primary micelles.

The formation of rodlike micelles as we understand them currently was first proposed by Debye and Anacker [41] based on light scattering measurements on solutions of cationic alkyl trimethyl ammonium bromides in the presence of added salt, KBr. They concluded that the dissymmetry measurements were not in agreement with the formation of spherical micelles and cylindrical lamellar micelles but agreed with micelles being very large and rodlike at high concentrations of added salt. They proposed that "the cross-section of the rod would be circular with the polar heads of the detergent lying on the periphery and the hydrocarbon tails filling the interior. The ends of such a rod would most certainly have to be rounded off with polar heads." They observed that the effect of salt on micelle growth was more significant, the longer the tail length of the surfactant. A number of studies that followed confirmed the formation of such large rodlike micelles, as summarized in the article by Anacker [42].

The polydispersed nature of rodlike micelles was first recognized by Scheraga and Backus [43]. They conducted flow birefringence measurements in solutions of cetyltrimethylammonium bromide (CTAB) and concluded that large asymmetrical aggregates are formed whose length increases with increasing salt concentration. More importantly, they found that a monodispersed aggregate model does not describe the experimental data and that aggregates must be highly polydispersed. They provided an interpretation for this size distribution by arguing that for spherical micelles formed at low salt concentrations, relatively high energies are involved (based on Debye's theory) if the micelle size is to be changed by more than two or three monomers. Hence, the small micelles should have narrow size distributions. In contrast, at high salt concentrations, the charged groups become shielded so that it does not require much work to be done to bring up more molecules. This would increase the mean micelle size and also lead to a high polydispersity. They also speculated as to the bending and flexibility of the micelles, although these features were not explored.

The existence of viscoelastic behavior in surfactant solutions was first discovered by Pilpel [44,45], based on rheological studies on aqueous solutions of sodium and potassium oleate in the presence of electrolytes. He interpreted the viscoelasticity as arising from a change in shape of the aggregates, with the small soap micelles formed at low electrolyte concentrations transforming to long interlinked cylinders at high electrolyte concentrations.

1.5.1 HALSEY MODEL

The first theoretical concept to describe rodlike micelles was proposed by Halsey [24]. As discussed previously, two factors that limit the growth of micelles and permit them to exist without phase separation had been proposed: one is the geometrical limit of chain length proposed by Hartley as the limiting radius of spherical micelles and the other is the repulsive interactions between the head groups at the micelle surface proposed in the Debye model (applied to the McBain–Harkins cylindrical portion of a lamellar micelle). Halsey [24] examined the growth of micelles avoiding both of these limitations, which resulted in finite stable aggregates without causing phase separation. Using Debye's approach, Halsey compared the energetics of one-dimensional growth of micelle as a disk and as a rod.

Halsey calculated the electrical potential at a small distance z from the end of a long rod of length $L = nz_0$, where z_0 is a characteristic distance between charges and w_e is the characteristic electrical energy appearing in the Debye model for electrostatic head group interactions. For large n, the potential will be

$$w_e \int_{z_0}^{nz_0} \frac{dz}{z} = w_e \ln n \qquad (1.21)$$

The total electrical work of forming a cylindrical micelle of size g is given by

$$W_e = \int_1^g w_e \ln n \, dn = w_e[g \ln g - g + 1] \qquad (1.22)$$

Comparing this expression for electrical energy of cylinders with the Debye expression (Equation 1.4) $W_e = w_e g^{3/2}$, for lamella, Halsey noted that the function $(g \ln g - g + 1)$ varies more slowly than $g^{3/2}$ as g increases. Therefore, he concluded that for large g, a rodlike micelle will correspond to lower energy and hence the more stable aggregate compared with a disk. He argued that because the surfactant molecules approaching one end of a long rod would feel only the near end, the rod length should have no effect on the total energy. This would make the length of the rod to be infinite. Because this is not the case, he proposed that the finite size of the micelles can be explained by the analogy between rodlike micelles and a one-dimensional gas (molecules fixed on a string like beads). Just as a one-dimensional gas does not condense whatever may be the nature of interactions, the rodlike micelles remain finite and do not cause phase separation. The micelles can be polydispersed analogous to the distribution of a linear polymer in equilibrium with the monomer. As for the other two smaller dimensions of the rodlike micelle, Halsey observed that one dimension must be limited by the length of the hydrocarbon chain and the second can either be limited by electrical forces (as in the Debye model) or by the length of the hydrocarbon chain (based on the structural description proposed by Hartley). Halesey's conclusions as to the large micelles being cylindrical, finite in size, polydispersed, and that such micelle formation is different from phase separation remain valid today.

1.6 STATISTICAL MECHANICAL THEORIES OF MICELLES

With the development of fundamental statistical mechanical models describing liquid state, such formalisms were also applied to describe the formation of micelles. Two of the significant contributions are due to Hoeve and Benson [25] and Poland and Scheraga [26,27] because, in both of their models, the free energy of formation of micelles was decomposed into many individual contributions identified in molecular terms similar to more recent predictive theories. These studies more clearly enumerated the various factors contributing to the free energy of micellization, including the hydrophobic effect associated with the transfer of surfactant tail from water to micelle and the surface energy associated with the exposed hydrophobic surface of the micelle.

1.6.1 HOEVE AND BENSON THEORY

In Hoeve and Benson's treatment [25], the micelles are thought to be aggregates of the hydrocarbon parts of the molecules, with the polar parts on the outside of the aggregate in contact with water. It was assumed that a spherical shape persists until the micelle becomes large enough that the radius of the sphere would exceed the maximum length of the hydrocarbon part of the molecule. When that occurs, the micelle is assumed to become flatter, leaving the polar parts fully hydrated and an oblate spherocylinder shape was assumed. The interior of the micelle is assumed to have properties similar to that of a liquid hydrocarbon. It is assumed that the system may be treated by classic partition functions, apart from contributions of vibrations, which are considered to be quantized and separable.

Aggregates of all sizes g are assumed to exist, with N_g being the number of aggregates of size g. For solutions of nonionic detergents in water, the most probable micelle size distribution is determined by minimizing the free energy F or maximizing the canonical partition function Q of the solution (note, $F = -kT \ln Q$). Hoeve and Benson obtained

$$\ln N_1 = \ln\left(\frac{Q_1 V}{\Lambda_1^3}\right) + \lambda, \quad \ln N_g = \ln\left(\frac{Q_g V}{\Lambda_g^3}\right) + \lambda g$$

$$\ln N_g = \ln\left(\frac{Q_g V}{\Lambda_g^3}\right) - g \ln\left(\frac{Q_1 V}{\Lambda_1^3}\right) + g \ln N_1 \tag{1.23}$$

where V is the system volume over which all the aggregates translate, Q_1 and Q_g are the internal partition functions for the unaggregated molecule and an aggregate of size g, Λ_1 and Λ_g are factors appearing in the external translational/rotational partition functions for the monomer and aggregate, and λ is the Lagrange multiplier that links the two equations providing an expression for the size distribution of aggregates. The translational/rotational terms Λ_1 and Λ_g are given by

$$\Lambda_g = \Lambda_g^{\text{trans}} \, \Lambda_g^{\text{rot}} = \left(\frac{h^2}{2\pi mgkT}\right)^{1/2} \left(\frac{h^2}{2\pi I_g kT}\right)^{1/2}$$

$$\Lambda_1 = \Lambda_1^{\text{trans}} \, \Lambda_1^{\text{rot}} = \left(\frac{h^2}{2\pi mkT}\right)^{1/2} \left(\frac{h^2}{2\pi I_1 kT}\right)^{1/2} \tag{1.24}$$

where m is the mass of a surfactant molecule, h is the Planck constant, and I_1 and I_g are the average moments of inertia of a monomer and aggregate of size g, respectively.

To evaluate Q_g for the aggregate, multiple contributions were considered. First, the assumption that the interior of a micelle is liquid in character was used. Hoeve and Benson suggested that because paraffinic hydrocarbons are not appreciably curled up in the liquid state, the contributions to the partition function of a hydrocarbon liquid are separable, and they wrote down the following liquid state contribution to Q_g.

$$Q_g^{\text{liq}} = \frac{1}{g!} \left[\left(\frac{2\pi mgkT}{h^2} \right)^{3/2} V_f \left(\frac{2\pi I_h kT}{h^2} \right)^{3/2} 8\pi^2 \gamma \right]^{g-1} q_h^g \quad (1.25)$$

Here, I_h is the moment of inertia of the hydrocarbon chain, q_h is the vibrational partition function for the chain, V_f is the free volume within the liquid phase where molecular translation occurs, and γ is the rotational free angle ratio accounting for the hindered rotation in the hydrocarbon liquid.

A second contribution comes from the recognition that the aggregate is a small liquid drop and therefore has a surface contribution. For a spherical drop of g molecules, with each molecule having a surface area A_H, the drop surface area is $g^{2/3}A_H$. The surface area A_H is calculated from the radius r_o of a hypothetical sphere whose volume is identical to that of the surfactant tail. Hoeve and Benson proposed the surface contribution as the product of the surface area and a characteristic interfacial energy σ between the drop and water.

$$Q_g^{\text{sur}} = \exp - \left(A_H g^{2/3} \sigma \right) = \exp - \left(4\pi r_o^2 g^{2/3} \sigma \right) \quad (1.26)$$

Hoeve and Benson then considered a third contribution to account for the constraint that in the micelle all ends of the hydrocarbon chains to which the polar heads are attached must be at the surface of the micelle, whereas in the liquid drop, the corresponding ends may occupy any position within the drop. The contribution of bringing the chain ends to the surface, resulting in head group crowding, was approximated as follows. Before bringing the ends to the surface, the free volume available for them is proportional to the micellar core volume $V_g = gv_o$, with v_o denoting the volume of a surfactant tail. After bringing the ends to the surface, the free volume is proportional to the free surface area available for them, which is assumed to be the difference between the surface area of the spherical micelle A_g and the space ga_P occupied by the g head groups already at the surface, that is, $A_g - ga_P$. Consequently, this contribution was written as proportional to the free volume restriction on micelle formation.

$$Q_g^{\text{crowd}} = C \left(\frac{A_g - ga_P}{V_g} \right) \quad (1.27)$$

Finally, a fourth contribution was recognized for nonionic surfactants with polar parts that are long polar chains (such as in the most common nonionic surfactant family with oligoethyleneoxide head groups), which strongly interact with the solvent and possess internal degrees of freedom. Using q_P, they denoted the contribution of the polar head due to internal vibrations and rotations and interactions with the solvent. Therefore,

$$Q_g^{\text{head}} = (q_P)^g \quad (1.28)$$

Combining all of these contributions, one can write

$$Q_g = Q_g^{\text{liq}} \ Q_g^{\text{sur}} \ Q_g^{\text{crowd}} \ Q_g^{\text{head}} \quad (1.29)$$

Following the same approach as for the micelle, Hoeve and Benson considered an expression for the partition function Q_1 of the monomer. In evaluating Q_1, the contributions from the polar part and the hydrophobic chain were separated, which is similar to what was done for the aggregate. Because the polar parts are assumed to remain in the water phase, q_P has the same contribution per molecule for micelles and for the single molecule. Similarly, the internal vibrational contribution q_h is also the same for the monomer as for the molecule in an aggregate. However, the contribution of the hydrocarbon part interacting with the surrounding water could not even be approximately evaluated. They recognized that structural effects exist in the water surrounding the hydrocarbon parts of the single molecules, and that the model of a liquid drop cannot be valid here. It was felt that only after gaining a better understanding of these effects could the contribution of the hydrocarbon part be quantitatively obtained. Keeping it as an unknown molecular partition function q_w, the overall partition function of the monomer was written as

$$Q_1 = q_w q_h q_P \tag{1.30}$$

By introducing the partition functions (Equations 1.29 and 1.30) in the expression for aggregate size distribution (Equation 1.23), Hoeve and Benson were able to identify some general features of the micellization of nonionic surfactants.

The illustrative size distribution curves they calculated for two different monomer concentrations are shown in Figure 1.5. The size distribution function was calculated showing a minimum and maximum in the size distribution and without the piecewise continuity introduced by Reich because of his postulate of a complete micelle. As a result, the maximum in the size distribution could be seen to change with increasing surfactant concentration, although only a little, and not fixed to one micelle size (g_{com} of the complete micelle) as in the Reich model. The cmc phenomenon is also evident because the amount of surfactant incorporated into the aggregates (excluding monomers) increases from approximately 10^{-8} to 2.64×10^{-4} and 21.2×10^{-4} for the three monomer concentrations shown in Figure 1.5. A small decrease in the monomer concentration close to the cmc makes the aggregate concentration close to zero. A small increase in the monomer concentration close to the cmc causes an order of magnitude increase in the amount of surfactants present as aggregates.

The work of Hoeve and Benson is important for many reasons. It was the first formal application of the statistical mechanical formulation to micelle formation, even though the need for such

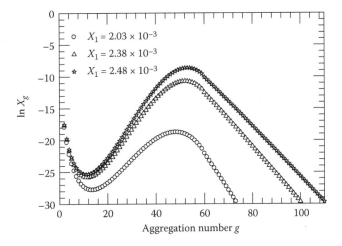

FIGURE 1.5 Size distribution of micelles calculated using the Hoeve and Benson model for a nonionic detergent at three different monomer concentrations. The parameters for the molecule used in the computations are identical to those used in the article by Hoeve and Benson [25].

an application was foreseen in the work of Ooshika, Reich, and Halsey. The formulation required the identification of multiple contributions to the partition function (or the free energy) and three important new contributions were identified compared with all previous studies that had identified van der Waals, electrical, and surface contributions. First, the treatment identified the importance of considering the molecular translational and rotational contributions, and these are part of the chain packing contributions we consider in current predictive theories. Second, by accounting for the crowding of head groups at the micelle surface for the nonionic surfactants, Hoeve and Benson were the first to propose a clear free energy contribution responsible for the finiteness of micelle size for nonionic surfactants. Third, Hoeve and Benson were the first to explicitly recognize that water structural changes will provide a critical contribution to the free energy and a more fundamental understanding of these structural changes is required before a quantitative predictive model of micelles can be developed. The subsequent work of Poland and Scheraga [26,27] and of Tanford [28–30] indeed emphasized detailed structural descriptions of these water structure changes and the quantitative and accurate estimation of the corresponding hydrophobic free energy.

Concerning ionic micelles, Hoeve and Benson noted that the micelles are highly charged and therefore the Debye–Hückel approximation is not adequate to describe ionic micelles. In applying the Poisson–Boltzmann equation, the challenge was to account for the fact that the counterions in the neighborhood of the micelle are crowded, even if the solution has a low concentration of ions. Another difficulty they pointed out related to the question of how "rough" the micelle surface is. Because the concentration of counterions is quite large in this region, the electrical free energy would be rather sensitive to the degree of roughness of the surface. These features relevant to the estimation of electrostatic interactions at the micelle surface remain unsatisfactorily explored even today and limit our ability to accurately predict aggregation properties in the presence of a variety of counterions, especially the counterion specificity in promoting a transition from spherical to rod-like micelles.

1.6.2 POLAND AND SCHERAGA THEORY

Poland and Scheraga [26,27] undertook the modeling of micelle formation as an illustration of the hydrophobic effect for which Nemethy and Scheraga [46–48] had developed a quantitative theory just a few years earlier. To simplify the calculations, instead of considering a size distribution of micelles, Poland and Scheraga [26] considered a system containing N surfactant molecules, present in the form of N_g micelles of size g. The equilibrium properties of the micelle were determined from the minimization of the free energy $F(g)$ of the system with respect to the micelle aggregation number g. Poland and Scheraga considered spherical micelles with polar heads on the surface and nonpolar tails in the interior partially coiled up (with some freedom of internal motion) and interacting with each other through hydrophobic bonds. The free energy of formation of these bonds was quantitatively estimated from the theory of hydrophobic bonding developed by Nemethy and Scheraga. They constructed a partition function Q_g for the micelle with g constituent molecules, by identifying an external contribution, an internal contribution, and a solvent interaction contribution.

$$F(g) = -kT \ln\left(\frac{Q_g^{N_g}}{N_g!}\right), \quad Q_g = Q_g^{\text{ext}} Q_g^{\text{int}} Q_g^{\text{sol}}$$

$$-\frac{F(g)}{NkT} = \frac{1}{g}\ln\frac{eQ_g}{N_g} = \frac{1}{g}\ln\frac{eQ_g^{\text{ext}}}{N_g} + \frac{1}{g}\ln Q_g^{\text{int}} + \frac{1}{g}\ln Q_g^{\text{sol}}$$

(1.31)

The external contribution is taken as the product of the classic partition functions for the translation and rotation of the micelle as a whole, similar to that in Hoeve and Benson's theory.

$$Q_g^{ext} = \left[\left(\frac{2\pi mgkT}{h^2} \right)^{3/2} V_f \left(\frac{2\pi I_h kT}{h^2} \right)^{3/2} 8\pi^2 \gamma \right]^{g-1} \tag{1.32}$$

Here, I_h is the moment of inertia of the hydrocarbon chain, q_h is the vibrational partition function for the chain, V_f is the free volume within the liquid phase where molecular translation occurs, and γ is the rotational free angle ratio accounting for the hindered rotation in the hydrocarbon liquid.

The internal partition function corresponds to the internal freedom of the micelle arising from the motions of the hydrocarbon tails. A monomer molecule in a micelle will make two large contributions to the internal partition function. One is the motion of the monomer molecule as a whole within the micelle and the other is the internal rotation (including complex vibrations and torsional oscillations) in the hydrocarbon tails. Because the latter is included in the hydrophobic bond energy per molecule and is accounted for as part of the solvent contribution discussed below, only the contribution of the former had to be included in Q_g^{int}. To account for the internal motion of the monomer in the micelle, they considered two approaches. In one, a free volume viewpoint is taken and the internal motion is treated as a translation; in the other, the micelle is treated as a lattice and the permutations of monomers in the lattice are calculated. From both approaches, they obtained essentially the same dependence of Q_g^{int} on g. Taking the free volume approach,

$$Q_g^{int} = \frac{1}{g!} \left[\left(\frac{2\pi mkT}{h^2} \right)^{3/2} V_f \right]^g, \quad V_f = V_{sh} \left(\frac{g-1}{g} \right) \tag{1.33}$$

In Equation 1.33, V_{sh} is the volume of a spherical shell within which all the head groups are constrained to translate and the factor $(g-1)/g$ represents the fact that the volume occupied by a molecule is not available to it for translation.

For calculating the solvent interaction partition function, they applied the results from the theory of hydrophobic bonding developed by Nemethy and Scheraga. In the monomeric state, the entire surface area of the surfactant tails comprises the hydrophobic surface exposed to water. In the micellar state, only the surface of the spherical micelle, excluding the space occupied by the head groups, comprises the hydrophobic surface exposed to water. Therefore, one can calculate the fraction θ_H of the total hydrocarbon surface involved in hydrophobic bonding, similar to that done by Reich. Poland and Scheraga expressed the solvent interaction contribution as the product of this fractional area of exposure θ_H involved in hydrophobic bonding and the hydrophobic bond energy ΔF_H per molecule. The hydrophobic bond energy per molecule had been estimated from Nemethy and Scheraga's theory of hydrophobic bonding as a function of temperature and the chain length of the molecule. The solvent interaction partition function is given by Equation 1.34. Here, for the hydrophobic bond energy, they used a temperature-dependent expression (which they established for amino acids) with the constants c_1, c_2, and c_3 taken to correspond to alanine–alanine bonds. The fractional area of exposure is approximated in the last step by neglecting the area a_P covered by the head group in comparison with the surface area A_H of a tail.

$$Q_g^{sol} = \exp\left[-\frac{g\Delta F_H}{RT} \theta_H \right], \quad \Delta F_H = c_1 + c_2 T + c_3 T^2$$

$$\theta_H = \frac{gA_H - g^{2/3} A_H}{gA_H - g a_P} = \left[\frac{A_H}{A_H - a_P} \right] \left[1 - g^{-1/3} \right] \approx \left[1 - g^{-1/3} \right] \tag{1.34}$$

Poland and Scheraga [27] also extended their treatment to ionic surfactants by including an electrostatic free energy for spherical micelles. They calculated this free energy by considering the Coulombic interactions between the charged head groups similar to the model of Debye, but applied to a spherical surface of radius R. As mentioned previously, the radius R is related to the aggregation number g through the relation $R = r_0 g^{1/3}$, where r_0 denotes the radius of a spherical hydrocarbon droplet whose volume is equal to that of a single surfactant tail. Therefore, the electrostatic partition function was represented as

$$Q_g^{\text{ele}} = \exp\left[\frac{ge^2}{2\varepsilon R}\right] = \exp\left[\frac{ge^2}{2\varepsilon r_0 g^{1/3}}\right] \qquad (1.35)$$

Combining the different free energy contributions, the overall dependence of the free energy $F(g)$ on the aggregation number has the form

$$\frac{F(g)}{NkT} = -\frac{1}{g}(\alpha - \ln X + 5\ln g) + \frac{\Delta F_H}{RT}\left(1 - g^{-1/3}\right) - \beta + \frac{1}{3}\ln g + \gamma g^{2/3} \qquad (1.36)$$

where α is extracted from the external free energy term (Equation 1.32), X is the surfactant concentration, β is extracted from the internal free energy term (Equation 1.33), and γ is extracted from the electrostatic energy term (Equation 1.35). Note that β and γ are relevant only for the micelle and will not appear in the free energy of a monomer.

Poland and Scheraga specified three conditions to obtain a stable micelle in a system containing N surfactant molecules: (C1) the free energy $F(g)$ of the solution of micelles must be an extremum ($\partial F/\partial g = 0$), (C2) the extremum must be a minimum ($\partial^2 F/\partial g^2 < 0$), and (C3) the free energy of the system of micelles $F(g)$ should be lower than the free energy of the system of monomers, $F(1)$. These three conditions lead to the following three constraints, respectively, on the hydrophobic bond energy, required for the formation of micelle of size g_0, where the last condition has been approximated taking into account that g_0 is much larger than 1. Furthermore, if the third condition is made into an equality, it provides the relation for the cmc.

$$\text{C1:} \quad \left(\frac{\partial F}{\partial g}\right) = 0 \Rightarrow \left(-\frac{\Delta F_H}{RT}\right) = g_0^{1/3}\left[\frac{3}{g_0}(\alpha - \ln X + 5\ln g_0 - 5) + 1\right] + 2\gamma g_0$$

$$\text{C2:} \quad \left(\frac{\partial^2 F}{\partial g^2}\right) < 0 \Rightarrow \left(-\frac{\Delta F_H}{RT}\right) < \frac{3}{2}\, g_0^{1/3}\left[1 + \frac{15}{g_0}\right] + 5\gamma g_0 \qquad (1.37)$$

$$\text{C3:} \quad F(g) < F(1) \Rightarrow \left(-\frac{\Delta F_H}{RT}\right) > \left[\alpha - \ln X + \frac{1}{3}\ln g_0 - \beta\right] + \gamma g_0^{2/3}$$

The conditions for micelle formation based on the above equilibrium requirements are shown in Figure 1.6 as a relation between the hydrophobic bond energy and the micelle size. The first constraint in Equation 1.37 implies that, for a given concentration, micelles will be formed when the values of ΔF_H and g_0 lie on the curve C1. The second constraint implies that micelles will form only for ΔF_H and g_0 values, which lie below the curve C2. Because the surfactant concentration fixes the curve C1, curves C1 and C2 determine the values of the hydrophobic bond energy and the corresponding micelle size. The second constraint imposes a restriction on the magnitude of ΔF_H for micelle formation to occur. If the hydrophobic bond strength is too large, small oligomers would be preferred to micelles. For lower hydrophobic bond energy, the formation of small oligomers is

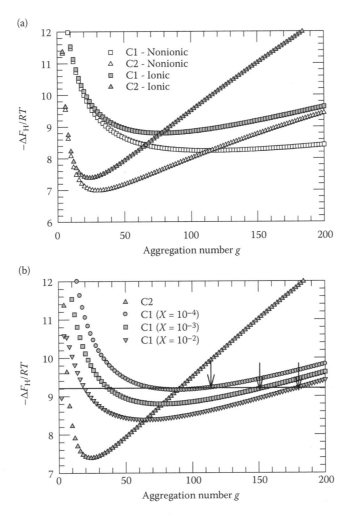

FIGURE 1.6 (a) Poland and Scheraga model predictions of conditions for micelle formation from ionic and nonionic surfactants at a surfactant concentration of $X = 10^{-3}$. Curves C1 and C2 are calculated using Equation 1.37 with the assumed model parameters of $\alpha = 1$, $\beta = 0.2$, and $\gamma = 0.003$. Micelle formation is allowed only for conditions below the curve C2. Equilibrium micelle size at the given concentration X can be found from the line C1 depending on the value of the hydrophobic bond energy of the surfactant. The leftward shift in the intersection of the curves C1 and C2 for the ionic surfactant imply that the equilibrium micelle size for the ionic surfactant will be smaller than that for the nonionic surfactant. (b) Shift in micelle size with a change in surfactant concentration for ionic surfactants. Curve C1 is calculated for three surfactant concentrations whereas curve C2 is independent of concentration. The arrows indicate the predicted equilibrium micelle size for the three concentrations, corresponding to the hydrophobic bond energy of 9.2 units. The micelle size increases with increasing surfactant concentration.

disallowed. Condition C3 indicates whether the concentration for which the curve C1 is constructed is above or below the cmc.

In the Poland and Scheraga model, the external, internal, and solvent interaction free energy contributions for the nonionic surfactants vary gradually when the aggregation number exceeds that of small oligomers, say 10. Correspondingly, the minimum in the free energy for nonionic surfactants is shallow resulting from a delicate balance between the slowly varying free energy contributions. In contrast, for ionic surfactants, the minimum is sharper because of the stronger dependence of the electrostatic energy on the micelle aggregation number. In the Poland and Scheraga treatment, the hydrophobic effect was fully recognized and the first attempt to employ a reasonable quantitative estimate was made. The residual interfacial contact between the hydrocarbon tail and water was accounted for in the solvent interaction term as in the Reich model rather than treating it as a surface free energy as was done in the Hoeve and Benson theory.

1.7 MICELLE SHAPE TRANSITIONS AND SIZE DISTRIBUTION

The free energy models constructed thus far have focused mainly on small micelles having either the lamellar-cylinder (McBain–Harkins micelle) or spherical (Hartley micelle) shape. Subsequent developments in the theory of micelle formation have attempted to construct a unified free energy model for the common aggregate shapes observed in surfactant solutions (Figure 1.7).

The small micelles are spherical in shape. When large rodlike micelles form, they can be visualized as having a cylindrical middle portion and parts of spheres as endcaps. The cylindrical middle and the spherical endcaps can have different diameters. When micelles can no longer pack into spheres (this happens for aggregation numbers for which a spherical aggregate will have a radius larger than the extended length of the surfactant tail), and if at the same time the rodlike micelles are not yet favored by equilibrium considerations, then small nonspherical globular aggregates form. Globular shapes such as prolate and oblate ellipsoids and shapes generated via ellipses of revolution have been proposed for these micelles. Some surfactants pack into a spherical bilayer structure called a vesicle, which encloses an aqueous cavity. In the outer and the inner layers of the vesicle, the surface area (in contact with water) per surfactant molecule and the number of surfactant molecules need not be equal to one another and the thicknesses of the inner and outer layers of the bilayer can also be different from one another.

The multiple equilibrium model of micellization (Equation 1.1) can be formally applied to aggregates of any shape and size. Correspondingly, the equilibrium condition corresponding to the minimum of the free energy of a solution made up of monomers, aggregates of all sizes and shapes, and water molecules can be represented in the form shown in Equation 1.38, which stipulates that the chemical potential of the singly dispersed surfactant molecule is equal to the chemical potential per molecule of an aggregate of any size and shape. In the multiple equilibrium description, each aggregate of a given size and shape is treated as a distinct chemical component characterized by a chemical potential. The conditions corresponding to the formation of aggregates are usually in the

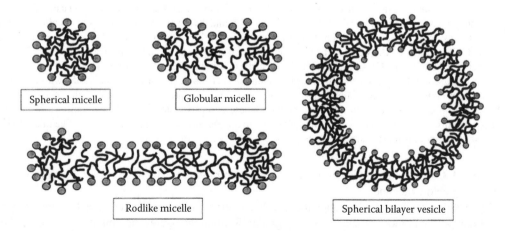

FIGURE 1.7 Schematic representation of surfactant aggregates in dilute aqueous solutions. The structures formed include spherical micelles, globular micelles, rodlike micelles with spherical endcaps, and spherical bilayer vesicles. One characteristic dimension in each of these aggregates is limited by the length of the surfactant tail.

realm of dilute solutions and, for these conditions, one can very simply relate the chemical potential of a component to the concentration of that component in solution, as shown in Equation 1.38.

$$\frac{\mu_g}{g} = \mu_1$$

(1.38)

$$\mu_g = \mu_g^o + kT \ln X_g, \quad \mu_1 = \mu_1^o + kT \ln X_1$$

Here, μ_g^o is the standard state chemical potential of the aggregate of size g having any shape, X_g is its mole fraction in solution, k is the Boltzmann constant, and T is the absolute temperature. The standard state of the solvent is defined as the pure solvent whereas the standard states of all the other species are taken to be infinitely dilute solution conditions. Combining the equilibrium condition with the concentration dependency of chemical potentials, we obtain the aggregate size distribution

$$X_g = X_1^g \exp-\left(\frac{\mu_g^o - g\mu_1^o}{kT}\right) = X_1^g \exp-\left(\frac{g\,\Delta\mu_g^o}{kT}\right)$$

(1.39)

Here, $\Delta\mu_g^o$ is the difference in the standard chemical potentials between a surfactant molecule present in an aggregate of size g and a singly dispersed surfactant in water. It is this free energy difference that is directly connected to the equilibrium constant K_g for micellization defined in Equation 1.1. To calculate the aggregate size distribution, we need an explicit equation for the standard state chemical potential difference $\Delta\mu_g^o$ or equivalently, the equilibrium constant for aggregation, K_g. Most of the theoretical studies in the last 30 years have focused on developing quantitatively accurate expressions for the dependence of this equilibrium constant on g.

Even in the absence of a free energy model for micellization, the thermodynamic relations provide many interesting results pertinent to the self-assembly process [49]. From the micelle size distribution, we can compute average aggregation numbers using the definitions

$$g_n = \frac{\sum gX_g}{\sum X_g}, \quad g_w = \frac{\sum g^2X_g}{\sum gX_g}, \quad g_z = \frac{\sum g^3X_g}{\sum g^2X_g}$$

(1.40)

where g_n, g_w, and g_z denote the number-average, the weight-average, and the z-average aggregation numbers, respectively, and the summations extend from 2 to ∞. The ratios g_w/g_n and g_z/g_w are unity for monodispersed systems and are equal to 2 and 3/2 for systems exhibiting very high polydispersity. Thus, either of these ratios can be used as an index of polydispersity. For a surfactant with any kind of head group, we can show [49] from the size distribution that the average aggregation numbers g_n and g_w depend on the concentration of the micellized surfactant (difference between the total surfactant concentration X_T and the cmc X_C) as follows:

$$\partial \ln g_n = \left(1 - \frac{g_n}{g_w}\right)\partial \ln \sum gX_g, \quad \partial \ln g_w = \left(\frac{g_z}{g_w} - 1\right)\partial \ln \sum gX_g$$

(1.41)

$$g_n \propto \left(\sum gX_g\right)^{\left(1 - \frac{g_n}{g_w}\right)}, \quad g_w \propto \left(\sum gX_g\right)^{\left(\frac{g_z}{g_w} - 1\right)}, \quad \sum gX_g = X_T - X_C$$

This equation states that the average aggregation numbers g_n and g_w must increase appreciably with increasing concentration of the micellized surfactant if the micelles are polydispersed; the average aggregation numbers must be virtually independent of the total surfactant concentration if the micelles are narrowly dispersed. Furthermore, Equation 1.41 shows that the exponent relating the average micelle size to the total surfactant concentration is a direct measure of the aggregate polydispersity. These are purely thermodynamic results independent of any free energy models for micellization. The conclusions of Debye as to the narrow distribution of small micelles and of Halsey on cylindrical micelles being polydispersed are consistent with these thermodynamic results, which are independent of any free energy model.

1.8   COOPERATIVE AND ANTICOOPERATIVE FREE ENERGY FUNCTION

Utilizing an increasing understanding of the energetic factors contributing to the micellization process, particularly with the recognition of hydrophobic interactions, Mukerjee was able to show how one part of the free energy function must promote the growth of the aggregate (the cooperative part) and another part of the free energy function must limit the growth and contribute to the finiteness of the aggregate (the anticooperative part) [36,50–53]. In developing his analysis, Mukerjee also discovered how very subtle changes in the two parts of the free energy function can affect the formation of spherical micelles versus rodlike micelles. As previously mentioned, Reich had already recognized the need for cooperativity in micelle formation for a micelle of some large enough aggregation number to form whereas the formation of dimers, trimers, and other small oligomers is prevented.

1.8.1   Representation of Cooperativity

Mukerjee started with the representation of micelle formation as a stepwise association process.

$$A_1 + A_{g-1} \overset{k_g}{\Leftrightarrow} A_g, \quad K_g = \prod_2^g k_j \tag{1.42}$$

where k_g is the stepwise association equilibrium constant for the formation of a g-mer from the combination of a $(g-1)$-mer with a monomer. It is differentiated from the monomer–micelle equilibrium constant K_g defined in Equation 1.1 and is related to it as shown above. The stepwise association equilibrium constant is also directly linked to the free energy change $\Delta\mu_g^o$ through the relation

$$-kT\ln k_g = g\Delta\mu_g^o - (g-1)\Delta\mu_{g-1}^o = \frac{d(g\Delta\mu_g^o)}{dg}; \quad -kT\ln K_g = g\Delta\mu_g^o \tag{1.43}$$

Thus, the overall monomer–micelle equilibrium constant K_g is directly related to the magnitude of the free energy of micellization $g\Delta\mu_g^o$ whereas the stepwise association equilibrium constant k_g is related to the dependence of this free energy on g.

If k_g increases with g, then the larger aggregates are favored over the smaller ones and the system is considered to exhibit positive cooperativity. If k_g decreases with g, then the formation of larger aggregates is increasingly disfavored and the system is said to exhibit negative or anticooperativity. When k_g is independent of g, the association is said to be continuous and noncooperative. In this case, polydispersed aggregates form and their size distribution is monotonically decreasing.

The shielding of the hydrophobic part of the micellar core from water becomes more and more effective with every incremental addition of a surfactant to the micelle. Thus, the incremental change in free energy due to the hydrophobic interactions becomes more negative with increasing size of the micelles. This has the tendency to increase k_g with increasing g. However, as g increases, the micellar surface becomes increasingly crowded with the polar head groups of the surfactants.

Consequently, for every incremental addition of a surfactant to the micelle, the repulsion between the polar head groups increases. This has the tendency to decrease k_g with increasing g.

At the initial stages of aggregation, that is, for relatively small values of g, the incremental change in the hydrophobic interactions is greater than that in the head group repulsions. Hence, there is an initial region of positive cooperativity in which k_g increases with increasing g. Beyond some critical aggregation number, the incremental change in the head group repulsions exceeds that in the hydrophobic interactions. Therefore, beyond a maximum value corresponding to a critical aggregation number, k_g begins to decrease with g, signaling a region of negative cooperativity. When large cylindrical micelles begin to form, the incremental change in the surface area of the micelle per amphiphile becomes a constant. As a result, the incremental addition of a surfactant molecule to the micelle alters neither the incremental changes in the attractive hydrophobic interactions nor the repulsive head group interactions. Consequently, k_g becomes independent of g, indicating a final region of noncooperativity.

1.8.2 COOPERATIVITY AND FORMATION OF CYLINDRICAL MICELLES

Mukerjee [36,51] proposed an empirical equation (Equation 1.44) for the functional dependence of K_g and k_g on g to show that subtle changes in the anticooperative region determine whether small micelles or large cylindrical micelles form.

$$-\ln K_g = \frac{g\Delta\mu_g^o}{kT} = 2\,(g-1)\ln(g-1)-0.02\,(g-1)^2+2.7896\,(g-1)$$

$$(1.44)$$

$$-\ln k_g = \frac{d(g\Delta\mu_g^o\,/\,kT)}{dg} = 2\ln(g-1)-0.04\,(g-1)+4.7896$$

Specifically, he considered two situations differing from one another in the value of g where the anticooperative region ends and the noncooperative region begins, as shown in Figure 1.8. Mukerjee showed that when the anticooperative region extends to the aggregation number 116 and beyond, only small spherical or globular micelles with narrow size distribution form. The average size does not

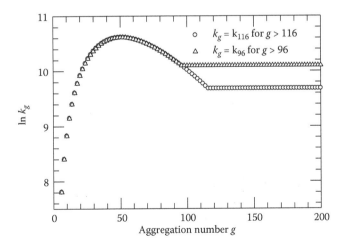

FIGURE 1.8 Schematic representation of the stepwise association equilibrium constant, showing the region of cooperativity and the region of anticooperativity. At large aggregation numbers, when cylindrical micelles form, the stepwise association constant becomes independent of size and is a constant, corresponding to a region of noncooperativity. The stepwise association equilibrium constant shown here is calculated using the empirical Equation 1.44 used by Mukerjee [36,51] to illustrate the phenomena.

change even when the total surfactant concentration is increased by orders of magnitude. In this case, the shallow free energy minimum and its range (determined by where the anticooperative region ends) are favorable enough for the stability of the smaller micelles and a transition to rods does not occur. On the other hand, when the anticooperative region ends at an aggregation number of 96, large polydispersed rodlike micelles form, which significantly change their average size with increasing surfactant concentration. In this case, the range of the shallow free energy minimum is not large enough to assure the stability of the small micelles and a transition to the large micelles occurs.

Mukerjee [50,51] was the first to treat the thermodynamics of rodlike aggregates by recognizing that two characteristic equilibrium constants are necessary to describe their formation. In the stepwise aggregation process, Mukerjee employed an equilibrium constant k_2 for the formation of a dimer that was different from the equilibrium constant k for the subsequent stepwise association for all aggregates larger than the dimer. From the equilibrium relations, Mukerjee showed how the weight average aggregation number g_w of the micelle is related to the concentration of the surfactant in solution and also the dependence of the cmc (X_C) on the equilibrium constants, as follows:

$$g_w = 2 \left(\frac{k}{k_2} \right)^{1/2} \left(\frac{X_T - X_1}{X_1} \right)^{1/2}, \quad X_C \approx X_1 = \frac{1}{k} \tag{1.45}$$

1.8.3 Alternate Representations of Cylindrical Micelles Formation

The approach pioneered by Mukerjee is equivalent to the thermodynamic treatments presented in later studies by Tausk and Overbeek [54], Israelachvili et al. [31], Missel et al. [55], and Nagarajan [56] to describe the transition from spherical to rodlike micelles. Following the treatment presented by Israelachvili et al., starting from the proposed structure for rodlike micelles shown in Figure 1.7 with a middle cylindrical part and quasi-spherical endcaps, one can identify two characteristic equilibrium constants associated with the molecules in the cylindrical part and those in the endcaps, respectively. The standard chemical potential of a rodlike micelle of size g with g_{cap} molecules in the two spherical endcaps and $(g - g_{cap})$ molecules in the cylindrical middle can be written as

$$\mu_g^o = (g - g_{cap}) \mu_{cyl}^o + g_{cap} \, \mu_{cap}^o \tag{1.46}$$

where μ_{cyl}^o and μ_{cap}^o are the standard chemical potentials of the molecules in the two regions of the rodlike aggregate, respectively. Introducing the above relation in the aggregate size distribution (Equation 1.39), we obtain

$$X_g = \left[X_1 \exp\left(\frac{\Delta\mu_{cyl}^o}{kT} \right) \right]^g \exp\left\{ g_{cap} \left(\frac{\Delta\mu_{cap}^o - \Delta\mu_{cyl}^o}{kT} \right) \right\} \tag{1.47}$$

where $\Delta\mu_{cyl}^o$ and $\Delta\mu_{cap}^o$ are the differences in the standard chemical potentials between a surfactant molecule in the cylindrical middle or the endcaps of the rodlike micelle and a singly dispersed surfactant molecule. Equation 1.47 can be rewritten as

$$X_g = \frac{1}{K} Y^g, \quad Y = \left[X_1 \exp\left(\frac{\Delta\mu_{cyl}^o}{kT} \right) \right], \quad K = \exp\left\{ g_{cap} \left(\frac{\Delta\mu_{cap}^o - \Delta\mu_{cyl}^o}{kT} \right) \right\} \tag{1.48}$$

where K is a measure of the free energy penalty for the molecules present in the spherical endcap compared with those in the cylindrical portion. The average aggregation numbers can be computed from Equation 1.40 by analytically summing the series functions:

$$g_n = g_{cap} + \left(\frac{Y}{1-Y}\right), \quad g_w = g_{cap} + \left(\frac{Y}{1-Y}\right)\left(1 + \frac{1}{Y + g_{cap}(1-Y)}\right) \qquad (1.49)$$

Equation 1.49 shows that for values of Y close to unity, very large aggregates are formed. The total concentration of surfactant present in the aggregated state is given by the expression:

$$\sum g X_g = \frac{g_{cap}}{K}\left(\frac{Y^{g_{cap}}}{1-Y}\right)\left(1 + \frac{Y}{g_{cap}(1-Y)}\right) = X_T - X_1 \qquad (1.50)$$

Here, X_{tot} and X_1 refer to the total amount of surfactant and the amount of surfactant present as singly dispersed molecules (practically, the cmc), so that the difference between them is the amount of surfactant in the micellar form. In the limit of Y close to unity and $g_{cap}(1-Y) \ll 1$, Equation 1.50 reduces to

$$\sum g X_g = \frac{1}{K}\left(\frac{1}{1-Y}\right)^2 = X_T - X_1 \qquad (1.51)$$

Introducing Equation 1.51 in Equation 1.49, the dependence of the average aggregation numbers on the surfactant concentration is obtained.

$$g_n = g_{cap} + \frac{1}{(1-Y)} = g_{cap} + [K(X_T - X_1)]^2$$

$$g_w = g_{cap} + \frac{2}{(1-Y)} = g_{cap} + 2[K(X_T - X_1)]^2 \qquad (1.52)$$

Because a realistic value for X_{tot} is less than 10^{-2} (that is, a surfactant concentration of 0.55 M or less), it is evident that K must be in the range of 10^8 to 10^{12} if large rodlike micelles ($g \sim 10^3$ to 10^5) are to form at physically realistic surfactant concentrations. The polydispersity index (g_w/g_n) goes from unity to two as the micelles grow from globules at low surfactant concentration to giant rodlike micelles at high surfactant concentrations. These results are identical to the conclusions reached by Mukerjee. These general thermodynamic results have since been used as the framework to evaluate the constant K, known in the literature as the sphere-to-rod transition parameter, based on either experimental measurements of aggregate growth with concentration [57,58] or by theoretical modeling of the free energy functions discussed in the following sections.

1.9   TANFORD'S PRINCIPLE OF OPPOSING FORCES

Tanford proposed the concept of opposing forces to formulate a quantitative expression for the standard free energy change on aggregation, incorporating quantitatively accurate expressions for the hydrophobic interactions [28–30]. Tanford proposed that the standard free energy change associated with the transfer of a surfactant from its infinitely dilute state in water to an aggregate of size g has three contributions:

$$\left(\frac{\Delta\mu_g^o}{kT}\right) = \left(\frac{\Delta\mu_g^o}{kT}\right)_{Tail} + \left(\frac{\Delta\mu_g^o}{kT}\right)_{Int} + \left(\frac{\Delta\mu_g^o}{kT}\right)_{Head} \qquad (1.53)$$

The first term $\left(\Delta\mu_g^{o}/kT\right)_{\text{Tail}}$ is a negative free energy contribution arising from the transfer of the tail from its unfavorable contact with water to the hydrocarbon-like environment of the aggregate core. The transfer free energy contribution depends on the surfactant tail but not on the aggregate shape or size. The second term $\left(\Delta\mu_g^{o}/kT\right)_{\text{Int}}$ provides a positive contribution to account for the fact that the entire surface area of the tail is not removed from water but there is still residual contact with water at the surface of the aggregate core. This is represented as the product of a contact free energy per unit area σ (or an interfacial free energy) and the surface area per molecule of the aggregate core, a. The third term $\left(\Delta\mu_g^{o}/kT\right)_{\text{Head}}$ provides another positive contribution representing the repulsive interactions between the head groups that crowd at the aggregate surface. The repulsions may be due to steric interactions (for all types of head groups) and also electrostatic interactions (dipole–dipole interactions for zwitterionic head groups and ion–ion repulsions for ionic head groups). Because the repulsion would increase if the head groups came close to one another, Tanford proposed an expression with an inverse dependence on a. Thus, the standard free energy change per molecule on aggregation proposed by Tanford has the form:

$$\left(\frac{\Delta\mu_g^{o}}{kT}\right) = \left(\frac{\Delta\mu_g^{o}}{kT}\right)_{\text{Tail}} + \left(\frac{\sigma}{kT}\right)a + \left(\frac{\alpha}{kT}\right)\frac{1}{a} \qquad (1.54)$$

where α is the head group repulsion parameter, k the Boltzmann constant, and T the temperature.

Starting from the free energy model of Tanford, the equilibrium aggregation behavior can be examined either by treating the surfactant solution as consisting of aggregates with a distribution of sizes or by treating the aggregate as constituting a pseudophase. If the aggregate is viewed as a pseudophase, in the sense of small systems thermodynamics, the equilibrium condition corresponds to a minimum in the standard free energy change per molecule, $\Delta\mu_g^{o}/kT$. The minimization can be done with respect to either the aggregation number g or the area per molecule a because they are dependent on one another through the geometrical relations given in Table 1.1.

One obtains in this manner, the equilibrium condition:

$$\frac{\partial}{\partial a}\left(\frac{\Delta\mu_g^{o}}{kT}\right) = 0 \Rightarrow \left(\frac{\sigma}{kT}\right) - \left(\frac{\alpha}{kT}\right)\frac{1}{a^2} = 0 \quad \text{at} \quad a = a_{\text{e}} = \left(\frac{\alpha}{\sigma}\right)^{1/2} \qquad (1.55)$$

TABLE 1.1
Geometrical Relations for Spherical and Cylindrical Micelles and Bilayers

Variable	Sphere	Cylinder	Bilayer
Volume of core $V = gv_0$	$4\pi R^3/3$	πR^2	$2R$
Surface area of core $A = ga$	$4\pi R^2$	$2\pi R$	2
Area per molecule a	$3v_0/R$	$2v_0/R$	v_0/R
Shape parameter $P = v_0/aR$	$1/3$	$1/2$	1
Largest aggregation number g_{max}	$4\pi\ell_0^3/3v_0$	$4\pi\ell_0^2/v_0$	$2\ell_0/v_0$
Aggregation number g	$g_{\text{max}}(3v_0/a\ell_0)^3$	$g_{\text{max}}(2v_0/a\ell_0)^2$	$g_{\text{max}}(v_0/a\ell_0)$

Note: Variables V, A, g, and g_{max} refer to the entire spherical aggregate, unit length of a cylinder, or unit area of a bilayer. R is the radius of spherical or cylindrical micelle or the half-bilayer thickness. v_0 and ℓ_0 are the volume and extended length of the surfactant tail. g_{max} is the largest aggregation number possible for the given geometry based on the constraint that the aggregate core is filled and the tail cannot stretch beyond its extended length.

The cmc (denoted as X_C in mole fraction units), in the pseudophase approximation, is obtained from the relation,

$$\ln X_C = \left(\frac{\Delta \mu_g^o}{kT} \right)_{\text{Tail}} + \left(\frac{\sigma}{kT} \right) a_e + \left(\frac{\alpha}{kT} \right) \frac{1}{a_e} \qquad (1.56)$$

In Tanford's free energy expression (Equation 1.54), the first contribution, the tail transfer free energy, is negative (Figure 1.9). Hence, this contribution is responsible for the aggregation to occur. It affects only the cmc (as shown by Equation 1.56) but not the equilibrium area a_e (as shown by Equation 1.55) or the size and shape of the aggregate. The second contribution, the free energy of residual contact between the aggregate core and water, is positive and decreases in magnitude as the area a decreases. A decrease in the area a corresponds to an increase in the aggregation number g, for all aggregate shapes, as shown in Table 1.1. Hence, this contribution promotes the growth of the aggregate. The third contribution, the free energy due to head group repulsions, is also positive and increases in magnitude if the area a decreases or the aggregation number g increases. Hence, this contribution is responsible for limiting the growth of aggregates to a finite size. Thus, Tanford's model clearly identifies why aggregates form, why they grow, and why they do not keep growing but remain finite in size.

Tanford's free energy expression is widely recognized in the micellar literature and is commonly referred to as the principle of opposing forces. Most theoretical work on micelles since 1973 has been very much influenced by Tanford's formalism. Indeed, we had used Tanford's model to propose a theoretical definition for the cmc as a transition point in the micelle size distribution [59,60], in contrast with the usual practical definition based on an observed change in the trend of any measured physical property as a function of the surfactant concentration. Also, we used the Tanford model to predict that double chain surfactant molecules form micelles as well as vesicles in solutions [34].

In formulating the free energy model, Tanford recognized that the state of the hydrocarbon tail of the surfactant within an aggregate is different from that of a similar hydrocarbon in a bulk liquid state because one end of the surfactant tail is constrained to remain at the aggregate–water interface whereas the entire tail has to be packed within a given aggregate shape maintaining liquid-like density in the aggregate core. He accounted for this empirically by making a correction to the estimate for the transfer free energy contribution. Because this correction was taken to be independent of the aggregate shape and size, it had no effect on the size and shape of the aggregates.

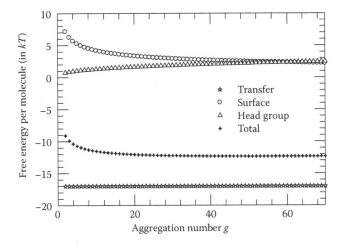

FIGURE 1.9 Contributions to the free energy of micelle formation based on Tanford's principle of opposing forces.

Furthermore, in representing the surface free energy contribution, Tanford did not use the value of the macroscopic hydrocarbon–water interfacial tension for the parameter σ appearing in Equation 1.54. Instead, he estimated σ as equal to the magnitude of the free energy change per unit area in transferring a hydrocarbon chain from liquid hydrocarbon to water. The sum of the transfer free energy and the surface free energy terms in Tanford's model is exactly equivalent to the hydrophobic effect calculated by Poland and Scheraga.

Finally, the head group repulsions were represented by a phenomenological expression with an inverse dependence on the area per molecule. A firm theoretical basis to compute this contribution for nonionic surfactants was not available, while the phenomenological expression provided only an approximation for electrostatic interaction energies for ionic and zwitterionic head groups. All these features have been improved upon in the recent predictive theories of micellization discussed in Section 1.11.

1.10   MOLECULAR PACKING MODEL FOR SELF-ASSEMBLY

Israelachvili et al. [31] proposed the concept of molecular packing parameter and demonstrated how the size and the shape of the aggregate at equilibrium can be predicted from molecular packing considerations. The molecular packing parameter is defined as $v_0/a\ell_0$, where v_0 and ℓ_0 are the volume and the length of the surfactant tail and a is the surface area of the hydrophobic core of the aggregate expressed per molecule in the aggregate (usually referred to as the area per molecule). If we consider a spherical micelle with a core radius R, made up of g molecules, then the volume of the core $V = gv_0 = 4\pi R^3/3$, the surface area of the core $A = ga = 4\pi R^2$, and hence $R = 3 v_0/a$, from simple geometrical relations (Table 1.1). If the micelle core is packed with surfactant tails without any empty space, then the radius R cannot exceed the extended length ℓ_0 of the tail. Introducing this constraint in the expression for R, one obtains the constraint on the molecular packing parameter, that is, $0 \leq v_0/a\ell_0 \leq 1/3$, for spherical micelles.

For spherical, cylindrical, or bilayer aggregates, the geometrical relations for the volume V and the surface area A are given in Table 1.1. The variables V, A, and g in Table 1.1 refer to the entire spherical aggregate, unit length of a cylindrical aggregate, or unit area of a bilayer aggregate, respectively, for the three shapes. These geometric relations, together with the constraint that at least one dimension of the aggregate (the radius of the sphere or the cylinder, or the half-bilayer thickness, all denoted by R) cannot exceed ℓ_0, lead to the following well-known [31] connection between the molecular packing parameter and the aggregate shape: $0 \leq v_0/a\ell_0 \leq 1/3$ for spheres, $1/3 \leq v_0/a\ell_0 \leq 1/2$ for cylinders, and $1/2 \leq v_0/a\ell_0 \leq 1$ for bilayers. Therefore, if we know the molecular packing parameter, the shape and size of the equilibrium aggregate can be readily identified, as shown above. This is the predictive sense in which the molecular packing parameter of Israelachvili et al. [31] has found significant use in the literature.

The notion of molecular packing into various aggregate shapes was recognized even in the earlier works of Tartar [61] and Tanford [28], as can be seen, for example, from Figure 9.1 of Tanford's classic monograph. However, only after this concept was explored thoroughly in the work of Israelachvili et al., taking the form of the packing parameter, did it evoke wide appreciation in the literature. For common surfactants, the ratio v_0/ℓ_0 is a constant independent of tail length, equal to 21 Å² for a single tail and 42 Å² for a double tail [28]. Consequently, only the area a_e reflects the specificity of the surfactant in the packing parameter $v_0/a_e\ell_0$.

The area per molecule a_e is a thermodynamic quantity obtained from equilibrium considerations of minimum free energy and is not a simple variable connected to the geometrical shape and size of the surfactant head group. Israelachvili et al. invoked the free energy model of Tanford [28] to estimate the equilibrium area per molecule, a_e.

$$\left(\frac{\Delta\mu_g^o}{kT}\right) = \left(\frac{\Delta\mu_g^o}{kT}\right)_{\text{Tail}} + \left(\frac{\sigma}{kT}\right)a + \left(\frac{\alpha}{kT}\right)\frac{1}{a}, \quad \frac{\partial}{\partial a}\left(\frac{\Delta\mu_g^o}{kT}\right) = 0 \text{ at } a_e = \left(\frac{\alpha}{\sigma}\right)^{1/2} \qquad (1.57)$$

The packing parameter $v_0/a_e\ell_0$ is readily calculated using the area a_e obtained from Equation 1.57. One can observe that a_e will be small and the packing parameter will be large if the head group interaction parameter α is small. The area a_e will increase and the packing parameter will decrease if the interfacial free energy per unit area σ decreases. These simple considerations allow one to predict many features of surfactant self-assembly as summarized below.

1.10.1 NONIONIC SURFACTANTS

For nonionic surfactants with oligoethylene oxide as the head group, the steric repulsions between the head groups can be expected to increase if the number of ethylene oxide units in a head group increases. Correspondingly, the head group parameter α will increase in magnitude if the number of ethylene oxide units in the head group increases. Therefore, when the number of ethylene oxide units is small, α is small, a_e is small, $v_0/a_e\ell_0$ is large, and bilayer aggregates (lamellae) are favored. For a larger number of ethylene oxide units, α increases, a_e increases, $v_0/a_e\ell_0$ decreases, and cylindrical micelles become possible. When the number of ethylene oxide units is further increased, $v_0/a_e\ell_0$ becomes small enough so that spherical micelles will form, with their aggregation number g decreasing with increasing ethylene oxide chain length. Furthermore, the increase in α and a_e for the nonionic surfactants with increasing ethylene oxide chain length gives rise to an increase in the cmc as the head group size increases. For a given ethylene oxide chain length and an equilibrium packing parameter corresponding to spherical micelles, increasing the chain length of the surfactant tail will cause an increase in the aggregation number of the micelles as predicted by the geometrical relation in Table 1.1.

1.10.2 IONIC AND ZWITTERIONIC SURFACTANTS

Comparing ionic and nonionic surfactants, the head group interaction parameter α will be larger for the ionic surfactants than for the nonionic surfactants because one has to also consider ionic repulsions between the head groups in the former case. Therefore, the equilibrium area per molecule, a_e will be larger and $v_0/a_e\ell_0$ will be smaller for the ionics compared with the nonionics. As a result, ionic surfactants would form aggregates of smaller aggregation number compared with nonionic surfactants of the same tail length. Furthermore, the increase in α and a_e for the ionic surfactants also gives rise to an increase in the cmc when ionic and nonionic surfactants of equal tail lengths are compared. The head group repulsions for zwitterionic surfactants are intermediate between those for ionic and nonionic surfactants. Therefore, both the cmc for zwitterionic surfactants and their aggregation numbers will be intermediate between those for ionic and nonionic surfactants of the same tail length.

1.10.3 SOLUTION CONDITIONS

For a given surfactant molecule, the head group repulsion can be modified by a change in the solution conditions. For example, adding salt to an ionic surfactant solution decreases electrostatic repulsions between ionic head groups; increasing the temperature for a nonionic surfactant molecule with ethylene oxide head group decreases steric repulsions between the nonionic head groups. Because the head group repulsion parameter α decreases, the equilibrium area per molecule, a_e, will decrease and the packing parameter $v_0/a_e\ell_0$ will increase. Thus, one can achieve a transition from spherical micelles to rodlike micelles and possibly to bilayer aggregates, by modifying solution conditions (adding salt to ionic surfactants and increasing the temperature for the nonionic surfactants) that control head group repulsions.

1.10.4 DOUBLE TAIL SURFACTANTS

If single tail and double tail surfactant molecules are compared, for the same equilibrium area per molecule, a_e, the double tail molecule will have a packing parameter $v_0/a_e\ell_0$ twice as large as that of the single tail molecule. Therefore, the double tail molecule can self-assemble to form bilayer vesicles whereas the corresponding single tail molecule aggregates into only spherical or globular micelles.

1.10.5 Polar Organic Solvents

If the solvent is changed from water to a mixed aqueous–organic solvent, then the interfacial tension parameter σ decreases. For a given surfactant, this would lead to an increase in the equilibrium area per molecule a_e, and hence, a decrease in $v_o/a_e\ell_o$. Therefore, upon the addition of a polar organic solvent to an aqueous surfactant solution, bilayers will transform into micelles, rodlike micelles into spherical micelles, and spherical micelles into those of smaller aggregation numbers including only small molecular clusters. Starting from phospholipid surfactants that have two hydrophobic tails per molecule, we can form bilayer vesicles, cylindrical micelles, or spherical micelles, respectively, if we use an aqueous–organic solvent mixture with increasing amounts of polar organic solvent in the solvent mixture.

All of the above predictions are in agreement with numerous experiments and are by now well established in the literature (see numerous experimental studies that have been included in the *Surfactants in Solution* series of books as well as in various journals). One can thus see evidence of the predictive power of the phenomenological free energy model and the molecular packing considerations, despite their remarkable simplicity. These concepts can be used to predict the self-assembly behavior of other novel amphiphiles and one example is the predictions made for bola amphiphiles that have a single hydrophobic chain separating two terminal polar head groups in each molecule [62]. The concept of a molecular packing parameter has been widely cited in the literature for chemistry, physics, and biology because it allows a simple and intuitive insight into the self-assembly phenomenon [63–65]. Without any doubt, the contributions of Tanford and of Israelachvili et al. have had a lasting effect on the surfactant literature over the past 35 years.

In the packing parameter model utilizing the free energy expression of Tanford, the surfactant tail has no recognizable role in determining the shape and size of the self-assembled structure. We have proposed a modified free energy model that takes into account tail-packing constraints inside the micelle core [66]. The inclusion of this contribution introduces an explicit as well as an implicit dependence of the aggregation properties on the tail length, in addition to the accepted dependence on the head group. We have shown that the tail length's effects are critical, especially for predicting when the transitions between aggregate shapes occur [67]. We revisit the packing parameter model in Section 1.11, taking advantage of the more refined free energy expressions we have developed as part of a predictive theory of surfactant self-assembly.

1.11 MOLECULAR SCALE PREDICTIVE THEORIES OF MICELLE FORMATION

Many of the general principles of surfactant aggregation have now been adequately clarified by the theoretical studies mentioned. Yet, these studies have not provided quantitative a priori predictions of aggregation behavior starting from the molecular structure of the surfactants and the solution conditions such as temperature, ionic strength, surfactant concentration, etc. For example, in the free energy model of Tanford, some contributions were estimated by utilizing experimental data available for micellar solutions. First, in writing the transfer free energy term, Tanford started with an expression for the transfer of a hydrocarbon chain from bulk water to a bulk hydrocarbon liquid. Then, recognizing that the tail conformation in the micelle interior is different from that in a bulk hydrocarbon liquid, he proposed an empirical correction to the free energy to ensure agreement with the experimental cmc data. Despite its empirical nature, this correction term had universality in the sense that it was not altered from one surfactant to another but was considered valid for all surfactants. This empirical expression assumed the chain conformation energy to have no influence on the aggregate size and shape. Second, the free energy contribution due to the electrostatic interactions between the polar head groups of ionic surfactants was estimated using the Debye–Hückel approximation in which an empirical correction was incorporated. A constant numerical factor of about half (0.46 being the best fit) in the Debye–Hückel expression was observed to be necessary to describe the experimental cmc data. Again, this empirical constant was not changed for different ionic surfactants but was used as a universal correction constant. Third, no explicit expression dependent on the head group was obvious in the Tanford model to describe head group interactions in nonionic surfactants. Furthermore, for the nonionic surfactants containing poly(oxyethylene) chain polar head groups, the interactions among the head groups may require a model different from one that could be used for the more compact head groups such as glucosides. Finally, Tanford treated the surface energy using estimates of hydrophobic bond energy per unit area rather than using the well-known hydrocarbon–water interfacial tension.

The free energy models developed in our work [66] and by the Blankschtein group [68] have attempted to eliminate these empirical features to allow for truly a priori predictions of the aggregation properties, given the surfactant molecular structure. We describe only our work in detail below but would direct the readers to explore many important articles from the Blankschtein group [69–76], whose approach is broadly similar to ours but differing in many important details.

In constructing the free energy model, we do not utilize any information derived from experiments on surfactant solutions to ensure that the predictions from the model remain truly a priori. To arrive at such a free energy expression, first we developed an analytical equation for the chain conformation free energy that is dependent on the size and the shape of the aggregates by borrowing results from the analysis of the chain conformation in copolymer melts [77]. This contribution allowed new predictions to be made that were overlooked by the empirical, shape-independent free energy correction term used earlier by Tanford and also in our earlier modeling [78–81]. Second, the electrostatic interactions in ionic micelles were computed using an approximate analytical solution to the Poisson–Boltzmann equation derived by Evans and Ninham [82], which incorporates curvature corrections. The use of this equation is found to provide a satisfactory estimate of the ionic interaction free energy and avoids the earlier use of the empirical coefficient of 0.46 in conjunction with the Debye–Hückel approximation. Third, the interactions among poly(oxyethylene) head groups in nonionic micelles was treated by taking into account the polymeric nature of the polar head group and considering the mixing and deformation free energies in the head group region in addition to the steric repulsions at the micellar core surface. Fourth, the surface energy is treated using macroscopic oil–water interfacial tension. Finally, the temperature dependences of all the free energy contributions are explicitly described, allowing one to calculate the temperature dependence of the aggregation behavior.

The details of the predictive theory are briefly presented below. More extensive discussions including comparisons with experimental data can be found in our previous publications [66,83–86]. We emphasize again that in developing the free energy model, we have not made use of any information obtained a posteriori from experimental aggregation data. Obviously, various molecular constants are involved in the calculations but they can be readily estimated from the molecular structural properties of surfactants.

1.11.1 Contributions to Free Energy Change on Aggregation

In our model [66], the standard free energy difference $\left(\Delta\mu_g^{\circ}/kT\right)$ between a surfactant molecule in an aggregate of size g and one in the singly dispersed state is decomposed into a number of contributions on the basis of molecular considerations. First, the hydrophobic tail of the surfactant is removed from contact with water and transferred to the aggregate core, which is like a hydrocarbon liquid. Second, the surfactant tail inside the aggregate core is subjected to packing constraints because of the requirements that the polar head group should remain at the aggregate–water interface and the micelle core should have a hydrocarbon liquid–like density. Third, the formation of the aggregate is associated with the creation of an interface between its hydrophobic domain and water.

Fourth, the surfactant head groups are brought to the aggregate surface giving rise to steric repulsions between them. Last, if the head groups are ionic or zwitterionic, then electrostatic repulsions between the head groups at the aggregate surface also arise. Explicit analytical expressions for each of these free energy contributions were developed [66] in terms of the molecular characteristics of the surfactant and they are briefly discussed in this section.

1.11.2 TRANSFER OF THE SURFACTANT TAIL

The contribution to the free energy from the transfer of surfactant tail from water to the micelle core is estimated by considering the core to be similar to a liquid hydrocarbon. This allows us to estimate the transfer free energy using compiled experimental data on the solubility of hydrocarbons in water [87,88]. Note that no information derived from experiments involving surfactants is used anywhere in our model. We have estimated [66] the group contributions to the transfer free energy for a methylene and a methyl group in an aliphatic tail to be

$$\left(\frac{\Delta\mu_g^o}{kT}\right)_{Tr} = 5.85\ln T + \frac{896}{T} - 36.15 - 0.0056\,T \text{ (for the } CH_2 \text{ group)}$$

$$\left(\frac{\Delta\mu_g^o}{kT}\right)_{Tr} = 3.38\ln T + \frac{4064}{T} - 44.13 + 0.02595\,T \text{ (for the } CH_3 \text{ group)}$$

$$(1.58)$$

Because the micelle core differs from a liquid hydrocarbon, a contribution to the free energy from this different state should be considered.

1.11.3 DEFORMATION OF THE SURFACTANT TAIL

In formulating his phenomenological free energy model, Tanford recognized that the state of the hydrocarbon tail of the surfactant within an aggregate is different from that of a similar hydrocarbon in bulk liquid state because one end of the surfactant tail is constrained to remain at the aggregate–water interface whereas the entire tail has to be packed within a given aggregate shape maintaining liquid-like density in the aggregate core (Figure 1.10). He accounted for this empirically by making a correction to the estimate for the transfer free energy contribution. Because this correction was taken to be independent of the aggregate shape and size, it could not describe any effect of the molecular packing constraints on the size and shape of the aggregates.

Detailed chain-packing models to estimate this free energy contribution $\left(\Delta\mu_g^o/kT\right)_{def}$ as a function of the aggregate shape and size have been developed following different approaches by Gruen [91–93], Dill and Flory [90,94–96], and Ben-Shaul et al. [97–99]. In our work, we have adapted the method suggested for block copolymers by Semenov [77] to obtain an analytical expression for the tail deformation energy [66]. For spherical micelles,

$$\left(\frac{\Delta\mu_g^o}{kT}\right)_{Def} = \left(\frac{9P\pi^2}{80}\right)\left(\frac{R^2}{NL^2}\right)$$

$$(1.59)$$

where P is the shape parameter defined in Table 1.1, R is the core radius, L is the segment length and, N is the number of segments in the tail ($N = \ell_o/L$, where ℓ_o is the extended length of the tail). As suggested by Dill and Flory [90,94], a segment is assumed to consist of 3.6 methylene groups (hence, $L = 0.46$ nm). L also represents the spacing between alkane molecules in the liquid state, namely, $L^2 = 0.21$ nm^2 is the cross-sectional area of the polymethylene chain. Equation 1.59 is also used for globular micelles and the spherical endcaps of rodlike micelles. For the cylindrical middle

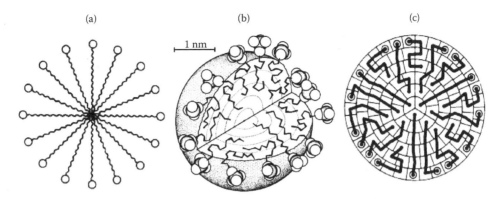

FIGURE 1.10 In the micelle, the methylene group of the surfactant tail that is attached to the polar head group is constrained to remain at the aggregate–water interface. The other end (the terminal methyl group) is free to occupy any position inside the aggregate as long as uniform liquidlike density is maintained in the aggregate core. In the classic picture of a Hartley micelle (a), these packing features are not recognized because the radial orientation of the tails is inconsistent with the requirement of uniform liquidlike density in the core. Gruen [89] presented the space-filling model for a micelle (b) consisting of approximately 55 molecules with dodecyl hydrophobic chain. Clearly, it is necessary for the tail to deform nonuniformly along its length to satisfy both the packing and the uniform density constraints. Dill and Flory [90] developed a lattice representation for the packing inside the micelle (c) meeting the liquidlike density requirement in the core. (a and c: Reprinted from K.A. Dill and P.J. Flory, *Proc. Natl. Acad. Sci. USA*, 78, 676–680, 1981. With permission. b: With kind permission from Spring Science + Business Media: *Progr. Colloid Polym. Sci.*, 70, 1985, 6–16, D.W.R. Gruen.)

of rodlike micelles, the coefficient 9 is replaced by 10, the radius R now represents the radius of the cylinder, and $P = 1/2$. For bilayers, the coefficient 9 is replaced by 10, the radius R represents the half-bilayer thickness, R and $P = 1$.

1.11.4 FORMATION OF AGGREGATE CORE-WATER INTERFACE

Micelle formation generates an interface between the hydrophobic core and the surrounding water medium. The free energy of formation of this interface is calculated as the product of the surface area in contact with water and the macroscopic interfacial tension σ_{agg} characteristic of the interface [66]:

$$\left(\frac{\Delta\mu_g^o}{kT}\right)_{int} = \left(\frac{\sigma_{agg}}{kT}\right)(a - a_o) \tag{1.60}$$

Here, a is the surface area of the micelle core per surfactant molecule, and a_o is the surface area per molecule shielded from contact with water by the polar head group of the surfactant. Expressions for the area per molecule a corresponding to different aggregate shapes are provided in Table 1.1. The area a_o depends on the extent to which the polar head group shields the cross-sectional area L^2 of the surfactant tail. If the head group cross-sectional area a_p is larger than L^2, the tail cross-section is shielded completely from contact with water and $a_o = L^2$. If a_p is smaller than L^2, then the head group shields only a part of the cross-sectional area of the tail from contact with water, and in this case, $a_o = a_p$. Thus, a_o is equal to the smaller of a_p and L^2. The aggregate core–water interfacial tension σ_{agg} is taken equal to the interfacial tension σ_{sw} between water (W) and the aliphatic hydrocarbon of the same molecular weight as the surfactant tail (S). The interfacial tension σ_{sw} can be calculated in terms of the surface tensions σ_s of the aliphatic surfactant tail and σ_w of water [100]:

$$\sigma_{SW} = \sigma_S + \sigma_W - 2.0 \; \psi \; (\sigma_S \sigma_W)^{1/2}$$

$$\sigma_S = 35.0 - 325 \; M^{-2/3} - 0.098 \, (T - 298) \tag{1.61}$$

$$\sigma_W = 72.0 - 0.16 \, (T - 298)$$

In Equation 1.18, ψ is a constant equal to 0.55 [100], M is the molecular weight of the surfactant tail, and the surface tensions are expressed in mN/m. Data from Ref. [101] were used to obtain the expression for σ_W.

We note that the use of macroscopic interfacial tension was proposed in the work of Hoeve and Benson [25]; in contrast, Poland and Scheraga [26] and Tanford [28–30] estimated σ as the transfer energy of the tail per unit area. The two characteristic energies per unit area differ by a factor of nearly 3.

1.11.5 HEAD GROUP INTERACTIONS—STERIC

Upon micelle formation, the surfactant head groups are brought to the micelle surface where they are crowded compared with the infinitely dilute state of the singly dispersed surfactant molecules. Figure 1.11 schematically depicts the three types of head groups having a polyoxyethylene chain, a zwitterionic head, and an ionic head, respectively. For nonionic surfactants, including those with a compact head group or an oligomeric chain head group (as in the case of polyoxyethylene head groups), the head group region has a higher local concentration for the hydrophilic groups. If the surfactant has a zwitterionic head group, the dipoles are oriented normal to the interface and stacked such that the poles of the dipoles are located on parallel surfaces. If the surfactant has an anionic or cationic head group, then the ionic groups are crowded at the surface. The counterions are distributed around the surface depending on their valence and the charge density at the surface.

In the case of nonionic surfactants, the crowding of the head groups at the micelle surface generates steric repulsions among the head groups. For compact head groups, by analogy with the repulsion term in the van der Waals equation of state, we have proposed [66,78,79] the expression

$$\left(\frac{\Delta\mu_g^o}{kT} \right)_{steric} = -\ln\left(1 - \frac{a_P}{a} \right) \tag{1.62}$$

where a_P is the cross-sectional area of the polar head group near the micellar surface. This equation is used for all aggregate shapes. We note that such an expression for calculating the steric repulsions appeared in the theory of Hoeve and Benson [25]. This approach to calculating the steric interactions is inadequate when the polar head groups are not compact such as in the case of nonionic surfactants with oligoethylene oxide head groups. An alternate treatment for head group interactions in

Nonionic   Zwitterionic   Ionic

FIGURE 1.11   Different types of head groups of classic surfactants.

such systems, considering the head groups to be polymer-like, has been developed by us [66] with mixed success in predicting the micellar properties.

1.11.6 HEAD GROUP INTERACTIONS—DIPOLAR

If the surfactant has a zwitterionic head group with a permanent dipole moment, then dipole–dipole interactions arise at the micelle surface. The dipoles are oriented normal to the interface and stacked such that the poles of the dipoles are located on parallel surfaces. The interaction free energy is estimated by considering that the poles of the dipoles generate an electrical capacitor and the distance between the planes of the capacitor is equal to the distance of charge separation d (or the dipole length) in the zwitterionic head group. Consequently, for spherical micelles, one obtains [66]

$$\left(\frac{\Delta\mu_g^o}{kT}\right)_{dipole} = \frac{2\pi e^2 R}{\varepsilon a_\delta kT}\left(\frac{d}{R+\delta+d}\right) \tag{1.63}$$

where e is the electronic charge, ε the dielectric constant of the solvent, R the radius of the spherical core, and δ the distance from the core surface to the place where the dipole is located. This equation is also employed for globular micelles and the endcaps of rodlike micelles. For the cylindrical part of the rodlike micelles, the capacitor model yields

$$\left(\frac{\Delta\mu_g^o}{kT}\right)_{dipole} = \frac{2\pi e^2 R}{\varepsilon a_\delta kT}\ln\left(1+\frac{d}{R+\delta}\right) \tag{1.64}$$

where R is now the radius of the cylindrical core of the micelle. The dielectric constant ε is taken to be that of pure water [101] and is calculated using the expression [66]

$$\varepsilon = 87.7 \exp\left[-0.0046\,(T-273)\right] \tag{1.65}$$

1.11.7 HEAD GROUP INTERACTIONS—IONIC

If the surfactant has an anionic or cationic head group, then ionic interactions arise at the micellar surface. An approximate analytical solution to the Poisson-Boltzmann equation derived by Evans and Ninham [82] is used in our calculations.

$$\left(\frac{\Delta\mu_g^o}{kT}\right)_{ionic} = 2\left[\ln\left(\frac{S}{2}+U\right)-\frac{2}{S}(U-1)-\frac{2C}{\kappa S}\ln\left(\frac{1}{2}+\frac{1}{2}U\right)\right] \tag{1.66}$$

$$U = \left[1+\left(\frac{S}{2}\right)^2\right]^{1/2}, \quad S = \frac{4\pi e^2}{\varepsilon\kappa a_\delta kT}$$

In Equation 1.66, the area per molecule a_δ is evaluated at a distance δ from the hydrophobic core surface (see Table 1.1) where the center of the counterion is located. The first two terms on the right-hand side of Equation 1.66 constitute the exact solution to the Poisson–Boltzmann equation for a planar geometry and the last term provides the curvature correction. The curvature-dependent factor C is given by [66]

$$C = \frac{2}{R_S + \delta}, \quad \frac{1}{R_C + \delta} \tag{1.67}$$

for spheres/spherical endcaps of spherocylinders and cylindrical middle part of spherocylinders, respectively. κ is the reciprocal Debye length and is related to the ionic strength of the solution through

$$\kappa = \left(\frac{8\pi n_o e^2}{\varepsilon kT} \right)^{1/2}, \quad n_o = \frac{(C_1 + C_{add})N_{Av}}{1,000} \tag{1.68}$$

In the above equation, n_o is the number of counterions in solution per cubic centimeter, C_1 is the molar concentration of the singly dispersed surfactant molecules, C_{add} is the molar concentration of the salt added to the surfactant solution, and N_{Av} is Avogadro's number.

1.11.8   HEAD GROUP INTERACTIONS FOR OLIGOMERIC HEAD GROUPS

For nonionic surfactants with polyoxyethylene chains as head groups, the calculation of the head group interactions using Equation 1.62 for the steric interaction energy becomes less satisfactory because it is difficult to define an area a_p characteristic of the oligomeric head groups without ambiguity. For sufficiently large polyoxyethylene chain lengths, it is more appropriate to treat the head group as a polymeric chain when estimating the free energy of head group interactions. We have developed a treatment in our free energy model [66] based on the following conceptual approach. In a singly dispersed surfactant molecule, the polyoxyethylene chain is viewed as an isolated free polymer coil swollen in water. In micelles, the polyoxyethylene chains present in the region surrounding the hydrophobic core (referred to as the shell or corona) can be viewed as forming a solution denser in polymer segments compared with the isolated polymer coil. The difference in the two states of polyoxyethylene provides a contribution to the free energy of aggregation, which is computed as the sum of the free energy of mixing of the polymer segments with water and the free energy of polymer chain deformation. As the polyoxyethylene chain length decreases, the use of polymer statistics becomes less satisfactory.

Two limiting models of micellar corona are considered. One assumes that the corona has a uniform concentration of polymer segments. The maintenance of such a uniform concentration in the corona is possible for curved aggregates only if the chains deform nonuniformly along the radial coordinate. The second model assumes a radial concentration gradient of chain segments in the corona consistent with the uniform deformation of the chain. The quantitative results from either of the models are affected by the polymer solution theory used for the calculation of the mixing free energies.

The mean-field approach of Flory [102] requires only the polyoxyethylene–water interaction parameter χ_{WE} for calculating the free energies, and because of its simplicity, it was used in our free energy model. However, it should be noted that composition and temperature dependencies must be assigned to χ_{WE} to describe the thermodynamic properties of aqueous polymer solutions. Such dependencies have not yet been satisfactorily established and, hence, our predictive calculations were performed assuming χ_{WE} to be a constant. For both uniform concentration and nonuniform concentration models, the detailed free energy expressions are described in ref. [66] and are not reproduced here.

1.11.9   ESTIMATION OF MOLECULAR CONSTANTS NEEDED FOR PREDICTIVE COMPUTATION

For predictive purposes, a few molecular constants are needed. The molecular constants associated with the surfactant tail are the volume v_o and the extended length ℓ_o of the tail. For the head groups, one needs the cross-sectional area a_p for all types of head groups, the distance δ from the core surface where the counterion is located in the case of ionic head groups, and the dipole length d and the distance δ from the core surface at which the dipole is located in the case of a zwitterionic head group. All these molecular constants can be estimated from the chemical structure of the surfactant molecule [66,85]. There are no free parameters and the calculations are completely predictive in nature.

The molecular volume v_0 of the surfactant tail containing n_C carbon atoms is calculated from the group contributions of $(n_C - 1)$ methylene groups and the terminal methyl group; the group molecular volumes are estimated from the density versus temperature data (T in K) available for aliphatic hydrocarbons [103].

$$v_0 = v(CH_3) + (n_C - 1)v(CH_2)$$

$$v(CH_3) = 54.6 + 0.124\,(T - 298)\,\text{Å}^3)$$ (1.69)

$$v(CH_2) = 26.9 + 0.0146\,(T - 298)\,\text{Å}^3)$$

For double-tailed surfactants, v_0 is calculated by accounting for both the tails. For the fluorocarbon tails, extensive volumetric data are not yet available to estimate the temperature dependence of the molecular volumes of CF_3 and CF_2 groups. Using the data available for 25°C, we estimate [66] that $v(CF_3) = 1.67\ v(CH_3)$ and $v(CF_2) = 1.44\ v(CH_2)$. The ratios between the volumes of the fluorocarbon and the hydrocarbon groups are assumed to be the same at all temperatures.

The extended length of the surfactant tail ℓ_0 at 298 K is calculated using a group contribution of 0.13 nm for the methylene group and 0.28 nm for the methyl group [28–30]. In the absence of any information, and given the small volumetric expansion of the surfactant tail over the range of temperatures of interest, the extended tail length ℓ_0 is considered as temperature independent. The small volumetric expansion of the surfactant tail is accounted for by small increases in the cross-sectional area of the surfactant tail. The extended length of the fluorocarbon chain is estimated using the same group contributions as for hydrocarbon tails, namely, 0.1265 nm for the CF_2 group and 0.2765 nm for the CF_3 group.

The head group area a_p is calculated as the cross-sectional area of the head group near the hydrophobic core–water interface as described here for some example head group structures. The glucoside head group in β-glucosides has a compact ring structure [53] with an approximate diameter of 0.7 nm, and hence, the effective cross-sectional area of the polar head group a_p is estimated as 0.40 nm². For sodium alkyl sulfates, the cross-sectional area of the polar group a_p has been estimated to be 0.17 nm². For the zwitterionic N-betaine head group, a_p has been estimated to be 0.30 nm². The area per molecule a_0 of the micellar core, which is shielded by the head group from having contact with water, is the smaller of a_p or L^2, as discussed previously.

For ionic surfactants, the molecular constant δ depends on the size of the ionic head group, the size of the hydrated counterion, and the proximity of the counterion to the charge on the surfactant ion. Visualizing that the sodium counterion is placed on top of the sulfate anion, we estimate $\delta = 0.545$ nm for sodium alkyl sulfates, and 0.385 nm for sodium alkyl sulfonates. For alkyl pyridinium bromide, the surfactant cation is very near the hydrophobic core, and we estimate $\delta = 0.22$ nm. For zwitterionic surfactants, we need the molecular constant d, which is the distance of separation of the charges on the dipole (or the dipole length), and also the constant δ, which is the distance from the hydrophobic core surface at which the dipole is located. From the head group molecular structure, we estimate $d = 0.5$ nm and $\delta = 0.07$ nm for N-alkyl betaines, and $d = 0.62$ nm and $\delta = 0.65$ nm for the lecithin head group.

1.11.10　INFLUENCE OF FREE ENERGY CONTRIBUTIONS ON AGGREGATION BEHAVIOR

For illustrative purposes, Figure 1.12 presents the calculated free energy contributions $\left(\Delta\mu_g^\circ / kT\right)$, expressed per molecule of surfactant for cetyl pyridinium bromide in water as a function of g.

Of all the contributions, only the transfer free energy of the surfactant tail is negative, and is responsible for the aggregated state of the surfactant being favored over the singly dispersed state.

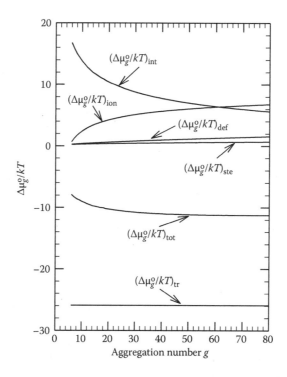

FIGURE 1.12 Contributions to the standard free energy difference between a surfactant molecule in the micelle and a surfactant monomer in water, calculated as a function of the aggregation number g of the micelle for cetyl pyridinium bromide in water at 25°C. Subscripts refer to the following: tot (total), tr (transfer free energy of tails), def (deformation free energy of tails), int (interfacial free energy), ste (head group steric interactions), and ion (head group ionic interactions). Refer to text for detailed discussion.

The transfer free energy contribution is independent of the micellar size and, hence, has no influence on the structural characteristics of the equilibrium aggregate. All the remaining free energy contributions to $\left(\Delta\mu_g^\circ/kT\right)$ are positive and depend on the aggregate size. It is clear from the geometrical relations for aggregates (see Table 1.1) that as the aggregation number g increases, the area per molecule a decreases. Consequently, the free energy of formation of the aggregate core–water interface decreases with increasing aggregation number. This free energy is thus responsible for the positive cooperativity, which favors the growth of aggregates to large sizes. All remaining free energy contributions (namely, the surfactant tail deformation energy, the steric repulsions between the head groups, the dipole–dipole interactions between zwitterionic head groups, and the ionic interactions between ionic head groups) increase with increasing aggregation number. These free energy contributions are, therefore, responsible for the negative cooperativity that limits the aggregates to finite sizes. All the free energy contributions, however, affect the magnitude of the cmc.

The calculated size distributions for cetylpyridinium bromide are presented in Figure 1.13 for two values of the molar concentration C_1 (= 55.55 X_1) of the singly dispersed surfactant. As expected, a small variation in C_1 gives rise to a large variation in the total aggregate concentration, C_{tot}. The average aggregation number is, however, practically the same at these two concentrations. This implies (see Equation 1.41) that the size dispersion of the aggregates is narrow, which can also be seen in Figure 1.13.

We have performed extensive predictive calculations of the cmc and aggregate size as a function of the tail length and head group type for anionic, cationic, nonionic, and zwitterionic surfactants. The parameter K representing the transition from spherical to rodlike micelles as a function of temperature and salt concentration, the dependence of cmc and aggregate size on the concentration

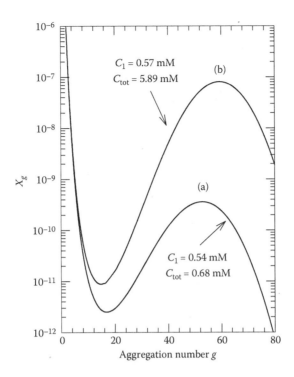

FIGURE 1.13 Calculated size distribution of cetyl pyridinium bromide aggregates in water at 25°C at two values of total surfactant concentration C_{tot} and singly dispersed surfactant concentration C_1. (a) $C_1 = 0.54$ mM; $C_{tot} = 0.68$ mM; (b) $C_1 = 0.57$ mM; $C_{tot} = 5.89$ mM.

of the added electrolyte, the dependence of cmc on temperature, and the formation of vesicles from double chain surfactants. These predicted results have been compared against available experimental data and show that reasonable predictions are possible using the detailed free energy model described above. The detailed computational results and comparison against experimental data can be found in refs. [66,84–86] and are not reproduced here.

1.11.11 MODIFIED PACKING PARAMETER MODEL ACCOUNTING FOR CHAIN LENGTH EFFECTS

As mentioned previously, the packing parameter model of Israelachvili et al. [31] used Tanford's free energy expression and consequently concluded that the surfactant head group is the only key variable determining the aggregation properties. However, the fee energy model should include the tail deformation free energy, which has an explicit dependency on the chain length of the surfactant tail. The expression developed for the tail deformation free energy takes into account the fact that the surfactant tail has to deform nonuniformly along its length to fill the aggregate core with uniform density. The packing free energy contribution (Equation 1.59) has the form

$$\left(\frac{\Delta \mu_g^o}{kT} \right)_{Def} = \left(\frac{3\pi^2}{80} \right) \left(\frac{R^2}{NL^2} \right), \ \left(\frac{5\pi^2}{80} \right) \left(\frac{R^2}{NL^2} \right), \ \left(\frac{10\pi^2}{80} \right) \left(\frac{R^2}{NL^2} \right) \tag{1.70}$$

for spheres, cylinders, and bilayers. In the above equation, L is a characteristic segment length that is taken to be 4.6 Å, as mentioned earlier, and N is the number of segments in a tail such that $NL^3 = v_o$. Because $R = 3\, v_o/a$, $2\, v_o/a$, and v_o/a for the three geometries (Table 1.1), the packing free energy contribution can be rewritten as

$$\left(\frac{\Delta\mu_g^o}{kT}\right)_{Def} = \frac{Q}{a^2}, \quad Q_{sph} = \left(\frac{27}{8}\right)v_oL, \quad Q_{cyl} = \left(\frac{20}{8}\right)v_oL, \quad Q_{bilayer} = \left(\frac{10}{8}\right)v_oL \qquad (1.71)$$

where the symbol Q is used to denote the coefficient of $1/a^2$ in the free energy expression and stands for Q_{sph}, Q_{cyl}, or $Q_{bilayer}$ depending on the aggregate shape. If we add this contribution to the Tanford model (Equation 1.54), the equilibrium area a_e given previously by Equation 1.55 is now obtained from

$$\left(\frac{\Delta\mu_g^o}{kT}\right) = \left(\frac{\Delta\mu_g^o}{kT}\right)_{Tail} + \left(\frac{\sigma}{kT}\right)a + \left(\frac{\alpha}{kT}\right)\frac{1}{a} + \frac{Q}{a^2}$$

$$\frac{\partial}{\partial a}\left(\frac{\Delta\mu_g^o}{kT}\right) = 0 \Rightarrow \left(\frac{\sigma}{kT}\right) - \left(\frac{\alpha}{kT}\right)\frac{1}{a^2} - \frac{2Q}{a^3} = 0 \quad \text{at} \quad a = a_e = \left(\frac{\alpha + 2QkT/a_e}{\sigma}\right)^{1/2}$$

$$(1.72)$$

Because the variable Q is dependent on the tail, the tail has direct influence over the equilibrium area a_e, when compared with Equation 1.55 based on the free energy model of Tanford.

To illustrate the influence of the tail for model predictions using the modified packing parameter model, we have chosen three different values for the head group interaction parameter (α/kT) keeping $(\sigma/kT) = 0.12$ Å^{-2}, consistent with σ being 50 mN/m. Q depends on the tail length and the shape of the aggregate as shown in Equation 1.71. The calculated results for three tail lengths of single tail surfactants are provided in Table 1.2. Shown in the table are the equilibrium area a_e and the molecular packing parameter $v_o/a_e\ell_o$ calculated based on Equation 1.55, ignoring the tail deformation free energy contribution and those based on Equation 1.72 accounting for this contribution. When using Equation 1.72 to calculate a_e, the expression for Q_{sph}, Q_{cyl}, or $Q_{bilayer}$ are introduced while calculating the areas corresponding to the sphere, cylinder, or the bilayer, respectively. In each case, a numerical solution is obtained and all valid solutions corresponding to the three shapes are listed in Table 1.2. A solution is valid only if the calculated area lies within the range allowed for the aggregate shape.

TABLE 1.2

Molecular Packing Model Accounting for Surfactant Tail Packing

	a_e (Å^2)	$v_o/a_e\ell_o$	a_e (Å^2)	$v_o/a_e\ell_o$	a_e (Å^2)	$v_o/a_e\ell_o$	a_e (Å^2)	$v_o/a_e\ell_o$
			Model Accounting for Surfactant Tail Packing					
n_C	Only Head Group Controls (Equation 1.55)		Spheres (Equation 1.72)		Cylinders (Equation 1.72)		Bilayers (Equation 1.72)	
For $\sigma/kT = 0.12$ Å^{-2}, $\alpha/kT = 120$ Å^2								
8	31.6	0.664			45.1	0.466	39.9	0.527
12	31.6	0.664			48.8	0.43		
16	31.6	0.664			51.9	0.404		
For $\sigma/kT = 0.12$ Å^{-2}, $\alpha/kT = 240$ Å^2								
8	44.7	0.47			53.6	0.392		
12	44.7	0.47			56.5	0.372		
16	44.7	0.47	62.45	0.336	59.1	0.356		
For $\sigma/kT = 0.12$ Å^{-2}, $\alpha/kT = 300$ Å^2								
8	50	0.42			57.5	0.365		
12	50	0.42	62.9	0.334	60.1	0.349		
16	50	0.42	65.7	0.320	62.5	0.336		

For example, if the area per molecule calculated assuming a spherical shape turns out to be 48.1 Å^2, such an area is not realizable with a spherical structure because the smallest area per molecule attainable for a sphere is 63 Å^2. Hence, this solution is not valid.

One can now compare the predictions of a_e and $v_0/a_e\ell_0$ obtained by ignoring the packing free energy and by accounting for the packing free energy. The inclusion of the packing free energy (Equation 1.72) results in the equilibrium area a_e being larger than that estimated from Equation 1.55, neglecting this contribution. For $\alpha/kT = 120$ Å^2, $a_e = 31.6$ Å^2 from Equation 1.55, hence $v_0/a_e\ell_0 = 0.664$, and correspondingly, bilayer structures are predicted, independent of the tail length. However, when Equation 1.72 is applied, one obtains $n_C = 8$, $a_e = 45.1$ Å^2 and $v_0/a_e\ell_0 = 0.466$, implying a cylindrical aggregate. In this case, there is also another solution $a_e = 39.9$ Å^2 and $v_0/a_e\ell_0 = 0.527$, implying a bilayer aggregate. For $n_C = 12$ or 16, only one valid solution exists corresponding to the formation of cylindrical aggregates. Similarly, for $\alpha/kT = 240$ Å^2, Equation 1.55 predicts $v_0/a_e\ell_0 = 0.47$ and correspondingly, large rodlike micelles for all three tail lengths. However, Equation 1.72 predicts $v_0/a_e\ell_0$ in the range of 0.33 to 0.39 depending on the tail length, indicating the formation of only smaller cylinders for $n_C = 8$ or 12 and near-spherical aggregates for $n_C = 16$. For $\alpha/kT = 300$ Å^2, when Equation 1.55 is used, we obtain $v_0/a_e\ell_0 = 0.42$, consistent with cylindrical micelles for all three tail lengths. However, Equation 1.72, for the same conditions, predicts $v_0/a_e\ell_0$ in the range 0.32 to 0.365 depending on the tail length, suggesting only spherical or small globular aggregates. In all cases, the predictions are significantly modified by the incorporation of the tail packing free energy contribution. The incorporation of the packing free energy leads to a more direct influence exerted by the surfactant tail on the equilibrium area per molecule a_e, hence on the packing parameter and the size and shape of the equilibrium aggregate.

1.12   EXTENDING THE THEORY OF MICELLIZATION

The predictive theories developed in our work and by the Blankschtein group have demonstrated that reasonably accurate a priori predictions of a variety of aggregation properties are now possible for different classes of surfactants. The theory of the formation of micelles developed with predictive capabilities has been extended to a wide range of self-assembly phenomena exhibited by surfactants. Keeping the general approach of the theory intact and by adding or modifying free energy contributions to the self-assembly process appropriately, the extended theoretical models have been developed for quantitative treatments of mixed micelles exhibiting ideal and a variety of nonideal mixing behavior [83]; the formation of vesicles from surfactants and surfactant mixtures [66,85]; the generation of coexisting partially miscible hydrocarbon and fluorocarbon micelles [83]; the formation of giant rodlike micelles from surfactants, surfactant mixtures, and surfactant-alcohol mixtures [86]; the aggregation behavior of gemini surfactants, which are equivalent to two single chain surfactants covalently connected to one another by a hydrophobic spacer chain near the head groups [104]; the solubilization of hydrocarbons in micelles [66]; the selective solubilization of hydrocarbons from their mixtures in surfactant micelles [66]; the aggregation of surfactant in polar solvents [105]; the micellization of surfactant in aqueous–organic binary mixed solvents [106]; the formation of surfactant aggregates at hydrophobic solid surfaces [107] and hydrophilic solid surfaces [108]; and the formation of droplet and bicontinuous type microemulsions [109].

For classic surfactants, it is possible to estimate the structural parameters that appear as molecular constants in the free energy model relatively easily as we have described previously. But for surfactants with either or both of the tail and head group having complex structures, these molecular constants are not easy to determine and the hydrophobic contributions cannot be estimated from the relations we have presented previously. To address such situations arising for complex hydrophobic and hydrophilic groups, in the last decade, Blankschtein has identified a clever way to combine computer simulation techniques with analytical theory to arrive at necessary structural constants and free energy estimates [110–115].

The time-consuming computer simulations are first used to get at a few essential molecular features not easily obtainable on phenomenological grounds and, with this information as input, analytical theory is used to arrive at quantitative predictions with minimal computational efforts. For example, in amphiphilic systems such as sterols or their derivatives, it is not easy to establish quantitative properties to describe the polar head and the nonpolar tail unambiguously. In some systems, the solubilizate molecule taken up by the micelle may not be a simple hydrophobe. In such cases, by conducting simulations of the amphiphile or the solubilizate at an oil–water interface, Blankschtein was able determine quantitative hydrophobicity and hydrophilicity values for the molecules that could be used as input to the analytical theory to make predictions of the self-assembly behavior. In another example, Blankschtein used computer simulations to get at the hydration of hydrophobic groups in water and in aggregates as a way to arrive at a better quantitative estimate of the hydrophobic interactions, which is then input to the analytical model.

Blankschtein has used this approach in yet another interesting example dealing with mixed micelle formation. Starting from a pure component micelle established by the analytic free energy model, he conducted computer simulations to quantify the free energy changes that arise when one surfactant molecule is replaced by another. Using this information as input, he is then able to use the analytic free energy model to make accurate predictions of mixed surfactant systems. This creative combination of computer simulations with the analytical free energy model is an important new addition to the literature and can prove to be very useful in solving self-assembly problems of complex molecules.

1.13 PROSPECTS

From the review of theoretical modeling described above, one can persuasively argue that quantitative understanding of surfactant self-assembly for classic surfactants with hydrocarbon tails, assembling into spherical and cylindrical micelles, is well established. Recent research has attempted to utilize the available free energy models and molecular packing considerations to address other emerging theoretical problems.

Many experimental studies have shown that ionic surfactants can form large wormlike micelles, which transform into branched micelles and interconnected network structures, exhibiting very interesting rheological properties. Andreev and Victorov [116] have developed a theory for such systems by constructing an analytical molecular-thermodynamic model for the free energy of branching portions of wormlike ionic micelles. In this model, the junction of three cylindrical aggregates is represented by a combination of portions of a torus and a bilayer and the free energy of formation of such a shape is written by following exactly the approach described earlier for spherical, cylindrical, and lamellar aggregates. The model correctly predicts the sequence of stable aggregate morphologies, including a narrow bicontinuous zone, as functions of hydrocarbon tail length, head group size, and solution ionic strength. It has been found that torus-like micelles form and often coexist with branched micellar systems. A free energy model using macroscopic bending elasticity constants has been developed by Bergström [117] to analyze the formation and stability of such micellar aggregates. The theory identifies the conditions for which the torus-like micelles are stable over bilayers. The theory also revealed that torus-like micelles, in general, are favored at the expense of long spherocylindrical micelles as a result of elimination of the unfavorable endcaps. Modeling of these systems using molecular thermodynamic free energy expressions and quantitative comparison of predictions against experimental data are interesting future areas of study.

Theorists are also challenged by novel developments in amphiphiles, especially those that combine biological or inorganic elements into classic surfactant or lipid structures. On the one hand, these novel amphiphiles are developed utilizing our current knowledge of surfactant self-assembly, best represented by the intuitive molecular packing model. At the same time, they also display novel aggregation phenomena that require new free energy models that remain to be developed. One example is the DNA-programmed lipid consisting of two alkyl hydrophobic tails, linked covalently to the 5′-termini of a single-stranded DNA (ssDNA) oligonucleotide that functions as hydrophilic head group synthesized in the Gianneschi laboratory [118]. Effectively, the DNA hybridization generates "new" surfactants and by manipulating the size, shape, and charge of the polar head group of the surfactant molecule via DNA hybridization and displacement cycles, one could achieve the desired aggregation patterns. The approach offers us a programmable, reversible trigger for nanostructure morphology with which informational, encoded nanostructures are made possible and the surfactant is selectively responsive, in a logical manner, to multiple input signals. Another example is the synthesis of anionic surfactants containing a purely inorganic multinuclear head group of the polytungstate type R-$[PW_{11}O_{39}]_3$ by Polarz and coworkers [119]. The self-assembly behavior of the polyoxometalate surfactants into micelles and lyotropic phases has been studied. A surfactant of this class can play the role of a surface-active agent, such as for emulsification as well as a catalyst for a chemical reaction, simultaneously.

A wide range of surfactants designed using peptides have been synthesized and their interesting aggregation properties have been examined. In a recent study by Gianneschi and coworkers [120], micelles were prepared from amphiphiles with a hydrophobic small polymer tail and multiple hydrophilic peptides as the head group. The peptides were selected to be substrates for cancer-associated proteins and were amenable to proteolysis. Because a specific peptide in the head group can be chemically modified by proteolysis using a specific enzyme, the head group properties of the amphiphile are altered and an enzyme-responsive switching of the morphology of the micelles is observed. Such multienzyme responsive surfactant systems can lead to novel materials capable of signaling specific patterns of multiple biochemical stimuli. The ability to program the nature of micelle responses to disease-associated enzymes, through peptide design, has implications for *in vivo* delivery and detection methods. Tirrell and coworkers [121] have investigated peptide amphiphiles that remain soluble under physiological conditions for their ability to self-assembly into micellar aggregates. Various classes of peptide amphiphiles were considered. The monomer design, secondary structure of the peptide, and the resulting micelle size and shape are related qualitatively to the competing forces between the hydrophobic tail and the peptide head group. Tirrell and coworkers [122] have also investigated wormlike micelles formed from peptide amphiphiles. The wormlike micelles resembling nanofibers are viewed as potential synthetic extracellular matrix materials for tissue engineering and regenerative medicine. They interpreted the neutron scattering and atomic force microscopy data as demonstrating the existence of transient spherical micelles in the early stage and subsequent micelle chain elongation by attachment of spherical micelles to the end of growing cylindrical micelles to form wormlike micelles in a process analogous to chain growth polymerization. One may note that such a mechanism for the formation of cylindrical micelles by attachment of spherical micelles was postulated at the early stages of micellization history [17] when cylindrical micelles were first observed. However, later studies of surfactant systems had shown that self-assembly into cylindrical aggregates can directly occur without requiring the preformation of spherical micelles. It will be of interest to compare and contrast the different mechanisms proposed for the peptide amphiphiles versus classic amphiphiles.

A general approach to dynamically constructing and destroying amphiphiles has recently been proposed by Montenegro et al. [123]. These dynamic amphiphiles have a charged head, a hydrophobic tail, and a dynamic connector or "bridge." The dynamic covalent bonds of bridges, formed with, for example, hydrazones, disulfides, or oximes, are weaker than common covalent bonds but stronger than noncovalent interactions, such as hydrogen bonds. These bridges can be made and destroyed easily. Because of the dynamic nature of their bridges, dynamic amphiphiles can be formed, modified, and destroyed *in situ*, depending on environmental conditions. Current efforts in extending such dynamic amphiphiles are focused on incorporating more unusual tails such as fullerenes, mesogenes, or fluorophiles. This library of amphiphiles is being screened for gene and siRNA delivery applications.

Existing theoretical models of micellization need to be extended to account for specific interactional terms connected with novel head groups such as peptides, oligonucleotides, polyoxymetalates, etc., and also to account for novel hydrophobic tails such as hydrophobic peptides, fullerenes, etc., to quantitatively describe the wide pattern of aggregation observed. A free energy model, similar to that for surfactants, has recently been developed by Semenov and Subbotin [124], for peptide amphiphiles made up entirely of amino acids–hydrophobic alanine (as tail) and hydrophilic aspartic acid (as head group). Such a theoretical approach to free energy calculations remains to be extended to the rich variety of peptide amphiphile structures as well as other classes of amphiphiles mentioned above.

The novel design of amphiphilic systems incorporating memory, responsiveness, and programmability and the ability to combine organic, inorganic, and biological moieties within a single amphiphile structure, makes this area a very interesting and fertile field of research for years to come. Most importantly, the novel amphiphiles have opened entirely new areas of applications in materials science and nanomedicine, far beyond the classic colloidal applications of conventional amphiphiles.

ACKNOWLEDGMENTS

Professor Ruckenstein has collaborated in many parts of the work discussed here as can be inferred from the cited references, and the author has benefited from numerous discussions with him. The predictive theory described in the paper was developed during the author's tenure at The Pennsylvania State University. I thank Natick Soldier Research, Development and Engineering Center for support during the preparation of this manuscript.

REFERENCES

1. J.W. McBain. General discussion on colloids and their viscosity. *Trans. Faraday Soc.*, **9**, 99–101 (1913).
2. J.W. McBain, E.C.V. Cornish and R.C. Bowden. Studies of the constitution of soap in solution: Sodium myristate and sodium laurate. *J. Chem. Soc., Trans.*, **101**, 2042–2056 (1912).
3. J.W. McBain and C.S. Salmon. Colloidal electrolytes: Soap solutions and their constitution. *Proc. R. Soc. Lond. A*, **97**, 44–65 (1920).
4. J.W. McBain. Solubilization and other factors in detergent action. In *Advances in Colloid Science*, E.O. Kramer (Ed.). Interscience Publishers, New York, pp.99-142 (1942).
5. J.W. McBain. *Colloid Science*. D.C. Heath and Company, San Francisco, CA (1950).
6. N.K. Adam. The structure of surface films on water. *J. Phys. Chem.*, **29**, 87–101 (1924).
7. W.D. Harkins, E.C.H. Davies and G.L. Clark. The orientation of molecules in the surfaces of liquids, the energy relations at surfaces, solubility, adsorption, emulsification, molecular association, and the effect of acids and bases on interfacial tension (Surface energy VI). *J. Am. Chem. Soc.*, **39**, 541–596 (1917).
8. W.D. Harkins, R.W. Mattoon and M.L. Corrin. Structure of soap micelles indicated by x-rays and the theory of molecular orientation. I. Aqueous solutions. *J. Am. Chem. Soc.*, **68**, 220–228 (1946).
9. R.W. Mattoon, R.S. Stearns and W.D. Harkins. Structure for soap micelles as indicated by a previously unrecognized x-ray diffraction band. *J. Chem. Phys.*, **15**, 209–210 (1947).
10. W.D. Harkins. A cylindrical model for the smallsoap micelle. *J. Chem. Phys.*, **16**, 156–157 (1948).
11. W.D. Harkins. Soap solutions: Salt, alcohol, micelles, rubber. *Scientific Monthly*, **70**, 220–228 (1950).
12. G.S. Hartley. *Aqueous Solutions of Paraffin Chain Salts*. Hermann et Cie, Paris (1936).
13. G.S. Hartley, B. Collie and C.S. Samis. Transport numbers of paraffin-chainsalts in aqueous solution. Part I.-Measurement of transport numbers of cetylpyridinium and cetyltrimethylammonium bromides and their interpretation in terms of micelle formation, with some data also forcetane sulphonic acid. *Trans. Faraday Soc.*, **32**, 795–815 (1936).
14. G.S. Hartley and D.F. Runnicles. Salt micelles from diffusion measurements: The determination of the size of paraffin-chain. *Proc. R. Soc. Lond. A*, **168**, 420–440 (1938).
15. G.S. Hartley. Organized structure in soap solutions. *Nature*, **163**, 767–768 (1949).
16. W. Philippoff. Micelles and x-rays. *J. Colloid Sci.*, **5**, 169–191 (1950).
17. W. Philippoff. Colloidal and polyelectrolytes. The micelle and swollen micelle. On soap micelles. *Discuss. Faraday Soc.*, **11**, 96–107 (1951).
18. P. Debye. Note on light scattering in soap solutions. *J. Colloid Sci.*, **3**, 407–409 (1948).
19. P. Debye. Light scattering in soap solutions. *J. Phys. Colloid Chem.*, **53**, 1–8 (1949).

20. P. Debye. Light scattering in soap solutions. *Ann. NY Acad. Sci.*, **51**, 575–592 (1949).

21. M.E. Hobbs. The effect of salts on the critical concentration, size and stability of soap micelles. *J. Phys. Colloid Chem.*, **55**, 675–683 (1951).

22. Y. Ooshika. A theory of critical micelle concentration of colloidal electrolyte solutions. *J. Colloid Sci.*, **9**, 254–262 (1954).

23. I. Reich. Factors responsible for the stability of detergent micelles. *J. Phys. Chem.*, **60**, 257–262 (1956).

24. J.D. Halsey, Jr. On the structure of micelles. *J. Phys. Chem.*, **57**, 87–89 (1953).

25. C.A.J. Hoeve and C.G. Benson. On the statistical mechanical theory of micelle formation in detergent solutions. *J. Phys. Chem.*, **61**, 1149–1158 (1957).

26. D.C. Poland and H.A. Scheraga. Hydrophobic bonding and micelle stability. *J. Phys. Chem.*, **69**, 2431–2442 (1965).

27. D.C. Poland and H.A. Scheraga. Hydrophobic bonding and micelle stability: The influence of ionic head groups. *J. Colloid Interface Sci.*, **21**, 273–283 (1966).

28. C. Tanford. *The Hydrophobic Effect*. Wiley, New York, 1973.

29. C. Tanford. Thermodynamics of micelle formation: Prediction of micelle size and size distribution. *Proc. Natl. Acad. Sci. USA*, **71**, 1811–1815 (1974).

30. C. Tanford. Theory of micelle formation in aqueous solutions. *J. Phys. Chem.*, **78**, 2469–2479 (1974).

31. J.N. Israelachvili, J.D. Mitchell and B.W. Ninham. Theory of self-assembly of hydrocarbon amphiphiles into micelles and bilayers. *J. Chem. Soc. Faraday Trans. II*, **72**, 1525–1568 (1976).

32. K.L. Mittal (Ed.). *Micellization, Solubilization and Microemulsions*, Vols. 1 and 2. Plenum Press, New York (1977).

33. G.S. Hartley. Micelles—Retrospect and prospect. In *Micellization, Solubilization and Microemulsions*, Vol. 1, K.L. Mittal (Ed.). Plenum Press, New York, pp. 23–43 (1977).

34. C. Tanford. Thermodynamics of micellization of simple amphiphiles. In *Micellization, Solubilization and Microemulsions*, Vol. 1, K.L. Mittal (Ed.). Plenum Press, New York, pp. 119–131 (1977).

35. R. Nagarajan and E. Ruckenstein. Thermodynamics of amphiphilar aggregation into micelles and vesicles. In *Micellization, Solubilization and Microemulsions*, Vol. 1, K.L. Mittal (Ed.). Plenum Press, New York, pp. 133–149 (1977).

36. P. Mukerjee. Size distribution of micelles: Monomer–micelle equilibrium, treatment of experimental molecular weight data, the sphere-to-rod transition and a general association model. In *Micellization, Solubilization and Microemulsions*, Vol. 1, K.L. Mittal (Ed.). Plenum Press, New York, pp. 171–194 (1977).

37. N.A. Mazer, M.C. Carey and G.B. Benedek. The size, shape and thermodynamics of sodium dodecyl sulfate (SDS) micelles using quasi-elastic light-scattering spectroscopy. In *Micellization, Solubilization and Microemulsions*, Vol. 1, K.L. Mittal (Ed.). Plenum Press, New York, pp. 359–381 (1977).

38. K. Hess, W. Phllippoff and H. Kiessig. Soap solutions. *Kolloid Z.*, **88**, 40–51 (1939).

39. K. Hess and J. Gundermann. Roentgen graphical tests on inactive and pouring colloid solutions (Evidence of the orientation from colloid parts in the flow through capillaries through the occurrence of fiber diagrams, hydration of colloid parts in the solution). *Ber. Dtsch. Chem. Ges*, **70**, 1800–1808 (1937).

40. M.L. Corrin. Interpretation of x-ray scattering from solutions of long chain electrolytes on the basis of a spherical micelle. *J. Chem. Phys.*, **16**, 844–845 (1948).

41. P. Debye and E.W. Anacker. Micelle shape from dissymmetry measurements. *J. Phys. Colloid Chem.*, **55**, 644–655 (1951).

42. E.W. Anacker. Micelle formation of cationic surfactants in aqueous media. In *Cationic Surfactants*, E. Jungermann (Ed.), Chap. 7. Marcel Dekker, New York, pp. 203-288 (1970) (and references therein).

43. H.A. Scheraga and J.K. Backus. Flow birefringence in solutions of n-hexadecyltrimethylammonium bromide. *J. Am. Chem. Soc.*, **73**, 5108–5112 (1951).

44. N. Pilpel. On gel formation in soaps. *J. Colloid Sci.*, **9**, 285–299 (1954).

45. N. Pilpel. Viscoelasticity in aqueous soap solutions. *J. Phys. Chem.*, **60**, 779–782 (1956).

46. G. Nemethy and H.A. Scheraga. Structure of water and hydrophobic bonding in proteins. 1. A model for thermodynamic properties of liquid water. *J. Chem. Phys.*, **36**, 3382–3400 (1962).

47. G. Nemethy and H.A. Scheraga. Structure of water and hydrophobic bonding in proteins. 2. Model for thermodynamic properties of aqueous solutions of hydrocarbons. *J. Chem. Phys.*, **36**, 3401–3417 (1962).

48. G. Nemethy and H.A. Scheraga. Structure of water and hydrophobic bonding in proteins. 3. Thermodynamic properties of hydrophobic bonds in proteins. *J. Phys. Chem.*, **66**, 1773–3417 (1962).

49. R. Nagarajan. Modeling solution entropy in the theory of micellization. *Colloids Surf. A*, **71**, 39–64 (1993).

50. P. Mukerjee. Size distribution of small and large micelles—Multiple equilibrium analysis. *J. Phys. Chem.*, **76**, 565–570 (1972).

51. P. Mukerjee. Micellar properties of drugs—Micellar and nonmicellar patterns of self-association of hydrophobic solutes of different molecular-structures—Monomer fraction, availability, and misuses of micellar hypothesis. *J. Pharm. Sci.*, **63**, 972–981 (1974).

52. P. Mukerjee. Hydrophobic and electrostatic interactions in ionic micelles. Problems in calculating monomer contributions to free energy. *J. Phys. Chem.*, **73**, 2054–2060 (1969).

53. P. Mukerjee. The nature of the association equilibria and hydrophobic bondingin aqueous solutions of association colloids. *Adv. Colloid Interface Sci.*, **1**, 241–275 (1967).

54. R.J.M. Tausk and J.Th.G. Overbeek. Physical chemical studies of short-chain lecithin homologues. IV. A simple model for the influence of salt and the alkyl chainlength on the micellar size. *Biophys. Chem.*, **2**, 175–179 (1974).

55. P.J. Missel, N.A. Mazer, G.B. Benedek, C.Y. Young and M.C. Carey. Thermodynamic analysis of the growth of sodium dodecyl sulfate micelles. *J. Phys. Chem.*, **84**, 1044–1057 (1980).

56. R. Nagarajan. Are large micelles rigid or flexible: A reinterpretation of viscosity data for micellar solutions. *J. Colloid Interface Sci.*, **90**, 477–486 (1982).

57. P.J. Missel, N.A. Mazer, G.B. Benedek, C.Y. Young and M.C. Carey. Influence of chain-length on the sphere-to-rod transition in alkyl sulfate micelles. *J. Phys. Chem.*, **87**, 1264–1277 (1983).

58. P.J. Missel, N.A. Mazer, M.C. Carey and G.B. Benedek. Influence of alkali-metal counterion identity on the sphere-to-rod transition in alkyl sulfate micelles. *J. Phys. Chem.*, **93**, 8354–8366 (1989).

59. E. Ruckenstein and R. Nagarajan. Critical micelle concentration—Transition point for micellar size distribution. *J. Phys. Chem.*, **79**, 2622–2626 (1975).

60. R. Nagarajan and E. Ruckenstein. Relation between the transition point in micellar size distribution, the cmc, and the cooperativity of micellization. *J. Colloid Interface Sci.*, **91**, 500–506 (1983).

61. H.V. Tartar. A theory of the structure of the micelles of normal paraffin chain salts in aqueous solution. *J. Phys. Chem.*, **59**, 1195–1199 (1955).

62. R. Nagarajan. Self-assembly of bola amphiphiles. *Chem. Eng. Comm.*, **55**, 251–273 (1987).

63. J.N. Israelachvili, D.J. Mitchell and B.W. Ninham. Theory of self-assembly of lipid bilayers and vesicles. *Biochim. Biophys. Acta*, **470**, 185–201 (1977).

64. J.N. Israelachvili, S. Marcelja and R.G. Horn. Physical principles of membrane organization. *Q. Rev. Biophys.*, **13**, 121–200 (1980).

65. J.N. Israelachvili. *Intermolecular and Surface Forces*, 1st ed. Academic Press, London (1985).

66. R. Nagarajan and E. Ruckenstein. Theory of surfactant self-assembly: A predictive molecular thermodynamic approach. *Langmuir*, **7**, 2934–2969 (1991).

67. R. Nagarajan. Molecular packing parameter and surfactant self-assembly: The neglected role of the surfactant tail. *Langmuir*, **18**, 31–38 (2002).

68. S. Puvvada and D. Blankschtein. Molecular-thermodynamic approach to predict micellization, phase-behavior and phase-separation of micellar solutions. 1. Application to nonionic surfactants. *J. Chem. Phys.*, **92**, 3710–3724 (1990).

69. S. Puvvada and D. Blankschtein. Thermodynamic description of micellization, phase-behavior, and phase-separation of aqueous-solutions of surfactant mixtures. *J. Phys. Chem.*, **96**, 5567–5579 (1992).

70. A. Naor, S. Puvvada and D. Blankschtein. An analytical expression for the free-energy of micellization. *J. Phys. Chem.*, **96**, 7830–7832 (1992).

71. P.K. Yuet and D. Blankschtein. Molecular-thermodynamic modeling of mixed cationic/anionic vesicles. *Langmuir*, **12**, 3802–3818 (1996).

72. N. Zoeller, L. Lue and D. Blankschtein. Statistical-thermodynamic framework to model nonionic micellar solutions. *Langmuir*, **13**, 5258–5275 (1997).

73. A. Shiloach and D. Blankschtein. Predicting micellar solution properties of binary surfactant mixtures. *Langmuir*, **14**, 1618–1636 (1998).

74. V. Srinivasan and D. Blankschtein. Effect of counterion binding on micellar solution behavior: 1. Molecular-thermodynamic theory of micellization of ionic surfactants. *Langmuir*, **19**, 9932–9945 (2003).

75. V. Srinivasan and D. Blankschtein. Effect of counterion binding on micellar solution behavior: 2. Prediction of micellar solution properties of ionic surfactant-electrolyte systems. *Langmuir*, **19**, 9946–9961 (2003).

76. V. Srinivasan and D. Blankschtein. Prediction of conformational characteristics and micellar solution properties of fluorocarbon surfactants. *Langmuir*, **21**, 1647–1660 (2005).

77. A.N. Semenov. Contribution to the theory of microphase layering in block-copolymer melts. *Soviet Phys. JETP*, **61**, 733–742 (1985).

78. R. Nagarajan and E. Ruckenstein. Critical micelle concentration: A transition point for micellar size distribution. A statistical thermodynamical approach. *J. Colloid Interface Sci.*, **60**, 221–231 (1977).

79. R. Nagarajan and E. Ruckenstein. Aggregation of amphiphiles as micelles or vesicles in aqueous media. *J. Colloid Interface Sci.*, **71**, 580–604 (1979).

80. R. Nagarajan. Molecular theory for mixed micelles. *Langmuir*, **1**, 331–341 (1985).

81. R. Nagarajan. Micellization, mixed micellization and solubilization—The role of interfacial interactions. *Adv. Colloid Interface Sci.*, **26**, 205–264 (1986).

82. D.F. Evans and B.W. Ninham. Ion binding and the hydrophobic effect. *J. Phys. Chem.*, **87**, 5025–5032 (1983).

83. R. Nagarajan. Micellization of binary surfactant mixtures: Theory. In *Mixed Surfactant Systems*, P.M. Holland and D.N. Rubingh (Eds.), Chap. 4, ACS Symposium Series 501. American Chemical Society, Washington, DC, pp. 54–95 (1992).

84. R. Nagarajan and E. Ruckenstein. Self-assembled systems. In *Equations of State for Fluids and Fluid Mixtures*, J.V. Sengers, R.F. Kayser, C.J. Peters and H.J. White Jr. (Eds.), Chap. 15. Elsevier Science, Amsterdam, pp. 589–749 (2000).

85. R. Nagarajan. Theory of micelle formation. Quantitative approach to predicting the micellar properties from surfactant molecular structure. In *Structure-Performance Relationships in Surfactants*, K. Esumi and M. Ueno (Eds.), Chap. 1. Marcel Dekker, New York, pp. 1–110 (2003).

86. R. Nagarajan. Molecular thermodynamics of giant micelles. In *Giant Micelles. Properties and Applications*, R. Zana and E. Kaler (Eds.), Chap. 1 Taylor and Francis, New York, pp. 1–40 (2007).

87. M.H. Abraham. Thermodynamics of solution of homologous series of solutes in water. *J. Chem. Soc. Faraday Trans. I*, **80**, 153–181 (1984).

88. M.H. Abraham and E. Matteoli. The temperature-variation of the hydrophobic effect. *J. Chem. Soc. Faraday Trans. I*, **84**, 1985–2000 (1988).

89. D.W.R. Gruen. The standard picture of ionic micelles. *Progr. Colloid Polym. Sci.*, **70**, 6–16 (1985).

90. K.A. Dill and P.J. Flory. Molecular organization in micelles and vesicles. *Proc. Natl. Acad. Sci. USA*, **78**, 676–680 (1981).

91. D.W.R. Gruen. Statistical mechanical model of the lipid bilayer above its phase-transition. *Biochim. Biophys. Acta*, **595**, 161–183 (1980).

92. D.W.R. Gruen. The packing of amphiphile chains in a small spherical micelle. *J. Colloid Interface Sci.*, **84**, 281–283 (1981).

93. D.W.R. Gruen. A model for the chains in amphiphilic aggregates. 1. Comparison with a molecular-dynamics simulation of a bilayer. *J. Phys. Chem.*, **89**, 146–153 (1985).

94. K.A. Dill and P.J. Flory. Interphases of chain molecules—Monolayers and lipid bilayer-membranes. *Proc. Natl. Acad. Sci. USA*, **77**, 3115–3119 (1980).

95. K.A. Dill and R.S. Cantor. Statistical thermodynamics of short-chain molecule interphases. 1. Theory. *Macromolecules*, **17**, 380–384 (1984).

96. K.A. Dill, D.E. Koppel, R.S. Cantor, J.D. Dill, D. Bendedouch and S.H. Chen. Molecular conformations in surfactant micelles. *Nature*, **309**, 42–45 (1984).

97. A. Ben-Shaul, I. Szleifer and W.M. Gelbart. Statistical thermodynamics of amphiphile chains in micelles. *Proc. Natl. Acad. Sci. USA*, **81**, 4601–4605 (1984).

98. A. Ben-Shaul, I. Szleifer and W.M. Gelbart. Chain organization and thermodynamics in micelles and bilayers. 1. Theory. *J. Chem. Phys.*, **83**, 3597–3611 (1985).

99. A. Ben-Shaul and W.M. Gelbart. Theory of chain packing in amphiphilic aggregates. *Ann. Rev. Phys. Chem.*, **36**, 179–211 (1985).

100. L.A. Girifalco and R.J. Good. A theory for the estimation of surface and interfacial energies. 1. Derivation and application to interfacial tension. *J. Phys. Chem.*, **61**, 904–909 (1957).

101. R.C. Weast. *CRC Handbook of Chemistry and Physics*, 60th ed. CRC Press, Boca Raton, FL (1980).

102. P.J. Flory. *Principles of Polymer Chemistry*. Cornell University Press, Ithaca, NY (1962).

103. T.E. Daubert and R.P. Danner. *Physical and Thermodynamic Properties of Pure Chemicals*. Design Institute for Physical Property Data, American Institute of Chemical Engineers, Hemisphere Publishing Corporation, New York (1988).

104. T.A. Camesano and R. Nagarajan. Micelle formation and CMC of gemini surfactants: A thermodynamic model. *Colloids Surf. A*, **167**, 165–177 (2000).

105. R. Nagarajan and C.C. Wang. Solution behavior of surfactants in ethylene glycol: Probing the existence of a CMC and of micellar aggregates. *J. Colloid Interface Sci.*, **178**, 471–482 (1996).

106. R. Nagarajan and C.C. Wang. Theory of surfactant aggregation in water/ethylene glycol mixed solvents. *Langmuir*, **16**, 5242–5251 (2000).

107. R.A. Johnson and R. Nagarajan. Modeling self-assembly of surfactants at solid–liquid interfaces. I. Hydrophobic surfaces. *Colloids Surf. A*, **167**, 31–46 (2000).

108. R.A. Johnson and R. Nagarajan. Modeling self-assembly of surfactants at solid–liquid interfaces. II. Hydrophilic surfaces. *Colloids Surf. A*, **167**, 21–30 (2000).

109. R. Nagarajan and E. Ruckenstein. Molecular theory of microemulsions. *Langmuir*, **16**, 6400–6415 (2000).

110. B.C. Stephenson, K.J. Beers and D. Blankschtein. Complementary use of simulations and molecular— Thermodynamic theory to model micellization. *Langmuir*, **22**, 1500–1513 (2006).

111. B.C. Stephenson, A. Goldsipe, K.J. Beers and D. Blankschtein. Quantifying the hydrophobic effect: 1. A computer simulation/molecular-thermodynamic model for the self-assembly of hydrophobic and amphiphilic solutes in aqueous solution. *J. Phys. Chem. B*, **111**, 1025–1044 (2007).

112. B.C. Stephenson, A. Goldsipe, K.J. Beers and D. Blankschtein. Quantifying the hydrophobic effect: 2. A computer simulation/molecular-thermodynamic model for the micellization of nonionic surfactants in aqueous solution. *J. Phys. Chem. B*, **111**, 1045–1062 (2007).

113. B.C. Stephenson, K.J. Beers and D. Blankschtein. Quantifying the hydrophobic effect: 3. A computer simulation/molecular-thermodynamic model for the micellization of ionic and zwitterionic surfactants in aqueous solution. *J. Phys. Chem. B*, **111**, 1063–1075 (2007).

114. B.C. Stephenson, K.A. Stafford, K.J. Beers and D. Blankschtein. Application of computer simulation free-energy methods to compute the free energy of micellization as a function of micelle composition. 1. Theory. *J. Phys. Chem. B*, **112**, 1634–1640 (2008).

115. B.C. Stephenson, K.A. Stafford, K.J. Beers and D. Blankschtein. Application of computer simulation free-energy methods to compute the free energy of micellization as a function of micelle composition. 2. Implementation. *J. Phys. Chem. B*, **112**, 1641–1656 (2008).

116. V.A. Andreev and A.I. Victorov. Molecular thermodynamics for micellar branching in solutions of ionic surfactants. *Langmuir*, **22**, 8298–8310 (2006).

117. L.M. Bergström. Thermodynamics and bending energetics of torus-like micelles. *J. Colloid Interface Sci.*, **327**, 191–197 (2008).

118. M.P. Thompson, M.-P. Chien, T.-H. Ku, A.M. Rush and N.C. Gianneschi. Smart lipids for programmable nanomaterials. *Nano Lett.*, **10**, 2690–2693 (2010).

119. S. Landsmann, C. Lizandara-Pueyo and S. Polarz. A new class of surfactants with multinuclear, inorganic head groups. *J. Am. Chem. Soc.*, **132**, 5315–5321 (2010).

120. T.-H. Ku, M.-P. Chien, M.P. Thompson, R.S. Sinkovits, N.H. Olson, T.S. Baker and N.C. Gianneschi. Controlling and switching the morphology of micellar nanoparticles with enzymes. *J. Am. Chem. Soc.*, **133**, 8392–8395 (2011).

121. A. Trent, R. Marullo, B. Lin, M. Black and M. Tirrell. Structural properties of soluble peptide amphiphile micelles. *Soft Matter*, **7**, 9572–9582 (2011).

122. T. Shimada, N. Sakamoto, R. Motokawa, S. Koizumi and M. Tirrell. Self-assembly process of peptide amphiphile worm-like micelles. *J. Phys. Chem. B*, **116**, 240–243 (2012).

123. J. Montenegro, E.-K. Bang, N. Sakai and S. Matile. Synthesis of an enlarged library of dynamic DNA activators with oxime, disulfide and hydrazone bridges. *Chem. Eur. J.*, **18**, 10436–10443 (2012).

124. A.N. Semenov and A.V. Subbotin. Theory of self-assembling structures of model oligopeptides. *Macromolecules*, **43**, 3487–3501 (2010).

2 Ionic Surfactants and Ion-Specific Effects
Adsorption, Micellization, and Thin Liquid Films

Radomir I. Slavchov, Stoyan I. Karakashev, and Ivan B. Ivanov

CONTENTS

2.1   INTRODUCTION

In 1887, F. Hofmeister published an article [1] in which the precipitation efficiency of different salts on proteins was investigated. Hofmeister found that the critical electrolyte concentration for protein precipitation exhibited regularity and ordered the ions by their efficiency; nowadays, this ordering is called the *Hofmeister series*. For a series of salts with the same cation and different anions, the precipitation efficiency increases (i.e., the precipitation concentration decreases) in the following sequence [2] ("Ac⁻" is acetate ion; F⁻ was added by us):

$$Ac^- > F^- > OH^- > Cl^- > Br^- > NO_3^- > BF_4^- > I^- > ClO_4^- \tag{2.1}$$

Similarly, for salts sharing the same anion, the Hofmeister series for the cations reads [2]:

$$Li^+ < Na^+ < K^+ < Rb^+ < Cs^+ < NH_4^+ < NMe_4^+ \tag{2.2}$$

("Me" stands for methyl group). These ion sequences in series (Equations 2.1 and 2.2) were found to be approximately independent of the protein, although the *direction* of the effect depends [3,4], among other factors, on the sign of the protein's net charge (in Hofmeister experiments, proteins were negatively charged). Since the work of Hofmeister, a large number of experimental studies have demonstrated similar regularities in various phenomena. The interested reader can find a collection of articles on the Hofmeister effect and related phenomena in two special issues of *Current Opinion in Colloid and Interface Science* in 2004 [5] and in perhaps the only book dedicated to this topic, *Specific Ion Effects* [2].

Earlier attempts for interpretation of the Hofmeister series and its effect on the interaction between proteins, macromolecules, or colloidal particles were qualitative and invoked mainly the ion size, the ion interaction with water, and the "hydration force" (see e.g., [5]). Ninham and coworkers were probably the first to advocate the role of the van der Waals forces for the interaction between ions in solution, for the adsorption of electrolytes, for the interaction between proteins or colloidal particles, etc. In a series of articles [2,6–15], Ninham and coworkers applied and tested Ninham's idea for explaining the adsorption of simple electrolytes. The initial results were encouraging [7], but the subsequent efforts to obtain good quantitative results [8–10] met with difficulties. Tavares et al. [16] studied theoretically the Hofmeister effect on the interaction of charged proteins. They calculated the purely electrostatic and the van der Waals contributions and found that the van der Waals interaction gives rise to a strong attractive force. No comparison with experimental data was carried out.

In this chapter, we will concentrate on the Hofmeister effect on the properties of ionic surfactant solutions and related phenomena. Warszynski and coworkers [17–19] accounted explicitly for the role of the ion-specific effects in the ionic surfactants' adsorption and determined their adsorption constants K_s. In ref. [17], they investigated the effects of Li⁺, Na⁺, NH₄⁺, K⁺, and Cs⁺ on the surface tension of anionic surfactant solutions, whereas ref. [18] was devoted to the anions. Aratono and coworkers [20–23] carried out numerous meticulous experimental studies, coupled with thermodynamic analysis, of the adsorption of a number of surfactants with various counterions at water|gas (W|G) and water|oil (W|O) interfaces. A large number of studies were dedicated to the ion-specific effects on the critical micelle concentration (cmc) of surfactants, see for example, refs. [24,25].

It is also worth mentioning two theories which did not address the Hofmeister effect, but played an important role in the theory of adsorption of ionic surfactants. The first one is the theory of Davies [26,27], who derived the equation of state (EOS) of dilute monolayers of ionic surfactants. The role of Davies' model in the theory of the adsorption of ionic surfactants is as significant as the ideal solution model is in the theory of solutions. The second theory belongs to Borwankar and Wasan [28], who proposed a simple approach for the derivation of the adsorption isotherm of ionic surfactants. We use both theories below.

Following Ninham's ideas, Ivanov et al. [29] proposed and tested experimentally a relatively simple theoretical model of the effect of the type of electrolyte on the adsorption constant of ionic surfactants. In the core of the theory is a quantity, u_{i0}, called *ion-specific adsorption energy*—it is equal to the van der Waals adsorption energy of an ion at the interface. It encompasses, in a single simple expression, all the major factors controlling ion-specific adsorption: the ion polarizability and ionization potential, the radius of the hydrated ion and the possible deformation of the hydration shell upon ion adsorption at the interface. The ion-specific adsorption energy u_{i0} turned out to be independent of the type of the surfactant, which allows using the calculated value for a given ion for the interpretation of adsorption data for different surfactants but with this same counterion. In ref. [30], the theory was successfully applied to other phenomena involving ionic surfactants: micellization, disjoining pressure of thin liquid films, and emulsion stability.

Our goal in this article is (i) to present concisely the results from refs. [29,30], related to some phenomena involving ionic surfactants and the Hofmeister effect, and to supplement them with more examples and new developments; (ii) to present a new EOS and a new adsorption isotherm for ionic surfactants, and to check them against surface/interfacial tension data; (iii) to discuss briefly the theories and experimental data of other authors, when they are related to the subject studied; and (iv) to outline some possible future developments (including extensions or refinements of the models presented) and several other phenomena to which the theory and the procedures presented here could be applied.

Practically all theories of adsorption of ionic surfactants (including the present one) are based on one or more theories of the nonionic ones. In several recent studies [31–34], it was shown that the most popular EOS of nonionic surfactants suffered from important shortcomings. Methods for overcoming these problems were suggested there and are discussed shortly in the present article.

The article is organized as follows: Section 2.2 is devoted to the adsorption of ionic surfactants from dilute surfactant solutions. The adsorption model used here is an extension of Henry's isotherm to ionic surfactants. That is why, in Section 2.2.1.1, we present briefly the theory developed in refs. [31,33] for the adsorption constants K_s, the adsorption thickness δ_a, and adsorption energy E_a. Sections 2.2.1.2 and 2.2.1.3 are devoted to the electrostatic and thermodynamic foundations of the theory of adsorption of ionic surfactants. In Section 2.2.1.4, the Davies adsorption isotherm for ionic surfactants [26,27], along with his expression for the surface potential, are derived and discussed. In Section 2.2.3, the difference in the adsorption behavior of ionic surfactants at W|G versus W|O interface is analyzed and the role of the spreading pressure and the liquid-expanded (LE) adsorbed layer is explained. The entire Section 2.2.3 is devoted to the counterion-specific effects on ionic surfactants adsorption, developed in ref. [29], more precisely—to the derivation of an equation for the central quantity of the theory—*the ion-specific adsorption energy u_{i0}*, and the development of a procedure for its theoretical calculation. The experimental verification of the theoretical results from Section 2.2 is given in Section 2.2.4. Davies model and the theory of the adsorption constant are analyzed by using numerous data for adsorption of ionic surfactants at W|G and W|O interfaces. The theory of the Hofmeister effect on the adsorption of ionic surfactants is compared with data for numerous individual counterions and their mixtures (Section 2.2.4.2). The main goal of Section 2.3 is the formulation and experimental verification of a new model of dense monolayers of ionic surfactants. Because it is an extension of the respective model for nonionic surfactants [31–33], in Sections 2.3.1.1 and 2.3.1.2, we present a rather detailed analysis and some new considerations about three basic quantities of the theory: the actual area of the molecule α, and the interaction constants B_{attr} and β. Section 2.3.1.3 presents the new surface EOS and adsorption isotherm of nonionic surfactants, derived previously in refs. [32,33]. They are generalized in Section 2.3.2.1 for nonlocalized adsorption of ionic surfactants with ion-specific effects. The verification of the theory with data for adsorption of ionic surfactants at W|G and W|O interfaces led to satisfactory results (Section 2.3.2.2) and allowed the formulation of several new electrostatic and ion-specific effects on the adsorption parameters (Section 2.3.2.3).

The generality of the theory of the counterion-specific effects to ionic surfactants is demonstrated in Section 2.4 by the analysis of other phenomena involving ionic surfactants that are different from adsorption [30], namely, the cmc and disjoining pressure in thin liquid films. A theory (based on the model of Shinoda et al. [25]) of the Hofmeister effect on the cmc is presented. The cmc of ionic surfactants in the presence of a mixture of counterions is also considered. Section 2.5 summarizes the main results obtained and discussed in this article and outlines the possibility for refinement and extension of the theory of the phenomena considered as well as for its application to other phenomena.

In the cases analyzed in the present article, remarkable agreement between theory and experiment was found and the results were in agreement with the Hofmeister series (2.1 and 2.2). Despite the diversity of the phenomena studied, all interpretations were based on a single parameter: the ion-specific adsorption energy u_{i0}. For a given counterion, a single value of u_{i0} was used to explain various phenomena; this value was calculated from the same equation without using adjustable parameters. The success of the theory in explaining or predicting quantitatively the ion-specific effects on several different phenomena involving ionic surfactants makes us believe that, at least for ionic surfactants, it is close to a "firmly based theory," which was missing according to Kunz et al. [15].

2.2 ADSORPTION OF IONIC SURFACTANTS FROM DILUTE SOLUTIONS

2.2.1 ADSORPTION IN THE ABSENCE OF ION-SPECIFIC EFFECTS

2.2.1.1 Henry's Adsorption Constant of Nonionic Surfactants: Adsorption Energy and Thickness

Before adsorption from dilute solutions of ionic surfactants, we will briefly discuss the basic theory of nonionic surfactants. The chemical potentials of a nonionic surfactant in an ideal solution of concentration C_s and in an ideal adsorbed monolayer of adsorption Γ_s are, respectively,

$$\mu^B = \mu_0^B + k_B T \ln C_s \tag{2.3}$$

$$\mu^S = \mu_0^S + k_B T \ln \Gamma_s \tag{2.4}$$

Here, superscripts "B" and "S" denote bulk and surface phase, and μ_0^S and μ_0^B are the corresponding standard chemical potentials. At equilibrium, the chemical potentials μ^B and μ^S must be equal. This leads to Henry's adsorption isotherm:

$$\Gamma_s = K_s C_s \tag{2.5}$$

where the adsorption constant K_s is defined by the relation

$$k_B T \ln K_s \equiv \mu_0^B - \mu_0^S \tag{2.6}$$

As it is obvious from the derivation, Henry's adsorption isotherm is valid only for adsorption layers consisting of noninteracting nonionic surfactant molecules, which is possible for dilute adsorption layers only.

The first step in our analysis will be the consideration of the dependence of K_s on the structure of the surfactant molecule and of the interface. The explicit expression for the adsorption constant K_s in terms of molecular parameters is usually written as

$$K_s = \delta_a \exp(E_a/k_B T) \qquad (2.7)$$

where δ_a is referred to as the "thickness of the adsorbed layer," and E_a is the adsorption energy. Davies and Rideal (Equation 4.2 in ref. [26]), proposed to use the length of the surfactant molecule for the thickness δ_a, an assumption adopted later by others (e.g., [35–37]). Davies and Rideal represented the adsorption energy E_a as

$$E_a = E_0 + u_{CH_2} n_C \qquad (2.8)$$

Here, n_C is the number of carbon atoms in the hydrophobic chain and u_{CH_2} is the (positive) free energy of transfer of a $-CH_2-$ group from the solution to the adsorption layer. E_0 is the n_C-independent part of E_a, which was ascribed in ref. [26] solely to the adsorption energy E_{head} of the hydrophilic head (cf. Equation 4.3 in ref. [26]). In fact, the assumptions of Davies and Rideal for both δ_a and E_a are not entirely correct. More rigorous treatment based on classic statistical thermodynamics was given in ref. [31], which will be presented in the section below. The results will be compared with experimental data for ionic surfactants in Section 2.2.4.1.

The essential contributions to the interaction potential of a surfactant molecule with the surface are the following (Figure 2.1):

1. When the cap of the hydrophobic chain touches the surface, a portion of the water–hydrophobic phase interface of area $\alpha_\perp$ disappears. The contribution of this process to $u(z)$ is modeled as a contact potential at $z = n_C l_{CH_2}$ (l_{CH_2} is length per $-CH_2-$ group):

$$u_{(i)}(z) = \begin{cases} \sigma_0 \alpha_\perp, & z < n_C l_{CH_2} \\ 0, & z > n_C l_{CH_2} \end{cases} \qquad (2.9)$$

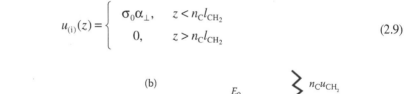

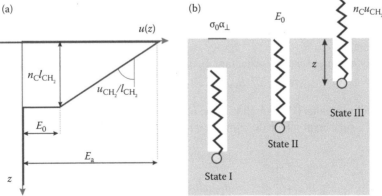

FIGURE 2.1 (a) Interaction potential $u(z)$ between a surfactant molecule and the interface as a function of the distance z between the ionic head and the surface (Equation 2.12). (b) According to the model we used, at distances $z > n_C l_{CH_2}$ (state I), there is no significant interaction. At $z = n_C l_{CH_2}$ (state II), energy is gained due to the disappearance of pure water surface of area $\alpha_\perp$, and the transfer energy of the cap of the $-CH_3$ group (Equation 2.14). At shorter distances (state III), there is a linear dependence of u on z related to the energy of transfer $n_C u_{CH_2}$ of the hydrocarbon chain from water to the gas phase (Equation 2.11).

where σ_0 is the surface tension of the pure water–hydrophobic phase interface. For W|G interface at 25°C, $\sigma_0 = 72$ mN/m, and for a typical W|O, $\sigma_0 \approx 50$ mN/m. The cross-sectional area of the hydrocarbon tail is $\alpha_\perp = \pi R_{CH_2}^2 \approx 21$ Å^2 (following from the value $R_{CH_2} = 2.6$ Å of the cross-sectional radius of the chain [38]). Thus, the energy $\sigma_0 \alpha_\perp$ is of the order of 10^{-20} J or approximately $2.5 \times k_B T$.

2. Let the free energy of transfer of a single –CH_2– group from the bulk solution to the adsorption layer be u_{CH_2}. For the energy u_{Me} of transfer of the –CH_3 group, we assume proportionality to the contact area of this group with water [39]. One can approximate the shape of –CH_3 as a cylinder with a cap. The lateral area of the cylinder is assumed equal to that of a –CH_2– group, $\alpha_\parallel$, and the cap area is assumed equal to the cross-sectional area $\alpha_\perp$ of the hydrocarbon tail. The values of the two areas are $\alpha_\perp = \pi R_{CH_2}^2 \approx 21$ Å^2 and $\alpha_\parallel = 2\pi R_{CH_2} l_{CH_2} \approx 21$ Å^2 (the values $R_{CH_2} = 2.6$ Å and $l_{CH_2} = 1.26$ Å were used [38]). Consequently, the two areas are equal and the energy corresponding to each of these areas is $u_{CH_2} \alpha_\parallel$. The energy pertaining to the cap can be represented as a contact potential with the same z-dependence as $u_{(i)}$ in Equation 2.9:

$$u_{(ii)}(z) = \begin{cases} u_{CH_2}, & z < n_C l_{CH_2} \\ 0, & z > n_C l_{CH_2} \end{cases} \tag{2.10}$$

The second part of u_{Me} (pertaining to the lateral area of –CH_3) is not included in Equation 2.10; it will be included in the next term, the potential $u_{(iii)}$. The contributions (1) and (2) were derived by Ivanov et al. [31] and independently by Kumpulainen et al. [40].

3. Assuming for simplicity that the carbon chain remains normal to the interface, one can model the hydrophobic energy due to –CH_2– adsorption (plus the lateral energy of the –CH_3 group) as a linear function of the distance z between the surfactant head and the interface:

$$u_{(iii)}(z) = u_{CH_2} z / l_{CH2}, \quad n_C l_{CH_2} > z > 0 \tag{2.11}$$

4. Although the hydrophilic head remains immersed in the hydrophilic phase, it also interacts with the interface. This interaction probably involves both short-range and long-range (such as van der Waals and electrostatic) forces. Because these forces are not yet fully understood, we will account for their contribution to the adsorption energy E_a by an empirical constant E_{head}.

5. One finally assumes that the surfactant cannot desorb into the hydrophobic phase, that is, $u(z) = \infty$ at $z < 0$.

Combining contributions (1) through (5), one obtains an approximate expression for the interaction potential of a surfactant molecule with the interface (see Figure 2.1):

$$u(z) = \begin{cases} \infty, & 0 > z; \\ -E_a + u_{CH_2} z / l_{CH2}, & n_C l_{CH2} > z > 0; \\ 0 & z > n_C l_{CH2}, \end{cases} \tag{2.12}$$

where the adsorption energy E_a is given by

$$E_a = E_{head} + \alpha_\perp \sigma_0 + u_{CH_2}(n_C + 1) \tag{2.13}$$

In Equation 2.12, the free energy of surfactant in the bulk solution is used as a reference state. Comparison of Equations 2.13 and 2.8 leads to an explicit expression for the empirical constant E_0 of Davies and Rideal [26]:

$$E_0 = E_{head} + u_{CH_2} + \alpha_\perp \sigma_0 \qquad (2.14)$$

It encompasses not only E_{head}, as assumed by Davies and Rideal, but also all the other contributions to E_a, unrelated to the adsorption of the $-CH_2-$ chain.

Our next goal is to derive a new expression for the "thickness of the adsorbed layer" δ_a, and the relation of δ_a and the adsorption energy E_a to the adsorption constant K_s. This can be done by statistical calculation of the adsorption Γ_s. Using Boltzmann distribution, the potential $u(z)$ (Equation 2.12) and the Gibbs' definition of adsorption as an excess [41], for an ideal monolayer, one can write:

$$\Gamma_s = C_s \int_0^{n_C l_{CH_2}} (e^{-u(z)/k_B T} - 1)\,dz = \frac{k_B T l_{CH_2}}{u_{CH_2}} e^{E_a/k_B T} \left(1 - e^{-n_C u_{CH_2}/k_B T} - \frac{n_C u_{CH_2}}{k_B T} e^{-E_a/k_B T} \right) C_s \qquad (2.15)$$

This is, in fact, a detailed expression of Henry's adsorption isotherm, $\Gamma_s = K_s C_s$. Because the exponents in the brackets are negligible, it yields Equation 2.7 for K_s. The comparison with Equation 2.7 shows that the adsorption layer thickness is

$$\delta_a = k_B T l_{CH_2}/u_{CH_2} \qquad (2.16)$$

Using Tanford's values for u_{CH_2} and l_{CH_2}, at 300 K, one obtains $\delta_a = 0.9$ and 1.2 Å for W|O and W|G interfaces, correspondingly. This is in contrast with the assumption that δ_a is of the order of the thickness of the adsorption layer [26,37]: indeed, for chain length $n_C = 12$, the ratio of the two thicknesses, $n_C l_{CH_2}$ and δ_a, as defined by Equation 2.16, is approximately 12.

2.2.1.2 Poisson–Boltzmann Equation and Electroneutrality: Gouy Equation

Consider an electrolyte solution positioned in the semi-space $z > 0$, as illustrated in Figure 2.2. Let each surfactant ion possess charge e_s (only monovalent surfactants will be considered, so that $e_s = \pm e_0$, where e_0 is the elementary charge). Consequently, the surface where the surfactants' heads are situated ($z = 0$, see Figure 2.2) has a surface charge density $e_s \Gamma_s$, due to the surfactant's adsorption in the adsorbed layer Γ_s. This surface charge and the ions in the diffuse layer create electrostatic potential $\phi(z)$ in the electrolyte solution, which is determined, at first approximation, by the Poisson–Boltzmann equation

$$\varepsilon \frac{d^2\phi}{dz^2} = -\sum_i e_i C_i \exp(-e_i \phi/k_B T) \qquad (2.17)$$

Here, ε is the absolute dielectric constant, e_i and C_i are the ith component charge, in units (C), and bulk molecular number concentration, in units (m^{-3}), k_B is Boltzmann constant, and T is absolute temperature. In this equation, the variables ϕ and $d\phi/dz \equiv -E$ (electric field) can be separated, by using the identity $2d^2\phi/dz^2 = d(E^2)/d\phi$. This leads to

$$d(E^2) = -\frac{2}{\varepsilon} \sum_i e_i C_i \exp(-e_i \phi/k_B T)\,d\phi \qquad (2.18)$$

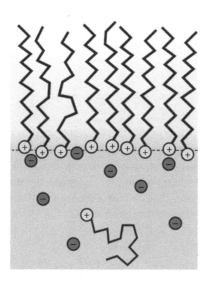

FIGURE 2.2 Structure of the adsorption layer. The adsorption of surfactant creates surface charge density at the interface. The surface charge is neutralized by the diffuse double layer.

The first integral of the Poisson–Boltzmann equation is obtained by integrating Equation 2.18 in limits $z = \infty$ to z, using as a first boundary condition, $E = 0$ and $\phi = 0$ at $z = \infty$:

$$E(z)^2 = \frac{2k_\mathrm{B}T}{\varepsilon} \sum_i C_i (\mathrm{e}^{-e_i\phi(z)/k_\mathrm{B}T} - 1) \tag{2.19}$$

The second boundary condition (at $z = 0$) is the Gauss condition for electroneutrality:

$$\varepsilon E\big|_{z=0} = e_\mathrm{s}\Gamma_\mathrm{s} \tag{2.20}$$

We will denote the surface potential $\phi(0)$ by ϕ^S. Setting $z = 0$ and $\phi = \phi^\mathrm{S}$ into Equation 2.19, and eliminating $E(z = 0)$ from the electroneutrality condition (Equation 2.20), the Gouy equation is obtained [42]:

$$\frac{\kappa_0^2}{4}\Gamma_\mathrm{s}^2 = \sum_i C_i \left(\mathrm{e}^{-e_i\phi^\mathrm{S}/k_\mathrm{B}T} - 1\right) \tag{2.21}$$

Here, the quantity

$$\kappa_0^2 \equiv 2e_0^2 \big/ \varepsilon k_\mathrm{B}T \tag{2.22}$$

is the concentration-independent part of the Debye parameter: $\kappa^2 \equiv \kappa_0^2 C_\mathrm{t}$ and $\kappa_0^2 \equiv 2 \times r_\mathrm{B}$, where $r_\mathrm{B} \equiv e_0^2 \big/ \varepsilon k_\mathrm{B}T$ is the so-called Bjerumm length. In the case of 1:1 electrolyte, the Gouy equation (Equation 2.21) simplifies to

$$\frac{\kappa_0}{4}\Gamma_\mathrm{s} = \sqrt{C_\mathrm{t}}\,\sinh(\Phi^\mathrm{S}/2) \tag{2.23}$$

Here, C_t is the total electrolyte concentration in units (m^{-3}) and

$$\Phi^S \equiv e_0 \left| \phi^S \right| / k_B T \tag{2.24}$$

is the absolute value of the dimensionless surface potential. At high surface potentials ($\Phi^S \gg 1$), a good approximation of the Gouy equation (Equation 2.23) is

$$\Gamma_s = \frac{2}{\kappa_0} \sqrt{C_t} \exp(\Phi^S/2) \tag{2.25}$$

In the case of 1:1 electrolyte, Equation 2.19 can be integrated analytically. First, we take the root of Equation 2.19:

$$\frac{d e_0 \phi(z)/k_B T}{dz} = -\kappa_0 \sqrt{C_t} \left(e^{e_0 \phi(z)/2 k_B T} - e^{-e_0 \phi(z)/2 k_B T} \right) \tag{2.26}$$

Direct integration of Equation 2.26 gives an explicit relation between z and φ:

$$\kappa_0 \sqrt{C_t} z/2 = \operatorname{arctanh}\left(e^{e_0 \phi/2 k_B T} \right) - \operatorname{arctanh}\left(e^{e_0 \phi^S/2 k_B T} \right) \tag{2.27}$$

2.2.1.3 Thermodynamics of the Diffuse Double Layer: Adsorption and Surface Tension

The ion distribution in the electric double layer depends on the local potential $\phi(z)$. The ion adsorption Γ_j^{DL} of any ion j in the diffuse layer can be calculated by using Gibbs' definition of adsorption as an excess [41]:

$$\Gamma_j^{DL} \equiv C_j \int_0^\infty \left(e^{-e_j \phi(z)/k_B T} - 1 \right) dz \tag{2.28}$$

where the superscript "DL" indicates adsorption of the counterions, coions, and surfactant ions in the double layer only. In principle, the *total* surfactant adsorption is a sum of Γ_s^{DL} and the surface concentration Γ_s (which is the adsorption in the adsorption layer, driven by hydrophobic forces). The surfactant ions in the diffuse layer are repelled by the interface because they have the same charge. Usually, the surface potential Φ^S is high, so that the surfactant concentration in the diffuse layer is close to zero. This leads to a relatively small negative adsorption of the order of $\Gamma_s^{DL} \sim -C_s/\kappa$. Because $\left| \Gamma_s^{DL} \right| \ll \Gamma_s$, the adsorption of surfactant in the diffuse layer can be neglected. The same refers to the coions. Hence, under these conditions, only the adsorption of the counterions in the diffuse layer is of importance.

To calculate the integrals defined by Equation 2.28, it is convenient to change the integration variable to φ, by using the relation $dz = d\phi/(d\phi/dz)$,

$$\Gamma_j^{DL} \equiv C_j \int_{\phi^S}^0 \frac{\exp(-e_j \phi/k_B T) - 1}{d\phi/dz} d\phi \tag{2.29}$$

By inserting here the expression (2.26) for $d\phi/dz$, one can obtain explicit formulae for the adsorptions Γ_j^{DL}. For 1:1 electrolyte at high surface potential Φ^S, the result for the adsorption of the counterion i reads

$$\Gamma_i^{\mathrm{DL}} = \frac{2C_i}{\kappa_0 \sqrt{C_\mathrm{t}}} \left(e^{\Phi^S/2} - 1 \right) \xrightarrow{\Phi^S \to \infty} \frac{2C_i}{\kappa_0 \sqrt{C_\mathrm{t}}} \exp(\Phi^S/2) \tag{2.30}$$

To calculate the surface tension, the Gibbs isotherm is used [41]. If only one counterion of concentration C_i is present in the system, and the bulk solution is assumed ideal, one has

$$d\sigma = -k_\mathrm{B}T\Gamma_\mathrm{s} d\ln C_\mathrm{s} - k_\mathrm{B}T\Gamma_i^{\mathrm{DL}} d\ln C_i \tag{2.31}$$

Because at high surface potential, the charge of the adsorbed layer is compensated only by the counterion in the diffuse layer, one has $\Gamma_i^{\mathrm{DL}} = \Gamma_\mathrm{s}$. Then, the Gibbs isotherm (2.31) simplifies to

$$d\sigma = -2k_\mathrm{B}T\,\Gamma_\mathrm{s}\,d\ln C \tag{2.32}$$

where C is the mean ionic activity of the surfactant [43–45], defined with

$$C = C_\mathrm{s}^{1/2} C_i^{1/2} \tag{2.33}$$

If the solution is not ideal, the mean ionic activity C in Equation 2.32 will include the mean activity coefficient γ:

$$C = \gamma C_\mathrm{s}^{1/2} C_i^{1/2} \tag{2.34}$$

2.2.1.4 Davies' Adsorption Isotherm

We now consider an ideal solution of ionic surfactant of concentration C_s in equilibrium with an "ideal" charged adsorbed monolayer with surface potential ϕ^S. The chemical potentials in the two states are

$$\mu^B = \mu_0^B + k_\mathrm{B}T \ln C_\mathrm{s}; \quad \mu^S = \mu_0^S + k_\mathrm{B}T \ln \Gamma_\mathrm{s} + e_\mathrm{s}\phi^S \tag{2.35}$$

The difference from the corresponding expressions for nonionic surfactants (Equations 2.3 and 2.4), is the presence of the additional electrostatic energy term $e_\mathrm{s}\phi^S$ in μ^S.

The condition for equilibrium between the surfactant molecules in the bulk solution and at the surface reads:

$$\mu_0^B + k_\mathrm{B}T \ln C_\mathrm{s} = \mu_0^S + k_\mathrm{B}T \ln \Gamma_\mathrm{s} + e_\mathrm{s}\phi^S \tag{2.36}$$

Introducing here the dimensionless potential Φ^S (Equation 2.24), one obtains

$$\Gamma_\mathrm{s} = K_\mathrm{s} C_\mathrm{s} \exp(-\Phi^S) \tag{2.37}$$

Equation 2.37 was first derived by Davies [26,27] (see also ref. [28]). The adsorption constant K_s in this equation is defined by Equation 2.6, but in this case, $\mu_0^B - \mu_0^S$ may contain electrostatic contributions.

The elimination of Γ_s from Equations 2.25 and 2.37 leads to an equation for the dependence of the surface potential Φ^S on the composition of the bulk solution [27]:

$$3\Phi^S = \ln \frac{\kappa_0^2 K_\mathrm{s}^2}{4} + \ln \frac{C_\mathrm{s}^2}{C_\mathrm{t}} = 6\ln \frac{\kappa_0 K_0}{2} + \ln \frac{C_\mathrm{s}^2}{C_\mathrm{t}} \tag{2.38}$$

Equation 2.38 shows that the surface potential Φ^S increases with C_s and K_s (due to the increased adsorption) and decreases with the total electrolyte concentration C_t (due to the additional screening effect of the electrolyte on the surface charge).

Inserting back the surface potential (2.38) into the isotherm (2.37), one obtains a generalization of Henry's isotherm for the adsorption of ionic surfactants:

$$\Gamma_s = K_0 C^{2/3} \tag{2.39}$$

where C is given by Equation 2.33, and K_0 is the adsorption constant of the ionic surfactant. It is related to Henry's constant K_s:

$$K_0 = \left(4K_s/\kappa_0^2\right)^{1/3} \tag{2.40}$$

The fact that according to Equation 2.39, Γ_s depends only on the mean ionic activity C is an explicit formulation of what is known as *salting-out effect* on ionic surfactant adsorption [44,45]. Equation 2.39 was first derived and confirmed by experimental data for $C_nH_{2n+1}SO_4^+$ at the W|O interface by Davies [27]. We will refer to this as Davies' isotherm. By using the procedure of Borwankar and Wasan [28], Ivanov et al. [29] derived Equation 2.39 and obtained the explicit expression (2.40) for K_0 (see also Section 2.3.2.1). According to Equation 2.40, K_0 should not depend on the electrolyte concentration, at least for moderate concentrations.

Substituting Equation 2.39 in the Gibbs isotherm (2.32) and integrating, one obtains the surface pressure isotherm:

$$\pi^S \equiv \sigma_0 - \sigma = 3k_BTK_0C^{2/3} \tag{2.41}$$

which is also due to Davies [27]. Comparison with Equation 2.39 shows that $\pi^S = 3k_BT\Gamma_s$. Because the surface pressure of an ideal layer of a nonionic surfactant is $k_BT\Gamma_s$, it follows that the contribution to π^S of the double layer at high surface potential is [26]

$$\pi_{el}^S = 2k_BT\Gamma_s \tag{2.42}$$

For the sake of simplicity, until now, the ever-present counterion-specific effects were disregarded. It will be shown in Section 2.2.3 that these effects modify the adsorption constant, leading, instead of K_0 in Equations 2.39 and 2.41, to a new constant, $K = K_0\exp(-u_{i0}/2k_BT)$, where u_{i0} is the counterion-specific adsorption energy (cf. Equation 2.56 below).

2.2.2 Adsorption Behavior at W|G versus W|O Interface: LE Layer and Spreading Pressure

Below, we analyze the experimental data for π^S versus the 2/3-power of the mean ionic activity $C^{2/3}$ at W|O and W|G interfaces for low and medium surfactant concentrations—data for sodium dodecyl sulfate, $C_{12}H_{25}SO_4Na$, at W|O and W|G interfaces in the presence of various concentrations of NaCl are shown in Figure 2.3. At W|G interface, the data exhibit two well-defined regions. At very low surface pressures (less than ca. 2–3 mN/m), a close-to-linear dependence without intercept is observed. At intermediate concentrations (up to cmc) and pressures, there is a second linear region, but with negative intercept. Denoting this intercept by π_0, one can write for this region instead of Equation 2.41:

$$\pi^S = \pi_0 + 3k_BTK^{LE}C^{2/3} \tag{2.43}$$

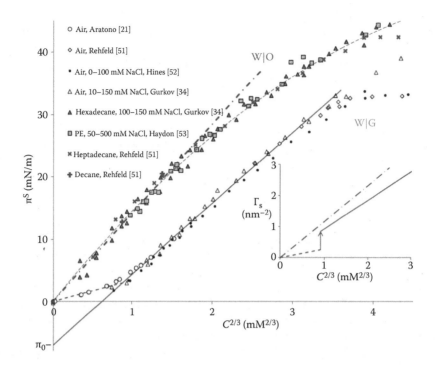

FIGURE 2.3 **(See color insert.)** Interfacial pressure π^S versus the 2/3-power of the mean activity $C^{2/3}$ for $C_{12}H_{25}SO_4Na$ solutions at different NaCl concentrations. (Data for the W|G surface from M. Aratono et al., *J. Colloid Interface Sci.* 98, 33–38, 1984; T.D. Gurkov et al., *Colloids Surf. A* 261, 29–38, 2005; S.J. Rehfeld, *J. Phys. Chem.* 71, 738–745, 1967; J.D. Hines, *J. Colloid Interface Sci.* 180, 488–492, 1996. With permission.) For the W|O interface, the oil is heptadecane, decane [51], hexadecane [34], and petroleum ether [53]. Solid line, data fit for the W|G in the range $C^{2/3}$ = 1.2 to 3 mM$^{2/3}$ (the LE region), according to Equation 2.43. The short, dashed line stands for W|G data in the range $C^{2/3}$ = 0 to 1 mM$^{2/3}$ (gaseous monolayer region). Dash-dot line, data fit of the W|O interface in the range $C^{2/3}$ = 0 to 1.8 mM$^{2/3}$ (Equation 2.41). Long dashed line, quadratic fit of W|O data in the range $C^{2/3}$ = 0 to cmc$^{2/3}$ (Equation 2.47). The adsorption parameters determined from these fits are listed in Table 2.1. Inset, the corresponding adsorption isotherms, $\Gamma_s(C^{2/3})$, calculated from Equation 2.39 with the adsorption parameters determined by the fits. The jump of Γ_s at C = 0.81 mM corresponds to a phase transition from a gaseous monolayer to LE state.

By analogy with Langmuir's treatment of noncharged monolayers [46], this behavior can be explained by assuming that the monolayer is in an LE state. In this state, the adsorbed hydrophobic tails form a very thin, but more or less dense, oil film spread onto the water phase. In contrast, in the first region, at lower concentrations, the surfactant molecules are in a gaseous state in which they are isolated from each other. The intersection point between the two lines probably corresponds to phase transition between the gaseous and the LE states (see Figure 2.3). Such a phase transition was also discussed by Aratono et al. [21], and observed with soluble nonionic surfactants by Kumpulainen et al. [40].

Langmuir's idea for the origin of π_0 (e.g., [47]) can be quantified as follows. Let σ_0^{WO} be the interfacial tension of the pure W|O interface, and let σ_0^{OG} be the oil–gas surface tension. The oil-like thin film formed by the adsorbed hydrophobic tails can be considered as a single "interface" (membrane) of interfacial tension $\sigma_0^M = \sigma_0^{WO} + \sigma_0^{OG}$, which for this system is the counterpart of the interfacial tension of the pure interface σ_0. The hydrophilic heads of the surfactant are "adsorbed" at the W|O interface of the thin film. If their adsorption is ideal, one can use Equation 2.41 with σ_0 replaced with σ_0^M to calculate σ:

$$\sigma = \sigma_0^{OG} + \sigma_0^{WO} - 3k_BTK^{LE}C^{2/3} \qquad (2.44)$$

However, by definition, the surface pressure π^S at the W|G interface is defined with respect to the pure W|G surface of tension σ_0^{WG}, that is, $\pi^S = \sigma_0^{WG} - \sigma$. Inserting Equation 2.44 into this definition, and comparing the result to Equation 2.43, one obtains

$$\pi_0 = \sigma_0^{WG} - \sigma_0^{WO} - \sigma_0^{OG} \tag{2.45}$$

According to these simple considerations, the intercept π_0 coincides with the spreading coefficient of a hydrocarbon on water [46]. Therefore, π_0 is referred to as *spreading pressure*. Langmuir's explanation of π_0 is confirmed by the data in Figure 2.3. Indeed, the spreading coefficient of dodecane on water is −6.4 mN/m (the values $\sigma_0^{WO} = 53.7$, $\sigma_0^{OG} = 25.3$, and $\sigma_0^{WG} = 72.6$ mN/m at 22°C were used [48]), versus $\pi_0 = -7$ mN/m determined from the data in Figure 2.3. However, this picture is oversimplified, and as will be shown in Section 2.2.4 below, π_0, in fact, depends on the counterion,* and probably on the surfactant ionic head. The reason for these dependences is not yet clear.

The situation is different with the adsorption at the W|O interface. In this case, there is no oil–gas interface. Then, in Equation 2.45, one must replace σ_0^{WG} with σ_0^{WO}, and σ_0^{OG} with σ_0^{OO} (the latter is of course zero). Thus, one finds $\pi_0 = 0$, in accordance with the data in Figure 2.3: indeed, the surface pressure at the W|O interface follows rather well the simple dependence of Davies (Equation 2.41), with no intercept, up to $C^{2/3} \approx 2$ mM$^{2/3}$ ($C \approx 3$ mM).

The following observations deserve additional attention:

1. In accordance with the salting-out effect and Equation 2.32, the surface tension σ depends on the mean activity C only, as defined by Equation 2.34. Indeed, regardless of the electrolyte concentration, all data fall on two master curves π^S versus $C^{2/3}$ (one for W|G and one for water–alkane interface). This is so only if activities rather than concentrations are used—this also follows from Equation 2.32. The activity coefficients γ in Figure 2.3 were calculated by the formula [43,49]

$$\lg \gamma = -\frac{A\sqrt{C_t}}{1 + B\sqrt{C_t}} + bC_t \tag{2.46}$$

   If the total electrolyte concentration $C_t = C_{el} + C_s$ is in units (M), the Debye constant is $A = 0.5108$ M$^{-1/2}$ (at 298.15 K); for the empirical constants B and b, we used the mean values $B = 1.25$ M$^{-1/2}$ and $b = 0.0083$ M^{-1} for all salts.

2. The dependence of π^S on $C^{2/3}$ is linear up to surface pressures of approximately 25 mN/m for W|O and 30 mN/m for W|G interfaces (the values of the slopes and the adsorption parameters are listed in Table 2.1). The respective values of $C^{2/3}$ are approximately 1.8 and 3.2 mM$^{2/3}$. The difference between the two systems is due to the larger second virial coefficient (larger repulsive interactions) at the W|O interface, which leads to earlier deviation from ideality. This is evident from the values of the respective second virial coefficients (for details, see Section 2.3.2.2 below).

3. At the W|G interface, the line drawn for the gaseous adsorbed layer is only tentative, as the validity of the Davies isotherm in this region is questionable due to the low potential and the possible effects of the charge discreteness. Phase transition from gaseous to LE state occurs at the W|G at $C = 0.81$ mM, corresponding to π^S of approximately 3.0 mN/m (Aratono et al. [21] found 0.83 mM and 3.9 mN/m, respectively). This corresponds to a transition from $\Gamma_s = 0.24$ nm^{-2} to $\Gamma_s = 0.78$ nm^{-2}, that is, from a less dense gaseous structure to a more dense liquid-like monolayer (see the Γ_s vs. $C^{2/3}$ plot in Figure 2.3).

---

* cf. Figure 2.12.

TABLE 2.1
Values of the Adsorption Parameters

	Linear Fit with Equation 2.43			Quadratic Fit (Equation 2.47)	
	$d\pi^S/dC^{2/3}$ (mN/mM$^{2/3}$m)	π_0 (mN/m)	K	π_0 (mN/m)	K
W|G, LE	11.4	−7.0	129	−9.1	156
W|O	13.5	0	161	0	178

Note: Determined from the data in Figure 2.3, according to Equation 2.47 for W\|O interface and Equation 2.43 for W\|G ($T = 25°$C).

4. A number of different procedures were tested to determine K, including linear regression on a few initial points (straight lines in Figure 2.3) and square polynomial fit (long dashed curve in Figure 2.3) with the equation

$$\pi^S = \pi_0 + 3k_B TKC^{2/3} + bC^{4/3} \tag{2.47}$$

which is, in fact, Equation 2.134 derived in Section 2.3 below. Although the linear regression model looks satisfactory, we prefer the result from the polynomial fit because it generally yields K values closer to the ones obtained by using more realistic models, such as those discussed in Section 2.3.

5. From the $\pi^S(C^{2/3})$ data and Equation 2.47, we found for the LE region $K^{LE} = 156$ (Table 2.1). The latter is very close to the value for adsorption at W\|O interface, $K^{WO} = 178$, which suggests that both processes are similar.

Davies also accounted for the cohesive (i.e., negative spreading) pressure of soluble ionic surfactants at the W\|G interface [27]. However, neither Langmuir [46] nor Davies [27] used the simple isotherm (Equation 2.44). Instead, Langmuir used a correction for steric repulsion between the

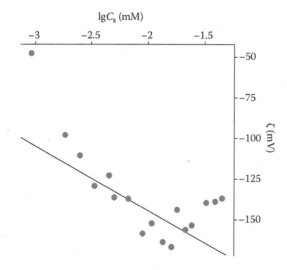

FIGURE 2.4 Dependence of the ζ potential (mV) at the water–hexadecane interface on $\lg C_s$ (mM) for $C_{12}H_{25}SO_4^-Na^+$ in the presence of 10 mM of NaCl [34]. Solid line, theoretical dependence (Equation 2.38) of the surface potential Φ^S (assumed equal to the ζ potential) on C_s, with no adjustable parameters ($K = 178$; see Table 2.1).

TABLE 2.2
Specific Adsorption Energies of the Ions Considered ($T = 298$ K)

Cation	R_b (Å)	n_w Equation 2.67	R_h (Å) Equation 2.68	N_w Equation 2.70	L_{ww} Equation 2.69 (m⁶J) $\times 10^{80}$	$\alpha_{p,i}$ (Å³)	I_i (J) $\times 10^{18}$	L_{wi} Equation 2.69 (m⁶J) $\times 10^{80}$	u_{i0}/k_BT Type I, Equation 2.65	u_{i0}/k_BT Type II, Equation 2.66
Li⁺	0.69[1]	5.22	2.41	0.13	41.5	0.03[1]	12.1[2]	11.5		−0.09
Na⁺	1.02[1]	3.53	2.18	0.40	134	0.15[3]	7.58[2]	53.1		−0.33
NH₄⁺	1.53[4]	2.35	2.14	1.36	453	1.64[1]	2.13[2]	378	−0.61	
K⁺	1.41[1]	2.55	2.12	1.07	354	0.79[3]	5.07[2]	253	−0.90	
Rb⁺	1.65[5]	2.18	2.17	1.71	568	1.4[1]	4.41[2]	431	−0.98	
NMe₄⁺	2.80[1]	1.29	2.94	8.36	2770	9.08[1]	2.43[2]	2220	−1.05	

Anion	R_b (Å)	n_w	R_h (Å)	N_w	L_{ww} (m⁶J) $\times 10^{80}$	$\alpha_{p,i}$ (Å³)	I_i (J) $\times 10^{18}$	L_{wi} Equation 2.69 (m⁶J) $\times 10^{80}$	u_{i0}/k_BT, Type I	u_{i0}/k_BT, Type II
Ac⁻	1.65[1]	2.18	2.17	1.71	568	5.50[1]	0.544[1]	545		−0.185
OH⁻	1.33[1]	2.71	2.11	0.90	297	2.04[1]	0.345[1]	134		−0.736
F⁻	1.33[1]	2.71	2.11	0.90	297	1.04[2]	0.545[1]	99.1		−0.891
Cl⁻	1.64[1]	2.20	2.17	1.68	557	3.59[2]	0.580[1]	359	−1.43	
Br⁻	1.95[1]	1.85	2.31	2.82	937	5.07[2]	0.540[1]	480	−2.32	
NO₃⁻	2.00[1]	1.80	2.33	3.05	1010	3.93[1]	0.631[1]	420	−2.83	
N₃⁻	1.95[1]	1.85	2.31	2.82	937	4.45[1]	0.444[1]	360	−2.93	
ClO₄⁻	2.40[1]	1.50	2.61	5.26	1750	5.25[2]	0.758[1]	642	−3.28	
BF₄⁻	2.30[1]	1.57	2.53	4.63	1540	2.80[1]	0.902[1]	388	−3.84	

Source: (1) From Marcus, Y. *Ion Properties*. Marcel Dekker, New York, 1997. (2) From Nikolskij, B.P. *Handbook of Chemistry*. Khimija, Leningrad, 1963 [in Russian]. (3) From Tavares, F.W. et al., *J. Phys. Chem. B* 108, 9228–9235, 2004. (4) Dietrich, B. et al., *J. Phys. Chem.* 91, 6600–6606, 1987. (5) Lide, D.R. (Ed.). *CRC Handbook of Chemistry and Physics* (83rd ed.). CRC Press, Boca Raton, FL, 2002.

Note: R_b, bare ion radius; n_w, hydration number (Equation 2.67); R_h, hydrated ion radius (Equation 2.68); N_w, number of water molecules in the ensemble, replaced by the ion upon adsorption (Equation 2.70); L_{ww}, London constant of this ensemble (Equation 2.69); $\alpha_{p,i}$, polarizability of the ion; I_i, second ionization potential of the cations and negative electron affinity of the anions; u_{i0}, ion-specific adsorption energy. Equation 2.66 for ions of type I (no deformation of the hydration shell) and Equation 2.65 for ions of type II (with deformation of the hydration shell). The ions in the table are ordered by increasing absolute values of u_{i0}. The sequence of both cations and anions is the same as in Hofmeister series (Equations 2.1 and 2.2), but for the cations this order corresponds to increasing efficiency as opposite to the series (Equation 2.2).

heads, whereas Davies introduced an empirical dependence of π_0 on Γ_s. We prefer to discuss the attractive and repulsive interactions between the adsorbed molecules separately in Section 2.3.

In Figure 2.4, the Davies model was tested further by comparison of the theoretical surface potential ϕ^S (Equation 2.38), with ζ potential measurements at the water–hexadecane interface [34]. All parameters in Equation 2.38 are known. In fact, the experimental data also involves ion-specific effects. Hence, the calculation of ϕ^S was performed with Equations 2.56 and 2.58 in Section 2.2.3 below, which accounts for the effect of the counterion on K and on ϕ^S; the value $u_{i0} = -0.34\,k_BT$ was used for Na^+ (Table 2.2). It turned out that the contribution of the ion-specific effect to ϕ^S is small. Taking into account the experimental difficulties (see also ref. [50]), the theoretical predictions seem adequate.

Another test of the model behind in Equations 2.41 and 2.43—the adequacy of the results for K (Equation 2.40)—will be given in Section 2.2.4.

2.2.3 COUNTERION-SPECIFIC EFFECTS ON THE ADSORPTION OF IONIC SURFACTANTS FROM DILUTE SOLUTIONS

2.2.3.1 Gouy Equation with Specific Interactions

We turn now to the theoretical treatment of the influence of the van der Waals interactions (which we consider as the most important specific interaction) of the counterions with the interface on the adsorption of a monovalent ionic surfactant. Toward this goal, an extended Poisson–Boltzmann equation involving both electrostatic and van der Waals potentials, is solved approximately. The ensuing generalized form of the Gouy equation, along with some thermodynamic considerations, will be used in the remaining parts of this section to account for the ion-specific effects on the adsorption and related phenomena with ionic surfactants.

Near the adsorbed layer of an ionic surfactant, the van der Waals forces between the counterion and the bulk phases lead to an increase of the local concentration of counterions in the diffuse double layer. The repulsive interactions disallow the counterion to approach the interface at a distance that is less than its radius R (bare or hydrated). Both interactions, repulsive and attractive, were modeled in a simple manner in ref. [29] by the following expression for the energy $u_i(z)$ of specific interaction between the ion and the interface:

$$u_i(z) = \frac{R_i^3}{(R_i + z)^3}\, u_{i0} \tag{2.48}$$

Here, R_i is the ionic radius (of the bare or of the hydrated ion, as will be discussed in Section 2.2.3.2) and u_{i0} is the van der Waals energy of an ion in the plane $z = 0$ situated at distance R_i from the interface (Figure 2.5). If one assumes that the van der Waals energy (Equation 2.48) and the electrostatic energy $e_i\phi^S$ are additive, the Boltzmann distribution will involve the sum of the two: $e_i\phi^S + u_i(z)$. The Poisson–Boltzmann equation (Equation 2.17) will then read:

$$\varepsilon\frac{d^2\phi}{dz^2} = -\sum_i e_i C_i e^{-(e_i\phi + u_i(z))/k_BT} \tag{2.49}$$

This equation can be integrated by analogy with the derivation of Gouy equation (Equation 2.21), by using $2d^2\phi/dz^2 = d(E^2)/d\phi$ and the Gauss condition (Equation 2.20). The result is

$$e_0^2\Gamma_s^2 = -2\varepsilon\sum_i e_i C_i e^{-u_{i0}/k_BT}\int_0^{\phi^S} e^{-e_i\phi/k_BT} e^{-(u_i(z)-u_{i0})/k_BT}\,d\phi \tag{2.50}$$

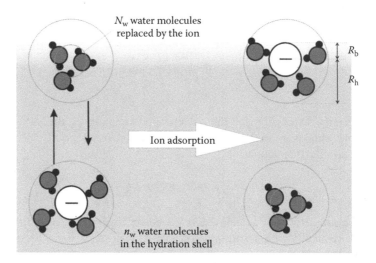

N_w water molecules
replaced by the ion

R_b

R_h

Ion adsorption

n_w water molecules
in the hydration shell

FIGURE 2.5 **(See color insert.)** Scheme of the process of adsorption of type I ions. Left, the ion is in the bulk. Right, the ion is at the interface. The n_w hydrating water molecules might be pushed away by the interface, so that the shortest distance of approach of the ion to the interface is the bare ion radius R_b. Upon adsorption, the ion exchanges position with an ensemble of N_w water molecules. For type II ions, the shortest distance of approach of the ion to the interface is the hydrated ion radius R_h.

At high surface potentials, only the counterions need to be taken into account in the sum in the right-hand side of this equation. This approximation is of crucial importance for the success of our theory because it simplifies all subsequent calculations. It can be also used in the case of ionized proteins and polymers (when $|\phi^S/k_BT| \gg 1$), but not for the adsorption of simple electrolytes. In the latter case, both cations and anions (whose properties are similar) have comparable participation in the diffuse layer whose local potential depends, in fact, on the small differences in their local concentrations [54]. With this approximation, the integrals on the right-hand side of Equation 2.50 can be evaluated by using an iterative procedure [29]. As zeroth iteration, one can use in the integrand the results for the case of the absent ion-specific effect obtained in Section 2.2.1.4, that is, one must set in Equation 2.50 $\phi^S = \phi_0^S$, where ϕ_0^S is given by Equation 2.38, and $z = z_0(\phi)$, given by Equation 2.27. Integration by parts then leads to

$$\int_0^{\phi_0^S} e^{-e_i\phi/k_BT} e^{-(u_i(z_0)-u_{i0})/k_BT}\, d\phi = -\frac{k_BT}{e_i} e^{\Phi_0^S}(1-F_u) \qquad (2.51)$$

where F_u stands for the expression

$$F_u = e^{-\Phi_0^S}\left[e^{u_{i0}/k_BT} + \int_0^{\phi_0^S} e^{-e_i\phi/k_BT}\, \frac{de^{-(u_i(z_0)-u_{i0})/k_BT}}{d\phi}\, d\phi \right] \qquad (2.52)$$

At high surface potentials, the value of F_u was found to be much smaller than unity [29], so that it can be neglected in Equation 2.51. Using this approximation, one substitutes Equation 2.51 into Equation 2.50 to obtain a generalization of the Gouy equation (Equation 2.21), accounting for the ion-specific effect:

$$\Gamma_s^2 = \frac{4}{\kappa_0^2}\sum_i C_i e^{-u_{i0}/k_BT} e^{\Phi_0^S} \qquad (2.53)$$

(the Debye parameter κ_0 is given by Equation 2.22). If only one counterion is present in the system, Equation 2.53 simplifies to

$$\Gamma_s^2 = \frac{4}{\kappa_0^2} C_t e^{-u_{i0}/k_B T} e^{\Phi_0^S} \tag{2.54}$$

Substituting here the expression for the zeroth approximation of the surface potential Φ_0^S (Equation 2.38), one obtains an extension of the Davies isotherm $\Gamma_{s0} = K_0 C^{2/3}$ (Equation 2.39), accounting for ion-specific interactions:

$$\Gamma_s = K_0 \, e^{-u_{i0}/2k_B T} C^{2/3} \equiv K C^{2/3} \tag{2.55}$$

Here, C is the mean activity (Equation 2.33). Based on Equation 2.55 and the expression in Equation 2.40 for the nonspecific adsorption constant K_0, one finds the expression for the ion-specific adsorption constant K:

$$K = K_0 e^{-u_{i0}/2k_B T} = \left(4K_s/\kappa_0^2\right)^{1/3} e^{-u_{i0}/2k_B T} \tag{2.56}$$

This procedure also allows for the determination of the first iteration of the surface potential Φ^S. To do so, the EOS (Equation 2.37) is used, with Γ_s given by Equation 2.55. After solving the result with respect to Φ^S, one obtains

$$\Phi^S = \frac{1}{3} \ln \frac{\kappa_0^2 K_s^2}{4} - \frac{1}{3} \ln \frac{C_s^2}{C_t} + \frac{u_{i0}}{2k_B T} \tag{2.57}$$

The comparison of Equations 2.55 and 2.57 with the respective zeroth (nonspecific) approximations for the surface potential Φ_0^S (Equation 2.38), and the adsorption Γ_{s0} (Equation 2.39), leads to

$$\Gamma_s = \Gamma_{s0} \exp(-u_{i0}/2k_B T), \quad \Phi^S = \Phi_0^S + u_{i0}/2k_B T \tag{2.58}$$

Note that the procedure is applied to the EOS (Equation 2.37) of an "ideal" monolayer. For other systems in which the surfactant molecules interact directly with each other (e.g., with van der Waals or steric forces), it is possible that other system parameters are affected (examples will be given in Sections 2.3 and 2.4).

In Equation 2.53, it is assumed that the surface charge density is due to the surfactant ions only (i.e., it is $e_s\Gamma_s$). In reality, the counterions can also penetrate into the adsorbed layer and in the empty spaces between the surfactant heads, but because of the relatively low values of the specific adsorption energies u_{i0} (cf. Table 2.2), they do not remain firmly bound (unlike the surfactant ions) to the interface. Hence, most common ions must be treated as part of the diffuse layer. This was proven directly by Shimamoto et al. [22], who studied experimentally, using total reflection X-ray absorption, the fine structure of ion distribution in the adsorbed and diffuse layers.

2.2.3.2 Specific Interaction between an Ion and the Interface

The adsorption potential of the counterion u_{i0} is related to a number of parameters, among them: the molecular or ion static polarizabilities, $\alpha_{p,w}$ and $\alpha_{p,i}$, and the ionization potentials I_w and I_i of the water molecule and the counterion, as well as the radii of the hydrated and bare ion (R_h and R_b, respectively). These parameters are not always available and even when they are, they are not very reliable. As shown in ref. [29], the adsorption energy u_{i0} depends strongly on the choice of the parameters and can vary by orders of magnitude. When several values of a given parameter were available in the literature, all were tested and the one providing the best coincidence with the experimental data was retained.

In ref. [29], the calculation of the energy u_{i0} was performed, using the London expression for the intermolecular potential u_{ij} between molecules of types i and j at a distance r_{ij} [39]:

$$u_{ij} = -L_{ij}/r_{ij}^6 \qquad (2.59)$$

where the London constant L_{ij} is related to the static polarizabilities $\alpha_{p,i}$ and $\alpha_{p,j}$ and the ionization potentials I_i and I_j of the interacting species:

$$L_{ij} = \frac{3\alpha_{p,i}\alpha_{p,j}}{2} \frac{I_i I_j}{I_i + I_j} \qquad (2.60)$$

Upon adsorption, the counterion displaces an ensemble of N_w water molecules (Figure 2.5). In the initial state (before adsorption), the ion is in the bulk and has energy u_i^B and the N_w water molecules are at the interface, with total energy u_w^S (subscript indices "i" and "w" stand for "ion" and "water," respectively, whereas superscript indices "S" and "B" stand for "surface" and "bulk," respectively; see Figure 2.5). In the final state (adsorbed ion), the ion and the water molecules have exchanged positions and their energies became u_i^S and u_w^B, respectively. Thus, the ion adsorption energy u_{i0}, which is equal to the change in the energy upon adsorption, is

$$u_{i0} = \left(u_i^S - u_i^B\right) - \left(u_w^S - u_w^B\right) \qquad (2.61)$$

The hydration shell of the large ions is loose because they have lower hydration numbers n_w and a larger area of bare ions. That is why it is assumed that when they are adsorbed, the hydration shell is deformed by the interface, as shown in Figure 2.5. Hence, they can approach the interface up to a distance equal to the radius R_b of the bare ion. We will refer to them as "type I ions." They correspond to the chaotropes of Collins [55]. Smaller ions ("type II") have denser adsorption shells, which cannot be rearranged upon adsorption, so that they will most probably remain immersed in water, along with their hydration layer. Therefore, they can approach the interface only to distances equal to R_h. Type II ions correspond to the cosmotropes of Collins [55].

We will first calculate the energy u_i^S of the type I ions. Toward this aim, the London potential (Equation 2.59) is integrated over the volume of the water phase excluding the hydration shell, with r_{ij} being the distance between the volume element $d^3r = rdrd\phi dz$ and the ion positioned at $r = 0, z = 0$ (that is, the integration is over $z > -R_b$ and $r^2 + z^2 < R_h^2$). The integration is performed in cylindrical coordinates:

$$u_i^S = -\int_{-R_b}^{R_h} \int_{\sqrt{R_h^2 - z^2}}^{\infty} \frac{L_{iw}\rho_w 2\pi r dr dz}{(r^2 + z^2)^3} - \int_{R_h}^{\infty} \int_0^{\infty} \frac{L_{iw}\rho_w 2\pi r dr dz}{(r^2 + z^2)^3} = -\frac{2\pi}{3} \frac{L_{iw}\rho_w}{R_h^3} \left(1 + \frac{3}{4} \frac{R_b}{R_h}\right) \qquad (2.62)$$

here ρ_w is the molecular number concentration of water. Similarly, the bulk energy of the ion is (integration in spherical coordinates)

$$u_i^B = -\int_{R_h}^{\infty} \frac{L_{iw}}{r_{iw}^6} \rho_w 4\pi r_{iw}^2 \, dr_{iw} = -\frac{4\pi}{3} \frac{L_{iw}\rho_w}{R_h^3} \qquad (2.63)$$

The respective energies of the ensemble of water molecules (assuming a sphere of radius R_h, or a part of it) are

$$u_w^S = -\frac{2\pi}{3} \frac{L_{ww}\rho_w}{R_h^3}\left(1 + \frac{3}{4}\frac{R_b}{R_h}\right); \ u_w^B = -\frac{4\pi}{3}\frac{L_{ww}\rho_w}{R_h^3} \tag{2.64}$$

Substituting Equations 2.62 through 2.64 into the expression (Equation 2.61) for u_{i0}, one obtains an explicit relation for the adsorption energy with the ionic properties of type I ions:

$$u_{i0} = \left(1 - \frac{3}{4}\frac{R_b}{R_h}\right)\frac{2\pi}{3}\frac{\rho_w}{R_h^3}(L_{iw} - L_{ww}) \tag{2.65}$$

To calculate u_{i0} for type II ions, one must set $R_h = R_b$ in Equation 2.65, which simplifies the expression to

$$u_{i0} = \frac{\pi}{6}\frac{\rho_w}{R_h^3}(L_{iw} - L_{ww}) \tag{2.66}$$

The values of the hydration number n_w and the radius R_h of the hydrated ion depend strongly on the method used for their determination and can vary widely (see, e.g., p. 143 in ref. [56]). It seems that more reasonable results can be obtained by model calculations, rather than experimentally. For monovalent ions, Marcus [57] found that the hydration number n_w can be represented by the empirical relation

$$n_w = A_v/R_i \tag{2.67}$$

where $A_v = 3.6$ Å for all ions. He further assumed that the hydrating n_w water molecules, considered as spheres with radius $R_w = 1.38$ Å and volume $v_w = 11$ Å^3, are smeared around the ion, forming a layer of thickness $R_h - R_b$ and volume:

$$n_w v_w = \frac{4\pi}{3}\left(R_h^3 - R_b^3\right) \tag{2.68}$$

The last relation is used to calculate R_h. The values of n_w and R_h calculated in this way [29,57,58] are shown in Table 2.2. Robinson and Stokes (Equation 9.27 in ref. [43]) used a similar approach, but with a water molecular volume of $v_w = 30$ Å^3, which follows from the density of water. They also used different values of the hydration number n_w, which were calculated from the ion diffusivity (see Table 11.10 in ref. [43]). Ivanov et al. [29] calculated the radius R_h using both sets of parameters; for the ions of interest, the results for R_h did not differ much from each other. Both sets of R_h, those calculated by the method of Marcus and by the method of Robinson and Stokes, differ however much from the often-quoted values (e.g., in ref. [39]) of $R_h = 3.8$, 3.6, and 3.3 Å for Li$^+$, Na$^+$, and K$^+$, respectively, and 3.5, 3.3, and 3.3 Å for F$^-$, Cl$^-$, and Br$^-$, respectively.

The London constants L_{iw} for the ion–water molecule interaction, and L_{ww} for the interaction of N_w water molecules with a single water molecule are calculated directly from Equation 2.60:

$$L_{iw} = \frac{3\alpha_{p,i}\alpha_{p,w}}{2}, \quad L_{ww} = \frac{3}{4}N_w\alpha_{p,w}^2 I_w \tag{2.69}$$

For the calculation of L_{ww}, the ensemble of N_w water molecules is regarded as a sphere with polarizability $N_w \alpha_{p,w}$ [29]. The number N_w was assumed to be equal to the ratio between the volume of the bare ion and the volume of one water molecule [13]:

$$N_w = R_b^3/R_w^3 \qquad (2.70)$$

where R_w is the radius of the water molecule. For the value of R_w, two possibilities were tested in ref. [29]: (i) the average volume per molecule (30 Å^3), based on the water density, yields $R_w = 1.93$ Å; and (ii) the actual volume of a water molecule, 11 Å^3, corresponds to $R_w = 1.38$ Å. Better agreement with the experimental data was obtained with the second option, $R_w = 1.38$ Å. The value of the static polarizability of water used was $\alpha_{p,w} = 1.48$ Å^3 and that of the ionization potential was $I_w = 2.02 \times 10^{-18}$ J [59].

The values of the ionization potentials of the ions I_i are also questionable. In ref. [29], the ionization potential in vacuum was used for halogen ions. For the alkaline ions, it was corrected for the hydration effect, although the correction was small. Later, we found some new data about the system parameters, which showed that the hydration correction was even smaller and hereafter it was disregarded. For the cations, we used the second ionization potential because the first one corresponds to the ionization of the respective atom, not ion. Because the anions have already accepted one extra electron, their ionization potential must be equal to the negative value of the electron affinity.

2.2.4 COMPARISON WITH THE EXPERIMENTS

In the theoretical parts of this section (Subsections 2.2.1 through 2.2.3), we showed that a complete theory of the adsorption constant K and the adsorption isotherm of dilute monolayers requires detailed, more rigorous treatment of several important effects (most of which are new): (i) contribution $\sigma_0 \alpha_\perp$ to the adsorption energy E_a (cf. Equation 2.13) due to the penetration of the adsorbing surfactant tail through the clean interface; (ii) correct theory and new expression (Equation 2.16) for the adsorption "thickness" δ_a; (iii) purely electrostatic (nonspecific) contribution to the adsorption constant K_0 (Equation 2.40); (iv) effect and nature of the spreading pressure π_0 in LE monolayers; and (v) ion-specific effect on the adsorption constant K (Equation 2.56); (vi) origin and calculation of the ion-specific adsorption energy u_{i0} (Equations 2.61, 2.65, and 2.66). It turned out that these effects must be properly accounted for to give adequate treatment and interpretation of the experimental data. Although our ultimate goal is to check our theory of the Hofmeister effect, cf. (v) and (vi), we were forced by the logic of the study to analyze first effects (i–iv) and to relegate the analysis of the ion-specific effect to the end of this subsection.

2.2.4.1 Experimental Verification of the Theory of Adsorption Constant K

Our expression for the adsorption energy E_a differs from Equation 4.3 of Davies and Rideal [26] with the presence of the new terms $\sigma_0 \alpha_\perp$ and the additional u_{CH_2}. As shown in Section 2.2.1.1 and in Figure 2.1, this term stems from the disappearance of area $\alpha_\perp$ of interfacial tension σ_0 when the surfactant molecule is adsorbed at the interface. To demonstrate the significance of this energy, we will analyze the data by Rehfeld [51] and Gillap et al. [62] for the adsorption of $C_{12}H_{25}SO_4Na$ at various W|O interfaces, where the oil phase is varied—this is a simple way to change the interfacial tension σ_0 of the *clean* surface without excessively affecting the other parameters of Equation 2.13, and allows direct observation of the expected effect of σ_0 on K.

Substituting the expression (Equation 2.7) for K_s into the definition (Equation 2.56) of K, and using the result (Equation 2.13) for the adsorption energy E_a, one obtains

$$\ln K + \frac{u_{i0}}{2k_BT} \equiv \ln K_0 = \text{const} + \frac{u_{CH_2}}{3k_BT}n_C + \frac{\alpha_\perp}{3k_BT}\sigma_0 \qquad (2.71)$$

where

$$\text{const} = \frac{E_{\text{head}}}{3k_B T} + \frac{u_{\text{CH}_2}}{3k_B T} + \frac{1}{3} \ln \frac{4\delta_a}{\kappa_0^2} \tag{2.72}$$

In Equation 2.71, we preferred to correct the experimental adsorption constant K with the term $u_{i0}/2k_B T$, standing for the ion-specific adsorption energy of Na$^+$ ion ($u_{i0} = -0.34 \times k_B T$; cf. Table 2.2), to obtain the counterion-independent constant K_0 (Equation 2.40).

The effect of the nature of the hydrophobic phase is twofold. First (and more important), the change of the oil will affect σ_0 in the last term in Equation 2.71. The interfacial tension σ_0 of the pure W|O interfaces in Rehfeld's experiments ranges from 31.3 mN/m for water–1-hexene to 53.2 mN/m for water–heptadecane. According to Equation 2.71, this corresponds to a difference of approximately 0.4 in the value of $\ln K_0$. The second effect of the hydrophobic phase is on the transfer energy u_{CH_2}, which also depends, to a certain extent, on the nature of the oil. According to Tanford [38], the energy u_{CH_2} for transfer of a $-\text{CH}_2-$ group from water to hydrocarbons does not differ significantly for alkanes and alkenes. Aveyard and Briscoe [63] found only a weak dependence of u_{CH_2} on the length of the alkanes. We could not find data for the aromatic and cyclic hydrocarbons used by Rehfeld, but the relatively good coincidence between theoretical dependence and experimental values depicted in Figures 2.6 and 2.7 suggests that this second effect is smaller. Therefore, for all systems considered, only the term $\sigma_0\alpha_\perp$ in Equation 2.71 will vary significantly with the nature of the oil.

Three typical surface pressure isotherms $\pi^S(C^{2/3})$ for oils of different interfacial tensions σ_0, based on the data of Rehfeld [51], are shown in Figure 2.6. From these data, the values of K were determined according to Equation 2.47 and corrected with $u_{i0}/2k_B T$ according to Equation 2.71 to obtain the counterion-independent quantity $\ln K_0$. The obtained $\ln K_0$ values for several different oils with the same surfactant $C_{12}H_{25}SO_4Na$ are plotted in Figure 2.7 in coordinates $\ln K_0$ versus σ_0.

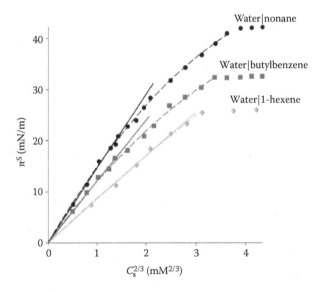

FIGURE 2.6 Surface pressure π^S versus 2/3-power of surfactant concentration $C_s^{2/3}$ for adsorption of $C_{12}H_{25}SO_4Na$ at the W|O interface for three typical oils with different interfacial tensions ($\sigma_0 = 50.9$, 40.1, and 31.3 mN/m for water–nonane, water–butylbenzene, and water–1-hexene, respectively). (Experimental data from S.J. Rehfeld, *J. Phys. Chem.* 71, 738–745, 1967. With permission.) $T = 25°C$. Solid lines, the linear dependence (Equation 2.39); dashed lines, fit with quadratic polynomial (Equation 2.47), up to cmc. From the polynomial dependences, the adsorption constants K, used in Figure 2.7, are determined.

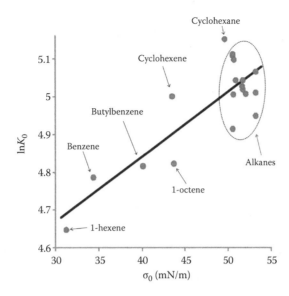

FIGURE 2.7 Dependence of $\ln K_0$ of $C_{12}H_{25}SO_4Na$ on the interfacial tensions σ_0 of the pure W|O interfaces. All points are determined from plots similar to those in Figure 2.6. The values of K were calculated from quadratic fit of π^S versus $C_s^{2/3}$ (Equation 2.47), and were corrected for the counterion effect according to Equation 2.71 to yield K_0. Solid line, theoretical dependence (Equation 2.71), with the theoretical slope $\alpha_\perp/3k_BT = 0.0172$ m/mN (Equation 2.71), corresponding to $\alpha_\perp = 21.2$ Å^2 [38]. The intercept is 4.16 ± 0.08.

The line is plotted according to Equation 2.71 with the theoretical slope $\alpha_\perp/3k_BT = 0.0172$ m/mN, corresponding to the value $\alpha_\perp = 21.2$ Å^2 given by Tanford [38] for the hydrocarbon chain cross-sectional area. The theoretical intercept, calculated using Equations 2.71 and 2.72 with values from the parameters given in Section 2.2.1.1, is 4.8. Its experimental value, according to Figure 2.7, is 4.16 ± 0.08, which is reasonably close to the theoretical prediction. These data confirm our theory of the adsorption energy and thickness (Equations 2.13 and 2.16).

We now turn to the dependence of the adsorption constant K on the number of carbon atoms n_C in the hydrophobic tail of homologous surfactants. It is obvious from Figure 2.8 that the addition of $-CH_2-$ groups leads to a strong increase of the surface pressure π^S, which is related, according to Equation 2.71, to the energy u_{CH_2} of transfer of a $-CH_2-$ group from water to the hydrophobic phase. Both Tanford [38] and Davies and Rideal (Table 4I in ref. [26]) cite different values of u_{CH_2} for W|O and W|G interfaces. For the water–alkane interface, Tanford gives $u_{CH_2}^{WO} = 5.75 \times 10^{-21}$ J versus Davies and Rideal's $u_{CH_2}^{WO} = 5.98 \times 10^{-21}$ J. For the W|G surface, Tanford gives $u_{CH_2}^{WO} = 4.35 \times 10^{-21}$ J. Davies and Rideal found that $u_{CH_2}^{WG}$ depends on the coverage: for dilute monolayers, $u_{CH_2}^{WG} = 4.17 \times 10^{-21}$ J, whereas for denser monolayers (90 Å^2 per molecule) $u_{CH_2}^{WG} = 4.85 \times 10^{-21}$ J.

We will analyze the effect of the chain length n_C of a homologous series of surfactants on the value of K_0 at different W|O interfaces (Figure 2.8a). Because the interfacial tensions σ_0 of the oils used in the experiment (especially of the nonsaturated ones) differ significantly, we transformed Equation 2.71 in such a way that this effect was eliminated. This was achieved by subtracting the term $\alpha_\perp\sigma_0/3k_BT$ and adding instead a term $\alpha_\perp\sigma_0^{alkane}/3k_BT$ referring to a typical alkane, for example, decane with $\sigma_0^{alkane} = 52$ mN/m. As a result, a new adsorption constant, K_0^{alkane}, independent of σ_0 was introduced:

$$\ln\ K_0^{alkane} \equiv \ln\ K_0 + \frac{\alpha_\perp}{3k_BT}\left(\sigma_0^{alkane} - \sigma_0\right) = \text{const} + \frac{\alpha_\perp}{3k_BT}\sigma_0^{alkane} + \frac{u_{CH_2}}{3k_BT}n_C \qquad (2.73)$$

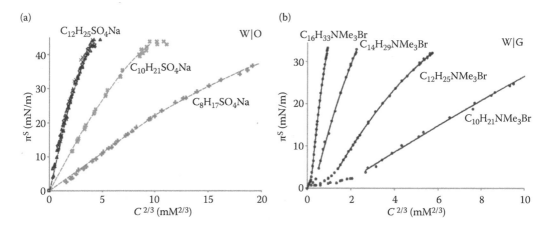

FIGURE 2.8 Dependence of the surface pressure π^S versus 2/3-power of the mean activity, $C^{2/3}$. (a) Alkylsulfates of different chain lengths at the W|O interface, with or without added NaCl. The data for $C_{12}H_{25}SO_4Na$, with various amounts of NaCl, are the same as in Figure 2.3; the data for $C_8H_{17}SO_4Na$ and $C_{10}H_{21}SO_4Na$ at the water–decane interface with the addition of 50 to 500 mM of NaCl [64]. The data were processed as in Figure 2.6: dashed lines are quadratic fits with Equation 2.47. (b) Alkyltrimethylammonium bromides of different chain lengths at the W|G interface. Data for $C_{10}H_{21}NMe_3Br$ with 0 to 10 mM NaBr [18]. (Data for $C_{12}H_{25}NMe_3Br$, $C_{14}H_{29}NMe_3Br$, and $C_{16}H_{33}NMe_3Br$ from Aratono, M. personal communication, 2010. With permission.) Solid lines correspond to quadratic fits (Equation 2.47). From the fits, the values of the adsorption constants K were obtained; the results are used in Figure 2.9.

K_0^{alkane} is the expected adsorption constant of a surfactant with n_C carbon atoms at the water–alkane interface. The data for adsorption of sodium alkylsulfates and alkyltrimethylammonium bromides at the W|O interface, plotted in Figure 2.9, confirms Equation 2.73. The value $u_{CH_2}^{WO} = 5.75 \times 10^{-21}$ J, quoted by Tanford, was used to draw the lines.

We now turn to the adsorption of homologous series of surfactants at the W|G surface in the LE region (Figure 2.8b). We determined the adsorption constant K for each particular system by fitting the experimental data $\pi^S(C^{2/3})$ with Equation 2.47. Because σ_0 is the same for all surfactants, the analysis can be carried out by using Equation 2.71. The calculated dependence of $\ln K_0$ on n_C for the LE monolayers at W|G is again linear, as shown in Figure 2.9, that is, Equation 2.71 is valid. Surprisingly, within experimental error, the slope is the same as for the W|O interface, that is, u_{CH_2} is again 5.75×10^{-21} J, which is considerably larger than the values quoted by Davies and Rideal, 4.17 to 4.85×10^{-21} J [26]. The coincidence of the W|G and W|O values of u_{CH_2} is certainly because the transfer of a –CH_2– group is from water to an LE adsorption layer (and not to a gas!), which is nearly equivalent to a transfer from water to oil. We found it difficult to estimate the corresponding energy of transfer from water to *gaseous* monolayers due to the insufficient and contradictory data for adsorption in this region.

In Figure 2.10, the surface pressure isotherms of four $C_{12}H_{25}NMe_3^+$ salts at the W|G surface are shown. Obviously, the counterion can drastically increase the surface activity of the surfactant ion, and the effect of the counterion follows the Hofmeister series (Equation 2.2). Similar curves were obtained (but not shown) for the other surfactants considered below—alkylsulfates and 1-dodecyl-4-dimethyl aminopyridinium ($C_{12}H_{25}PyrNMe_2^+$) halogenides.

Our aim now is to demonstrate how the results in Figure 2.10 can be explained quantitatively with the model developed in Section 2.3. Equation 2.56 can be presented in logarithmic form:

$$\ln K = \ln K_0 - \frac{1}{2}\frac{u_{i0}}{k_B T} \tag{2.74}$$

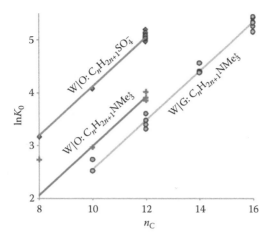

FIGURE 2.9 Dependence of the logarithm of the adsorption constant $\ln K_0$ on the number of carbon atoms n_C in the hydrocarbon chain. Circles, data for alkylsulfates at the W|O interface; the values of K_0 were calculated from $\pi^S\left(C_s^{2/3}\right)$ dependences similar to those shown in Figure 2.8. The data for $C_{12}H_{25}SO_4Na$ are from the same sources as in Figure 2.7; data for $C_8H_{17}SO_4Na$ and $C_{10}H_{21}SO_4Na$ with 0 to 500 mM of NaCl, and the oil is either decane [64] or other alkanes [62]. Correction for σ_0 was made, and all data were reduced to $\sigma_0 = 52$ mN/m according to Equation 2.73. Crosses, adsorption of alkyltrimethylammonium salts at the W|O interface. Data for $C_8H_{17}NMe_3^+$ and $C_{10}H_{21}NMe_3^+$ with decane in the presence of 0 to 500 mM NaCl [64]; data for $C_{12}H_{25}NMe_3Br$ and Cl with hexane, hexadecane [65], and petroleum ether and 0 to 500 mM NaCl [53]. Diamonds, data for the adsorption of alkyltrimethylammonium bromides and chlorides at the W|G surface, with various amounts of salt [18,65,66]. Lines, the theoretical dependence (Equation 2.71) with the value $u_{CH_2} = 5.75 \times 10^{-21}$ J given by Tanford [38]. The intercepts were determined as fitting parameters: −0.55 for alkylsulfates at the W|O interface, −1.63 for alkyltrimethylammonium salts at the W|O interface, and −2.12 at the W|G interface.

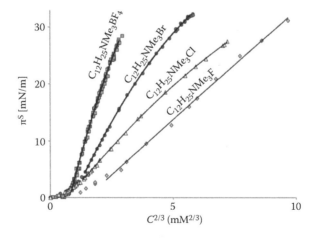

FIGURE 2.10 Surface pressure π^S versus $C^{2/3}$ (2/3-power of the mean activity) for $C_{12}H_{25}NMe_3^+$ salts at the W|G surface. The surface pressure at a given concentration increases in accordance with Hofmeister series $\left(BF_4^- > Br^- > Cl^- > F^-\right)$. Data for $C_{12}H_{25}NMe_3BF_4$ with the addition of 0 to 15 mM NaBF$_4$ [20]; $C_{12}H_{25}NMe_3Br^-$ and Cl$^-$ are without additives [20,65]; the data for $C_{12}H_{25}NMe_3F$ are obtained with 100 mM NaF added to 0 to 15 mM $C_{12}H_{25}NMe_3Br$ solutions [68]. $T = 23°C$ to $25°C$. The lines are quadratic fits. The results were used for calculation of the respective adsorption constants K, used in Figure 2.11.

This suggests that the model can be checked by plotting $\ln K$ versus u_{i0}. We found scarce experimental data for surface tension of the same ionic surfactant with different counterions (with or without added salt). To expand the databases to the full set of counterions of interest, we used the relation (Equation 2.71) between K_0 and the number of carbon atoms n_C. We were able to recalculate, from the value of K of any surfactant from a homologous series, the adsorption constant K_{12} for a surfactant with $n_C = 12$:

$$\ln K_{12} = \ln K - \frac{u_{CH_2}}{3k_B T}(n_C - 12) \tag{2.75}$$

where $u_{CH_2} = 5.75 \times 10^{-21}$ J as determined above. For a homologous series of surfactants with the same ionic head at the same interface and temperature, this standard constant K_{12} should depend on u_{i0} only (Equations 2.71 and 2.72).

The specific adsorption energies u_{i0} of the ions were taken from Table 2.2. The values of K were found by curve-fitting with the quadratic dependence (Equation 2.47). They were then used to calculate the nonspecific adsorption constant K_0 for each of the surfactant ions through Equation 2.74. In agreement with the theory, the obtained values of $\ln K_0$ are the same for a given surfactant ion with any considered counterion. In some cases, adsorption data for a mixture of two counterions only were available, for example, refs. [18,67,68]. Then, the value of K_0, which is unique for the two counterions, was determined by using a procedure described in Section 2.2.4.2; after that, K was recalculated for the prevailing ion from Equation 2.56, $K = K_0 \exp(-u_{i0}/2k_B T)$.

The dependence of the adsorption constant on the adsorption energy u_{i0} of the counterion is illustrated in Figure 2.11, where $\ln K_{12}$ versus $-u_{i0}/k_B T$ is plotted for three different surfactant ions. The lines are drawn according to Equation 2.74 by using the theoretical slope 1/2 and the average values of $\ln K_0$ for each surfactant ion: 4.80 ± 0.13 for $C_{12}H_{25}SO_4^-$, 3.74 ± 0.02 for $C_{12}H_{25}PyrNMe_2^-$, and $\ln K_0 = 3.28 \pm 0.10$ for $C_{12}H_{25}NMe_3^+$. The theory describes the data adequately. Note that the values of the adsorption energy u_{i0} listed in Table 2.2 were obtained without using any free adjustable parameter.

Below, we arranged the counterions according to the experimental values of the adsorption constants K_{12} (Figure 2.11), that is, according to their "adsorption efficiency" for a given surfactant head group:

$$Ac^- < F^- < Cl^- < Br^- \lesssim NO_3^- < BF_4^- \quad \text{for cationic surfactants} \tag{2.76}$$

$$Li^+ < Na^+ < NH_4^+ < Rb^+ \lesssim K^+ < NMe_4^+ \quad \text{for anionic surfactants} \tag{2.77}$$

Because of the relation (Equation 2.74) between K and u_{i0}, the order will remain the same if one uses as criterion the absolute value of u_{i0} (see Table 2.2). The cation series (Equation 2.77) follows the Hofmeister series (Equation 2.2), with one exception—the NH_4^+ ion. In contrast, although the sequence (Equation 2.76) of the anions is the same as the Hofmeister series (Equation 2.1), the signs "<" and ">" are opposite, that is, the "adsorption efficiency" in Equation 2.76 increases whereas the "precipitation efficiency" in Equation 2.1 decreases from left to right. A possible reason for the coincidence of the two series of cations is that Hofmeister worked only with negatively charged proteins, similar to interfaces with adsorbed anionic surfactants, corresponding to Equation 2.77. The situation with the anions is the opposite—Hofmeister's proteins were negatively charged, whereas the interface with adsorbed cationic surfactants is positive. This explanation is in agreement with the finding of Schwierz et al. [4], who argued that the relative order of anions may reverse depending on the charge of the

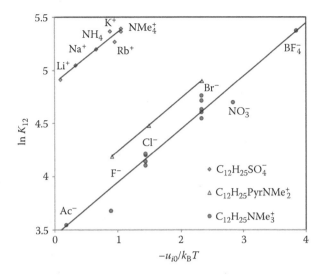

FIGURE 2.11 Dependence of the adsorption constant K_{12} on the ion adsorption energy $-u_{i0}/k_B T$ of surfactants with three different head groups at the W|G surface. The values of K_{12} were determined from $\pi^s(C^{2/3})$ data as those in Figure 2.10. For this plot, we used the calculated values of u_{i0} (Table 2.2), corresponding to the model presented in Sections 2.2.3.1 and 2.2.3.2. Lines, comparison with the theoretical dependence with fixed slope 1/2 (Equation 2.74). Sources: $C_{12}H_{25}SO_4^-$ with Li^+ alone [67,69] and Li^+ with added 1 mM NH_4^+ [67]; for Na^+ (Figure 2.3); NH_4^+ and NMe_4^+ stand for 5 to 10 mM NH_4^+ or NMe_4^+ with 1 to 3 mM Li^+ [67]; K^+ and Rb^+ are without added salt [69]; $T = 23°C$ to $33°C$. K values for $C_{12}H_{25}PyrNMe_2^+$ halogenides are calculated from Koelsch's data [70] (no added salt or 100 mM NaF/NaCl/NaBr added to $C_{12}H_{25}PyrNMe_2Br$ at room temperature). $C_{12}H_{25}NMe_3^+$ is from Aratono (BF_4^- with the addition of 0–10 mM $NaBF_4$, NaCl, NaBr [20,21,65] and Bergeron [66]). Data for $C_{10}H_{21}NMe_3^+$ (Br^- with 0–10 mM NaCl [18,66]), $C_{14}H_{29}NMe_3^+$ (Cl^- and Br^- [65,66]), and $C_{16}H_{33}NMe_3^+$ (Cl^- and Br^- in the presence of 0–100 mM salt [18,65,66]; 10–100 mM Ac^-, NO_3^- or F^- in the presence of 0–0.5 mM Br^- [18]) are also used in this figure—the corresponding adsorption constants K of these surfactants were reduced to the standard constant K_{12} of $C_{12}H_{25}NMe_3^+$ through Equation 2.75. All measurements with $C_nH_{2n+1}NMe_3^+$ have been performed at $T = 20°C$ to $25°C$.

surface, with $I^- > Cl^- > F^-$ on positively charged surfaces but $F^- > Cl^- > I^-$ on negatively charged surfaces (see also ref. [3]).

The Hofmeister effect also influences the other parameters in the surface pressure isotherm (Equation 2.47). We found a strong correlation between the spreading pressure π_0 of the surfactant and its counterion. In Figure 2.12, the spreading pressure of $C_nH_{2n+1}NMe_3^+$ salts is plotted against u_{i0}. The dependence is close to linear; the spreading pressure increases in absolute value with $-u_{i0}$.

The good coincidence between the theoretical dependence Equation 2.74 with the theoretical values of u_{i0} (Table 2.2) and experiment, demonstrated above, suggests that the effect of the type of counterions on the adsorption constant K is due not only to steric reasons, related to ion size, as it is sometimes assumed [71–73], but is also due to van der Waals interactions.

There are at least three effects that we observed and partially explained but we believe that they still need additional in-depth analysis and clarification: (i) the factors determining the value of the nonspecific adsorption constant K_0 of the surfactant; (ii) the value and nature of the adsorption constant of a surfactant at W|G surface in *gaseous* monolayers; and (iii) the reason for the dependence of the spreading pressure π_0 on the ion-specific adsorption energy u_{i0} (Figure 2.12). We have some preliminary ideas and calculations but we lack, for the time being, enough reliable data to check and improve them. Hence, we postpone these issues for future studies.

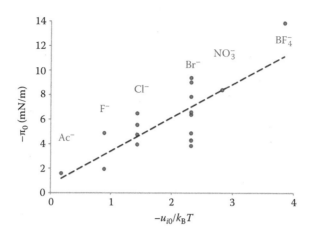

FIGURE 2.12 Hofmeister effect on Langmuir's spreading pressure π_0 of $C_nH_{2n+1}NMe_3^+$ salts at the W|G surface: dependence of $-\pi_0$ on $-u_{i0}/k_BT$. The spreading pressure is equivalent to the intercept of the quadratic fits of $\pi^S(C^{2/3})$ (Figure 2.10). π_0 shows strong correlation with the ion adsorption energy u_{i0}. Data sources as in Figure 2.11.

2.2.4.2 Adsorption in the Presence of a Mixture of Counterions

Often, ionic surfactants are used with a mixture of counterions, for example, refs. [18,20]. Well-studied experimentally are systems with two counterions, one of which came with the surfactant and the second one with the salt additive (e.g., $C_{12}H_{25}SO_4^-Na^+$ in the presence of KCl additive). To derive an analogue of the isotherm (Equation 2.55); for a mixture of two counterions, one must use the multi-ion form of the generalized Gouy equation (Equation 2.53) instead of its one-counterion version (Equation 2.54). By analogy with the derivation of Equations 2.55 through 2.57, the expression for the zeroth approximation for the potential Φ_0^S (Equation 2.38), is inserted into the Gouy equation (Equation 2.53) to give:

$$\Gamma_s^2 = \left(\frac{4K_s}{\kappa_0^2}\right)^{2/3} \sum_i C_i e^{-u_{i0}/k_BT} \frac{C_s^{2/3}}{C_t^{1/3}} \tag{2.78}$$

After taking the square root of this equation and making some rearrangements, a generalization of the adsorption isotherm (Equation 2.55) for ion mixtures is obtained:

$$\Gamma_s = K_0 \left(\sum_i x_i e^{-u_{i0}/k_BT}\right)^{1/2} C_t^{1/3} C_s^{1/3} \tag{2.79}$$

where K_0 is given by Equation 2.40, C_t is the total electrolyte concentration, and $x_i = C_i/C_t$ is the fraction of ith counterions from all counterions. For the most common case of two counterions, Equation 2.79 can be rewritten as

$$\Gamma_s = K_0 \left(x_1 e^{-u_{10}/k_BT} + x_2 e^{-u_{20}/k_BT}\right)^{1/2} C_t^{1/3} C_s^{1/3} \tag{2.80}$$

We proceed now to the calculation of the surface pressure π^s. If the adsorption of the coion is neglected, Gibbs isotherm reads:

$$d\pi^s = \Gamma_1 d\mu_1 + \Gamma_2 d\mu_2 + \Gamma_s d\mu_s \qquad (2.81)$$

If the bulk surfactant solution is ideal, Equation 2.81 yields $C_s(\partial \pi^S/\partial C_s)_{C_1,C_2} = k_B T \Gamma_s$. This equation can be integrated, after substituting into it Γ_s from Equation 2.80. The result is:

$$\pi^S = 3k_B T K_0 \left(\frac{C_1}{C_t} e^{-u_{10}/k_B T} + \frac{C_2}{C_t} e^{-u_{20}/k_B T} \right)^{1/2} C_t^{1/3} C_s^{1/3} + \pi_0(C_1, C_2) \qquad (2.82)$$

The integration constant π_0 is Langmuir's spreading pressure. In the case of the W|O interface, it is obvious that $\pi_0 = 0$ if the small effect of the salt itself on σ_0 [6,74] is neglected. If only one counterion is present in the system, then $C_1 = C_t$ and $C_2 = 0$, and the bracket in the above equation simplifies to $\exp(-u_{10}/2k_B T)$, and Equation 2.82 becomes identical to the one-counterion surface tension isotherm (Equation 2.43). Equation 2.82 is compared in Figure 2.13 to experimental data from ref. [18] for solutions of $C_{16}H_{33}NMe_3X$ and NaX, where X stands for Cl$^-$ and Br$^-$. For these two ions and the data in the figure, we assumed that the dependence of the spreading pressure π_0 on C_1 and C_2 was negligible.

For the W|O interface, where $\pi_0 = 0$, one can determine the adsorptions Γ_1 and Γ_2 of the two counterions from the surface pressure isotherm (Equation 2.82). From Gibbs isotherm (Equation 2.81), it follows that

$$\Gamma_1 = \frac{C_1}{k_B T} \left(\frac{\partial \pi^S}{\partial C_1} \right)_{C_2,C_s} \quad \text{and} \quad \Gamma_2 = \frac{C_2}{k_B T} \left(\frac{\partial \pi^S}{\partial C_2} \right)_{C_1,C_s} \qquad (2.83)$$

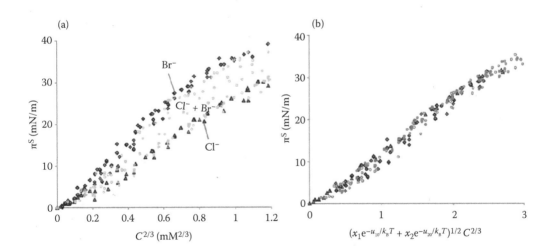

FIGURE 2.13 **(See color insert.)** (a) Surface pressure π^S versus 2/3-power of the mean activity $C^{2/3}$ for $C_{16}H_{33}NMe_3Cl$ with or without added NaCl (diamonds); $C_{16}H_{33}NMe_3Br$ with or without NaBr (triangles); $C_{16}H_{33}NMe_3^+$ in the presence of both ions Cl$^-$ and Br$^-$ (empty circles) at the W|G surface. The data for a single counterion, Cl$^-$ or Br$^-$, falls on separate master curves according to the salting-out effect (Equation 2.32), with slopes $3k_B T K_{Br}$ and $3k_B T K_{Cl}$ correspondingly. In contrast, the data for the counterion mixtures are dispersed between these two curves. (b) Drawn in coordinates π^S versus $\left(e^{-u_{10}/k_B T} x_1 + e^{-u_{20}/k_B T} x_2 \right)^{1/2} C_t^{1/3} C_s^{1/3}$, all data fall on a single master curve with slope $3k_B T K_0$, according to Equation 2.80. (Data from Para, G. et al., *Adv. Colloid Interface Sci.* 122, 39–55, 2006.)

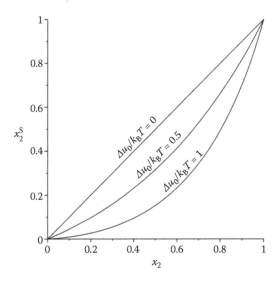

FIGURE 2.14 Interface composition $x_2^S = \Gamma_2/\Gamma_s$ as a function of the bulk composition $x_2 = C_2/C_t$ at different values of the difference $\Delta u_0 = (u_{20} - u_{10})/k_BT$. The surface is enriched with the ion having larger (in absolute value) ion adsorption energy u_{i0}.

Here, substituting π^S from Equation 2.82 with $\pi_0 = 0$, one can obtain the adsorptions of both counterions and thereby, together with Equation 2.80, the composition of the interface in terms of molar parts $x_i^S \equiv \Gamma_i/\Gamma_s$:

$$x_2^S = 1 - x_1^S = x_2 \frac{1 + \dfrac{1}{2} x_1 \left[1 - e^{(u_{20} - u_{10})/k_BT} \right]}{x_2 + x_1 e^{(u_{20} - u_{10})/k_BT}} \tag{2.84}$$

The dependence of the surface composition x_2^S on the bulk composition x_2 at three values of the difference $(u_{20} - u_{10})/k_BT$ is illustrated in Figure 2.14.

2.3   EOS AND ADSORPTION ISOTHERM OF DENSE MONOLAYERS

2.3.1   NONLOCALIZED ADSORPTION OF NONIONIC SURFACTANTS

Most theories of the EOS and adsorption isotherms of ionic surfactants are obtained by simply extending theories of nonionic surfactants to account for the electrostatic interaction. This is usually done by using the Gouy equation (Equation 2.21), for example, see refs. [28,36,75]. Our analysis revealed that most of the problems encountered in the theory of adsorption of ionic surfactants stem, in fact, from certain drawbacks of the respective theories of nonionic surfactants. These problems were dealt with in refs. [31–33]; for the readers' convenience, we will present in this section a brief account of some of the findings from these articles.

The two most widely used EOS for nonionic surfactants are those of Langmuir–Frumkin (usually called the Frumkin equation) and the two-dimensional (2-D) equation of van der Waals (also known as the Volmer–De Boer equation):

1. Frumkin EOS [76,77],

$$\frac{\pi^S}{k_BT} = -\frac{1}{\alpha_L} \ln(1 - \alpha_L \Gamma_s) - B_{attr} \Gamma_s^2 \tag{2.85}$$

Here, α_L is the area parameter of Langmuir's model (we will show in Section 2.3.1.1 that α_L is different from the cross-sectional area of the molecule); B_{attr} is the attractive part of the second virial coefficient accounting for the long-ranged attractive interactions between the adsorbed molecules (see Section 2.3.1.2).

2. Van der Waals EOS [78,79],

$$\frac{\pi^S}{k_B T} = \frac{\Gamma_s}{1 - \alpha_V \Gamma_s} - B_{attr}\Gamma_s^2 \qquad (2.86)$$

Here, α_V is the area parameter of Volmer's model, also different from the cross-sectional area of the molecule.

A third important EOS for hard discs (unfortunately, not much used in the literature) was derived by Helfand et al. [80]. If their result is modified with the same attractive term $-B_{attr}\Gamma_s^2$ as in Equations 2.85 and 2.86, the following equation is obtained:

3. Modified Helfand–Frisch–Lebowitz (HFL) EOS [31,33,80],

$$\frac{\pi^S}{k_B T} = \frac{\Gamma_s}{\left(1 - \alpha\Gamma_s\right)^2} - B_{attr}\Gamma_s^2 \qquad (2.87)$$

The first terms on the right-hand sides of Equations 2.85 through 2.87 refer to hard core interactions only and were proposed by Langmuir [76], Volmer [78], and Helfand et al. [80], respectively. The attractive term $-B_{attr}\Gamma_s^2$ was introduced into the Langmuir isotherm by analogy with the three-dimensional van der Waals EOS by Frumkin [77], to make it applicable to surfactants with large hydrophobic tails. De Boer [79] did the same for Volmer EOS, and in refs. [31,33], it was done for the EOS of HFL. One must keep in mind that an additive term, proportional to Γ_s^2, strictly speaking, can account only for the contribution of binary collisions, that is, it is correct only for low surface concentrations, whereas the hard core parts are valid (in the framework of the respective model) for any surface concentration.

Let us discuss the physical model behind these equations. The Langmuir EOS can be applied for adsorption on a 2-D lattice of adsorption centers if the following conditions hold [81]: (i) each center can be occupied only by one molecule, (ii) the molecules cannot exchange positions over the surface and jump from center to center, and (iii) the adsorbed molecules do not interact with each other neither by attractive nor by repulsive forces. The first two limitations must also apply to Frumkin EOS (Equation 2.85). Quite obviously, the combination of these conditions cannot be realized with a fluid adsorbed layer—despite this, the Langmuir and Frumkin EOSs have been widely applied to such systems. One possible reason is the fact that the integration of Langmuir EOS with Gibbs adsorption isotherm to eliminate Γ_s leads exactly to the empirical Szyszkowski equation [82], which describes well the dependence of the surface tension σ on the surfactant concentration C_s for low-molecular weight nonionic surfactants.

Tonks [83] has shown that the equation of Volmer is rigorous for delocalized adsorption of solid rods at a line. However, such a one-dimensional (1-D) model is hardly applicable to a 2-D fluid interface. A more realistic model for the adsorbed monolayer of nonionic surfactant would be a 2-D system of hard discs. Reiss et al. [84] developed a very astute procedure (which they called *scaled particle theory*) for treating systems of hard core particles. They solved exactly several problems, related to hard particles, but it turned out that the 2-D case (hard discs at interfaces) had in principle no exact analytical solution [85]. Nevertheless, Helfand et al. [80] succeeded in deriving an almost exact simple 2-D EOS for nonattracting hard discs (the original HFL equation is Equation 2.87 with

$B_{attr} = 0$). It is impossible to present in a concise manner their theory and we refer the interested reader to their original article [80]. Rusanov [86] obtained their result using a totally different approach.

The respective adsorption isotherms can be obtained by integrating the Gibbs equation for nonionic surfactant ($d\pi^S = k_B T \Gamma_s d\ln C_s$): by substituting in it π^S from either Equations 2.85, 2.86, or 2.87, and integrating with respect to Γ_s, one finds the corresponding adsorption isotherms for nonionic surfactants*:

$$\text{1. Langmuir–Frumkin: } K_s C_s = \frac{\Gamma_s}{1-\alpha_L \Gamma_s} \exp(-2B_{attr}\Gamma_s) \qquad (2.88)$$

$$\text{2. van der Waals: } K_s C_s = \frac{\Gamma_s}{1-\alpha_V \Gamma_s} \exp\left(\frac{\alpha_V \Gamma_s}{1-\alpha_V \Gamma_s} - 2B_{attr}\Gamma_s\right) \qquad (2.89)$$

$$\text{3. Modified HFL: } K_s C_s = \frac{\Gamma_s}{1-\alpha \Gamma_s} \exp\left[\frac{\alpha \Gamma_s(3-2\alpha\Gamma_s)}{(1-\alpha\Gamma_s)^2} - 2B_{attr}\Gamma_s\right] \qquad (2.90)$$

The isotherm, following from the original HFL EOS, can be obtained by setting $B_{attr} = 0$ in Equation 2.90. Equations 2.88 through 2.90 will be extended in Section 2.3.2 to ionic surfactants.

2.3.1.1 The Area per Molecule α

The reliability of the hard core parts of Equations 2.85 through 2.87 (with $B_{attr} = 0$) can be checked by comparing their virial expansions in terms of Γ_s with the exact virial EOS, obtained numerically [80], namely,

$$\pi^S/k_B T = \Gamma_s + 2\alpha\Gamma_s^2 + 3.128\alpha^2\Gamma_s^3 + 4.262\alpha^3\Gamma_s^4 + 4.95\alpha^4\Gamma_s^5 + \dots \qquad (2.91)$$

This expansion can be compared with the virial expansion of the hard core part of the HFL EOS (Equation 2.87):

$$\pi^S/k_B T = \Gamma_s + 2\alpha\Gamma_s^2 + 3\alpha^2\Gamma_s^3 + 4\alpha^3\Gamma_s^4 + 5\alpha^4\Gamma_s^5 + \dots \qquad (2.92)$$

Equation 2.92 nearly coincides with the exact expansion (Equation 2.91). This means that HFL EOS (Equation 2.87) with $B_{attr} = 0$, is a very good approximation for the nonlocalized adsorption of rigid discs both at low and high surface coverages. In contrast, the expansions of the hard core parts of Langmuir and Volmer EOS (Equations 2.85 and 2.86), are in obvious disagreement with Equation 2.91:

$$\text{Langmuir: } \pi^S/k_B T = \Gamma_s + \alpha_L\Gamma_s^2/2 + \alpha_L^2\Gamma_s^3/3 + \alpha_L^3\Gamma_s^4/4 + \alpha_L^4\Gamma_s^5/5 + \dots, \qquad (2.93)$$

$$\text{Volmer: } \pi^S/k_B T = \Gamma_s + \alpha_V\Gamma_s^2 + \alpha_V^2\Gamma_s^3 + \alpha_V^3\Gamma_s^4 + \alpha_V^4\Gamma_s^5 + \dots \qquad (2.94)$$

---

* Helfand et al. derived only the EOS with $B_{attr} = 0$ (Equation 2.87). The adsorption isotherms (Equation 2.90), based on this equation, and its modification for interacting surfactants (Equation 2.87), were also derived and extended to ionic surfactants in refs. [29,31–33]. However, because these isotherms are virtually straightforward consequences of Equation 2.87, we prefer to call all of them modified adsorption isotherms of HFL.

This could be expected because their physical bases are very different from nonlocalized adsorption at fluid interfaces. The above conclusions are visualized in Figure 2.15, where the curves correspond to the indicated EOS and the points refer to the exact numerical values. The plot of the HFL equation almost coincides with the numerical values, which confirms that it is nearly exact. The Langmuir and Volmer EOS are in substantial error compared with the exact numerical solution.

It is interesting to find out how the values of α_L and α_V are related to the actual area α used in the HFL model. To answer this question, Ivanov et al. [31] compared the ratio $\pi^S/k_B T \Gamma_s$, predicted by the three equations. Because the left-hand sides of the three EOS (Equations 2.85 through 2.87) are model-independent and equal, so must be the right ones. After dividing the three equations by Γ_s (for easier calculations) and setting the so-obtained right-hand side equal, one arrives at the following conditions for the identity of Langmuir and HFL EOS:

$$-\frac{1}{\alpha_L \Gamma_s} \ln(1 - \alpha_L \Gamma_s) = \frac{1}{(1 - \alpha \Gamma_s)^2} \tag{2.95}$$

Similarly, for Volmer and HFL EOS, one has:

$$\frac{1}{1 - \alpha_V \Gamma_s} = \frac{1}{(1 - \alpha \Gamma_s)^2} \tag{2.96}$$

Equation 2.96 can be easily solved for $\alpha_V(\Gamma_s)$:

$$\alpha_V/\alpha = 2 - \alpha \Gamma_s \tag{2.97}$$

In contrast, the relation between α_L and α can be found only by numerically solving Equation 2.95. The asymptotic behavior of the solution for α_L at $\Gamma_s \to 0$ is

$$\alpha_L/\alpha = 4 - \frac{14}{3} \alpha \Gamma_s + \dots \tag{2.98}$$

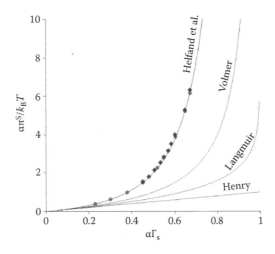

FIGURE 2.15 Dependence of the dimensionless surface pressure $\alpha \pi^S/k_B T$ on the surface coverage $\alpha \Gamma_s$: comparison of the hard core parts of HFL, Volmer, and Langmuir EOS (Equations 2.85 through 2.87; solid lines) with Henry's EOS ($\pi^S/k_B T = \Gamma_s$) and exact numerical calculations for delocalized adsorption of rigid discs at fluid interface (points). Circles, Monte Carlo calculations [80]; diamonds, dynamic calculations [80]. For all lines, α denotes the area per molecule (α, α_V, or α_L) corresponding to the plotted EOS.

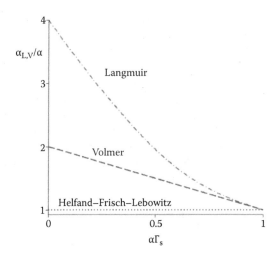

FIGURE 2.16 Dependence of the ratio of the area parameters α_L and α_V to the actual molecular area $\alpha = \pi d^2/4$ (d – molecular diameter) on the surface coverage $\alpha \Gamma_s$ for the EOS of Langmuir (dash-dot line) and Volmer (dashed line; cf. Equations 2.95 through 2.97). For comparison, we have plotted the dotted line, which presents the analogous ratio (which is unity) for the molecular area of the HFL model.

whereas at $\alpha \Gamma_s \to 1$ it is $\alpha_L/\alpha = 2 - \alpha \Gamma_s + \dots$ The obtained dependences of α_L and α_V on $\alpha \Gamma_s$ are shown in Figure 2.16. Rusanov [87] had ideas similar to ours about the dependence of the minimum area per molecule on the adsorption Γ_s. By using a different approach, he found the limiting values 4 and 1 at $\alpha \Gamma_s \to 0$ and $\alpha \Gamma_s \to 1$, respectively, for the Langmuir model. For the initial slope at $\Gamma_s \to 0$, however, he obtained -6.616 instead of our value $-14/3$ (see Equation 2.98).

The results from this section lead to the conclusion that the area per molecule, determined by using a model isotherm whose hard core part is different from the HFL equation, may not be a true physical constant, but may depend on the model used and the adsorption Γ_s.

2.3.1.2 The Interaction Parameter β

At low coverages, EOS can be expanded into virial series with respect to the powers of Γ_s:

$$\frac{\pi^s}{k_B T} = \Gamma_s + B_2 \Gamma_s^2 + B_3 \Gamma_s^2 + \dots \tag{2.99}$$

The second virial coefficient B_2 for hard particles of diameter d (cross-sectional area $\alpha = \pi d^2/4$) with attractive potential $u(r)$ between two particles at a distance r is (cf. e.g., ref. [81]):

$$B_2 = -\pi \int_0^\infty \left(e^{-u/k_B T} - 1 \right) r \, dr = 2\alpha - \pi \int_d^\infty \left(e^{-u/k_B T} - 1 \right) r \, dr \tag{2.100}$$

The second integral in this equation is, in fact, B_{attr} in Equations 2.85 through 2.87. Its value depends on the model used for $u(r)$, as will be demonstrated below with two examples.

In the first example, it is assumed that $u(r)$ is small and long-ranged. If $|u/k_B T| \ll 1$, then the exponent in Equation 2.100 can be expanded into series up to the linear term. If in addition $u(r)$ is represented by the London potential, $u = -L/r^6$ (L is London interaction constant), then B_{attr} is:

$$B_{attr} \equiv \pi \int_d^\infty \left(e^{-u/k_B T} - 1 \right) r \, dr \approx -\frac{\pi}{k_B T} \int_d^\infty u(r) r \, dr = \frac{\pi L}{4 k_B T d^4} \tag{2.101}$$

B_{attr} has dimension of area, so it seems natural to scale it with the area of the molecule α. The result is $B_{attr}/\alpha = u_c/k_BT$, where the contact potential $u_c = L/d^6$ is the absolute value of the attraction energy at contact between the molecules (u at $r = d$). We will call the parameter $\beta \equiv u_c/k_BT$ the *attraction constant*. Consequently, Equation 2.100 reads:

$$B_2 = 2\alpha - B_{attr} = 2\alpha - \alpha\beta \tag{2.102}$$

This result for B_2 can also be obtained by expanding the modified HFL equation (Equation 2.87) into a series in terms of $\alpha\Gamma_s$ and using $B_{attr} = \alpha\beta$.

The second example, which we will use in Section 2.3.1.3, is the short-ranged "sticky potential," introduced by Baxter [88]:

$$u(r) = \begin{cases} \infty, & r > d; \\ -E, & d < r < d(1+\lambda); \\ 0, & r > d(1+\lambda). \end{cases} \tag{2.103}$$

It is represented in Figure 2.17 by a solid line: up to the minimum distance d between the molecules the potential energy u is $+\infty$; then, in the potential well between d and $d(1 + \lambda)$, where $\lambda < 1$, it remains constant and equal to $-E(\lambda)$; after that it becomes zero. When two particles are in the attractive potential well and $\lambda \to 0$, the particles "stick" (hence, the name of the potential). It is assumed that under these conditions, $E \to \infty$ but in such a way that the attractive potential well gives finite contribution to B_{attr}. Baxter proposed an attraction potential depending logarithmically on λ. We will use the following expression (slightly different from Baxter's) for E:

$$E = k_BT \ln (u_c/k_BT\lambda) \tag{2.104}$$

In the limit $\lambda \to 0$, the energy $E \to \infty$. Here, u_c is still an undetermined coefficient. Substituting the expression (Equation 2.104) for E into Equation 2.101, and performing the integration (by keeping λ constant), one finds an expression that gives the evolution of B_{attr} with λ; in the limit $\lambda \to 0$, it leads to the final expression for B_{attr}:

$$B_{attr} = 4\alpha\left(\frac{u_C}{k_BT} - \lambda\right)\left(1 + \frac{\lambda}{2}\right) \xrightarrow{\lambda \to 0} 4\alpha\frac{u_c}{k_BT} \tag{2.105}$$

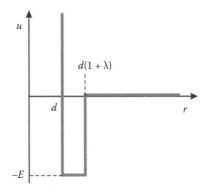

FIGURE 2.17 Scheme of the sticky potential (Equation 2.103; solid line).

Thus, for sticky potential, one has

$$B_{attr} = 4\alpha\beta; \quad B_2 = 2\alpha - 4\alpha\beta, \quad \text{where} \quad \beta = u_c/k_BT \tag{2.106}$$

The limiting transition $\lambda \to 0$ affects only the well width λd. The "moment of sticking" corresponds to $\lambda = 0$; hence, it is natural to assume that u_c is again the absolute value of the contact energy—for example, for discs attracting each other with the London potential, u_c is given again by $u_c = L/d^6$.

Unlike B_2 (and hence, B_{attr}), which depends on the model used for $u(r)$, the attractive constant $\beta = u_c/k_BT$ depends only on the nature of the interacting molecules and its value can be considered as a measure of the strength of this interaction—this allows easier comparison between different surfactants. The dependence of B_{attr} on the model used for $u(r)$ is reflected by the difference in the numerical coefficients in the expressions in Equations 2.102 and 2.106 for the cases of small, long-ranged potential and sticky potential, respectively.

2.3.1.3 New EOS and Adsorption Isotherm for Nonionic Surfactants

We checked the reliability of the basic EOS, discussed in the beginning of this section, by using experimental interfacial tension data for $C_{12}H_{25}SO_4Na$ [51], $C_{12}H_{25}NMe_3Cl$ [65,89], and $C_{12}H_{25}NMe_3Br$ [65] at the W|O interface. We considered the adsorption at the water–hexane interface because for W|O systems, β must be positive and small [39], and this fact can be used as a criterion for the correctness of a given isotherm. Ionic surfactants were selected because the nonionics are usually soluble in oil. We processed the data using three adsorption isotherms (Equations 2.88 through 2.90), based on the three basic EOS with attraction: Frumkin (Equations 2.85), van der Waals (Equations 2.86), and the modified HFL equations (Equations 2.87). The results for α and β, where $\beta = B_{attr}/\alpha$ as given by Equation 2.102, are summarized in Table 2.3 (the data processing is described in Section 2.3.2.2).

All three isotherms were fitted with high-precision (coefficient of determination R_d^2 higher than 0.999), but led to very different results for α and β. Thus, we reach a very important conclusion: the good fit of the data does not guarantee that the adsorption model is correct. The data for α confirm our conclusions in Section 2.3.1.1, namely, that the value of the area per molecule α strongly depends on the model used, and only the HFL model leads to realistic values, of the order of 20 to 25 Å^2. The most troubling result, however, is related to the values of β, which theoretically is expected to have positive value close to zero [39]. Instead, β was found negative for all isotherms. Because we had no doubt about the validity of the HFL model with $\beta = 0$, we concluded that the reason for the poor results was the way the attraction between the surfactant molecules was accounted for in Equations 2.85 through 2.87.

This gave us the impetus to derive in ref. [32] a new, hopefully more correct, EOS and adsorption isotherm whose hard core repulsion part is the same as in the original HFL model (see Equations

---

TABLE 2.3

Adsorption Parameters α and β, Obtained by Curve-Fitting of the Interfacial Tension Data for the Water–Hexane Interface by Using the Corresponding Models (Frumkin, van der Waals, and Modified HFL)

Model	$C_{12}H_{25}SO_4Na$ [51]		$C_{12}H_{25}NMe_3Cl$ [65,89]		$C_{12}H_{25}NMe_3Br$ [65]	
	α [Å^2]	β	α [Å^2]	β	α [Å^2]	β
Frumkin: Equations 2.85 and 2.88	41.4	−1.7	48.1	−2.2	42.6	−1.7
van der Waals: Equations 2.86 and 2.89	30.0	−3.2	35.3	−4.3	31.5	−3.0
Modified HFL: Equations 2.87 and 2.90	20.6	−1.3	24.4	−2.0	21.9	−1.0

2.87 and 2.90 with $B_{attr} = 0$). The new isotherm was thoroughly checked with experimental data for several nonionic surfactants and performed quite well [33]. Here, we will present a brief account of this model, and in Section 2.3.2 below, it will be extended to ionic surfactants.

To find some clues for the derivation for the 2-D case, first the 1-D case of nonlocalized adsorption was solved. The 1-D system considered consists of N_s rods, each of length d, strung on a straight thread of total length ℓ (which is the 1-D counterpart of the surface area A), interacting with a sticky potential. The convenient thermodynamic potential for this problem is the Gibbs isothermic–isobaric potential for N_s molecules: $G = N_s\mu_s$ (Chapters 1 and 2 in ref. [81]). The partition function Δ_p for the system under consideration is:

$$\Delta_p = \int_0^\infty Z\exp(-\ell\pi^\ell/k_BT)\,d\ell, \quad G = -k_BTN_s\ln\Delta_p \tag{2.107}$$

where Z is the canonical partition function of the adsorbate, and π^ℓ is the line pressure (J/m). The integration over ℓ in Equation 2.107 corresponds to an increase of the system length ℓ from 0 to ∞ at constant temperature, pressure π^ℓ, and number of adsorbed molecules N_s. From the Gibbs–Duhem equation of this system, one can calculate the adsorption Γ_s:

$$N_s d\mu_s = \ell d\pi^\ell \rightarrow \frac{1}{\Gamma_s} = \frac{\ell}{N_s} = \frac{d\mu_s}{d\pi^\ell} = -k_BT\left(\frac{\partial\ln\Delta_p}{\partial\pi^\ell}\right)_{N_s} \tag{2.108}$$

The last equation leads to the EOS. This approach is rigorous but rather complicated because of the necessity to know the partition function Z.

Hemmer and Stell [90] simplified the required calculation significantly. They argued that if the intermolecular potential is short enough, it will act only between neighboring molecules. Hence, they replaced the partition function Z in Equation 2.107 with $\exp(-u/k_BT)$, where u is the short range intermolecular potential (Equation 2.103). Because ℓ is variable, it was replaced in Equation 2.107 by the distance r between the interacting molecules. Thus, Equation 2.107 was simplified to [90]:

$$\Delta_p = \int_0^\infty e^{-u(r)/k_BT}\,e^{-\pi^\ell r/k_BT}\,dr \tag{2.109}$$

This equation was used in ref. [32] to derive a new EOS by inserting the expression for $u(r)$ (Equation 2.103). One can easily solve the obtained integral to find:

$$\ln\Delta_p = \ln\frac{k_BT}{\pi^\ell} - \frac{\pi^\ell d}{k_BT} + \ln\left[e^{-\lambda\pi^\ell d/k_BT}\left(1 - \frac{\beta}{\lambda}\right) + \frac{\beta}{\lambda}\right] \tag{2.110}$$

where $\beta = u_c/k_BT$. From Equations 2.110 and 2.108, one has:

$$\frac{1}{\Gamma_s d} = \frac{k_BT}{\pi^\ell d} + 1 + \frac{\lambda e^{-\lambda\pi^\ell d/K_BT}\left(1 - \dfrac{\beta}{\lambda}\right)}{e^{-\lambda\pi^\ell d/k_BT}\left(1 - \dfrac{\beta}{\lambda}\right) + \dfrac{\beta}{\lambda}} \tag{2.111}$$

In the limit $\lambda \to 0$, Equation 2.111 yields:

$$\frac{1}{\Gamma_s d} = \frac{k_B T}{\pi^\ell d} + 1 - \frac{\beta}{1 + \beta \pi^\ell d / k_B T} \tag{2.112}$$

This is a quadratic equation with respect to π^ℓ; solving it, one obtains the EOS $\pi^\ell(\Gamma_s)$:

$$\frac{\pi^\ell}{k_B T} = \frac{R_\beta^{1D} - 1}{2 \beta d}, \quad \text{where} \quad R_\beta^{1D} = \sqrt{1 + 4\beta \frac{\Gamma_s d}{1 - \Gamma_s d}} \tag{2.113}$$

This result was obtained by Gurkov and Ivanov [32] and also by Tutschka and Cuesta [91]. The latter authors used a different method and Equation 2.113 for them was an intermediate result while solving a different problem.

By multiplying Equation 2.113 by $1 + R_\beta^{1D}$, one can transform it to

$$\frac{\pi^\ell}{k_B T} = \frac{\Gamma_s}{1 - \Gamma_s d} \times \frac{2}{1 + R_\beta^{1D}} \tag{2.114}$$

The first factor in the right-hand side is the hard core part of the new EOS, which is the same as the hard core part of Volmer EOS (Equation 2.86).

To derive the 2-D analogue of Equation 2.114, a heuristic approach was used in ref. [33]. It was based on a useful procedure for deriving an EOS, proposed by Hemmer and Stell [90], which accounts rigorously for the attraction between hard spheres up to terms of the order of B_{attr}^2. This procedure was modified in ref. [33] for the case of a 2-D fluid composed of attracting hard core particles. The respective result for π^S is:

$$\pi^S = \pi_{hc}^S + \frac{B_{attr}}{B_{hc}} \left(\pi_{hc}^S - \Gamma_s \frac{\partial \pi_{hc}^S}{\partial \Gamma_s} \right) \tag{2.115}$$

where the subscript "hc" denotes "hard core." If π_{hc}^S is given by HFL EOS (Equation 2.87) with $B_{attr} = 0$, then $B_{hc} = 2\alpha$ (Equation 2.106). For sticky potential, $B_{attr} = 4\alpha\beta$, so that Equation 2.115 yields:

$$\frac{\pi^S}{k_B T} = \frac{\Gamma_s}{(1 - \alpha \Gamma_s)^2} \left(1 - 4\beta \frac{\alpha \Gamma_s}{1 - \alpha \Gamma_s} \right) \tag{2.116}$$

On the other hand, Equation 2.114 can be expanded in power series in β:

$$\frac{\pi^\ell}{k_B T} \xrightarrow{\beta \to 0} \frac{\Gamma_s}{1 - \Gamma_s d} \left(1 - \beta \frac{\Gamma_s d}{1 - \Gamma_s d} + \dots \right) \tag{2.117}$$

A comparison of Equations 2.116 and 2.117 reveals that they have an analogous structure: the respective hard core terms, $\Gamma_s/(1 - \Gamma_s d)$ for the Volmer and $\Gamma_s/(1 - \alpha \Gamma_s)^2$ for HFL equations, are multiplied by the same function of β and Γ_s, with the only difference being the numerical coefficients 1 and 4, respectively. In the case of Equation 2.117, this function stems from the expansion of the factor containing R_β^{1D}, it is not difficult to realize that the two functions will become identical if the numerical coefficient in R_β^{1D} (Equation 2.113), is changed from 4 to 16. Hence, it was hypothesized that the analogue of the 1-D EOS (Equation 2.113) for 2-D adsorption must read:

$$\frac{\pi^S}{k_B T} = \frac{\Gamma_s}{(1 - \alpha\Gamma_s)^2} \times \frac{2}{1 + R_\beta}, \quad R_\beta = \sqrt{1 + 16\beta \frac{\alpha\Gamma_s}{1 - \alpha\Gamma_s}} \quad (2.118)$$

The corresponding adsorption isotherm was obtained by substituting π^S from Equation 2.118 into Gibbs isotherm ($d\pi^S = k_B T\Gamma_s d\ln C_s$) and integrating. The result was

$$K_s C_s = \frac{\Gamma_s}{(1 - \alpha\Gamma_s)}\left(\frac{2}{1 + R_\beta}\right)^{\frac{1+8\beta}{4\beta}} \times \exp\left[\frac{\alpha\Gamma_s(4 - 3\alpha\Gamma_s)}{(1 - \alpha\Gamma_s)^2} \times \frac{2}{1 + R_\beta}\right] \quad (2.119)$$

The new adsorption isotherm (Equation 2.119) was checked numerically in ref. [33] by direct calculation of the third virial coefficient—it turned out to be very close to the one obtained by the expansion in series up to Γ_s^3 of Equation 2.118. This gives hope that Equations 2.118 and 2.119 will work reasonably well at least for small β and moderate degrees of coverage $\alpha\Gamma_s$. Note that the expansions in β used above are much more general and precise than the virial expansions in Γ_s with the same number of terms. The new adsorption isotherm was confirmed in ref. [33] by extensive analysis of the adsorption data obtained with dimethyl alkylphosphine oxides in refs. [92] and [93], and with aliphatic acids in refs. [94] and [95].

2.3.2 Dense Adsorption Layers of Ionic Surfactants: Ion-Specific Effects

2.3.2.1 Equations of State and Adsorption Isotherms of Ionic Surfactants

We will now modify the EOS of Equations 2.85 through 2.87 and 2.118, and the corresponding adsorption isotherms Equations 2.88 through 2.90 and 2.119, to make them applicable to ionic surfactants in the presence of ion-specific effects. A formal thermodynamic derivation similar to the one in Section 2.2.1.4 is possible, but we will instead use the two-stage adsorption procedure of Borwankar and Wasan [28], which better reveals the pitfalls of the derivation. These authors accounted for the electrostatic and nonelectrostatic contributions to the adsorption Γ_s, by assuming that (i) when the surfactant ion is in the bulk of the solution, it is under the action of the electrostatic potential $\phi(z)$ only, and its concentration C_s^S at the subsurface $z = 0$ is determined by Boltzmann distribution with the surface potential ϕ^S: $C_s^S = C_s \exp(-e_s\phi^S/k_B T)$; and (ii) After the ionic head has reached the surface $z = 0$, the hydrophobic tail is adsorbed. They also assumed that the adsorption constant K_s of process (ii) was the same as for a nonionic surfactant with the same tail.

We now apply their procedure to all adsorption isotherms for nonionic surfactants. According to assumption (i) above, the concentration of surfactant ions at $z = 0$ will be C_s^S. If applied to the equilibrium between the subsurface and the surface (with C_s^S instead of C_s), all isotherms can be written in the general form

$$K_s C_s^S \equiv K_s C_s \exp(-\Phi^S) = \Gamma_s \gamma^S(\alpha\Gamma_s, \beta) \quad (2.120)$$

Here, γ^S is the surface activity coefficient of the ionic surfactant, which depends on the adsorption model: γ^S can be easily deduced by setting the right-hand side of Equation 2.120 to be equal to the right-hand side of the adsorption isotherms, Equations 2.88 through 2.90 and 2.119. We will use Equation 2.120 to obtain the adsorption isotherm $\Gamma_{s0}(C)$ of ionic surfactants in the absence of ion-specific effects, which will be used as a zeroth approximation afterward. We will denote by subscript "0" the quantities pertaining to the case of absent ion-specific effects—these are Γ_{s0}, K_0, and α_0. Substituting in Equation 2.120 the factor $\exp(-\Phi^S)$ from the Gouy equation (Equation 2.25), after elementary algebra (see also Equations 2.33, 2.40, and 2.56), one obtains the nonspecific adsorption isotherm of an ionic surfactant:

$$K_0 C^{2/3} = [\gamma^S(\alpha_0\Gamma_{s0}, \beta)]^{1/3}\Gamma_{s0} \tag{2.121}$$

K_0 is defined by Equation 2.40 and C is mean activity (Equation 2.34). If one sets $\gamma^S = 1$, Equation 2.121 simplifies to Davies isotherm (Equation 2.39).

To obtain the first iteration of the adsorption isotherm of dense monolayers in the presence of ion-specific effects, we will use the first equation (Equation 2.58), which is a direct corollary of the generalized Gouy equation (Equation 2.54). We write it now as

$$\Gamma_{s0} = \Gamma_s \exp(u_{i0}/2k_BT) \tag{2.122}$$

Substituting Γ_{s0} from Equation 2.122 into the adsorption isotherm (Equation 2.121), one obtains

$$KC^{2/3} = [\gamma^S(\alpha\Gamma_s, \beta)]^{1/3}\Gamma_s \tag{2.123}$$

Here, K is the ion-specific constant given by Equation 2.56. We have defined the ion-specific molecular area α as

$$\alpha = \alpha_0\exp(u_{i0}/2k_BT) \tag{2.124}$$

This equation is the quantitative formulation of the Hofmeister effect on the effective area per molecule α, which will be discussed in Section 2.3.2.3. It reveals that the higher the absolute value of the counterion adsorption energy u_{i0} is, the smaller the molecular area α. That α may depend on the nature of the counterion was inferred by Goddard and coworkers based on their data obtained with insoluble nonadecylbenzene sulfonate monolayers with different cations [96] or docosyltrimethylammonium monolayers with different anions [97].

By applying the result (Equation 2.123) to the adsorption isotherms (Equations 2.88 through 2.90) and using Equation 2.102, $B_{attr} = \alpha\beta$, one obtains the respective isotherms of ionic surfactants:

1. Langmuir–Frumkin:

$$KC^{2/3} = \frac{\Gamma_s}{(1-\alpha_L\Gamma_s)^{1/3}} \exp(-2\beta\alpha_L\Gamma_s/3) \tag{2.125}$$

2. van der Waals:

$$KC^{2/3} = \frac{\Gamma_s}{(1-\alpha_V\Gamma_s)^{1/3}} \exp\left[\frac{\alpha_V\Gamma_s}{3(1-\alpha_V\Gamma_s)} - \frac{2}{3}\beta\alpha_V\Gamma_s\right] \tag{2.126}$$

3. modified HFL:

$$KC^{2/3} = \frac{\Gamma_s}{(1-\alpha\Gamma_s)^{1/3}} \exp\left[\frac{\alpha\Gamma_s(3-2\alpha\Gamma_s)}{3(1-\alpha\Gamma_s)^2} - \frac{2}{3}\beta\alpha\Gamma_s\right] \tag{2.127}$$

Setting in the last equation $\beta = 0$, one obtains:

4. HFL:

$$KC^{2/3} = \frac{\Gamma_s}{(1-\alpha\Gamma_s)^{1/3}} \exp\left[\frac{\alpha\Gamma_s(3-2\alpha\Gamma_s)}{3(1-\alpha\Gamma_s)^2}\right] \tag{2.128}$$

5. The new isotherm (Equation 2.119) leads to:

$$KC^{2/3} = \frac{\Gamma_s}{(1-\alpha\Gamma_s)^{1/3}} \left(\frac{2}{1+R_\beta}\right)^{\frac{1+8\beta}{12\beta}} \times \exp\left[\frac{\alpha\Gamma_s(4-3\alpha\Gamma_s)}{3(1-\alpha\Gamma_s)^2} \times \frac{2}{1+R_\beta}\right] \tag{2.129}$$

with R_β given by Equation 2.118.

Next, we proceed to the surface pressure π^S. For high surface potentials, Φ^S, this task is easy because the contribution of the electrical double layer to the surface pressure π^S is $2k_BT\Gamma_s$ (as we already showed in Section 2.1.3). For dense monolayers, this can be proven rigorously in the following way. As a starting point, we use the Gibbs isotherm (Equation 2.32). To obtain the convenient exponent $C^{2/3}$, we will rewrite it as

$$d\pi^S = 3k_BT\Gamma_s d\ln C^{2/3} \tag{2.130}$$

Substituting $C^{2/3}$ from Equation 2.123, one obtains

$$d\pi^S = k_BT\Gamma_s d\ln\gamma^S\Gamma_s + 2k_BTd\Gamma_s \tag{2.131}$$

The first term in this equation is the same for both ionic and nonionic surfactants. The second one refers to ionic surfactants only. Equation 2.131 can be written in integral form:

$$\pi^S = \pi_0^S(\Gamma_s) + 2k_BT\Gamma_s \tag{2.132}$$

where $\pi_0^S(\Gamma_s)$ is the expression for the EOS of a nonionic surfactant. Hence, it is enough to add a term $2k_BT\Gamma_s$ to the right-hand side of EOS (Equations 2.85 through 2.87 and 2.118) for nonionic surfactants to obtain the corresponding EOS for ionic surfactants. The EOS (Equation 2.132) determines the dependence of π^S on the adsorption Γ_s and, together with the adsorption isotherm (Equation 2.123), it parametrically defines the dependence of π^S on C_s.

By expanding in series of Γ_s any of the adsorption isotherms (Equations 2.125 through 2.129), and inverting the series, one obtains:

$$\Gamma_s = KC^{2/3} - \frac{2}{3}B_2K^2C^{4/3} + \dots, \tag{2.133}$$

in which $B_{attr} = 4\alpha\beta$ for Equation 2.129 and $B_{attr} = \alpha\beta$ for the other isotherms. The corresponding dependence of π^S on C follows directly from the integration of Gibbs isotherm (Equation 2.130):

$$\pi^S/k_BT = 3KC^{2/3} - B_2K^2C^{4/3} + \dots \tag{2.134}$$

This is the virial expansion of $\pi^S(C)$. We used this equation to determine K in Section 2.4.

2.3.2.2 Experimental Results and Analysis

We already showed that Frumkin, van der Waals, and the modified HFL models yield some unreliable results (cf. Table 2.3). That is why we will perform analysis of the experimental data for $\pi^S(C)$ only with two models, each comprising an EOS $\pi^S(\Gamma_s)$ and an adsorption isotherm $\Gamma_s(C)$—these are the HFL model (Equations 2.87 and 2.128 with $B_{attr} = 0$) and the new model (Equations 2.118 and 2.129). In both EOS, $2k_B T\Gamma_s$ is added according to Equation 2.132.

Several factors complicate the determination of the parameters K, α, β and π_0 of the two models from experimental data. First of all, this is the number of fitting parameters (four for the new model for W|G; three for HFL model for W|O interface). This leads to a broad minimum of the merit function—the dispersion of $\pi(C^{2/3})$. The insufficient experimental information (small number of available data for different surfactants and interfaces) makes the problem even worse.

A second problem is that the experimental data $\pi^S(C)$ are usually known with rather high errors. The main reasons for these errors are [52,98]: (i) extremely pure water, salt, and surfactant and careful formulation of the experimental conditions are essential for reliable results; and (ii) careful calibration is needed for all types of tensiometric methods. The last factor usually affects the absolute value of the surface tension σ, but not the difference between its two subsequent values. Often, quite different experimental curves for the same system can be made to coincide by suitably shifting them in the vertical direction. An example with data for $C_{12}H_{25}SO_4Na$ at the water–hexane interface is shown in Figure 2.18. The data of Aratono [99] are 3.5 mN/m lower than the results of Rehfeld [51], and 5.5 mN/m higher than the results of Gillap et al. [62]. The original three sets of data differ in Figure 2.18a but coincide when suitably shifted in Figure 2.18b. This problem makes it difficult to determine exactly the spreading pressure π_0, especially when the authors have not measured the interfacial tension of the pure interface (at zero surfactant concentration).

A third problem is the assessment of the reliability of the obtained parameters K, α, β, and π_0. In the case of multiparametric nonlinear models (such as the HFL defined with Equations 2.128 and 2.87, and the new model defined with Equations 2.118 and 2.129), this assessment requires detailed and cumbersome numerical analysis of the merit function (similar to the one in ref. [33]).

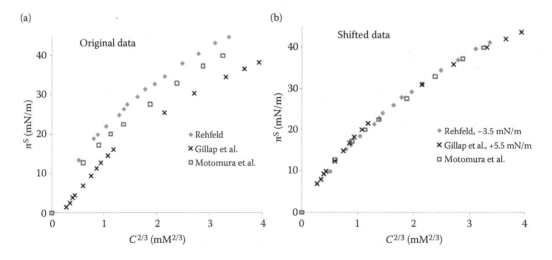

FIGURE 2.18 **(See color insert.)** Surface tension versus $C^{2/3}$ for $C_{12}H_{25}SO_4Na$ at water–hexane interface. (a) The three sets differ in absolute values, which we attribute to incorrect calibration of the tensiometers. (b) When suitably shifted, all data coincide. (Data from Rehfeld, S.J., *J. Phys. Chem.* 71, 738–745, 1967; Gillap, W.R. et al., *J. Colloid Interface Sci.* 26, 232–236, 1968; Motomura, K. et al., *J. Colloid Interface Sci.* 67, 247–254, 1978.)

These three problems will be tackled in a future publication in preparation, in which the verification of the adsorption theory will be performed on more systems. Instead of such a detailed approach, which cannot be presented concisely in this chapter, we will use a simpler procedure for illustrative purposes that yields, we believe, reasonable results.

This procedure involves the following steps:

1. Determination of Γ_s from the experimental dependence $\pi^S(C^{2/3})$ by numerical differentiation. To calculate Γ_s, we followed Rehfeld [51] who proposed to smooth the experimental data $\pi^S(C)$ with a quadratic regression model.

$$\pi^S = b_0 + b_1 \ln C^{2/3} + b_2 \ln^2 C^{2/3} \tag{2.135}$$

where the coefficients b_i are determined from the curve fitting. This dependence is further used to calculate the adsorption Γ_s according to Gibbs isotherm (Equation 2.130):

$$\Gamma_s = \frac{C^{2/3}}{3k_B T}\frac{d\pi^S}{dC^{2/3}} = \frac{1}{3k_B T}(b_1 + 2b_2 \ln C^{2/3}) \tag{2.136}$$

It turned out that the dependence (Equation 2.136) approximates these adsorption isotherms well in the moderate to high surfactant concentration region.* The approximation fails at $C \to 0$ because Equation 2.136 yields $\Gamma_s \to -\infty$, but this region is unimportant for the concentration ranges considered below.

2. The values of Γ_s found by procedure (1) plotted versus $C^{2/3}$ represent numerically the "experimental" adsorption isotherm $C^{2/3}(\Gamma_s)$. Similarly, the values of π^S from the experimental curves $\pi^S(C^{2/3})$ plotted versus the respective calculated values of Γ_s yield the "experimental" EOS. As an example, the "experimental" $\pi^S(\Gamma_s)$ points for $C_{12}H_{25}NMe_3Br$ at W|G and W|O interface are presented in Figure 2.19a. The processing of the so-obtained EOS and adsorption isotherm is easy for two reasons. First, the fitting procedure is performed with explicit theoretical functions $\pi_{th}(\Gamma_s; \alpha, \beta, \pi_0)$ and $\Gamma_{th}(\Gamma_s; K, \alpha, \beta)$. These are Equations 2.129 and 2.118, the latter with added $2k_B T\Gamma_s$. Second, the EOS and the adsorption isotherm both involve one less parameter—Equation 2.118 for $\pi^S(\Gamma_s)$ does not contain K, whereas Equation 2.129 for $C^{2/3}(\Gamma_s)$ does not contain π_0. We chose to use only the $\pi^S(\Gamma_s)$ data, as they are less sensitive to the failure of Rehfeld's procedure (Equation 2.136) at $C \to 0$. In summary, the parameters α, β, and π_0 are determined from the curve-fitting of $\pi^S(\Gamma_s)$, obtained with Rehfeld's procedure (Equations 2.135 and 2.136), with the theoretical models Equations 2.129, 2.118, and 2.132. This procedure was applied with the new adsorption model at the W|G interface and with the HFL model at the water–hexane interface.

3. With α, β, and π_0 obtained from (2), there is only one undetermined parameter left: the adsorption constant K. To determine this, the theoretical adsorption isotherm and EOS (Equations 2.129 and 2.118), were solved numerically with fixed values of α, β, and π_0 as obtained from (2). This leads to the theoretical relation $\pi_{th}(C^{2/3}; K)$. It was compared with the original experimental data π^S versus $C^{2/3}$, to determine the value of K (the simplex method was used for minimization of the dispersion of π^S vs. $C^{2/3}$). This step is illustrated in Figure 2.19b, where the points are experimental data for $C_{12}H_{25}NMe_3Br$ at the W|G and W|O interfaces and the fitting curves are shown.

---

* The results, presented in Table 2.3 and discussed in Section 2.3.1.3, were obtained by calculating the values of Γ_s using the above procedure. These data for Γ_s were then used for the fits of the adsorption isotherms of Frumkin, van der Waals, and the modified HFL (Equations 2.125 through 2.128).

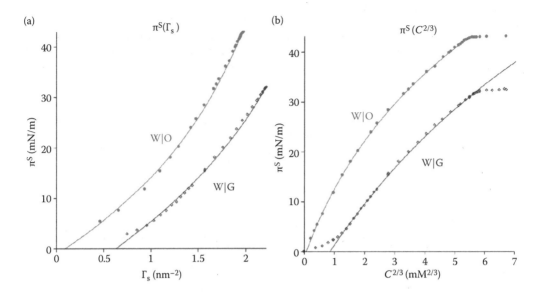

FIGURE 2.19 Illustration of the curve-fitting procedure used for determination of the adsorption parameters of $C_{12}H_{25}NMe_3Br$ at $W|G$ and $W|O$ interfaces [65]. (a) EOS, π^S versus Γ_s. "Experimental" values of Γ_s were obtained using Rehfeld's procedure (Equation 2.136). The lines are fits based on HFL EOS ($W|O$ interface; Equations 2.87 and 2.132), and the new EOS ($W|G$; Equation 2.118). The parameters α, β, and π_0 given in Table 2.3 were obtained from these fits. (b) π^S versus $C^{2/3}$. The points are experimental data [65]. The lines are fits with HFL (Equations 2.87 and 2.128), and the new model (Equations 2.118 and 2.129). The values of α, β, and π_0 are those determined from π^S versus Γ_s in (a), and the values of K in Table 2.3 are determined as fitting parameters.

The surfactants we chose to study using this procedure at the water–hexane and $W|G$ interfaces were $C_{12}H_{25}SO_4Na$ [51,52], $C_{12}H_{25}NMe_3Cl$, and $C_{12}H_{25}NMe_3Br$ [65,21,89]. The reason for this choice was that such a combination of surfactants allowed us to study the main effects that we were expecting: the effects of the ionic head and the counterion on the adsorption parameters. For this reason, we chose the same hydrophobic tail (dodecyl), and the same oil (hexane, the only one for which we found adsorption data for the three surfactants). For the $W|G$ surface, we used only the data for the LE state, that is, those after the kink in the $W|G$ curves $\pi^S(C^{2/3})$ in Figure 2.19b.

The results obtained for the adsorption parameters (K, α, β, π_0) are shown in Table 2.4. The calculated values of the adsorption constant K and the excess pressure π_0 in the LE state can be compared with those obtained from Figures 2.3 and 2.10 after direct quadratic regression analysis of the experimental data $\pi^S(C^{2/3})$, see Equation 2.134, which does not involve any model for the intermolecular interaction. The differences for the K values are less than 4% and those for the π_0 values less than 15%. The standard errors for α vary between 2.5% and 5%, whereas those for β vary between 20% and 30%. The relative values of the errors are in qualitative agreement with the numerical analysis, performed in ref. [33], which revealed that the errors were smallest for α, increased for K, and were even larger for β.

However, we met problems with fitting the data for the water–hexane interface with the new model (Equations 2.118 and 2.129). We obtained reasonable data for α and β, but the errors in the obtained values of K were large. Similar problems appeared in ref. [33] with the alkyldimethylphosphine oxide compounds with short chain length at the $W|G$ surface—because they exhibited very small values of β (~0.1), the fitting problems were then ascribed to the fact that the new adsorption isotherm (Equation 2.129) contains a power $(1 + 8\beta)/4\beta$, which diverges at $\beta \to 0$ and thus probably makes the fit uncertain for small β. Similar small values of β were then obtained

TABLE 2.4
Parameters of the Adsorption

Surfactant, Interface	Quadratic Regression (Equation 2.134)			HFL Model (Equations 2.128 and 2.87)			New Model (Equations 2.118 and 2.129)				Nonspecific Parameters (W|G)		
	π_0 (mN/m)	K	B_2 (Å²)	π_0 (mN/m)	K	α (Å²)	π_0 (mN/m)	K	α (Å²)	β	u_{i0}/k_BT	K_0	α_0 (Å²)
$C_{12}H_{25}SO_4Na$, water–hexane [51]	5.9	196	21.3	3.45	266	22.9							
$C_{12}H_{25}SO_4Na$, W|G [52]	−9.1	156	12.0				−10.5	155	20.3	0.58	−0.33	131	24.0
$C_{12}H_{25}NMe_3Cl$, water–hexane [21,65]	0	131	30.4	1.94	127	29.6							
$C_{12}H_{25}NMe_3Cl$, W|G [89]	−4.0	63.3	20.4				−4.2	60.6	31.2	0.65	−1.43	29.6	63.8
$C_{12}H_{25}NMe_3Br$, water–hexane [65]	0.26	142	20.8	−1.14	187	23.5							
$C_{12}H_{25}NMe_3Br$, W|G [65]	−9.0	117	19.2				−8.2	113	23.3	0.33	−2.32	35.4	74.3

Note: The parameters for the W\|O interface were obtained by curve-fitting with the HFL model (Equations 2.128, 2.87, and 2.132), and those for W\|G with the new model (Equations 2.118, 2.129, and 2.132).

for the water–hexane interface from the new isotherm (Equation 2.129) and we believe that the reason for the small values obtained for K is the same. We are currently working to resolve this problem. That is why we decided to confine ourselves to the water–hexane interface with data obtained from the HFL model only (Equations 2.128 and 2.87), with $B_{attr} = 0$. The results obtained with the HFL model are systematically larger for K and smaller for α (with 2–4 Å^2) than those obtained (but not shown here) with the new model. However, both of these follow the same qualitative trends as expected on physical grounds (see Section 2.2.3): decrease of K and increase of α with the decrease of $-u_{i0}/k_B T$. It is impossible to decide without detailed numerical analysis (planned for the future), which one of the two models leads to more reliable results for the W|O interface because, when applying the HFL isotherm, one forces the system to have $\beta = 0$, which might not always be true.

An interesting and surprising effect, related to the parameters α and β, is exhibited by the $\Gamma_s(C^{2/3})$ plots for $C_{12}H_{25}NMe_3Br$ at W|G and W|O interfaces in Figure 2.20: the adsorption Γ_s at W|G at small concentrations of C is smaller than that at the W|O interface. However, as C increases Γ_s at W|G increases faster than Γ_s at the W|O interface and the two adsorption curves intersect (the same type of behavior was found for $C_{12}H_{25}NMe_3Cl$ and $C_{12}H_{25}SO_4Na$, as well as with other surfactants studied; for lack of space, we do not reproduce the respective plots here). Because the K values do not change with C, the reason must be sought in the role of the other parameters, α and β, more precisely, in the second virial coefficient, which encompasses both factors: $B_2 = 2\alpha(1 - 2\beta)$, cf. Equation 2.106. According to the data in Table 2.3, $\beta \approx 0.5$ at the W|G interface so that $B_2 \ll 2\alpha$ for the three surfactants. For the W|O interface, $B_2 = 2\alpha$ is larger because $\beta = 0$. This means that the repulsive forces between the adsorbed surfactant ions will be larger in the latter case, thus hindering the increase of the adsorption. The effect of B_2 on the adsorption isotherm is also visible from the plots of $\pi^S(C^{2/3})$ in Figure 2.3: whereas the plots for the W|G systems are linear almost up to cmc, those for the W|O interface exhibit noticeable curvature.

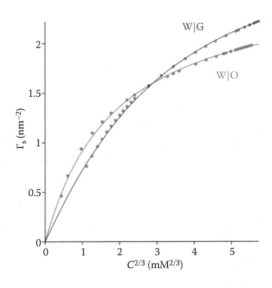

FIGURE 2.20 Illustration of the difference in the adsorptions Γ_s versus mean surfactant activity $C^{2/3}$ for $C_{12}H_{25}NMe_3Br$ at W|G and W|O interface. "Experimental" values of Γ_s were obtained using Rehfeld's procedure (Equation 2.136). The lines are theoretical predictions based on the HFL model (W|O), (Equations 2.128), and the new model (W|G; Equations 2.129), with parameters as in Table 2.3.

2.3.2.3 Electrostatic and Ion-Specific Effects on the Adsorption Parameters K, α, and β

The data in Table 2.3 exhibit a marked dependence of the area per molecule α on the ion-specific adsorption energy $-u_{i0}$, in agreement with Equation 2.124. We applied Equation 2.124 to calculate the values of α_0 for $C_{12}H_{25}SO_4Na$, $C_{12}H_{25}NMe_3Cl$, and $C_{12}H_{25}NMe_3Br$ at the W|G surface by using the values of $-u_{i0}/k_BT$ for the counterions Na^+, Cl^-, and Br^- (cf. Table 2.3). If the theory and the calculations are correct, the values of α_0, obtained from Equation 2.124, must coincide for different counterions, if the surfactant ion and everything else is the same. Indeed, the values of α for $C_{12}H_{25}NMe_3Cl$ and $C_{12}H_{25}NMe_3Br$ differ significantly (31.2 vs. 23.3 Å², respectively), whereas the α_0's are closer (63.8 vs. 74.3 Å²) as required by the theory. The small difference of approximately 10 Å² is probably due to experimental errors. Despite the close values of the α's, the α_0 is much smaller for $C_{12}H_{25}SO_4Na$ than for $C_{12}H_{25}NMe_3Cl$ and Br (because $-u_{i0}/k_BT$ is also smaller). Unfortunately, for the time being, we cannot perform similar calculations for adsorption at the W|O interface because a complete theory of the ion-specific adsorption energy u_{i0}^{WO} at such interface is lacking. Still, the same qualitative trend of α for $C_{12}H_{25}NMe_3Cl$ and Br at W|O interface is obeyed: $\alpha_{Cl} > \alpha_{Br}$.

There should be no direct influence of the ion-specific effect on the attraction constant β because it depends only on the van der Waals interaction energy. However, it can be affected indirectly by the ion-specific effect through its dependence on α (e.g., see Equation 33 in ref. [33]). It is conceptually easy to account for this dependence but such effort is hardly worthwhile, first, because it involves trivial but lengthy calculations, and second, because of the large errors in the β values.

The observed strong ion-specific effect on the area per molecule α is an indication of electrostatic repulsion between the surfactant ions, which is dampened by the counterions. This suggests that there is also probably some electrostatic effect on the values of the experimentally determined area α. Indeed, the hard core cross-sectional areas of $C_{12}H_{25}NMe_3Cl$ and Br must be almost equal but the values obtained for the respective α's differ significantly. To check this hypothesis, we analyzed the available data for adsorption of the same surfactant, $C_{12}H_{25}SO_4Na$, at similar W|O interfaces but at different concentrations of added NaCl. Because the surfactant and all ions are the same, there should be no ion-specific effect. In Figure 2.21, $\ln\alpha$ is plotted versus the square of the Debye length, $1/\kappa^2 \equiv \varepsilon k_BT/2e_0^2C_1$, because we hypothesized that at these relatively low salt concentrations, the area α must be related to the square of the Debye length, $1/\kappa$. The good linearity and the reasonable value of $\alpha = 19.3$ Å² at $1/\kappa = 0$ (i.e., in the absence of electrostatic effect) seem to confirm our hypothesis.

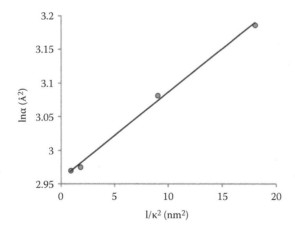

FIGURE 2.21 Dependence of the area per molecule α on the Debye length $1/\kappa$. Plot of $\ln \alpha$ versus $1/\kappa^2$ for $C_{12}H_{25}SO_4Na$ adsorption at W|O interface at various electrolyte concentrations: $C_{el} = 0$ for heptadecane [51], 0.05 M and 0.1 M for petroleum ether [53], and 0.01 M for hexadecane [34]. For the system, without added electrolyte, the value of C_s at $\pi^S = 20$ mN/m was used for the calculation of κ. Line, linear regression of the data. The intercept $\ln\alpha_0 = 2.96$ corresponds to $\alpha_0 = 19.3$ Å².

Unfortunately, in the absence of theory and of more experimental data, it is impossible to interpret the obtained value for the slope. Levine et al. [100] and Warszynski et al. [7] have proposed theoretical models of this effect but, because of their complexity and lack of experimental confirmation, we will not discuss them.

The relation between the adsorption constants K and K_0 for the three surfactants under consideration at the W|G surface confirms the findings in Section 2.4.1 (Table 2.3). The difference in K between $C_{12}H_{25}NMe_3Cl$ and $C_{12}H_{25}NMe_3Br$ is significant (60.6 vs. 113, respectively), but the values of K_0, calculated from Equation 2.56, are much closer (29.6 vs. 35.4). They are also close to the value $K_0 = 26.6$ from the intercept of the respective line in Figure 2.11—according to the theory, this should be so for identical surfactant ions. The nonspecific adsorption constant K_0 for $C_{12}H_{25}SO_4Na$ at the W|G surface was found to be $K_0 = 131$ (cf. Table 2.4), in fair agreement with the value 122 from Figure 2.11. This value is larger than the constants K_0 for the two $C_{12}H_{25}NMe_3^+$ salts, certainly because of the larger value of the adsorption constant K_s for dodecylsulfate surfactants (Equation 2.40; Figure 2.11). At the same time, the ratio $K/K_0 = 1.18$ is smaller for $C_{12}H_{25}SO_4Na$ than it is for the two $C_{12}H_{25}NMe_3^+$ surfactants (2.04 for $C_{12}H_{25}NMe_3Cl$ and 3.19 for $C_{12}H_{25}NMe_3Br$) because of the much smaller absolute value of the ion-specific adsorption energy u_{i0} of $C_{12}H_{25}SO_4Na$.

We believe that the results in this section confirm the self-consistency of our theory and calculation procedures.

2.4 HOFMEISTER EFFECT ON cmc OF IONIC SURFACTANTS AND DISJOINING PRESSURE IN FOAM FILMS

It was shown in Section 2.2.3 that the ion-specific energy u_{i0} depends only on the ion and its interaction with the bulk phases and not on the type of the surfactant. This makes it possible to use the same value of u_{i0} for a given counterion for the interpretation of data obtained with various surfactants in different phenomena. To check the latter hypothesis, we attempted in ref. [30] to interpret two phenomena that were rather different from the adsorption of surfactant: the cmc of the ionic surfactant and the disjoining pressure in thin liquid films. In this section, we give a brief account of the obtained results.

2.4.1 EFFECT OF COUNTERION ON THE CMC OF IONIC SURFACTANTS

A number of properties of micellar solutions of ionic surfactants show correlation to Hofmeister series. For example, the degree of binding [101], micelle aggregation number and shape [102], clouding point [103], enthalpy of micellization [104], and viscosity of micellar solutions [105]. The "classic" example is, of course, the cmc; the counterion effect on cmc was extensively investigated experimentally [25,106]. Correlation to ion size and polarizability was observed [24].

In this section, the cmc of surfactant solutions will be explicitly related to the counterion adsorption energy u_{i0}. Only the case of ionic surfactants that are 1:1 electrolytes will be considered. The ion-specific effects will be investigated by extending the semiempirical approach of Shinoda et al. [25], which predicts the cmc of ionic surfactants in the absence of a Hofmeister effect. Shinoda's approach is based on Gouy theory, and successfully explains the experimentally observed dependence of cmc on the electrolyte concentration C_{el}, known as Corrin–Harkins equation [107]:

$$\text{lncmc} = \text{const} - K_g \ln(\text{cmc} + C_{el}) \qquad (2.137)$$

Following Shinoda et al. [25], we assume that the micellar solution can be regarded as consisting of bulk solution of monomers (indexed with superscript "B") and of micellar pseudo-phase (superscript "M"). The monomer bulk solution of concentration C_s is assumed to be ideal, and the ionic surfactant to be totally dissociated. In such case, the chemical potential of the surfactant monomers will be $\mu^B = \mu_0^B + k_B T \ln C_s$ (Equation 2.3). Next, we assume, as Shinoda did, that the electrostatic contribution to the chemical potential μ^M of the surfactant ion in the micelle is equal to the electrostatic work $E_{el} = K_g e_s \phi^S$ for transferring the surfactant (monovalent) ion into the micelle, that is,

$$\mu^M = \mu_0^M + K_g e_s \phi^S \tag{2.138}$$

Here, K_g is an empirical correction coefficient, whose physical meaning was discussed by Shinoda, who conjectured that $0 < K_g < 1$ with the argument that "it is logical to assume that every monomer introduces charge smaller than 1 into the micelle, due to the counterions that accompany it" [25]. Similar approaches were used by other authors, such as Nagarajan [108] and Rao and Ruckenstein [109]. The experimental values of K_g typically lay in the range between 0.4 and 0.6, usually close to 1/2 [25].

The condition for chemical equilibrium is the equality of the potentials defined by Equations 2.3 and 2.138. If this equality is solved for the surfactant concentration C_s in the monomer solution, the following relation is obtained:

$$\ln C_s = \frac{\mu_0^M - \mu_0^S}{k_B T} + K_g \Phi^S \tag{2.139}$$

where the dimensionless potential Φ^S is given by Equation 2.24. To determine C_s and Φ^S, we also need the electroneutrality condition, that is, the Gouy equation. The curvature effect can be readily taken into account [110]; however, following Shinoda and for the sake of simplicity, we will neglect this effect. Therefore, we will use the Gouy equation for flat surfaces with ion-specific effects included (Equation 2.50).

Let Γ_s be the number of surfactant molecules per unit area of the micelle, that is, it is the micellar aggregation number divided by the area of the micelle. Once again following Shinoda, we assume that Γ_s is a constant, independent of the salt concentration. This assumption, in which it was assumed that the ion-specific effect modifies both the adsorption and the surface potential while the surfactant and the electrolyte concentrations remain constant, makes the iterative procedure used in Section 2.2.3.1 inapplicable. On the contrary, in the case of micelles, the adsorption Γ_s is constant whereas the ion-specific effect changes only Φ^S and thereby, through the equilibrium condition (Equation 2.139), the surfactant concentration C_s, that is, the cmc. This modifies the calculation as follows. The Gouy equation (Equation 2.50) written for one surfactant ion and one counterion reads:

$$e_0^2 \Gamma_s^2 / 2\varepsilon = e_0 C_t e^{-u_{i0}/k_B T} \int_0^{\phi^S} e^{e_0 \phi/k_B T} e^{-(u(z)-u_{i0})/k_B T} \, d\phi \tag{2.140}$$

When analyzing Equation 2.50 to obtain Equation 2.53, it was shown that the result was proportional to $\exp(\Phi_0^S)$. We will make a similar approximation about Equation 2.140. However, in this case, no iteration is possible because Γ_s is assumed constant. Therefore, to obtain meaningful results, one must keep, as the upper limit of the integral in Equation 2.140, the true surface potential Φ^S corresponding to a given counterion concentration. Then, the integral must be replaced by $-k_B T \exp(\Phi^S)/e_0$, rather than by $-k_B T \exp(\Phi_0^S)/e_i$, as in Equation 2.51 (where F_u is negligible). Then, Equation 2.140 yields:

$$\Gamma_s^2 = \frac{4}{\kappa_0^2} C_t e^{-u_{i0}/k_B T} e^{\Phi^S} \tag{2.141}$$

which is the counterpart of Equation 2.53. Thus, Equation 2.141 yields the following expression for Φ^S:

$$\Phi^S = \ln \frac{\kappa_0^2 \Gamma_s^2}{4C_t} + \frac{u_{i0}}{k_B T} \qquad (2.142)$$

Eliminating Φ^S from the chemical equilibrium condition (Equation 2.139) and the generalized Gouy equation (Equation 2.142), one obtains an equation for C_s:

$$\ln C_s + K_g \ln C_t = (1 + K_g) \ln C_0 + K_g \frac{u_{i0}}{k_B T} \qquad (2.143)$$

where C_0 stands for the standard cmc:

$$\ln C_0 = \frac{\mu_0^M - \mu_0^S}{(1 + K_g) k_B T} + \frac{K_g}{1 + K_g} \ln \frac{\kappa_0^2 \Gamma_s^2}{4} \qquad (2.144)$$

C_0 is the hypothetical cmc of the ionic surfactant in the absence of ion-specific effects. An expression similar to (Equation 2.144) was derived by Shinoda et al. [25]. If the ion-specific adsorption energy u_{i0} is set equal to zero, Equation 2.143 is reduced to the Corrin–Harkins equation (Equation 2.137), with explicit expression for the constant there.

In the absence of added electrolyte, the total electrolyte concentration C_t is equal to the surfactant concentration C_s, and Equation 2.143 can be solved for C_s (i.e., for cmc)

$$\ln C_s = \ln C_0 + \frac{K_g}{1 + K_g} \frac{u_{i0}}{k_B T} \qquad (2.145)$$

This simple equation explicitly accounts for the Hofmeister effect on the cmc of ionic surfactants in the absence of added salt (the cmc in the presence of added salt with the same counterion is described in Equation 2.143). For the typical case where $K_g = 1/2$, one finds the dependence of cmc on the counterion $C_s = C_0 \exp(u_{i0}/3k_B T)$, where the values of u_{i0} are those listed in Table 2.2.

Equation 2.143 was generalized for the important case of different counterions of the surfactant and the added electrolyte [30]. We will consider once again only the case of 1:1 electrolytes. We must use the more general form of the Gouy equation for a mixture of counterions (Equation 2.53) where Φ_0^S must be replaced by Φ^S. For a micellar solution with only two monovalent counterions, the result reads:

$$\left(C_1 e^{-u_{10}/k_B T} + C_2 e^{-u_{20}/k_B T} \right) e^{\Phi^S} = \frac{\kappa_0^2}{4} \Gamma_s^2 \qquad (2.146)$$

which is the analogue of Equation 2.142 for the case of two counterions. Usually, one of the counterions is introduced into the solution with the ionic surfactant (so that $C_1 \equiv C_s$) and the other one is introduced with the added electrolyte ($C_2 \equiv C_{el}$). Eliminating Φ^S from Equation 2.146 and the condition for chemical equilibrium from Equation 2.139, one obtains

$$\ln C_s = (1 + K_g) \ln C_0 - K_g \ln \left(C_s e^{-u_{s0}/k_B T} + C_{el} e^{-u_{el0}/k_B T} \right) \qquad (2.147)$$

where C_0 is defined by Equation 2.144; u_{s0} and u_{el0} are, respectively, the adsorption energies at the micellar interface of the counterions stemming from the surfactant ("s") and the added electrolyte ("el"). Equation 2.147 is a generalization of Equation 2.143 and the Corrin–Harkins equation (Equation 2.137) for the case of two different monovalent counterions.

Experimental results for the ion-specific effect on cmc are summarized in ref. [30]. Only data for 1:1 electrolytes were used. The theory was checked with experimental data for alkyltrimethylammonium salts (Figure 2.22 through Figure 2.25) and dodecylsulfate salts. The data for cmc of the homologous series of alkyltrimethyl ammonium salts $(C_{10}H_{21} - C_{18}H_{37})NMe_3^+$ in the absence of added salts [17,25,106,111–126] are presented in Figure 2.22. They were obtained with data from 62 measurements with 17 different surfactants. The data are plotted as $\ln C_s$ versus $-u_{i0}/k_B T$ according to Equation 2.145. Because a complete theory of u_{i0} at the W|O interface is still absent, we assumed that for micelles one can use the data for u_{i0} at the W|G interface from Table 2.2. One of the reasons for doing so was the fact that the average density of the hydrophobic chains inside the micelle is probably lower than it is in a typical oil phase. This assumption is supported by the fact that the calculated free energy per unit area due to the W|O interface of micelles is less than half of the typical value for σ_0 of the water–alkane interface [39]. The other reason was that our preliminary

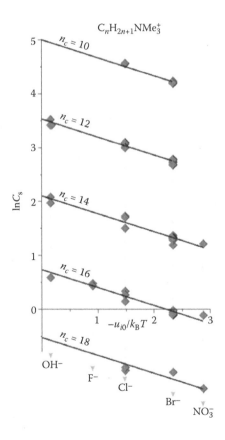

FIGURE 2.22 Dependence of $\ln C_s$ (mM) on the counterion adsorption energy $-u_{i0}/k_B T$ for different hydrocarbon chain lengths n_C of the surface active ion $C_nH_{2n+1}NMe_3^+$ [17,25,106,111–126]; $T = 25°C$ to $30°C$. The black parallel lines are the theoretical dependences according to Equation 2.145 with $K_g = 1/2$. The slope is consequently $K_g/(1 + K_g) = 1/3$ for all lines. The values of $\ln C_0$, obtained from the intercepts, are used in Figure 2.23. (Reprinted from *Adv. Colloid Interface Sci.*, 168, Ivanov, I.B., R.I. Slavchov, E.S. Basheva, D. Sidzhakova, and S.I. Karakashev, Hofmeister effect on micellization, thin films and emulsion stability, 93–104. Copyright 2011, with permission from Elsevier.)

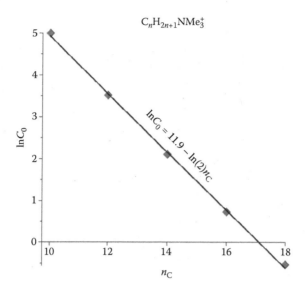

FIGURE 2.23 Dependence of $\ln C_0$ (mM) on the carbon chain length n_C of the surface active ion $C_nH_{2n+1}NMe_3^+$. The values of C_0 were obtained from the intercept of the cmc dependence on the adsorption energy, $\ln C_s(u_{i0})$, shown in Figure 2.22. The value of the slope of the linear dependence $\ln C_0 = A + B\,n_C$ is $B = -\ln 2$ [25,127]. (Reprinted from *Adv. Colloid Interface Sci.*, 168, Ivanov, I.B., R.I. Slavchov, E.S. Basheva, D. Sidzhakova, and S.I. Karakashev, Hofmeister effect on micellization, thin films and emulsion stability, 93–104. Copyright 2011, with permission from Elsevier.)

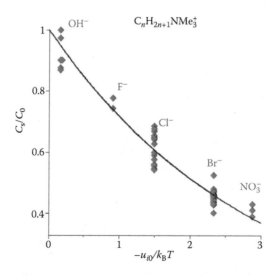

FIGURE 2.24 Normalized cmc, C_s/C_0, versus the counterion adsorption energy $-u_{i0}/k_BT$ for $C_nH_{2n+1}NMe_3^+$ ion ($n_C = 10 - 18$) with different anions. In agreement with Equation 2.145, the normalized cmc does not depend on the carbon chain length n_C—that is why all experimental points in Figure 2.24 now fall on a single curve. (Reprinted from *Adv. Colloid Interface Sci.*, 168, Ivanov, I.B., R.I. Slavchov, E.S. Basheva, D. Sidzhakova, and S.I. Karakashev, Hofmeister effect on micellization, thin films and emulsion stability, 93–104. Copyright 2011, with permission from Elsevier.)

$$C_{14}H_{29}NMe_3^+$$

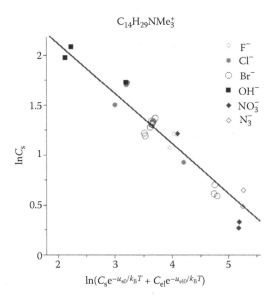

FIGURE 2.25 Dependence of $\ln C_s$ on the added salt concentration and the ion adsorption energy u_{i0}, according to Equation 2.145. The points are experimental data for $C_{14}H_{29}NMe_3^+$ ions with different anions and anion mixtures, with and without added salt. The symbol "F⁻" refers to $C_{14}H_{29}NMe_3Br$ with the addition of NaF [128]; "Cl⁻" to $C_{14}H_{29}NMe_3Cl$ alone [106,113–115,23], or $C_{14}H_{29}NMe_3Br$ with the addition of NaCl [128]; "Br⁻" to $C_{14}H_{29}NMe_3Br$ alone [112,114,117,121], or $C_{14}H_{29}NMe_3Br$ with the addition of bromides [128,112]; "OH⁻" to $C_{14}H_{29}NMe_3OH$ alone [120,121]; "NO_3^-" to $C_{14}H_{29}NMe_3NO_3$ alone [119], or $C_{14}H_{29}NMe_3Br$ with $NaNO_3$ [128]; "N_3^-" to $C_{14}H_{29}NMe_3Br$ with the addition of NaN_3 [128]. The line is the theoretical dependence (Equation 2.145), with slope $K_g = 1/2$. $T = 25°C$ to $30°C$. (Reprinted from *Adv. Colloid Interface Sci.*, 168, Ivanov, I.B., R.I. Slavchov, E.S. Basheva, D. Sidzhakova, and S.I. Karakashev, Hofmeister effect on micellization, thin films and emulsion stability, 93–104. Copyright 2011, with permission from Elsevier.)

estimates of u_{i0} for W|O showed that it was only approximately 10% higher than it was at the W|G interface. Finally, the good agreement obtained below between the experimental data and the theoretical equations (Equations 2.143, 2.145, and 2.147) suggests that such an assumption is probably legitimate.

The cmc data in Figure 2.22 follows the Hofmeister series (Equation 2.2). The linear dependence from Equation 2.145 is obeyed, within experimental accuracy. The slope is fixed to 1/3, corresponding to $K_g = 1/2$. The intercept is the only fitting parameter, and gives the standard cmc, C_0. The value of C_0 depends on the structure of the surface active ion, but not on the counterion (Equation 2.144). The C_0 values, determined from Figure 2.22, are presented in Figure 2.23 as a function of the carbon chain length, n_C. The known linear dependence $\ln C_0 = A + Bn_C$ is obeyed. Our value of the slope B coincides with the known value $B = -\ln 2$, valid for all monovalent nonbranched ionic surfactants [25,27]. The value of the intercept, obtained as fitting parameter, namely, $A = 11.9$, refers to the whole alkyltrimethylammonium salts homologous series. The above values of A and B, along with knowledge of the counterion adsorption energy u_{i0}Table 2.2 for a list of u_{i0} values), allow the prediction of the cmc of any alkyltrimethylammonium salt.

The data from Figure 2.22 are plotted in Figure 2.24 as C_s/C_0 versus $-u_{i0}/k_BT$. All data fall on the same curve; indeed, according to Equation 2.145, the ratio C_s/C_0 for a given surfactant head depends only on u_{i0}/k_BT. The effect of the counterion is quite large—its change can shift cmc by approximately ±50%.

The data for cmc of $C_{14}H_{29}NMe_3^+$ in the presence of a mixture of counterions are presented in Figure 2.25. Typically, the Br⁻ counterion comes with the surfactant, and the second counterion is

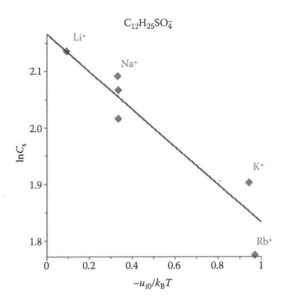

FIGURE 2.26 Dependence of cmc of $C_{12}H_{25}SO_4^-$ salts on the counterion used: $\ln C_s$ (mM) versus the ion adsorption energy $-u_{i0}/k_BT$ according to Equation 2.145. Data from refs. [69,128]; $T = 30°C$ to $33°C$. (Reprinted from *Adv. Colloid Interface Sci.*, 168, Ivanov, I.B., R.I. Slavchov, E.S. Basheva, D. Sidzhakova, and S.I. Karakashev, Hofmeister effect on micellization, thin films and emulsion stability, 93–104. Copyright 2011, with permission from Elsevier.)

added with the salt. The concentration of both ions was commensurable. The data compared well with our Equation 2.147 with a slope of $K_g = 1/2$.

Other ionic surfactants exhibit similar behavior [30]. This is illustrated in Figure 2.26, where data for the cmc of the anionic $C_{12}H_{25}SO_4^-$ with different cations are shown. The model (Equation 2.145) describes the data well within experimental error. Convincing results were also obtained for cmc of $C_{12}H_{25}SO_4^-$ in the presence of a mixture of cations [30]. Our model with dodecylammonium $C_{12}H_{25}NH_3^+$ and dodecanoate $C_{12}H_{25}COO^-$ salts with various counterions [25,106] was less successful, probably due to hydrolysis: the data followed the linear dependence of $\ln C_s$ on u_{i0}/k_BT, as required by Equation 2.147, but the value of K_g was significantly smaller than $K_g = 1/2$.

The order of decrease of cmc of the surfactants shown in Figures 2.24 and 2.26 is:

$$OH^- < F^- < Cl^- < Br^- \lesssim NO_3^- \quad \text{for} \quad C_nH_{2n+1}NMe_3^+ \tag{2.148}$$

$$Li^+ < Na^+ < K^+ < Rb^+ \quad \text{for} \quad C_{12}H_{25}SO_4^- \tag{2.149}$$

The above sequences correspond to the order of increase of the adsorption constant K_{12} (Equations 2.76 and 2.77).

2.4.2 Ion-Specific Effect on the Disjoining Pressure of Foam Films Stabilized with Ionic Surfactants

Although the adsorption energies of the counterions, u_{i0}, are not very large, they may significantly change the electrostatic component Π_{el} of the disjoining pressure Π of a thin film. The theory of the electrostatic disjoining pressure has been developed by many authors, above all by Churaev and

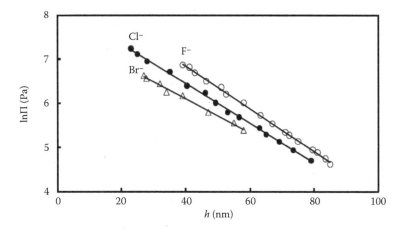

FIGURE 2.27 Plot of $\ln\Pi$ (Pa) versus h (nm) for foam films stabilized with $C_{16}H_{33}NMe_3Br$ and NaX (X = F⁻, Cl⁻, Br⁻). (Reprinted from *Adv. Colloid Interface Sci.*, 168, Ivanov, I.B., R.I. Slavchov, E.S. Basheva, D. Sidzhakova, and S.I. Karakashev, Hofmeister effect on micellization, thin films and emulsion stability, 93–104. Copyright 2011, with permission from Elsevier.)

associates [129]. According to their theory, which neglects the ion-specific effects, if the electrostatic potential in the middle of the film is low (this is equivalent to either $\kappa h \gg 1$ or $\Phi_0^S \ll 1$), the electrostatic disjoining pressure, Π_{el} is given by the following expression:

$$\Pi_{el} = 64 k_B T C_t \tanh^2\left(\Phi_0^S/4\right)\exp(-\kappa h) \equiv \Pi_0 \exp(-\kappa h) \tag{2.150}$$

where C_t is the total ion concentration. Because during the derivation of Equation 2.150 in ref. [129] no assumptions about the surface potential were made, we decided that to account for the ion-specific effects, it should be sufficient to merely replace Φ_0^S in Equation 2.150 with Φ^S from Equation 2.57.

The equation obtained was tested in ref. [30] by measuring the disjoining pressure Π in foam films stabilized with* 0.01 mM $C_{16}H_{33}NMe_3Br$ with the addition of 0.09 mM of one of the salts NaX (X = F⁻, Cl⁻, Br⁻). The films were formed in a thin film pressure balance by using the Mysels–Jones porous plate technique [130,131], and the thickness h was measured interferometrically. Because the films are rather thick, one can disregard the contribution of the van der Waals disjoining pressure. This permits identifying Π with Π_{el}, and therefore, Equation 2.150 can be used for the calculation of Π, if Φ_0^S is replaced with Φ^S. The lines in Figure 2.27 obey Equation 2.150, which in logarithmic form reads:

$$\ln\Pi_{el} = \ln\Pi_0 - \kappa h \tag{2.151}$$

These lines are shifted, but close to parallel, which means that the ion-specific effect affects mainly Π_0, which can be calculated from the intercepts. Thus, from Equation 2.150 (with Φ^S instead of Φ_0^S) one can calculate Φ^S for the three systems. Equation 2.57 and the second equation (Equation 2.58) suggest to plot the experimental Φ^S versus the dimensionless ion adsorption potential$-u_{i0}/k_B T$. This is done in Figure 2.28. The relatively good linearity and the value of the experimental slope

---

* A mistake was made in Section 4 of ref. [30], where the cited values of the surfactant and salt concentrations are a hundred times larger than the true values given here.

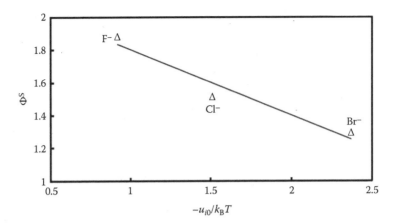

FIGURE 2.28 Combined surface potential Φ^S versus $-u_{i0}/k_B T$ for foam films stabilized with $C_{16}H_{33}NMe_3Br$ and NaX (where X = F⁻, Cl⁻, Br⁻). The potential Φ^S was determined from the line intercepts in Figure 2.27 and Equations 2.150 and 2.151. The intercept yields the purely electrostatic potential, $\Phi_0^S = 2.2$ (Equation 2.58). The absolute value of the slope is 0.4, which is close to the theoretical value 1/2 in Equation 2.58. (Reprinted from *Adv. Colloid Interface Sci.*, 168, Ivanov, I.B., R.I. Slavchov, E.S. Basheva, D. Sidzhakova, and S.I. Karakashev, Hofmeister effect on micellization, thin films and emulsion stability, 93–104. Copyright 2011, with permission from Elsevier.)

(0.4), which is close to the theoretical slope (1/2; cf. Equation 2.58), seem to confirm the role of the ion-specific effect on the disjoining pressure. It was also confirmed qualitatively by experimental studies on the effect of counterions on the stability of emulsion drops stabilized by ionic surfactants [30]. The efficiency of the counterions in decreasing the surface potential Φ^S follows the order

$$F⁻ < Cl⁻ < Br⁻ \qquad\qquad (2.152)$$

This sequence is similar to the respective series (Equations 2.76 and 2.148) based on the adsorption constant K and the values of the cmc.

2.5 SUMMARY AND CONCLUSIONS

Our analysis of Davies' adsorption isotherm (Equations 2.39) revealed that (i) when neutral electrolyte has been added, it remains valid but instead of surfactant concentration, C_s, one must use mean ion activity C (Equation 2.34); and (ii) the nonspecific adsorption constant K_0 of ionic surfactants must be independent of C and is closely related to a constant K_s accounting for the field-independent contribution to K_0. The theoretical expression (Equation 2.7) for Henry's adsorption constant was corrected in two respects. First, a statistical derivation showed that the "thickness of the adsorbed layer" δ_a is determined by the transfer energy u_{CH_2} of a $-CH_2-$ group from water to the interface, rather than being equal to the extended length (~20 Å) of the surfactant tail as suggested by Davies and Rideal [26,27]. Second, the expression for the adsorption energy E_a was corrected by introducing a new term, $\sigma_0 \alpha_\perp$, accounting for the disappearance of area $\alpha_\perp$, equal to the cross-sectional area of the surfactant chain, when the latter penetrates the pure interface of interfacial tension σ_0. The last effect was checked by plotting $\ln K_0$ of $C_{12}H_{25}SO_4Na$ versus the interfacial tensions σ_0 of the pure W|O interfaces for several oils—the plot in Figure 2.7 was linear with a slope equal to the theoretical one.

The other theoretical conclusions and the surface tension isotherms (Equations 2.41 and 2.44) were confirmed in Section 2.2.4 by using numerous data for the adsorption of ionic surfactants. It was found that at the W|G surface, the dependence of the surface pressure π^S on $C^{2/3}$ has a kink at intermediate concentrations, followed by a linear region with a negative intercept π_0 (Figure 2.3). This intercept is due to the hydrophobic tails forming an oil-like LE film on the water phase. There is no intercept at the W|O interface, which is due to the absence of such an LE film. The dependence of $\ln K_0$ on the number of carbon atoms n_C in the hydrophobic chain revealed that the free energies u_{CH_2} for transfer of one $-CH_2-$ group to both interfaces are equal, that is, the oil-like LE adsorbed layer behaves as an oil phase.

Section 2.2.3 is devoted to the counterion-specific effects on the adsorption of ionic surfactants. The generalized Poisson–Boltzmann equation (involving the London potential; Equation 2.49) was integrated analytically. This led to simple relations between the ion-specific adsorption energy, u_{i0}, the ion-specific quantities (surface potential Φ^S and adsorption constant K) and the respective non-specific ones (Φ_0^S and K_0; cf. Equations 2.56 and 2.57). Equations 2.65 and 2.66 for u_{i0} encompass the most important effects, determining the occurrence of the ion-specific phenomena: radii of the bare and hydrated ions, R_b and R_h, possible deformation of the hydration shells at the interface for large ions, polarizabilities α_p and ionization potentials I of the counterions and of the pure bulk phase. Table 2.2 summarizes the values of $u_{i0}/k_B T$ for a number of ions, calculated without using any adjustable parameter. They were tested in Figure 2.11 against the values of K from the plots of π^S versus $C^{2/3}$ for several surfactants and counterions. All data fall well within the lines $\ln K$ versus $-u_{i0}/k_B T$ drawn with the theoretical slope 1/2.

The analysis of the popular models for adsorption of surfactants at fluid interfaces (those of Frumkin [77] and van der Waals [79]) revealed that they are flawed in the definitions of all constants involved: the adsorption constant K, the area per molecule α, and the attraction constant β [31–33]. Hence, in ref. [32], new EOS and adsorption isotherms for nonionic surfactants were derived (Equations 2.118 and 2.119). Their hard core part is based on the practically exact HFL model for hard discs [80] (Equation 2.87). The attraction interaction is accounted for by the sticky potential (Equation 2.103) of Baxter [88]. In the present article, these results were used to derive new analytical EOS and adsorption isotherms for ionic surfactants, accounting also for ion-specific effects (Equations 2.129, 2.118, and 2.132). They were tested by adsorption data for three surfactants ($C_{12}H_{25}SO_4Na$, $C_{12}H_{25}NMe_3Cl$, and Br) on two interfaces: W|G and water–hexane. The results obtained by this new model were encouraging (cf. Table 2.4 and the comments thereafter). Moreover, some interesting new electrostatic and ion-specific effects were found. It turned out that similar to the adsorption constant K, the area per molecule α exhibits strong ion-specific effects (cf. Equation 2.124): its nonspecific values α_0 at W|G for $C_{12}H_{25}NMe_3^+Cl^-$ and Br^- are about two to three times larger than the respective α's determined by the fit of the adsorption data with the adsorption isotherm (Table 2.4). However, because the repulsion is dampened by the ion-specific effect, the experimental values of the α's are smaller than the α_0's and those of the K's are larger than the K_0's. More importantly, the α_0's and K_0's are almost equal for the two counterions, Cl^- and Br^-, as they should be according to the theory. We consider these results as strong confirmation of our theory. The electrostatic nature of α was confirmed by plotting $\ln\alpha$ versus the square of the Debye length $1/\kappa^2$ for different electrolyte concentrations (cf. Figure 2.21 and the related discussion). Another surprising effect is exhibited by the comparison of the adsorption Γ_s of the same surfactant at W|O and W|G interfaces (Figure 2.20). At small concentrations C, the adsorption Γ_s at W|G is smaller than that at W|O, but as C increases, the adsorption at W|G becomes larger. This leads to an intersection of the two curves $\Gamma_s(C)$. We attributed this effect to the second virial coefficient B_2 (leading to repulsion), which is smaller at the W|G interface than at the W|O interface.

When deriving the expressions (Equations 2.65 and 2.66) for u_{i0}, no assumptions were made about the phenomenon studied or the nature of the ionic surfactants. Hence, we decided to use the present theory of the ion-specific effects and the calculated values of u_{i0} from Table 2.2 to interpret two other phenomena involving ionic surfactants: the cmc and the disjoining pressure of thin films. In Section 2.4, Shinoda's theory of the cmc of ionic surfactants was generalized to account for the specific effect of the counterion (Equation 2.143). The new theory was confirmed by experimental data for cmc of several surfactants with different counterions (Figures 2.22 and 2.26)—in agreement with Equation 2.145, the plots of ln cmc versus $-u_{i0}/k_BT$ were linear with theoretical slope $-1/3$. A model for a mixture of several counterions is proposed (Equation 2.147), and confirmed with experimental data (Figure 2.25). Finally, the Derjaguin-Landau-Verwey-Overbeek (DLVO) theory of the electrostatic disjoining pressure Π_{el} of thin liquid films was also extended to account for ion-specific effects. From the experimental curves $\Pi_{el}(h)$ (where h is the film thickness) for films stabilized with $C_{16}H_{33}NMe_3Br$ and with the addition of excess amounts of the salts NaX (X = F⁻, Cl⁻, Br⁻), the values of the surface potential Φ^S were calculated and plotted versus $-u_{i0}/k_BT$ of the respective counterions (Figure 2.28). Although we deal with three points only, the interpolation gave a line with slope -0.4, slightly different from the theoretical value $-1/2$ (Equation 2.57).

Based on all these results, we may conclude that the simple theory of the counterion-specific effects, presented above, is surprisingly efficient and universal—it has thus far been successfully applied to several different phenomena and led not only to correct interpretations (in all cases, quantitative) of the considered effects but also helped us predict and explain several new effects that have not been observed previously. An important general result was that (with one exception—the NH_4^+ ion) in all studied phenomena, the sequence of the ions was ordered by efficiency following Hofmeister's series (Equations 2.1 and 2.2) but the ordering of the anions was in reverse (compare Equations 2.76 and 2.1). A similar reversal of the Hofmeister series was attributed in refs. [3,4] to the sign of the interfacial charge. It is worthwhile to remind the readers that the theory is based on a single quantity, the ion-specific adsorption energy u_{i0}, which was calculated from Equations 2.65 and 2.66 without using any adjustable parameters. That is why we hope that u_{i0} is (at least for ionic surfactants and, possibly, for proteins) the main quantity involved in Hofmeister effects.

2.6 PROSPECTS

As mentioned above, the theory of the ion-specific adsorption energy u_{i0} in Section 2.2.3 and its experimental verification suggest that it depends solely on the nature of the counterion and, possibly, of the hydrophobic phase. Our preliminary unpublished calculations show that, at least for W|O interfaces, the latter effect is quite small. This makes us believe that this study is a good basis for further investigations of ion-specific effects. Part of these investigations should lead to a better understanding of several new effects, described above. These are (i) the dependence of the intercept π_0 of $\pi^S(C^{2/3})$ on the ion adsorption energy u_{i0} and its relation to the structure of the oil-like layer formed by the hydrocarbon tails of the surfactant (Figure 2.12); (ii) the role of the surfactant head on the value of the nonspecific constant K_0 of the surfactant ion; and (iii) the electrostatic effect, mentioned above, on the area per molecule α. We are planning to clarify these issues in the near future.

There are several other phenomena, which seem closely related to the ion-specific adsorption energy u_{i0}. We hope that they could be tackled without need for significant modifications of the present theory. These tasks are (i) the application of the present theory to a study of the ion-specific effect on the electrostatic interaction of proteins; this can be done in at least three ways: by interpreting data for the second virial coefficient, by investigating spread protein layers, or by studying thin liquid films stabilized by proteins at low ionic strength. In every case, one must vary at least the counterion, pH, and salt concentrations; (ii) Application of the theory to interpret the ion-specific effects on the micellar aggregation number and the transition from spherical to cylindrical micelles [102]; and (iii) More detailed analysis of the ion-specific effects on the DLVO theory and use of the results to modify the coalescence theory of drops and bubbles.

As the van der Waals forces are ever present, so are the ion-specific effects. This is confirmed by a short list of phenomena in which ion-specific effects have been revealed and in which our theory could possibly be applied: surface tension of electrolyte solutions [2,6–8,54], microemulsions [14] and vesicles [132], and properties of lipid monolayers [96,97]. Our approach might be applicable even for systems not involving surfactants or lipids. Indeed, Parsons et al. [133] recently successfully calculated the interaction energy between silica and alumina particles using a theory whose main features are similar to our theory of the ion-selective adsorption energy u_{i0}, developed in ref. [29] and used in the present article.

LIST OF SYMBOLS AND ABBREVIATIONS

B_2	second virial coefficient ($B_2 = B_{hc} + B_{attr}$)
B_{attr}	attractive part of the second virial coefficient
B_{hc}	hard core (repulsive) part of the second virial coefficient
C_{el}	concentration of added electrolyte in the bulk
C_s	surfactant concentration in the bulk
C_t	total electrolyte concentration, $C_t \equiv C_s + C_{el}$
C	mean ion activity $\left(C = \gamma C_t^{1/2} C_s^{1/2}\right)$
d	rod length (one-dimensional adsorption of solid rods)
E_a	adsorption energy
E_{head}	energy of transfer of the surfactant's head from the bulk to the interface
e_0	elementary charge
e_i	charge of the ith component
I	ionization potential
k_B	Boltzmann constant
K	adsorption constant of an ionic surfactant ($\Gamma_s = KC^{2/3}$; $K = K_0 e^{-u_{i0}/2k_B T}$)
K_0	adsorption constant in the absence of ion-specific effect, $K_0^3 \equiv 4K_s/\kappa_0^2$
K_s	Henry's constant $\left(\Gamma_s = K_s C_s^S\right)$
L_{ij}	London constant
N_w	number of water molecules displaced by an ion upon adsorption
n_w	number of water molecules in the hydration shell of an ion
u_{i0}	specific adsorption energy of the counterion
u_{CH_2}	free energy for transfer of –CH2– from water to the hydrophobic phase
x_i	composition of salt mixture, $x_i \equiv C_i/C_t$
x_i^S	surface composition of salt mixture, $x_i^S \equiv \Gamma_i/\Gamma_s$
z	Cartesian coordinate
α	area of a molecule
α_L	area of a molecule in isotherms, based on Langmuir's model
α_p	static polarizability
α_V	area of a molecule in isotherms, based on Volmer's model
β	attraction parameter
γ	activity coefficient
Γ_s	surfactant adsorption
Γ_i^{DL}	adsorption of an ion in the diffuse double layer
δ_a	adsorption thickness
ε	absolute dielectric constant
κ	reciprocal Debye radius
κ_0	concentration-independent part of the Debye parameter, $\kappa_0^2 = 2e_0^2/\varepsilon k_B T$
μ	chemical potential
μ^S	chemical potential at the interface (function of Γ_s)
σ	interfacial tension
σ_0	tension of a pure interface (in the absence of surfactant and salt)
π_0	spreading (or negative cohesive) pressure
π^S	surface pressure, $\sigma_0 - \sigma$
Π	disjoining pressure

ϕ	electrostatic potential
ϕ^s	surface potential
Φ^S	the absolute value of the dimensionless surface potential, $\Phi^S \equiv e_0 \lvert \phi^s \rvert / k_B T$
Ac^-	acetate ion, CH_3COO^-
cmc	critical micelle concentration
EOS	equation of state
HFL	Helfand–Frisch–Lebowitz model
LE	liquid-expanded adsorption layer
Me	methyl group, $-CH_3$
W|G	water–gas interface
W|O	water–oil interface

ACKNOWLEDGMENTS

This work was supported by the Bulgarian National Science Fund Grants DDVU 02/12 and DDVU 02/54. We are indebted to Prof. Makoto Aratono and Prof. Piotr Warszynski for providing some of their experimental data used in this chapter, and to Dora Dimitrova for obtaining and processing some experimental data.

REFERENCES

1. F. Hofmeister. About regularities in the protein precipitating effects of salts and the relation of these effects with the physiological behaviour of salts. *Arch. Exp. Pathol. Pharmakol.*, **24**, 247–260 (1887).
2. W. Kunz (Ed.). *Specific Ion Effects.* World Scientific Publishing Co., New York (2011).
3. M. Boström, F.W. Tavares, S. Finet, F. Skouri-Panet, A. Tardieu and B.W. Ninham. Why forces between proteins follow different Hofmeister series for pH above and below pI. *Biophys. Chem.*, **117**, 217–224 (2005).
4. N. Schwierz, D. Horinek and R.R. Netz. Reversed anionic Hofmeister series: The interplay of surface charge and surface polarity. *Langmuir*, **26**, 7370–7379 (2010).
5. Issues 1 and 2 of *Curr. Opin. Colloid Interface Sci.*, **9**(1–2), 1–198 (2004).
6. B.W. Ninham and V. Yaminsky. Ion binding and ionspecificity: The Hofmeister effect and Onsager and Lifshitz theories. *Langmuir*, **13**, 2097–2108 (1997).
7. M. Boström, D.R.M. Williams and B.W. Ninham. Surface tension of electrolytes: Specific ion effects explained by dispersion forces. *Langmuir*, **17**, 4475–4478 (2001).
8. M. Boström, W. Kunz and B.W. Ninham. Hofmeister effects in surface tension of aqueous electrolyte solution. *Langmuir*, **21**, 2619–2623 (2005).
9. L.A. Moreira, M. Boström, B.W. Ninham, E.C. Biscaia and F.W. Tavares. Hofmeister effects: Why protein charge, pH titration and protein precipitation depend on the choice of background salt solution. *Colloids Surf. A*, **282–283**, 457–463 (2006).
10. M. Boström and B.W. Ninham. Contributions from dispersion and Born self free energies to the solvation energies of salt solutions. *J. Phys. Chem. B*, **108**, 12593–12595 (2004).
11. M. Boström and B.W. Ninham. Dispersion self-free energies and interaction free energies of finite-sized ions in salt solutions. *Langmuir*, **20**, 7569–7574 (2004).
12. M. Boström, D.R.M. Williams and B.W. Ninham. Specific ion effects: Why DLVO theory fails for biology and colloid systems. *Phys. Rev. Lett.*, **87**, 168103/1-4 (2001).
13. W. Kunz, L. Belloni, O. Bernard and B.W. Ninham. Osmotic coefficients and surface tensions of aqueous electrolyte solutions: Role of dispersion forces. *J. Phys. Chem. B*, **108**, 2398 (2004).
14. S. Murgia, M. Monduzzi and B.W. Ninham. Hofmeister effects in cationic microemulsions. *Curr. Opin. Colloid Interface Sci.*, **9**, 102–106 (2004).
15. W. Kunz, P. Lo Nostro and B.W. Ninham. The present state of affairs with Hofmeister effects. *Curr. Opin. Colloid Interface Sci.*, **9**, 1–18 (2004).
16. F.W. Tavares, D. Bratko, H.W. Blanch and J.M. Prausnitz. Ion-specific effects in the colloid–colloid or protein–protein potential of mean force: Role of salt–macroion van der Waals interactions. *J. Phys. Chem. B*, **108**, 9228–9235 (2004).

17. P. Warszynski, K. Lunkenheimer and G. Czichocki. Effect of counterions on the adsorption of ionic surfactants at fluid–fluid interfaces. *Langmuir*, **18**, 2506–2514 (2002).

18. G. Para, E. Jarek and P. Warszynski. The Hofmeister series effect in adsorption of cationic surfactants—Theoretical description and experimental results. *Adv. Colloid Interface Sci.*, **122**, 39–55 (2006).

19. G. Para, E. Jarek and P. Warszynski. The surface tension of aqueous solutions of cetyltrimethylammonium cationic surfactants in presence of bromide and chloride counterions. *Colloids Surf. A*, **261**, 65–73 (2005).

20. H.H. Li, Y. Imai, M. Yamanaka, Y. Hayami, T. Takiue, H. Matsubara and M. Aratono. Specific counterion effect on the adsorbed film of cationic surfactant mixtures at the air/water interface. *J. Colloid Interface Sci.*, **359**, 189–193 (2011).

21. M. Aratono, S. Uryu, Y. Hayami, K. Motomura and R. Matuura. Phase transition in the adsorbed films at water/air interface. *J. Colloid Interface Sci.*, **98**, 33–38 (1984).

22. K. Shimamoto, A. Onohara, H. Takumi, I. Watanabe, H. Tanida, H. Matsubara, T. Takiue and M. Aratono. Miscibility and distribution of counterions of imidazolium ionic liquid mixtures at the air/water surface. *Langmuir*, **25**, 9954–9959 (2009).

23. Y. Hayami, H. Ichikawa, A. Someya, M. Aratono and K. Motomura. Thermodynamic study on the adsorption and micelle formation of long chain alkyltrimethylammonium chlorides. *Colloid Polym. Sci.*, **276**, 595–600 (1998).

24. E.D. Goddard, O. Harva and T.G. Jones. The effect of univalent cations on the critical micelle concentration of sodium dodecyl sulfate. *Trans. Faraday Soc.*, **49**, 980–984 (1953).

25. K. Shinoda, T. Nakagawa, B.-I. Tamamushi and T. Isemuta. *Colloidal Surfactants, Some Physicochemical Properties*. Academic Press, New York (1963).

26. J.T. Davies and E. Rideal. *Interfacial Phenomena*. Academic Press, New York (1963).

27. J.T. Davies. Adsorption of long-chain ions I. *Proc. R. Soc. London, Ser. A*, **245**, 417–428 (1958).

28. R.P. Borwankar and D.T. Wasan. Equilibrium and dynamics of adsorption of surfactants at fluid-fluid interfaces. *Chem. Eng. Sci.*, **43**, 1323–1337 (1988).

29. I.B. Ivanov, K.G. Marinova, K.D. Danov, D. Dimitrova, K.P. Ananthapadmanabhan and A. Lips. Role of the counterions on the adsorption of ionic surfactants. *Adv. Colloid Interface Sci.*, **134–135**, 105–124 (2007).

30. I.B. Ivanov, R.I. Slavchov, E.S. Basheva, D. Sidzhakova and S.I. Karakashev. Hofmeister effect on micellization, thin films and emulsion stability. *Adv. Colloid Interface Sci.*, **168**, 93–104 (2011).

31. I.B. Ivanov, K.P. Ananthapadmanabhan and A. Lips. Adsorption and structure of the adsorbed layer of ionic surfactants. *Adv. Colloid Interface Sci.*, **123–126**, 189–212 (2006).

32. T.D. Gurkov and I.B. Ivanov. Layers of nonionic surfactants on fluid interfaces; Adsorption and interactions in the frames of a statistical model. In Proceedings of the 4th World Congress on Emulsions, Lyon, France; Paper No. 2.1, p. 509 (2006).

33. I.B. Ivanov, K.D. Danov, D. Dimitrova, M. Boyanov, K.P. Ananthapadmanabhan and A. Lips. Equations of state and adsorption isotherms of low molecular non-ionic surfactants. *Colloids Surf. A*, **354**, 118–133 (2010).

34. T.D. Gurkov, D.T. Dimitrova, K.G. Marinova, C. Bilke-Crause, C. Gerber and I.B. Ivanov. Ionic surfactants on fluid interfaces: Determination of the adsorption; role of the salt and the type of the hydrophobic phase. *Colloids Surf. A*, **261**, 29–38 (2005).

35. E.D. Shchukin, A.V. Pertsov and E.A. Amelina. *Colloid Chemistry*. Univ. Press, Moscow (1982) [in Russian]. Elsevier, Amsterdam (2001) [in English].

36. P.A. Kralchevsky, K.D. Danov, G. Broze and A. Mehreteab. Thermodynamics of ionic surfactant adsorption with account for the counterion binding: Effect of salts of various valency. *Langmuir*, **15**, 2351–2365 (1999).

37. K.D. Danov, P.A. Kralchevsky, K.P. Ananthapadmanabhan and A. Lips. Interpretation of surface-tension isotherms of n-alkanoic (fatty) acids by means of the van der Waals model. *J. Colloid Interface Sci.*, **300**, 809–813 (2006).

38. C. Tanford. *The Hydrophobic Effect*. Wiley, New York (1980).

39. J.N. Israelachvili. *Intermolecular and Surface Forces*. Academic Press, New York (2011).

40. A.J. Kumpulainen, C.M. Persson, J.C. Eriksson, E.C. Tyrode and C.M. Johnson. Soluble monolayers of *n*-decyl glucopyranoside and *n*-decyl maltopyranoside. Phase changes in the gaseous to the liquid-expanded range. *Langmuir*, **21**, 305–315 (2005).

41. J.W. Gibbs. *The Collected Works of J. W. Gibbs*, Vol. I. Longmans, Green, New York (1931).

42. L.G. Gouy. Sur la constitution de la charge électrique à la surface d'un électrolyte. *J. Phys. Radium*, **9**, 457–468 (1910).

43. R.A. Robinson and R.H. Stokes. *Electrolyte Solutions*. Butterworth Scientific Publications, London (1959).

44. E.H. Lucassen-Reynders. Surface equation of state for ionized surfactants. *J. Phys. Chem.*, **70**, 1777–1785 (1966).

45. E.H. Lucassen-Reynders. Adsorption at fluid interfaces. In *Anionic Surfactants: Physical Chemistry of Surfactant Action*, E.H. Lucassen-Reynders (Ed.), Chap. 1, Surfactant Science Series, Vol. 11. Marcel Dekker, New York (1981).

46. I. Langmuir. Oil lenses on water and the nature of monomolecular expanded films. *J. Chem. Phys.*, **1**, 756–776 (1933).

47. N.K. Adam. *The Physics and Chemistry of Surfaces*. Clarendon Press, Oxford (1941).

48. A. Goebel and K. Lunkenheimer. Interfacial tension of the water/n-alkane interface. *Langmuir*, **13**, 369–372 (1997).

49. E.A. Guggenheim. Specific thermodynamic properties of aqueous solutions of strong electrolytes. *Philos. Mag.*, **19**, 588–643 (1935).

50. K. Wojciechowski, A. Bitner, P. Warszynski and M. Zubrowska. The Hofmeister effect in zeta potentials of CTAB-stabilised toluene-in-water emulsions. *Colloids Surf. A*, **376**, 122–126 (2011).

51. S.J. Rehfeld. Adsorption of sodium dodecyl sulfate at various hydrocarbon-water interfaces. *J. Phys. Chem.*, **71**, 738–745 (1967).

52. J.D. Hines. The preparation of surface chemically pure sodium n-dodecyl sulfate by foam fractionation. *J. Colloid Interface Sci.*, **180**, 488–492 (1996).

53. D.A. Haydon and F.H. Taylor. On adsorption at the oil/water interface and the calculation of electrical potentials in the aqueous surface phase I. *Phil. Trans. R. Soc. Lond. A*, **252**, 225–248 (1960).

54. R.I. Slavchov, J.K. Novev, T.V. Peshkova and N.A. Grozev. Surface tension and surface $\Delta\chi$-potential of concentrated Z+:Z− electrolyte solutions. *J. Colloid Interface Sci.* **403**, 113–126 (2013).

55. K.D. Collins. Charge density-dependent strength of hydration and biological structure. *Biophysical*, **72**, 65–76 (1997).

56. N.A. Izmailov. *Electrochemistry of Solutions*, 3rd ed. Khimia, Moscow (1976) [in Russian].

57. Y. Marcus. *Ion Properties*. Marcel Dekker, New York (1997).

58. Y. Marcus. Thermodynamics of ion hydration and its interpretation in terms of a common model. *Pure. Appl. Chem.*, **59**, 1093–1101 (1987).

59. B.P. Nikolskij. *Handbook of Chemistry*. Khimija, Leningrad (1963) [in Russian].

60. B. Dietrich, J.-P. Kintzinger, J.-M. Lehn, B. Metz and A. Zahidi. Stability, molecular dynamics in solution, and x-ray structure of the ammonium cryptate [NH4+.cntnd.2.2.2]hexafluorophosphate. *J. Phys. Chem.*, **91**, 6600–6606 (1987).

61. D.R. Lide (Ed.). *CRC Handbook of Chemistry and Physics*, 83rd ed. CRC Press, Boca Raton, FL (2002).

62. W.R. Gillap, N.D. Weiner and M. Gibaldi. Effect of hydrocarbon chain length on adsorption of sodium alkyl sulfates at oil/water interfaces. *J. Colloid Interface Sci.*, **26**, 232–236 (1968).

63. R. Aveyard and B.J. Briscoe. Adsorption of n-alkanols at alkane/water interfaces. *J. Chem. Soc., Faraday Trans.*, **168**, 478–491 (1972).

64. D.A. Haydon and F.H. Taylor. Adsorption of sodium octyl and decyl sulphates and octyl and decyl trimethylammonium bromides at the decane-water interface. *Trans. Faraday Soc.*, **58**, 1233–1250 (1962).

65. M. Aratono. Personal communication (2010).

66. V. Bergeron. Disjoining pressures and film stability of alkyltrimethylammonium bromide foam films. *Langmuir*, **13**, 3474–3482 (1997).

67. C. Das and B. Das. Effect of tetraalkylammonium salts on the micellar behavior of lithium dodecyl sulfate: A conductometric and tensiometric study. *J. Mol. Liq.*, **137**, 152–158 (2008).

68. D. Dimitrova. Personal communication (2011).

69. J.R. Lu, A. Marrocco, T. Su, R.K. Thomas and J. Penfold. Adsorption of dodecyl sulfate surfactants with monovalent metal counterions at the air-water interface studied by neutron reflection and surface tension. *J. Colloid Interface Sci.*, **158**, 303–316 (1993).

70. P. Koelsch and H. Motschmann. Varying the counterions at a charged interface. *Langmuir*, **21**, 3436–3442 (2005).

71. J. Rogers and J.H. Schulman. A Mechanism of the Selective Flotation of Soluble Salts. In: Proceedings of the Second International Congress of Surface Activity, vol. III: Electrical phenomena, solid-liquid interface, ed. Schulman J.H. London: Butterworth, 243, 1957.

72. D.F. Sears and J.H. Schulman. Influence of water structures on the surface pressure, surface potential, and area of soap monolayers of lithium, sodium, potassium, and calcium. *J. Phys. Chem.*, **68**, 3529–3534 (1964).

73. I. Weil. Surface concentration and the Gibbs adsorption law. The effect of the alkali metal cations on surface behavior. *J. Phys. Chem.*, **70**, 133–140 (1969).

74. L. Onsager and N.N.T. Samaras. The surface tension of Debye-Hückel electrolytes. *J. Chem. Phys.*, **2**, 528–536 (1934).

75. V.V. Kalinin and C.J. Radke. An ion-binding model for ionic surfactant adsorption at aqueous-fluid interfaces. *Colloids Surf. A*, **114**, 337–350 (1996).

76. I. Langmuir. The adsorption of gases on plane surfaces of glass, mica and platinum. *J. Am. Chem. Soc.*, **40**, 1361–1403 (1918).

77. A. Frumkin. Electrocapillary curve of higher aliphatic acids and the state equation of the surface layer. *Z. Phys. Chem.*, **116**, 466 (1925).

78. M. Volmer. Thermodynamische Folgerungen aus der Zustandsgleichung ftir adsorbierte. Stoffe. *Z. Phys. Chem.*, **115**, 253–260 (1925).

79. J.H. De Boer. *The Dynamical Character of Adsorption*. Clarendon Press, Oxford (1953).

80. E. Helfand, H.L. Frisch and J.L. Lebowitz. Theory of the two- and one-dimensional rigid sphere fluids. *J. Chem. Phys.*, **34**, 1037–1042 (1961).

81. T.L. Hill. *An Introduction to Statistical Thermodynamics*. Addison-Wesley, Reading, MA (1962).

82. A.W. Adamson and A.P. Gast. *Physical Chemistry of Surfaces*, 6th ed. Wiley, New York (1997).

83. L. Tonks. The complete equation of state of one, two and three-dimensional gases of hard elastic spheres. *Phys. Rev.*, **50**, 955–963 (1936).

84. H. Reiss, H.L. Frisch and J.L. Lebowitz. Statistical mechanics of rigid spheres. *J. Chem. Phys.*, **31**, 369–380 (1959).

85. F. Lado. Equation of state of the hard disk fluid from approximate integral equations. *J. Chem. Phys.*, **49**, 3092–3096 (1968).

86. A.I. Rusanov. The essence of the new approach to the equation of the monolayer state. *Colloid J.*, **69**, 131–143 (2007).

87. A.I. Rusanov. Equation of state and phase transitions in surface monolayer. *Colloids Surf. A*, **239**, 105–111 (2004).

88. R.J. Baxter. Percus-Yevick equation for hard spheres with surface adhesion. *J. Chem. Phys.*, **49**, 2770–2774 (1968).

89. M. Yamanaka, M. Aratono, K. Motomura and R. Matuura. Effect of pressure on the adsorption and micelle formation of aqueous dodecyltrimethylammonium chloride solution-hexane system. *Colloid Polym. Sci.*, **262**, 338–341 (1984).

90. P.C. Hemmer and G. Stell. Strong short-range attractions in a fluid. *J. Chem. Phys.*, **93**, 8220 (1990).

91. C. Tutschka and J.A. Cuesta. Overcomplete free energy functional for D=1 particle systems with next neighbor interactions. *J. Stat. Phys.*, **111**, 1125–1148 (2003).

92. A.V. Makievski and D.O. Grigoriev. Adsorption of alkyl dimethyl phosphine oxides at the solution/air interface. *Colloids Surf. A*, **143**, 233–242 (1998).

93. P. Warszynski and K. Lunkenheimer. Influence of conformational free energy of hydrocarbon chains on adsorption of nonionic surfactants at the air/solution interface. *J. Phys. Chem. B*, **103**, 4404–4411 (1999).

94. L. Malysa, R. Miller and K. Lunkenheimer. Relationship between foam stability and surface elasticity forces: Fatty acid solutions. *Colloids Surf.*, **53**, 47–62 (1991).

95. K. Lunkenheimer and R. Hirte. Another approach to a surface equation of state. *J. Phys. Chem.*, **96**, 8683–8686 (1992).

96. E.D. Goddard and H.C. Kung. Counterion effects in long-chain sulfonate monolayers. *J. Colloid Interface Sci.*, **37**, 585–594 (1971).

97. E.D. Goddard, O. Kao and H.C. Kung. Counterion effects in charged monolayers. *J. Colloid Interface Sci.*, **27**, 616–624 (1968).

98. G. Loglio, P. Pandolfini, L. Liggieri, A.V. Makievski and F. Ravera. Determination of interfacial properties by the pendant drop tensiometry: Optimisation of experimental and calculation procedures. In *Bubble and Drop Interfaces*, R. Miller and L. Liggieri (Eds.), Chap. 2, Book Series: Progress in Colloid and Interface Science, Vol. 2. Brill, Leiden (2011).

99. K. Motomura, M. Aratono, N. Matubayasi and R. Matuura. Thermodynamic studies on adsorption at interfaces III. Sodium dodecyl sulfate at water | hexane interface. *J. Colloid Interface Sci.*, **67**, 247–254 (1978).

100. S. Levine, K. Robinson, G.M. Bell and J. Mingins. The discreteness-of-charge effect at charged aqueous interfaces I. *J. Electroanal. Chem.*, **38**, 253–269 (1972).

101. L. Gaillon, J. Lelievre and R. Gaboriaud. Counterion effects in aqueous solutions of cationic surfactants: Electromotive force measurements and thermodynamic model. *J. Colloid Interface Sci.*, **213**, 287–297 (1999).

102. J.V. Joshi, V.K. Aswal, P. Bahadur and P.S. Goyal. Role of counterion of the surfactant molecule on the micellar structure in aqueous solution. *Curr. Sci.*, **83**, 47–49 (2002).

103. M. Benrraou, B.L. Bales and R. Zana. Effect of the nature of the counterion on the properties of anionic surfactants 1. *J. Phys. Chem. B*, **107**, 13432–13440 (2003).

104. M.H. Ropers, G. Czichocki and G. Brezesinski. Counterion effect on the thermodynamics of micellization of alkyl sulfates. *J. Phys. Chem. B*, **107**, 5281–5288 (2003).

105. S. Kumar, D. Sharmav and Kabir-ud-Din. Effect of additives on the clouding behavior of sodium dodecyl sulfate plus tetra-n-butylammonium bromide system. *J. Surfactants Deterg.*, **7**, 271–275 (2004).

106. M.J. Rosen. *Surfactants and Interfacial Phenomena*, 3rd ed. Wiley, New York (2004).

107. M.L. Corrin and W.D. Harkins. The effect of salts on the critical concentration for the formation of micelles in colloidal electrolytes. *J. Am. Chem. Soc.*, **69**, 683–688 (1947).

108. R. Nagarajan. Micellization, mixed micellization and solubilization—The role of interfacial interactions. *Adv. Colloid Interface Sci.*, **26**, 205–264 (1986).

109. I.V. Rao and E. Ruckenstein. Micellization behavior in the presence of alcohols. *J. Colloid Interface Sci.*, **113**, 375–387 (1986).

110. V. Srinivasan and D. Blankschtein. Effect of counterion binding on micellar solution behavior 1. *Langmuir*, **19**, 9932–9945 (2003).

111. A. Jakubowska. Interactions of different counterions with cationic and anionic surfactants. *J. Colloid Interface Sci.*, **346**, 398–404 (2010).

112. R. Zielinski. Micelle formation in aqueous NaBr solutions of alkyltrimethylammonium bromides. *Pol. J. Chem.*, **72**, 127–136 (1998).

113. A. Malliaris, J. Lang and R. Zana. Micellar aggregation numbers at high surfactant concentration. *J. Colloid Interface Sci.*, **110**, 237–242 (1986).

114. T.M. Perger and M. Bester-Rogac. Thermodynamics of micelle formation of alkyltrimethylammonium chlorides from high performance electric conductivity measurements. *J. Colloid Interface Sci.*, **313**, 288–295 (2007).

115. S. Durand-Vidal, M. Jardat, V. Dahirel, O. Bernard, K. Perrigaud and P. Turq. Determining the radius and the apparent charge of a micelle from electrical conductivity measurements by using a transport theory: Explicit equations for practical use. *J. Phys. Chem. B*, **110**, 15542–15547 (2006).

116. B.L. Bales and R. Zana. Characterization of micelles of quaternary ammonium surfactants, as reaction media I. *J. Phys. Chem. B*, **106**, 1926–1939 (2002).

117. D.F. Evans and P.J. Wightman. Micelle formation above 100-degrees-C. *J. Colloid Interface Sci.*, **86**, 515–524 (1982).

118. P.F. Grieger and C.A. Kraus. Properties of electrolytic solutions. 35. Conductance of some long chain salts in methanol-water mixtures. *J. Am. Chem. Soc.*, **70**, 3803–3811 (1948).

119. A. Gonzalez-Perez, J. Czapkiewicz, J.L. Del Castillo and J.R. Rodriguez. Micellar properties of tetradecyltrimethylammonium nitrate in aqueous solutions at various temperatures and in water–benzyl alcohol mixtures at 25 degrees C. *Colloid Polym. Sci.*, **282**, 1359–1364 (2004).

120. S. Hashimoto, J.K. Thomas, D.F. Evans, S. Mukherjee and B.W. Ninham. Unusual behavior of hydroxide surfactants. *J. Colloid Interface Sci.*, **95**, 594–596 (1983).

121. P. Lianos and R. Zana. Use of pyrene excimer formation to study the effect of NaCl on the structure of sodium dodecyl-sulfate micelles. *J. Phys. Chem.*, **84**, 3339–3341 (1980).

122. S.K. Mehta, K.K. Bhasin, R. Chauhan and S. Dham. Effect of temperature on critical micelle concentration and thermodynamic behavior of dodecyldimethylethylammonium bromide and dodecyltrimethylammonium chloride in aqueous media. *Colloids Surf. A*, **255**, 153–157 (2005).

123. H.B. Klevens. Critical micelle concentrations as determined by refraction. *J. Phys. Colloid Chem.*, **52**, 130–148 (1948).

124. J. Osugi, M. Sato and N. Ifuku. Micelle formation of cationic detergent solution at high pressures. *Rev. Phys. Chem. Jpn.*, **35**, 32 (1965).

125. H. Okuda, T. Imae and S. Ikeda. The adsorption of cetyltrimethylammonium bromide on aqueous surfaces of sodium-bromide solutions. *Colloids Surf.*, **27**, 187–200 (1987).

126. A.P. Brady and S. Ross. The measurement of foam stability. *J. Am. Chem. Soc.*, **66**, 1348–1356 (1944).

127. A.B. Scott, H.V. Tartar and E.C. Lingafelter. Electrolytic properties of aqueous solutions of octyltrimethylammonium octanesulfonate and decyltrimethylammonium decanesulfonate. *J. Am. Chem. Soc.*, **65**, 698–701 (1943).

128. K. Maiti, D. Mitra, S. Guha and S.P. Moulik. Salt effect on self-aggregation of sodium dodecylsulfate (SDS) and tetradecyltrimethylammonium bromide (TTAB): Physicochemical correlation and assessment in the light of Hofmeister (lyotropic) effect. *J. Mol. Liq.*, **146**, 44–51 (2009).

129. N.V. Churaev, B.V. Derjagiun and V.M. Muller. *Surface Forces*. Springer, New York (1987).
130. K.J. Mysels and M.N. Jones. Direct measurement of variation of double-layer repulsion with distance. *Discuss. Faraday Soc.*, **42**, 42–50 (1966).
131. V. Bergeron and C.J. Radke. Equilibrium measurements of oscillatory disjoining pressures in aqueous foam films. *Langmuir*, **8**, 3020–3026 (1992).
132. M. Gradzielski. Vesicle gel—Phase behavior and processes of formation. *Curr. Opin. Colloid. Interface Sci.*, **9**, 149–153 (2004).
133. D.F. Parsons, M. Boström, T.J. Maceina, A. Salis and B.W. Ninham. Why direct or reversed hofmeister series? Interplay of hydration, non-electrostatic potentials, and ion size. *Langmuir*, **26**, 3323–3328 (2010).

Part II

Surfactants at Solid—Liquid Interfaces

3 Wettability of Solid-Supported Lipid Layers

Emil Chibowski, Malgorzata Jurak, and Lucyna Holysz

CONTENTS

3.1 INTRODUCTION

Phospholipids, which belong to a group of lipids, besides the principal role they play in biological membranes, are also biosurfactants [l], especially those with shorter hydrocarbon chain(s). They can stabilize or destabilize dispersed systems, such as emulsions or suspensions. Lecithin of natural origin is commonly used in food processing, pharmaceutical products, cosmetics, and others. The stability of the dispersed systems is generally determined by electrostatic repulsive forces (zeta potential) and attractive London dispersion forces. In the case of long-chain surfactants, steric stabilization can also take place. However, the principal role they play is that of constituents of the living cell membrane, where they form bilayers. Therefore, the model biological membranes, which are monolayers, bilayers, and multilayers of lipids and phospholipids, have been extensively investigated for several decades, not only with respect to the membrane properties but also with respect to many practical applications. In this chapter, we will briefly discuss the historical background of these investigations, the preparation methods for the solid-supported phospholipid layers, and the methods for their investigations, with special emphasis on the advanced techniques applied nowadays. Finally, studies on the wettability of such layers that we have recently carried out will be described. Some of the results will also be compared with the atomic force microscopy (AFM) images of the investigated layers. Some future research we are planning to carry out will also be mentioned.

3.1.1 HISTORICAL BACKGROUND

3.1.1.1 The Langmuir Films

The history of Langmuir films can be said to begin with Benjamin Franklin, who in 1773 observed that the waves on a pond became calm on about half-an-acre of its surface when he dropped a teaspoon of oil on it [2]. Although he did not realize it, he probably for the first time had created a monolayer oil film on the water's surface. Then, it took a century before Lord Rayleigh explained this phenomenon in a "scientific way." In a simple way, he calculated from the oil volume used and the pond surface area that the thickness of the oleic acid film was 1.6 nm. Then, Agnes Pockels conducted some experiments in her kitchen sink using a thread in the shape of a loop placed on the water surface, and found that the films of different oils spread on the water's surface could be compressed by up to 0.2 nm^2, whereas the surface tension of water did not change. With help from Lord Rayleigh, in 1891, she published her findings in *Nature* [3]. In the next step, Irving Langmuir confirmed the results of Pockels and concluded that the molecules in the films were oriented vertically irrespective of the chain length of molecules.

Thus, the monolayer films on a liquid surface are formed by insoluble amphiphilic molecules whose polar heads are directed toward the aqueous phase whereas the hydrophobic tails protrude from the surface, preferring to interact with air, contrary to the hydrophilic heads that prefer to interact with polar water molecules (Figure 3.1). As a result, a reduction in the surface free energy (surface tension) of water takes place.

Among the molecules that form a monolayer on the water's surface, there are different kinds of biomolecules. Moreover, a mixed monolayer on the surface can be formed by two or more different compounds, whose structure, packing, and mutual interactions are drastically different from those in a single-component monolayer, and thus its properties can be tuned. The monolayers spread onto the water's surface are nowadays commonly called *Langmuir films*. The basic characterization of the Langmuir films is the isotherm of the film pressure π versus surface area A occupied by one molecule in the film. The film pressure increases during its compression, that is, the reduction of the surface area occupied by the film molecules. This causes a film transition from the gaseous (G), through the liquid-expanded (LE), and the liquid-condensed (LC), and finally to the solid-like

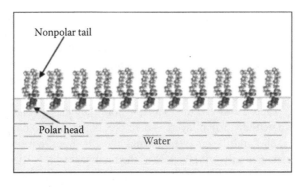

FIGURE 3.1 The scheme of a Langmuir film spread on water surface.

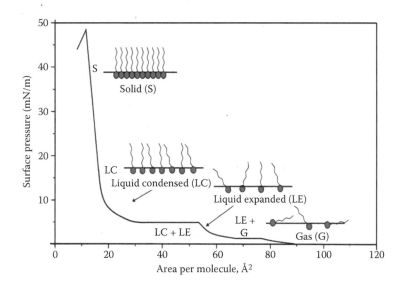

FIGURE 3.2 The scheme of different states of a Langmuir monolayer.

states—upon further compression, the collapse of the film can occur [4]. Figure 3.2 shows schematically different states of a monolayer film and the corresponding π versus A curve.

3.1.1.2   The Langmuir–Blodgett Films

This term designates the monolayer, bilayer, or multilayer films of organic molecules transferred from the liquid surface (the Langmuir film) onto a solid support. The name for films of this kind originates, of course, from the names of Irving Langmuir and Katharine Blodgett, who invented the method of their preparation and developed scientific foundations for the investigation of these films [1]. Katharine Blodgett was a coworker of Irving Langmuir who at the beginning of their joint investigations hired her as his assistant. The Langmuir–Blodgett (LB) films are characterized by well-defined thickness, packing, and topography, which greatly depend on the pressure at which the Langmuir film is transferred onto the solid substrate. From a biological point of view, the monolayers and bilayers of lipid and protein molecules are most attractive because these molecules are the principal constituents of biological membranes of living cells and such films are good models to study their properties. It should also be mentioned that other methods are now known to produce the solid-supported films of lipids or proteins, like the spreading of a solution onto a solid surface [5–9], spin-coating [10–13], or vesicle fusion [14–22]. These and other methods will be described in the next section.

3.1.1.3   The Structure of Biological Membrane

A model of a biological membrane is shown in Figure 3.3. Native biological membranes are composed of a lipid bilayer, which consists of phospholipids (the most abundant), glycolipids, sterols (cholesterol), and proteins.

In the bilayer, the polar groups of phospholipids are directed outward and the hydrocarbon tails are "inside" the membrane, where they mutually interact by attractive nonpolar London dispersion forces. The length of the hydrocarbon chains is 16 to 20 carbon atoms, and they may be saturated or unsaturated. Therefore, the thickness of the bilayer is approximately 5 nm. The length and saturation/unsaturation of the tails determine the membrane's fluidity. The presence of unsaturated chains causes larger fluidity of the membrane, that is, it is packed less tightly. Singer and Nicolson

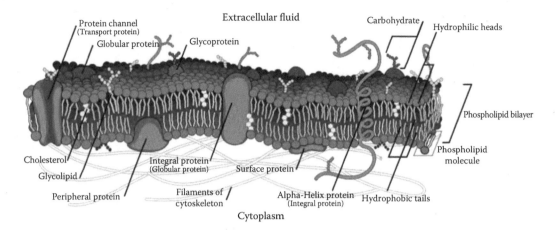

FIGURE 3.3 **(See color insert.)** A model of the biological cell membrane. (From http://en.wikipedia.org/wiki/Cell_membrane.)

[24] proposed a fluid mosaic model of the biological membranes (now commonly accepted), which can be considered as a two-dimensional (2-D) liquid, in which all the lipid and protein molecules can diffuse more or less easily. However, the membranes contain different microdomains of protein–protein complexes, pickets and fences formed by the actin-based cytoskeleton, and lipid rafts, which are microdomains enriched in cholesterol. Under physiological conditions the molecules in the membrane are in the liquid crystalline state. The structure of the cell membrane causes its low permeability and prevents polar solutes (ions, amino or nucleic acids, and others) from diffusing easily across the membrane. However, a "controlled transmembrane movement" takes place through special channels (pores or gates).

It is difficult to investigate the processes occurring in such complex biological systems; therefore, it is necessary to build simpler models, such as the solid-supported lipid films. Figure 3.4 shows the scheme of two principal phospholipids constituting biological membranes, that is, 1,2-dipalmitoyl-*sn*-glycero-3-phosphocholine (DPPC), and 1,2-dioleoyl-*sn*-glycero-3-phosphocholine (DOPC).

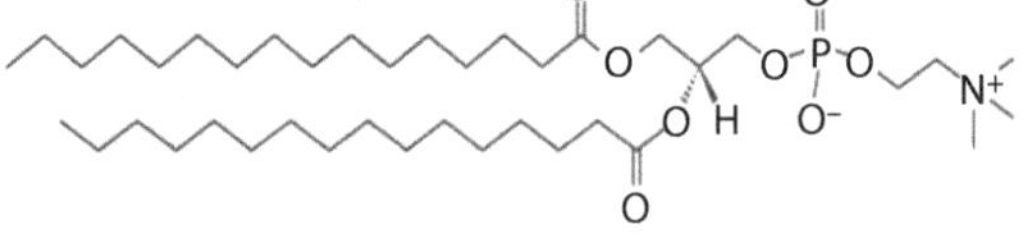

DPPC, cross-sectional molecular area S = 55 $Å^2$

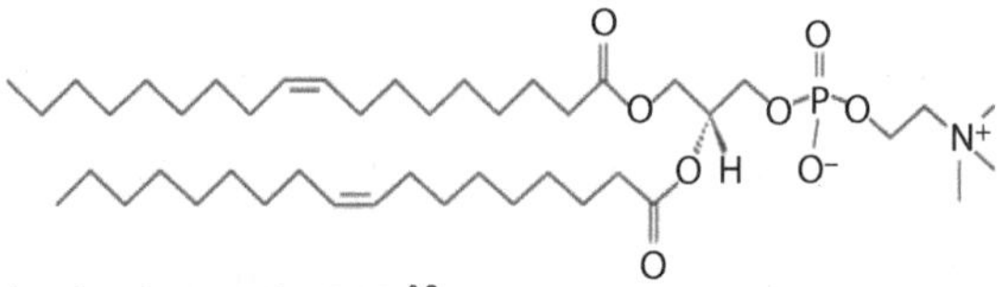

DOPC, cross-sectional molecular area S = 87.9 $Å^2$

FIGURE 3.4 The schemes of DPPC and DOPC phospholipid molecules.

3.1.2 Fabrication of Solid-Supported Lipid Layers

As mentioned above, solid-supported lipid layers have been developed as models of cell membranes to study the properties of biological membrane systems, as well as for potential biotechnological applications. The most popular biomimetic systems used for such purposes are solid-supported lipid monolayers and bilayers [8–12,25–27], polymer-cushioned lipid bilayers [28], hybrid bilayers [29,30], tethered lipid bilayers [31], suspended lipid bilayers, or supported vesicular layers [14–18]. These model membranes can be prepared through a few different routes: direct spreading from an organic solution [5–9], spin-coating [10–13], vesicle fusion [14–22], and the LB or Langmuir–Schaefer (LS) techniques [19,27,32–36]. Other techniques of preparation that can be applied include spray-coating, dip-coating, self-assembly, adsorption from solution [12,25,26], electrodeposition [19], or thermal evaporation [30].

The spreading of organic lipid solutions on solid surfaces is a simple, fast, practical, and efficient method to prepare thin lipid films [8,9,37]. Under suitable conditions, depending on the concentration of solution used, this method yields multilayers with a thickness of a few hundred to several thousand bilayers, but it is practically impossible to obtain just several bilayers and control their number. Moreover, to obtain homogeneous films, the solid support should be completely wettable by the organic solution. Thus, the use of an appropriate solvent is crucial for producing homogeneous films [8].

Spin-coating is a cheap and fast method to produce stacks of a well-controlled number of bilayers (1–30) on a solid support [10,11]. An excess amount of the lipid dissolved in an organic solvent is placed on the solid support, which is then rotated at a high speed to spread the solution by centrifugal force. Some authors maintain that the spin-coating allows the preparation of highly ordered structures of a well-defined thickness, which can be controlled by the kind and size of the support surface, by varying the rotation speed and time, the concentration of the lipid solution, the kind of solvent, and the amount of the solution deposited onto the solid support [10–12,38,39]. The main disadvantage of this technique is its being limited by the choice of solvent.

Another method for depositing supported lipid layers on both flat and textured surfaces is the adsorption of vesicles (liposomes) from an aqueous suspension and their subsequent fusion on the support surface [17,40]. This technique, first reported by Brian and McConnell [15], is commonly applied nowadays due to its simplicity [14,17,18,40–42]. The first stage involves the vesicles' adhesion and their rupture, and the second involves the fusion of the bilayer patches, which leads to the formation of a homogeneous membrane. It is a relatively simple and safe technique for the formation of multicomponent films composed of many kinds of lipids and proteins. This is because the vesicles do not come into contact with the air–water interface, which obviates possible denaturation of hydrophobic proteins. In addition, all the components are fully hydrated during the process of layer preparation [40]. Factors affecting the fusion process are properties of the solid support, salt concentration, pH, ionic strength, temperature, osmotic pressure and size of vesicles, and surface charge [14,22,41]. The vesicles deposition process can be carried out both on flat and textured surfaces, including even colloidal particles, which act as a rigid skeleton for the membrane and control its structure [20,43]. This is an advantage with regard to the LB technique, which is effective only on flat surfaces.

Moreover, instead of lipid vesicles, micellar solutions of mixtures of surfactants and lipids can be used for the spontaneous formation of self-assembled structures [44]. The driving force for this process results from the difference in solubility of the components. The application of surfactant micelles does not alter the packing or the conformation of the lipid and does not leave a residue of surfactant in the bilayer [44].

The LB technique is one of the most useful methods to prepare highly organized 2-D lipid structures on solid supports with molecular level precision. This procedure, as mentioned above, was first described by Langmuir and effectively developed by Blodgett [45–47]. It relies on the successive transfer of monolayers onto a flat solid support (glass, quartz, silicon, mica, gold, and others) by withdrawing the immersed plate that passes through the monolayer compressed to a given surface pressure present on the liquid subphase [48–50]. The dipping can be repeated to create well-ordered multilayer films with defined orientation of the molecules. This method ensures precise control of the layer thickness and its uniform deposition. However, there are several parameters that exert significant influence on the LB film transfer: these include the nature of the spread monolayer, composition and pH of the subphase, temperature, surface pressure, deposition rate, type of solid support, and in the case of multilayer transfer, period of storage between the successive cycles [51,52].

An alternative method for transferring a monomolecular film onto a hydrophobic solid support is the horizontal lift technique described by Langmuir and Schaefer. The method involves horizontal touching of the Langmuir film present on the subphase surface with the solid support already possessing a preformed monolayer [48,49,53]. Tamm and McConnell [27] first conducted the lipid deposition, by the combination of the LB and LS methods, a technique that is often used nowadays [19]. On the other hand, the incorporation of any proteins into the lipid bilayer cannot be easily carried out by these techniques because of their irreversible denaturation [16]. But a combination of the LB with the vesicle fusion technique [16,19], in which vesicle fusion follows prior deposition of the phospholipid monolayer, allows the fabrication of asymmetric bilayers and incorporation of transmembrane proteins [16]. However, these methods are not suitable for the preparation of more than three bilayers, which would have uniformly ordered and defect-free surfaces. Strong interactions occur because of close contact between the lipid membrane and the solid support [11]. This affects the function of the protein and its lateral mobility. Moreover, such films can be easily damaged [11]. To overcome these problems and to simultaneously separate the membrane from the solid support, various ways to anchor planar lipid layers via chemical tethers have been developed [28,29,31,54]. Larger distances between the membrane and support can be obtained by placing a polymer film (polymer cushion) [28,31,54], by the self-assembly of chemically modified lipids, or by direct fusion of spacer lipids [29]. The unique advantage of such a hybrid bilayer membrane is its high mechanical stability. However, this additional monolayer of the phospholipid is often more crystalline than the first internal monolayer and this can change the specific membrane functions. Another drawback of these combined methods is that certain coupling molecules are associated specifically with a given kind of support, for example, thiols for gold surfaces or silanes for quartz and mica surfaces [29].

3.1.3 Current and Prospective Advanced Techniques for the Investigation of Solid-Supported Lipid Layers

The assembly of biomimetic membranes on a solid surface possesses attractive possibilities of being probed by various surface sensitive analytical techniques based on different principles, for example, optical techniques, such as fluorescence microscopy [55], fluorescence recovery after photobleaching (FRAP) [27,54,56], surface plasmon resonance (SPR) spectroscopy [30,57], ellipsometry [58,59], optical waveguide light mode spectroscopy (OWLS) [60], chiral second harmonic generation spectroscopy (SHG) [61], or sum frequency generation spectroscopy (SFG) [62]; scanning probe techniques, such as AFM [11,58,63–66], and electroacoustic techniques, such as quartz crystal microbalance (QCM) [59]. Moreover, the QCM is a simple and useful method for measuring mass changes [59]. It can be extended by performing direct monitoring of energy dissipation (QCM-D), considered a useful indicator of the viscoelastic properties of the deposited layer, and the real-time kinetics of the different processes [67]. The QCM-D technique has already been used in combination with optical techniques, both for separate optical measurements, for example, by SPR [57] and by OWLS [60], and recently in combination with optical reflectometry [68]. Another powerful method is electrochemical impedance spectroscopy (EIS). Initially used for the determination of electrical properties of the membrane, such as its resistivity and the double-layer capacitance, it is now applied for the investigation of electrode processes and complex interfaces, especially in combination with other complementary techniques, such as QCM-D [69]. An analysis of the system response provides information about the interface, its structure, and the reactions taking place.

Moreover, QCM [59,67], SPR [30,59], surface-enhanced Raman spectroscopy (SERS), and infrared refection–absorption spectroscopy (IRRAS) permit monitoring of the lipid film fabrication and the binding of biomolecules [29]. Ellipsometry is often employed for measuring the thickness of the deposited and fused lipid films [58,59]. X-ray photoelectron spectrometry (XPS) has been successfully applied to determine the quantitative chemical composition, chemical bonding, thickness, and molecular organization of the supported lipid films [70]. Time-of-fight secondary ion mass spectrometry (TOF-SIMS) allows a detailed analysis of the morphology of biomolecules on lipid surfaces and submicrometer imaging of the domains [71].

Much of the research is focused on the microscopic structure of the lipid films. The most general and widely utilized technique is fluorescence microscopy [55]. It gives a wealth of information about the structures on the micrometer scale, but because of the resolution limit, it provides little information about the submicrometer structure. To obtain higher resolution information, many of the newer scanning probe techniques can be employed, such as AFM [11,58,63–66], the related frictional force microscopy [72], and more recently, near-field scanning optical microscopy (NSOM) [66,73]. With these techniques, a detailed view of the stability, morphology, and submicrometer length structure of the lipid films can be obtained [72]. However, it is quite difficult to record changes in the thin film structure by AFM measurements, even in the noncontact mode in short times. By comparison, a host of acoustic-based and optical-based sensor techniques, including SPR [30], ellipsometry [58,59], reflectometry [74], and QCM-D monitoring [67,74] can be used to quantitatively track changes in film thickness and mass without introducing any external marker. Therefore, by combining the information gained through these techniques, one can identify interesting kinetic processes taking place in the model membranes.

Neutron and x-ray scattering have proven to be two of the most powerful techniques for structure determination, which are capable of providing dynamic and structural information [75]. In addition, x-ray and neutron reflectometry (XR, NR) have emerged as powerful surface/interface probes used to characterize the fine structure of the surface, interface, and thin films as well [76,77]. The most common approach is specular reflectivity, which measures the scattering density profile perpendicular to the surface with adsorbed film [78].

3.2 STUDIES OF WETTABILITY OF LIPID LAYERS

3.2.1 Surface Free Energy

To determine the wetting properties of a solid surface, the contact angle of a water droplet settled on the surface is most often measured. In fact, contact angle is the only directly measured quantity that reflects the interactions (forces) acting at solid/liquid/gas interfaces. It can be said that the contact angle is a "visible balance" of the interacting forces along the three phase contact line. In the case of a polar liquid such as water, despite London dispersion forces, which are always present, polar Lewis acid–base interactions are also present. They are the electron–donor and electron–acceptor interactions originating from water's ability to form hydrogen bonds, both as the donor and acceptor of electrons.

However, to evaluate the polar interactions of a given solid surface, contact angles of another polar liquid and one more apolar have to be measured. Then, applying the theoretical approach proposed by van Oss et al. [79,80], one can calculate the components of the surface free energy. In this approach, called the Lifshitz–van der Waals/acid–base (LWAB), it is assumed that the surface free energy is the sum of the apolar Lifshitz–van der Waals component γ_S^{LW} (which is actually the London dispersion component γ_S^d as determined experimentally) and the polar Lewis acid–base γ_S^{AB} component (the electron–donor γ_S^- and electron–acceptor γ_S^+ parameters):

$$\gamma_S = \gamma_S^{LW} + \gamma_S^{AB} = \gamma_S^{LW} + 2\left(\gamma_S^- \gamma_S^+\right)^{1/2} \tag{3.1}$$

Then, the work of adhesion W_A of a liquid to the solid surface can be expressed in the following way:

$$W_A = \gamma_L(1+\cos\theta_a) = 2\left(\gamma_S^{LW}\gamma_L^{LW}\right)^{1/2} + 2\left(\gamma_S^+\gamma_L^-\right)^{1/2} + 2\left(\gamma_S^-\gamma_L^+\right)^{1/2} \qquad (3.2)$$

where the subscripts "S" and "L" mean solid and liquid, respectively, and θ_a is the advancing contact angle of the probe liquid.

Polar water and formamide as well as apolar diiodomethane, whose surface tension components are known, are used as the probe liquids. Then, three equations similar to Equation 3.2 can be solved simultaneously and the solid surface free energy components can be determined. Other approaches used to determine the surface free energy of solids can be found in a review published by Etzler [81].

Moreover, the apparent surface free energy can also be evaluated from the contact angle hysteresis (CAH) using the model proposed by Chibowski [82–84]. The CAH approach relates total apparent surface free energy of the solid γ_S to the surface tension of the probe liquid γ_L used and its CAH, which is defined as the difference between the advancing θ_a and receding θ_r contact angles:

$$\gamma_S = \frac{\gamma_L(1+\cos\theta_a)^2}{(2+\cos\theta_r+\cos\theta_a)} \qquad (3.3)$$

Using this approach, the surface free energy of a solid can be evaluated from the measured contact angles of only one probe liquid, whose surface tension is known. The free energy of a solid surface calculated from the CAH of water droplet involves total apparent apolar and polar interactions. However, when evaluated from the diiodomethane CAH, it represents almost totally the apolar interaction of the tested surface.

In this article, we focus mostly on the contact angles of water and diiodomethane; we will also provide a few examples of the calculated surface free energy components of the investigated lipid layers. The results of surface free energy determination of these systems can be found in our previously published articles [6,7,13,33–35,84] as well as in the entry of the *Encyclopedia of Surface and Colloid Science* [85].

3.2.2   Contact Angles on Solid-Supported Lipid Layers

3.2.2.1   Contact Angles on DPPC Layers Formed by Aqueous Solution Spreading

The advancing and receding contact angles of polar water and apolar diiodomethane on initially bare surfaces of solid supports (glass, mica, and poly[methyl methacrylate] [PMMA]), subsequently covered by up to five consecutive statistical DPPC monolayers, are plotted in Figures 3.5 and 3.6, respectively. The layers were deposited by spreading from aqueous or methanol solutions. The statistical monolayer number was calculated from a cross-section of the lipid molecule, which was determined from the π versus A isotherm and the geometric surface area of solid support. As can be seen, the contact angle values measured on the DPPC layers depend on the kind of support used and the number of statistical monolayers. The greatest changes in contact angle occur on the first two monolayers deposited on glass and mica (Figures 3.5 and 3.6), and the smallest ones take place on the successive monolayers spread on PMMA. Generally, the contact angles of polar water (Figure 3.5) are smaller than those of diiodomethane (Figure 3.6), which is an apolar liquid whose nature of interactions are almost totally London dispersive. Therefore, the contact angles of diiodomethane provide information about the strength of dispersion interactions of the surface. The highest advancing contact angles of water and simultaneously the lowest advancing contact angles of diiodomethane were obtained on the films produced on PMMA, which was found to be a monopolar (Lewis base)

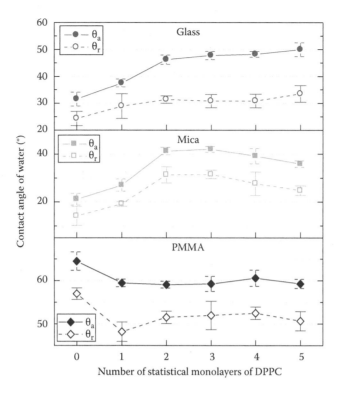

FIGURE 3.5 Advancing and receding contact angles of water depending on the number of statistical mono-layers of DPPC deposited on different supports. The layers were obtained by spreading aqueous (on glass and mica) or methanol (on PMMA) solutions of DPPC.

support [82,86]. Keeping in mind the strong polar nature of water, its relatively high contact angles clearly demonstrate the apolar nature of the films on PMMA. Moreover, the changes in contact angles measured for water and diiodomethane droplets on DPPC layers deposited on bipolar (Lewis acid–base) surfaces of glass and mica (Figures 3.5 and 3.6) are similar for each liquid. These contact angles are higher than those on the respective bare support surfaces and they are dependent on the thickness of the DPPC film. On the other hand, contact angles of water on the layers deposited on PMMA decrease relative to those on bare PMMA surface. In all systems, CAH occurs (Figures 3.5 and 3.6) and that of water droplets even amounts to approximately 10°. Much smaller CAH appears if diiodomethane is used as the probe liquid. The larger CAH in the case of water droplets may be a consequence of water penetration inside the DPPC layer structure when the three phase contact line recedes, that is, during the receding contact angle measurements.

Summing up, the DPPC monolayers on glass and mica supports make these surfaces more hydrophobic. This is a result of the molecules being arranged in such a way that the hydrocarbon chains are directed outward and the polar heads toward the support. Then, a further increase in their contact angles occurs probably due to the increased number of apolar tails of DPPC molecules being accessible for polar molecules of the probe liquids. This is because the films obtained by the solution spreading technique are not densely packed and their molecules may aggregate to form islets with the hydrocarbon chains sticking out, and patches of uncovered support surface may be present. In the case of the weakly polar PMMA surface, the situation is somehow different because the presence of the DPPC layer causes a small increase in hydrophobicity (Figure 3.5).

The difference between water contact angles measured on DPPC layers deposited on PMMA and those measured on glass or mica indicates that the structure of the DPPC films on these solid surfaces must be different. Hence, one may conclude that the layers' wettability is determined both

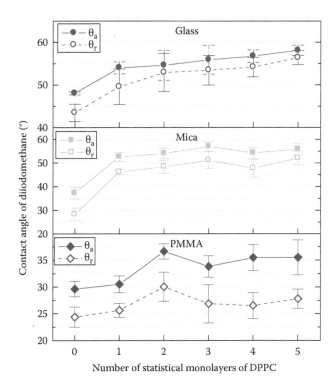

FIGURE 3.6 Advancing and receding contact angles of diiodomethane depending on the number of statistical monolayers of DPPC deposited on different supports. The layers were obtained by spreading aqueous (on glass and mica) or methanol (on PMMA) solutions of DPPC.

by the orientation of DPPC molecules in the layers and by their homogeneity. In the literature, the presence of a uniform monolayer or bilayer as well as the presence of patches of the molecules in other cases was also reported [6,9,27,87].

3.2.2.2 Contact Angles on DPPC and DOPC Layers Formed by Chloroform Solution Spreading

It was found that the structure of the deposited phospholipid film can also be affected by the kind of solvent used for the preparation of DPPC solution, which is reflected in the contact angle changes. Figures 3.7 and 3.8 illustrate the values of advancing and receding contact angles of water and diiodomethane on DPPC layers fabricated on glass by spreading of the chloroformic solution [7]. Generally, in the case of water droplets, they are a few degrees higher (compare Figures 3.5 and 3.7) and, in the case of diiodomethane, a few degrees lower (compare Figures 3.6 and 3.8) than those obtained by spreading DPPC dispersed in water. This suggests that the layers produced by the spreading of organic solvent solution are slightly more hydrophobic. Moreover, the wetting properties of the phospholipid films deposited on glass depend on the kind of phospholipid used for covering the support and the layer thickness (Figures 3.7 and 3.8). The contact angle increases after the deposition of the first monolayer of DPPC or DOPC relative to bare glass surface and, for consecutive layers, the contact angle regularly fluctuates from layer to layer in a narrow range, depending on the type of phospholipid, which is more visible on DPPC layers. The smaller contact angles of water suggest that more polar heads of the molecules are directed outward from the surface, depending on the number of layers. The behavior of the molecules at the solid support is determined by many factors, among others, the affinity for the solid surface and its physical state and their structure and organization [7]. At the experiment temperature, saturated DPPC (C16) exists as a gel phase, whereas DOPC, whose molecule has two unsaturated chains (C18), exists in a disordered liquid crystal phase [88]. Therefore, the surface

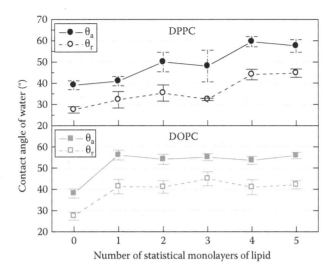

FIGURE 3.7 Advancing and receding contact angles of water depending on the number of statistical mono-layers of DPPC and DOPC deposited on glass by spreading their solutions in chloroform. (From Golabek, M., and L. Holysz, *Appl. Surface Sci.,* 256, 5463–5469, 2010.)

area occupied by a DOPC molecule must be larger than that occupied by DPPC [89]. Hence, DOPC layers on glass are slightly more hydrophobic. This is manifested in higher contact angles of water, immediately observable on one statistical monolayer. The contact angles are practically the same up to five statistical monolayers, whereas on DPPC layers, they fluctuate but progressively increase, reaching (on the fourth monolayer) the value already obtained on one DPPC monolayer (Figure 3.7). These results also show that the interactions occurring across the solid (film)/liquid interface are reflected in the changes in contact angles of apolar and polar liquids (Figures 3.7 and 3.8) [7].

Thus, it appears that the presence of cholesterol (Chol) in the DPPC solution, which is spread onto the glass surface, causes significant changes in the wetting properties of the films formed. The advancing and receding contact angles of water and diiodomethane, measured on pure DPPC, Chol,

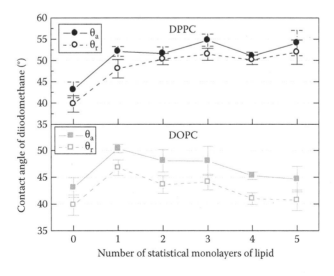

FIGURE 3.8 Advancing and receding contact angles of diiodomethane depending on the number of statis-tical monolayers of DPPC and DOPC deposited on glass by spreading their solutions in chloroform. (From Golabek, M., and L. Holysz, *Appl. Surface Sci.,* 256, 5463–5469, 2010.)

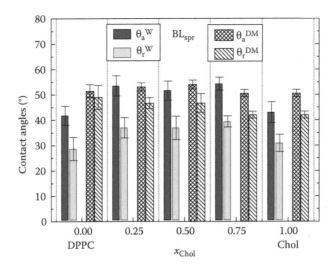

FIGURE 3.9 Advancing and receding contact angles of water and diiodomethane on DPPC/Chol bilayers deposited on glass by spreading (BL_{spr}) their solutions in chloroform. (From Golabek, M. et al., *Colloids Surf. A* 391, 150–157, 2011.)

and mixed DPPC/Chol bilayers, as a function of Chol molar fraction, are plotted in Figure 3.9 [84]. Unexpectedly, the water contact angles on two-component bilayers are bigger than those on pure DPPC and Chol bilayers, regardless of the binary system composition (Figure 3.9). The effect of Chol on the conformational and orientational ordering of the hydrocarbon chains of DPPC in the mixed monolayers [90] can be depicted by the contact angle values. It is commonly known as the condensation effect of cholesterol [35,91,92], and is connected with a reduction in the mean area per molecule in mixed DPPC/Chol layers, which reflects the tighter molecular organization and the larger strength of interactions between molecules [93,94]. The condensing effect promotes a larger rigidity of the layer [93], which prevents the penetration of the probe liquid into the film's interior.

However, curiously enough, the contact angles of water measured on the films, which have been dipped for 5 min in water and dried (Figure 3.10) are much greater than those on the layers not in

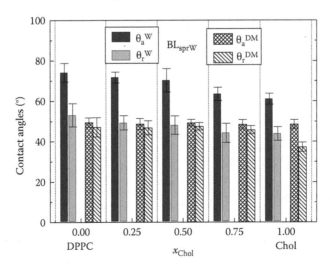

FIGURE 3.10 Advancing and receding contact angles of water and diiodomethane on DPPC/Chol bilayers deposited on glass by spreading and contacting for 5 min with water (BL_{sprW}). (From Golabek, M. et al., *Colloids Surf. A* 391, 150–157, 2011.)

contact with water (Figure 3.9). In the case of the pure DPPC bilayer, the difference amounts to about 32° (Figures 3.9 and 3.10) and the contact angle decreases smoothly with increasing Chol content in the bilayer, from 74° to only 17° for the pure Chol bilayer (Figure 3.10) [84]. This means that the studied films in contact with water undergo some reorganization, making the film structure more hydrophobic. Moreover, the contact of lipid bilayers with water causes an increase of water's CAH (Figures 3.9 and 3.10), which indicates easier penetration of water droplets into the film structure during the receding contact angles measurements. Interestingly, here the contact angles of diiodomethane are smaller than those on the bilayers before they are in contact with water and are practically independent of the bilayer composition (Figure 3.10) [84].

3.2.2.3 Contact Angles on DPPC Layers on Various Supports Formed by Spin-Coating

In this section, the results of contact angle measurements on solid-supported lipid layers obtained by spin-coating technique are presented. The advancing and receding contact angles measured for droplets of water and diiodomethane on the DPPC layers versus the concentrations of phospholipid solutions used to cover the solid support surface are shown in Figures 3.11 and 3.12, respectively. A distinct maximum in changes in the contact angle of water can be observed if 0.5 to 0.75 mg/mL of the DPPC solution is applied for the deposition (Figure 3.11). Then, the measured contact angles decrease upon the increase in the surface coverage. Regardless of the kind of the solid support at the solution concentration of 2 to 2.5 mg/mL, they reach practically the same value, which is even smaller than that of the respective bare solid surface. Further increases in the DPPC solution concentration did not influence the contact angle values (not presented). This suggests that these values characterize the wettability of DPPC molecules only. Despite the similar trend in the contact angle changes measured on the layers covering these three solid supports, the contact angle values are different, especially those on silicon and mica (Figure 3.11). This must be due to the specificity

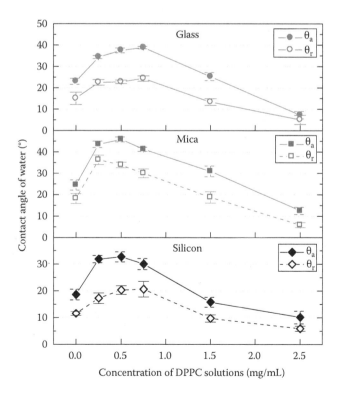

FIGURE 3.11 Advancing and receding contact angles of water on DPPC layers deposited on glass, mica, and silicon using the spin-coating method, depending on the phospholipid concentration.

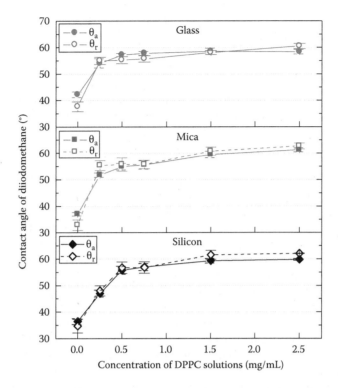

FIGURE 3.12 Advancing and receding contact angles of diiodomethane on DPPC layers deposited on glass, mica, and silicon using the spin-coating method, depending on the phospholipid concentration.

of the supports and different energetic properties of their surfaces [13], which affect the structure, packing, and organization of the lipids. Furthermore, the water CAH, determined on DPPC films deposited on these three solids, is clearly seen and amounts to approximately 10°, except for the thicker layer for which it is only a few degrees. In contrast, in the case of diiodomethane, practically no CAH appears (Figure 3.12).

Despite the fact that the wetting properties of these solid-supported DPPC layers depend on the technique used for their preparation, to some extent, the values of contact angles measured on them can be correlated with the layer thickness. Namely, in the case of DPPC films produced by aqueous or chloroform solution spreading, on the first statistical monolayer, the advancing contact angle of water on glass is equivalent to 37.5° (Figure 3.5) and 41.0° (Figure 3.7), respectively. However, in the case of mica support, a similar contact angle of water at 41.3° (Figure 3.5) was measured on two statistical DPPC monolayers. On the other hand, the maximum value of the water advancing contact angle on the DPPC layers obtained by spin-coating method on glass is 38.8°, and on mica it is 45.8° (Figure 3.11). Accordingly, one may conclude that the thickness of the films obtained from 0.5 mg DPPC/mL solution by the spin-coating technique corresponds to one to two statistical monolayers prepared by spreading of the DPPC solution at its appropriate concentration (Figures 3.5, 3.7, and 3.11). However, the contact angles of water measured on the DPPC monolayers and bilayers transferred by LB and LS techniques on glass are 53.8° and 53.1°, respectively (Figure 3.13) [84]. Within experimental error, these values are comparable to those determined for two statistical monolayers on glass obtained by spreading DPPC from the chloroform solution, that is, 50.0°. Taking into account possible differences in the films' structure resulting from different packing and ordering of the molecules, which depend on the preparation method, the reproducibility of the contact angle values in the repeated experiments is quite satisfactory.

The increase in the values of contact angles is mostly caused by the reduced polar interactions of the probe liquid with the DPPC film, in which the lipid headgroups are oriented toward the

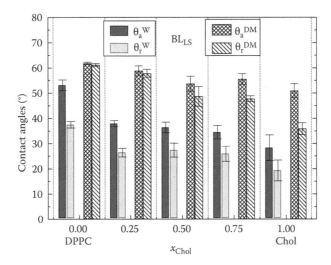

FIGURE 3.13 Advancing and receding contact angles of water and diiodomethane on the DPPC/Chol bilayer deposited on glass using the LS method (BL$_{LS}$). (From Golabek, M. et al., *Colloids Surf. A* 391, 150–157, 2011.)

hydrophilic solid surface whereas the hydrocarbon chains are directed outward to the air environment. If the polar support is only partially covered with lipid, the monolayer structure is not uniform and patches of bare solid support may occur. Hence, the measured contact angles are lower. However, when a thicker lipid film is deposited onto the substrate, its structure is more uniform and densely packed, which results in the values of the contact angle being higher than those on the bare surface. On the other hand, it is likely that when the solid surface is fully covered by DPPC molecules (at 2.5 mg/mL), its wettability is determined by DPPC molecules alone. Thus, the contact angle changes appearing with an increase in concentration of the coating solution are due to the DPPC film's structural organization or reorganization (Figures 3.11 and 3.12).

Additionally, the film structure images obtained using the AFM technique (not shown here; see ref. [13]) proved that depending on the method and procedure of deposition, even on the same support, homogeneous or fractional DPPC films can be fabricated [13].

3.2.2.4 Contact Angles on Phospholipid Layers Deposited Using the LB/LS Techniques

Both the monolayer and the bilayer of DPPC deposited on glass surfaces by the LB/LS technique cause a significant increase in the water contact angle value, from about 39.1° on bare glass surface to approximately 53.8° and 53.1° on the surface with the monolayer or the bilayer, respectively (Figure 3.14) [84]. The advancing contact angles of diiodomethane are higher than those of water and they are 62.4° and 61.8° for the monolayer and the bilayer, respectively (Figure 3.13). The incorporation of Chol molecules into DPPC films provokes changes in the contact angles, which gradually decrease upon an increase in Chol content in the mixed bilayer (Figure 3.13). This effect is most pronounced in the case of water contact angles, which decrease from 53° (pure DPPC bilayer) to 30° (pure Chol bilayer), and their hysteresis is 8° to 15°. However, the contact angles of apolar diiodomethane decrease only slightly with an increase in the Chol molar fraction (within ca. 5°). Moreover, practically no hysteresis of the contact angles occurs. These decreasing water contact angles show that with increasing content of Chol in the bilayer, its hydrophilicity increases, but the diiodomethane contact angles prove that the London dispersion forces do not change much (Figure 3.13). Only at a higher Chol content does visible CAH occur. These contact angles of diiodomethane (Figure 3.13) are approximately 15° higher than those on the DPPC bilayers prepared by the spreading method (Figure 3.9) and they are higher by approximately 20° than those measured on the bilayers in contact with water for 5 min (Figure 3.10) [84].

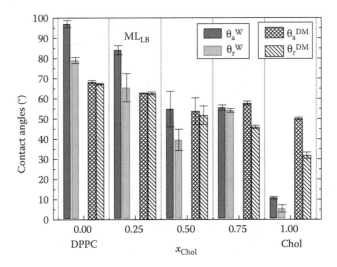

FIGURE 3.14 Advancing and receding contact angles of water and diiodomethane on DPPC/Chol monolayers deposited on mica using the LB method (ML$_{LB}$) as a function of cholesterol molar ratio. (From Chibowski, E., and M. Jurak, *Colloids Surf. A* 383, 56–60, 2011.)

Investigations carried out by other authors [58,63,95,96] showed that the role of solid support in the lipid membrane formation is crucial. It affects the distribution of lipid molecules in the film [58], influences the membrane structure and phase behavior [95], and the solid roughness is important for the interactions between the membrane and the solid surface [63]. For example, it was found that the lipid layers on silicon oxide are less stable than those on the mica support [63].

Our studies show that DPPC monolayers and bilayers deposited on the surface of mica are characterized by higher values of contact angles than those on DPPC layers on glass. For water, they are 97.0° and 89.6°, respectively, and for diiodomethane droplets, they are 68.2° and 67.5°, respectively (Figures 3.14 through 3.17). The large values of contact angles on the DPPC monolayer confirm the dense packing of molecules on mica whose surface is strongly hydrophilic [34]. The freshly cleaved mica surface possesses high surface charge density (3.2×10^7 e/cm²) [97], which facilitates strong interactions between phospholipid molecule and solid support, leading to a compact monolayer formation. In the case of DPPC bilayer on mica (Figures 3.15 and 3.17), the water contact angles are slightly smaller than those on the monolayer (Figures 3.14 and 3.16), which indicates that, in the bilayer, more DPPC molecules can interact with polar molecules of water by hydrogen bonds. However, on DPPC layers deposited on glasswater contact angles are the same for the monolayer and bilayer within experimental error. This shows that depending on the kind of solid support, reorganization in the DPPC bilayer may occur. The carefully prepared phospholipid bilayer, while exposed to air right after being transferred from water, can self-reorganize, exposing the acyl chains outward [98], and form flat domains and vesicles on the homogenous DPPC monolayer [70].

Moreover, the contact angle measured on a less condensed DPPC layer transferred at 10 mN/m on mica surface is smaller than that on the layer deposited at 35 mN/m, and they are 54.3° and 70.3° (for water) and 63.8° and 64.6° (for diiodomethane) on the monolayer and bilayer, respectively. The more loosely packed structure fosters the penetration of polar molecules of the probe liquid into the layer interior [34], thus causing a decrease in the contact angle values. It is worth mentioning that the contact angles measured on the DPPC monolayer on mica transferred at 10 mN/m are comparable to those obtained on the DPPC monolayer on glass but transferred at 35 mN/m (Figure 3.13). The similar contact angle values suggest similar packing of DPPC molecules in these two monolayers, despite differences in the film's pressure upon their transferring.

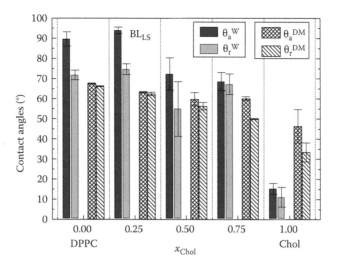

FIGURE 3.15 Advancing and receding contact angles of water and diiodomethane on DPPC/Chol bilayers deposited on mica using the LS method (BL_{LS}) as a function of cholesterol molar ratio.

Furthermore, the presence of Chol (Figures 3.14 and 3.15) or DOPC (Figures 3.16 and 3.17) in the DPPC layer induces a decrease in contact angle values as a function of Chol or DOPC molar fraction. In the case of binary DPPC/Chol monolayers and bilayers, a similar trend in contact angle changes occurs, that is, water contact angles are higher than those of diiodomethane, except for the layers of pure Chol and mixed DPPC/Chol$_{ML}$, if the Chol content was larger than DPPC (Figures 3.14 and 3.15). It is exactly the opposite of the contact angles on these layers on glass, where smaller water contact angles than those measured for diiodomethane droplets can be observed (Figure 3.13). Generally, the layers on mica are characterized by higher contact angles because of their tight structure. The more condensed structure of the films makes the penetration of liquid molecules inside the layer more difficult and it results in a higher contact angle of water, which principally interacts by the dispersion forces. It is obvious that the properties of the solid support and mutual interactions between

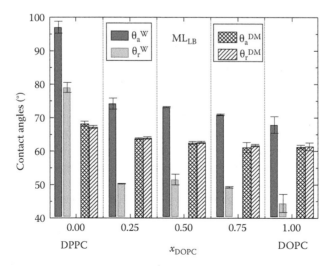

FIGURE 3.16 Advancing and receding contact angles of water and diiodomethane on DPPC/DOPC monolayers deposited on mica using the LB method (ML_{LB}) at different DOPC molar ratios. (From Chibowski, E., and M. Jurak, *Colloids Surf. A* 383, 56–60, 2011.)

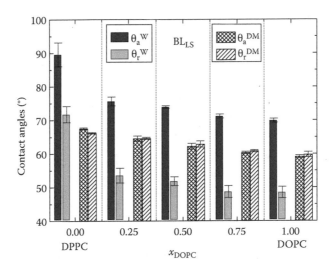

FIGURE 3.17 Advancing and receding contact angles of water and diiodomethane on DPPC/DOPC bilayers deposited on mica using the LS method (BL$_{LS}$) at different DOPC molar ratios.

the molecules are particularly responsible for the compactness of the layer. It is worth noticing that in the DPPC/Chol system, an attraction between molecules exists and is known as the condensation effect of cholesterol [35,91,92]. Some authors attribute this effect to a stable binary complex formation as a consequence of the existence of a strong cohesive force between molecules, especially at a small content of Chol (20%–25%) [35,91,92,99]. However, the patched film may be formed at the larger Chol molar ratios [91,92]. Depending on the kind of solid support used, the deposited layer is more or less condensed, which explains the differences in the values of contact angles.

The physical properties of the natural membranes are strongly affected by the kind of lipid present. At low cholesterol content (x_{Chol} = 0.25), its molecules interfere with the tightly packed DPPC molecules, which is required for the formation of orderly packed film. On the other hand, incorporating unsaturated phospholipid DOPC containing *cis*-double bonds to saturated phospholipid DPPC disrupts the tight packing of the DPPC membranes. This is because a phase separation occurs, resulting in the coexistence of spatially separated regions, namely, the condensed (ordered) DPPC-rich domains and a fluid (disordered) DOPC matrix [34]. Consequently, an expansion of the mean area per molecule with increasing DOPC content follows. Therefore, the resulting layer is less packed and can be more permeable for water and other small molecules [35]. It should be taken into account that the size of the diiodomethane molecule is more than three times larger than that of water [35]. Therefore, water molecules can easily penetrate into the lipid films, especially due to the presence of polar heads of phospholipids in the layer structure. This is reflected in the advancing and receding contact angle values and their hystereses. The highest water contact angle is obtained on the pure DPPC monolayer (Figures 3.14 through 3.17) and the smallest one on the Chol monolayer. Generally, on DPPC/Chol (Figures 3.14 and 3.15) smaller contact angles of water are found than on the DPPC/DOPC monolayer (Figures 3.16 and 3.17), except for the layers at x_{Chol} = 0.25, for which high advancing contact angles (Figures 3.14 and 3.15) point to the nonpolar character of the film. Moreover, the contact angles of diiodomethane change less than those of water, and the diiodomethane CAH is very small. It has to be emphasized that during the receding contact angle measurement, when the three phase contact line is receded, the liquid can penetrate into the film surface. Hence, larger CAH of water can be observed.

To sum up, the differences in the contact angles indicate that these values are determined by the kind of solid support, the film's homogeneity and composition, as well as the ordering and packing of molecules.

3.2.3 THE EFFECT OF ENZYMES ON CHANGES IN THE WETTABILITY OF PHOSPHOLIPID LAYERS

The monolayers and bilayers of phospholipids deposited on mica support using the LB/LS technique were found to be good models to study the membrane hydrolysis catalyzed by phospholipases due to their well-defined structures. Phospholipases are enzymes that are activated at the interface, whose activity strongly depends on the morphology and physicochemical properties of the phospholipid substrate. Several methods have been employed for monitoring the kinetics of enzymatic hydrolysis of DPPC layers under defined conditions, such as in monolayer studies at the air–water interface [100], electrochemical methods [101], fluorescence spectroscopy [102], fluorescence microscopy [103], and AFM [104]. In our studies, the advancing and receding contact angles on DPPC layers under phospholipase A_2 (PLA$_2$) and phospholipase C (PLC) were measured. The changes in advancing and receding contact angles of water and diiodomethane on the DPPC monolayer and bilayer caused by PLA$_2$ treatment at 0.02 units/mL and pH 8.9 are shown in Figures 3.18 and 3.19, respectively. As defined, one unit corresponds to the amount of enzyme that hydrolyzes 1 µmol of 3-sn-phosphatidylcholine per minute at pH 8.0 and 37°C. A drastic decrease in the contact angle values occurs within the first minute and then a gradual decrease of contact angles during 2 h can be observed, but water contact angles are much lower than those of diiodomethane. This behavior can be interpreted as an increase in the layer's surface hydrophilicity. The changes for both the monolayer and the bilayer run in a similar way, but the contact angles determined for the bilayer are slightly higher than those on the monolayer. It is evident that hydrolysis of the monolayer and the bilayer starts in the first minutes after the layer's contact with PLA$_2$, and is completed within 2 h, which is confirmed by the convergence of contact angle values after this time. PLA$_2$ hydrolyzes the sn-2 ester linkage of DPPC molecules to release palmitic acid and lysophosphatidylcholine. The latter product of hydrolysis is water-soluble and may leave the interface for the bulk phase but the palmitic acid, due to its affinity for the lipid layer, can first accumulate in the DPPC membrane [105]. During this stage, hydrogen bonds can be formed between the probe liquid and the polar products of hydrolysis, which are responsible for an increase in the electron–donor interactions γ_S^- displayed in Figure 3.20 [36]. Moreover, after the first 5 min of the hydrolysis, the electron–donor interactions γ_S^- of the bilayer are significantly stronger than those of the monolayer, 80 and 55 mJ/m^2, respectively (Figure 3.20) [36]. This can be a result of different packing and orientation of the

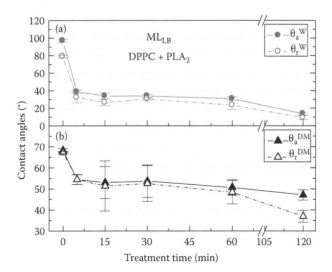

FIGURE 3.18 Changes in advancing and receding contact angles of water (a) and diiodomethane (b) on DPPC monolayer deposited on mica caused by enzyme PLA$_2$ treatment at 0.02 units/mL and pH 8.9. The monolayer was deposited using the LB method (ML$_{LB}$).

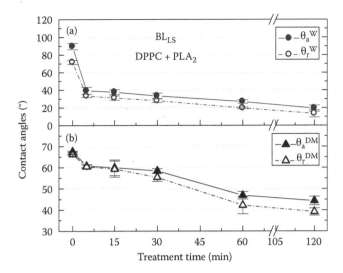

FIGURE 3.19 Changes in advancing and receding contact angles of water (a) and diiodomethane (b) on DPPC bilayer deposited on mica caused by enzyme PLA_2 treatment at 0.02 units/mL and pH 8.9. The bilayer was deposited using the LS method (BL_{LS}).

polar products of hydrolysis in the membrane. Product accumulation can lead to phase segregation (formation of the product domains inside the membrane), which shows that the phospholipid molecules are not well packed. Such loose molecular organization, the presence of hydrolysis product domains, and the loss of materials as a result of dissolution provide the best conditions for enzymes to penetrate and provide easier access for the enzyme to the phospholipid molecules. It should accelerate the activation and catalytic activity of enzymes [104]. Moreover, the negatively charged palmitic acid and positively charged PLA_2 activate the initial step of enzyme binding to

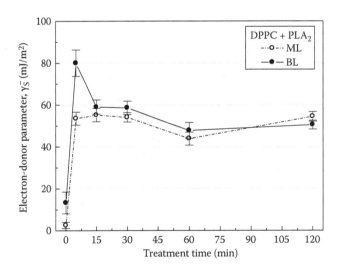

FIGURE 3.20 Changes in the electron–donor parameter of DPPC bilayer deposited on mica resulting from enzyme PLA_2 treatment at 0.02 units/mL and pH 8.9. The monolayer (ML) was deposited using the Langmuir–Blodgett method and bilayer (BL) using the LS method. (From Jurak, M., and E. Chibowski, *Appl. Surf. Sci.* 256, 6304–6312, 2010.)

the membrane via electrostatic interactions [106]. With the progress of catalysis, the domains of hydrolysis products grow, causing vast morphological changes in the membrane. This provokes the instability of its structure, dissolution of the products into the bulk phase, and finally total degradation of the membrane. The decrease in γ_S^- interactions (Figure 3.20) points to the dissolution of these products during a longer period of catalysis. Therefore, it seems likely that after accomplishment of the monolayer and the bilayer degradation, the measured contact angles reach the same value (compare Figures 3.18 and 3.19).

During the hydrolysis of the DPPC layer, patches of bare surface of mica appear too. They are accessible to probe liquids during the contact angle measurements [36,107], and thus contribute to the resulting values. Therefore, it may generally be said that the changes in contact angle result from structural changes in the membrane structure, which are caused by PLA$_2$'s actions.

Thus, the influence of PLC enzyme on the wetting properties of DPPC bilayers on mica is depicted in Figure 3.21 via the contact angle changes. PLC is a water-soluble enzyme, which catalyzes the hydrolysis of the ester bond in the C-3 position of the DPPC molecule. Here, the hydrolysis products are water-insoluble dipalmitoylglycerol and water-soluble phosphocholine [36]. Because the phosphocholine is released to the solution during the enzymatic reaction, morphological changes in the bilayer structure mainly originate from the dipalmitoylglycerol and unreacted DPPC [4], which undergo a rearrangement at the interface. Redistribution of the molecules results in visible changes in the contact angle values (Figure 3.21), which are significant only during the first 15 min. Accordingly, the drastic increase in the electron–donor interactions γ_S^- within the first few minutes (Figure 3.22) can be ascribed to domain formation in the membrane structure and possible molecular rearrangement of the dipalmitoylglycerol. Consequently, the slight decrease in these interactions is connected with the partial dissolution of the domains of hydrolysis products into the buffer solution [36].

It is believed that the difference between these two enzymes (PLA$_2$ and PLC), as for their effect on the wettability of the DPPC layers, principally lies in the properties of emergent hydrolysis products and the ordering of the phospholipid chains, which plays a major role in the enzyme binding and its activity at lipid interfaces.

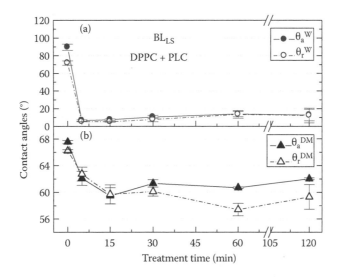

FIGURE 3.21 Changes in advancing and receding contact angles of water (a) and diiodomethane (b) on DPPC bilayer on mica caused by enzyme PLC treatment. The bilayer was deposited using the LS method (BL$_{LS}$).

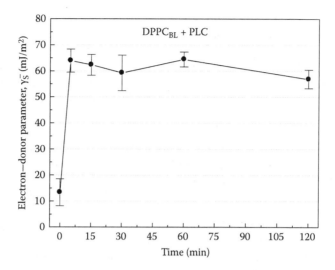

FIGURE 3.22 Changes in the electron–donor parameter of DPPC bilayer deposited on mica resulting from enzyme PLC treatment at 0.02 units/mL and pH 8.0. The bilayer (BL) was deposited using the LS method. (From Jurak, M., and E. Chibowski, *Appl. Surf. Sci.* 256, 6304–6312, 2010.)

3.2.4 PERSPECTIVES

In our opinion, future work needs to be done to connect knowledge of wetting properties with the utility of nanoscale advanced methods allowing direct visualization of the coexisting phases in the monolayers and the bilayers prepared from binary and ternary lipid mixtures, as well as tracking the changes in membrane structure caused by enzyme action. It would be important to have insight into the interactions occurring between particular components of the membrane, which seem to be complex and are not yet well understood. Moreover, new applications of such solid-supported lipid layers involve investigations of the membrane rafts. They are small, heterogeneous sterol-enriched or sphingolipid-enriched domains, which play functional roles in modulating cellular processes. Despite many biophysical techniques having been used to characterize rafts, progress in this area is still limited. Future work should be focused on examining binary and ternary lipid mixtures comprising phosphatidylcholine (PC), sphingomyelin (SM), and cholesterol (Chol), which can form coexisting phases in which the ordered liquid film structures would be analogous to the membrane rafts. Therefore, systematic studies are required to gain deeper insight into the mechanisms of the raft's formation and the enzymatic hydrolysis taking place in such membranes as well.

3.3 CONCLUSIONS

Based on the results of the advancing and receding contact angles measured on the solid-supported lipid layers, it may be concluded that:

- Macroscopic contact angles provide interesting information about the wetting properties of layers, which can also be useful for practical purposes
- The hydrophobic/hydrophilic character and the structure of solid-supported lipid layers depend on the method of their preparation and surface properties of the solid support
- Nevertheless, the layers also demonstrate some common features, which are observed if one calculates their surface free energy changes from the measured contact angles
- The most hydrophobic surface of the phospholipid film is that of DPPC deposited on glass or mica whereas the cholesterol layer deposited on mica seemed to be the most hydrophilic

- Enzymes affect the layer's wettability, which is manifested as a drastic decrease of water contact angle within the first 5 to 10 min, which means that the hydrophilicity of the layer increases significantly. The enzymatic effect depends on the kind of enzyme used for the hydrolysis process
- The increased hydrophilic character of the layer surface results from the appearance of polar products of the hydrolysis reaction. It is manifested as a sharp increase in the electron–donor component of the film surface free energy. These changes may also increase the activity of the enzyme and finally break down the layer structure
- The abrupt and gradual changes in the contact angles, and the surface free energy component calculated from them, allow tracking changes in the membrane structure, ordering, and interactions as a function of the layer composition, thickness, and the kind of the solid support or enzyme action. This provides an opportunity to use the hydrolysis processes to obtain the surfaces with defined hydrophobic/hydrophilic properties

ACKNOWLEDGMENTS

We thank Dr. Kash Mittal for our long-term coöperation and for helping us attend many symposia. We gratefully acknowledge the financial support from the Polish Ministry of Science and Higher Education, project no. N N204 272839.

REFERENCES

1. P. Rahman and E. Gapke. Production, characterization and applications of biosurfactants—Review. *Biotechnology*, **7**, 360–370 (2008).
2. A.W. Adamson and A.P. Gast. *Physical Chemistry of Surfaces*, 6th ed. Wiley, New York (1997).
3. A. Pockels. Letters to the editor: Surface tension. *Nature*, **43**, 437–439 (1891).
4. Q. He and J. Li. Hydrolysis characterization of phospholipid monolayers catalyzed by different phospholipases at the air–water interface. *Adv. Colloid Interface Sci.*, **131**, 91–98 (2007).
5. Y.-L. Lee and Ch.-Y. Chen. Surface wettability and platelet adhesion studies on Langmuir–Blodgett films. *Appl. Surf. Sci.*, **207**, 51–62 (2003).
6. M. Jurak and E. Chibowski. Topography and surface free energy of DPPC layers deposited on a glass, mica, or PMMA support. *Langmuir*, **22**, 7226–7234 (2006).
7. M. Golabek (Miroslaw) and L. Holysz. Changes in wetting and energetic properties of glass caused by deposition of different lipid layers. *Appl. Surf. Sci.*, **256**, 5463–5469 (2010).
8. M. Seul and M.J. Sammon. Preparation of surfactant multilayer films on solid substrates by deposition from organic solution. *Thin Solid Films*, **185**, 287–305 (1990).
9. T. Spangenberg, N.F. de Mello, T.B. Creczynski-Pasa, A.A. Pasa and H. Niehus. AFM in-situ characterization of supported phospholipid layers formed by solution spreading. *Phys. Stat. Sol. (a)*, **201**, 857–860 (2004).
10. U. Mennicke and T. Salditt. Preparation of solid-supported lipid bilayers by spin-coating. *Langmuir*, **18**, 8172–8177 (2002).
11. G. Pompeo, M. Girasole, A. Cricenti, F. Cattaruzza, A. Flamini, T. Prosperi, J. Generosic and A. Congiu Castellano. AFM characterization of solid-supported lipid multilayers prepared by spin-coating. *Biochim. Biophys. Acta*, **1712**, 29–36 (2005).
12. M. Rappolt, H. Amenitsch, J. Strancar, C.V. Teixeira, M. Kriechbaum, G. Pabsta, M. Majerowicz and P. Laggnera. Phospholipid mesophases at solid interfaces: In-situ X-ray diffraction and spin-label studies. *Adv. Colloid Interface Sci.*, **111**, 63–77 (2004).
13. M. Jurak and E. Chibowski. Wettability and topography of phospholipids DPPC multilayers deposited by spin-coating on glass, silicon, and mica slides. *Langmuir*, **23**, 10156–10163 (2007).
14. I. Reviakine and A. Brisson. Formation of supported phospholipid bilayers from unilamellar vesicles investigated by atomic force microscopy. *Langmuir*, **16**, 1806–1815 (2000).
15. A.A. Brian and H.M. McConnell. Allogeneic stimulation of cytotoxic T cells by supported planar membranes. *Proc. Natl. Acad. Sci. USA*, **81**, 6159–6163 (1984).

16. E. Kalb, S. Frey and L.K. Tamm. Formation of supported planar bilayers by fusion of vesicles to supported phospholipid monolayers. *Biochim. Biophys. Acta*, **1103**, 307–317 (1992).

17. Z.V. Leonenko, A. Carnini and D.T. Cramb. Supported planar bilayer formation by vesicle fusion: The interaction of phospholipid vesicles with surfaces and the eject of gramicidin on bilayer properties using atomic force microscopy. *Biochim. Biophys. Acta*, **1509**, 131–147 (2000).

18. Z.V. Leonenko, E. Finot, H. Ma, T.E.S. Dahms and D.T. Cramb. Investigation of temperature-induced phase transitions in DOPC and DPPC phospholipid bilayers using temperature-controlled scanning force microscopy. *Biophys. J.*, **86**, 3783–3793 (2004).

19. M. Li, M. Chen, E. Sheepwash, Ch.L. Brosseau, H. Li, B. Pettinger, H. Gruler and J. Lipkowski. AFM studies of solid-supported lipid bilayers formed at a Au(111) electrode surface using vesicle fusion and a combination of Langmuir–Blodgett and Langmuir–Schaefer techniques. *Langmuir*, **24**, 10313–10323 (2008).

20. S. Bucak, Ch. Wang, P.E. Laibinis and T.A. Hatton. Dynamics of supported lipid bilayer deposition from vesicle suspensions. *J. Colloid Interface Sci.*, **348**, 608–614 (2010).

21. P. Lenz, C.M. Ajo-Franklin and S.G. Boxer. Patterned supported lipid bilayers and monolayers on poly(dimethylsiloxane). *Langmuir*, **20**, 11092–11099 (2004).

22. R.P. Richter, R. Bérat and A.R. Brisson. Formation of solid-supported lipid bilayers: An integrated view. *Langmuir*, **22**, 3497–3505 (2006).

23. Available at http://en.wikipedia.org/wiki/Cell_membrane.

24. S.J. Singer and G.L. Nicolson. The fluid mosaic model of the structure of cell membranes. *Science*, **175**, 720–731 (1972).

25. J. Katsaras. Adsorbed to a rigid substrate, dimyristoylphosphatidylcholine multibilayers attain full hydration in all mesophases. *Biophys. J.*, **75**, 2157–2162 (1998).

26. R. Tero, M. Takizawa, Y.J. Li, M. Yamazaki and T. Urisu. Deposition of phospholipid layers on SiO_2 surface modified by alkyl-SAM islands. *Appl. Surf. Sci.*, **238**, 218–222 (2004).

27. L.K. Tamm and H.M. McConnell. Supported phospholipid bilayers. *Biophys. J.*, **47**, 105–113 (1985).

28. H.L. Smith, M.S. Jablin, A. Vidyasagar, J. Saiz, E. Watkins, R. Toomey, A.J. Hurd and J. Majewski. Model lipid membranes on a tunable polymer cushion. *Phys. Rev. Lett.*, **102**, 228102–228104 (2009).

29. A.L. Plant. Supported hybrid bilayer membranes as rugged cell membrane mimics. *Langmuir*, **15**, 5128–5135 (1999).

30. C. Rossi, J. Homand, C. Bauche, H. Hamdi, D. Ladant and J. Chopineau. Differential mechanisms for calcium-dependent protein/membrane association as evidenced from SPR-binding studies on supported biomimetic membranes. *Biochemistry*, **42**, 15273–15283 (2003).

31. C. Rossi and J. Chopineau. Biomimetic tethered lipid membranes designed for membrane-protein interaction studies. *Eur. Biophys. J.*, **36**, 955–965 (2007).

32. G. Zhavnerko and G. Marletta. Developing Langmuir–Blodgett strategies towards practical devices. *Mater. Sci. Eng. B*, **169**, 43–48 (2010).

33. E. Chibowski, L. Holysz and M. Jurak. Effect of a lipolytic enzyme on wettability and topography of phospholipid layers deposited on solid support. *Colloids Surf. A*, **321**, 131–136 (2008).

34. M. Jurak and E. Chibowski. Surface free energy and topography of mixed lipid layers on mica. *Colloids Surf. B*, **75**, 165–174 (2010).

35. E. Chibowski and M. Jurak. Interaction energy of model lipid membranes with water and diiodomethane. *Colloids Surf. A*, **383**, 56–60 (2011).

36. M. Jurak and E. Chibowski. Influence of (phospho)lipases on properties of mica supported phospholipids layers. *Appl. Surf. Sci.*, **256**, 6304–6312 (2010).

37. N. Kuèerka, Y. Liu, N. Chu, H.I. Petrache, S. Tristram-Nagle and J.F. Nagle. Structure of fully hydrated fluid phase DMPC and DLPC lipid bilayers using X-Ray scattering from oriented multilamellar arrays and from unilamellar vesicles. *Biophys. J.*, **88**, 2626–2637 (2005).

38. A.C. Simonsen and L.A. Bagatolli. Structure of spin-coated lipid films and domain formation in supported membranes formed by hydration. *Langmuir*, **20**, 9720–9728 (2004).

39. E. ten Grotenhuis, W.J.M. van der Kemp, J.G. Blok, J.C. van Miltenburg and J.P. van der Eerden. Scanning force microscopy of cholesterol multilayers prepared with the spin-coating technique. *Colloids Surf. B*, **6**, 209–218 (1996).

40. L.J. Johnston. Nanoscale imaging of domains in supported lipid membranes. *Langmuir*, **23**, 5886–5895 (2007).

41. I. Reviakine, A. Simon and A. Brisson. Effect of Ca^{2+} on the morphology of mixed DPPC-DOPS supported phospholipid bilayers. *Langmuir*, **16**, 1473–1477 (2000).

42. A. Charrier and F. Thibaudau. Main phase transitions in supported lipid single-bilayer. *Biophys. J.*, **89**, 1094–1101 (2005).

43. A.L. Troutier and C. Ladaviere. An overview of lipid membrane supported by colloidal particles. *Adv. Colloid Interface Sci.*, **133**, 1–27 (2007).

44. C. Lee, H. Wacklin and C.D. Bain. Changes in molecular composition and packing during lipid membrane reconstitution from phospholipid–surfactant micelles. *Soft Matter*, **5**, 568–575 (2009).

45. K.B. Blodgett. Monomolecular films of fatty acids on glass. *J. Am. Chem. Soc.*, **56**, 495–495 (1934).

46. K.B. Blodgett. Films built by depositing successive monomolecular layers on a solid surface. *J. Am. Chem. Soc.*, **57**, 1007–1007 (1935).

47. K.B. Blodgett and I. Langmuir. Built-up films of barium stearate and their optical properties. *Phys. Rev.*, **51**, 964–982 (1937).

48. G.L. Gaines, Jr. *Insoluble Monolayers at Liquid-Gas Interfaces*. Interscience Publishers, New York (1966).

49. J. Lyklema. *Fundamentals of Interface and Colloid Science*, Vol. III. Academic Press, London (2000).

50. A. Chyla. *Langmuir–Blodgett films and Their Application to Molecular Electronics, No. 2*. Scientific Papers of the Institute of Theoretical Chemistry of the Wroclaw University of Technology, Monographs Series, Wroclaw, Poland (2004).

51. J.B. Peng, M. Prakash, R. Macdonald, P. Dutta and J.B. Ketterson. Formation of multilayers of dipalmitoylphosphatidylcholine using the Langmuir–Blodgett technique. *Langmuir*, **3**, 1096–1097 (1987).

52. A. Girard-Egrot, R.M. Morelis and P.R. Coulet. Direct influence of the interaction between the first layer and a hydrophilic substrate on the transition from Y- to Z-type transfer during deposition of phospholipids Langmuir–Blodgett films. *Langmuir*, **12**, 778–783 (1996).

53. I. Langmuir and V.J. Schaefer. Activities of urease and pepsin monolayers. *J. Am. Chem. Soc.*, **60**, 1351–1360 (1938).

54. L. Renner, T. Osaki, S. Chiantia, P. Schwille, T. Pompe and C. Werner. Supported lipid bilayers on spacious and pH-responsive polymer cushions with varied hydrophilicity. *J. Phys. Chem. B*, **112**, 6373–6378 (2008).

55. J.T. Groves, N. Ulman and S.G. Boxer. Micropatterning of fluid lipid bilayers on solid supports. *Science*, **275**, 651–653 (1997).

56. T.H. Anderson, Y. Min, K.L. Weirich, H. Zeng, D. Fygenson and J.N. Israelachvili. Formation of supported bilayers on silica substrates. *Langmuir*, **25**, 6997–7005 (2009).

57. E. Reimhult, M. Zäch, F. Höök and B. Kasemo. A multitechnique study of liposome adsorption on Au and lipid bilayer formation on SiO_2. *Langmuir*, **22**, 3313–3319 (2006).

58. R.P. Richter and A.R. Brisson. Following the formation of supported lipid bilayers on mica: A study combining AFM, QCM-D, and ellipsometry. *Biophys. J.*, **88**, 3422–3433 (2005).

59. P.-Å. Ohlsson, T. Tjärnhage, E. Herbai, S. Löfås and G. Puu. Liposome and proteoliposome fusion onto solid substrates, studied using atomic force microscopy, quartz crystal microbalance and surface plasmon resonance. Biological activities of incorporated components. *Bioelectrochem. Bioenerg.*, **38**, 137–148 (1995).

60. D. Thid, M. Bally, K. Holm, S. Chessari, S. Tosatti, M. Textor and J. Gold. Issues of ligand accessibility and mobility in initial cell attachment. *Langmuir*, **23**, 11693–11704 (2007).

61. J.C. Conboy and M.A. Kriech. Measuring melittin binding to planar supported lipid bilayer by chiral second harmonic generation. *Anal. Chim. Acta*, **496**, 143–153 (2003).

62. X. Chen and Z. Chen. SFG studies on interactions between antimicrobial peptides and supported lipid bilayers. *Biochim. Biophys. Acta*, **1758**, 1257–1273 (2006).

63. H.M. Seeger, A. Di Cerbo, A. Alessandrini and P. Facci. Supported lipid bilayers on mica and silicon oxide: Comparison of the main phase transition behavior. *J. Phys. Chem. B*, **114**, 8926–8933 (2010).

64. L. Picas, C. Suárez-Germà, M.T. Montero and J. Hernández-Borrell. Force spectroscopy study of Langmuir–Blodgett asymmetric bilayers of phosphatidylethanolamine and phosphatidylglycerol. *J. Phys. Chem. B*, **114**, 3543–3549 (2010).

65. H.-J. Butt, R. Berger, E. Bonaccurso, Y. Chen and J. Wang. Impact of atomic force microscopy on interface and colloid science. *Adv. Colloid Interface Sci.*, **133**, 91–104 (2007).

66. C.W. Hollars and R.C. Dunn. Submicron structure in L-α-dipalmitoylphosphatidylcholine monolayers and bilayers probed with confocal, atomic force, and near-field microscopy. *Biophys. J.*, **75**, 342–353 (1998).

67. H.-M. Cho, D.Y. Cho, J.Y. Jeon, S.Y. Hwang, I.S. Ahn, J. Choo and E.K. Lee. Fabrication of protein-anchoring surface by modification of SiO_2 with liposomal bilayer. *Colloids Surf. B*, **75**, 209–213 (2010).

68. M. Edvardsson, S. Svedhem, G. Wang, R. Richter, M. Rodahl and B. Kasemo. A new QCM-D and reflectometry instrument—Applications to supported lipid structures and their biomolecular interactions. *Anal. Chem.*, **81**, 349–361 (2009).

69. E. Briand, M. Zäch, S. Svedhem, B. Kasemo and S. Petronis. Combined QCM-D and EIS study of supported lipid bilayer formation and interaction with pore-forming peptides. *Analyst*, **135**, 343–350 (2010).

70. J.M. Solletti, M. Botreau, F. Sommer, W.L. Brunat, S. Kasas, T.M. Duc and M.R. Celio. Elaboration and characterization of phospholipid Langmuir–Blodgett films. *Langmuir*, **12**, 5379–5386 (1996).

71. M.J. Baker, L. Zheng, N. Winograd, N.P. Lockyer and J.C. Vickerman. Mass spectral imaging of glycophospholipids, cholesterol, and glycophorin A in model cell membranes. *Langmuir*, **24**, 11803–11810 (2008).

72. B. Bhushan, J.N. Israelachvili and U. Landman. Nanotribology: Friction, wear and lubrication at the atomic scale. *Nature*, **374**, 607–616 (1995).

73. A. Ianoul, P. Burgos, Z. Lu, R.S. Taylor and L.J. Johnson. Phase separation in supported phospholipid bilayers visualized by near-field scanning optical microscopy in aqueous solution. *Langmuir*, **19**, 9246–9254 (2003).

74. N.-J. Cho, G. Wang, M. Edvardsson, J.S. Glenn, F. Hook and C.W. Frank. Alpha-helical peptide-induced vesicle rupture revealing new insight into the vesicle fusion process as monitored in situ by quartz crystal microbalance-dissipation and reflectometry. *Anal. Chem.*, **81**, 4752–4761 (2009).

75. G. Pabst, N. Kuĉerka, M.-P. Nieh, M.C. Rheinstädter and J. Katsaras. Applications of neutron and X-ray scattering to the study of biologically relevant model membranes. *Chem. Phys. Lipids*, **163**, 460–479 (2010).

76. D.J. McGillivray, G. Valincius, D.J. Vanderah, W. Febo-Ayala, J.T. Woodward, F. Heinrich, J.J. Kasianowicz and M. Losche. Molecular-scale structural and functional characterization of sparsely tethered bilayer membranes. *Biointerphases*, **2**, 21–33 (2007).

77. M. Broniatowski, M. Flasinski, J. Majewski and P. Dynarowicz-Latka. X-ray grazing incidence diffraction and Langmuir monolayer study of the interaction of beta-cyclodextrin with model lipid membranes. *J. Colloid Interface Sci.*, **348**, 511–521 (2010).

78. A. Junghans and I. Köper. Structural analysis of tethered bilayer lipid membranes. *Langmuir*, **26**, 11035–11040 (2010).

79. C.J. van Oss, M.K. Chaudhury and R.J. Good. Interfacial Lifshitz–van der Waals and polar interactions in macroscopic systems. *Chem. Rev.*, **88**, 927–941 (1988).

80. C.J. van Oss. Acid–base interfacial interactions in aqueous media. *Colloids Surf. A*, **78**, 1–49 (1993).

81. F.M. Etzler. Characterization of surface free energies and surface chemistry of solids. In *Contact Angle, Wettability and Adhesion*, Vol. 3, K.L. Mittal (Ed.). VSP, Utrecht, pp. 219–264 (2003).

82. E. Chibowski, A. Ontiveros-Ortega and R. Perea-Carpio. On the interpretation of contact angle hysteresis. *J. Adhesion Sci. Technol.*, **16**, 1367–1404 (2002).

83. E. Chibowski. Surface free energy of a solid from contact angle hysteresis. *Adv. Colloid Interface Sci.*, **103**, 149–172 (2005).

84. M. Golabek, M. Jurak, L. Holysz and E. Chibowski. The energetic and topography changes of mixed lipid bilayers deposited on glass. *Colloids Surf. A*, **391**, 150–157 (2011).

85. E. Chibowski, M. Jurak and L. Holysz. Preparation, investigation techniques and surface free energy of solid supported phospholipid layers. In *The Encyclopedia of Surface and Colloid Science*, 2nd ed. Taylor & Francis Group p.1–22 (2012), Boca Raton, Florida, USA, ISBN: 0-8493-9615-8; eISBN: 0-8493-9614-x (2012), p. 1–22, doi: 10.1081/E-ESCS-120047339.

86. H. Radelcuk, L. Holysz and E. Chibowski. Comparison of the Lifshitz–van der Waals/acid-base and contact angle hysteresis approaches for determination of solid surface free energy. *J. Adhesion Sci. Technol.*, **16**, 1547–1568 (2002).

87. B. Cross, A. Steinberger, C. Cotton-Bizonne, J.P. Rieu and E. Charlaix. Boundary flow of water on supported phospholipid films. *Europhys. Lett.*, **73**, 390–395 (2006).

88. M. Hishida, H. Seto and K. Yoshikawa. Smooth/rough layering in liquid-crystalline/gel state of dry phospholipid film, in relation to its ability to generate giant vesicles. *Chem. Phys. Lett.*, **411**, 267–272 (2005).

89. D. Ghosh and J. Tinoco. Monolayer interactions of individual lecithins with natural sterols. *Biochim. Biophys. Acta B*, **266**, 41–49 (1972).

90. C. Ohe, T. Sasaki, M. Noi, Y. Goto and K. Itoh. Sum frequency generation spectroscopic study of the condensation effect of cholesterol on a lipid monolayer. *Anal. Bioanal. Chem.*, **388**, 73–79 (2007).

91. M. Kodama, O. Shibata, S. Nakamura, S. Lee and G. Sugihara. A monolayer study on three binary mixed systems of dipalmitoylphosphatidyl choline with cholesterol, cholestanol and stigmasterol. *Colloids Surf. B*, **33**, 211–226 (2004).

92. Y. Su, Q. Li, L. Chen and Z. Yu. Condensation effect of cholesterol, stigmasterol, and sitosterol on dipalmitoylphosphatidylcholine in molecular monolayers. *Colloids Surf. A*, **293**, 123–129 (2007).

93. K. Gong, S.-S. Feng, M.L. Go and P.H. Soew. Effects of pH on the stability and compressibility of DPPC/cholesterol monolayers at the air–water interface. *Colloids Surf. A*, **207**, 113–125 (2002).

94. M. Alwarawrah, J. Dai and J. Huang. A molecular view of the cholesterol condensing effect in DOPC lipid bilayers. *J. Phys. Chem. B*, **114**, 7516–7523 (2010).

95. C. Xing and R. Faller. Density imbalances and free energy of lipid transfer in supported lipid bilayers. *J. Chem. Phys.*, **131**, 175104–175107 (2009).

96. L.K. Tamm. Lateral diffusion and fluorescence microscope studies on a monoclonal antibody specifically bound to supported phospholipid bilayers. *Biochemistry*, **27**, 1450–1457 (1988).

97. G. Qi, Y. Yang, H. Yan, L. Guan, Y. Li, X. Qiu and Ch. Wang. Quantifying surface charge density by using an electric force microscope with a referential structure. *J. Phys. Chem. C*, **113**, 204–207 (2009).

98. I.S. Costin and G.T. Barnes. Two-component monolayers. II. Surface pressure–area relations for the octadecanol–docosyl sulphate system. *J. Colloid Interface Sci.*, **51**, 106–121 (1975).

99. P. Dynarowicz-Łątka and K. Kita. Molecular interaction in mixed monolayers at the air water interface. *Adv. Colloid Interface Sci.*, **79**, 1–17 (1999).

100. S. Ransac, H. Moreau, C. Riviere and R. Verger. Monolayer techniques for studying phospholipase kinetics. *Method Enzymol.*, **197**, 49–65 (1991).

101. S. Chen and H.D. Abruna. Enzymatic activity of a phospholipase A_2: An electrochemical approach. *Langmuir*, **13**, 5969–5973 (1997).

102. M.K. Jain and B.P. Maliwal. Spectroscopic properties of the states of pig pancreatic phospholipase A_2 at interfaces and their possible molecular origin. *Biochemistry*, **32**, 11838–11846 (1993).

103. D.W. Grainger, A. Reichert, H. Ringsdorf and C. Salesse. Hydrolytic action of phospholipase A_2 in monolayer phase transition region: Direct observation of enzyme domain formation using fluorescence microscopy. *Biochim. Biophys. Acta*, **1023**, 365–379 (1990).

104. K. Balashev, N.J. DiNardo, T.H. Callisen, A. Svendsen and T. Bjørnholm. Atomic force microscope visualization of lipid bilayer degradation due to action of phospholipase A_2 and Humicola lanuginosa lipase. *Biochim. Biophys. Acta*, **1768**, 90–99 (2007).

105. L.K. Nielsen, J. Risbo, T.H. Callisen and T. Bjørnholm. Lag-burst kinetics in phospholipase A_2 hydrolysis of DPPC bilayers visualized by atomic force microscopy. *Biochim. Biophys. Acta*, **1420**, 266–271 (1999).

106. J.B. Henshaw, C.A. Olsen, A.R. Farnbach, K.H. Nielson and J.D. Bell. Definition of the specific roles of lysolecithin and palmitic acid in altering the susceptibility of dipalmitoylphosphatidylcholine bilayers to phospholipase A_2. *Biochemistry*, **37**, 10709–10721 (1998).

107. M. Jurak and E. Chibowski. Zeta potential and surface free energy changes of solid-supported phospholipid (DPPC) layers caused by the enzyme phospholipase A_2 (PLA$_2$). *Adsorption*, **15**, 211–219 (2009).

4 Surfactant Adsorption Layers at Liquid Interfaces

Reinhard Miller, Valentin B. Fainerman, Vincent Pradines,
Volodymyr I. Kovalchuk, Nina M. Kovalchuk,
Eugene V. Aksenenko, Libero Liggieri, Francesca Ravera,
Giuseppe Loglio, Altynay Sharipova, Yuri Vysotsky,
Dieter Vollhardt, Nenad Mucic, Rainer Wüstneck,
Jürgen Krägel, and Aliyar Javadi

CONTENTS

4.1 INTRODUCTION

Surfactants are an enormously important group of chemicals and have been known for thousands of years. The earliest surfactant-like compounds were, of course, not synthetic but of natural origin, found by careful observations of situations happening in daily life. Although several thousand years passed without real progress, modern chemistry accelerated this development, and the time of understanding the first studied natural surface active compounds such as fatty acids and fatty alcohols until the most modern special surfactants of these days has been less than a century.

Surfactants are important in many technological fields and are used in daily life, such as detergents, wetting agents, emulsifiers, foaming agents, and dispersants. For efficient applications, the interfacial properties of surfactants must be known on a quantitative level. Among the many important properties, nonequilibrium interfacial behavior is the most essential. It is impossible to provide a complete overview on surfactants in a single book chapter, and the same problem persists even for an entire book. In 1987, Martin Schick tried to summarize only the knowledge on the physical chemistry of nonionic surfactants, and he filled more than a 1000 pages [1]. Stache and Kosswig [2] published a small encyclopedia (a volume of 500 pages) about all the surfactants known up to 1990. Among the many other books and various aspects of surfactants we can mention here, only three have been very subjectively selected and focus on special aspects, namely, the chemistry, the main properties of surfactants in bulk and at interfaces, and the huge variety of application fields [3–5]. The book by Rosen has even been updated a third time.This contribution is not devoted to the chemistry nor does it focus on the various applications, rather it deals with the state of art of the theoretical and experimental characterization of surfactants at liquid interfaces. We will show that the theoretical models are very advanced and even mixed surfactant adsorbed layers can be handled quantitatively. In addition to classic adsorption models (Langmuir or Frumkin), new specific ideas have recently been proposed, such as different coexisting orientations of adsorbed molecules, two-dimensional aggregation at the interface, and interfacial intrinsic compressibility. These new models help explain some peculiarities observed at water–oil interfaces where the oil molecules can be embedded in the interfacial layer. Based on these equilibrium thermodynamic models, the dynamics of adsorption as well as the dilational rheology can be determined quantitatively. Here, in particular, the theory for the viscoelasticity of adsorption layers for mixed surfactant solutions represents the present state of the art, whereas again the particular role of solvent molecules included into the interfacial layer is still not clear.

In addition to the theoretical basis, a number of excellent experimental tools have also been developed recently, to a large extent even as commercial instruments that provide reliable sets of data for the quantitative analysis of many interfacial properties, such as dynamic interfacial tensions, dilational elasticity, and viscosity. The dynamics of adsorption layers is most frequently studied by capillary pressure techniques (CPT), from which the maximum bubble pressure tensiometry (BPT) represents the world champion for studies at shortest adsorption times. An equivalent technique, the capillary pressure tensiometry, can also be applied to liquid–liquid interfaces to gather experimental data for adsorption times, which were not accessible until very recently. The method of choice most frequently used now is surely the drop profile analysis tensiometry (PAT), which was introduced by Neumann (he called it axisymmetric drop shape analysis—ADSA) as a professional tool for interfacial studies only in 1983 [6]. This method is now the most reliable and versatile tool for characterizing liquid interfaces. However, it was recently discussed in detail that a single method was unable to cover the requirements of a quantitative characterization of a surfactant in the whole concentration range [7]. In addition, it was shown that not every surfactant can adequately be investigated due to the limitations of the available instruments. Nowadays, it seems that the most efficient set of instruments for the quantitative analysis of surfactants at liquid interfaces is the combination of bubble pressure/capillary pressure tensiometry and drop/bubble profile analysis tensiometry.

4.2 HISTORY OF SURFACTANTS

The use of natural compounds containing surfactants originated in the early stages of civilization. Examples include smashed lupine peas, utilized in the pharaoh's Egypt as a detergent, and the crushed roots and leaves of soapwort, utilized by the ancient Greeks to clean fabrics. Also, the unconscious use of protein surfactants in various processes related to food preparation, in particular as related to milk.

The first chemicals similar to soaps were already mentioned in about 2500 BC cuneiform Sumeric scripts found in Ur [8]. Accordingly, soaps were obtained by boiling vegetable oils with potash. Some recipes were even more elaborate, making use of resins and salt. Papyri from the Ptolemaic period (~300 BC) report of taxes paid for in terms of castor oil, which was boiled with soda to obtain a primitive soap [9]. This knowledge was then transferred to the Roman civilization. Galen (~AD 150) mentions the production of soap by Germans and Gauls [10]. However, the Latin term "sapo"—from which the word "soap" originates—is used by the Roman natural philosopher

Pliny to indicate an ointment produced in Gallia from ashes and animal fat. It is not clear, however, whether that was really used as a modern soap. In addition, the finding of soap among the toiletries from the Pompei excavations is rather questionable. In the late Roman age, we find instead clear references to the use of soap for hair washing. From the Middle Ages until the modern ages, soap was regularly produced in Europe, even on a large scale, mainly based on the utilization of animal fats in the northern countries and of olive oil in the south [11].

The word *surfactant* is derived from *surface active agent*, and was created only recently. The first synthetic surfactants were sulfate oils and short-chain alkyl sulfonate naphthalenes, which appeared between the late nineteenth and the early twentieth century [12]. A surfactant molecule, in its simplest structure (Figure 4.1), consists of a hydrophobic and a hydrophilic part, that is, typically a hydrophilic head and a hydrophobic tail.

As mentioned previously, there are a huge number of different surfactants known today. Soaps were the first detergents described and used by man, and fatty acids and fatty alcohols are additional surface active molecules systematically studied by surface scientists (see Figure 4.2). Alkyl sulfate (Figure 4.3) is possibly the first synthetic detergent synthesized, produced, and applied on a large scale.

The chemist Heinrich Bertsch (Figure 4.4) described its synthesis and filed a great number of patents for its application as the first synthetic detergent. In 1932, he invented the first synthetic washing powder, FEWA (which stands for *Für Eure Wäsche Alles* or "all for your laundry"). This product, originally based mainly on sodium lauryl (dodecyl) sulfate (SDS), is still produced by Henkel.

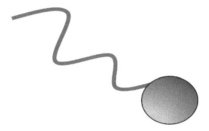

FIGURE 4.1   Scheme of a typical surfactant molecule.

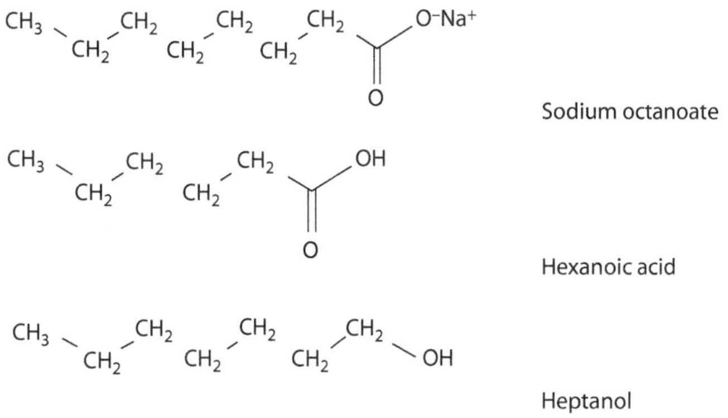

$$CH_3-CH_2-CH_2-CH_2-CH_2-CH_2-CH_2-C(=O)-O^-Na^+$$

Sodium octanoate

$$CH_3-CH_2-CH_2-CH_2-CH_2-C(=O)-OH$$

Hexanoic acid

$$CH_3-CH_2-CH_2-CH_2-CH_2-CH_2-CH_2-OH$$

Heptanol

FIGURE 4.2   Chemical structure of the first natural surfactants.

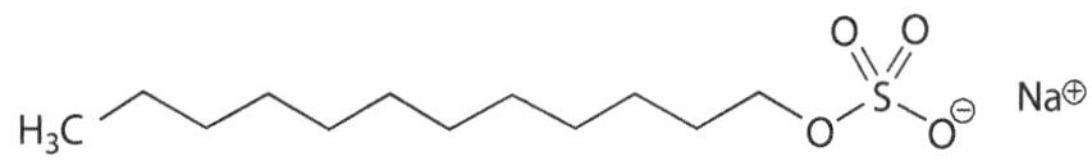

$$H_3C\text{---}(CH_2)_n\text{---}O\text{---}S(=O)_2\text{---}O^\ominus \quad Na^\oplus$$

FIGURE 4.3   Chemical structure of sodium dodecyl sulfate (SDS).

FIGURE 4.4   Heinrich Gottlob Bertsch (Born, January 11, 1897 in Rosenfeld, Germany; Died, March 19, 1981 in Berlin, Germany). (Courtesy of G. Czichocki, Berlin, Germany.)

Meanwhile, we have thousands of synthetic surfactants and it is impossible to summarize them briefly. In a recent monograph, a systematization and guidelines for their synthesis were published along with methods of their characterization and fields of application [3].

4.3   EQUATIONS OF STATE/ADSORPTION ISOTHERMS

As mentioned previously, dynamic interfacial properties of surfactants are the most important characteristics; however, without knowing the thermodynamic quantities, they do not make sense. The adsorption isotherm and equation of state of a surfactant is an important baseline indispensable for a quantitative analysis of a liquid interface. Such a water–air or water–oil interface covered by a self-assembled (adsorbed) surfactant layer is schematically shown in Figure 4.5.

Before analytical relationships between the main interfacial parameters were developed, a general but more qualitative rule was discovered, known now as the Traube rule. The chemist Isidor Traube (Figure 4.6) discovered a linear increase in surface activity of members in a homologous series of amphiphiles (for example, fatty alcohols). He showed that a given decrease in surface tension is obtained for any two neighbor members of a homologous series at concentrations in which the ratio is more or less constant [13]. As a rough rule, we can say that the surface activity of a surfactant is increased by one order of magnitude when two CH_2 groups are added to the hydrophobic chain.

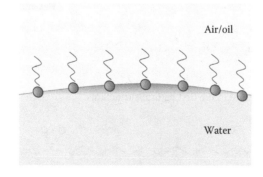

FIGURE 4.5   Schematic of a water–oil interfacial layer covered by surfactant molecules.

FIGURE 4.6 Isidor Traube (Born, March 31, 1860 in Hildesheim, Germany; Died, October 27, 1943 in Edinburgh, UK). (From Edsall, J.T., *Proc. Am. Philos. Soc.* 129, 371–406, 1985.)

This simple rule immediately entails a question regarding the term "surface activity." which is defined in the literature in various ways. A measure for classifying the surface activity of a surfactant can be its critical micelle concentration (cmc), it could also be the surface tension decrease at a given concentration, or the minimum surface tension that can be reached. Also, the number of molecules that can be adsorbed at a unit surface area can serve as a measure for the surface activity. All the parameters mentioned are somehow linked with each other and have their pros and cons. For the discussion here, it is not important to choose one or the other definition.

4.3.1 THE HENRY ADSORPTION MODEL

The chemist William Henry (Figure 4.7) found that gas solubility in a liquid is a linear function of its pressure [14]. The linear relationship between solubility and pressure of a solute is also applicable to the adsorption of a gas as a function of its partial pressure and the adsorption of surfactants at liquid interfaces, which has the form

$$\Gamma = Kc \qquad (4.1)$$

FIGURE 4.7 William Henry (Born, December 12, 1775 in Manchester, UK; Died, September 2, 1836 in Pendlebury, UK). (Available at http://de.wikipedia.org/wiki/William_Henry_(Chemiker).)

where K is the Henry constant and Γ is the adsorbed amount at the bulk concentration c. The coefficient K can be used as a reasonable measure for the surface activity.

This linear adsorption isotherm has of course only a rather limited range of applications and reflects the situation in an interfacial layer only at very low coverage, let us say when the adsorbed surfactant molecules behave as an ideal gas.

4.3.2 The Langmuir/von Szyszkowski Adsorption Model

Von Szyszkowski [15], in his studies on fatty acids, tried to find a trend in the change in the surface tension isotherm with varying the alkyl chain length. He demonstrated that an equation of the form

$$\frac{\gamma_0 - \gamma}{\gamma_0} = B \ln\left[1 + \frac{c}{A}\right] \tag{4.2}$$

described the measured values, that is, the dependence of surface tension γ as a function of bulk concentration c, very well (γ_0 is the surface tension of pure water). In the early work of von Szyszkowski (Figure 4.8), the parameters A and B were empirical constants because the entire equation was of an empirical nature.

Langmuir (Figure 4.9) gave these empirical coefficients physical meaning. Langmuir derived an adsorption isotherm on the basis of physical principles having the following form [16]:

$$\Gamma = \Gamma_{max} \frac{c/a}{1 + c/a} \tag{4.3}$$

where Γ is the adsorbed amount at the bulk concentration c of the surfactant. The coefficient a is called the Langmuir equilibrium constant and often serves as a measure of the surface activity of a surfactant. In terms of the Langmuir adsorption model, the parameter a corresponds to the concentration at which half of the surface is covered by surfactant molecules $\Gamma = \Gamma_{max}/2$, whereas Γ_{max} stands for the maximum number of adsorbed molecules per unit surface area.

Irving Langmuir was awarded the Nobel Prize in Chemistry in 1932 for his work on surface chemistry, and later, an American Chemical Society journal was named after him. When we set $B = RT\,\Gamma_{max}/\gamma_0$ and $A = a$ from Equation 4.2, we obtain an equation

FIGURE 4.8 Bohdan von Szyszkowski (Born, June 20, 1873 in Trybuchy, Russia; Died, August 13, 1931 in Myślenice, Poland). (Photo courtesy of M. Nowakowska, Krakow, Poland.)

FIGURE 4.9 Irving Langmuir (Born, January 31, 1881 in Brooklyn, NY; Died, August 16, 1957 in Woods Hole, MA). (Available at http://de.wikipedia.org/wiki/Irving_Langmuir.)

$$\gamma = \gamma_0 - RT\Gamma_{max} \ln(1 + c/a) \tag{4.4}$$

that can be directly transferred into Equation 4.3 by using the Gibbs fundamental adsorption equation in its simplest form

$$\Gamma = -\frac{1}{RT}\frac{d\gamma}{d\ln c} \tag{4.5}$$

In this way, the empirical parameters introduced by von Szyszkowski attain a physical meaning. The Henry constant is also obtained from the Langmuir model given by Equation 4.3 for small surfactant concentrations as $K = \Gamma_{max}/a$. Hence, the Langmuir equilibrium constant a can also be used as a measure for the surface activity of a surfactant.

With respect to the adsorption of surfactants, Gibbs has played a central role (Figure 4.10). In fact, he introduced a consistent thermodynamic model for interfaces, which has been the basis of all subsequent developments in modern surface science. Such a model provided, in particular,

FIGURE 4.10 Josiah Willard Gibbs (Born, February 11, 1839 in New Haven, CT; Died, April 28, 1903 New Haven, CT). (Available at http://de.wikipedia.org/wiki/Josiah_Willard_Gibbs.)

the formal definition of the concept of adsorption and its relation with measurable quantities. The fundamental equation (Equation 4.5) combines adsorption Γ, interfacial tension γ, and the bulk concentration of the surfactant c.

This equation is rarely applied directly to experimental data; however, it can help in transferring theoretical relationships into equations containing experimentally accessible quantities. For example, Equation 4.4 can be transferred into the following equation of state:

$$\gamma = \gamma_0 + RT\Gamma_{max} \ln\left[1 - \frac{\Gamma}{\Gamma_{max}}\right] \tag{4.6}$$

4.3.3 The Adsorption Model of Frumkin

Analyzing the experimental data obtained from surface tension as a function of bulk concentration $\gamma(c)$, Frumkin (Figure 4.11) found out from his measurements with fatty acids that with increasing chain length of the surfactant molecule the deviation of the results from the Langmuir model regularly increased [17].

To compensate for this deviation from the Langmuir model, Frumkin assumed that the mutual interaction between adsorbed molecules was important. This interaction is of a hydrophobic nature and should increase with the alkyl chain length of the surfactant molecules. To take this effect into consideration, he added, empirically, a term to the equation of state (Equation 4.6) and obtained

$$\gamma = \gamma_0 + RT\Gamma_{max} \ln\left[1 - \frac{\Gamma}{\Gamma_{max}}\right] + a_{int}\left[\frac{\Gamma}{\Gamma_{max}}\right]^2 \tag{4.7}$$

Here, a_{int} is the interaction constant between the adsorbed molecules. Although this term was originally added empirically, it was later shown by Lucassen-Reynders that it actually results from a clear thermodynamic approach [18].

FIGURE 4.11 Alexander N. Frumkin (Born, October 24, 1895 in Kishinev (today Chişinău), Moldavia; Died, May 27, 1976 in Tula, Russia). (From Frumkin, A., *Z. Phys. Chem. (Leipzig)* 116, 466–484, 1925.)

In addition to the given classic models, there have been many more attempts that cannot be summarized here, in particular due to their limited range of application. However, some very new models developed recently should be discussed here, as they are generalizations of the models given above and continue the line from the model of Henry via the Langmuir and Frumkin models to the so-called reorientation or surface aggregation models. Actually, they reduce to the Frumkin or Langmuir model under simplified conditions.

4.3.4 THE ADSORPTION MODELS WITH REORIENTATION AND AGGREGATION

The reorientation model [19,20] has been developed only recently and contains two molar areas of an adsorbed molecule, ω_1 required at an unoccupied interface and ω_2 as the minimum area needed at a covered interface. The physical picture of such change in the molar area in an adsorption layer is schematically shown in Figure 4.12.

A change from a more tilted orientation in a diluted state to a more upright orientation obviously leads to a smaller area in which a molecule can occupy at the interface. The basic equations developed for this model are based on the work of Butler [21] and his definition of the chemical potentials for a bulk solution and an interfacial layer. This leads to the following model equations:

$$\frac{\Pi\omega}{RT} = -\ln(1 - \Gamma\omega)\text{(equation of state)} \qquad (4.8)$$

$$bc = \frac{\Gamma_2\omega}{(1 - \Gamma\omega)^{\omega_2/\omega}} \text{ adsorption isotherm} \qquad (4.9)$$

whereas the mean molar area averaged over all possible adsorption states is given by

$$\omega = \frac{\omega_1\Gamma_1 + \omega_2\Gamma_2}{\Gamma_1 + \Gamma_2} \qquad (4.10)$$

Here, $\Pi = \gamma_0 - \gamma$ is the surface pressure, $\Gamma = \Gamma_1 + \Gamma_2$ is the total adsorption, and the constant b is the adsorption constant. For $\omega_1 = \omega$, this model turns into the Langmuir model and we obtain $b = 1/a$. The reorientation model can be extended by taking into consideration the nonideality of entropy and enthalpy [22] and it was shown in many articles that it is superior to the Langmuir and Frumkin models, in particular for surfactants of small mutual interaction at the interface, that is, typically for molecules that require a rather large molar area. Prominent examples are the oxyethylated alcohols C_nEO_m.

The aggregation model [23] was also developed recently. It contains only one molar area of an adsorbed molecule at a covered interface (ω_1), as it is the case for the Langmuir and Frumkin models. In addition, we have Γ_c as the critical surface aggregation concentration and n as the surface

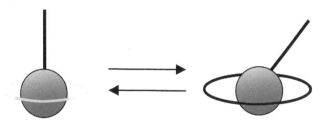

FIGURE 4.12 Schematic for the molecular change in orientation of surfactant and the corresponding molar area at the interface.

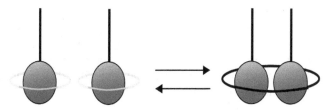

FIGURE 4.13 Schematic for the molecular aggregation of surfactant and the corresponding molar area at the interface.

aggregation number. The schematic given in Figure 4.13 shows how the formation of very small two-dimensional aggregates (like dimers) requires less space in the interfacial layer than two single monomers.

The basic equations for this model are

$$\frac{\Pi\omega}{RT} = -\ln\left\{1 - \Gamma_1\omega\left[1 + (\Gamma_1/\Gamma_c)^{n-1}\right]\right\} \text{ (equation of state)} \tag{4.11}$$

$$bc = \frac{\Gamma_1\omega}{\left\{1 - \Gamma_1\omega\left[1 + (\Gamma_1/\Gamma_c)^{n-1}\right]\right\}^{\omega_1/\omega}} \text{ (adsorption isotherm)} \tag{4.12}$$

The mean molar area can be obtained from a weighted average

$$\omega = \omega_1 \frac{1 + n(\Gamma_1/\Gamma_c)^{n-1}}{1 + (\Gamma_1/\Gamma_c)^{n-1}} \tag{4.13}$$

The aggregation model can be seen as a generalization of the Langmuir model and describes systems that are also well described by the Frumkin model. Actually, the reason for the formation of small aggregates is the strong attraction between adsorbed molecules.

4.3.5 ADSORPTION MODELS FOR SURFACTANT MIXTURES

In most applications, surfactant mixtures are used to achieve the desired behavior of the involved interfaces. For a long time, a quantitative understanding of mixed adsorption layers was not possible and only rough estimations existed. The first quantitative model based on the known properties of the components of the mixture was proposed in 2001 [24]. The set of equations given below for a binary mixture refers to a generalized Frumkin model, including extra effects of the nonideality of entropy:

$$\Pi = -\frac{RT}{\omega_0}\left[\ln(1 - \theta_1 - \theta_2) + \theta_1\left(1 - \frac{1}{n_1}\right)\right.$$
$$\left. + \theta_2\left(1 - \frac{1}{n_2}\right) + a_1\theta_1^2 + a_2\theta_2^2 + 2a_{12}\theta_1\theta_2\right] \tag{4.14}$$

$$b_1c_1 = \frac{\theta_1}{(1 - \theta_1 - \theta_2)^{n_1}}\exp(-2a_1\theta_1 - 2a_{12}\theta_2)$$
$$\exp\left[(1 - n_1)\left(a_1\theta_1^2 + a_2\theta_2^2 + 2a_{12}\theta_1\theta_2\right)\right] \tag{4.15}$$

$$b_2 c_2 = \frac{\theta_2}{(1 - \theta_1 - \theta_2)^{n_2}} \exp(-2a_2\theta_2 - 2a_{12}\theta_1)$$
$$\exp\left[(1 - n_2)\left(a_1\theta_1^2 + a_2\theta_2^2 + 2a_{12}\theta_1\theta_2\right)\right] \tag{4.16}$$

θ is the surface coverage. Simplifications lead to a relationship that is easily applied to experimental data

$$\exp\overline{\Pi} = \exp\overline{\Pi}_1 + \exp\overline{\Pi}_2 - 1 \tag{4.17}$$

where $\overline{\Pi} = \Pi\omega/RT$, $\overline{\Pi}_1 = \Pi_1\omega/RT$, and $\overline{\Pi}_2 = \Pi_2\omega/RT$ are the dimensionless surface pressures of the mixture and the individual solutions of components 1 and 2, respectively, at the same surfactant concentrations c_i as in the mixture. The average molar area ω is given by

$$\omega = \frac{\omega_1\Gamma_1 + \omega_2\Gamma_2}{\Gamma_1 + \Gamma_2} \tag{4.18}$$

where the index now refers to the surfactant. The parameters n_i are the relative molar areas of the compounds $n_i = \omega_i/\omega$, whereas the coefficients a_i are the interaction parameters of molecules of component i among each other, whereas a_{12} is the cross-interaction coefficient. Although the system of equations looks quite bulky, it contains the parameters of the single compounds and the only additional parameter is a_{12}.

As an example of this, on a first glance rather complicated model, the data for a mixed surfactant system will be briefly discussed here. In Figure 4.14, the isotherms for the two rather different surfactants are shown: the nonionic surfactant $C_{10}EO_5$ and the anionic SDS. In addition, the measured surface tension isotherms at different mixing ratios are shown. The solid lines are dependencies

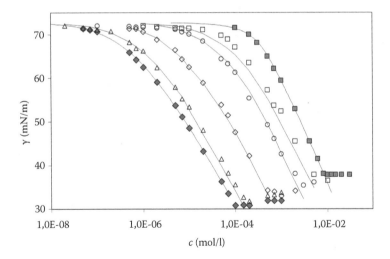

FIGURE 4.14 Surface tension isotherms for $C_{10}EO_5$ (◆), SDS in 0.01 M NaCl (■), and their mixtures with ratios 1:1 (△), 1:10 (◇), 1:100 (○), and 1:500 (□). Theoretical isotherms for individual solutions calculated from Equations 4.15 and 4.16: for $C_{10}EO_5$ $\omega_1 = 3.06 \times 10^5$ m²/mol, $a_1 = -0.7$, and $b_1 = 3.84 \times 10^6$ l/mol; for SDS in 0.01 M NaCl, $\omega_2 = 2.24 \times 10^5$ m²/mol, $a_2 = 0.95$, and $b_2 = 1.135 \times 10^3$ l/mol, $a_{12} = 0$; solid lines—theoretical calculations for mixtures performed with Equations 4.14 through 4.18. (Data reported in Fainerman, V.B., and R. Miller. Adsorption isotherms at liquid interfaces. In *Encyclopedia of Surface and Colloid Science*, Somasundaran, P., and A. Hubbard (Eds.), 2nd ed, Vol. 1, Taylor & Francis, Boca Raton, 1–15, 2009.)

calculated as best fits to the data for the two individual surfactants and calculated for the mixture with the above-mentioned model using the same surfactant parameters. The isotherms for the mixtures are perfectly described by this model without considering any additional parameter, as the mutual interaction parameter was assumed to be zero, $a_{12} = 0$.

It was shown that this model for surfactant mixtures also works for mixtures of polymers (in particular proteins) with surfactants [25,26].

4.4   ADSORPTION KINETICS OF SURFACTANTS AT LIQUID INTERFACES

The adsorption of surfactants is, of course, a process in time. When a fresh surface is created, molecules have to move toward the interface and become adsorbed. In 1907, Milner [27] was the first to propose that transport from the bulk was responsible for this dynamic character of adsorption. In 1937, Bond and Puls [28] qualitatively proposed the idea about a diffusion-controlled adsorption model. Working with ionic surfactants, Doss [29] claimed that the effect of the electrical double layer should also be considered.

In 1946, the first quantitative diffusion controlled model was published by Ward and Tordai [30]. The result of their derivation was an integral equation

$$\Gamma(t) = 2c_0 \sqrt{\frac{Dt}{\pi}} - 2\sqrt{\frac{D}{\pi}} \int_0^{\sqrt{t}} c(0, t - \tau) d\sqrt{\tau} \tag{4.19}$$

that combines two time functions, the adsorption at the interface $\Gamma(t)$ and the so-called subsurface concentration $c(0,t)$. The solution of this Volterra integral equation of the second kind (with a weak singularity at $t = 0$) requires an additional relationship. There is no analytical solution for Equation 4.19 and it took more than 30 years until Miller and Lunkenheimer [31], in 1978, proposed a procedure for numerically solving the diffusion-controlled adsorption model of Ward and Tordai. As an additional relationship necessary for any type of solution, the respective adsorption isotherm was used, assuming that it was also valid under dynamic conditions, that is, that there was a permanently local equilibrium between the surface and the subsurface.

For complicated types of adsorption isotherms, there is obviously no possibility to find an analytical solution of the integral equation (Equation 4.19); however, for a linear Henry isotherm, Sutherland derived an analytical solution [32] that can be used as a good approximation at low surface coverage. For a Langmuir isotherm, Ziller and Miller [33] developed an approximation via a particular transformation to solve Equation 4.19 for a Langmuir isotherm with a polynomial, the coefficient of which had been tabulated for all values of c/a. For other adsorption isotherms, no analytical solutions are available and therefore either numerical procedures [34] or approximations are applied, such as the short time approximation by Hansen [35] and the long time approximation by Rillaerts and Joos [36], the range of applicability of which was accurately analyzed by Makievski et al. [37]. The short time approximation reads

$$\Gamma(t) = \frac{\gamma(t)}{RT} = 2c_0 \sqrt{\frac{Dt}{\pi}} \tag{4.20}$$

where $\Delta\gamma(t) = \gamma_0 - \gamma(t)$ is the change in interfacial tension with time. The long time approximation, independent of the assumption of any adsorption isotherm, reads

$$\gamma(t) = \frac{RT\Gamma(t)^2}{c_0\sqrt{\pi Dt}} \tag{4.21}$$

The function $\Gamma(t)$ can be replaced by a $\gamma(t)$ relationship derived from a suitable adsorption model and, in this way, also provides access to experimentally available data.

The models discussed thus far assume only transport by diffusion as the process that controls the rate of adsorption. It was Baret in 1969 who proposed a kinetically controlled adsorption mechanism [38], that is, a mechanism that considers an additional rate constant describing the transfer of molecules from the subsurface to the interface or back. There are various equations used in the literature for this purpose. In his proposal, Baret used a rate equation, which at equilibrium, becomes the Langmuir adsorption isotherm:

$$\frac{\Gamma(t)}{dt} = k_{ads}c_0(1 - \Gamma(t)/\Gamma_\infty) - k_{des}\Gamma(t)/\Gamma_\infty \qquad (4.22)$$

Baret also proposed a combined model in which the bulk concentration c_0 is replaced by the subsurface concentration $c(0,t)$, which leads to a combined integro-differential equation. An alternative approach for kinetically controlled adsorption, valid for every adsorption isotherm, was then developed by Ravera et al. [39] and Liggieri et al. [40]. Such an approach results in a generalized Ward–Tordai equation with an effective diffusion coefficient depending on the rate constant and on the adsorption.

The first researchers who proposed a quantitative model for mixed surfactant solutions were Krotov and Rusanov [41], whereas Fainerman considered the effect of micelles on adsorption kinetics for the first time in ref. [42]. A much more complex approach was only recently published by Danov et al. [43].

There are more firsts in the area of adsorption kinetics, such as the quantification of electrical double layer effects [44], the quantitative description of the adsorption kinetics of proteins [45], or protein–surfactant mixtures [46]. We will not go into the details of these more sophisticated models here.

Some examples of the above models are shown in Figures 4.15 and 4.16, and illustrate the remarkable effect of the isotherm on the course of the adsorption process. Impressively, Figure 4.15 shows how the interaction coefficient a_{int} changes the adsorption process.

Each adsorption model has similar peculiarities, as shown for the example of the aggregation model in Figure 4.16. The increasing aggregation number n slows down the adsorption process,

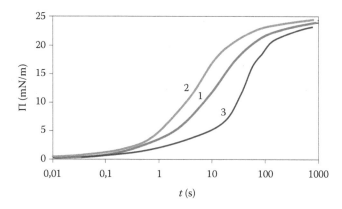

FIGURE 4.15 Dynamic surface pressure calculated for the Frumkin isotherm; $c_0 = 5 \times 10^{-8}$ mol/cm³, $D = 6 \times 10^{-10}$ m²/s, $\omega = 1.5 \times 10^5$ m²/mol, $b = 5 \times 10^6$ cm³/mol, $a_{int} = 0$ (1); Langmuir isotherm; $a_{int} = -1.5$ (2); $a_{int} = +1.5$ (3). (With kind permission from Springer Science+Business Media: *Colloid Polym. Sci.*, On the theory of adsorption kinetics of ionic surfactants at fluid interfaces. 1. The effect of electric DL under quasi-equilibrium conditions. 261, 1983, 335–339, Dukhin, S.S., R. Miller, and G. Kretzschmar.)

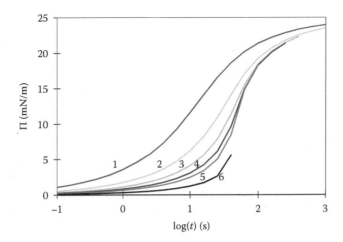

FIGURE 4.16 Dynamic surface pressure of adsorption monolayer for the formation of *n*-mers ($\Gamma_c < 10^{-9}$ mol/m²): curve 1, Langmuir isotherm ($n = 1$); curves 2 to 6 for $n = 2, 3, 4, 5$, and 10, respectively. (Reprinted from *Colloids Surf. A*, 183, Miller, R., E.V. Aksenenko, V.B. Fainerman, and U. Pison, 381–390, Copyright 2001, with permission from Elsevier.)

analogous with the Frumkin model (Figure 4.15) in which a larger attraction (positive values of a_{int}) also decelerates the adsorption rate.

4.5 ADSORPTION KINETICS OF MIXED SURFACTANT SOLUTIONS

As mentioned above, Krotov and Rusanov were the first to discuss the simultaneous adsorption process of more than one surfactant; however, they restricted their discussion to a simplified Langmuir adsorption model due to the lack of any better model [47].

The generic description of the diffusion-controlled adsorption of surfactants from mixed solutions was proposed for the first time in a quantitative model by Miller and coworkers [47]. It is

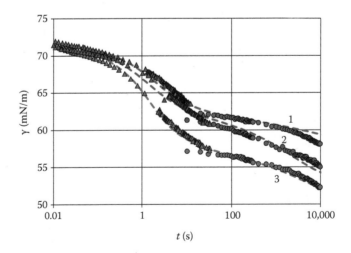

FIGURE 4.17 Dynamic surface tensions of mixtures of C_{10}DMPO and C_{14}DMPO: concentration ratio $c_{10}/c_{14} = 10^{-7}$ mol/cm³/10^{-9} mol/cm³ (1), 10^{-7} mol/cm³/3×10^{-9} mol/cm³ (2), 2×10^{-7} mol/cm³/3×10^{-9} mol/cm³ (3); dotted lines: calculated for $D = 1 \times 10^{-6}$ (1), 3×10^{-6} (2), 2×10^{-6} (3) cm²/s, ▲ – BPA1, ● – PAT1. (Reprinted from *Colloid Polym. Sci.*, 259, Miller, R., On the solution of diffusion controlled adsorption kinetics for any adsorption isotherm, 375–381, Copyright 1981, with permission from Elsevier.)

actually a set of Ward and Tordai equations, linked with each other through a quantitative adsorption model for surfactant mixtures containing r compounds:

$$\Gamma_i(t) = 2\sqrt{\frac{D_i}{\pi}} \left[c_{oi}\sqrt{t} - \int_0^{\sqrt{t}} c_i(0, t-\tau)d\sqrt{\tau} \right]; \ i = 1, 2, \ldots, r \qquad (4.23)$$

Hence, quantitative solutions of Equation 4.23 require quite sophisticated numerical algorithms to consider suitable adsorption models, such as the one given by Equations 4.14 through 4.18. Experimental data for a mixture of two nonionic surfactants, $C_{10}DMPO$ and $C_{14}DMPO$, and three mixing ratios are shown along with theoretical calculations with the model mentioned in Figure 4.17.

The agreement between experimental data and theory is excellent and underlines first that the adsorption of this type of surfactants is diffusion controlled, and second that the mixture can be quantitatively understood using the proposed model.

4.6 EXPERIMENTAL TOOLS AVAILABLE

A recent review was dedicated to the question on the optimum selection of experimental tools for the study of adsorption kinetics and dilational rheology, respectively [7]. It was shown that a combination of maximum bubble pressure analyzer (BPA) and drop profile analysis tensiometry (PAT) provides the most efficient set of instruments to study the dynamics of adsorption in a time range between 10^{-4} up to 10^5 s, approximately 28 h and nine orders of magnitude. Regarding the dilational rheology, a combination of oscillating drops or bubbles based on profile tensiometry (0.001–0.2 Hz) and capillary pressure tensiometry (0.1–100 Hz) provides experimental possibilities for studies in a frequency range of five orders of magnitude. For these studies, oscillating drop and bubble tensiometers are available, such as PAT and oscillating drop and bubble analyzer (ODBA) or equivalent apparatuses. As this chapter actually deals with the development of knowledge on surfactants at liquid interfaces, the part about experimental tools is kept short and reference to recent detailed descriptions is made.

4.6.1 MAXIMUM BUBBLE PRESSURE TENSIOMETRY

The most efficient instrument for studies of the short time adsorption of surfactants at liquid–gas interfaces is the maximum bubble pressure tensiometer. As an example, the BPA1 of SINTERFACE Technologies, Berlin, Germany, was described by Fainerman et al. [48,49]. For the investigation of the interface between two immiscible liquids, most of the static methods can be used; however, they provide no access to short time data, as explained further below.

The application of the BPA1 to a micellar solution of a surfactant was discussed recently in ref. [50] for the nonionic surfactant $C_{14}EO_8$ (Figure 4.18). It was shown that it is possible to derive rate constants of micellar kinetics from dynamic interfacial tensions. This is because the disintegration of a micelle serves as an additional source of individual surfactant molecules ready to adsorb.

Therefore, the rate of adsorption from a micellar solution is larger than from a solution of a concentration exactly at the critical micelle concentration cmc. Note that bubble pressure tensiometry can provide data even in the time range of milliseconds and shorter, as one can see from Figure 4.18.

4.6.2 DROP AND BUBBLE PROFILE ANALYSIS TENSIOMETRY

PAT, pioneered by Neumann and his group as axisymmetric drop shape analysis (ADSA) [6], has been established as the most efficient measurement technique in surface science laboratories. The actual state of the art on PAT was described in detail in recent book chapters by Loglio et al. [51] and

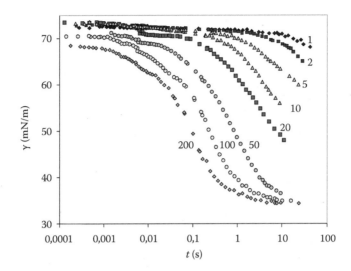

FIGURE 4.18 Dynamic surface tension for $C_{14}EO_8$ solutions as a function of the effective surface lifetime. The surfactant concentrations are given as multiples of the cmc. (Reprinted with permission from Fainerman, V.B., A.V. Makievski, and R. Miller. Accurate analysis of the bubble formation process in maximum bubble pressure tensiometry, *Rev. Sci. Instrum.*, 75, 213–221 Copyright 2004, American Institute of Physics.)

Zholob et al. [52]. The method is based on the Young–Laplace equation, which accurately describes the shape of axisymmetric drops and bubbles. Hence, if one can determine the exact profile coordinates of a pendent drop or buoyant bubble and compare it with the theoretical profile calculated by the Young–Laplace equation of capillarity, the best fit is obtained for the right interfacial tension, assuming the density difference and gravitational acceleration are correctly known.

Although in the past, PAT was an improvised instrument run in only a few laboratories and a workstation was required for the extraction of shape coordinates and Laplace fitting, today all instruments can be run online with PCs or even laptops. This makes PAT the workhorse of surface science laboratories.

As emphasized in the literature, drop profile tensiometry is applicable to both study the adsorption kinetics in a time range from seconds to many hours and to slow the interfacial perturbations used to generate interfacial relaxation processes. In ref. [53], the first transient surface relaxation experiments with the drop PAT were proposed. In Figure 4.19, an example for a surfactant solution is shown.

Several years later, Benjamins et al. [54] used the same methodology for generating slow harmonic drop surface area oscillations and extracted the dilational rheology from the response function.

4.6.3 Capillary Pressure Tensiometry

The capillary pressure tensiometry (CPT) was first used for equilibrium surface tension measurement by Passerone et al. [55], and then it was developed for adsorption kinetic studies [56,57]. This method was further refined for much more dynamic interfacial situations. This step required a quantitative consideration of the additional "disturbing" effects of hydrodynamics [58]. The present state of the art allows acquiring experimental data at adsorption times of a few milliseconds. This methodology is also the most suitable one for interfacial studies of systems where both liquid phases have a similar density or even for any liquid system under weightlessness. An overview of such microgravity investigations was given recently in ref. [59].

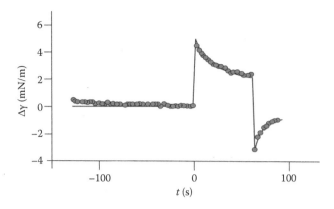

FIGURE 4.19 Surface tension response $\Delta\gamma(t)$ for a trapezoidal surface area perturbation of an aqueous $C_{12}DMPO$ solution drop of 1.67×10^{-5} mol/L. Line, best fit to a diffusion controlled relaxation model. (According to Loglio, G., P. Pandolfini, L. Liggieri, A.V. Makievski, and F. Ravera. Determination of interfacial properties by the pendant drop tensiometry: Optimisation of experimental and calculation procedures. In *Bubble and Drop Interfaces* (*Progress in Colloid and Interface Science*), Vol. 2, edited by R. Miller and L. Liggieri, Brill, Leiden, 7–38, 2011.)

4.6.4 OSCILLATING DROPS AND BUBBLES FOR STUDIES OF THE DILATIONAL RHEOLOGY

Capillary pressure setups have also been applied to drop and bubble oscillations, a methodology for the determination of the dilational viscoelasticity of liquid interfacial layers. The first experiments with oscillating bubbles had already been proposed in 1970 by Kretzschmar and Lunkenheimer [60]; however, their apparatus did not contain the electric pressure sensors that are now available. In recent times, only the CPT has been significantly refined for studies of the adsorption dynamics and dilational viscoelasticity to allow efficient measurements. Although the apparatus of Russev et al. [61] was only theoretical, the option of oscillations was up to 100 Hz, but experimentally, this was limited to approximately 1 Hz. However, the most powerful drop and bubble oscillation apparatus ODBA (SINTERFACE Technologies) can actually reach frequencies that are much higher than 30 Hz (even 100 Hz) [62]. Figure 4.20 shows an example of a high-frequency study on an aqueous

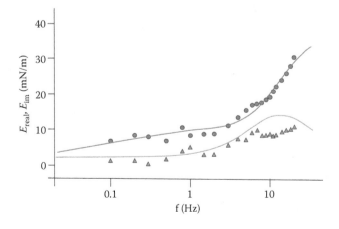

FIGURE 4.20 Real (●) and imaginary (▲) part of the dilational viscoelasticity modulus for the water–hexane interface with an aqueous phase containing 1% of silica nanoparticles and 2×10^{-4} mol/L CTAB. (According to Kretzschmar, G., and K. Lunkenheimer, *Ber. Bunsenges. Phys. Chem.*, 74, 1064–1071, 1970.)

solution of nanoparticles mixed with cetyltrimethylammonium bromide (CTAB) at the water–hexane interface.

The quantitative analysis of such experiments is not trivial and requires an exact consideration of the hydrodynamics in the drop and in the measuring cell, as shown in detail elsewhere [63,64].

4.7 OUTLOOK

The state of the art of surfactants at liquid interfaces is quite advanced. Most of the investigations have been done at the water–air interface and we can say that there is a quantitative understanding of the dynamics as well as of the equilibrium properties of surfactant adsorption layers. This also holds partially true for dilational rheology; however, the accuracy of the respective experiments can be significantly improved. This development is important because dilational viscoelasticity is a property that is extremely sensitive to peculiarities in adsorption layers. Despite an excellent knowledge on surfactant adsorption layers of single and even mixed systems, there is thus far no reliable correlation with the properties of foam films and, consequently, of foams. These interrelations remain a great challenge in surface science.

Although adsorption layers at the water–air interface are well studied, knowledge of surfactant layers at interfaces between two liquids, such as at the water–oil interface, is still rather limited. For these systems, the partitioning of surfactant between the liquid phases can significantly influence the behavior under dynamic conditions. These effects are qualitatively and quantitatively understood for single diluted surfactants. The adsorption phenomena in concentrated surfactant solutions (above the cmc) or solutions of surfactant mixtures are instead nearly unexplored, while being important for application purposes. The effect of the nature of an oil on the properties of the corresponding interfacial layers is very complex, and only the first steps have been taken toward a quantitative understanding. For example, it was shown for the homologous series of alkyl trimethylammonium bromides (C_nTAB) that, at the water–hexane interface, the shorter chain surfactants C_{10} and C_{12} allow hexane molecules to be embedded into the surface layer, whereas the longer chains C_{14} and C_{16} interact strongly with each other and squeeze the hexane progressively out of the interfacial layer [65]. Obviously, the direct interaction between the hydrophobic parts of the surfactant molecules with the oil molecules depends on the alkyl chain length of both molecules, and that the surface coverage is of immanent importance. Systematic studies on more homologous series and at the interface with different oils are pending.

Very recently, it was shown that experiments at the aqueous surfactant solution interface to an alkane vapor represent a new possibility of molecular understanding of interfacial surfactant–oil interaction [66]. Effects of the alkane chain length on the vapor pressure and, in turn, on the strength and rate of interfacial layer penetration have been observed.

Finally, an important emerging topic concerns the experimental investigation and modeling of mixed microparticle/nanoparticle surfactant layers at liquid interfaces, which have a practical effect on the stability control of emulsions and foams in different industrial and natural processes. The mesoscopic nature of these layers raises new theoretical challenges because even the elementary concepts of surface tension or adsorption need a fundamental reconsideration.

ACKNOWLEDGMENTS

This research was financially supported by the European Space Agency (PASTA), the Deutsche Luft- und Raumfahrt (DLR 50WM1129), and the Deutsche Forschungsgemeinschaft SPP 1506 (Mi418/18-2). Financial assistance from the Bundesministerium für Bildung, Wissenschaft, Forschung und Technologie (BMBF) and the Ukrainian Ministry of Education and Science (common project UKR 10/039) is also gratefully acknowledged. The work was supported by COST actions CM1101 and MP1106.

REFERENCES

1. M.J. Schick (Ed.). *Nonionic Surfactants: Physical Chemistry*. Surfactant Science Series, Vol. 23. Marcel Dekker, New York (1987).
2. H. Stache and K. Kosswig (Eds.). *Tensid-Taschenbuch*, 3rd ed. Hanser Verlag, Munich (1990).
3. D. Möbius, R. Miller and V.B. Fainerman (Eds.). *Surfactants: Chemistry, Interfacial Properties, Applications*. Studies in Interface Science, Vol. 13. Elsevier, Amsterdam (2001).
4. K. Holmberg, B. Jönsson, B. Kronberg and B. Lindman. *Surfactants and Polymers in Aqueous Solution*, 2nd ed. Wiley, Chichester (2002).
5. M.J. Rosen. *Surfactants and Interfacial Phenomena*, 3rd ed. Wiley, Hoboken (2004).
6. Y. Rotenberg, L. Boruvka and A.W. Neumann. Determination of surface tension and contact angle from the shapes of axisymmetric fluid interfaces. *J. Colloid Interface Sci.*, **93**, 169–183 (1983).
7. N. Mucic, A. Javadi, N.M. Kovalchuk and R. Miller. Dynamics of interfacial layers—Experimental feasibilities for adsorption kinetics and dilational rheology. *Adv. Colloid Interface Sci.*, **168**, 167–178 (2011).
8. C. Singer, E.J. Holmyard, A.R. Hall and T.I. Williams (Eds.). *A History of Techonology*, Vol. 1, Chap. 11. Clarendon Press, Oxford (1954).
9. R.J. Forbes. *Studies in Ancient Technology*, Vol. 4, Chap. 2. Brill, Leiden (1964).
10. J.R. Partington. *A History of Greek Fire and Gun Powder*, Chap. 7. JHU Press, Baltimore, MD (1999).
11. C. Singer, E.J. Holmyard, A.R. Hall and T.I. Williams (Eds.). *A History of Techonology*, Vol. 2, Chap. 10. Clarendon Press, Oxford (1956).
12. D. Myers. *Surfactant Science and Technology*. Wiley-Interscience, Hoboken p. 3 (2006).
13. I. Traube. Über die Capillaritätsconstanten organischer Stoffe in wässerigen Lösungen. *Liebigs Ann. Chem.*, **265**, 27–55 (1891).
14. W. Henry. Experiments on the quantity of gases absorbed by water, at different temperatures, and under different pressures. *Phil. Trans. R. Soc. London*, **93**, 29–42 (1803).
15. B. von Szyszkowski. Experimentelle Studien über kapillare Eigenschaften der wäßrigen Lösungen von Fettsäuren. *Z. Phys. Chem. (Leipzig)*, **64**, 385–414 (1908).
16. I. Langmuir. The constitution and fundamental properties of solids and liquids. II. Liquids. *J. Am. Chem. Soc.*, **39**, 1848–1907 (1917).
17. A. Frumkin. Die Kapillarkurve der höheren Fettsäuren und die Zustandsgleichung der Oberflächenschicht. *Z. Phys. Chem. (Leipzig)*, **116**, 466–484 (1925).
18. E.H. Lucassen-Reynders. Interactions in mixed monolayers I. Assessment of interaction between surfactants. *J. Colloid Interface Sci.*, **42**, 554–562 (1973).
19. V.B. Fainerman, R. Miller, R. Wüstneck and A.V. Makievski. Adsorption isotherm and surface tension equation for a surfactant with changing partial molar area. 1. Ideal surface layer. *J. Phys. Chem.*, **100**, 7669–7675 (1996).
20. V.B. Fainerman, R. Miller and R. Wüstneck. Adsorption isotherm and surface tension equation for a surfactant with changing partial molar area. 2. Non-ideal surface layer. *J. Phys. Chem.*, **101**, 6479–6483 (1997).
21. J.A.V. Butler. The thermodynamics of the surfaces of solutions. *Proc. R. Soc. Ser. A*, **138**, 348–375 (1932).
22. V.B. Fainerman and R. Miller. Adsorption isotherms at liquid interfaces. In *Encyclopedia of Surface and Colloid Science*, 2nd ed., Vol. 1, P. Somasundaran and A. Hubbard (Eds.). Taylor & Francis, Boca Raton, FL, pp. 1–15 (2009).
23. V.B. Fainerman and R. Miller. Surface tension isotherms for surfactant adsorption layers including surface aggregation. *Langmuir*, **12**, 6011–6014 (1996).
24. V.B. Fainerman, R. Wüstneck and R. Miller. Surface tension of mixed surfactant solutions. *Tenside Surfact. Deterg.*, **38**, 224–229 (2001).
25. V.B. Fainerman, S.A. Zholob, M.E. Leser, M. Michel and R. Miller. Models of two-dimensional solution assuming the internal compressibility of adsorbed molecules: A comparative analysis. *J. Phys. Chem.*, **108**, 16780–16785 (2004).
26. R. Miller, V.B. Fainerman, M.E. Leser and M. Michel. Surface tension of mixed non-ionic surfactant/protein solutions: Comparison of a simple theoretical model with experiments. *Colloids Surf. A*, **233**, 39–42 (2004).
27. S.R. Milner. On surface concentration and the formation of liquid films. *Phil. Mag.*, **13**, 96–101 (1907).
28. W.N. Bond and H.O. Puls. The change of surface tension with time. *Phil. Mag.*, **24**, 864–888 (1937).
29. K.S.G. Doss. Alterung der Oberflächen von Lösungen. IV. Über die Natur der Potentialschranke, welche die Anreicherung der Moleküle des gelösten (1). *Kolloid Z.*, **86**, 205–213 (1939).

30. A.F.H. Ward and L. Tordai. Time-dependence of boundary tensions of solutions. *J. Phys. Chem.*, **14**, 453–461 (1946).

31. R. Miller and K. Lunkenheimer. Zur Adsorptionskinetik an fluiden Phasengrenzen. Eine numerische Lösung für den diffusionskontrollierten Adsorptionsvorgang. *Z. Phys. Chem. (Leipzig)*, **259**, 863–868 (1978).

32. K.L. Sutherland. The kinetics of adsorption at liquid surfaces. *Austr. J. Sci. Res.*, **A5**, 683–696 (1952).

33. M. Ziller and R. Miller. On the solution of diffusion controlled adsorption kinetics by means of orthogonal collocation. *Colloid Polym. Sci.*, **264**, 611–615 (1986).

34. R. Miller. On the solution of diffusion controlled adsorption kinetics for any adsorption isotherm. *Colloid Polym. Sci.*, **259**, 375–381 (1981).

35. R.S. Hansen. Diffusion and the kinetics of adsorption of aliphatic acids and alcohols at the water-air interface. *J. Colloid Sci.*, **16**, 549–560 (1961).

36. E. Rillaerts and P. Joos. Measurement of the dynamic surface tension and surface dilational viscosity of adsorbed mixed monolayers. *J. Colloid Interface Sci.*, **88**, 1–7 (1982).

37. A.V. Makievski, V.B. Fainerman, R. Miller, M. Bree, L. Liggieri and F. Ravera. Determination of equilibrium surface tension values by extrapolation via long time approximations. *Colloids Surf. A*, **122**, 269–273 (1997).

38. J.F. Baret. Theoretical model for an interface allowing a kinetic study of adsorption. *J. Colloid Interface Sci.*, **30**, 1–12 (1969).

39. F. Ravera, L. Liggieri and A. Steinchen. Sorption kinetics considered as a renormalised diffusion process. *J. Colloid Interface Sci.*, **156**, 109–116 (1993).

40. L. Liggieri, F. Ravera and A. Passerone. A diffusion-based approach to mixed adsorption kinetics. *Colloids Surf. A*, **114**, 351–359 (1996).

41. V.V. Krotov and A.I. Rusanov. On the adsorption kinetics of surfactants in liquid solutions II. The case of surfactants mixture. *Kolloidn. Zh.*, **39**, 58–65 (1977).

42. V.B. Fainerman. Kinetics of adsorption of surfactants from micellar solutions (theory). *Kolloidn. Zh.*, **43**, 94–100 (1981).

43. K.D. Danov, P.A. Kralchevsky, N.D. Denkov, K.P. Ananthapadmanabhan and A. Lips. Mass transport in micellar surfactant solutions: 2. Theoretical modeling of adsorption at a quiescent interface. *Adv. Colloid Interface Sci.*, **119**, 17–33 (2006).

44. S.S. Dukhin, R. Miller and G. Kretzschmar. On the theory of adsorption kinetics of ionic surfactants at fluid interfaces. 1. The effect of electric DL under quasi-equilibrium conditions. *Colloid Polym. Sci.*, **261**, 335–339 (1983).

45. R. Miller, E.V. Aksenenko, V.B. Fainerman and U. Pison. Kinetics of adsorption of globular proteins at liquid/fluid interfaces. *Colloids Surf. A*, **183**, 381–390 (2001).

46. V.B. Fainerman, S.A. Zholob, M. Leser, M. Michel and R. Miller. Competitive adsorption from mixed non-ionic surfactant/protein solutions. *J. Colloid Interface Sci.*, **274**, 496–501 (2004).

47. R. Miller, A.V. Makievski, C. Frese, J. Krägel, E.V. Aksenenko and V.B. Fainerman. Adsorption kinetics of surfactant mixtures at the aqueous solution—Air interface. *Tenside Surfact. Deterg.*, **40**, 256–259 (2003).

48. V.B. Fainerman, A.V. Makievski and R. Miller. Accurate analysis of the bubble formation process in maximum bubble pressure tensiometry. *Rev. Sci. Instrum.*, **75**, 213–221 (2004).

49. V.B. Fainerman and R. Miller. Maximum bubble pressure tensiometry: Theory, analysis of experimental constraints and applications. In *Bubble and Drop Interfaces*, Vol. 2, in Progress in Colloid and Interface Science, R. Miller and L. Liggieri (Eds.). Brill, Leiden, pp. 73–118 (2011).

50. V.B. Fainerman, V.D. Mys, A.V. Makievski, J.T. Petkov and R. Miller. Dynamic surface tension of micellar solutions in the millisecond and sub–millisecond time range. *J. Colloid Interface Sci.*, **302**, 40–46 (2006).

51. G. Loglio, P. Pandolfini, L. Liggieri, A.V. Makievski and F. Ravera. Determination of interfacial properties by the pendant drop tensiometry: Optimisation of experimental and calculation procedures. In *Bubble and Drop Interfaces*, Vol. 2, in Progress in Colloid and Interface Science, R. Miller and L. Liggieri (Eds.). Brill, Leiden, pp. 7–38 (2011).

52. S.A. Zholob, A.V. Makievski, R. Miller and V.B. Fainerman. Advances in calculation methods for the determination of surface tensions in drop profile analysis tensiometry. In *Bubble and Drop Interfaces*, Vol. 2, in Progress in Colloid and Interface Science, R. Miller and L. Liggieri (Eds.). Brill, Leiden, pp. 49–74 (2011).

53. R. Miller, R. Sedev, K.-H. Schano, Ch. Ng and A.W. Neumann. Relaxations of surfactant adsorption layers at solution/air interfaces measured by using ADSA. *Colloids Surf. A*, **69**, 209–216 (1993).

54. J. Benjamins, A. Cagna and E.H. Lucassen-Reynders. Viscoelastic properties of triacylglycerol/water interfaces covered by proteins. *Colloids Surf. A*, **114**, 245–254 (1996).

55. A. Passerone, L. Liggieri, N. Rando, F. Ravera and E. Ricci. A new experimental method for the measurement of the interfacial tensions between immiscible fluids at zero Bond number. *J. Colloid Interface Sci.*, **146**, 152–162 (1991).

56. L. Liggieri, F. Ravera and A. Passerone. Dynamic interfacial tension measurements by a capillary pressure method. *J. Colloid Interface Sci.*, **169**, 226–237 (1995).

57. C.A. MacLeod and C.J. Radke. A growing drop technique for measuring dynamic interfacial tension. *J. Colloid Interface Sci.*, **160**, 435–448 (1993).

58. A. Javadi, J. Krägel, P. Pandolfini, G. Loglio, V.I. Kovalchuk, E.V. Aksenenko, F. Ravera, L. Liggieri and R. Miller. Short time dynamic interfacial tension as studied by the growing drop capillary pressure technique. *Colloids Surf. A*, **365**, 62–69 (2010).

59. V.I. Kovalchuk, F. Ravera, L. Liggieri, G. Loglio, P. Pandolfini, A.V. Makievski, S. Vincent-Bonnieu, J. Krägel, A. Javadi and R. Miller. Capillary pressure studies under low gravity conditions. *Adv. Colloid Interface Sci.*, **161**, 102–114 (2010).

60. G. Kretzschmar and K. Lunkenheimer. Untersuchungen zur Bestimmung der Elastizität von Adsorptionsschichten löslicher grenzflächenaktiver Stoffe. *Ber. Bunsenges. Phys. Chem.*, **74**, 1064–1071 (1970).

61. S.C. Russev, N. Alexandrov, K.G. Marinova, K.D. Danov, N.D. Denkov, L. Lyutov, V. Vulchev and C. Bilke-Krause. Instrument and methods for surface dilational rheology measurements. *Rev. Sci. Instrum.*, **79**, 104102-1–104102-10 (2008).

62. A. Javadi, J. Krägel, A.V. Makievski, N.M. Kovalchuk, V.I. Kovalchuk, N. Mucic, G. Loglio, P. Pandolfini, M. Karbaschi and R. Miller. Capillary pressure technique for measuring the dynamic interfacial tensions and dilational rheology of liquid interfacial layers. *Colloids Surf. A*, **407**, 159–168 (2012).

63. F. Ravera, G. Loglio, P. Pandolfini, E. Santini and L. Liggieri. Determination of the dilational viscoelasticity by the oscillating drop/bubble method in a capillary pressure tensiometer. *Colloids Surf. A*, **365**, 2–13 (2010).

64. F. Ravera, G. Loglio and V.I. Kovalchuk. Interfacial dilational rheology by oscillating bubble/drop methods. *Curr. Opin. Colloid Interface Sci.*, **15**, 217–228 (2010).

65. V. Pradines, V.B. Fainerman, E.V. Aksenenko, J. Krägel, N. Mucic and R. Miller. Alkyltrimethylammonium bromides adsorption at liquid/fluid interfaces in the presence of neutral phosphate buffer. *Colloids Surf. A*, **371**, 22–28 (2010).

66. A. Javadi, N. Moradi, M. Karbaschi, V.B. Fainerman, H. Möhwald and R. Miller. Alkane vapor and surfactants co-adsorption on aqueous solution interfaces. *Colloids Surf. A*, **391**, 19–24 (2011).

67. J.T. Edsall. Isidor Traube: Physical chemist, biochemist, colloid chemist and controversialist. *Proc. Am. Philos. Soc.*, **129**, 371–406 (1985).

68. Ya. M. Kolotyrkin, O.A. Petrii and A.M. Skundin. *Russ. J. Electrochem.*, **31**, 709–712 (1995).

69. R. Wüstneck, R. Miller, J. Kriwanek and H.-R. Holzbauer. Quantification of synergistic interaction between different surfactants using a generalized Frumkin-Damaskin adsorption isotherm. *Langmuir*, **10**, 3738–3742 (1994).

70. R. Miller. On the solution of diffusion controlled adsorption kinetics for any adsorption isotherm. *Colloid Polym. Sci.*, **259**, 375–381 (1981).

71. E.V. Aksenenko, V.B. Fainerman and R. Miller. Dynamics of surfactant adsorption from solution considering aggregation within the adsorption layer. *J. Phys. Chem.*, **102**, 6025–6028 (1998).

72. G. Loglio, P. Pandolfini, R. Miller, A.V. Makievski, J. Krägel, F. Ravera and B.A. Noskov. Perturbation-response relationship in liquid interfacial systems: Non-linearity assessment by frequency-domain analysis. *Colloids Surf. A*, **261**, 57–63 (2005).

73. F. Ravera, M. Ferrari, L. Liggieri, G. Loglio, E. Santini and A. Zanobini. Liquid-liquid interfacial properties of mixed nanoparticle-surfactant systems. *Colloids Surf. A*, **323**, 99–108 (2008).

5 Wetting and Spreading by Aqueous Surfactant Solutions

Natalia Ivanova and Victor M. Starov

CONTENTS

5.1 INTRODUCTION

Both wetting and dewetting play an important role in many natural and technological processes. In a number of applications, surface wettability is macroscopically described by the equilibrium or, more frequently, by a static advancing contact angle. This description is mostly used to describe wetting properties of liquids on smooth, chemically homogeneous surfaces and pure liquids excluding adsorption and evaporation effects. However, such an approach is not sufficient for the description of a host of technological processes because the kinetic aspects should also be considered. In a number of applications, dynamic wetting and dewetting processes are of crucial importance. The spreading velocity is often an important criterion based on which the efficiency of surface-active substances (surfactants) can be estimated.

The dynamic behavior of a pure liquid on an ideal solid surface can often be successfully described by the equilibrium contact angle (or rather static advancing/receding contact angles), the dynamic (time dependent) contact angle, as well as the spreading velocity. Such approaches [1,2] lead to the spreading force $\gamma_{LV}[\cos \theta_0 - \cos \theta(t)]$, where θ_0 is the static advancing or static receding contact angle. Note that in the case of spreading, θ_0 should be selected as a static advancing contact angle, whereas in the case of dewetting, θ_0 should be selected as a static receding contact angle. Energy is required for expanding the solid–liquid interface, and the energy will dissipate due to viscous shear in the liquid. According to the molecular kinetic theory [3], however, displacements are due to the surface diffusion at the three-phase contact (TPC) line as a possible reason for the spreading force. This approach completely neglects viscous dissipation in the liquid.

In general, when a liquid drop is placed on a solid surface, either it spreads over the surface, that is, it completely wets the surface (Figure 5.1a), or it forms a finite contact angle with the surface. If the contact angle is between 0° and 90°, the situation is referred to as partial wetting (Figure 5.1b).

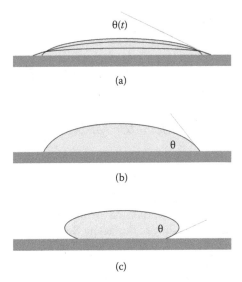

FIGURE 5.1 Different wetting situations. (a) Complete wetting case: a droplet completely spreads out and only a dynamic contact angle can be measured, which tends to zero; (b) partial wetting case: the final static advancing contact angle is between 0° and 90°; (c) nonwetting case: contact angle is larger than 90°.

However, if the contact angle is larger than 90°, the liquid does not wet the surface and the situation is referred to as nonwetting (Figure 5.1c). A more detailed description of advancing, receding, Young's contact angles, as well as problems of experimental and theoretical verification of equilibrium contact angle have been recently provided [4,7]. A reduction of water surface tension by the adsorption of surfactant molecules on the water–vapor interface and adsorption of surfactant molecules on solid–liquid and solid–vapor interfaces alters the nonwetting behavior of aqueous solutions on hydrophobic substrates into a partial or even complete wetting behavior. Surfactants have been used for a long time and their influence on surface wettability is well known and widely used. However, employing surfactants to enhance spreading complicates the wetting process through time-dependent diffusion, adsorption processes at the interfaces involved, and Marangoni phenomenon. The same processes are important in the case of water penetration into hydrophobic porous media. Aqueous surfactant solutions can spontaneously penetrate into hydrophobic porous substrates and the penetration rate depends on both the surfactant type and its concentration. Both the liquid–vapor interfacial tension γ_{LV} and the contact angle of moving meniscus θ_A (advancing contact angle) become concentration dependent.

The major process determining the penetration of aqueous surfactant solutions into hydrophobic porous media or spreading over highly hydrophobic substrates seems to be the adsorption of surfactant molecules onto a bare hydrophobic substrate in front of the moving TPC line. The latter process results in a partial hydrophilization of the hydrophobic surface in front of the meniscus or drop, which causes spontaneous imbibition or spreading.

Pure water does not spontaneously penetrate into hydrophobic capillaries and shows the advancing contact angle larger than 90°. The latter means that water can only be forced into the capillary either by an excess pressure, or by adding surface-active agents. Let us consider in more detail the very beginning of the imbibition process into a hydrophobic capillary, when a surfactant solution touches the capillary inlet. The advancing contact angle at this moment is larger than 90° and the liquid cannot penetrate into the capillary. Solid–liquid and liquid–vapor interfacial tensions do not vary with time in the initial stage because the adsorption of surfactant molecules on these surfaces is a fast process compared with the imbibition rate. Only solid–vapor interfacial tension γ_{SV} can vary. If the adsorption of surfactant molecules at the bare hydrophobic surface, Γ_{SV}, in the vicinity of the

TPC line takes place, the solid–vapor interfacial tension increases with time. After Γ_{SV} reaches some critical value, Γ_{SV}^c, the advancing contact angle reaches 90° and the spontaneous imbibition process can start at higher surfactant concentrations. The latter consideration shows that there is a critical bulk concentration, C^*, below which Γ_{SV} remains below its critical value Γ_{SV}^c and the spontaneous imbibition process does not take place.

The excess free energy Φ of the droplet on a solid substrate (see Figure 5.2) is:

$$\Phi = \gamma_{LV}S + P_eV + \pi R^2 (\gamma_{SL} - \gamma_{SV}) \tag{5.1}$$

where S is the area of the liquid–vapor interface; $P_e = P_a - P_1$ is the excess pressure inside the liquid, P_a and P_1 are the ambient air pressure and pressure inside the liquid, respectively; R is the droplet base radius; γ_{LV}, γ_{SL}, and γ_{SV} are the liquid–vapor, solid–liquid, and solid–vapor interfacial tensions, respectively. The last term on the right-hand side of Equation 5.1 gives the difference between the energy of the surface covered by the liquid drop and the energy of the same solid surface without the droplet. Equation 5.1 shows that the excess free energy decreases if (a) the liquid–vapor interfacial tension decreases, (b) the solid–liquid interfacial tension decreases, and (c) the solid–vapor interfacial tension increases [17]. The last very important conclusion is often overlooked.

In the absence of surfactants, the drop forms a contact angle above 90° with a hydrophobic substrate. In the presence of surfactants, the following three transfer processes take place from the liquid onto all three interfaces: surfactant adsorption at both (i) the inner solid–liquid interface and (ii) the liquid–vapor interface, and (iii) transfer of surfactant molecules from the drop onto the solid–vapor interface in front of the drop on the bare hydrophobic substrate. As mentioned above, all three processes lead to a decrease of the excess free energy of the system. However, adsorption processes (i) and (ii) result in a decrease of corresponding interfacial tensions γ_{SV} and γ_{LV}, but the transfer of surfactant molecules onto the solid–vapor interface in front of the drop results in an increase of a local free energy; however, the total free energy of the system decreases according to Equation 5.1 [17]. That is, surfactant molecule transfer (iii) takes place via a relatively high potential barrier and, hence, occurs considerably more slowly than adsorption processes (i) and (ii). Hence, processes (i) and (ii) are "fast" processes compared with the third process (iii).

Despite the enormous technological importance of the spreading of aqueous surfactant solutions over solid surfaces, information on possible spreading mechanisms is limited in the literature. Specific surface forces come into play in the vicinity of the TPC line. The appearance of disjoining/conjoining pressure in the vicinity of the TPC line is a manifestation of the surface forces action. Actually, the consideration of events in the vicinity of the TPC line is impossible without knowledge of disjoining/conjoining pressure [7]. Disjoining/conjoining pressure isotherms in the presence of surfactants are well investigated in the case of free liquid films [5], much less is known in the case of liquid films on solid substrates [6]. This is the reason that we are currently not able to give an answer to how surfactant molecules are transferred to the vicinity of the TPC line. The region between a bulk liquid (meniscus or droplet) and equilibrium thin liquid film in front is referred to as the transition zone [7]. Inside the transition zone, surface forces and capillary forces (disjoining/conjoining Derjaguin's pressure) are equally important. In the case of aqueous surfactant solutions, our knowledge of the transition zone from the meniscus to thin films in front is still very limited.

FIGURE 5.2 Schematic of a droplet placed on a solid surface; γ_{LV}, γ_{SL}, and γ_{SV} are liquid–vapor, solid–liquid, and solid–vapor interfacial tensions, respectively, at the TPC line; R is the radius of the droplet base. The droplet is small enough and the gravity action can be neglected.

It was shown in ref. [7] that the well-known Young's equation for the equilibrium contact angle is an empirical one and should be replaced with the Derjaguin–Frumkin equation. The latter equation expresses the equilibrium contact angle via surface forces acting in the TPC line vicinity. However, in view of our limited knowledge regarding the surfactant behavior in the vicinity of moving TPC lines, we use Young's equation for describing spreading processes over hydrophobic surfaces. Our hypothesis on surfactant adsorption at a bare hydrophobic substrate in front of the moving meniscus allows us to develop some theoretical predictions, which are in reasonable agreement with known experimental data in the literature.

The situation has even been less investigated in the case of simultaneous spreading and imbibition into porous substrates. We present some theoretical and experimental investigations of the process, which should be considered as a first step in this direction.

We also consider a much better theoretically understood process of flow caused by the surface tension gradient (Marangoni flow). We show that the flow caused by the deposition of a small droplet of a surfactant solution on the surface of a thin aqueous film is governed by the surface tension gradient only. This process recently became a powerful tool for investigating the phenomenon of superspreading.

5.2 SPREADING OF AQUEOUS SURFACTANT SOLUTIONS ON HYDROPHOBIC SURFACES

Most of the natural (e.g., plant leaves; refs. [8,9]) and artificial surfaces (e.g., polymers; ref. [10]) are poorly wetted or are not wetted at all by water and aqueous solutions (see Figure 5.3). These surfaces are referred to as low-energy surfaces or hydrophobic surfaces. However, various technological processes require aqueous solutions to spread out on originally hydrophobic surfaces [11–16].

According to Equation 5.1, a direct way to promote the spreading of aqueous droplets over hydrophobic surfaces is to reduce the interfacial tensions γ_{LV} and γ_{SL} or increase γ_{SV} through the addition of surfactants, which are amphiphilic surface active molecules capable of adsorption at all three interfaces involved [16–19].

Surfactants are widely used to facilitate the wetting of surfaces and spreading water-based formulations in diverse industrial applications such as agrochemical, pharmaceutical, home care products, cosmetics, and coatings [11–16]. However, the mechanism behind the spreading of aqueous surfactant solutions over hydrophobic substrates is more complicated in comparison to the spreading of pure liquids and has not been completely understood yet. The reason is that the spreading dynamic of surfactant solutions is affected by (i) a time-dependent adsorption/desorption of surfactant molecules at all interfaces (i.e., liquid–vapor, solid–liquid, and solid–vapor) involved, drastically changing the interfacial tensions and energy balance at the moving TPC line; (ii) the resulting interfacial tension

(a)   (b)

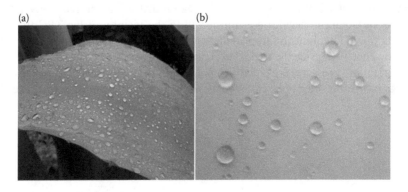

FIGURE 5.3 Rain droplets on the leaf of an Agavaceae plant (a). Water droplets deposited from a sprayer on the surface of a polypropylene film (b). (Redrawn from Ivanova, N.A., and V.M. Starov, *Curr. Opin. Colloid Interface Sci.*, 16, 285, 2011.)

gradients and Marangoni flow as a consequence; and (iii) the disjoining/conjoining (Derjaguin's) pressure gradient. Very little is known about this pressure in the case of nonwetting.

The adsorption processes as well as the changing interfacial tensions, in turn, are influenced by the concentration of surfactants in the bulk of the droplet, which reflects in the spreading behavior (see for example, Figure 5.4; ref. [17]). Figure 5.4 shows the spreading kinetics of aqueous solutions of trisiloxane surfactant TEO_8 (see below) on Teflon AF depending on concentration of TEO_8. At the lowest concentrations, droplets spread very slowly and a delay in spreading is detected (the upper curve in Figure 5.4). Increasing the concentration decreases the initial contact angle of droplets due to the fast adsorption kinetics of molecules on the liquid–vapor interface and enhances spreading, allowing reaching a final quasi-equilibrium contact angle much faster.

It is important to note that trisiloxane surfactants (organosilicone surfactants) on moderately hydrophobic surfaces exhibit "superspreading," which means a rapid spreading of droplets of these surfactants solutions to a final contact angle close to zero [25–27]. The mechanism responsible for superspreading is still debated, and a comprehensive review concerning this problem is presented by Venzmer [20].

The character of spreading can be predetermined by the surface free energy of solids and the pattern of surface chemical groups on the solid surface, onto which the deposition of surfactant solutions occurs. Let us consider, as an example, two hydrophobic surfaces: (i) polyethylene (PE) having only relatively "weak" nonpolar methylene $-CH_2$ groups at its solid–vapor interface. PE exhibits the microscopic contact angle of pure water, θ_w, in the range $\theta_w = 92°$ [21] to $103.5°$ [22] depending on the formation method; and (ii) poly(methylmethacrylate) (PMMA) with more hydrophobic nonpolar methyl $-CH_3$ groups (note that $-CH_3$ groups are more nonpolar as compared with $-CH_2$ groups; ref. [23]). However, PMMA has, on its surface, comparatively hydrophilic (polar) ester groups. These ester groups, despite the presence of highly nonpolar $-CH_3$ groups, decrease the values of θ_w down to $80°$ [10] or even $68°$ [24].

The nature and structural composition of surfactant molecules influences the spreading behavior as well. Let us consider, for example, the high-performance wetting agents such as nonionic ethoxylated alcohols ($C_m EO_n$ for short) with $-CH_2$ hydrophobic tails and ethoxylated organosilicone surfactants (known as trisiloxane surfactants, TEO_n) [25–27] with silicone-based hydrophobic tails screened by

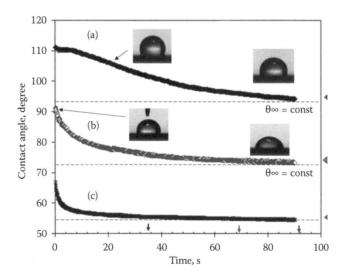

FIGURE 5.4 Evolution of advancing contact angles of droplets of aqueous trisiloxane solution (TEO_8) at various concentrations inside the droplet: (a) $C < cac$; (b) $cac < C < cwc$; and (c) $C > cwc$ on Teflon AF–coated silicon wafers. (Redrawn from Ivanova, N.A., and V.M. Starov, *Curr. Opin. Colloid Interface Sci.*, 16, 285, 2011.)

–CH$_3$ groups. Both surfactants have identical hydrophilic polyoxyethylene (EO) chains; however, they exhibit different character and power of spreading on identical hydrophobic surfaces [28,29].

5.2.1 BIMODAL KINETICS OF SPREADING OF SURFACTANT SOLUTIONS: THE CLASSIC POWER LAW

Analysis of the recent literature [20,22,30–38] shows that spreading of aqueous surfactant solutions on hydrophobic surfaces, as well as imbibition of surfactant solutions into hydrophobic capillaries, shows mostly bimodal kinetics: two stages of spreading/imbibition with different rates controlled by the surfactant adsorption mode. The latter, in turn, is determined by the concentration of surfactant, and the nature of both substrate and surfactant. In some cases, even three stages of spreading/imbibition were detected [33,35,38].

One of the widely used methods to analyze the mechanism of spreading of surfactant solutions is to fit the experimental data by the power law $R \sim t^n$. This method is widely used to analyze wetting by pure liquids. Briefly, the relationship between the spreading/power exponent, n, and the corresponding mechanism is as follows. In the case of small droplets and complete spreading, $n = 0.1$ (Tanner's law), showing the capillary-driven spreading regime [39], or $n = 0.14$, the same regime but based on a molecular approach to the dissipation of energy [40]. Note that the exponent $n = 0.1$ is in excellent agreement with experimental data in the case of complete wetting or spreading over both smooth solid substrates [41] and porous substrates saturated with the same liquid [42]. Marangoni forces (surface tension gradients) are responsible for spreading with $n = 0.25$ [43,44]. For relatively large droplets, the gravity-driven spreading dominates, yielding $n = 0.125$.

von Bahr and coworkers [36,37] detected two-stage spreading dynamics of nonionic polyoxyethylene alcohol surfactants (C_mEO_6, where m is the length of hydrophobic tail, 6 is the number of ethylene oxide units in the hydrophilic head) solutions on hydrophobically modified glass and gold surfaces. The first stage was very fast (30 ms; ref. [37]) with the moving edge as $n = 0.5$ (referred to as a nondiffusive regime; ref. [36]), after this short period, the spreading process slowed down relaxing toward the final contact angle. The first stage, according to the authors, is due to different factors such as inertia, capillarity, relaxation of balance of interfacial tensions [36,37], and the lifetime of the droplet [36], but the slow stage is determined by the diffusion of surfactant monomers from bulk solution to the liquid–vapor interface to be adsorbed on that interface. The authors also noticed that adsorption at the solid–liquid interface leads to a decrease of the spreading rate during the second stage, especially at concentrations below the critical micelle concentration (cmc).

Dutschk and coworkers [30–32] found that the first stage (fast spreading) of C_mEO_5 surfactants on different polymers (Parafilm M [Pechiney Plastic Packaging, Chicago, USA] and polypropylene, etc.) lasted less than 1 s with the radius evolving according to $n = 0.5$ but the second stage was much slower and is characterized by spreading exponent $n \ll 0.1$, which is consistent with the results obtained in refs. [36,37]. To explain the extremely low spreading rate, they used the theory proposed by Starov and coworkers [17,18]. According to this theory, the spreading in the slow stage is governed by a slow adsorption of surfactant monomers in front of the moving edge onto the solid–vapor interface making it hydrophilic (γ_{SV} increasing) [17,18]. The characteristic time (τ) of transfer of surfactant molecules onto the solid–vapor interface decreases with increasing bulk concentration of surfactant, whereas above the cmc, τ should level off and reach its lowest value. Figure 5.5a shows excellent agreement between theory and experiments [31]. This mechanism, first proposed by Churaev et al. [34] and theoretically developed by Starov and coworkers [17,18], was later confirmed by direct experimental observation of adsorption of molecules in front of the moving contact line by Garoff's group [45]. Note that the adsorption of surfactant molecules in front of the moving TPC line results in a decrease of the total excess free energy of the system; however, the solid–vapor interfacial tension increases locally [17,18]. The latter means that the mentioned process takes place via a potential barrier, that is, it is much slower as compared with other adsorption processes.

Drelich et al. [21] have shown that in the case of spreading aqueous solutions of $C_{12}EO_n$ over hydrophobic toner and polypropylene, the first stage occurs for a long period, up to a few minutes

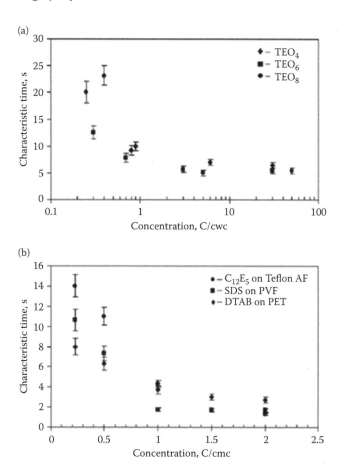

FIGURE 5.5 Dependencies of the characteristic time of surfactant transfer on a bare hydrophobic substrate in front of the TPC line on bulk concentration of surfactants during the slow stage of spreading: (a) trisiloxane surfactants (TEO_4, TEO_6, and TEO_8) on PTFE AF–coated silicon wafers; (b) conventional ionic sodium dodecyl sulfate (SDS), dodecyltrimethylammonium bromide (DTAB), and nonionic ($C_{12}EO_5$) surfactants on different polymers: polyvinylfluoride (PVF), polyethyleneterephthalate (PET), and Teflon AF. (a. Redrawn from Starov, V. et al., *Adv. Colloid Interface Sci.*, 161, 153, 2010. b. Redrawn from Dutschk, V., and B. Breitzke, *Tenside Surf. Det.*, 42, 1, 2005.)

after deposition. During the first stage, the contact angle significantly decreased, and afterward a slower relaxation was observed. That is, the first stage was found to last much longer as compared with those reported in refs. [30–32,36,37], although the type of surfactants and contact angle of water on solid substrates (close to 90°) used by all groups of researchers were quite similar. Probably, the reason for the difference could be due to the difference in chemical compositions of substrates used by different authors. In some cases, polymer substrates were used with hydrophobic groups, whereas in other cases, the gold surfaces coated by organosulfur monolayers with inclusions of OH polar groups were used, which could affect the spreading rate.

Svitova et al. [38] reported two stages of complete spreading of $C_{12}EO_3$ surfactants and trisiloxanes TEO_8 and TEO_{12} on graphite. For polyoxyethylene alcohols at $C > cac$ (cac is the critical aggregation concentration; critical wetting concentration > cac) the first and second stages were described by the power laws with $n = 0.1$ and $n = 0.4$, respectively. In the case of trisiloxanes at $C > cwc$ the authors found higher spreading exponents: $n = 0.2$ and $n = 0.5$, respectively. Only one-stage partial wetting was observed for both types of surfactants at $C < cmc/cwc$.

Ivanova et al. [19] studied spreading kinetics of trisiloxane surfactants and polyoxyethylene alcohol surfactant solutions on Teflon AF. Teflon AF has very low surface energy (15 mJ/m², according to Sigma Aldrich data). It was found that only partial wetting or nonwetting takes place and the spreading in this case can be referred to as "microspreading." At concentrations above cac/cwc, two stages were observed: a short fast stage fitted using a power law with $n \leq 0.05$, followed by a relatively longer stage, which is the slow motion of the TPC line toward the final position. At low concentrations below cac, only the slow stage of spreading occurs, which is similar to the results from Svitova et al. [38], which were obtained on moderately hydrophobic graphite. Moreover, in some cases at low surfactant concentrations, a delay in spreading up to a few seconds was detected [19]. The second stage was fitted according to Starov's theory [18] and reasonably well agreements were obtained (see Figure 5.5b). However, the mechanism for the first stage was not explained in ref. [19]. Recently, Starov et al. [17] suggested that the first stage is likely determined by the surface tension relaxation at the liquid–vapor interface immediately after deposition of the droplet.

Some authors do not report the bimodal kinetics of spreading of the surfactants studied, but they provide the evolution of the spreading exponents with concentration [46,47], which is useful for the identification of the driving forces responsible for spreading in a certain concentration range of surfactants. Zhang and Han [46] studied the spreading of complex surfactants, glucosamide-based trisiloxanes solutions, on polystyrene substrate. They examined the spreading kinetics over 2 s after the droplet was deposited on the substrate. It was found that for some surfactants, the exponent n increases from approximately 0.1 to 0.25, showing that at low concentrations, the spreading is driven by capillary forces, and at high concentrations, the Marangoni forces dominate. Gemini-like glucosamide surfactants do not show noticeable spreading capability, the spreading exponent in this case hardly reaches $n = 0.1$ at the highest concentrations [46]. Rafaï et al. [47] showed that the power exponent for trisiloxane surfactant, TEO_8, spreading over poly(ethylene terephthalate) increases when the bulk surfactant concentration increases until a plateau value of $n = 1$ is reached for the highest concentration. Marangoni forces (i.e., surface tension gradient) were considered to be responsible for the spreading over the whole range of concentrations: the onset of radial surface tension gradient corresponds to $n = 0.25$, but if the surface tension gradient is rapidly established over the height of the droplet at very high concentrations then it yields $n = 1$. The radius of the droplets of anionic surfactant solutions increases due to the capillary forces ($n = 0.1$) at the highest concentrations, but at low concentrations ($C < cmc$), these solutions do not spread at all over poly(ethylene terephthalate) [47].

This brief overview demonstrates that the character of spreading is strongly affected by the concentration of surfactant because different mechanisms of spreading are observed at $C < cac/cmc$ and at $C > cac/cmc$. The competition between the rate of adsorption of surfactant molecules and rate of depletion of molecules at the expanding interface due to their adsorption at the solid–liquid or due to the slow diffusion to the interface during the spreading reflects in the changing spreading rate of droplets.

Attempts to fit the data on spreading of surfactant solutions by the power law showed that the fitting with only one exponent does not work. In some cases, the power exponent is so small that the mechanism behind spreading can hardly be related to well-determined mechanisms corresponding to certain values of n. In many cases, spreading occurs in a crossover mode and is controlled by competition between different mechanisms that drive the spreading.

5.2.2 INFLUENCE OF SURFACE ENERGY OF HYDROPHOBIC SOLID SUBSTRATE

In the early 1960s, Zisman [23] introduced the concept of critical surface tension, γ_c, as a parameter that characterizes a solid surface. This parameter shows whether liquid with a known liquid–vapor interfacial tension, γ_{LV}, will wet a solid surface or not. His method involved the measuring of contact angles of a series of liquids with known γ_{LV} on the given solid surface. The critical surface tension is defined by a linear extrapolation of $\cos \theta$ versus γ_{LV} plot to the intersection with the line $\cos \theta = 1$. According to Zisman's plot, liquids with $\gamma_{LV} < \gamma_c$ spread out over the solid substrate. The surface tension of pure water ($\gamma_{LV} = 72.8$ mJ/m²) is much higher than the critical surface tension values of hydrophobic materials (γ_c < 50 mJ/m² at 20°C; ref. [10]). Hence, spreading of water does not occur on these materials.

The spreading and wetting behavior of surfactant solutions, with respect to the relationship between the surface energy of substrate and the surface energy of the liquid–vapor interface of solutions, has been studied by many authors [23,28–33,48–51]. In the literature, a solid surface is usually characterized either by the surface energy/critical surface tension or by the contact angle of the droplet of pure water on the surface. Typically, surface energies of polymers do not exceed 50 mJ/m² [23]. Therefore, according to Zisman's rule, aqueous solutions of surfactants should promote spreading if surfactants are capable of reducing the surface tension of water to less than the critical surface tension of the substrate. This conclusion is supported by Dutschk and coworkers [30–32], who have shown that the solutions of some ionic surfactants whose liquid–vapor surface tension at the cmc is close to 40 mJ/m² does not spread at all compared with polymers whose critical surface tension is less than 30 mJ/m² and are characterized by θ_w > 90° (such as polypropylene, Parafilm, Teflon AF). At the same time, nonionic surfactants that decrease the surface tension of water down to approximately 30 mJ/m² [25] exhibit partial wetting on these polymer substrates. The same trend was observed by Rafaï et al. [47] for anionic surfactant solutions at high concentrations on poly(ethylene terephthalate).

Radulovic et al. [49] have undertaken a comparative analysis of spreading exponents for aqueous solutions of Silwet L-77 (De Sangosse Ltd., Cambridge, UK), which is a commercially available trisiloxane surfactant, at 0.1 wt% on hydrophobic substrates with different degrees of hydrophobicity that was interpreted in terms of contact angles of pure water droplets, θ_w, on these substrates. The authors found a nearly linear increase in exponent n from roughly 0.1 to 0.8 with a decrease of pure water contact angle from 118° (Teflon AF–coated Si wafers; ref. [19]) to 70° (poly(ethylene terephthalate); ref. [47]). Detailed studies have been done on the effect of the surface energy of substrate for a series of trisiloxane homologues [50] as well as for trisiloxane surfactants, polyoxyethylene alcohols, and ionic surfactants [28,29,51].

Wagner et al. [50] have shown that the spreading of trisiloxane homologues (γ_{LV} = 20 mJ/m² at C = cac) is favorable on nonpolar surfaces and on slightly polar but still hydrophobic surfaces with a moderate surface energy of 30 to 40 mJ/m². However, the lowest and highest surface energies of solid substrates suppress the wetting capability of all these surfactants. A similar trend was observed by Stoebe et al. [28] for trisiloxanes on gold-coated substrates covered by organosulfur monolayers with different degrees of hydrophobicity varying from 57° to 112° in terms of θ_w. Polyoxyethylene alcohols and ionic surfactant solutions [29,51], regardless of surfactant concentration and the length of ethylene oxide (EO) chains, exhibit sharp maximum in the spreading rate on slightly hydrophobic organosulfur monolayers with θ_w = 57°, containing mostly polar OH groups. On substrates with only CH₃ groups (θ_w = 112°) or a small amount OH groups (θ_w = 93°) solutions do not spread at all.

5.2.3 THE ROLE OF EQUILIBRIUM AND DYNAMIC SURFACE TENSIONS AT LIQUID–VAPOR INTERFACE

It looks like it is possible to conclude from the above paragraph as well as from Young's equation (cos θ = (γ_{SV} − γ_{SL})/γ_{LV}) that the lower the liquid–vapor interfacial tension, γ_{LV}, the better the spreading capability of the solution in terms of spreading rate/complete wetting, contact angle values, and more hydrophobic surfaces can be wetted by this solution. However, despite this view, it has been shown that the lowest equilibrium γ_{LV} value does not guarantee fast spreading or "superspreading" [25,26] over hydrophobic surfaces. It is known that trisiloxane surfactants (having "umbrella-like" structure of hydrophobic part of the molecule; ref. [25]) with different numbers of EO groups, varying from four to nine groups, reduce the surface tension of water down to the equilibrium value of approximately 20 to 22 mJ/m² [25,27,52,53]. Ivanova et al. [52] have shown that despite this, these trisiloxanes demonstrate different capabilities to wet substrates with relatively high surface energy (30–37 mJ/m²). A more hydrophobic trisiloxane with short EO chain (EO₄) does not spread completely, but demonstrates a partial wetting and reaches a final contact angle on hydrophobic polystyrene ($\theta_w \approx 89°$) and polypropylene ($\theta_w = 97°$), even at $C > $ cwc [52].

Sieverding et al. [54] compared spreading performance of some commercially available organo-silicone surfactants used in agrochemical industry, such as Silwet L-77, Break-Thru S233, and Break-Thru S240 surfactants (EVONIK Industries, Essen, Germany). All these surfactants have nearly the same equilibrium and dynamic surface tensions (DSTs) varying in the range of 24.9 to 23.6 mJ/m², but demonstrate considerable differences in spreading capability: Silwet L-77 and Break-Thru S233 behave as real "superspreaders" on polypropylene surfaces, but Break-Thru S240 does not promote spreading of water on this substrate.

Another type of surfactant that is capable of reducing the liquid–vapor surface tension of water to 16 to 18 mJ/m² is the fluorocarbon surfactant [55,56]. Such low interfacial tension gives grounds for expecting the lowest contact angles and excellent wetting behavior of these surfactant solutions on hydrophobic surfaces. However, studies by Ananthapadmanabhan et al. [56] and by Tiberg and Cazabat [57] have shown that these solutions have much lower spreading performance on polypro-pylene and parafilm surfaces [56] and on octadecyltrichlorosilane monolayers [57] compared with trisiloxane surfactants.

Ananthapadmanabhan et al. [56] pointed out that the turbidity of solutions (the presence of dis-persed phase in solutions) at certain concentrations influences spreading, rather than the lowest equilibrium surface tension.

Trying to explain the lowest spreading exponent ($n < 0.1$), which was found in the case of glu-cosamide, a Gemini-like surfactant, Zhang and Han [46] suggested that the DST at the liquid–vapor interface reflects the spreading dynamics. Indeed, despite the fact that all surfactants studied reached essentially equal equilibrium values of γ_{LV} at a concentration above cac, the Gemini-like surfactant exhibits a much slower decrease of γ_{LV} down to $\gamma_{LV} \approx \gamma_c$ compared with other surfactants, which according to the authors, does not allow the development of enough Marangoni forces over the surface of the droplet to spread it over the substrate. According to Drelich et al. [21] the kinet-ics of adsorption of surfactants on the liquid–vapor interface might have an effect on the contact angle relaxation at the beginning of the spreading process. This is consistent with the experiments by von Bahr et al. [36] on the influence of the so-called lifetime of the droplet, that is, the time after the droplet was formed on the syringe tip, on the spreading characteristics. These results were also related to the liquid–vapor interfacial tension relaxation. The authors above found that the direct deposition (zero lifetime) of a droplet results in a spreading delay and an initial contact angle close to the contact angle of pure water. Increasing the lifetime to tens of seconds leads to a rather low initial contact angle of the droplet and a very fast approach to reaching the final contact angle com-pared with the zero lifetime case. The abovementioned behavior depends on the concentration and the rate of diffusion/adsorption of the surfactant at the liquid–vapor interfaces.

Starov et al. [17], based on theoretical considerations, have shown that adsorption at a liquid–vapor interface (or the relaxation of liquid–vapor interfacial tension) does not favor spreading at low concentrations, that is, when an initial contact angle of a droplet is more than 90° [17], but it does at higher concentrations.

In some cases, studies of DST of surfactant solutions can help in discovering new interesting properties of these surfactants related to their spreading performance [53,58,59]. Kumar et al. [58] measured DST of trisiloxane solutions to check whether the adsorption dynamics fulfilled the con-ditions necessary for the high rate of spreading. It seems that the Frumkin equation fit the DST data well only at relatively low concentrations, but not for the higher concentrations. Hence, the diffusion of monomers cannot provide the necessary amount of surfactants to maintain the spreading rate featured at high concentrations. Based on their DST data and analysis, the authors suggested that direct adsorption of bulk aggregates on interfaces is capable of maintaining the necessary flux of surfactants to the interface of the spreading droplet [58].

More recently, Ritacco et al. [53] carried out detailed measurements of DST for these surfactants at very short times (<1 s). It was discovered that in the case of trisiloxanes (with relatively long EO_n

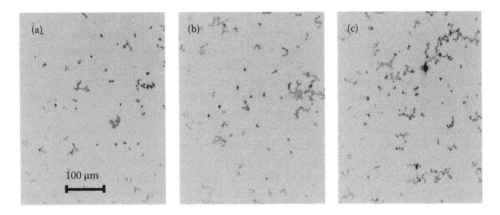

FIGURE 5.6 BAM images (inverted color) of surface aggregates for solutions TEO_6 (a), TEO_7 (b), and TEO_8 (c) at a concentration of 0.002 mol/m³. (Redrawn from Ivanova, N.A., and V.M. Starov, *Curr. Opin. Colloid Interface Sci.*, 16, 285, 2011.)

chains, $n = 6 - 9$ possessing "superspreading" behavior) two inflection points on DST curves were present [53]. Using the Brewster angle microscopy (BAM), Ritacco et al. [53] directly observed the appearance of aggregates on the liquid–vapor interface for these trisiloxanes in a range of concentrations (see Figure 5.6). This suggests that the surfactant molecules are present at the liquid–vapor interface in two states: as monomers and as surface aggregates [53,59]. The latter could act as reservoirs of surfactant monomers in the course of spreading, which confirms the conclusions reached by Kumar et al. [58].

The above discussion shows that DST behavior plays an important role in the dynamic spreading and wetting processes. For more detailed information on the basics of DST and its relation to diffusion and adsorption processes, refer to comprehensive reviews by Eastoe and Dalton [60] and Chang and Frances [61].

5.2.4 SPREADING CHARACTERISTICS VERSUS THE LENGTH OF THE POLYOXYETHYLENE CHAIN

The role of the length of polyoxyethylene chains of amphiphilic surfactant molecules in spreading behavior cannot be underestimated. On the one hand, this hydrophilic moiety provides solubility of surfactants in water phase. However, on the other hand, polyoxyethylene chains influence the arrangement of surfactant molecules in the adsorbed layers at air–liquid and solid–liquid interfaces and the shape of molecular aggregates formed in bulk aqueous phase. As a consequence, the length affects the dynamic and the equilibrium surface tension values of surfactant solutions.

In the case of trisiloxanes with an "umbrella-like" shape [25], Ruckenstein [62] theoretically argued that molecules with moderately long EO_{7-8} chains are capable of stimulating high spreading capability of water on hydrophobic surfaces. Increasing the EO length leads to a lowering of the adsorbability of surfactants at interfaces because of strong attraction interaction with water and thus reduces spreading efficiency. On the contrary, Gentle and Snow [27] found experimentally that adsorption of trisiloxanes does not vary with the EO_n length within the range of 4 to 16, but does decrease when $n > 16$. However, the theoretical arguments [62] are consistent with the results reported in refs. [19,28,50], where it was found that trisiloxanes at EO_{5-8} showed the highest spreading rate on hydrophobic surfaces as compared with slightly soluble short EO chains and highly soluble long EO chains. It was also shown by Wagner et al. [50] that on low hydrophobic (slightly polar) surfaces, trisiloxanes with relatively longer EO chains show faster spreading. Wagner et al. [50] have also shown that the concentration of surfactants is important for the spreading behavior. The latter behavior was associated with the bulk phase condition of trisiloxane solution that, in turn, depends on the length of the EO chain and temperature.

In the case of ethoxylated alcohols, Stoebe et al. [29] found a decreasing spreading rate of $C_{12}EO_{3-8}$ homologues on hydrophobic surfaces with an increase in the length of EO chain; however, nothing like this was observed in the case of hydrophilic surfaces. It was suggested that the latter is a result of the complex interplay between the adsorption capability of molecules at the solid and the aggregation in the bulk, which is determined by optimal hydrophilic/hydrophobic balance. However, AFM studies [63] of the relationship between surface aggregation on graphite and bulk aggregation of these $C_{12}EO_n$ surfactants have shown that at the highest surfactant concentrations, $n = 5$ to 10 EO chains formed hemicylindrical aggregates on graphite, whereas in the bulk solution, spherical and rod-like micelles were present. This means that there are no significant changes in the shapes of aggregates and consequently in the interplay mentioned above within the selected range of $n = 5$ to 8 EO chains. Nevertheless, the difference was found for very short EO [38,63] and very long EO [63] chains as compared with EO_{5-10} chains. Adsorption of bilayers at the surface and the formation of lamellae in the bulk phase were observed for the EO_3 surfactant, because bending of these molecules is thermodynamically unfavorable, and aggregation at the surface was exhibited by the EO_{23} surfactant.

Ivanova et al. [64] studied the wetting capability of $C_{10}EO_n$ surfactant solutions at concentrations ranging from 0.1 to 4 cmc on three polymeric surfaces: Teflon AF (117°), Parafilm (106°), and polypropylene (97°). It was shown that regardless of the hydrophobicity of surfaces, the final quasi-equilibrium advancing contact angle increases as the number of EO units increases for all three substrates used. This implies that the common trend of the final contact angle behavior of $C_{10}EO_n$ surfactants is maintained until the surfaces are characterized by $\theta_w > 90°$.

5.2.5 INFLUENCE OF OTHER FACTORS

It is known that instability to hydrolysis is a weak point of organosilicone surfactants [25]. Despite the high spreading performance, these surfactants lose their efficiency very quickly after preparation [25,65,66]. Recently, a comparative study of the aging processes of hydrocarbon and trisiloxane surfactants in water in terms of wetting power was undertaken by Radulovic et al. [65]. It was found that unlike conventional surfactants (e.g., Triton X-100), commercial trisiloxane Silwet L-77 gradually lost its wetting capability over a few days. Figure 5.7 is an illustration of the aging process of Silwet L-77 solution [65]: a droplet increases its final contact angle day by day over 10 days due to hydrolysis, which causes the degradation of surfactant molecules. However, the concentration was found to play a crucial role in maintaining the spreading capability: very dilute solutions lost their spreading ability over just 24 h, whereas very concentrated (0.1 wt%) solutions were more viable and their spreading kinetics was changed only slightly over several days.

As mentioned previously, the spreading behavior depends on the presence of dispersed phase, type, and size of aggregates in the bulk solution [25,28,29,46,67,70] and the phase behavior of surfactant solutions [25,68,69]. Wagner et al. [68] observed increasing spreading capability of trisiloxane solutions when solutions were close to the transition region between the two-phase state and the lamellar phase state, which depends on temperature and concentration. Zhu et al. [67] showed that the size of aggregates in the bulk of the solution determines the spreading rate in the following way: the smaller the aggregates, the faster the spreading. Zhang and Han [46] found that, for glucosamide-based surfactants, the existence of large bulk aggregates at low concentrations favored

| 1 day | 6 days | 8 days | 10 days |

FIGURE 5.7 Wetting performance of aged 0.1 wt% Silwet L-77 solution after 1 (53°), 6 (72°), 8 (91°), and 10 (103°) days on Teflon AF surface. (Redrawn from Radulovic, J. et al., *Chem. Eng. Sci.*, 65, 5251, 2010.)

inducing Marangoni forces and increased the spreading area during the initial spreading stage. According to a number of reports [25,28,29,70], the presence of lamellae or bilayer aggregates in trisiloxane solutions and their direct "unzipping" adsorption at the interfaces promote rapid spreading of these solutions.

Humidity of the ambient atmosphere and preadsorbed water layers on hydrophobic solid surfaces have been reported by some authors [67,70] as important factors inducing "superspreading" behavior of trisiloxane surfactant solutions. However, a weak influence of humidity on the spreading dynamics of these solutions was shown by Rafaï et al. [47] and Lin et al. [71]. It is possible to state, despite these contradictory conclusions, that the evaporation rate of spreading droplets depends on the surrounding humidity and hence, at least the contact angle values could be affected by low humidity because evaporation mostly occurs at the edge of the spreading droplets [72].

5.3 SPONTANEOUS INCREASE AND IMBIBITION OF SURFACTANT SOLUTIONS INTO HYDROPHOBIC CAPILLARIES

In the case of the partial wetting when the advancing contact angle takes on values of $0 < \theta_A < \pi/2$ the penetration of the liquid into a horizontal or vertical (at short times) capillary is described by the following equation [73]:

$$z(t) = \left(\frac{R\gamma_{LV} \cos\theta_A}{2\eta} t \right)^{1/2}$$

(5.2)

where the subscript A indicates the advancing contact angle, $z(t)$ is the length of the part of the capillary filled with the liquid, R is the capillary radius, γ is the liquid–vapor interfacial tension, η is the dynamic liquid viscosity, and t is time.

In the case of a vertical capillary, the law of liquid penetration corresponding to the long time limit, when the liquid increases to a stationary level $z_\infty = 2\gamma \cos\theta_A/\rho gR$ where gravity balances the capillary force, is given by the equation [74]

$$z(t) = z_\infty \left[1 - \exp\left(-\frac{\rho gR^2}{8\eta z_\infty} t \right) \right]$$

(5.3)

where ρ is liquid density and g is gravity acceleration.

When the advancing contact angle $\theta_A > \pi/2$, pure water does not penetrate spontaneously into the hydrophobic capillaries. In this case, the liquid can be forced into the capillaries by applying external forces such as pressure, electric field (electrowetting) [75], or thermocapillary forces [76].

However, as found in refs. [34,77–79], water containing surfactant molecules penetrates spontaneously into hydrophobic capillaries. It was shown that adsorption of the surfactant molecules onto the capillary walls ahead of the moving meniscus makes the penetration of surfactant solutions into the hydrophobic capillaries possible. It is assumed that surfactant molecules adsorption onto the solid–liquid and liquid–vapor interfaces proceeds much faster compared with the characteristic timescale of penetration. The latter means that the interfacial tensions γ_{SL} and γ_{LV} near the meniscus do not change appreciably over time because adsorption processes are sufficiently fast. Indeed, experiments on the spontaneous capillary imbibition of Syntamide-5 (Gamma Chemical, Dzerzhinsk, Russia) aqueous solutions into horizontal hydrophobic capillaries [34,78,79] (Figure 5.8) have shown that characteristic timescales of the imbibition and increase are approximately 100 s and 10^5 s, respectively, whereas a timescale estimation of diffusion kinetics in the capillary cross-section is only $R^2/D \approx 0.1$ s ($R \sim 10$ μm and $D \sim 10^{-5}$ cm^2/s) [77,80].

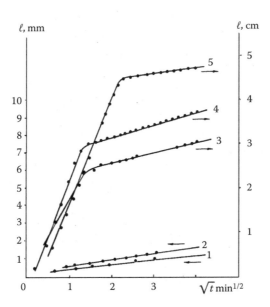

FIGURE 5.8 Evolution of the imbibition length, ℓ, with time, t (min) for aqueous solutions of Syntamide-5 in a horizontal hydrophobized quartz capillary with $R = 16$ μm at various surfactant concentrations, C, in the bulk: below cmc (left scale) (1) $C = 0.05\%$; (2) $C = 0.1\%$; above cmc (right scale) (3) $C = 0.4\%$; (4) $C = 0.5\%$; (5) $C = 1\%$. (Redrawn from Churaev, N.V. et al., *Colloid Polym. Sci.*, 259, 747, 1981.)

Adsorption on the solid–vapor interface leads to an increase of its interfacial tension with time and, consequently, to a decrease of the advancing contact angle, which becomes $\theta_A < \pi/2$ at some critical adsorption of surfactant corresponding to the critical bulk concentration, C^* [34,77,79]. Spontaneous imbibition into the hydrophobic capillary commences when the concentration of surfactant in solution C is above C^*, that is, at $C > C^*$. A theoretical analysis of the spontaneous imbibition of surfactant solutions into hydrophobic capillaries, taking into account the convective transfer, the surface diffusion of surfactant molecules, and the adsorption of molecules on a hydrophobic surface, was reported in refs. [34,79,80].

The mechanism of spontaneous imbibition is assumed to be controlled by surfactant concentration C_m close to the moving meniscus, which is below the C. Hence, the expression $\gamma_{LV} \cos \theta_A$ in Equation 5.2 should depend on the concentration C_m. It was shown in refs. [34,77–80] that the dependency $\varphi(C_m) = \gamma_{LV} \cos \theta_A$ can be approximated as

$$\varphi(C_m) = \alpha(C_m - C^*) \tag{5.4}$$

where the coefficient α was calculated by Starov [77]. According to Equation 5.4, $\varphi(C_m)$ should be a linear function of concentration at $C_m > C^*$, which is in agreement with experimental results [34,79]. In case the surfactant concentration C is below the cmc, the imbibition rate according to Equations 5.2 and 5.4 can be represented as follows [34,79]

$$z(t) = \left(\frac{R\alpha(C_m - C^*)}{2\mu} t \right)^{1/2} \tag{5.5}$$

The surfactant concentration C_m near the moving meniscus remains constant with time ($C_m =$ const) and its value depends on the diffusion coefficient and adsorption characteristics of surfactant molecules.

If the surfactant concentration C is above the cmc, the adsorption of surfactant molecules on the solid–vapor interface near the meniscus leads to the disintegration of micelles, decreasing the surfactant concentration C_m from $C_m = C_0 >$ cmc to $C_m <$ cmc. At the moment when the $C_m =$ cmc, the imbibition rate is changed and a further decrease of C_m causes the appearance of a zone free of micelles behind the meniscus. After this, the concentration C_m no longer changes and remains below the cmc, and the second slower stage of the process starts. The micelle front moves $\sim t^{1/2}$, but moves slower than the meniscus. It seems from the above analysis that the imbibition process under $C >$ cmc conditions should be characterized by two stages. Indeed, the experiments on the penetration of aqueous solutions of Syntamide-5 [34,78] into the hydrophobic capillaries has clearly shown the existence of two kinetic stages. The second stage is slower than the first one and the only individual surfactant molecules move with the meniscus. Note that during the second stage, the spontaneous imbibition rate is only slightly higher than that in the case when the concentration of surfactant C is below the cmc.

Unlike the case of the spontaneous imbibition of surfactant solutions into a flat hydrophobic capillary, in which the concentration of surfactant C_m near the meniscus is constant, the spontaneous increase into a vertical hydrophobic capillary is controlled by a different mechanism because the concentration C_m cannot remain constant. Spontaneous capillary rise experiments with Syntamide-5 solutions in a vertical hydrophobized quartz capillary [78] showed that the time evolution of the increase satisfies Equation 5.2 at the initial stage of the process. However, the slopes of experimental dependencies $z(\sqrt{t})$ according to Equation 5.2 are so small that they correspond to the θ_A values being only a few seconds less than $\pi/2$. At such θ_A values, the capillary rise would be expected to stop when the liquid has reached a level of $z_{max} = 10^{-3}$ cm. However, it does not stop at this level and goes up to a height of 3 to 4 cm. The explanation for this effect is that the meniscus increases following the surface diffusion front of surfactant molecules, which hydrophilize the walls of the capillary in front of the moving meniscus [34,77]. It was shown that the position of the solution meniscus can be described by the following equation:

$$z(t) = \frac{2\alpha}{\rho g h}(C_m - C^*) \tag{5.6}$$

which is valid only if $C_m(t)$ increases with time and, hence, the concentration of surfactant $C_m(t)$ near the meniscus in the case of spontaneous capillary rise into a hydrophobic capillary does not remain constant. The maximum level of the capillary rise z_{max} is reached after the concentration C_m becomes equal to the concentration at the capillary inlet C. Then, the increase stops at the maximum height determined as follows:

$$z_{max} = \frac{2\alpha}{\rho g h}(C - C^*) \tag{5.7}$$

This means that the experimental data presented by Churaev et al. [34] corresponds to only $z(t) \ll z_{max}$, that is, the initial stage of the capillary rise as shown in Figure 5.9. This figure shows that the experimental data deviate from the straight line in accordance with theoretical arguments presented by Starov [77]. These theoretical predictions were also confirmed in the study by Tiberg et al. [35], in which the spontaneous increase of $C_{10}E_6$ and $C_{14}E_6$ aqueous solutions into hydrophobic cylindrical capillaries was reported. Tiberg et al. [35] showed that during the initial stage of the process, the position of the moving meniscus, $z(t)$, was described by $z(t) \sim \sqrt{t}$ as predicted by Churaev et al. [34]. In the final stage, the meniscus moves toward the equilibrium level as predicted by Starov [77]. It was also observed by Tiberg et al. [35] that at the same concentrations, $C_{10}E_6$ solutions increase faster than $C_{14}E_6$ solutions. The velocity of the capillary rise was also considered in the study by Bain [81].

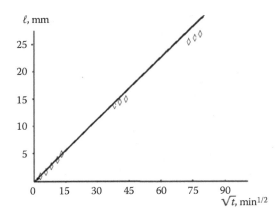

FIGURE 5.9 Spontaneous capillary rise in a vertical hydrophobized quartz capillary (R = 11 μm), Syntamide-5 surfactant solution (C = 0.1%). Time evolution of the imbibition length. Symbols, experimental points; solid line, according to the theory for a horizontal capillary. (Redrawn from V.M. Starov, *J. Colloid Interface Sci.*, 270, 180, 2004.)

5.4 SPREADING OF SURFACTANT SOLUTIONS OVER THIN AQUEOUS LAYERS: INFLUENCE OF SOLUBILITY AND MICELLE DISINTEGRATION

Thin liquid films can be found in many engineering, geological, and biophysics environments. Their applications are significant in many coating processes [82,83] and physiological applications [84]. The presence of nonuniform temperature or surface active compounds across thin liquid films leads to the formation of shear stresses, also known as Marangoni gradients, at the liquid–vapor interface. These gradients cause mass transfer on and in the liquid layer due to surface tension nonuniformity. Marangoni stresses transport the liquid from areas of low surface tension to areas of high surface tension (flow generation) and, in doing so, also deform the interface resulting in height variations (deformation and even possible instability of liquid films). In this section, we restrict our discussion to the influence of surfactants on thin liquid films.

An understanding of Marangoni-induced flows is important as these can be beneficial or detrimental in different applications. Surfactants are normally present in a healthy mammalian lung to reduce surface tension forces, which keeps the lungs compliant and prevents collapse of the small airways during exhalation. However, most prematurely born babies do not produce adequate amounts of these surfactants, which leads to respiratory distress syndrome. This condition is treated by surfactant replacement therapy in which surfactants are introduced into the lungs. These surfactants spread by other forces in the large to medium pulmonary airways. In small airways, surface tension gradients dominate and Marangoni flow transports the surfactant to the distal regions of the lung [85,86].

A problem in paint coating processes, in which paint is dried by solvent evaporation, is that the nonuniformity of evaporation leads to Marangoni stresses that cause deformation in the film and, hence, permanent defects on the paint's surface [87,88]. Another application of the Marangoni effect is its use for drying silicon wafers after a wet processing step during the manufacturing of integrated circuits. An alcohol vapor is blown through a moving nozzle over the wet wafer surface and the subsequent Marangoni effect will cause the liquid on the wafer to pull itself off the surface—effectively leaving a dry wafer surface [89].

Spreading of surfactant solutions on thin liquid films was reviewed by Afsar-Siddiqui et al. [90] a few years ago. Below, we summarize the progress made in recent years. A moving circular wave front forms after a small droplet of aqueous surfactant solution is deposited on a thin aqueous layer as illustrated in Figure 5.10. The time evolution of the moving front radius was monitored [91], in

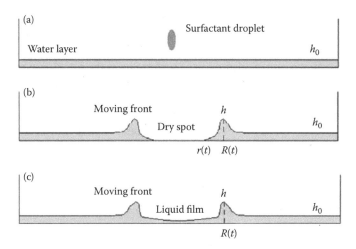

FIGURE 5.10 (a) A small droplet of aqueous surfactant solution is deposited on top of a thin aqueous layer of thickness h_0; (b) dry spot formation in the center; $R(t)$ is the radius of the circular moving front, $r(t)$ is the radius of the dry spot in the center, and H is the height of the moving front; (c) the same as in case b without a dry spot formation in the center. (Redrawn from Lee, K.S., and V.M. Starov, *J. Colloid Interface Sci.*, 314, 631, 2007.)

which an experimental methodology was designed to investigate the influence of the Marangoni force on spreading of surfactant solutions over thin aqueous layers. Surfactants with different solubilities at concentrations above the cmc were considered. In all cases [91], two stages of the front motion were observed: the first fast stage, followed by a slower second stage. Both the first and the second stages of spreading considerably depend on the solubility of surfactants: the higher the solubility, the slower the stages. If the solubility is high enough, then during the second stage, the front reaches the final position and does not move any further. The lower the solubility, the higher the exponent in the spreading law $R(t) = \text{const} \cdot t^n$ during the first stage and for low-solubility surfactants, it reaches approximately 0.75. Moreover, it was shown that the formation of a dry spot in the center is determined by the speed of the first stage: the higher this speed, the lower the probability for a dry spot formation. Hence, the dry spot forms in the case of soluble surfactants and does not form in the case of insoluble surfactants.

The observations by Lee and Starov [91] differed from earlier theoretical approaches [92], thus the influence of surfactant solubility and disintegration of micelles were incorporated to improve on the previous theoretical model. According to the theoretical predictions of Lee and Starov [91], low-solubility surfactant produced both faster first and second stages. Low-solubility surfactant (Tween 20, Acros Organics UK) produced a power law exponent of 0.73 ± 0.01, which was closest to the maximum attainable exponent of 0.75 predicted theoretically [91]. For the highly soluble surfactant, dodecyl trimethylammonium bromide (DTAB), the solubility was most significant during the second stage, where the spreading front reached its final position and did not move any further in agreement with the theoretical predictions [91].

5.5 CONCLUSIONS

Wetting and spreading phenomena occurring in the presence of nonionic surfactants on hydrophobic solids and thin aqueous layers are reviewed. Spontaneous increase and penetration of aqueous surfactant solutions into hydrophobized capillaries are discussed. Special attention is paid to wetting/spreading/penetration of nonionic hydrocarbon and organosilicone surfactants that are widely used in industrial and scientific applications. It is shown that the spreading of aqueous surfactant solutions on hydrophobic solids and penetration of the solutions into hydrophobic capillaries exhibit mostly two-stage dynamics with different rates: a first fast stage, followed by a second slower stage. The reason for two-stage dynamics is that the spreading of surfactant solutions is determined by the competition between the rate of adsorption of surfactants at the liquid–vapor, liquid–solid, and solid–vapor interfaces and the rate of depletion of molecules due to expansion of these interfaces in the course of spreading. The slow spreading stage is determined by the adsorption of surfactant monomers on the hydrophobic solid–vapor interface in front of the moving TPC line. Because the surfactant molecule transfer takes place via a relatively high potential barrier this occurs considerably slower compared with adsorption processes on the liquid–vapor and the liquid–solid interfaces.

In the case of spreading over thin aqueous layers, the rates of both stages are influenced by the solubility of surfactants and the rate of disintegration of micelles: the higher the solubility is, the slower both stages are.

The phase diagram and structure of surfactant aggregates in the bulk of the solution, the physicochemical properties of substrates, and a number of other properties could strongly influence the spreading and wetting dynamics of surfactant solutions.

The area of spreading and wetting phenomena is vast and thus it is impossible to provide a comprehensive review in one chapter. This is the reason that the authors concentrated mostly on their own research interests. The reader is referred to the following books [7,16,23,25] and recent reviews [17,20,93] for more information.

ACKNOWLEDGMENTS

We acknowledge the support from Engineering and Physical Sciences Research Council, UK (grant EP/D077869/1), MULTIFLOW project, EU Marie Curie and PASTA project, and the European Space Agency.

REFERENCES

1. R.G. Cox. The dynamics of the spreading of liquids on a solid surface. Part 1. Viscous flow. *J. Fluid Mech.*, **168**, 169 (1986).
2. O.V. Voinov. Hydrodynamics of wetting. *Fluid Dyn.*, **11**, 714 (1976).
3. T.D. Blake and J.M. Haynes. Kinetics of liquid–liquid displacement. *J. Colloid Interface Sci.*, **30**, 421 (1969).
4. E. Chibowski. On some relations between advancing, receding and Young's contact angles. *Adv. Colloid Interface Sci.*, **133**, 51 (2007).
5. D. Exerowa and P. Krugliakov. *Foam and Foam Films: Theory, Experiment, Application.* Studies in Interface Science, Vol. 5. Elsevier, New York (1988).
6. N. Churaev and Z. Zorin. Wetting films. *Adv. Colloid Interface Sci.*, **40**, 109 (1992).
7. V. Starov, M. Velarde and C. Radke. *Wetting and Spreading Dynamics.* Surfactant Sciences Series, Vol. 138. Taylor & Francis, Boca Raton, FL (2007).
8. A. Otten and S. Herminghaus. How plants keep dry: A physicist's point of view. *Langmuir*, **20**, 2405 (2004).
9. C. Muller and M. Reiderer. Plant surface properties in chemical ecology. *J. Chem. Ecol.*, **31**, 2621 (2005).
10. S. Wu. *Polymer Interface and Adhesion.* Marcel Dekker, New York, p. 88 (1982).
11. J.W. Adams. Silicones in the coating industry: Flow, leveling and defoaming. In *Surface Phenomena and Additives in Water-Based Coatings and Printing Technology*, M.K. Sharma (Ed.). Plenum Press, New York, p. 23 (1991).
12. L.L. Schramm. *Emulsions, Foams and Suspensions. Fundamentals and Applications.* Wiley-VCH, Weinheim, Germany, pp. 233–344 (2005).
13. H. Gecol, J.F. Scamehorn, S.D. Christian, B.P. Grady and F. Riddell. Use of surfactants to remove water based inks from plastic films. *Colloids Surf. A*, **189**, 55 (2001).
14. D. Penner, R. Burow and F. Roggenbuck. Use of organosilicone surfactants as agrichemical adjuvants. In *Silicone Surfactants*, R.M. Hill, (Ed.). Marcel Dekker, New York, pp. 241–258 (1999).
15. R.M. Hill. Silicone surfactants–new developments. *Curr. Opin. Colloid Interface Sci.*, **7**, 255 (2002).

16. T.F. Tadros. *Applied Surfactants: Principles and Applications*. Wiley-VCH, Weinheim, Germany (2005).
17. V. Starov, N. Ivanova and R.G. Rubio. Why do aqueous surfactant solutions spread over hydrophobic substrates? *Adv. Colloid Interface Sci.*, **161**, 153 (2010).
18. V.M. Starov, S.R. Kosvintsev and M.G. Velarde. Spreading of surfactant solutions over hydrophobic substrates. *J. Colloid Interface Sci.*, **227**, 185 (2000).
19. N. Ivanova, V. Starov, D. Johnson, N. Hilal and R. Rubio. Spreading of aqueous solutions of trisiloxanes and conventional surfactants over PTFE AF coated Siliconwafers. *Langmuir*, **25**, 3564 (2009).
20. J. Venzmer. Superspreading—20 years of physicochemical research. *Curr. Opin. Colloid Interface Sci.*, **16**, 335 (2011).
21. J. Drelich, R. Zahn, J.D. Miller and J.K. Borchardt. Contact angle relaxation for ethoxylated alcohol solution on hydrophobic surfaces. In *Contact Angle, Wettability and Adhesion*, Vol. 2, K.L. Mittal (Ed.). VSP, Utrecht, pp. 253–264.
22. C.W. Extrand. Water contact angles and hysteresis of polyamide surfaces. *J. Colloid Interface Sci.*, **248**, 136 (2002).
23. W. Zisman. Contact Angle. Wettability and Adhesion. *Adv. Chem. Series*, Vol. 43. American Chemical Society, Washington, DC (1964).
24. Y. Ma, X. Cao, X. Feng, Y. Ma and H. Zou. Fabrication of super-hydrophobic film from PMMA with intrinsic water contact angle below 90°. *Polymer*, **48**, 7455 (2007).
25. R.M. Hill. Siloxane surfactants. In *Silicone Surfactants*, R.M. Hill (Ed.). Marcel Dekker, New York, pp. 1–45 (1999).
26. R.M. Hill. Superspreading. *Curr. Opin. Colloid Interface Sci.*, **3**, 247 (1998).
27. T.E. Gentle and S.A. Snow. Adsorption of small silicone polyether surfactants at the air/water interface. *Langmuir*, **11**, 2905 (1995).
28. T. Stoebe, Z. Lin, R.M. Hill, M.D. Ward and H.T. Davis. Surfactant-enhanced spreading. *Langmuir*, **12**, 337 (1996).
29. T. Stoebe, R.M. Hill, M.D. Ward and H.T. Davis. Enhanced spreading of aqueous films containing ethoxylated alcohol surfactants on solid substrates. *Langmuir*, **13**, 7270 (1997).
30. V. Dutschk, B. Breitzke and K. Grundke. Wetting of aqueous surfactant solutions on polymer surfaces. *Tenside Surf. Det.*, **40**, 250 (2003).
31. V. Dutschk and B. Breitzke. Spreading characteristics of aqueous surfactant solutions on polymer surfaces. *Tenside Surf. Det.*, **42**, 1 (2005).
32. V. Dutschk, K.G. Sabbatovskiy, M. Stolz, K. Grundke and V.M. Rudoy. Unusual wetting dynamics of aqueous surfactant solutions on polymer surfaces. *J. Colloid Interface Sci.*, **267**, 456 (2003).
33. C.C. Wei Ping, N.A. Ivanova, V.M. Starov, N. Hilal and D. Johnson. Spreading behaviour of aqueous trisiloxane solutions over hydrophobic polymer substrates. *Colloid J.*, **71**, 391 (2009).
34. N.V. Churaev, G.M. Martynov, V.M. Starov and Z.M. Zorin. Some features of capillary imbition of surfactant solutions. *Colloid Polym. Sci.*, **259**, 747 (1981).
35. F. Tiberg, B. Zhmud, K. Hallstensson and M. von Bahr. Capillary rise of surfactant solutions. *Phys. Chem. Chem. Phys.*, **2**, 5189 (2000).
36. M. von Bahr, F. Tiberg and B. Zhmud. Spreading dynamics of surfactant solutions. *Langmuir*, **15**, 7069 (1999).
37. M. von Bahr, F. Tiberg and V. Yaminsky. Spreading dynamics of liquids and surfactant solutions on partially wettable hydrophobic substrates. *Colloids Surf. A*, **193**, 85 (2001).
38. T. Svitova, R.M. Hill and C.J. Radke. Adsorption layer structures and spreading behavior of aqueous non-ionic surfactants on graphite. *Colloids Surf. A*, **183–185**, 607 (2001).
39. L.H. Tanner. The spreading of silicone oil drops on horizontal surfaces. *J. Phys. D.*, **12**, 1473 (1979).
40. T.D. Blake and J.M. Haynes. Kinetics of liquid/liquid displacement. *J. Colloid Interface Sci.*, **30**, 421 (1969).
41. V.M. Starov, V.V. Kalinin and J.-D. Chen. Spreading of liquid drops over dry surfaces. *Adv. Colloid Interface Sci.*, **50**, 187 (1994).
42. V.M. Starov, S.R. Kosvintsev, V.D. Sobolev, M.G. Velarde and S.A. Zhdanov. Spreading of liquid drops over saturated porous layers. *J. Colloid Interface Sci.*, **246**, 372 (2002).
43. D.P. Gaver and J.B. Grotberg. The dynamics of a localized surfactant on a thin-film. *J. Fluid Mech.*, **213**, 127 (1990).
44. O.E. Jensen and J.B. Grotberg. Insoluble surfactant spreading on a thin viscous film: Shock evolution and film rupture. *J. Fluid Mech.*, **240**, 259 (1992).
45. N. Kumar, K. Varanasi, R.D. Tilton and S. Garoff. Surfactant self-assembly ahead of the contact line on a hydrophobic surface and its implications for wetting. *Langmuir*, **19**, 5366 (2003).

46. Y. Zhang and F. Han. The spreading behaviour and spreading mechanism of new glucosamide-based trisiloxane on polystyrene surfaces. *J. Colloid Interface Sci.*, **337**, 211 (2009).

47. S. Rafaï, D. Sarker, V. Bergeron, J. Meunier and D. Bonn. Spreading: Aqueous surfactant drops spreading on hydrophobic surfaces. *Langmuir*, **18**, 10486 (2002).

48. J. Radulovic, K. Sefiane and M.E.R. Shanahan. On the effect of pH on spreading of surfactant solutions on hydrophobic surfaces. *J. Colloid Interface Sci.*, **332**, 497 (2009).

49. J. Radulovic, K. Sefiane and M.E.R. Shanahan. Spreading and wetting behavior of trisiloxanes. *J. Bionic. Eng.*, **6**, 341 (2009).

50. R. Wagner, Y. Wu, H.V. Berlepsch and L. Perepelittchenko. Silicon-modified surfactants and wetting: IV. Spreading behavior of trisiloxane surfactants on energetically different solid surfaces. *Appl. Organometal. Chem.*, **14**, 177 (2000).

51. T. Stoebe, R.M. Hill, M.D. Ward and H.T. Davis. Enhanced spreading of aqueous films containing ionic surfactants on solids substrates. *Langmuir*, **13**, 7276–7281 (1997).

52. N. Ivanova, V. Starov, R. Rubio, H. Ritacco, N. Hilal and D. Johnson. Critical wetting concentrations of trisiloxane surfactants. *Colloids Surf. A*, **354**, 143 (2010).

53. H.A. Ritacco, F. Ortega, R.G. Rubio, N. Ivanova and V.M. Starov. Equilibrium and dynamic surface properties of trisiloxane aqueous solutions. Part 1. Experimental results. *Colloids Surf. A*, **365**, 199 (2010).

54. E. Sieverding, G.D. Humble and I. Fleute-Schlachter. A new herbicide adjuvant based on a non-super-spreading trisiloxane surfactant. *J. Plant Dis. Prot.*, **20**, 1005 (2006).

55. E.G. Schwarz and W.G. Reid. Surface-active agents—Their behavior and industrial use. *Indus. Eng. Chem.*, **56**, 26 (1994).

56. K.P. Ananthapadmanabhan, E.D. Goddard and P. Chandar. A study of the solution, interfacial and wetting properties of silicone surfactants. *Colloids Surf. A*, **44**, 281 (1990).

57. F. Tiberg and A.-M. Cazabat. Self-assembly and spreading of non-ionic trisiloxane surfactants. *Europhys. Lett.*, **25**, 205 (1994).

58. N. Kumar, A. Couzis and C. Maldarelli. Measurement of the kinetic rate constants for the adsorption of superspreading trisiloxanes to an air/aqueous interface and the relevance of these measurements to the mechanism of superspreading. *J. Colloid Interface Sci.*, **267**, 272 (2003).

59. H.A. Ritacco, V.B. Fainerman, F. Ortega, R.G. Rubio, N. Ivanova and V.M. Starov. Equilibrium and dynamic surface properties of trisiloxane aqueous solutions. Part 2. Theory and comparison with experiment. *Colloids Surf. A*, **365**, 204 (2010).

60. J. Eastoe and J.S. Dalton. Dynamic surface tension and adsorption mechanisms of surfactants at the air-water interface. *Adv. Colloid Interface Sci.*, **85**, 103 (2000).

61. C.-H. Chang and E.I. Frances. Adsorption dynamics of surfactants at the air/water interface: A critical review of mathematical models, data, and mechanisms. *Colloids Surf. A*, **100**, 1 (1995).

62. E. Ruckenstein. Effect of short-range interactions on spreading. *J. Colloid Interface Sci.*, **179**, 136 (1996).

63. H.N. Patrick, G.G. Warr, S. Manne and I.A. Aksay. Self-assembly structures of nonionic surfactants at graphite/solution interfaces. *Langmuir*, **13**, 4349 (1997).

64. N.A. Ivanova, Zh. Zhantenova and V.M. Starov. Wetting dynamics of polyoxyethylene alkyl ethers and trisiloxanes in respect of polyoxyethylene chains and properties of substrates. *Colloids Surf. A*, **413**, 307 (2012).

65. J. Radulovic, K. Sefiane and M.E.R. Shanahan. Ageing of trisiloxane solutions. *Chem. Eng. Sci.*, **65**, 5251 (2010).

66. Zh. Peng, Q. Wu, T. Caic, H. Gaob and K. Chena. Syntheses and properties of hydrolysis resistant twin-tail trisiloxane surfactants. *Colloids Surf. A*, **342**, 127 (2009).

67. S. Zhu, W.G. Miller, L.E. Scriven and H.T. Davis. Superspreading of water-silicone surfactant on hydrophobic surfaces. *Colloids Surf. A*, **90**, 63 (1994).

68. R. Wagner, Y. Wu, G. Czichocki, H.V. Berlepsch, F. Rexin and L. Perepelittchenko. Silicon-modified surfactants and wetting: II. Temperature-dependent spreading behaviour of oligoethylene glycol derivatives of heptamethyltrisiloxane. *Appl. Organometal. Chem.*, **13**, 201 (1999).

69. M. He, R.M. Hill, Z. Lin, L.E. Scriven and H.T. Davis. Comparison of the liquid crystal phase behaviour of four trisiloxane superwetter surfactants. *J. Phys. Chem.*, **97**, 8820 (1993).

70. J. Venzmer and S.P. Wilkowski. Trisiloxane surfactants—Mechanisms of spreading and wetting. In *Pesticide Formulation and Application Systems*, Vol. 18, ASTM STP 1347, J.D. Nalewaja, G.R. Goss and R.S. Tann (Eds.). American Society for Testing and Materials, West Conshohocken, Philadelphia, USA, pp. 140–151 (1998).

71. Z. Lin, R.M. Hill, T. Davis and M.D. Ward. Determination of wetting velocities of surfactant super-spreaders with the quartz crystal microbalance. *Langmuir*, **10**, 4060 (1994).

72. V. Starov and K. Sefiane. On evaporation rate and interfacial temperature of volatile sessile drops. *Colloids Surf. A*, **333**, 170 (2009).

73. E.W. Washburn. The dynamics of capillary flow. *Phys. Rev.*, **17**, 273 (1921).

74. B.V. Zhmud, F. Tiberg and K. Halstensson. Dynamics of capillary rise. *J. Colloid Interface Sci.*, **228**, 263 (2000).

75. T. Laurent, L. Thierry and N. Liviu. Dynamic spreading of a liquid finger driven by electrowetting: Theory and experimental validation. *J. Appl. Phys.*, **101**, 044907 (2007).

76. T.S. Sammarco and M.A. Burns. Thermocapillary pumping of discrete drops in microfabricated analyses devices. *AIChE J.*, **45**, 350 (1999).

77. V.M. Starov. Spontaneous rise of surfactant solutions into vertical hydrophobic capillaries. *J. Colloid Interface Sci.*, **270**, 180 (2004).

78. N. Churaev and Z. Zorin. Penetration of aqueous surfactant solutions into thin hydrophobized capillaries. *Colloids Surf. A*, **100**, 131 (1995).

79. P.P. Zolotarev, V.M. Starov and N.V. Churaev. Kinetics of the absorption of solutions of surfactants in capillaries. 2. Theory. *Colloid J.*, **38**, 895 (1976).

80. V.M. Starov. Surfactant solutions and porous substrates: Spreading and imbibitions. *Adv. Colloid Interface Sci.*, **111**, 3 (2004).

81. C.D. Bain. Penetration of surfactant solutions into hydrophobic capillaries. *Phys. Chem. Chem. Phys.*, **7**, 3048 (2005).

82. L.W. Schwartz, D.E. Weidner and R.R. Eley. Surface-tension-gradient effects on flow in thin coating layers. *Polym. Mater. Sci. Eng.*, **73**, 490 (1995).

83. J. Patzer, J. Fuchs and E.P. Hoffer. Surface tension effect in bubble-jet printing. *Proc. SPIE*, **167**, 2413 (1995).

84. H.P. Lee. Progress and trends in ink-jet printing technology. *J. Imaging Sci. Technol.*, **42**, 49 (1998).

85. D.L. Shapiro and R.H. Notter (Eds.). *Surfactant Replacement Therapy*. A.R. Liss, New York (1989).

86. W.-T. Tsai and L.-Y. Liu. Transport of exogenous surfactants on a thin viscous film within an axisymmetric airway. *Colloids Surf. A*, **234**, 51 (2004).

87. G.P. Bierwagen. Surface defects and surface flows in coatings. *Prog. Org. Coat.*, **19**, 59 (1991).

88. V.R. Gundabala and A.F. Routh. Thinning of drying latex films due to surfactant. *J. Colloid Interface Sci.*, **303**, 306 (2006).

89. I.-S. Chang and J.-H. Kim. Development of clean technology in wafer drying processes. *J. Clean. Prod.*, **9**, 227 (2001).

90. A.B. Afsar-Siddiqui, P.F. Luckham and O.K. Matar. The spreading of surfactant solutions on thin liquid films. *Adv. Colloid Interface Sci.*, **106**, 183 (2003).

91. K.S. Lee and V.M. Starov. Spreading of surfactant solutions over thin aqueous layers: Influence of solubility and micelles disintegration. *J. Colloid Interface Sci.*, **314**, 631 (2007).

92. V.M. Starov, A. de Ryck and M.G. Velarde. On the Spreading of an Insoluble Surfactant over a Thin Viscous Liquid Layer. *J. Colloid Interface Sci.*, **190**, 104 (1997).

93. N.A. Ivanova and V.M. Starov. Wetting of low free energy surfaces by aqueous surfactant solutions. *Curr. Opin. Colloid Interface Sci.*, **16**, 285 (2011).

6 Wetting Instabilities in Langmuir–Blodgett Film Deposition

Volodymyr I. Kovalchuk, Emiliy K. Zholkovskiy,
Mykola P. Bondarenko, and Dieter Vollhardt

CONTENTS

6.1 INTRODUCTION

Surfactants (surface active agents) are widely spread in nature, in industry, and in our everyday life. From colloid and interface science, we know that surfactants can most frequently be found at the interfaces between two contacting media. By modifying the surface properties, the properties of the whole system or process are modified. Therefore, the number of surfactant applications are too numerous. One such application is in the formation of multilayer films and coatings.

The method to produce thin molecular films with highly ordered layer structure and precisely controlled thickness was developed by Irving Langmuir and Katharine Blodgett in the 1930s and, since that time, it has continued to attract scientific attention. Such films composed of oriented amphiphilic molecules, and named Langmuir–Blodgett (LB) films, have been studied as antireflection coatings, simple models of biological membranes, substrates for biologically active molecules, thin light-conductive films, hydrophobic/hydrophilic substrates for wetting studies, model systems in tribology, systems with specific optical and electronic properties, and so on.

In recent times, with the help of modern techniques, the internal structure and peculiarities of morphology of LB films have been studied in detail. These studies confirmed the existence of various internal defects in many LB films. Under certain experimental conditions, striped structures have been observed, which were also initially considered as "defects." It was later realized that such striped structures could be used to create patterned surfaces with specific physical and chemical properties. Surface patterning has recently been developed into an important branch of nanoscience due to a large number of applications in which patterned surfaces with nanosized features could be used.

Microcontact printing, various lithographic and other techniques, based on the so-called top-down strategy, demonstrate a substantial progress in producing patterns with enhanced spatial resolution. These techniques, however, have become more and more complicated and more expensive.

In this respect, the bottom-up approach based on the formation of transfer-induced regular patterns during wetting/dewetting processes attracts more and more attention. This simple, low-cost, and highly efficient method is a promising alternative to the top-down approach. It allows a rapid (in a few minutes) covering of macroscopic areas, as large as several square centimeters, using regular patterns with lateral resolution in the nanometer range. This method allows a broad spectrum of patterns with various geometric, physical, and chemical properties. Also, an advantage of this method is that it can be used for deposition and patterning of very different materials including fragile (e.g., organic) materials.

Various patterned structures can be formed spontaneously at the three-phase contact line during the dewetting process, in particular, the LB transfer process. The structures are formed due to instabilities of the wetting meniscus, which generate inhomogeneous transfer of the material to the substrate surface. The meniscus dynamics strongly depends on the interactions in the region close to the contact line, as well as on different physical and chemical processes induced due to contact line motion. In particular, molecular, electrical, and other interactions; hydrodynamic flows, mass transfer processes, etc., can all affect the meniscus dynamics. Understanding these complicated interactions and processes is an important task that has to be solved to understand the effect of various experimental parameters on the properties of the patterned structures obtained.

6.2 EARLY STUDIES ON LB FILMS DEPOSITION AND THEIR FURTHER DEVELOPMENTS

Agnes Pockels was the first to realize that a simple trough with moving barriers was a very effective tool for studying the properties of free liquid interfaces [1]. Later, Irving Langmuir invented a much more sophisticated trough suitable for precise quantitative experiments and gave an explanation as to what happens at liquid interfaces on a molecular level [2]. Langmuir showed how adsorbed molecules are arranged at liquid interfaces and how molecular interactions are reflected in the properties of monolayers. By careful analysis of the available experimental data, he showed that amphiphilic substances at the air–water interface can form monomolecular layers in which the molecules have a specific orientation with hydrophilic groups oriented toward the aqueous subphase and hydrophobic parts exposed to air. This allowed him to estimate the size of hydrophilic groups and the length of hydrophobic chains in typical surfactant molecules. He also showed that CH_2 groups in the alkyl chains are arranged not in a line but in a zigzag manner. The discoveries made by Langmuir were an important stage in understanding the molecular nature of substances and brought him the Nobel Prize in Chemistry.

According to Langmuir, the monolayers formed at aqueous interfaces can be considered as two-dimensional (2-D) fluid phases, which can be easily compressed and expanded. Compression or expansion of the monolayer in a trough by simultaneous monitoring of the changes in interfacial tension can provide important quantitative information about molecular interactions and arrangements. In particular, 2-D phase transitions can be observed within interfacial layers by their compression/expansion. It seems that thermodynamic methods are also applicable to 2-D surface phases similar to three-dimensional phases.

In the early experiments of Langmuir and other scientists, it was shown that amphiphilic monolayers could be removed from a liquid subphase due to adhesion at the surface of a solid body when the last is withdrawn through a liquid interface containing the monolayer. Later, Katharine Blodgett showed that monolayers formed at aqueous subphase could be transferred to a solid substrate in a regular manner, forming multilayer films with a stepwise increase of thickness at each submersion/ withdrawal cycle of the substrate [3,4]. Now, such multilayer films deposited at a solid substrate are called LB films, in contrast to Langmuir films, which are monolayers floating at a liquid surface.

The technique of multilayer films production developed by Langmuir and Blodgett allows flexible control of the films' properties by choosing suitable monolayer material, substrate, and operative conditions [5–8]. This explains the high interest in such films in science and technology [9,10].

Many potential applications have been proposed for LB films, in particular, in molecular optics, electronic devices, and sensor applications. Also, LB films have proven to be useful model systems for many studies in different fields of science.

Since Langmuir's famous experiments, the trough with movable barriers and film balance (or Wilhelmy plate) remain indispensable instruments for studies of monomolecular films at liquid interfaces. The surface pressure–area (Π–A) isotherm represents a specific signature of each monolayer, which is determined by the chemical structure of the molecules forming the monolayer. Gaines [11], in his 1966 book, summarized the experimental results for many of the insoluble monolayers. Since that time, the amount of experimental results in this area has increased manifolds.

Very generalized Π–A isotherms (Figure 6.1) are helpful to understand the phase behavior of amphiphilic monolayers [12]. Three characteristic types of generalized Π–A isotherms representative for a certain temperature region ($T_1 < T_2 < T_3$) can be measured. Overlooking some of the details, two single phases of Langmuir monolayers can generally be distinguished, namely, the fluid phase (G) (gaseous, liquid-expanded [LE]) and the condensed phase (C) (liquid-condensed [LC], solid [S]). The experimental Π–A isotherms can indicate a further condensed/condensed phase transition according to a small plateau region (first order) or a kink in the steep part of the isotherm.

The most interesting type T_2 of Π–A isotherms shows a horizontal or nonhorizontal plateau region after a break point in the isotherm characteristic of a first-order main phase transition from the fluid phase to the condensed phase. The plateau represents the main fluid/condensed phase transition region, for example, L-α-dipalmitoylphosphatidylcholine (DPPC) monolayer spread on water at 20°C. The lower the temperature, the lower the surface pressure of the plateau region and the more extended the molecular area of the plateau region is. The dashed line with an apex for the fluid/condensed coexistence region indicates that with decreasing temperature, the surface pressure of the fluid/condensed coexistence (plateau) region also decreases but the area range for the coexisting phases becomes more extended.

At temperatures low enough and with increasing alkyl chain length of the amphiphiles, the two-phase coexistence region already exists at a surface pressure of $\Pi \sim 0$, that is, this is the general isotherm type T_1, for example, arachidic acid monolayer spread on 0.25 mmol/L CdCl$_2$ aqueous solution of pH 5.7 at 20°C.

On the other hand, at high enough temperatures, a continuous increase of the surface pressure with decreasing area per molecule suggests that for this isotherm type T_3, phase transition to the condensed phases does not occur over the whole area range. This third isotherm type T_3, which

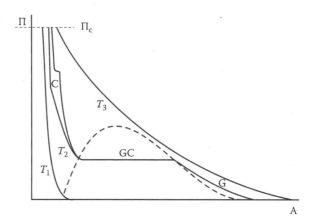

FIGURE 6.1 Generalized Π–A isotherms with representative examples for three temperatures $T_1 < T_2 < T_3$. Symbols of 2-D phases: G, fluid phase (gaseous and/or LE); C, condensed (LC and/or S); GC, fluid/condensed coexistence region; Π_c, collapse pressure. See the text for an explanation of the phase characteristics.

indicates a fluid phase (G) over the whole surface pressure range up to the collapse pressure Π_c, is not of interest for studies of the condensed phase characteristics of monolayers and LB films.

During the last few decades, powerful new techniques, such as Brewster angle microscopy, fluorescence microscopy, synchrotron x-ray diffraction and scattering, neutron reflection, and others have become available. By combining these techniques with Π–A isotherm measurements, they allow much deeper insight into the processes involved in Langmuir monolayers. They have allowed, in particular, the study of the texture and morphology of monolayers and their changes during 2-D phase transitions induced by the monolayer compression or expansion. The size and shape of the condensed phase domains have been deeply investigated depending on the chemical structure of the molecules in the monolayer [12].

The properties and the state of a monolayer being transferred to a solid substrate are very important for the stability of the deposition process and for the properties of the LB films obtained. On the other hand, how a monolayer is deposited depends crucially on the interactions of the molecules forming the monolayer with the substrate surface. Due to these interactions, the local monolayer state (structure, composition, and interfacial energies) in the three-phase line region are modified with respect to the flat water surface far from the substrate [13–18]. Because of the relatively short range of the interfacial forces, these interactions manifest themselves only within a close vicinity to the three-phase contact line. A variety of effects can be associated with such interactions, in particular, the complex behavior of the dynamic contact angle; changes in the LB film composition, structure, and morphology; changes in the monolayer transfer ratio; and the existence of a maximum deposition rate. Such effects are explained through molecular interactions and reorganization [19–26], hydrodynamic processes [24,25,27–30], adsorption–desorption processes [29,30], phase transitions [13–16], specific ion exchange processes, and electrostatic interactions [17]. Understanding these effects is necessary for improving the quality of LB coverage and for the production of LB films with the required specific properties.

In this respect, the contact angle formed during the substrate withdrawal seems to be very significant. For successful ("dry") deposition, large contact angles are necessary for rapid expulsion of water from the three-phase contact zone. Langmuir called such angles as "zipper angles" [8]. At small or zero contact angles, the entrainment of the water film does not allow stable multilayer formation.

The rate of water drainage defines the maximum deposition rate of the monolayer, U_{max} [28,31]. For faster water drainage and, correspondingly, larger maximum deposition rate, stronger adhesion between the monolayer and the substrate surface is necessary. As follows from general thermodynamic relationships, the larger contact angles (on the side of the aqueous phase; Figure 6.2) correspond to larger work of adhesion of the monolayer [32,33]. For the withdrawal mode, the contact angle is related to the work of adhesion of the hydrophilic heads of the monolayer to the hydrophilic head groups of the previously deposited monolayer as

$$W = \gamma(1 - \cos \theta_e) \tag{6.1}$$

where γ is the surface tension, and θ_e is the equilibrium contact angle, measured through the aqueous phase (Figure 6.2). Note, that the work of adhesion in Equation 6.1 is related to the monolayer adhesion at the substrate surface [33] and not to the adhesion of water.

The contact angle formed during the Langmuir monolayer deposition defines the hydrodynamics within the meniscus region and the maximum deposition rate, U_{max} (Figure 6.2). Using lubrication approximation, de Gennes showed that steady state deposition can be realized only if the velocity of the substrate movement is lower than a certain threshold [28]. For small contact angles, this threshold is given by the equation

$$U_{max} = \frac{\gamma \theta_e^3}{36\sqrt{3}\eta l} \tag{6.2}$$

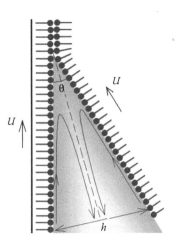

FIGURE 6.2 **(See color insert.)** Convective flow in the wedgelike region near the contact line between the floating monolayer and the substrate surface. U is the velocity at the surfaces, θ is the contact angle, h is the local thickness of the film formed between the substrate surface and the floating monolayer. (Reprinted from *Adv. Colloid Interface Sci.,* 168, Kovalchuk, V.I., E.K. Zholkovskiy, M.P. Bondarenko, V.M. Starov and D. Vollhardt. Concentration polarization effect at the deposition of charged Langmuir monolayers, 114–123. Copyright 2011, with permission from Elsevier.)

where η is the dynamic viscosity of the liquid, and $l \approx 12$, is a parameter that can be considered approximately as a constant. The de Gennes equation shows that the maximum velocity strongly depends on the equilibrium contact angle. If the substrate velocity exceeds U_{max}, the water film is entrained between the substrate and the deposited monolayer [28,31]. The maximum deposition speed is also predicted by a combined molecular–hydrodynamic approach proposed by J.G. Petrov and G.P. Petrov [29,30].

At constant withdrawal speed, a stationary acute contact angle is usually formed between the floating monolayer and the substrate surface, which is smaller than the static (equilibrium) contact angle, $\theta < \theta_e$ [20,22,23,25–27,33,35]. When the monolayer or the substrate surface are charged, then both the work of adhesion and the contact angle depend on the ionic subphase composition near the three-phase contact line, in particular on the subphase pH [20,22,23,25,26,35]. As will be shown further, during the monolayer transfer to the substrate, the local ion concentration will change. This should result in the variation of contact angle, work of adhesion, and maximum deposition rate.

Petrov and coauthors [31] experimentally determined the maximum deposition rate for monolayers of arachidic acid and cadmium arachidate at different ionic compositions in the subphase. They found that arachidic acid monolayers at pH 2 can be deposited at a relatively high substrate velocity of 1.3 cm/s. Cadmium arachidate monolayers at pH 6.8 in the presence of 3×10^{-4} M Cd^{2+} ions can also be deposited at a sufficiently high velocity of 0.6 cm/s. However, arachidic acid monolayers at pH = 6.8 in the absence of Cd^{2+} ions cannot be deposited even at small substrate velocities such as 0.005 cm/s. The authors explained these results as the difference in adhesion (reactivities), that is, interaction forces between the monolayer head groups and the solid surface for different monolayer states that depend on the subphase composition.

Taking this into account, one can summarize that monolayer state, work of adhesion, contact angle, and maximum deposition rate are interrelated. All of them are dependent on the subphase composition. Changes in the local subphase and monolayer composition induced within the meniscus region during the deposition process should be accompanied by changes in all these characteristics. For example, a change in the monolayer composition from cadmium salt to acid form can lead to a decrease in the work of adhesion at higher pH, and this results in a decrease of the contact angle. The latter should lead to a decrease in the maximum deposition rate, which can become lower than the velocity of the substrate movement. Such dependence of the characteristics of the process on the local subphase composition can lead to the appearance of a feedback within this dynamic system because the local subphase composition is, in turn, dependent on the contact angle, as will be discussed below. As a consequence, these effects can result in meniscus instability and auto-oscillations, which are responsible for pattern formations within deposited monolayers.

6.3 STRIPED PATTERN FORMATION IN LIPID MONOLAYERS

The interactions of the material forming the monolayer with the substrate surface have recently been intensively studied because of their importance in the formation of patterned interfaces during the Langmuir wetting process. One of the most studied examples of such patterns are the stripes formed by DPPC monolayers by their deposition onto a SiO$_2$ surface [13,14]. The regular stripes running *parallel* to the three-phase contact line (normal to the dipping direction) are created during continuous dynamic Langmuir wetting of lipid monolayers. The presence of patterned interfaces can be proved either by using a fluorescence microscope (by doping the lipid with a small amount of an amphiphilic fluorescing dye) or by taking an atomic force microscopy (AFM) image of the deposited pure DPPC monolayer. The on-line observations show that the striped patterns are not a result of rearrangements after the transfer but are formed during the deposition process. The process is accompanied by meniscus height and contact angle oscillations, which are a result of the meniscus instability.

The AFM data show that the height difference in the stripes is approximately 0.7 to 1.0 nm. Such height difference may be interpreted as alternating LC and LE phases within the deposited LB films. It is also important that the striped films are deposited in a narrow surface pressure range slightly below the pressure of the main LE/LC phase transition of DPPC monolayers (which is 4.4 mN/m at $T = 20°C$; Figure 6.1). This allowed the authors to conclude that the striped patterns are a consequence of the substrate-mediated monolayer condensation at the contact line induced due to the interaction between the lipid molecules and the substrate surface [13–16]. Such substrate-mediated condensation can be directly observed in experiments with a static meniscus at slowly increasing surface pressure [13].

The mechanism of pattern formation proposed in the study by Spratte et al. [14] assumes the existence of a feedback in the system being considered. During the substrate's continuous upward motion, the local monolayer state and the composition in the vicinity of the three-phase line change due to its interaction with the substrate's surface. The speed of the monolayer deposition is assumed to be dependent on the monolayer state (i.e., it is different for LE and LC modes) and, therefore, it can be switched during the deposition. Due to changes in the deposition speed, the three-phase line can rise up and relax downward periodically. Accordingly, the contact angle will also change and this should, in turn, influence the monolayer interaction with the substrate surface, the surface energies, the deposition speed, and so on. As a result, the DPPC molecules are deposited in stripes with more or less condensed 2-D phases.

The stripes width and periodicity depend on deposition speed, surface pressure, substrate preparation, and admixture (e.g., dye) concentration [14]. The stripes' width decreases from approximately 6 μm to 1 μm and the distance between them decreases from 80 μm to 1 to 2 μm, when the deposition speed increases from 0.5 to 4 μm/s. The width and the distance decrease with increasing surface pressure. It was later demonstrated by Gleiche et al. [36] that DPPC monolayers can be deposited onto a mica substrate with a withdrawal speed of 1000 μm/s, giving striped patterns with deposition-free channels of approximately 200 nm in width separated by LC phase stripes of 800 nm in width (Figure 6.3). The thinnest channels obtained with this method under proper conditions were approximately 100 nm in width at a channel density of 20,000/cm. The stripes and channels in such patterns demonstrate a high contrast in the wetting properties.

More detailed information on the dependency of DPPC stripes width and periodicity on surface pressure, deposition speed, and admixture (dye) concentration can be found in refs. [37,38]. Similar experimental results concerning the stripes' and channels' widths were obtained in a recent study by Köpf et al. [39], in which their variations with temperature were also documented. Investigation of the influence of substrate treatment on the DPPC pattern formation was performed by Hirtz et al. [40]. It has been shown that the substrate treatment influences not only the widths of stripes and channels but also the transfer pressure range for patterning.

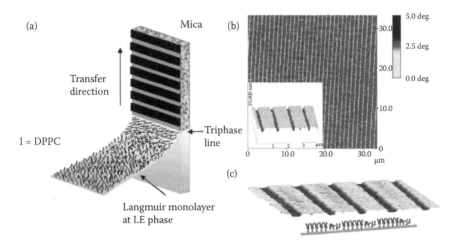

FIGURE 6.3 **(See color insert.)** Scheme of stripe formation during LB deposition of a pure DPPC monolayer onto the substrate surface (a); dynamic scanning force microscope (SFM) image of the patterned surface (b); and arrangement of DPPC molecules within the deposited film (c). (Reprinted with permission from Chen, X., S. Lenhert, M. Hirtz, N. Lu, H. Fuchs, and L. Chi. Langmuir–Blodgett patterning: A bottom-up way to build mesostructures over large areas. *Acc. Chem. Res.*, 40, 393–401. Copyright 2007. American Chemical Society.)

In refs. [18,39,41], a mathematical model was developed to describe the formation of regular stripe patterns during the transfer of surfactant monolayers onto solid substrates. The model takes into account interactions of the surfactant molecules with the substrate and a first-order phase transition within monolayers. The results of numerical simulations are in accord with the experimental data concerning the temperature variation of the width of stripes and channels. The model also allows for the prediction of a switching between the stripe patterns with orientation parallel and perpendicular to the contact line. There is, however, a question as to how this model is relevant to particular experimental systems because the model assumes complete evaporation of the solvent from the deposited film and, therefore, unidirectional flow within the meniscus, although it is usually assumed that the solvent expelled from the contact line due to strong monolayer adhesion produces a circular flow in typical LB systems [28–31]. Another question arises from the assumption of constant boundary conditions at the border with the reservoir, which cannot be relevant for systems in which meniscus oscillations play a significant role [14]. Nevertheless, the model is certainly useful for a deeper understanding of the complex phenomenon being considered.

The increased interest in pattern formation, in particular, also by deposition of lipid monolayers, is explained by a variety of applications for which such patterned surfaces can be used. Gleiche et al. [36] have shown that these structured surfaces can be used as a template with high adsorption selectivity due to different wetting conditions in the hydrophilic channels and the hydrophobic stripes. In particular, materials such as gold clusters stabilized by an organic ligand shell, dye molecules, and $FeCl_3$ can be selectively adsorbed in the channels [36].

According to Lenhert et al. [42], the patterned LB films can be converted into topographical features on the hard material of the substrate. In this case, the structured monolayer is used as a resist by selective chemical etching. Such prepared surfaces can be used as masters for nanoimprinting or the topographically induced alignment of biological cells. This methodology has been called Langmuir–Blodgett lithography [42,43]. Alternatively, the physisorbed LB pattern can be converted into a chemisorbed pattern by selective deposition of a self-assembled monolayer (SAM). The obtained SAM patterns can be used as secondary resists to selectively protect the substrate surface from etching [42,44]. The template-directed self-assembly can also be used for the fabrication of chemically patterned surfaces by selective deposition of two different silanes terminated with different functional groups onto patterned LB films [45].

Zhang et al. [46] demonstrated that patterned DPPC monolayers can be used as a mask to form aminosilane stripes that can be subsequently used to obtain regular arrays of metallic wires with uniform width and separation in a range of 100 nm to more than 1 μm. In particular, the arrays of straight submicrometer-wide copper wires over a silicon substrate were fabricated by this so-called template-assisted electrodeposition method.

To extend the range of properties, mixed Langmuir monolayers can be used to pattern the substrates. In particular, using mixed DPPC monolayers with a dye allows one to obtain luminescent stripe patterns with controlled submicrometer widths over large areas [38]. Such patterns are formed due to substrate-induced microphase separation. Chen et al. [47] examined the mixtures of DPPC with some other lipids. It was found that stripe width and periodicity in the mixed patterning films were smaller than those in stripes formed by pure DPPC monolayers at the same transfer conditions. But the lipids affect the pattern formation differently, probably due to their different miscibilities and diffusion coefficients in the monolayers.

Pattern formation by LB deposition of mixed phospholipid monolayers was also studied by Moraille and Badia [48,49]. The floating monolayers consisted of binary mixtures of DPPC and L-α-dilauroylphosphatidylcholine (DLPC) with different molar ratios. Under certain experimental conditions, these mixtures yield highly regular nanoscale patterns of parallel stripes or lines at clean mica substrates. Not only monolayer but also *bilayer* patterned films can be obtained due to condensation of phospholipids over the solidlike stripe domains of the underlying monolayer. In contrast to the previously discussed pure DPPC patterns, the mixed DPPC/DLPC patterns are formed at much higher surface pressures and result not from a phase transition but from the separation of DPPC and DLPC phases during the deposition.

A bilayer configuration of patterned films formed by the deposition of mixed phospholipid monolayers shows higher stability in aqueous environments [49], which can be important for potential applications. It is also important that in bilayer patterns, hydrophilic head groups are exposed to the outside instead of hydrophobic alkyl chains. This offers wider opportunities for further modifications of such patterned films.

Phase-separated LB films, produced from mixed phospholipid monolayers, can serve as surface templates for the selective and patterned deposition of macromolecules on a submicrometer scale [50,51]. For example, the difference in alkyl-chain packing densities and surface energies between the disordered LE phase of DLPC and the ordered LC phase of DPPC can be used for the selective adsorption of proteins (human serum albumin and human γ globulin) on LE stripes of phase-separated LB monolayers. In an article by Moraille and Badia [52], it was shown that by using surface patterns of the biologically active and inactive enantiomers within the mixed lipid bilayers, it was possible to spatially direct a stereospecific lipolytic enzyme reaction. By using phospholipids functionalized with metal-reactive –SSCH$_3$ groups, it is possible to produce periodic arrays of solid-supported metal nanostructures (such as Au, Ag, or Cu-coated stripes) because of the spatially directed adsorption of thermally evaporated metal atoms [53].

By varying the conditions of the LB deposition (transfer speed, surface pressure) it is possible to control the orientation of the stripes in patterned films with respect to the transfer direction (Figure 6.4). In particular, it is possible to obtain not only *horizontal* DPPC stripes as those considered above (Figure 6.4a) but also the patterns with *vertical* DPPC stripes oriented parallel to the dipping direction (Figure 6.4c) [37,42,47,54]. The probable mechanism behind the formation of vertical stripes is the so-called fingering instability driven by local surface tension gradients (Marangoni effect) formed in the vicinity of the three-phase line due to local variations of temperature or surface concentration [55]. At intermediate surface pressures and low transfer velocities, two types of stripes, vertical and horizontal, can appear simultaneously forming *crossed-stripe* structures (grids), such as those in Figure 6.4b [37,38,42,47,54]. In a study by Chen et al. [47], an approximate diagram was obtained showing which type of the stripe structure forms depending on the substrate velocity and surface pressure.

Pignataro et al. [56] obtained vertical stripes within dimyristoyl-phosphatidylcholine (DMPC) monolayers on mica (transferred at a low temperature of 10°C). In contrast to DPPC, the DMPC

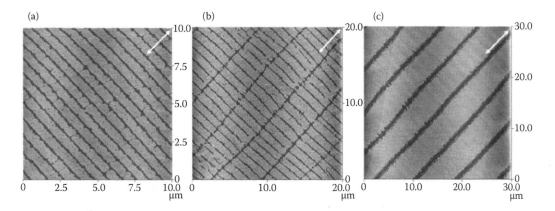

FIGURE 6.4 (See color insert.) AFM images of striped structures of pure DPPC formed under different transfer conditions (velocity and surface pressure): (a) horizontal stripes (60 mm/min and 3 mN/m); (b) crossed stripes (40 mm/min and 3 mN/m); and (c) vertical stripes (10 mm/min and 3 mN/m). The arrows show the transfer direction. (Reprinted with permission from Chen, X., S. Lenhert, M. Hirtz, N. Lu, H. Fuchs, and L. Chi. Langmuir–Blodgett patterning: A bottom-up way to build mesostructures over large areas. *Acc. Chem. Res.*, 40, 393–401. Copyright 2007. American Chemical Society.)

stripes were obtained at surface pressures above the main phase transition pressure, that is, in the LC region of the surface pressure–area isotherm. These patterns were obtained only in the range of substrate speeds between 3 and 10 mm/min.

Similar vertical stripes were also observed after the deposition of mixed lipid/lipopolymer mono-layers on glass substrates [57]. The authors verified that these stripes are the result of the dynamic phase separation of lipids and lipopolymers through the transfer. The stripes were formed at a much higher surface pressure (30 mN/m) than in the pure DPPC system. With increasing transfer velocity, the mean distance between the individual stripes decreases and the stripes become branched (Figure 6.5). Also, the increase in the polymer chain length leads to the formation of branched stripes. As demonstrated in the article by Purrucker et al. [57], the lipid/lipopolymer striped patterns

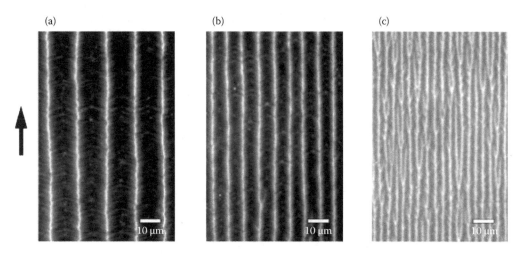

FIGURE 6.5 Fluorescence images of striped structures formed within the mixed lipid/lipopolymer mono-layers under different transfer velocities: (a) 50 μm/min; (b) 80 μm/min; and (c) 180 μm/s (surface pressures, 30 mN/m). The arrow shows the transfer direction. (Reprinted with permission from Purrucker, O., A. Förtig, K. Lüdtke, R. Jordan, and M. Tanaka. Confinement of transmembrane cell receptors in tunable stripe micro-patterns. *J. Am. Chem. Soc.*, 127, 1258–1264. Copyright 2005. American Chemical Society.)

can be used as self-assembled templates to confine cell receptor proteins that are incorporated preferably into the lipopolymer-rich regions. Thus, such patterned surfaces can be used, for example, for geometrical control of cell adhesion.

6.4 STRIPED PATTERNS FORMED BY CHARGED LANGMUIR MONOLAYERS

Mahnke et al. [58] observed regular patterns of horizontally oriented stripes (parallel to the meniscus) formed in bilayer LB films of arachidic acid. The arachidic acid monolayers were deposited onto hydrophobized microscopic glass slides from a subphase containing cadmium salt. The main distinction between these patterns and those formed by phospholipid monolayers is their dependence on the ionic composition within the subphase. The regular stripe patterns were obtained only in a narrow pH region at about pH 5.7 in the presence of 0.25 mmol/L of $CdCl_2$. The striped films were deposited at a surface pressure of approximately 30 mN/m, where the arachidic acid monolayer is in a condensed state (Figure 6.1). Thus, in this case, the substrate-mediated monolayer condensation cannot be the cause of pattern formation. Under similar conditions, similarly striped structures in arachidic acid/cadmium arachidate LB films were also observed by Kurnaz and Schwartz [59].

It is well known [17,60–62] that with a pH increase in the presence of approximately 0.25 mmol/L of $CdCl_2$, the composition of the floating monolayer changes continuously from arachidic acid to cadmium arachidate. Thus, under the subphase conditions in a pH region of around 5.7, as used in ref. [58], the floating monolayer should be a mixture of arachidic acid and cadmium arachidate. As shown by Mahnke et al. [58], during the deposition process, the stripe patterns were only preformed, that is, the deposited films retained a uniform thickness, although arachidic acid and cadmium arachidate have already been separated in the stripes. Generally, the usual procedure of skeletonization allows the removal of soluble components from a two-component monolayer, which consists of a soluble and an insoluble component, by a solvent. The insoluble component remains at the local position as a skeleton. In the present case, the arachidic acid was removed by dipping the substrate into a vessel with cyclohexane, whereas the cadmium salt remains within the deposited film as a skeleton.

The presence of grooves in the films after skeletonization shows that under these particular conditions, arachidic acid and cadmium arachidate are deposited not as a homogeneous mixture but are separated in stripes. The arachidic acid was removed by the skeletonization process and the remaining grooves indicate the position of the stripes of the removed arachidic acid. The presence of stripes within the deposited films after skeletonization was evidenced by AFM images and phase shift interference microscopy images (Figure 6.6). The depth of the grooves was approximately equal to the bilayer thickness. It was found that the distance between the main defect lines increases

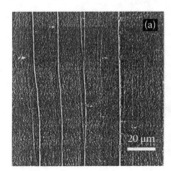 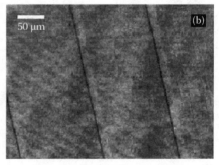

FIGURE 6.6 AFM image (a) and phase shift interference microscopy image (b) of striped structures in arachidic acid/cadmium arachidate bilayers after skeletonization in cyclohexane. (According to Mahnke, J., D. Vollhardt, K.W. Stöckelhuber, K. Meine, and H.J. Schulze. Regular stripe patterns in skeletonized Langmuir–Blodgett films of arachidic acid. *Langmuir* 15, 8220–8224, 1999. With permission.)

from 75 to approximately 300 μm with the increase of the dipping rate in the range between 3 and 8 mm/min.

The strong dependency of the stripe patterns on the subphase pH and the change of the stripe's composition from the cadmium arachidate to the arachidic acid form demonstrate the basic role of the counterion binding at the monolayer and, therefore, the importance of the electrical interactions in the meniscus region for the deposition process.

The fatty acid molecules can dissociate in presence of aqueous subphase, according to the reaction:

$$RH \rightarrow R^- + H^+ \tag{6.3}$$

where RH and R⁻ denote the nondissociated and dissociated fatty acids, respectively. In the presence of bivalent metal ions (M^{2+}) within the subphase, the dissociated fatty acid can form complexes with these ions:

$$2R^- + M^{2+} \rightarrow R_2M \tag{6.4}$$

$$R^- + M^{2+} \rightarrow RM^+ \tag{6.5}$$

There are different opinions in the literature as to which is the stoichiometry of complexes of bivalent metal ions with fatty acids in Langmuir monolayers: either only the positively charged 1:1 complex RM⁺ [60], or the neutral 1:2 complex R_2M [61], or both of them [63] can form. The chemical equilibriums at the monolayer surface determine the amounts of dissociated fatty acid (R⁻) and charged complex (RM⁺) and, accordingly, the surface charge density [17,60–62]:

$$\sigma = -F\left(X_{R^-} - X_{RM^+}\right) \tag{6.6}$$

where X_{R^-} and X_{RM^+} are the surface molar concentrations of the respective species and F is the Faraday constant. Thus, the surface charge density σ will depend on the local pH and the bivalent metal ions concentration within the subphase. The surface charge density can be calculated if the respective equilibrium (binding) constants are known [17,60–62].

Because of both the floating monolayer and the substrate surface (covered by the previously deposited monolayer) are charged, they interact electrostatically. Electrostatic repulsion decreases the monolayer adhesion to the substrate and, hence, the contact angle. During the formation of patterns, the composition of the deposited monolayer changes periodically from cadmium arachidate to arachidic acid and back, and thus, contact angle oscillations should be expected during the deposition process. Indeed, meniscus auto-oscillations were directly observed during the deposition of patterned films [58].

According to Equation 6.2, the maximum monolayer transfer rate decreases with the decrease of the contact angle. The substrate velocities in ref. [58] were in the range 0.005 to 0.013 cm/s. Such velocities are smaller than the maximum transfer rate determined by Petrov et al. [31] for cadmium arachidate monolayers, and are larger than that observed for arachidic acid monolayers (see Section 6.2). Thus, during the deposition of an arachidic acid stripe in ref. [58], the deposition rate was (obviously) very low, the contact line was apparently pinned at the substrate surface, and the meniscus rose up with the substrate. In turn, during the deposition of a cadmium arachidate stripe, the maximum deposition rate was higher than the substrate velocity and the meniscus slowed down.

Thus, the interaction of the floating monolayer with the substrate surface also plays a crucial role in the case of fatty acid monolayers. However, in this case, electrostatic interactions become much more important. The interaction energy changes with the monolayer composition, which is closely related to the local ionic composition within the subphase.

At the surface of a charged monolayer, an electrical double layer (EDL) is formed in the aqueous subphase. Within the meniscus region close to the three-phase contact line, the EDLs formed at the two interfaces overlap (Figure 6.7) [17]. Because of this overlap, the local electrical potential near the contact line should be *larger* than in the regions without EDL interaction. At the same time, the local surface charge density should be *smaller* here. The charge density decreases near the contact line due to enhanced counterions binding at higher local electrical potentials [17]. The decrease of charge density is an important precondition for good monolayer adhesion and successful deposition.

During the deposition process, the monolayer at the aqueous interface passes through the region of the EDL overlap where it binds counterions from solution. Due to counterions binding at the contact line, the local surface charge density decreases to zero, and the monolayer is deposited onto the substrate surface already in an electroneutral state. This requires, however, a sufficiently fast transfer of counterions toward this region. Although the kinetics of ion redistribution within an EDL is rather fast, it can seem insufficient at higher deposition rates, especially in the case of small contact angles [64–67]. This can lead to incomplete neutralization of the monolayer charge, poor adhesion of the monolayer, and entrainment of thin water film during the deposition.

It has been shown that under equilibrium conditions, the local monolayer composition (defined by the acid to salt ratio) near the contact line changes insignificantly [17]. However, the change in the composition can be much larger under dynamic conditions. This change in the monolayer composition is explained by the change in the local subphase composition induced due to the monolayer transfer, as considered further.

When the floating monolayer and the substrate surface begin to move, they produce a circular convective flow of the solution within the meniscus (Figure 6.2). The ions located in EDL closer to the moving interfaces are advected toward the contact line, whereas those located in the central part of the meniscus cross-section are advected back from the contact line, toward the bulk solution. As the counterions have higher concentrations near the charged interfaces than in the outer part of the EDL, their average convective flux in each meniscus cross-section is directed toward the contact line [64–67]. For the coions, which have smaller concentrations near the interfaces, the average convective flux moves in the opposite direction.

It is also important that the degree of diffuse layer overlap is different in different cross-sections of the meniscus. The closer the cross-section is to the contact line, the stronger is the overlap. Accordingly, the total convective flux of each ion in the given cross-section depends on the local distance h between the moving interfaces (Figure 6.2). The convective fluxes of ions decrease toward the contact line [64–67]. Thus, there is a difference in the convective ion fluxes in different cross-sections, which results in the ion's redistribution within the meniscus region after the beginning of

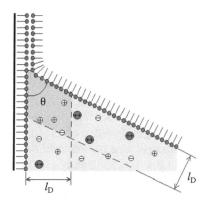

FIGURE 6.7   (See color insert.) Diffuse double layers overlapping in the meniscus region; l_D is the double layer thickness (Debye length). (According to Kovalchuk, V.I. et al., *J. Phys. Chem. B* 105, 9254–9265, 2001. With permission.)

the monolayer transfer. There is also a difference between the number of counterions removed with the deposited film and those brought to the contact line from the bulk by convective fluxes [64–67]. As a result, after a certain relaxation time, ion concentration and electrical potential profiles are formed that induce local electrodiffusion ion fluxes within the solution, in addition to the convective fluxes [64–69].

When the electrodiffusion ion fluxes fully develop, they can compensate for the difference in the convective fluxes in different cross-sections, and a steady-state transfer regime is established (Figure 6.8, curves 1 and 2). When the substrate is suddenly stopped, relaxation of the concentration profiles toward the equilibrium begins (Figure 6.8, curves 3 and 4). These relaxation processes should be rather long, in the order of several minutes or tens of minutes, because they are limited by the ion's diffusion on the distances of the meniscus size (1–2 mm) [68,69]. Such formation of concentration and electrical potential profiles is typical for concentration polarization effects in electrolyte solutions.

The meniscus relaxation after stopping the deposition process has been observed in a number of experimental studies on LB deposition of charged monolayers [20,33,35,70,71]. The typical relaxation times are 10 min or longer. Such a slow relaxation cannot be explained by hydrodynamic relaxation, which should be much faster. But it can be explained by the slow diffusion relaxation of the ion concentration within the meniscus region, which has much longer characteristic times [68,69]. The ion redistribution mechanism is additionally confirmed by the observed strong dependency of the relaxation time on the solution pH (the relaxation becomes slower with increasing ionization of the monolayer) [35].

The slow meniscus relaxation can also be a consequence of possible dissolution of the deposited monolayer material [70] and the following partial hydrophobization of the substrate surface. However, such a process would lead to the lower stability of the formed LB films and to the transformation of the Y type into X type deposition. Therefore, in contrast to ion redistribution, this cannot be a general mechanism for all types of films.

The change in the local concentrations of counterions within the subphase near the contact line should influence their binding at the monolayer surface. This can lead to variations of the monolayer composition and respective changes in the work of adhesion and contact angle during the deposition

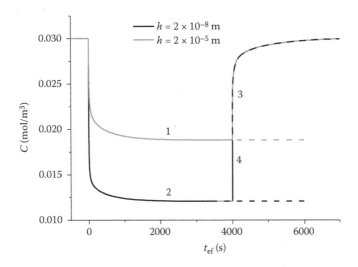

FIGURE 6.8 **(See color insert.)** Relaxation of the local electrolyte concentration at two distances from the contact line (corresponding to $h = 2 \times 10^{-8}$ m and 2×10^{-5} m) after the beginning (1,2) and stopping (3,4) of the monolayer transfer (the initial concentration $C_0 = 0.03$ mol/m^3, the ratio of the velocity to the contact angle in radians $U/\theta = 1.5$ mm/s, $t_{ef} = t\,\theta^2$ is the effective time). (According to Kovalchuk, V.I. et al., *J. Phys. Chem. B*, 115, 1999–2005, 2011. With permission.)

process. However, the ion redistribution within the meniscus depends on the contact angle. The smaller the contact angle, the slower the counterions transfer toward the contact line, that is, the stronger the concentration polarization in the solution. Thus, there is a feedback between the local subphase composition, monolayer adhesion, and contact angle—the local subphase composition influences the monolayer adhesion and contact angle and vice versa. This feedback can lead to periodic meniscus instability and formation of stripes of different ionic compositions within the deposited monolayer [64].

With increasing deposition speed, the ion concentration and electrical potential profiles formed within the meniscus region become sharper and the concentrations of potential-determining counterions at the contact line become smaller [65–67]. At a certain critical speed, the concentrations of counterions can become so small, and the increasing surface charge density so large, that the work of adhesion decreases to zero. Then, a thin water film will be entrained by the substrate. From the above consideration, it is clear that this critical speed will depend on the initial ion concentration in the solution. The respective model [67] shows that for fatty acid monolayers, the critical speed should be larger at smaller pH and at larger bivalent metal ions concentration. This conclusion is in a good agreement with the experimental data of Petrov and coworkers [31,72,73]. A similar behavior is also expected for other charged monolayers. In particular, the behavior of positively charged monolayers of fatty amines in the presence of bivalent counterions [31,33,72,73] is also in good agreement with the proposed model.

It is well known that electrostatic interactions strongly depend on the presence of indifferent electrolytes in solution. The same is also true in the case of the charged monolayers deposition. In contrast to potential-determining counterions, the indifferent counterions are not removed from the subphase with the deposited monolayer, but they are advected toward the contact line with the convective flux within the subphase. Thus, they will accumulate near the contact line. Due to their accumulation, they replace the potential-determining counterions in the EDL overlap region. Because of this, the local surface charge density and electrostatic repulsion will increase [74]. Accordingly, the maximum deposition speed will decrease. This conclusion is also in a good agreement with the experimental data, which show that in the presence of indifferent counterions (such as sodium), the maximum deposition speed is much smaller or there is no deposition at all [21,31,63,72].

In the study by Bondarenko et al. [74], it was shown that because of accumulation, even very small concentrations of indifferent counterions (e.g., 1 µM), can have a considerable effect on the deposition process. Unfortunately, very often the presence of admixtures of indifferent counterions is not carefully controlled in the experiments. On the other hand, the addition of small controlled amounts of indifferent electrolytes can be used to modify the deposited film properties.

This analysis shows that the aqueous subphase takes part in the mechanism of pattern formation by LB deposition of charged monolayers. The type and amount of different counterions within the subphase are very important in the interaction of the monolayer with the substrate. But due to insufficiently fast redistribution of the ions, the local environment near the contact line is different from the bulk solution. These changes in local subphase composition are responsible for both the meniscus oscillations and the composition changes within the deposited monolayer.

6.5 PATTERN FORMATION IN POLYMER AND PARTICLE SYSTEMS

The formation of patterned structures by dewetting from a receding meniscus in polymer and nanoparticle systems has been intensively studied in the last 10 years or so. Most of these patterned structures are not real 2-D structures (in the sense of monolayer or bilayer molecular/particle films). However, in many ways, they resemble the patterns formed by LB transfer of monolayers.

Various highly ordered periodic structures are formed due to fingering instabilities, Rayleigh–Benard convection cells and other mechanisms at the dewetting front of an evaporating polymer solution ([75] and references therein). For example, Yabu and Shimomura [75] obtained regular periodic dot, stripe, and ladder (grid) patterns depending on the polymer concentration in solution

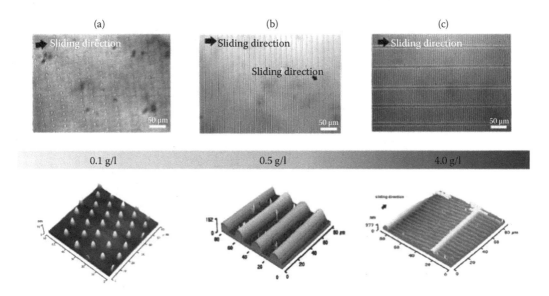

FIGURE 6.9 Polymer patterns formed by dewetting from polystyrene solutions of different concentrations (a) 0.1 g/l, (b) 0.5 g/l, (c) 4.0 g/l, visualized by optical (upper row) and atomic force (lower row) microscopy. (According to Yabu, H., and M. Shimomura, *Adv. Funct. Mater.*, 15, 575–581, 2005. With permission.)

(Figure 6.9). The ladder (grid) patterns (Figure 6.9c) resemble those formed by DPPC monolayers (cf. Figure 6.4b). Such ladder (grid) patterns are formed, supposedly, by the combination of two mechanisms: stick-slip behavior of the meniscus and fingering instability.

Huang et al. [76] have recently demonstrated that microparticles and nanoparticles can be arranged into lines of single particle thickness using the LB transfer procedure (Figure 6.10). The particle lines oriented parallel to the meniscus were obtained during the transfer of a particle Langmuir monolayer, prepared on the water's surface, to a substrate with a medium water contact angle. During the mono-layer transfer, a typical stick-slip motion of the meniscus was observed. Colloidal particles with a wide range of sizes and materials can be used to prepare the line patterns [76]. The particle density within the lines depends on the concentration of particles in the monolayer and the substrate speed.

As shown in the study by Huang et al. [76], for the deposition of single particle lines, diluted mono-layers (at surface pressures of <0.5 mN/m) and water contact angles of more than 20° are necessary. At higher surface pressures, thicker (multiparticle) lines are formed. A decrease of the water contact angle of less than 10° leads to a transition from horizontal to vertical (perpendicular to the meniscus) micrometer-scale particle stripes [77,78]. The authors presume that fingering instability is responsible for the formation of these vertical stripe patterns. A similar transition from horizontally to vertically oriented stripe patterns is also observed for phospholipid (DPPC) monolayers (Figure 6.4).

For the formation of patterned structures in polymer and particle systems, evaporation of the solvent at the dewetting front usually plays a much more important role than by deposition of ordi-nary surfactant monolayers. Evaporation of the solvent induces local surface tension gradients and the related thermal or solutal (caused by surface concentration gradients) Marangoni convection. Microscopic convective flows generated by evaporation modify the pattern formation mechanism. The rate of evaporation also determines the local thickness of thin liquid film formed at the contact line (or precursor film) and, therefore, influences the interactions in this zone. Evaporation by itself can cause the motion of the three-phase contact line with respect to the substrate surface, so that evaporation-driven patterns can also be formed on motionless substrates.

Shimomura and Sawadaishi [79–81] observed periodical stripes of ultra-fine particles running parallel to the receding direction of the solution edge. The stripes were obtained by dewetting of aqueous colloidal dispersions. At a small dispersion concentration, the thickness of stripes was

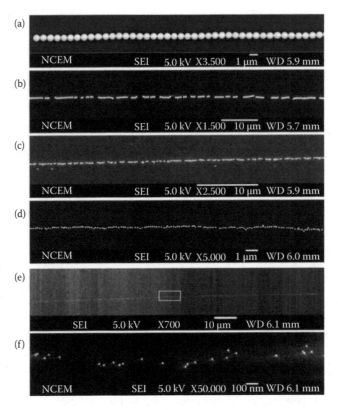

FIGURE 6.10 SEM images of single particle lines deposited on a partially wettable substrate using LB transfer procedure: (a) 0.9 μm, (b) 450 nm, and (c) 160 nm SiO₂ spheres, (d) 50 nm Ag nanocubes, and (e, f) 7 nm Pt nanoparticles (f is a higher magnification image of e). (According to Tao, A.R. et al., *Acc. Chem. Res.* 41, 1662–1673. 2008. With permission.)

equivalent to one monolayer of the particles. The periodicity of the stripes was dependent on the temperature of the substrate, which controls the evaporation rate of the solvent. By increasing the dispersion concentration, stripe patterns oriented perpendicular to the receding direction were observed, which were formed presumably by the stick–slip motion of the meniscus [80]. In the study by Sawadaishi and Shimomura [81], self-supporting assemblies of silica particles were obtained that could be removed from the substrate.

Both types of regular periodical particle stripes, running either perpendicular or parallel to the withdrawal direction, are often observed by dip-coating of colloidal suspensions on chemically and physically homogeneous substrate surfaces [80,82–85]. Horizontal stripes running perpendicular to the withdrawal direction are formed according to the stick–slip mechanism. In this case, however, the contribution from capillary interactions of particles at the contact line also becomes significant for the formation of patterns [82–84].

Vertical stripes running parallel to the withdrawal direction are formed due to the fingering instability mechanism [80,85]. It should be noted, however, that the fingering instability results in the formation of irregular, nonperiodical (e.g., dendritic) patterns, which occur more often in particle and polymer systems [86–89]. Frequently, such structures have hierarchically ordered branching resembling that in Figure 6.5c.

Spontaneously formed structures are also observed by the evaporation of liquid drops containing polymers or particles [90–94]. These structures, such as rings or dendritic structures, are very similar to those considered here, and many of these are governed by similar mechanisms. They are, however, not the topic of the present review, and therefore, are not considered here.

6.6 SUMMARY AND OUTLOOK

This chapter shows how the method proposed by Langmuir and Blodgett has been developed into a broad scientific field, a significant novel part of which is LB patterning. The review does not cover all results in this field, it only tries to give an insight into the main systems used in LB patterning and the main types of patterns. Currently, the real number of systems and patterns is much larger and, no doubt, it will be extended in the near future. More complicated patterns can be prepared by secondary processing of the structures obtained during wetting/dewetting processes.

The results considered here show that the structure type (stripes, grid, etc.) and its geometrical characteristics (e.g., periodicity) strongly depend on the material used for deposition, the substrate wetting properties, solvent properties, and the conditions of the deposition process (substrate speed, surface tension, rate of evaporation, and temperature). The effect of all these factors in each particular system is not yet completely clear because the mechanisms leading to meniscus instability and pattern formation are very complicated.

One of the open questions is the role of electrical interactions and electrokinetic relaxation in the vicinity of the contact line. This question is very important for understanding the formation of patterns in case of charged amphiphilic monolayers. Charged monolayers are a promising object for LB patterning because they have a number of important advantages. For example, by using charged monolayers, the transfer-induced patterns with a spatially periodic ionic composition (e.g., stripes containing different counterions) can be produced. Different types of ions incorporated into the patterned films will impart new physical or chemical properties (or both) to such films. The deposition process can also be controlled by regulation of the amounts of different counterions and coions within the subphase that influences the monolayer composition, the surface charge density, and the EDL thickness. Studies of electrical interactions and electrokinetic relaxation will also be important for polymer and particle systems.

ACKNOWLEDGMENTS

Financial assistance by the Bundesministerium für Bildung, Wissenschaft, Forschung und Technologie (BMBF) and the Ukrainian Ministry of Education and Science (common project UKR 10/039) and National Ukrainian Academy of Sciences (project 64/10-H) is gratefully acknowledged. This work was supported by COST D43 Action.

REFERENCES

1. A. Pockels. Surface tension. *Nature*, **43**, 437–439 (1891).
2. I. Langmuir. The constitution and fundamental properties of solids and liquids. II. Liquids. *J. Am. Chem. Soc.*, **39**, 1848–1906 (1917).
3. K.B. Blodgett. Monomolecular films of fatty acids on glass. *J. Am. Chem. Soc.*, **56**, 495 (1934).
4. K.B. Blodgett. Films built by depositing successive monomolecular layers on a solid surface. *J. Am. Chem. Soc.*, **57**, 1007–1022 (1935).
5. I. Langmuir and K.B. Blodgett. Über einige neue Methoden zur Untersuchung von monomolekularen Filmen. *Kolloid Z.*, **73**, 257–263 (1935).
6. K.B. Blodgett and I. Langmuir. Built-up films of barium stearate and their optical properties. *Phys. Rev.*, **51**, 964–982 (1937).
7. K.B. Blodgett. Properties of built-up films of barium stearate. *J. Phys. Chem.* **41**, 975–984 (1937).
8. I. Langmuir. Overturning and anchoring of monolayers. *Science*, **87**, 493–500 (1938).
9. G. Roberts (Ed.). *Langmuir–Blodgett Films*. Plenum Press, New York (1990).
10. U. Abraham. *An Introduction to Ultrathin Organic Films*. Academic Press, San Diego, CA (1991).
11. G.L. Gaines, Jr. *Insoluble Monolayers at Liquid–Gas Interfaces*. Wiley-Interscience, New York (1966).
12. D. Vollhardt. Morphology and phase behaviour of monolayers. *Adv. Colloid Interface Sci.*, **64**, 143–171 (1996).
13. H. Riegler and K. Spratte. Structural changes in lipid monolayers during the Langmuir–Blodgett transfer due to substrate/monolayer interactions. *Thin Solid Films*, **210/211**, 9–12 (1992).

14. K. Spratte, L.F. Chi and H. Riegler. Physisorption instabilities during dynamic Langmuir wetting. *Europhys. Lett.*, **25**, 211–217 (1994).
15. K. Spratte and H. Riegler. Steady state morphology and composition of mixed monomolecular films (Langmuir monolayers) at theair/water interface in the vicinity of the three-phaseline: Model calculations and experiments. *Langmuir*, **10**, 3161–3173 (1994).
16. K. Graf and H. Riegler. Molecular adhesion interactions between Langmuir monolayers and solid substrates. *Colloids Surf. A*, **131**, 215–224 (1998).
17. V.I. Kovalchuk, E.K. Zholkovskij, M.P. Bondarenko and D. Vollhardt. Dissociation of fatty acid and counterion binding at the Langmuir monolayer deposition: Theoretical considerations. *J. Phys. Chem. B.*, **105**, 9254–9265 (2001).
18. M.H. Köpf, S.V. Gurevich, R. Friedrich and L. Chi. Pattern formation in monolayer transfer systems with substrate-mediated condensation. *Langmuir*, **26**, 10444–10447 (2010).
19. J.F. Stephens. Mechanisms of formation of multilayers by the Langmuir–Blodgett techniques. *J. Colloid Interface Sci.*, **38**, 557–566 (1972).
20. G.L. Gaines, Jr. Contact angles during monolayer deposition. *J. Colloid Interface Sci.*, **59**, 438–446 (1977).
21. G. Veale and I.R. Peterson. Novel effects of counterions on Langmuir films of 22-tricosenoic acid. *J. Colloid Interface Sci.*, **103**, 178–189 (1985).
22. R.D. Neuman. Molecular reorientation in monolayers at the paraffin-water interface. *J. Colloid Interface Sci.*, **63**, 106–112 (1978).
23. R.D. Neuman and J.W. Swanson. Multilayers deposition of stearic acid-calcium stearate monomolecular films. *J. Colloid Interface Sci.*, **74**, 244–259 (1980).
24. M.R. Buhaenko, J.W. Goodwin, R.M. Richardson and M.F. Daniel. The influence of shear viscosity of spread monolayers on the Langmuir–Blodgett process. *Thin Solid Films*, **134**, 217–226 (1985).
25. M.R. Buhaenko and R.M. Richardson. Measurements of the forces of emersion and immersion and contact angles during Langmuir–Blodgett deposition. *Thin Solid Films*, **159**, 231–238 (1988).
26. R. Aveyard, B.P. Binks, P.D.I. Fletcher and X. Ye. Dynamic contact angles and deposition efficiency for transfer of docosanoic acid onto mica from $CdCl_2$ subphases as a function of pH. *Thin Solid Films*, **210/211**, 36–38 (1992).
27. J.B. Peng, B.M. Abraham, P. Dutta and J.B. Ketterson. Contact angle of lead stearate-covered water on mica during the deposition of Langmuir–Blodgett assemblies. *Thin Solid Films*, **134**, 187–193 (1985).
28. P.G. de Gennes. Deposition of Langmuir–Blodgett layers. *Colloid Polym. Sci.*, **264**, 463–465 (1986).
29. P.G. Petrov. Dynamics of deposition of Langmuir–Blodgett multilayers. *J. Chem. Soc., Faraday Trans.*, **93**, 295–302 (1997).
30. J.G. Petrov and G.P. Petrov. Molecular-hydrodynamic description of Langmuir–Blodgett deposition. *Langmuir*, **14**, 2490–2496 (1998).
31. J.G. Petrov, H. Kuhn and D. Möbius. Three-phase contact line motion in the deposition of spread monolayers. *J. Colloid Interface Sci.*, **73**, 66–75 (1980).
32. J.H. Clint and T. Walker. Interaction energies between layers of alkyl and partially fluorinated alkyl chains in Langmuir–Blodgett multilayers. *J. Colloid Interface Sci.*, **47**, 172–185 (1974).
33. J.G. Petrov and A. Angelova. Interaction free energies in Langmuir–Blodgett multilayers of docosylammonium phosphate. *Langmuir*, **8**, 3109–3115 (1992).
34. V.I. Kovalchuk, E.K. Zholkovskiy, M.P. Bondarenko, V.M. Starov and D. Vollhardt. Concentration polarization effect at the deposition of charged Langmuir monolayers. *Adv. Colloid Interface Sci.*, **168**, 114–123 (2011).
35. R. Aveyard, B.P. Binks, P.D.I. Fletcher and X. Ye. Contact angles and transfer ratios measured during the Langmuir–Blodgett deposition of docosanoic acid onto mica from $CdCl_2$ subphases. *Colloids Surf. A*, **94**, 279–289 (1995).
36. M. Gleiche, L.F. Chi and H. Fuchs. Nanoscopic channel lattices with controlled anisotropic wetting. *Nature*, **403**, 173–175 (2000).
37. S. Lenhert, M. Gleiche, H. Fuchs and L. Chi. Mechanism of regular pattern formation in reactive dewetting. *ChemPhysChem*, **6**, 2495–2498 (2005).
38. X. Chen, M. Hirtz, H. Fuchs and L. Chi. Self-organized patterning: Regular and spatially tunable luminescent submicrometer stripes over large areas. *Adv. Mater.*, **17**, 2881–2885 (2005).
39. M.H. Köpf, H. Harder, J. Reiche and S. Santer. Impact of temperature on the LB patterning of DPPC on mica. *Langmuir*, **27**, 12354–12360 (2011).
40. M. Hirtz, H. Fuchs and L. Chi. Influence of substrate treatment on self-organized pattern formation by Langmuir–Blodgett transfer. *J. Phys. Chem. B*, **112**, 824–827 (2008).

41. M.H. Köpf, S.V. Gurevich and R. Friedrich. Thin film dynamics with surfactant phase transition. *Europhys. Lett.*, **86**, 66003 (2009).

42. S. Lenhert, L. Zhang, J. Mueller, H.P. Wiesmann, G. Erker, H. Fuchs and L. Chi. Self-organised complex patterning: Langmuir–Blodgett lithography. *Adv. Mater.*, **16**, 619–624 (2004).

43. S. Lenhert, M.-B. Meier, U. Meyer, L. Chi and H.P. Wiesmann. Osteoblast alignment, elongation and migration on grooved polystyrene surfaces patterned by Langmuir–Blodgett lithography. *Biomaterials*, **26**, 563–570 (2005).

44. X. Wu, S. Lenhert, L. Chi and H. Fuchs. Interface interaction controlled transport of CdTe nanoparticles in the microcontact printing process. *Langmuir*, **22**, 7807–7811 (2006).

45. N. Lu, M. Gleiche, J. Zheng, S. Lenhert, B. Xu, L. Chi and H. Fuchs. Fabrication of chemically patterned surfaces based on template-directed self-assembly. *Adv. Mater.*, **14**, 1812–1814 (2002).

46. M. Zhang, S. Lenhert, M. Wang, L. Chi, N. Lu, H. Fuchs and N. Ming. Regular arrays of copper wires formed by template-assisted electrodeposition. *Adv. Mater.*, **16**, 409–413 (2004).

47. X. Chen, N. Lu, H. Zhang, M. Hirtz, L. Wu, H. Fuchs and L. Chi. Langmuir–Blodgett patterning of phospholipid microstripes: Effect of the second component. *J. Phys. Chem. B*, **110**, 8039–8046 (2006).

48. P. Moraille and A. Badia. Highly parallel, nanoscale stripe morphology in mixed phospholipid monolayers formed by Langmuir–Blodgett transfer. *Langmuir*, **18**, 4414–4419 (2002).

49. P. Moraille and A. Badia. Nanoscale stripe patterns in phospholipid bilayers formed by the Langmuir–Blodgett technique. *Langmuir*, **19**, 8041–8049 (2003).

50. P. Moraille and A. Badia. Spatially directed protein adsorption by using a novel, nanoscale surface template. *Angew. Chem. Int. Ed.*, **41**, 4303–4306 (2002).

51. A. Badia, P. Moraille, N.Y.W. Tang and M.E. Randlett. Nanostructured phospholipid membranes. *Int. J. Nanotechnol.*, **5**, 1371–1395 (2008).

52. P. Moraille and A. Badia. Enzymatic lithography of phospholipid bilayer films by stereoselective hydrolysis. *J. Am. Chem. Soc.*, **127**, 6546–6547 (2005).

53. N.Y.-W. Tang and A. Badia. Self-patterned mixed phospholipid monolayers for the spatially selective deposition of metals. *Langmuir*, **26**, 17058–17067 (2010).

54. X. Chen, S. Lenhert, M. Hirtz, N. Lu, H. Fuchs and L. Chi. Langmuir–Blodgett patterning: A bottom–up way to build mesostructures over large areas. *Acc. Chem. Res.*, **40**, 393–401 (2007).

55. O. Karthaus, L. Gråsjö, N. Maruyama and M. Shimomura. Formation of ordered mesoscopic polymer arrays by dewetting. *Chaos*, **9**, 308–314 (1999).

56. B. Pignataro, L. Sardone and G. Marletta. From micro- to nanometric scale patterning by Langmuir–Blodgett technique. *Mater. Sci. Eng. C*, **22**, 177–181 (2002).

57. O. Purrucker, A. Förtig, K. Lüdtke, R. Jordan and M. Tanaka. Confinement of transmembrane cell receptors in tunable stripe micropatterns. *J. Am. Chem. Soc.*, **127**, 1258–1264 (2005).

58. J. Mahnke, D. Vollhardt, K.W. Stöckelhuber, K. Meine and H.J. Schulze. Regular stripe patterns in skeletonized Langmuir–Blodgett films of arachidic acid. *Langmuir*, **15**, 8220–8224 (1999).

59. M.L. Kurnaz and D.K. Schwartz. Morphology of microphase separation in arachidic acid/cadmium arachidate Langmuir–Blodgett multilayers. *J. Phys. Chem.*, **100**, 11113–11119 (1996).

60. J.M. Bloch and W. Yun. Condensation of monovalent and divalent metal ions on a Langmuir monolayer. *Phys. Rev. A*, **41**, 844–862 (1990).

61. D.J. Ahn and E.I. Franses. Ion adsorption and ion exchange in ultrathin films of fatty acids. *AIChE J.*, **40**, 1046–1054 (1994).

62. M.E. Díaz, R.L. Cerro, F.J. Montes and M.A. Gálan. Non-ideal adsorption of divalent cations on a Langmuir monolayer: A theoretical model for predicting the composition of the resulting Langmuir–Blodgett films. *Chem. Eng. J.*, **131**, 155–162 (2007).

63. H. Hasmonay, M. Vincent and M. Dupeyrat. Composition and transfer mechanism of Langmuir–Blodgett multilayers of stearates. *Thin Solid Films*, **68**, 21–31 (1980).

64. V.I. Kovalchuk, M.P. Bondarenko, E.K. Zholkovskiy and D. Vollhardt. Mechanism of meniscus oscillations and stripe pattern formation in Langmuir–Blodgett films. *J. Phys. Chem. B*, **107**, 3486–3495 (2003).

65. V.I. Kovalchuk, E.K. Zholkovskij, M.P. Bondarenko and D. Vollhardt. Ion redistribution near the polar groups in the Langmuir wetting process. *J. Adhesion*, **80**, 851–870 (2004).

66. V.I. Kovalchuk, E.K. Zholkovskiy, M.P. Bondarenko and D. Vollhardt. Concentration polarization at the Langmuir monolayer deposition: Theoretical considerations. *J. Phys. Chem. B*, **108**, 13449–13455 (2004).

67. M.P. Bondarenko, E.K. Zholkovskiy, V.I. Kovalchuk and D. Vollhardt. Distributions of ionic concentrations and electric field around the three phase contact at high rates of Langmuir–Blodgett deposition. *J. Phys. Chem. B*, **110**, 1843–1855 (2006).

68. M.P. Bondarenko, V.I. Kovalchuk, E.K. Zholkovskiy and D. Vollhardt. Transient processes at the deposition of charged Langmuir monolayers. *Colloids Surf. A*, **354**, 226–233 (2010).

69. V.I. Kovalchuk, M.P. Bondarenko, E.K. Zholkovskiy, V.M. Starov and D. Vollhardt. Ions redistribution and meniscus relaxation during Langmuir wetting process. *J. Phys. Chem. B*, **115**, 1999–2005 (2011).

70. J.B. Peng, S. He, P. Dutta and J.B. Ketterson. Measurement of contact angle relaxation during the deposition of Langmuir–Blodgett films of cadmium stearate and valinomycin. *Thin Solid Films*, **202**, 351–357 (1991).

71. J.G. Petrov, A. Angelova and D. Möbius. Effect of the electrostatic and structural surface forces on the contact angles in Langmuir–Blodgett systems of cationic surfactants. 2. Static advancing and receding contact angles during deposition of mixed monolayers of methyl arachidate and dimethyldioctadecylammonium bromide. *Langmuir*, **8**, 206–212 (1992).

72. J.G. Petrov. Application of Blodgett multilayers for studying the role of the three phase contact line dynamic behaviours in dynamic wetting. *Z. Phys. Chem. (Leipzig)*, **266**, 706–712 (1985).

73. J.G. Petrov. Dependence of the maximum speed of wetting on the interaction in the three-phase contact zone. *Colloids Surf.*, **17**, 283–294 (1986).

74. M.P. Bondarenko, V.I. Kovalchuk, E.K. Zholkovskiy and D. Vollhardt. Concentration polarization at Langmuir monolayer deposition: The role of indifferent electrolytes. *J. Phys. Chem. B*, **111**, 1684–1692 (2007).

75. H. Yabu and M. Shimomura. Preparation of self-organized mesoscale polymer patterns on a solid substrate: Continuous pattern formation from a receding meniscus. *Adv. Funct. Mater.*, **15**, 575–581 (2005).

76. J. Huang, A.R. Tao, S. Connor, R. He and P. Yang. A general method for assembling single colloidal particle lines. *Nano Lett.*, **6**, 524–529 (2006).

77. A.R. Tao, J. Huang and A.P. Yang. Langmuir–Blodgettry of nanocrystals and nanowires. *Acc. Chem. Res.*, **41**, 1662–1673 (2008).

78. J. Huang, F. Kim, A.R. Tao, S. Connor and P. Yang. Spontaneous formation of nanoparticle stripe patterns through dewetting. *Nature Mater.*, **4**, 896–900 (2005).

79. M. Shimomura and T. Sawadaishi. Bottom-up strategy of materials fabrication: A new trend in nanotechnology of soft materials. *Curr. Opin. Colloid Interface Sci.*, **6**, 11–16 (2001).

80. T. Sawadaishi and M. Shimomura. Two-dimensional patterns of ultra-fine particles prepared by self-organization. *Colloids Surf. A*, **257–258**, 71–74 (2005).

81. T. Sawadaishi and M. Shimomura. Control of structures of two–dimensional patterns of nanoparticles by dissipative process. *Mol. Cryst. Liq. Cryst.*, **464**, 227(809)–231(813) (2007).

82. M. Ghosh, F. Fan and K.J. Stebe. Spontaneous pattern formation by dip coating of colloidal suspensions on homogeneous surfaces. *Langmuir*, **23**, 2180–2183 (2007).

83. J.A. Lee, K. Reibel, M.A. Snyder, L.E. Scriven and M. Tsapatsis. Geometric model describing the banded morphology of particle films formed by convective assembly. *ChemPhysChem.* **10**, 2116–2122 (2009).

84. S. Watanabe, K. Inukai, S. Mizuta and M.T. Miyahara. Mechanism for stripe pattern formation on hydrophilic surfaces by using convective self-assembly. *Langmuir*, **25**, 7287–7295 (2009).

85. Y. Cai and B.-M. Zhang Newby. Marangoni flow-induced self-assembly of hexagonal and stripelike nanoparticle patterns. *J. Am. Chem. Soc.*, **130**, 6076 (2008).

86. J. Matsui, T. Suzuki, T. Mikayama, A. Aoki and T. Miyashita. Self-organized dendritic patterns in the polymer Langmuir–Blodgett film. *Thin Solid Films*, **519**, 1998–2002 (2011).

87. Y. Kondo, H. Fukuoka, S. Nakano, K. Hayashi, T. Tsukagoshi, M. Matsumoto and N. Yoshino. Formation of wormlike aggregates of fluorocarbon-hydrocarbon hybrid surfactant by Langmuir–Blodgett transfer and alignment of gold nanoparticles. *Langmuir*, **23**, 5857–5860 (2007).

88. D.M. Kuncicky, R.R. Naik and O.D. Velev. Rapid deposition and long-range alignment of nanocoatings and arrays of electrically conductive wires from tobacco mosaic virus. *Small*, **2**, 1462–1466 (2006).

89. K. Demidenok, V. Bocharova, M. Stamm, E. Jähne, H.-J.P. Adler and A. Kiriy. One-dimensional SAMs of (12-Pyrrol-1-yl-dodecyl)-phosphonic acid templated by polyelectrolyte molecules. *Langmuir*, **23**, 9287–9292 (2007).

90. M. Byun, J. Wang and Z. Lin. Massively ordered microstructures composed of magnetic nanoparticles. *J. Phys.: Condens. Matter*, **21**, 264014 (2009).

91. J.R. Moffat, K. Sefiane and M.E.R. Shanahan. Effect of TiO_2 nanoparticles on contact line stick-slip behavior of volatile drops. *J. Phys. Chem. B*, **113**, 8860–8866 (2009).

92. M. Byun, S.W. Hong, F. Qiu, Q. Zou and Z. Lin. Evaporative organization of hierarchically structured polymer blend rings. *Macromolecules*, **41**, 9312–9317 (2008).

93. S.W. Hong, W. Jeong, H. Ko, M.R. Kessler, V.V. Tsukruk and Z. Lin. Directed self-assembly of gradient concentric carbon nanotube rings. *Adv. Funct. Mater.*, **18**, 2114–2122 (2008).

94. J.J. Diao and M.G. Xia. A particle transport study of vertical evaporation-driven colloidal deposition by the coffee-ring theory. *Colloids Surf. A*, **338**, 167–170 (2009).

7 Interfacial Studies of Coffee-Based Beverages
From Flavor Perception to Biofuels

*Michele Ferrari, Francesca Ravera,
Libero Liggieri, and Luciano Navarini*

CONTENTS

7.1 INTRODUCTION

Coffee is a beverage that is consumed daily by a large proportion of the human population worldwide (approximately 70%–80%). Physicochemical studies related to the surface properties of coffee-based beverages allow us to investigate aspects of the role of some classes of coffee constituents, such as lipids and other surface-active compounds, and also to obtain information about their relationship with consumer perception.

The worldwide success of *espresso* coffee beverage [1–4] seems to be based on the greater sensory satisfaction it gives to the consumer when compared with coffees prepared with other brewing methods. This coffee-brewing technique enhances several surface phenomena, such as foam and emulsion formation and stabilization, which strongly affect the organoleptic beverage properties related to the good wetting properties of the beverage on oral mucosa.

The different contributions of the components of commercially available coffee blends, well known as Arabica and Robusta from *Coffea arabica* and *Coffea canephora*, respectively, have been studied using dynamic experimental techniques such as maximum bubble pressure, whereas other techniques such as drop shape analysis can be used to characterize the air–coffee beverage interface. These two species are among those commercially used from the more than 60 known *Coffea* species (a genus belonging to the Rubiaceae family).

Different coffee-based beverages have been investigated ranging from espresso to soluble coffee. The presence and the possible role of some natural surface-active chemical components such as lipids on tensiometric behavior are consistent with a system having good wetting properties to oral cavity surfaces. In addition, the liquid–liquid interfacial properties of coffee oils extracted from roasted coffee have been investigated by studying samples obtained from different botanical species and geographical origins, as well as from roasted coffee blends and decaffeinated roasted coffee, and comparing them with common vegetable oils. The presence of coffee oil left in the powders after the beverage extraction can open new perspectives. In fact, coffee waste can be exploited as a renewable source of energy at low cost. This means that coffee grounds can be used in the future for alternative fuel production.

7.2 COFFEE: PREPARATION METHODS

Comparing the preparation methods can be helpful to obtain information about the composition of the corresponding beverage: the different conditions will result in a wide range of beverages, the different coffee fractions influenced mostly by the working water temperature or pressure (or both). It is well known that variations in the preparation variables (e.g., raw material, coffee/water ratio, temperature and pressure, percolation time, and beverage volume) can dramatically alter the "cup quality," not only in terms of taste and flavor (chemical composition) but also in terms of the characteristics of different phases present in the beverage (foam, emulsion, suspension, and solution) and their sensory effects (visual aspects and mouthfeel) and thus the physical properties.

Drip brew, or filter coffee, is a method for brewing coffee that involves pouring water over roasted, ground coffee beans contained on a filter. Water seeps through the coffee, absorbing its extractable fraction solely under gravity, and then passes through the bottom of the filter. The used coffee grounds are retained in the filter with the liquid falling (dripping) into a collecting vessel such as a carafe or pot [1,2]. The resulting beverage is close to a "true solution" of coffee compounds (typical total solids content in brew is equal to 13.0 g/L) [2], with a very limited presence of heterophases (very low lipid content and solid fine particles content). The beverage's physical properties are close to those of water at the same temperature.

Instant/soluble coffee is industrially prepared after roasting and grinding steps by extraction using water at a high temperature with a typical yield of 40% to 45%. Several extraction processes are used and described in the literature [3]. The aqueous extract is normally concentrated (by either evaporation or freeze concentration) before the drying process. The final product is dissolved directly in a cup of hot water (typically 2 g in 100 mL) in amounts based on personal preference. The resulting beverage is a solution and does not contain water-insoluble solids or oils. Like drip coffee, its physical properties are close to those of water at the same temperature.

Espresso brewing requires specialized equipment that can heat water to a temperature between 92°C and 94°C and pressurize it to 9 ± 2 bar [4]. The process is applied (percolation time) until the beverage volume in the cup meets the personal preference of the consumer or the regional traditions (or both). For example, in Italy, the volume range is 15 mL (*ristretto*) to 50 mL (*lungo*), with a typical volume of 25 to 30 mL for regular espresso. In Figure 7.1, the typical chemical composition of regular espresso coffee beverage is reported.

Unlike drip, filtered, or instant coffee, espresso is characterized by the presence of several coexisting phases, foams, emulsions, and suspensions in addition to the solution of coffee compounds. The beverage's physical properties are modulated by the percolation time: *ristretto espresso* coffee (low percolation time) shows higher total solids, viscosity, and density than *lungo espresso* coffee (high percolation time). At longer percolation times, the beverage's physical properties approach those of water at the same temperature.

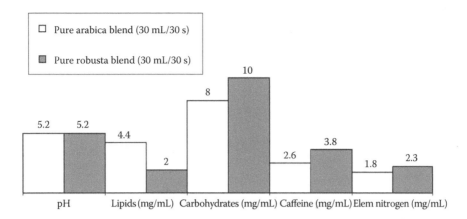

FIGURE 7.1 Comparison of espresso compositions from two different pure blends. (Reprinted from *Food Hydrocolloids*, 18, Navarini, L., M. Ferrari, F. Suggi Liverani, L. Liggieri, and F. Ravera. Dynamic tensiometric characterization of espresso coffee beverage, 387–393. Copyright 2004, with permission from Elsevier.)

7.3 FLAVOR PERCEPTION

The sensory description of complex systems, such as foods, includes the definition of certain standards to produce reliable measurements, and thus different scientific skills are mandatory. Despite a larger body of knowledge about food's bulk properties, the study of interfacial properties is more involved in the technological aspects than in sensory perception. Characterizing the perception of foods and beverages through their flavor, texture, and mouthfeel is a dynamic process that includes many surface physicochemical parameters, and evaluating their properties in the mouth has to take into account the associated physiological activities such as chewing, swallowing, and salivation.

7.3.1 INFLUENCING FACTORS

Food and beverage perception involves receptors embedded in the mouth's sensorial parts, such as the tongue, palate, and periodontal membrane, as well as muscles and tendons. Food psychorheology is a discipline aimed at understanding the sensory perception of all the physical properties of foods using advanced techniques and sensory methods [5]. The perception of food flavor is also related to the sensitive olfactory receptors. Although the transport of flavor molecules from the oral cavity to the olfactory region occurs in the gas phase, transport may also take place through a surface film. In fact, surface-active, nonvolatile flavor molecules can rapidly spread over the mucous layer in the nasal cavity. In addition, some saliva proteins exhibit a high monolayer spreading pressure and certain flavor molecules can reach the olfactory receptors even at extremely low vapor pressure [6].

Three factors—food properties, in-mouth environment, and psychosocial effects—govern the perception of flavor (Figure 7.2) [7–9]. Food properties are only related to the food itself, and include taste and aroma-active compounds, their interactions with the food matrix and the texture, color, and appearance of the matrix. These interactions also include the in-mouth environment and the activities inside the oral cavity, which are a complex set of ongoing events related to many variables such as saliva production, oral microbiota, breathing, mastication, and swallowing as well as anatomical aspects such as the oropharyngeal cavity size and the related mucous layer. Finally, psychosocial effects are all those daily life motivations that persuade someone to have a drink of coffee, including cognitive aspects like an individual's culture and memories, that person's mood or expectations, and effects on alertness and attention.

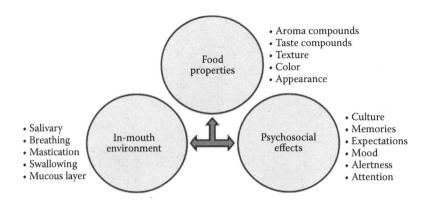

FIGURE 7.2 Groups of factors influencing flavor perception. (With kind permission from Springer Science+Business Media: *Flavours and Fragrances: Chemistry, Bioprocessing and Sustainability*, 1st ed., edited by R.G. Berger, 2007.)

7.3.2 Role of Saliva

Generally, we tend to compare the physicochemical properties of beverages to water (the main constituent) forgetting that a more appropriate model is our saliva. This bodily fluid is present as a thin film (varying from 72 to 100 μm in thickness) covering the surface area of our mouth (~220 cm²) [10]. The presence of thin adsorbed salivary films is also discussed in dentistry to assess the nature and significance of oral interfacial events, explaining the reason for a strong adhesion to healthy tooth surfaces *in vivo* [11].

Saliva facilitates important functions such as tasting, chewing, and swallowing, which are part of one's daily routine. Saliva is both a solvent and lubricant and also provides resistance to infection. Physically, saliva ensures self-cleaning of the oral cavity, avoids the degenerative changes by wetting the oral mucosa, sustains antibacterial activity by stabilizing lysozyme, thiocyanate, and immunoglobulins, and the buffering capacity of its bicarbonate content protects the teeth from an excess of acidic substances. Finally, saliva aids in the digestion of starch via the enzyme ptyalin (amylase) [12].

Saliva can be classified as

- Whole saliva: the combined secretion products of parotid, submandibular, sublingual, and minor salivary glands (labial, palatal, lingual, buccal, and glands of the lips) and
- Glandular saliva: the fluid secreted from a single type of salivary gland, before mixing takes place in the mouth

Several studies on saliva composition were carried out as part of a search for artificial substitutes [13]. Human whole saliva has a protein content of approximately 0.5 to 3.0 mg/mL. Salivary proteins may be produced by glands (e.g., α-amylase, lysozyme, mucins, proline-rich proteins), or derived from serum (e.g., albumin) or from immune cells (e.g., immunoglobulins). A significant amount of protein-bound carbohydrates exists in the saliva. Whole saliva contains approximately 10 to 100 μg/mL of lipids (glycolipids, free fatty acids, triglycerides, cholesteryl ester, and cholesterol) and a somewhat lower portion of phospholipids. Large fractions of salivary lipids are associated with proteins, especially mucins and proline-rich proteins. Many other substances are present in saliva, including nucleic acids, hormones, amino acids and their derivatives (urea, lactate, citrate, vitamins), and the chemical constituents of foods, toothpastes, and dental material [14].

The main macromolecular components of saliva are high–molecular weight glycoproteins that constitute the mucins. Mucins are present in two major populations known as MUC5B (old name,

MG1) and MUC7 (old name, MG2) [15]. MG1 (>1000 kDa) is composed of approximately 15% protein and 78% carbohydrate and it has an oligomeric structure in which mucin monomers are linked by disulfide bonds. MG2 (120–150 kDa) is composed of up to approximately 30% protein and 70% carbohydrates and is composed of single monomeric molecules [15].

The protein component of saliva largely determines its surface-active properties, which are very different from those of pure water. The surface tension of saliva (whole, stimulated) and its dynamic surface tension are reported in several studies (Figures 7.3 and 7.4) [16–18].

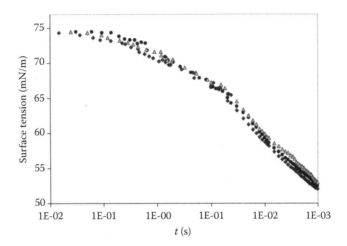

FIGURE 7.3 Dynamic tensiometric profile of human saliva collected from girls in the age group of 7 to 15 years. The volunteers did not suffer from any acute inflammation in the oral cavity nor chronic somatic diseases. The measurements obtained in this age group are quite similar. (Reprinted from *Colloids Surfaces B: Biointerfaces*, 74, Kazakov, V.N., A.A. Udod, I.I. Zinkovych, V.B. Fainerman, and R. Miller. Dynamic surface tension of saliva: General relationships and application in medical diagnostics, 457–461. Copyright 2009, with permission from Elsevier.)

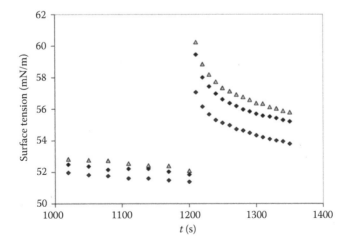

FIGURE 7.4 The same as in Figure 7.2 for times ranging for more than 1000 s. The relaxation process being used after the interfacial layer compression indicates the presence of long time adsorption components like higher molecular weight biopolymers. (Reprinted from *Colloids Surfaces B: Biointerfaces*, 74, Kazakov, V.N., A.A. Udod, I.I. Zinkovych, V.B. Fainerman, and R. Miller. Dynamic surface tension of saliva: General relationships and application in medical diagnostics, 457–461. Copyright 2009, with permission from Elsevier.)

TABLE 7.1
Surface Tension of Human Saliva

Time (s)	Temperature (°C)	Surface Tension (mN/m)	Reference
30	37	56.2 ± 2.1	[16]
300	37	49.5 ± 2.0	[16]
600	37	48.5 ± 1.6	[16]
1	25	68.7 ± 2.9	[17]
100	25	58.9 ± 4.7	[17]
∞	25	44.9 ± 3.8	[17]
Not reported	37	53.1 ± 2.7	[18]

Note: The values show the adsorption effect at longer time of higher molecular weight components. Discrepancies can arise as a function of age and population sample.

The dynamic surface tension decreases with time in line with expectation for protein systems. Table 7.1 shows the measured values (differences can be related to age or population sample).

Contact angle measurements on oral cavity surfaces illustrate the different wetting behaviors between saliva and water [19]: in particular, the contact angle at 60 s of whole saliva on human enamel *in vitro* is close to 63°, which is higher than that of water (~47°), whereas the contact angle at 30 s of whole saliva on human mucosa *in vivo* is close to 63°, which is lower than that of water (~70°).

Viscosity and rheological properties have been the topic of several studies as well. At the shear rate estimated at approximately 50 s^{-1} and a temperature of 37°C, the value of the viscosity of whole stimulated saliva was reported to be 1.9 ± 0.2 mPa·s [16]. Using a different technique (capillary viscosimetry), other authors [20] reported that stimulated whole saliva at 37°C values of relative viscosity (relative to water) were, on average, equivalent to 1.5 ranging from 1.24 ± 0.10 to 2.01 ± 0.47 mPa·s. The difference in the viscosity between stimulated and unstimulated whole saliva was investigated by Rantonen and Meurman [21]. They found that the viscosities of these two types of saliva differ not only in their absolute values but also in their responses to changes in salivary flow rates, and with respect to viscosity variation, the stimulated saliva was more viscosimetrically stable.

The rheological properties of saliva are typical of a non-Newtonian pseudoplastic fluid (viscosity decreasing with increasing shear rate), but were very different from those of water (Newtonian fluid: viscosity independent of shear rate) [22]. Only at extremely low shear rates (<10^{-4} s^{-1}) or at high frequencies (>100 Hz in the case of oscillatory flow), does saliva behave like a Newtonian fluid. Mellema et al. [23] report the non-Newtonian behavior of saliva and its viscoelasticity both as a bulk liquid and its surface. In fact, surface area changes and the consequent relaxation due to the diffusional exchange with the bulk can be considered as the mechanisms responsible for the surface tension variation, allowing surface rheology studies to be performed [24,25]. In the study by Mellema et al. [23], they provide numerous data on three different types of stimulated salivas: whole, submandibular, and parotid with the viscosity values of bulk liquids varying from 2.0 to 0.93 mPa·s under different flow conditions.

The viscoelasticity of total and glandular unstimulated salivas decreased in the following order: sublingual > palatal > whole ~ submandibular > parotid [22]. Other saliva properties that differ from water at 25°C are (a) the refractive index (whole saliva, 1.3333 ± 0.0001; submandibular saliva, 1.3333 ± 0.0002; parotid saliva, 1.3338 ± 0.0003); (b) density (whole and submandibular saliva, 999.3 kg m^{-3}; parotid saliva, 1010.5 kg m^{-3}); and (c) pH (whole saliva, 7.5 ± 0.1; submandibular saliva, 7.4 ± 0.5; parotid saliva, 7.9 ± 0.6) [23]. The pH of whole stimulated saliva is reported to be 7.4 [16]. However, values below 7 are closer for saliva in the mouth after its secretion. A value of 6.8 seems appropriate [13].

These studies demonstrate that a complex system like human saliva and its interaction with beverages within our oral cavity cannot be mimicked by water. Indeed, the components of saliva interact with aroma compounds causing a change in the release and perceived flavor by affecting the partition of aroma compounds between the liquid and gas phases [26]. It has been suggested that nonvolatile surface-active flavor molecules spread rapidly over the mucous layer in the nasal cavity [6]. Once again, the different surface properties of saliva in comparison to those of water have to be considered in discussing the sensory perception of beverages.

Viscosity also plays an important role in the mouthfeel of a beverage. Szczesniak [27] postulated that because a human being takes his or her equilibrium body state as the norm, the viscosity of saliva at rest should be taken as the norm (0 point) for viscosity judgments. Thus, beverages called "thin" should, when mixed with saliva in the mouth, exhibit a viscosity lower than saliva, whereas those called "thick" should have a viscosity greater than saliva, so that, in this regard, even the saliva flow rate can play an important role in flavor and texture perception.

Some studies show little correlation between the threshold of perception of primary flavors and saliva flow rate [28,29], but do show a correlation between the saliva flow of parotid or salivary glands, and the perception of texture and mouthfeel. For example, in a taste test, Maier [30] found a good correlation between perceived acidity and titratable acidity of coffee beverages using a pH end point close to the mean pH of the judges' saliva.

The most evident effect induced by food and beverage consumption is probably astringency, where saliva plays an important role if the response is induced by the consumption of organic acids. The rheology of the saliva is affected by both the proton concentration and the organic coniugated bases that sequester free calcium ions in saliva. Lowering the pH alters the rheological properties of saliva resulting in a decrease of its viscoelastic performance and lubricating ability [31].

7.3.3 ESPRESSO PERCEPTION

Espresso coffee can be defined as a polyphasic beverage, prepared from roasted and ground coffee and water alone, constituted by a foam layer of small bubbles with a particular tiger-tail pattern, on top of an emulsion of microscopic oil droplets in an aqueous solution of sugars, acids, protein-like material and caffeine, with dispersed gas bubbles and solid particles [32,33]. Thus, espresso coffee is a complex system, in which interfacial phenomena that occur in foam and emulsion formation and stabilization influence sensory properties such as visual aspect, aftertaste, and after-flavor (the sensation perceived for up to 15 min after having swallowed and emptied the mouth) and both body and mouthfeel. Foam layer and emulsified lipid fraction make espresso peculiar as compared with other coffee preparation methods. The foam layer plays a crucial role in providing the consumer the possibility to assess the preparation's quality by visual inspection, whereas time-dependent sensory properties are very important to prolong the pleasure of aroma perception after drinking.

For the latter, the sensory satisfaction is thus related to time-dependent wetting properties of the beverage on oral mucosa modulated by the drink's interfacial properties and transport phenomena that, in the mouth environment are, in fact, mediated by the water signal, considered like a marker for the regularity and the stability of breathing rhythm (in-mouth experience).

With the aim of studying this signal, nose space spectra (Figure 7.5) of the air exhaled during drinking of espresso coffee by an experienced coffee taster have been recorded by proton transfer reaction–mass spectrometry (PTR-MS) showing, after the first sip, an initial increase of the signal followed by a rapid decrease of the concentration of the compounds [7].

This behavior is influenced by the temperature dependence of the water–air partition coefficient, the dilution of coffee with saliva, the interaction of saliva components, and the adsorption and diffusion into the mucous layer [26,34].

To further increase complexity, it has to be stressed that the foam layer and aroma perception are interconnected, as recently discussed in studying a nontraditional espresso system. In particular, it has been found that the foam stability (in combination with the aroma's volatility properties)

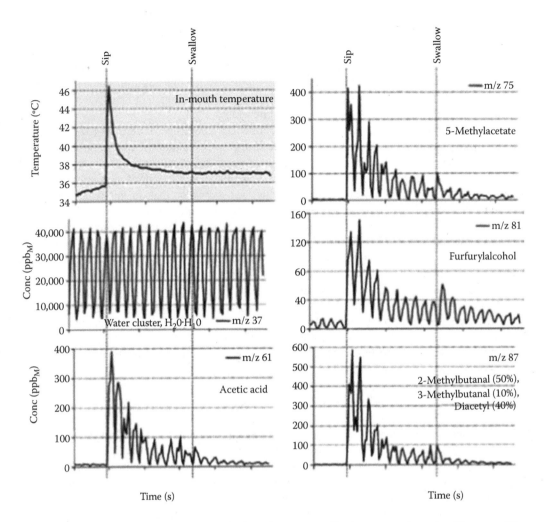

FIGURE 7.5 Nose space spectra during drinking of espresso coffee by an experienced coffee taster. Time evolution of physical parameters and some chemical components concentration during the drink consumption. (With kind permission from Springer Science+Business Media: *Flavours and Fragrances: Chemistry, Bioprocessing and Sustainability*, 1st ed., edited by R.G. Berger, 2007.)

influences the volatile release profile above the cup: 6 min after the beginning of the extraction, a low-stability foam layer provides higher above-the-cup volatile concentrations than a high-stability foam layer, suggesting that in addition to visual aspect, the foam layer acts as a barrier that entraps volatiles [35].

Moreover, the comparison between collapsed foam and the corresponding traditional (and non-traditional) espresso coffee bulk beverage reveals a quantitatively different chemical composition. In particular, the collapsed foam is richer in total lipids and in some selected sensory-relevant compounds compared with the corresponding bulk beverage, indicating that the foam layer is also very important from a taste point of view [36].

7.4 SURFACTANTS IN COFFEE AND RELATED FOAMING PROPERTIES

Despite the importance attributed to espresso coffee foam (also known as *crema*) by both experts and consumers, scientific literature on this subject is rather scarce. Very recently, this literature has been reviewed by Illy and Navarini [37]. The gas phase of the crema consists of carbon dioxide and gases formed in the Maillard reaction during coffee roasting and entrapped within the coffee cell structure, air, water vapor from the percolation process, and volatile aroma compounds that are released from the liquid into the gas phase. The formation of crema is affected by several factors. Carbon dioxide content and thermodynamic parameters during espresso coffee extraction are the main factors, which are involved in the proposed foam formation mechanism [38]. The presence of surface-active compounds makes the existence of a stable foam frequently possible.

The chemical nature of natural coffee surfactants has not yet been fully clarified. In the first chemical characterization of a pure Arabica blend espresso coffee brew, some classes of surfactants were quantitatively determined but, at that time, these were not recognized as surface-active agents [32]. In particular, proteins and lipids (the latter were determined as triglycerides, diglycerides, and fatty acids). The same investigation evidenced the different foaming behaviors of pure Arabica and pure Robusta blend espresso, suggesting that within the foam, the gas bubbles/lipid droplets balance governs foam formation and stability. At a given roasting degree, high carbon dioxide content and low lipid content leads to abundant but quickly vanishing foam, as observed on average for pure Robusta espresso. A similar behavior is observed with pure Arabica when bicarbonate-rich water is used to prepare traditional espresso coffee [39].

It has also been suggested that other classes of complex molecules such as glycolipids or glycoproteins might be involved in the formation and stabilization of the foam [40]. Foamability and foam stability of espresso coffee were determined as a function of the roasting degree along with other dependent variables, like total solids, pH, fat, protein, and carbohydrate contents [41].

The principal components analysis clearly indicated a very high correlation between foamability and protein content, and between foam stability and the fractions containing high–molecular weight polysaccharides, namely, a mixture of galactomannan and arabinogalactan (or better, complexes between polysaccharides, protein, and phenolic compounds caused by the roasting process, probably products of Maillard reactions and nonenzymatic browning) [41]. Moreover, a strong negative correlation between fat and foamability was put in evidence [42].

Among the possible surface-active coffee compounds, caffeine has been investigated: in the concentration range typical for espresso coffee, caffeine shows appreciable surface activity in water that is enhanced by the presence of sucrose [43].

Depending on the bean's postharvest processing method—dry method (DI), semidry method (DP), or wet method (W)—different amounts of soluble high–molecular weight materials can be found in coffee beverages. The polysaccharide fractions isolated from the beverages showed different total sugar and protein contents but nevertheless physicochemical properties such as viscosity and surface tension showed similar values. Beans processed using the wet method showed the highest yield in the polysaccharide fraction, whereas the beverage showed the highest mannose content, probably due to a higher soluble galactomannan content [44]. This finding suggests that the influence of coffee polysaccharides on beverage surface tension seems to be negligible.

The first foaming fraction from ground and roasted coffee was isolated by Navarini et al. [45] and characterized by the same group [46,47]. The addition of ammonium sulfate to the coffee extract (defatted dark roast Arabica blend extracted by solid–liquid extraction with water at 90°C) produced a precipitate that was redissolved, dialyzed, and freeze-dried.

This fraction, also known as total foaming fraction (TFF) redissolved in water, heavily foams on shaking. Isolation of TFF from pure Arabica and from pure Robusta revealed no significant differences in yield, elemental analysis, <Mw>, foam volume, and surface tension of a 0.4% w/v aqueous solution [47]. Different foaming behaviors between Arabica and Robusta may therefore reflect differences in other chemical compounds, for instance, lipids and carbon dioxide precursors (carbohydrates) [48].

A further difference between the foams obtained from pure Arabica espresso and that of pure Robusta is the special visual pattern known as "tiger skin" or "tiger tail" not observed in the latter. This effect, appreciated by consumers, is attributed to the presence in the foam of very fine coffee grounds along with cell wall fragments and it may reflect the different cellular structures between the two coffee species and thus the different behaviors when subjected to grinding [49].

However, in addition to the visual aspect, the solid particles in the Arabica espresso foam may be considered an additional class of surfactants, which can play a role in the foam's stabilization. No detailed studies have been focused at disclosing the role played by solid particles in espresso coffee foam. Optical microscopy has been performed for preliminary observation. In pure Arabica regular espresso freshly prepared, bubbles seem to be covered by solid particles (size <50 μm) [37]. The particle surface seems to mostly remain on the external side, suggesting a contact angle (through the aqueous phase) of less than 90° [50].

The particles in the dry espresso foam (pure Arabica), are clearly located in the plateau border, suggesting the tendency to be unattached. In fact, during drainage, unattached particles predominantly follow the net motion of the liquid [51]. This observation strongly suggests the possible stabilizing role of solid particles in the *crema*.

7.5 INTERFACIAL STUDIES ON ESPRESSO AND COMPARISON WITH OTHER COFFEE-BASED BEVERAGES

The contact of water with roasted coffee solids is the crucial step in producing a beverage. In fact, when instant coffee is not used, ground roasted coffee can be placed in contact with water resulting in three different extraction methodologies: decoction, infusion, or pressure [2]. Boiled coffee (very popular in Nordic countries) and Turkish/Greek coffee (served in all Mediterranean countries from Slovenia to Morocco) are, for example, prepared by the decoction method, whereas filtered coffee (also known as drip coffee), is obtained by the infusion method. Espresso coffee is prepared by the pressure method, and for this type of preparation, as mentioned above, thermodynamics and coffee chemistry create the conditions to produce a foam layer. The presence of several amphipathic solutes in coffee beverages, including low–molecular weight and high–molecular weight surfactants, has stimulated the study of interfacial properties in an effort to evidence possible differences between espresso and other types of coffee preparations.

7.5.1 DYNAMIC AND EQUILIBRIUM PROPERTIES

Maximum bubble pressure and pendant drop techniques have been successfully used to characterize espresso coffee beverage prepared following a standard procedure [33]. Surface tension values, even at relatively long times, are far from reaching a plateau (Figure 7.6), and this is particularly evident for Arabica espresso. In fact, the interfacial adsorption kinetics profile shows that the Robusta sample tends to a steady state more rapidly than the Arabica sample. Arabica has a higher content of lipids, which significantly lowers the surface tension of the water medium [33], suggesting that Arabica espresso coffee beverage is an effective "wetting agent" for the oral cavity [52].

Espresso coffee preparation has also been characterized over a wide range of total solids (from *ristretto* to *lungo* beverages) to ascertain the effect of prolonging the extraction on the surface properties, and on its dynamic surface behavior compared with other types of coffee beverages. In all cases, espresso coffee has a surface tension that is lower than other coffees (soluble coffee and drip coffee). The pressure applied to prepare espresso is very effective in extracting surface-active compounds that are not extracted by filter coffee preparation. Once again, emulsified lipids may well explain the observed differences because the lipid content in espresso (2–2.5 g/L) is remarkably higher than that in both instant and drip filter coffees (tens of milligrams per liter) [53].

The influence of coffee postharvest processing methods on beverage surface properties has been recently investigated. The same coffee lots were processed using dry, semidry, and wet methods. The corresponding beverages obtained from roasted products had identical surface tensions from all three methods [44].

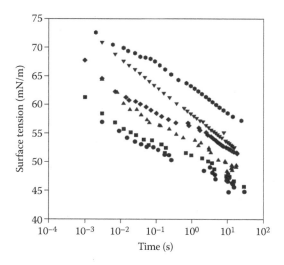

FIGURE 7.6 Dynamic surface tension of espresso coffee (●, *ristretto*; ■, *regular*; ▲, *lungo*) and of other types of coffee preparations: soluble coffee (⬤ c = 10 g/L, ◆ c = 40 g/L and, ▼ drip coffee). (Reprinted from *Food Hydrocolloids*, 21, Ferrari, M., L. Navarini, L. Liggieri, F. Ravera, and F. Suggi Liverani. Interfacial properties of coffee-based beverages, 1374–1378, Copyright 2007, with permission from Elsevier.)

7.5.2 CONTACT ANGLE MEASUREMENT

The wetting action of espresso coffee beverages on hydrophobic surfaces is higher than that of the other coffee preparations. It is also higher than that of human whole stimulated saliva, with values obtained for human saliva showing how slight its contribution is by how little it decreases the wetting angle. The spreading kinetics of espresso coffee determined from changes in the contact angle at 45°C for up to 300 s approaches a plateau (Figure 7.7), suggesting that times longer than 5 min are necessary to measure an equilibrium contact angle [52]. The hypothesis that coffee, particularly espresso coffee, is effective at wetting the human oral cavity is corroborated by the data reported in literature.

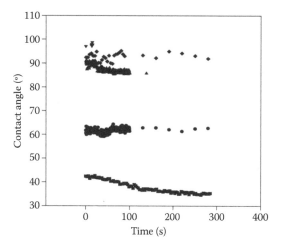

FIGURE 7.7 Spreading kinetics of espresso as determined from differences in contact angle and different coffee-based preparations at 20°C: ▲, drip coffee; ●, espresso coffee; ■, espresso at 45°C compared with ▼ water and ◆ saliva. (Reprinted from *Food Hydrocolloids*, 21, Ferrari, M., L. Navarini, L. Liggieri, F. Ravera, and F. Suggi Liverani. Interfacial properties of coffee-based beverages, 1374–1378. Copyright 2007, with permission from Elsevier.)

Coffee affects the mouth surfaces in a number of ways. Coffee is a stain, one of the most chromogenic staining solutions for denture materials and is usually used as a reference beverage in color stability studies [54,55]. Coffee significantly reduces the adhesion of *Streptococcus mutans* to dental surfaces (enamel and dentine) [56]. *S. mutans* is a major cause of human dental caries. Coffees contain components that interfere *in vitro* with *S. mutans* traits, for example, biofilm formation, which is relevant to cariogenesis [57]. Conceivably, interfacial interactions between coffee and the human oral cavity may be competitive with respect to those of saliva. For example, the long-lasting aftertaste after swallowing espresso could be a consequence of these interactions.

7.6 STUDIES ON COFFEE OILS

The two commercially relevant coffee species, Arabica and Robusta, contain between 7% and 17% w/w lipid fraction (known as coffee oil) with an average content of Arabica significantly higher (15%) than that of Robusta (10%). Coffee oil is composed mainly of triacylglycerols as well as fatty acids in similar quantities to those found in common edible vegetable oils. The relatively large unsaponifiable fraction is rich in diterpenes, contains sterols and tocopherols, and is very important as a chemical marker for authentication and traceability purposes. Both green and roasted coffee oils are used industrially, the former in the cosmetics industry and the latter in food applications such as flavorings [58]. In addition to the role played by coffee lipids in the foaming and surface properties of coffee-based beverages, emulsified lipids are relevant in differentiating:

- Foaming properties of espresso prepared from different botanical coffee species (Arabica and Robusta)
- Surface properties of bulk espresso prepared from different botanical coffee species (Arabica and Robusta), and
- Surface properties of coffees prepared by different methods

7.6.1 INTERFACIAL PROPERTIES

In coffee brews like espresso, emulsified lipids play a relevant role in the organoleptic properties of the beverage. Therefore, the study of the interfacial properties of coffee oil (the coffee lipid fraction) is needed to obtain information about the emulsifiability of the coffee lipid fraction. The liquid–liquid interfacial properties of coffee oils extracted from roasted coffee were recently investigated. Figure 7.8 characterizes the oils from roasted coffee of different botanical species and geographical origins as well as from roasted coffee blends and decaffeinated roasted coffee [59].

The presence of efficient surface-active compounds in roasted coffee oil is observed (see compression effect) in both the adsorption kinetics at the interface and the interfacial tension (IT) values that are similar to (and in some cases, remarkably lower than) those of other edible vegetable oils like sunflower and corn. No remarkable differences have been observed in comparing *Coffea* species and origins, even when blending and decaffeination processes showed a nonnegligible influence on IT values.

The interfacial properties of green coffee oil are similar to those of roasted coffee oil, suggesting that thermally induced modifications of coffee lipids as well as liposoluble Maillard reaction products play only a negligible role in the interfacial properties of roasted coffee oils. The surface-active compounds such as free fatty acids, diglycerides, and other unsaponifiable compounds present in the green coffee oil are not affected by the roasting process. Decaffeinated coffee oil (both green and roasted) has higher IT values than those of corresponding regular coffee oils, indicating that the decaffeination process removes other liposoluble surfactants in addition to caffeine (Table 7.2).

Given the interfacial behavior of roasted coffee oil, the combination of pressure, temperature, and rheological constraint with a low liquid/liquid IT value of coffee oil, the espresso brewing method provides a very efficient emulsification of the coffee lipids present in the starting coffee

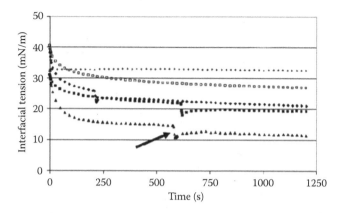

FIGURE 7.8   Interfacial tension (mN/m) as a function of time of corn oil and selected roasted coffee oil samples: (+) corn oil, (□) deca, (◆) aind, (■) rind, (▲) blend 1. The arrow indicates compression of the interfacial layer. (Reprinted from *Colloids Surfaces A*, 365, Ferrari, M., F. Ravera, E. De Angelis, F. Suggi Liverani, and L. Navarini. Interfacial properties of coffee oils, 79–82. Copyright 2010, with permission from Elsevier.)

TABLE 7.2
Equilibrium Interfacial Tension of Roasted Coffee Oil Samples

Sample	Oil Content (% w/w)	IT (mN/m)
Arabica (India)	13.5 ± 0.1	19.8 ± 1.5
Arabica (Brazil) 1	16.5 ± 0.4	17.0 ± 3.3
Arabica (Brazil) 2	16.4 ± 0.3	19.1 ± 0.1
Arabica (Blend) 1	15.2 ± 0.2	12.4 ± 1.4
Arabica (Blend) 2	15.8 ± 0.2	15.6 ± 0.4
Decaffeinated	15.5 ± 0.3	26.5 ± 0.3
Robusta (India)	9.2 ± 0.1	19.2 ± 0.1
Robusta (Brazil)	7.8 ± 0.1	20.9 ± 0.1

Note: The differences can depend on the composition of the oil phase, the botanical species, and on industrial processes.

dose, even for the very short brewing times (25–30 s) needed to prepare a cup of espresso. On the other hand, if sufficient pressure is not applied, as in other popular coffee preparation methods, the emulsion phase cannot be formed in the beverage despite the good emulsifiability of the coffee lipids. The similarities of interfacial properties of coffee oils with other vegetable oils allowed us to consider their use in alternative fuel production.

7.7   BIOFUELS FROM COFFEE OILS

Biodiesel is the name applied to fuels manufactured by the esterification of renewable oils, fats, and fatty acids. Coffee waste is a potential renewable energy source due to its oils and grounds content. Recently, coffee oil from roasted coffee waste had been suggested as a potential raw material for biodiesel production [60–62].

7.7.1   Chemistry and Exploitation of Coffee Waste

Preliminary estimates indicate that, in the United States alone, preparation of biodiesel and pellets from the grounds from waste generated by coffee shops can produce a profit of more than $8

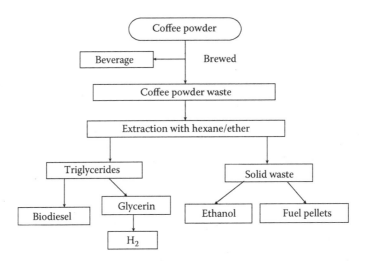

FIGURE 7.9 Schematic of the biodiesel production process from spent coffee grounds. (Reprinted from *Bioresource Technol.*, 99, Oliveira, L.S., A.S. Franca, R.S. Camargosm, and V.P. Ferraz. Coffee oil as a potential feedstock for biodiesel production, 3244–3250. Copyright 2008, with permission from Elsevier.)

million/year [63]. The oil and biodiesel were found to be stable for more than 1 month without any observable physical changes, so the biodiesel from spent coffee grounds is a strong candidate as an alternative to diesel.

Currently, spent coffee grounds are used primarily in gardens as compost for plants. Ideal coffee grounds for the soil have a C/N ratio (by weight) of 20:1. The C/N ratio after the oil extraction process showed that there was an insignificant change in this ratio before (19.8:1) and after (15.7:1) the extraction process. Thus, processed coffee grounds can be used as compost for the garden and as fuel pellets (1 lb pellet ≡ 8691 Btu of energy). In fact, after the oil is extracted from the grounds, the grounds can be pelletized immediately. In one example, approximately 15% of the oil obtained from spent coffee grounds was converted to biodiesel with a 100% yield. Gas chromatography-mass spectrometry and high-performance liquid chromatography analyses indicated that the coffee biodiesel consisted of both saturated (51.4%) and unsaturated (48.6%) esters.

Figure 7.9 shows a schematic of the biodiesel production process from spent coffee grounds. This biodiesel can be used industrially as an alternative to diesel, adding approximately 340 million gallons of biodiesel to the world's fuel supply [63].

7.8 SUMMARY

Coffee-drinking is a human routine that, especially for espresso coffee, is accomplished quickly, simply, and often taken for granted. However, if we look at coffee with the eye of a physical chemist, coffee's complexity is extremely surprising, particularly in terms of the complex interfacial interactions between the beverage and the drinker's mouth.

Coffee contains a number of surface-active substances and the type of preparation strongly influences the heterophasic nature of the beverage. The organoleptic characteristics depend not only on the water-soluble compounds but also on the foams and emulsions present in the drink (an almost unique characteristic of espresso). The interfacial properties of the coffee contribute significantly to the sensory perception and combine with those of the saliva to give the coffee drinker the full experience on the receptors present in the oral cavity. Among the many constituents of coffee, lipids are particularly interesting. Coffee oil not only contains liposoluble volatile aroma compounds but also some of its surface-active components make it easily emulsifiable, and these two characteristics make it particularly efficient in its interaction with the oral cavity in terms of aroma and aftertaste.

Only a small fraction of the oil is extracted during coffee preparation regardless of the method, and the bulk of the oil remains in the spent coffee grounds. Thus, spent coffee grounds may prove to be a viable source of biodiesel, a second surprising aspect of coffee.

Savoring a good cup is what coffee is about after all, but this pleasure is actually the result of a complex interplay between chemistry, physics, humans, and their friends.

REFERENCES

1. A. Peters. Brewing makes the difference. In Proceedings of the 14th International Scientific Colloquium on Coffee, San Francisco, USA, ASIC Paris, France, pp. 97–106 (1991).
2. M. Petracco. Beverage preparation: Brewing trends for the new millennium. In *Coffee: Recent Developments*, R.J. Clarke and O.G. Vitzthum (Eds.). Blackwell Science, London (2001).
3. R.J. Clarke. Technology III: Instant coffee. In *Coffee: Recent Developments*, R.J. Clarke and O.G. Vitzthum (Eds.). Blackwell Science, London (2001).
4. E. Illy. The complexity of coffee. *Sci. Am.*, **6**, 92–98 (2002).
5. J.X. Guinard and R. Mazzucchelli. The sensory perception of texture and mouthfeel. *Trends Food Sci. Technol.*, **7**, 213–219 (1996).
6. M. Larsson and K. Larsson. Neglected aspects of food flavor perception. *Colloids Surf. A*, **123–124**, 651–655 (1997).
7. R.G. Berger. *Flavours and Fragrances: Chemistry, Bioprocessing and Sustainability*. Springer Verlag, Berlin (2007).
8. E.D. Capaldi (Ed.). *Why We Eat What We Eat: The Psychology of Eating*. Am. Psychol. Assoc., Washington, DC (1996).
9. P. Rozin. *Toward a Psychology of Food Choice (Danone Chair Monographs)*. Inst. Danone, Brussels (1998).
10. C. Dawes. How much saliva is enough for avoidance of xerostomia. *Caries Res.*, **38**, 236–240 (2004).
11. P.-O. Glantz. Interfacial phenomena in the oral cavity. *Colloids Surf. A*, **123–124**, 657–670 (1997).
12. A. Woda. *Fisiologia del Sistema Stomatognatico*. Masson Italia Editori, Milano (in Italian) (1984).
13. J.-Y. Gal, Y. Fovet and M. Adib-Yadzi. About a synthetic saliva for *in vitro* studies. *Talanta*, **53**, 1103–1115 (2001).
14. T.K. Fábián, P. Fejerdy and P. Csermely. Chemical biology of saliva in health and disease. In *Wiley Encyclopedia of Chemical Biology*, T.P. Begley (Ed.). John Wiley & Sons, New York (2008).
15. C. Wickström, C. Christersson, J.R. Davies and I. Carlstedt. Macromolecular organization of saliva: Identification of insoluble MUC5B assemblies and non-mucin proteins in the gel phase. *Biochem J.*, **351**, 421–428 (2000).
16. C.E. Christersson, L. Lindh and T. Arnebrant. Film forming properties and viscosities of saliva substitutes and human whole saliva. *Eur. J. Oral Sci.*, **108**, 418–425 (2000).
17. V.N. Kazakov, A.A. Udod, I.I. Zinkovych, V.B. Fainerman and R. Miller. Dynamic surface tension of saliva: General relationships and application in medical diagnostics. *Colloids Surf. B Biointerfaces*, **74**, 457–461 (2009).
18. P.O. Glantz. The surface tension of saliva. *Odontol. Revy.*, **9**, 119–127 (1970).
19. A. Vissink, H.P. de Jong, H.J. Busscher, J. Arends and E.J. Gravenmade. Wetting properties of human saliva and saliva substitutes. *J. Dental. Res.*, **65**, 1121–1124 (1986).
20. M.E. Ortega Pantaleon, M. Calzado Suarez and M. Perez Marques. Evaluacion del flujo y viscosidad salival y su relacion con el indice de caries. *Medisan*, **2**, 33–39 (in Spanish) (1998).
21. P.J.F. Rantonen and J.H. Meurman. Viscosity of whole saliva. *Acta Odontol. Scand.*, **56**, 210–214 (1998).
22. W.A. Van Der Reijden, E.C.I. Veerman and A.V. Nieuw Amerongen. Shear rate dependent viscoelastic behavior of human glandular salivas. *Biorheology*, **30**, 141–152 (1993).
23. J. Mellema, H.J. Holterman, H.A. Waterman and C. Blom. Rheological aspects of mucin-containing solutions and saliva substitutes. *Biorheology*, **29**, 231–249 (1992).
24. L. Liggieri, M. Ferrari, D. Mondelli and F. Ravera. Surface rheology as a tool for the investigation of processes internal to surfactant adsorption layers. *Faraday Discuss.*, **129**, 125–140 (2005).
25. F. Ravera, M. Ferrari, E. Santini and L. Liggieri. Influence of surface processes on the dilational viscoelasticity of surfactant solutions. *Adv. Colloid Interface Sci.*, **117**(1–3), 75–100 (2005).
26. A.J. Taylor. Release and transport of flavors *in vivo*: Physicochemical, physiological, and perceptual considerations. *Compr. Rev. Food Sci. F.*, **1**, 45–57 (2002).
27. A.S. Szczesniak. Classification of mouthfeel characteristics of beverages. In *Food Texture and Rheology*, P. Sherman (Ed.). Academic Press, London, pp. 1–20 (1979).

28. J.X. Guinard, C. Zoumas-Morse, C. Walchak and H. Simpson. Relation between saliva flow and flavor release from chewing gum. *Physiol. Behav.*, **61**, 591–596 (1997).

29. J.X. Guinard, C. Zoumas-Morse and C. Walchak. Relation between parotid saliva flow and composition and the perception of gustatory trigeminal stimuli in foods. *Physiol. Behav.*, **63**, 109–118 (1997).

30. H.G. Maier. The acids of coffee. In Proceedings of the 12th International Scientific ASIC Colloquium on Coffee, Montreux, Switzerland, ASIC, Paris, France, pp. 229–237 (1987).

31. R.A. Sowalski and A.C. Noble. Comparison of the effects of concentration, pH and anions on astringency and sourness of organic acids. *Chem. Senses*, **23**, 343–349 (1998).

32. M. Petracco. Physico-chemical and structural characterization of espresso coffee brew. In Proceedings of the 13th International Scientific Colloquium on Coffee, Paipa, Colombia, ASIC, Paris, pp. 246–261 (1989).

33. L. Navarini, M. Ferrari, F. Suggi Liverani, L. Liggieri and F. Ravera. Dynamic tensiometric characterization of espresso coffee beverage. *Food Hydrocolloid*, **18**, 387–393 (2004).

34. A.J. Taylor and R.S.T. Linforth. Techniques for measuring volatile release *in vivo* during consumption of food. In *Flavour Release*, D.D. Roberts, A.J. Taylor and A.J. Tayim (Eds.). American Chemical Society, Washington, DC, pp. 8–21 (2000).

35. S. Dold, C. Lindinger, E. Kolodziejczyk, P. Pollien, S. Ali, J.C. Germain, S. Garcia Perin, N. Pineau, B. Folmer, K.-H. Engel, D. Barron and C. Hartmann. Influence of foam structure on the release kinetics of volatiles from espresso coffee prior to consumption. *J. Agric. Food Chem.*, **59**, 11196–11203 (2011).

36. L. Navarini, V. Lonzarich, S. Colomban, D. Rivetti, E. De Angelis, N. D'Amelio and F. Suggi Liverani. Chemical characterization of espresso coffee foam: Influence of the extraction method. In Proceedings of the 23rd International Colloquium on the Chemistry of Coffee, Bali, Indonesia, ASIC, Paris, France, pp. 208–216 (2010).

37. E. Illy and L. Navarini. Neglected food bubbles: The espresso coffee foam. *Food Biophys.*, **6**, 335–348 (2011).

38. L. Navarini, M. Barnaba and F. Suggi Liverani. Physicochemical characterization of espresso coffee foam. In Proceedings of the 21st International Colloquium on the Chemistry of Coffee, Montpellier, France, ASIC, Paris, France, pp. 320–327 (2006).

39. L. Navarini and D. Rivetti. Water quality for espresso coffee. *Food. Chem.*, **122**, 424–428 (2010).

40. A. Illy and R. Viani (Eds.). *Espresso Coffee: The Chemistry of Quality*. Academic Press, London (1995).

41. F.M. Nunes, M.A. Coimbra, A.C. Duarte and L. Delgadillo. Foamability, foam stability, and chemical composition of espresso coffee as affected by the degree of roast. *J. Agric. Food Chem.*, **45**, 3238–3243 (1997).

42. F.M. Nunes and M.A. Coimbra. Influence of polysaccharide composition in foam stability of espresso coffee. *Carbohydr. Polym.*, **37**, 283–285 (1998).

43. V. Aroulmoji, F. Hutteau, M. Mathlouthi and D.N. Rutledge. Hydration properties and the role of water in taste modalities of sucrose, caffeine and sucrose–caffeine mixtures. *J Agric Food Chem.*, **49**, 4039–4045 (2001).

44. A. Tarzia, M.B. dos Santos Scholz and C.L. de Oliveira Petkowicz. Influence of the postharvest processing method on polysaccharides and coffee beverages. *Int. J. Food Sci. Technol.*, **45**, 2167–2175 (2010).

45. L. Navarini, R. Gilli, V. Gombac, A. Abatangelo, M. Bosco and R. Toffanin. Polysaccharides from hot water extracts of roasted *Coffea arabica* beans: Isolation and characterization. *Carbohydr. Polym.*, **40**, 71–81 (1999).

46. L. Navarini and M. Petracco. Heavily foaming coffee fraction and process for its manufacture. European Patent Application, EP 1 021 957 A1 (1999).

47. M. Petracco, L. Navarini, A. Abatangelo, V. Gombac, E. D'Agnolo and F. Zanetti. Isolation and characterization of a foaming fraction from hot water extracts of roasted coffee. In Proceedings of the 18th International Scientific Colloquium on Coffee, Helsinki, Finland, ASIC, Paris, pp. 95–105 (1999).

48. B.A. Anderson, E. Shimoni, R. Liardon and T.P. Labuza. The diffusion kinetics of carbon dioxide in fresh roasted and ground coffee. *J. Food Eng.*, **59**, 71–78 (2003).

49. B.A. Anderson. The diffusion kinetics of carbon dioxide in fresh roasted coffee. Thesis (M.S.), University of Minnesota, Minneapolis, USA (2000).

50. B.P. Binks. Particles as surfactants—Similarities and differences. *Curr. Opin. Colloid Interface Sci.*, **7**, 21–41 (2002).

51. S.J. Neethling, H.T. Lee and J.J. Cilliers. The dispersion of particles within foams. In Proceedings of the XXI Int. Congress of Technical and Applied Mechanics, Warsaw, Poland (2004).

52. M. Ferrari, L. Navarini, L. Liggieri, F. Ravera and F. Suggi Liverani. Interfacial properties of coffee-based beverages. *Food Hydrocolloid*, **21**, 1374–1378 (2007).

53. W.M.N. Ratnayake, R. Hollywood, E. O'Grady and B. Stavric. Lipid content and composition of coffee brews prepared by different methods. *Food Chem Toxicol.*, **31**, 263–269 (1993).

54. T. Koksal and I. Dikbas. Color stability of different denture teeth materials against various staining agents. *Dent. Mater. J.*, **27**, 139–144 (2008).

55. N. Padiyar and P. Kaurani. Color stability: An important physical property of esthetic restorative materials. *Int. J. Clin. Dent. Sci.*, **1**, 81–84 (2010).

56. L.D. de Oliveira, E.H. da Silva Brandão, L.F. Landucci, C.Y. Koga-Ito and A.O. Cardoso Jorge. Effects of *Coffea arabica* on *Streptococcus mutans* adherence to dental enamel and dentine. *Braz. J. Oral Sci.*, **6**, 1438–1441 (2007).

57. M. Stauder, A. Papetti, D. Mascherpa, A.M. Schito, G. Gazzani, C. Pruzzo and M. Daglia. Antiadhesion and antibiofilm activities of high molecular weight coffee components against *Streptococcus mutans*. *J. Agric. Food Chem.*, **58**, 11662–11666 (2010).

58. T.A. Lucon Wagemaker, C.R. Limonta Carvalho, N. Borlina Maia, S.R. Baggio and O. Guerreiro Filho. Sun protection factor, content and composition of lipid fraction of green coffee beans. *Ind. Crop. Prod.*, **33**, 469–473 (2011).

59. M. Ferrari, F. Ravera, E. De Angelis, F. Suggi Liverani and L. Navarini. Interfacial properties of coffee oils. *Colloids Surf. A*, **365**, 79–82 (2010).

60. L.S. Oliveira, A.S. Franca, R.S. Camargos and V.P. Ferraz. Coffee oil as a potential feedstock for biodiesel production. *Bioresour. Technol.*, **99**, 3244–3250 (2008).

61. A.A. Nunes, A.S. Franca and L.S. Oliveira. Activated carbons from waste biomass: An alternative use for biodiesel production from solid residues. *Bioresour. Technol.*, **100**, 1786–1792 (2009).

62. L.S. Oliveira, A.S. Franca, J.C.F. Mendonca and M.C. Barros-Junior. Proximate composition and fatty acids profile of green and roasted defective coffee beans. *Lebensm. Wiss. Technol.*, **39**, 235–239 (2006).

63. N. Kondamudi, S.K. Mohapatra and M. Misra. Spent coffee grounds as a versatile source of green energy. *J. Agric. Food Chem.*, **56**, 11757–11760 (2008).

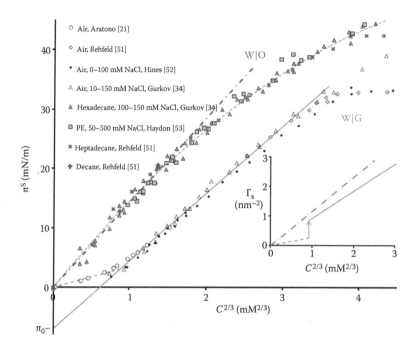

FIGURE 2.3 Interfacial pressure π^S versus the 2/3-power of the mean activity $C^{2/3}$ for $C_{12}H_{25}SO_4Na$ solutions at different NaCl concentrations. (Data for the W|G surface from M. Aratono et al., *J. Colloid Interface Sci.* 98, 33–38, 1984; T.D. Gurkov et al., *Colloids Surf. A* 261, 29–38, 2005; S.J. Rehfeld, *J. Phys. Chem.* 71, 738–745, 1967; J.D. Hines, *J. Colloid Interface Sci.* 180, 488–492, 1996. With permission.) For the W|O interface, the oil is heptadecane, decane [51], hexadecane [34], and petroleum ether [53]. Solid line, data fit for the W|G in the range $C^{2/3} = 1.2$ to 3 mM$^{2/3}$ (the LE region), according to Equation 2.43. The short, dashed line stands for W|G data in the range $C^{2/3} = 0$ to 1 mM$^{2/3}$ (gaseous monolayer region). Dash-dot line, data fit of the W|O interface in the range $C^{2/3} = 0$ to 1.8 mM$^{2/3}$ (Equation 2.41). Long dashed line, quadratic fit of W|O data in the range $C^{2/3} = 0$ to cmc $^{2/3}$ (Equation 2.47). The adsorption parameters determined from these fits are listed in Table 2.1. Inset, the corresponding adsorption isotherms, $\Gamma_s(C^{2/3})$, calculated from Equation 2.39 with the adsorption parameters determined by the fits. The jump of Γ_s at $C = 0.81$ mM corresponds to a phase transition from a gaseous monolayer to LE state.

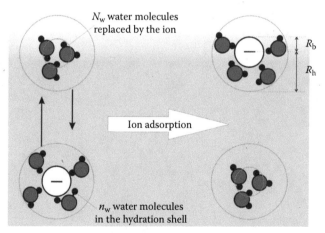

FIGURE 2.5 Scheme of the process of adsorption of type I ions. Left, the ion is in the bulk. Right, the ion is at the interface. The n_w hydrating water molecules might be pushed away by the interface, so that the shortest distance of approach of the ion to the interface is the bare ion radius R_b. Upon adsorption, the ion exchanges position with an ensemble of N_w water molecules. For type II ions, the shortest distance of approach of the ion to the interface is the hydrated ion radius R_h.

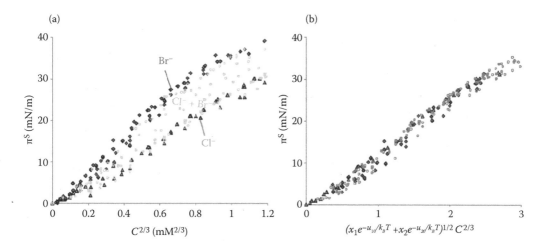

FIGURE 2.13 (a) Surface pressure π^S versus 2/3-power of the mean activity $C^{2/3}$ for $C_{16}H_{33}NMe_3Cl$ with or without added NaCl (diamonds); $C_{16}H_{33}NMe_3Br$ with or without NaBr (triangles); $C_{16}H_{33}NMe_3^+$ in the presence of both ions Cl^- and Br^- (empty circles) at the W|G surface. The data for a single counterion, Cl^- or Br^-, falls on separate master curves according to the salting-out effect (Equation 2.32), with slopes $3k_BTK_{Br}$ and $3k_BTK_{Cl}$ correspondingly. In contrast, the data for the counterion mixtures are dispersed between these two curves. (b) Drawn in coordinates π^S versus $\left(e^{-u_{10}/k_BT}x_1 + e^{-u_{20}/k_BT}x_2\right)^{1/2} C_t^{1/3}C_s^{1/3}$, all data fall on a single master curve with slope $3k_BTK_0$, according to Equation 2.80. (Data from Para, G. et al., *Adv. Colloid Interface Sci.* 122, 39–55, 2006.)

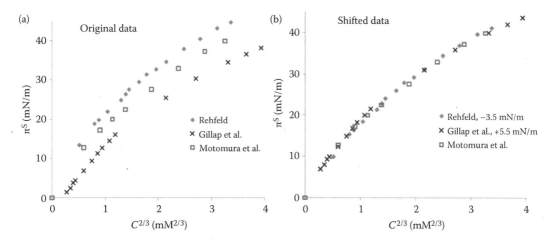

FIGURE 2.18 Surface tension versus $C^{2/3}$ for $C_{12}H_{25}SO_4Na$ at water–hexane interface. (a) The three sets differ in absolute values, which we attribute to incorrect calibration of the tensiometers. (b) When suitably shifted, all data coincide. (Data from Rehfeld, S.J., *J. Phys. Chem.* 71, 738–745, 1967; Gillap, W.R. et al., *J. Colloid Interface Sci.* 26, 232–236, 1968; Motomura, K. et al., *J. Colloid Interface Sci.* 67, 247–254, 1978.)

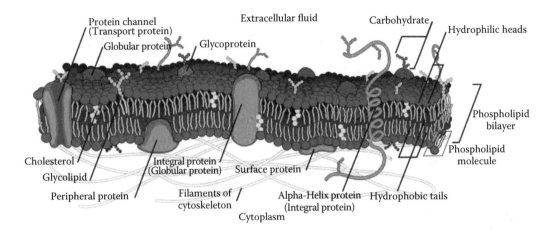

FIGURE 3.3 A model of the biological cell membrane. (From http://en.wikipedia.org/wiki/Cell_membrane.)

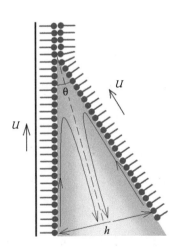

FIGURE 6.2 Convective flow in the wedgelike region near the contact line between the floating monolayer and the substrate surface. U is the velocity at the surfaces, θ is the contact angle, h is the local thickness of the film formed between the substrate surface and the floating monolayer. (Reprinted from *Adv. Colloid Interface Sci.*, 168, Kovalchuk, V.I., E.K. Zholkovskiy, M.P. Bondarenko, V.M. Starov and D. Vollhardt. Concentration polarization effect at the deposition of charged Langmuir monolayers, 114–123. Copyright 2011, with permission from Elsevier.)

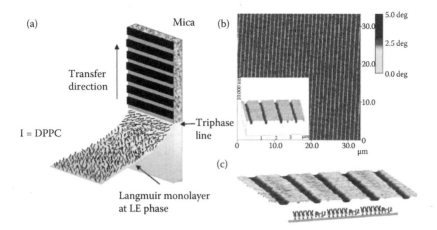

FIGURE 6.3 Scheme of stripe formation during LB deposition of a pure DPPC monolayer onto the substrate surface (a); dynamic scanning force microscope (SFM) image of the patterned surface (b); and arrangement of DPPC molecules within the deposited film (c). (Reprinted with permission from Chen, X., S. Lenhert, M. Hirtz, N. Lu, H. Fuchs, and L. Chi. Langmuir–Blodgett patterning: A bottom-up way to build mesostructures over large areas. *Acc. Chem. Res.*, 40, 393–401. Copyright 2007. American Chemical Society.)

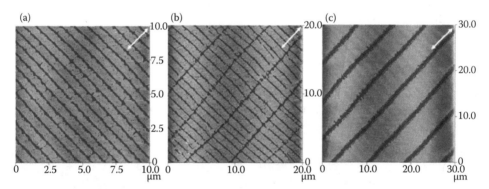

FIGURE 6.4 AFM images of striped structures of pure DPPC formed under different transfer conditions (velocity and surface pressure): (a) horizontal stripes (60 mm/min and 3 mN/m); (b) crossed stripes (40 mm/min and 3 mN/m); and (c) vertical stripes (10 mm/min and 3 mN/m). The arrows show the transfer direction. (Reprinted with permission from Chen, X., S. Lenhert, M. Hirtz, N. Lu, H. Fuchs, and L. Chi. Langmuir–Blodgett patterning: A bottom-up way to build mesostructures over large areas. *Acc. Chem. Res.*, 40, 393–401. Copyright 2007. American Chemical Society.)

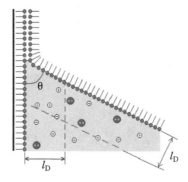

FIGURE 6.7 Diffuse double layers overlapping in the meniscus region; l_D is the double layer thickness (Debye length). (According to Kovalchuk, V.I. et al., *J. Phys. Chem. B* 105, 9254–9265, 2001. With permission.)

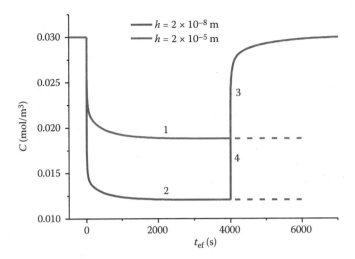

FIGURE 6.8 Relaxation of the local electrolyte concentration at two distances from the contact line (corresponding to $h = 2 \times 10^{-8}$ m and 2×10^{-5} m) after the beginning (1,2) and stopping (3,4) of the monolayer transfer (the initial concentration $C_0 = 0.03$ mol/m³, the ratio of the velocity to the contact angle in radians $U/\theta = 1.5$ mm/s, $t_{ef} = t\,\theta^2$ is the effective time). (According to Kovalchuk, V.I. et al., *J. Phys. Chem. B*, 115, 1999–2005, 2011. With permission.)

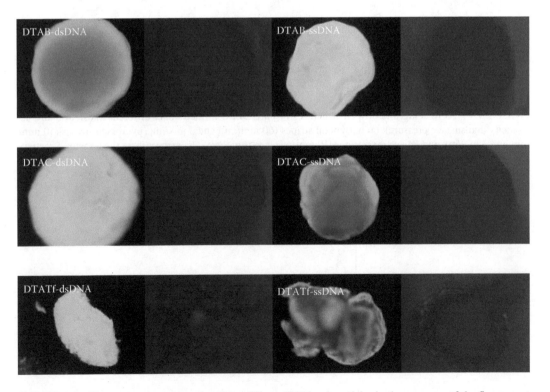

FIGURE 8.6 Fluorescence micrographs of the different DNA gel particles in the presence of the fluorescent dyes, AO (left) and NR (right) at 25°C. (Adapted from Morán, M.C. et al., *Soft Matter* 8, 3200–3211, 2012.)

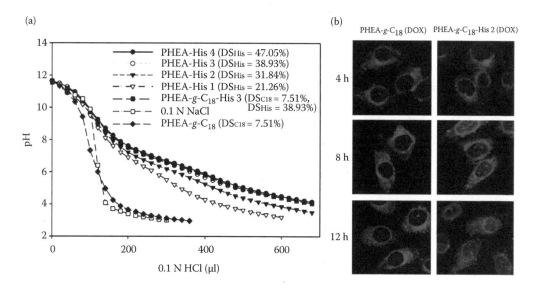

FIGURE 9.9 (a) Acid–base titration profile of 0.1 N NaCl, PHEA-g-C$_{18}$, PHEA-His series, and PHEA-g-C$_{18}$-His 3; and (b) confocal laser scanning microscopy images of HeLa cells incubated with PHEA-g-C$_{18}$ (DOX) and PHEA-g-C$_{18}$-His 2 (DOX) with incubation time. (From Yang, S.R. et al., *J. Control. Release* 114, 60–68, 2006. With permission.)

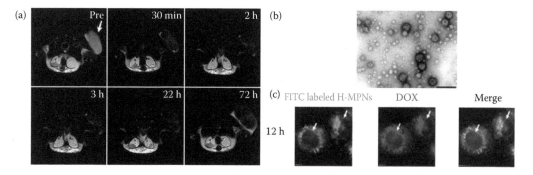

FIGURE 9.11 (a) *In vivo* T2-weighted MR images taken postinjection of PDMAEA-coated magnetic nanoparticles into a mouse bearing a CT26 tumor on its back. (From Lee, H.J., *Chem. Commun.*, 46, 3559, 2010. With permission.) TEM image of (b) H-MPNs and (c) confocal laser scanning microscopic images of HeLa cells treated with FITC-labeled H-MPNs (DOX) with 12-h incubation time. Scale bar represents 200 nm. (From Yang, H.-M. et al., *Chem. Commun.* 47, 5322–5324, 2011. With permission.)

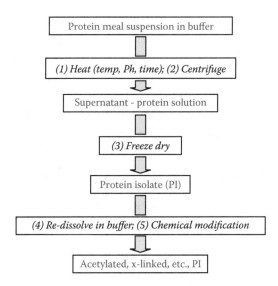

FIGURE 11.9 Flowchart for the isolation and chemical modification of plant proteins.

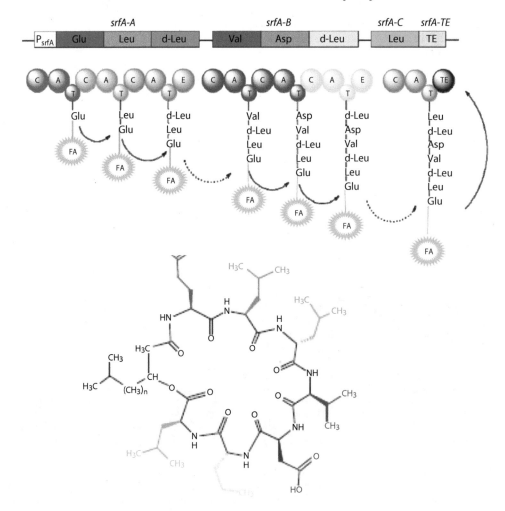

FIGURE 12.4 Surfactin is a lipopeptide that is produced naturally by *B. subtilis*. It is composed of a fatty acid chain that is 12 to 17 carbons in length and has seven amino acid cyclic peptides.

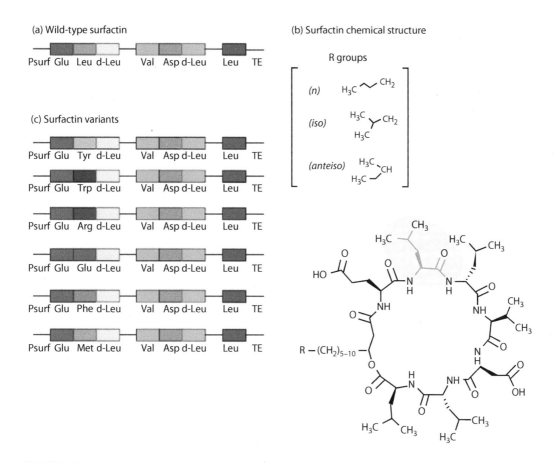

(a) Wild-type surfactin

Psurf Glu Leu d-Leu Val Asp d-Leu Leu TE

(b) Surfactin chemical structure

R groups

(n)

(iso)

(anteiso)

(c) Surfactin variants

Psurf Glu Tyr d-Leu Val Asp d-Leu Leu TE

Psurf Glu Trp d-Leu Val Asp d-Leu Leu TE

Psurf Glu Arg d-Leu Val Asp d-Leu Leu TE

Psurf Glu Glu d-Leu Val Asp d-Leu Leu TE

Psurf Glu Phe d-Leu Val Asp d-Leu Leu TE

Psurf Glu Met d-Leu Val Asp d-Leu Leu TE

$R - (CH_2)_{5-10}$

FIGURE 12.6 (a) Wild-type surfactin synthase operon; (b) chemical structure of wild-type surfactin; (c) six of the surfactin analogues in Modular's collection have substitutions at position 2 of the cyclic amino acid.

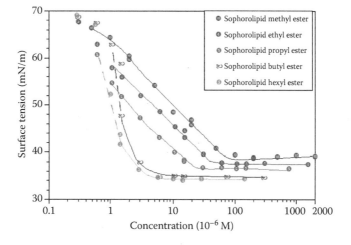

FIGURE 12.7 Surface tension of esters of sophorolipids. (From Zhang, L. et al., *Colloids Surf. A* 240, 75–82, 2004. With permission.)

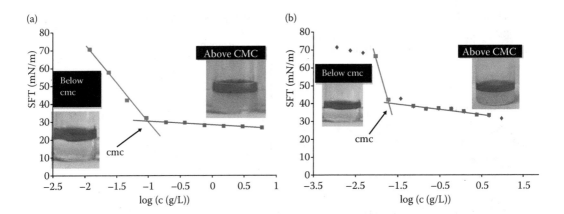

FIGURE 12.10 cmc values of (a) FA-Glu and (b) surfactin. SFT on the y axis indicates surface tension, and "c" on the x axis indicates concentration. Pictures inserted in the two graphs indicate that below the cmc values (left picture in the graphs), oil is not dispersed, whereas above the cmc values (right picture in the graphs), oil is dispersed in the surfactant solution. SFT indicates surface tension (mM/m).

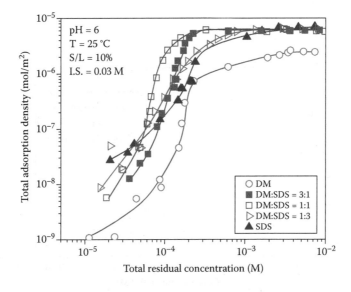

FIGURE 12.11 Adsorption of DM, SDS, and DM/SDS mixtures on alumina. IS indicates ionic strength, and S/L indicates the percentage solids of alumina in suspension. (From Zhang, L., and P. Somasundaran, *Colloid Interface Sci.* 302, 20, 2006. With permission.)

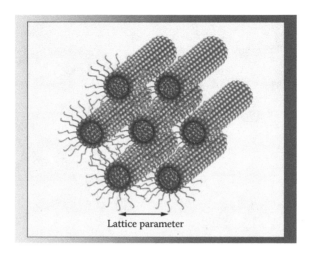

FIGURE 13.4 Schematic presentation of H_{II} mesophase showing the packing of water-filled rods surrounded by lipid layers.

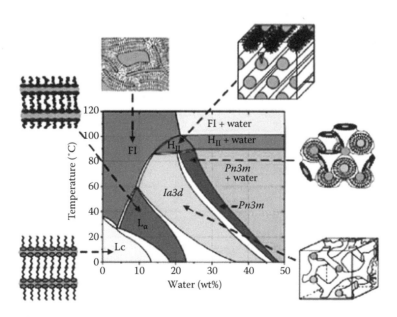

FIGURE 13.6 Temperature-composition phase diagram of the monoolein/water system (up to 50 wt% water). A cartoon representation of the various phase states is included in which shaded zones represent water. The mesophases are as follows: L_c, crystalline lamellar; L_α, lamellar; *Ia3d*, gyroid inverted bicontinuous cubic; *Pn3m*, primitive inverted bicontinuous cubic; H_{II}, inverted hexagonal; and F1, reverse micelles fluid phases. (From V. Cherezov et al., *J. Mol. Biol.*, 357, 1605–1618, 2006.)

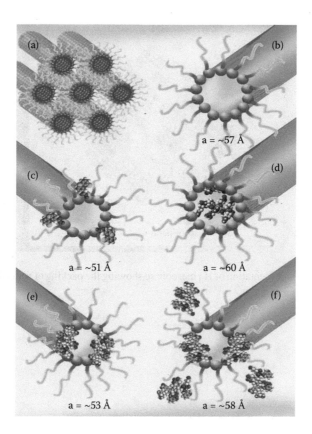

FIGURE 13.9 (a) Schematic illustration of the H_{II} mesophase general structure. The water populates the inner region of the cylinders; the GMO tails as well as the TAG point outward. (b) A single blank, empty system of the H_{II} mesophase, (c) Na-DFC-loaded system. The drug populates the interfacial area and causes channel shrinkage. (d) NONA-loaded H_{II} mesophase. The peptide populates the aqueous channels and swells them. (e, f) PEN-loaded systems. At low PEN concentration (*E*), it is adsorbed on the GMO headgroups and causes channel shrinkage. With increasing PEN concentration (f), it populates an additional hosting region at the interface; a = lattice parameter. (Reprinted with permission from M. Cohen-Avrahami et al., D. Libster, A. Aserin and N. Garti. Sodium diclofenac and cell-penetrating peptides embedded in H_{II} mesophases: Physical characterization and delivery. *J. Phys. Chem. B*, 115, 10189–10197. Copyright 2011 American Chemical Society.)

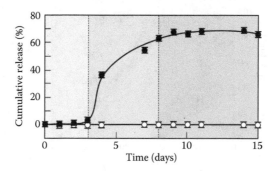

FIGURE 13.13 The release profile of DNA from the two columnar H_{II} mesophases considered: Release from the nonionic system (filled circle, GMO/tricaprylin/water/ascorbic acid) and release from the cationic-based columnar hexagonal phase (open circle, GMO/oleyl amine/tricaprylin/water/ascorbic acid) at room temperature. (I. Amar-Yuli, J. Adamcik, S. Blau, A. Aserin and R. Mezzenga. Controlled embedment and release of DNA from lipidic reverse columnar hexagonal mesophases. *Soft Matter*, 7, 8162–8168, 2011. Reproduced by permission of The Royal Society of Chemistry.)

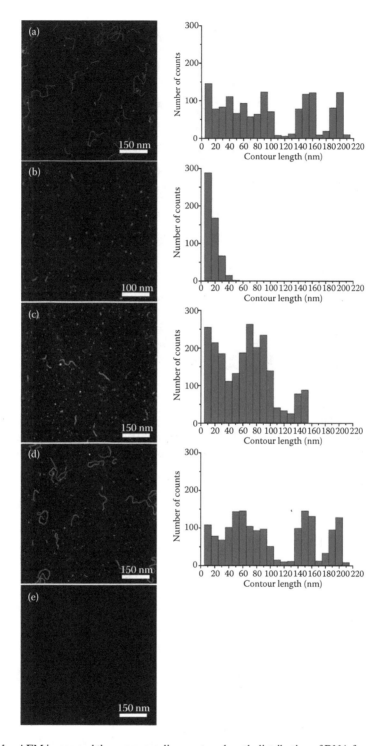

FIGURE 13.14 AFM image and the corresponding contour length distribution of DNA fragments: (a) before incorporating DNA into the H_{II} nonionic LLC system; (b) after 3 days release in the excess water environment; (c) after 4 days release; and (d) after 15 days release at room temperature. (e) AFM image after 15 days of DNA release from the cation-based H_{II} LLC system. (I. Amar-Yuli, J. Adamcik, S. Blau, A. Aserin and R. Mezzenga. Controlled embedment and release of DNA from lipidic reverse columnar hexagonal mesophases. *Soft Matter*, 7, 8162–8168, 2011. Reproduced by permission of The Royal Society of Chemistry.)

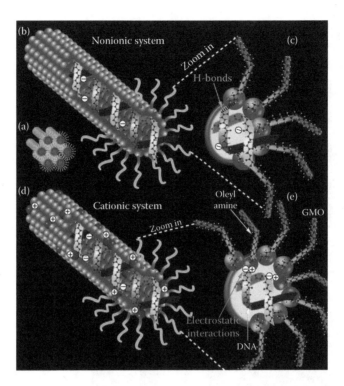

FIGURE 13.15 Schematic summary illustrating a typical columnar hexagonal phase (a), the DNA entrapment and interactions with the surfactants forming the nonionic (b, c) and cationic (d, e) H_{II} systems. The molecular ratio between all the components is not to scale in the image. (I. Amar-Yuli, J. Adamcik, S. Blau, A. Aserin and R. Mezzenga. Controlled embedment and release of DNA from lipidic reverse columnar hexagonal mesophases. *Soft Matter*, 7, 8162–8168, 2011. Reproduced by permission of The Royal Society of Chemistry.)

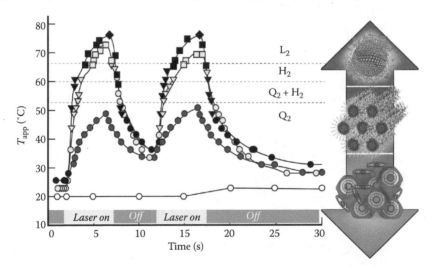

FIGURE 13.17 Effect of laser irradiation on apparent temperature of the phytantriol + water matrix (T_{app}) with change in GNR concentration. GNR = 0 nmol/L are white symbols, 0.3 nmol/L blue symbols, 1.5 nmol/L yellow symbols, and 3 nmol/L black symbols. Circles indicate v_2 phase, triangles indicate $v_2 + H_2$, squares indicate $H_2 + L_2$, and diamonds indicate L_2. The cartoon on the right indicates the type of phase structure present with increasing temperature. Note: $H_2 = H_{II}$. (Reprinted with permission from W.K. Fong, T. Hanley, B. Thierry, N. Kirby and B.J. Boyd. Plasmonic nanorods provide reversible control over nanostructure of self-assembled drug delivery materials. *Langmuir*, 26, 6136–6139. Copyright 2010 American Chemical Society.)

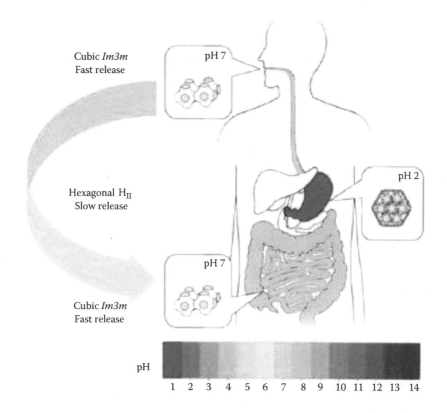

Cubic *Im3m*
Fast release

Hexagonal H$_{II}$
Slow release

Cubic *Im3m*
Fast release

pH 7

pH 2

pH 7

pH

1 2 3 4 5 6 7 8 9 10 11 12 13 14

FIGURE 13.19 Schematics of the proposed pH-responsive drug delivery strategy across the gastrointestinal tract. At pH 7, the drug nanocarrier has a reverse bicontinuous cubic phase (*Im3m*) symmetry; at pH 2, in the stomach, the symmetry of the mesophase changes to reverse hexagonal phase, slowing down diffusion and preventing the premature release of the drug; and at neutral pH in the intestine, the symmetry reverts to bicontinuous cubic phase (*Im3m*), triggering the release of the active ingredients loaded in the mesophase. (Reprinted with permission from R. Negrini and R. Mezzenga. pH-responsive lyotropic liquid crystals for controlled drug delivery. *Langmuir*, 27, 5296–5303. Copyright 2011 American Chemical Society.)

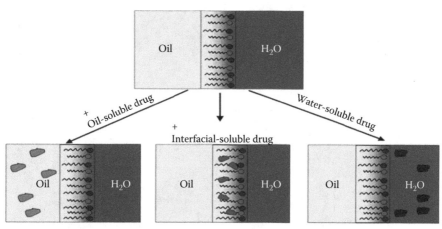

Oil   H$_2$O

Oil-soluble drug

+
Interfacial-soluble drug

Water-soluble drug

Oil   H$_2$O   Oil   H$_2$O   Oil   H$_2$O

Microemulsions can solubilize oil-soluble, water-soluble, and interface soluble drugs in a single phase liquid.

FIGURE 14.2 Three invisible compartments of microemulsions: oil, water, and the interfacial region.

By controlling the stability of the microemulsion, one can change the release rate of the drug molecules.

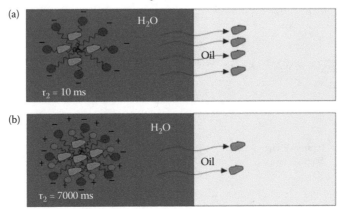

FIGURE 14.3 Inherent stability of a microemulsion droplet and the flux of the drug from the droplet to the surrounding and the mixed micelles shown in the lower part of the figure have inherently greater stability. (a) Less stable micelle and (b) more stable micelle.

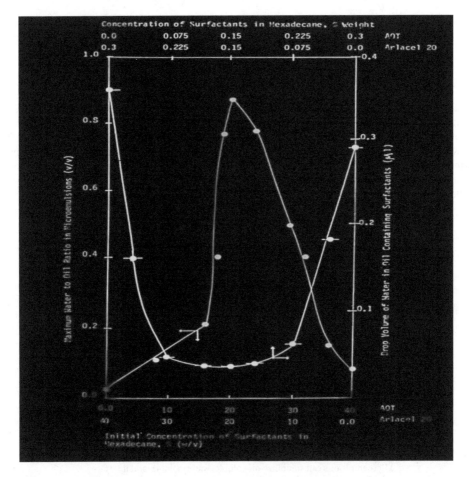

FIGURE 14.7 Solubilization of water in hexadecane + AOT + Arlacel-20 mixture. The maximum solubilization of water occurs when the drop-volume is at a minimum or when the interfacial tension between oil and water is at a minimum in the presence of AOT and Arlacel-20.

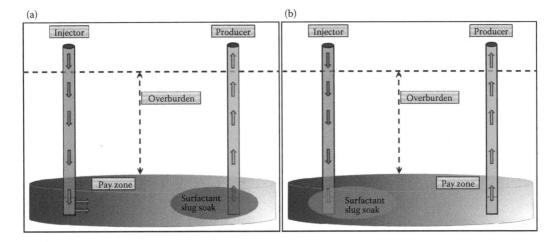

FIGURE 19.3 Surfactant-slug enhanced waterflooding process: (a) surfactant-soaked production zone; (b) surfactant-soaked injection zone. (Adapted from P.M. Mwangi. An experimental study of surfactant enhanced waterflooding. Masters' Thesis, Louisiana State University, 2010.)

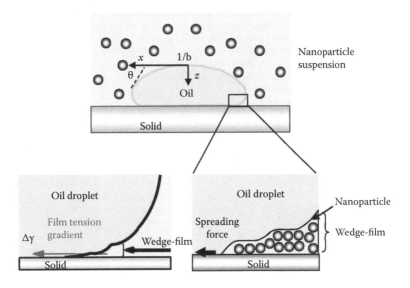

FIGURE 20.5 Schematic diagram of nanoparticle structuring in wedge-shaped thin aqueous film spreading under an oil drop during roll-up. (Reprinted with permission from Kondiparty, K., A. Nikolov, S. Wu, and D.T. Wasan. Wetting and spreading of nanofluids on solid surfaces driven by the structural disjoining pressure: Statics analysis and experiments. *Langmuir* 27, 3324–3335. Copyright 2011 American Chemical Society.)

Part III

Polymeric Surfactants and
Polymer/Surfactant Mixtures

8 DNA Release from Cross-Linked DNA Gels and DNA Gel Particles

M. Carmen Morán, Diana Costa, Maria da Graça Miguel, and Björn Lindman

CONTENTS

8.1 INTRODUCTION

DNA is an amphiphilic polymer that contains hydrophobic bases and carbohydrates and phosphate groups, which are hydrophilic. The bases promote an association between DNA molecules, whereas the hydrophilic parts oppose self-assembly. The double helix of DNA is its most important self-assembly structure. Mixtures of DNA and cationic surfactants show a strong associative behavior. In solution, DNA molecules present an extended conformation (coil), whereas at high concentrations of surfactant, the DNA molecules undergo compaction. This transition can be followed by fluorescence microscopy (FM). Complexes between DNA and cationic surfactants have attracted an increased interest lately due to the possibility of using these systems for gene transfection [1]. The nature of both the DNA and the surfactant [2–4] has been found to influence the phase separation limits and also the structure of the DNA-surfactant complexes. Surfactants are not the only class of molecules interacting strongly with DNA. For instance, the binding of positively charged proteins, as well as poly-L-lysine and poly-L-arginine, multivalent ions and multivalent polyamines, such as spermidine or spermine are known to condense large DNA coils [5,6], resulting in (associative) phase separation in sufficiently concentrated solutions.

Polymer gels that respond to changes in the surrounding environment, often referred to as responsive gels, have attracted much interest in the last few years [7–9]. The environmental conditions include changes in different parameters such as pH, solvent composition, ionic strength, temperature, pressure, buffer composition, chemicals, and surfactants. Complexes of polyelectrolyte gels with oppositely charged surfactants were intensively studied in the last decade [10–13] due to their interesting practical applications. Phase coexistence has been observed during volume transitions of polyelectrolyte gels after the uptake of surfactant ions from the solution [14,15]. Here, the transition to the collapsed state is promoted by the favorable electrostatic interaction between the polyion and surfactant micelles. Hydrogels, both natural and synthetic, are well suited for biomedical applications because of their tissue compatibility, arising from its high water content and its soft and rubbery consistency. Their flexibility in tailoring physicochemical properties, such as permeability and swelling, the ability to load drugs and release them in a controlled fashion, for times ranging from minutes to years, are also relevant issues [16]. To design appropriate delivery systems, the mesh size of swollen hydrogels, and the degree of swelling or polymer degradation can be tailored leading to adequate rates of solute diffusion and controlled/sustained release profiles. For most biomedical requirements, degradable hydrogels are preferred rather than nondegradable ones; in particular if they degrade on clinically relevant timescales under relatively mild conditions and thus eliminate the need for additional surgeries to remove the implanted systems. Biodegradable hydrogels seem to be a more challenging strategy for DNA release, once one can readily control the release rate by modulating the network structure by adjusting cross-linker density [17].

During the last few years, we have paid considerable attention to different aspects of covalent DNA gels, concerning their preparation, their characterization, the study of swelling isotherms in the presence of different cosolutes (such as surfactants), and their light-induced disruption [17,18–22]. In addition, a general understanding of interactions between DNA and oppositely charged agents has given us a basis for developing novel DNA gel particles [23]. The association strength, which is tuned by varying the chemical structure of the cationic cosolute, determines the spatial homogeneity of the gelation process, creating DNA reservoir devices and DNA matrix devices that can be designed to release DNA [23–29]. The particle morphology and size, swelling/dissolution behavior, degree of DNA entrapment and DNA release responses as a function of the nature of the cationic agent have been discussed [23–30]. Current directions on the preparation of covalent DNA gels [22] and DNA gel particles [31], including the DNA release, swelling behavior, photodegradation, cytotoxicity, and hemocompatibility assessments of these systems have been reviewed in this chapter.

8.2 COVALENTLY CROSS-LINKED DNA GELS

The swelling behavior and solute release mechanisms from pDNA hydrogels are relevant biological issues and this information is enormously important as a basic tool for the development of pDNA carriers that will used therapeutically. The most relevant aspect is the possibility of controlled release of pDNA. In therapeutic applications, purified pDNA should be employed to deliver the desired genetic information into the cells and to induce the production of relevant proteins. The acquired knowledge on this carrier is crucial as a model for progress in the design of systems for the delivery of drugs, proteins, genes, and ultimately, for developing gene therapies in the treatment of cancer. This issue has attracted considerable attention in recent years and it is already under investigation by our group. We are also interested in studying a strategically dual drug/DNA carrier for improving cancer therapy.

8.2.1 GEL FORMATION AND CHARACTERIZATION

Plasmid DNA (pDNA) hydrogels can be synthesized by cross-linking pDNA with ethylene glycol diglycidyl ether (EGDE), which is a bifunctional cross-linker with an epoxide structure. The reaction mechanism of gel formation seems to involve the guanine nitrogen atom at position seven (N-7),

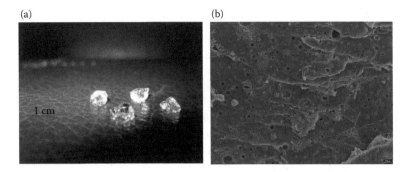

FIGURE 8.1 Picture (a) and scanning electron micrograph (b) of pDNA gels cross-linked with 0.2 wt% EGDE. (Adapted from Costa, D. et al., *Colloids Surf. B*, 92, 106–112, 2012.)

which attacks the more substituted carbon of the epoxide or the least hindered end of the epoxide. The reaction is of nucleophilic substitution type (S_N2 reaction). Gels with 0.2 wt% and 0.5 wt% cross-linkers were prepared [32]. All pDNA gels are clear and transparent (Figure 8.1a). Scanning electron micrography revealed a characteristic three-dimensional coral-like structure with small cavities confined by perforated membranes. In addition, they exhibit a spongy structure (Figure 8.1b).

8.2.2 Swelling Behavior of pDNA Gels Induced by Cosolutes

In pure water, charged networks swell due to the osmotic pressure from counterions, which originates from their translational entropy. The volume changes of gels are associated with osmotic effects rather than dehydration. The swelling is restricted by the cross-link density of the network and the deswelling by the volume occupied by the polymer network. The extent of the collapse is affected by the molecular parameters of the network such as cross-linking density, hydrophobic/hydrophilic balance of the polymer network, its charged groups content, its flexibility, and its ability to interact with additional solutes. In pure water, the degree of swelling for 0.2 wt% EGDE pDNA gels is considerably larger than that of corresponding 0.5 wt% EGDE gels [32]. An increase in the level of cross-linking agent leads to a larger resistance of the osmotic swelling force and thus to a reduced equilibrium swelling.

Upon the addition of a monovalent electrolyte, there is a progressive contraction of the gels. When gels preswollen in 1 mmol/L of NaOH are placed into salt solutions at different concentrations, they shrink due to the screening effect of the salt and mainly due to the concentration difference of mobile ions inside the gel and the external solution governed by the Donnan equilibrium [32]. In the presence of salt, the difference between ion concentrations inside and outside the gel is reduced. Consequently, the driving force for swelling decreases gradually with increasing salt concentration. The nature of the monovalent counterion has only a moderate effect on the deswelling of pDNA gels. The deswelling upon the addition of the divalent salts $CaCl_2$, $MgCl_2$, and $SrCl_2$ occurs at considerably lower salt concentrations and it seems to be more pronounced compared with the monovalent metal ions [32]. Qualitatively, we expect this observation because in order to maintain electroneutrality, adding one divalent cation to the gel requires the exchange of two monovalent cations. The swelling behavior of pDNA gels was also studied in the presence of polycationic species, which included chitosan, spermine (Spm), spermidine (Spd), and lysozyme [32]. The polyamines spermine (Spm^{4+}) and spermidine (Spd^{3+}) are essentially fully protonated at physiological pH. Chitosan is the most efficient cosolute in promoting gel collapse, followed by the polyamines and finally lysozyme. The higher charge of spermine compared with spermidine evidently makes spermine bind more strongly to pDNA with a concomitant higher gel collapse potential. A considerably larger concentration of lysozyme was needed to reach a corresponding collapse of the gel compared with the other polycationic species. A contribution to this difference can also be the fact that a linear polymer penetrates more easily into the network compared with a globular one.

8.2.3 Swelling Behavior of pDNA Gels Induced by Surfactants

In the bulk phase and at low concentrations, the surfactant molecules are dissolved as unimers, whereas at higher surfactant concentrations, a self-assembly into aggregates occurs. For the single-chain surfactants, the aggregates formed in this self-assembly are commonly spherical micelles with micelle formation starting at a well-defined concentration, the critical micelle concentration (cmc). In the presence of an oppositely charged polyelectrolyte, the micelle formation of an ionic surfactant is strongly facilitated, leading to a considerable lowering of the cmc; the cmc in the presence of a polymer is often referred to as the critical association concentration (cac). The stabilization of micelles due to an oppositely charged polyelectrolyte is mainly an entropic effect caused by a release of counterions. Therefore, the critical aggregation concentration for the surfactant in the presence of pDNA, cac, is considerably lower than the cmc. After the immersion of the swollen pDNA gels in the solutions of the oppositely charged surfactants, the surfactant ions migrate into the network and replace the network counterions, which are released. Adsorption of a considerable amount of C_nTA^+ ions leads to a transition of the swollen network to the collapsed state. The main reason for this transition is thus the aggregation of surfactant ions within the pDNA gel due to hydrophobic interactions between their hydrocarbon chains. Figure 8.2 shows swelling isotherms for pDNA gels upon the addition of a number of cationic surfactants. As can be seen, the surfactants have no effect at lower concentrations but there is a marked deswelling at higher concentrations, which becomes more important the longer the surfactant alkyl chain is. We note that the concentration for onset of deswelling varies by orders of magnitude between different surfactants [32]. We also note that the plateau value, obtained at high surfactant concentrations, is lower with longer alkyl chain lengths. The pronounced chain length dependence directly suggests a dominant role of surfactant self-assembly. These results for the different alkyl chain lengths confirm that the deswelling occurs below the normal cmc of the surfactant. The cmc values for $C_{16}TAB$, $C_{14}TAB$, $C_{12}TAB$, and C_8TAB are 0.9, 2.3, 15, and 144 mmol/L, respectively. We found that the surfactants induce volume changes starting at a certain rather well-defined concentration of cac ~ 0.015 mmol/L for $C_{16}TAB$, cac ~ 0.045 mmol/L for $C_{14}TAB$, cac ~ 0.08 mmol/L for $C_{12}TAB$, and cac ~ 1 mmol/L for

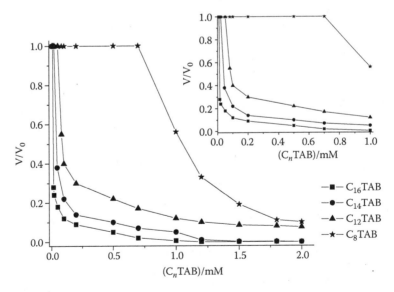

FIGURE 8.2 Swelling isotherms (V/V_0) for pDNA gels (0.2 wt% cross-linker) immersed in solutions of the cationic surfactants $C_{16}TAB$, $C_{14}TAB$, $C_{12}TAB$, and C_8TAB. The low concentration range is also represented in detail (see inset). Temperature 25°C, pH 8.5. The data were obtained by calculating the average of three independent experiments. The respective errors were determined and were less than 0.05. (Adapted from Costa, D. et al., *Colloids Surf. B*, 92, 106–112, 2012.)

C_8TAB. We also investigated the reversibility of the swelling process through the addition of an anionic surfactant to pDNA gels collapsed by a cationic surfactant. This study concerned the addition of different concentrations of a cationic surfactant ($C_{16}TAB$) followed, after gel collapse, by the addition of different concentrations of an anionic surfactant, sodium dodecyl sulfate (SDS). Because the relative swelling ratio, V/V_0, returned to between 90% and 100% of the initial state, the swelling of pDNA gels seems to be reversible [32]. The interaction between the two surfactants is stronger than that between a cationic surfactant and pDNA. The dynamic deswelling/swelling process could be useful in the control of the release rate of solutes from gels via "on–off" switching.

8.2.4 Photodegradation of pDNA Gels

Ethylene glycol ethers do not persist in the environment nor does it bioaccumulate in tissues, and are "practically nontoxic" to aquatic organisms; these ethers photooxidize in the presence of sunlight [33]. The degradation upon exposure to ultraviolet light (photooxidation) of EGDE leads to the removal of the chemical cross-links and can allow the release of the constituent network polymer, inducing changes in gel weight, mechanical properties, mesh size, porosity, and in the degree of swelling [22]. Thus, to demonstrate the pDNA gel disruption, experiments on pDNA release were performed after the gels had been irradiated with light (400 nm), and in the dark conditions, for pDNA gels cross-linked with 0.2 wt% and 0.5 wt% EGDE, as illustrated in Figure 8.3a. After irradiation, both gels underwent disruption leading to the release of pDNA with time. For 0.2 wt% and 0.5 wt% EGDE cross-linked gels, pDNA release behavior presents a short time lag in the first 24 h, after which the release gradually increases until a plateau is reached after approximately 400 h of photodegradation. The initial time lag may be related to the number of cross-links that have to be degraded to permit the release of pDNA. After irradiation, and at maximum release, pDNA gels cross-linked with 0.2 wt% EGDE released 87.8% of pDNA, whereas gels prepared from 0.5 wt% EGDE released 74.7% of pDNA in approximately 18 days. Not all the cross-linked pDNA is released, probably because the inhomogeneous distribution of the cross-linker leads to the existence of very concentrated cross-linker/DNA regions, from where the release of DNA is more difficult. Additionally, it has been noted that the extent of pDNA release is quite dependent on cross-linker density, as found previously for similar systems [17]: the higher the cross-linker concentration used in the gel preparation, the lower the amount of pDNA released. A completely different behavior was found for release studies performed in dark conditions (Figure 8.3b). In the absence of ultraviolet

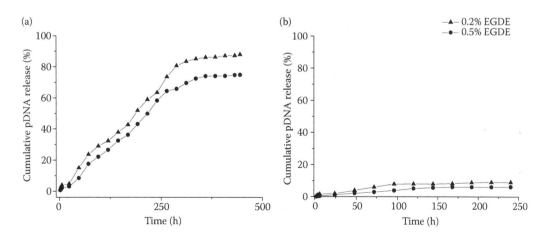

FIGURE 8.3 Cumulative release of pDNA from cross-linked pDNA gels with 0.2 wt% (▲) and 0.5 wt% (●) EGDE, as a function of time. Studies were performed after the irradiation of gels with light (400 nm) (a) and in the dark conditions (b). (Adapted from Costa, D. et al., *Langmuir* 27, 13780–13789, 2011.)

light irradiation, and for both gel types, minimal amounts of pDNA are released (<8%). In each gel system, this approximately corresponds to the free pDNA in the network, that is, the non–cross-linked pDNA. At the time of preparation, the relation between the pDNA concentration and the EGDE concentration allows the gel to maintain some of the chains in a non–cross-linked state.

8.2.5 pDNA Gel Biocompatibility

In vitro tests for biocompatibility focus on the cytotoxicity aspects of the polymeric material in the presence of live host cells. The MTT assay is a colorimetric method that allows the quantification of cell growth and proliferation and involves the reduction of MTT into purple formazan crystals in the presence of mitochondrial dehydrogenase. These formazan crystals are dissolved and analyzed by measuring absorbance at 570 nm. The amount of colored crystals is directly proportional to the number of live cells. The MTT assays were performed 1 and 2 days after cells were seeded on top of 0.2% and 0.5% EGDE pDNA gels. As shown in Figure 8.4, both pDNA hydrogels are nontoxic to cells because each formulation promoted dehydrogenase activity, with 0.2% EGDE gels being less toxic. After 1 day, the fibroblasts were found to be 96% and 74% viable relative to control cells for 0.2% and 0.5% EGDE, respectively. After 2 days of incubation with gels, the cell viability percentage decreased slightly, but the difference was not statistically different from the results obtained after 1 day. We can then conclude from the obtained data that the toxicity of these hydrogels is sufficiently low on human fibroblasts because no significant decrease in cell viability was detected in the interaction of cells with pDNA gels, even for prolonged periods. The presented formulation does not have an acute cytotoxic effect, and thus, this system should not elicit an inflammatory response that can ultimately result in failure to achieve normal cell growth and function.

Our ongoing research on pDNA-based carriers clearly shows the relevance of parameters such as the cross-linker density in the formulation cytotoxicity, cell internalization, and *in vitro* transfection. The codelivery of genes and drugs from biocompatible pDNA microgels is also under study, and preliminary results *in vitro* studies clearly show the advantage of this combined action, which will really be the future challenge in chemotherapy and gene delivery for cancer treatment.

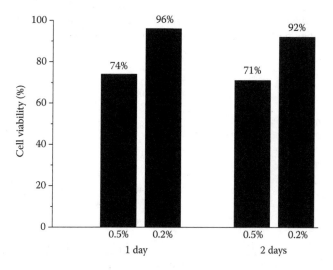

FIGURE 8.4 Cytotoxicity profiles of 0.2 wt% and 0.5 wt% EGDE pDNA gels on human fibroblasts after 1 and 2 days of incubation as measured by MTT assay. Percent viability is expressed relative to control cells: 0.5 wt%, 0.5 wt% EGDE cross-linked pDNA gel; 0.2 wt%, 0.2 wt% EGDE cross-linked pDNA gel. The data were obtained by calculating the average of three experiments. The respective errors were determined and were less than 0.05. (Adapted from Costa, D. et al., *Colloids Surf. B*, 92, 106–112, 2012.)

8.3 DNA GEL PARTICLES

For an oppositely charged polyelectrolyte–surfactant pair, the simplest type of phase separation involves the formation of two phases. If there are strong attractive interactions between the two components, the formation of one phase concentrated in both polymer and surfactant and another diluted in both components, that is, an associative phase separation, would be expected. The driving force for this strong association is the electrostatic interaction between the two components, as given by the entropic increase due to the release of the respective counterions. A general understanding of the interactions between DNA and oppositely charged agents has provided a basis for developing novel DNA gel particles. However, to date, little is known about the influence of the respective counterions on this oppositely charged polyelectrolyte–surfactant pair.

The contribution of the counterion entropy to the free energy of association depends on the detailed geometries and charge distributions of the separated and bound macro-ions. The apparent degree of counterion dissociation, α, also called the degree of micelle ionization, is an important parameter in the physical description of aqueous surfactant solutions [34,35]. Although ionic surfactants are strong electrolytes below the cmc, that is, fully ionized, the charge density on the micellar surface is so high that a fraction, $1 - \alpha$, of the counterions condense onto the surface and reduce the net charge; α is typically in the range of 0.2 to 0.5. This parameter significantly affects the surface properties of surfactants, such as the cmc, micellar size, reduction of interfacial (or surface) tension, etc. [35]. Colloidal properties, such as substrate binding efficiencies, transport properties, and phase transitions (e.g., from spheres to rodlike structures), may also show significant dependencies on α [36].

The effect of different counterions on the formation and properties of the DNA gel particles was examined by mixing DNA [either single-stranded DNA (ssDNA) or double-stranded DNA (dsDNA)] with the single chain surfactants dodecyltrimethylammonium chloride (DTAC), dodecyltrimethylammonium bromide (DTAB), dodecyltrimethylammonium hydrogen sulfate (DTAHs), and dodecyltrimethylammonium trifluoromethane sulfonate (DTATf). In particular, we used the hydrogen sulfate and trifluoromethane sulfonate anions and two halides (chloride and bromide) as counterions for this surfactant. Effects on the morphology of the particles obtained, the encapsulation of DNA and its release, as well as the hemocompatibility of these particles are presented using the counterion structure and the DNA conformation as controlling parameters. Analysis of the data indicates that the degree of counterion dissociation from the surfactant micelles and the polar/hydrophobic character of the counterion are important parameters in the final properties of the particles. The stronger interaction with amphiphiles for ssDNA than for dsDNA suggests the important role of both flexibility and hydrophobicity interactions.

8.3.1 GEL FORMATION AND CHARACTERIZATION

Particles were prepared at a charge ratio between DNA and cationic agent equal to 1, $R = $ [DNA]/[S$^+$], where [S$^+$] is the concentration of the corresponding surfactant. In all cases, the DNA concentration was set to 60 mmol/L. This DNA concentration was chosen because it produces high-viscosity solutions, which makes it an appropriate system for the preparation of stable DNA gel particles [23–29,31].

Particles were prepared by dropwise addition of DNA solutions to the surfactant solutions, equilibrated at 25°C or 45°C. Because of the relatively high viscosity of the DNA solution, mixing of the two solutions is not instantaneous. Therefore, before the two solutions can mix, the surfactant diffuses into the polyelectrolyte phase and forms a gel shell at the interface, stabilizing the particles. This is the general behavior observed for DNA placed in DTAB, DTAC, and DTATf solutions. However, in the case of DTAHs, DNA drops broke quickly on contact with the DTAHs solution, and the formation of the corresponding DNA gel particles did not take place. A similar behavior was observed in the case of particles prepared with denatured DNA. Changes in the pH values of the corresponding surfactant solutions could explain the behavior observed. Although the pH of the solutions containing DTAB, DTAC, and DTATf were in the 5.9 to 6.5 range, very low pH (≈1.8) was determined in the case of DTAHs. In this case, acidic conditions may play a role in the protonation of the DNA bases [37], contributing negatively to the polymer–surfactant interaction.

The characterization of micelles of DTAC and DTAB surfactants as reaction media showed significant differences in their degree of dissociation from the micelle: approximately 26% for bromide and approximately 37% for chloride in experiments performed at 25°C [38,39]. Recent studies showed degrees of triflate dissociation ranging from 0.13 to 0.15 for temperatures between 38°C and 47°C [40].

On the basis of these values, a good correlation between the degree of counterion dissociation for these three surfactants and the corresponding degree of DNA entrapment can be established. The degree of DNA entrapment is expressed through the loading efficiency (LE) and loading capacity (LC) values (Figure 8.5). Loading efficiency (LE) is calculated by comparing the amount of DNA

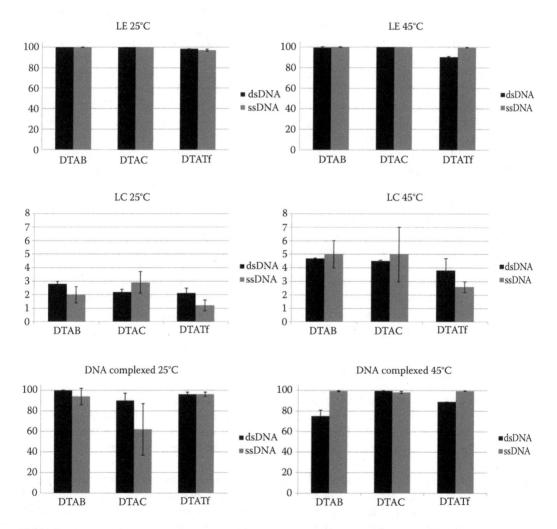

FIGURE 8.5 Characterization of the DNA gel particles with respect to DNA LE, LC, and DNA complexed as a function of the surfactant counterion, temperature, and secondary structure of the DNA. Complexed DNA is related to the amounts of DNA in the supernatant solutions and the skins derived from the particles, after particles were magnetically stirred overnight. All values were measured in triplicate and are given as average and standard deviation. (Adapted from Morán, M.C. et al., *Soft Matter* 8, 3200–3211, 2012.)

included in the particles with the total amount during particle formation. LC measures the amount of DNA entrapped in the particles as a function of their weight. The LE and the LC for the different formulations depends on the surfactant used.

Using DTAC and DTAB, the LE values found were higher than 99% for the two temperatures studied. In the case of DTATf, experiments carried out at 25°C showed LE values ranging between 97% and 98%. However, increasing the temperature to 45°C increases LE for the DTATf–ssDNA systems (>99%).

Although there are no differences between the LE values of the DTAC and DTAB systems, the most limited dissociation for the triflate counterion could explain the lower efficiency observed in the DTATf systems.

The determination of the entrapped DNA as a function of the weight of the particles showed LC values ranging from 1.2% to 2.9% for experiments carried out at 25°C. The lowest LC values were obtained for the DTATf–ssDNA system. Surfactants containing chloride or bromide as the corresponding counterions produce similar LC values. When particle formation takes place at 45°C, LC values are approximately twice as much as those obtained at lower temperatures.

An indication of the structural characteristics of these DNA particles can be obtained from the amount of DNA that was released when the break-up of the particles was induced mechanically. The percentages of DNA complexed are summarized in Figure 8.5. These values suggest that, by using these three surfactants, most of the DNA is complexed during the particle formation. The formation of these fully collapsed particles is consistent with the formation when using other surfactants with an identical hydrophobic contribution [28].

8.3.2 MORPHOLOGICAL CHARACTERIZATION OF THE DNA GEL PARTICLES

FM using the fluorescence dye, acridine orange (AO), was used to confirm the presence of DNA and to assess the secondary structure of the nucleic acid in the particles. AO (excitation, 500 nm; emission, 526 nm) intercalates into dsDNA as a monomer, whereas it binds to single-stranded DNA as an aggregate. Upon excitation, the monomeric AO bound to dsDNA fluoresces green, with an emission maximum at 530 nm. The aggregated AO on single-stranded DNA fluoresces red, with an emission at approximately 640 nm [41,42].

Figure 8.6 shows fluorescence micrographs of individual particles. FM images of freshly prepared particles using AO as staining agent (left panels) showed green emission, independently of the initial secondary structure of the DNA. The absence of red emission in the particles containing denatured DNA suggests that the accessibility of free DNA to the dye is hindered. The morphologies seen are consistent with the data on DNA distribution described above (Figure 8.5).

Similar studies were carried out using Nile red (NR) as staining agent. This molecule shows a solvatochromic behavior. In polar media, a red shift in the emission maximum can be observed, together with fluorescence quenching, due to the capacity of NR to establish hydrogen bonds with protic solvents [43]. As a consequence, the NR emission in water is weaker, with an emission maximum at 660 nm [44].

Some features of water distribution in these systems can be deduced from the FM studies using NR for staining (Figure 8.6). Particles formed with surfactants containing the two halides as counterions showed almost no emission of the dye. However, particles formed with the surfactant DTATf revealed an increase in its emission. These results suggest that, in the latter case, the dye NR remains less exposed to water in this system. It has been found that the dye NR is very sensitive to the local polarity (dielectric constant of the microenvironment) and can be used as a probe for hydrophobic surfaces in proteins. In a polar environment, NR has a low-fluorescence quantum yield, whereas in more hydrophobic environments, its quantum yield increases and its emission maximum becomes

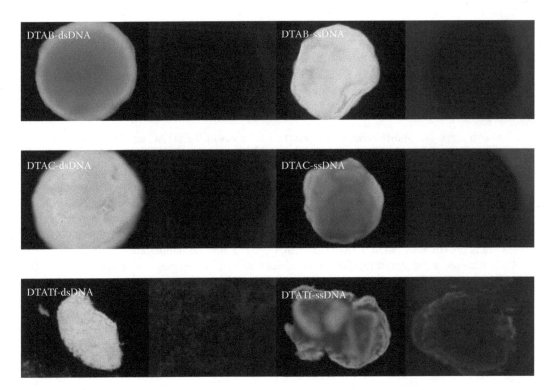

FIGURE 8.6 **(See color insert.)** Fluorescence micrographs of the different DNA gel particles in the presence of the fluorescent dyes, AO (left) and NR (right) at 25°C. (Adapted from Morán, M.C. et al., *Soft Matter* 8, 3200–3211, 2012.)

progressively blue-shifted. This property has been used for probing nonpolar sites in biomolecules (proteins and lipids) [45].

It is also based on the α values, as well as the character of the triflate anion, which is more hydrophobic than the anions bromide and chloride, that triflate ions in DTATf–DNA complexes could be expected to provide a more hydrophobic environment for the dye NR than halide counterions. Triflate ions, because they have three resonance forms, promote significant water structuring; the extent of water organization has been shown to be responsible for lipid head group dehydration [46,47]. The observed changes in NR emission argue in favor of these conclusions.

8.3.3 SWELLING KINETICS

Gels are considered to have great potential as drug reservoirs. Loaded drugs are released by diffusion from the gels or by erosion. Hence, the release mechanism can be controlled by swelling or dissolution of the gels. Figure 8.7 shows the relative weight (RW) ratio of the different gel particles after exposure to a Tris–HCl pH 7.4 buffer solution.

Swelling experiments carried out with the different DTA–DNA particles demonstrated that the RW depends on both the counterion of the surfactant and on the secondary structure of the nucleic acid. Although, in general, the degree of swelling seems to be higher when particles are prepared at 45°C, the degree of swelling for DTA–ssDNA systems (RW ratio, 3–13, using the maximum points as estimate) is higher than that using native DNA (RW, 3–5) for both temperatures. In addition, in the case of particles prepared with denatured DNA, it was found that the degree of swelling increased in the sequence DTATf < DTAB < DTAC.

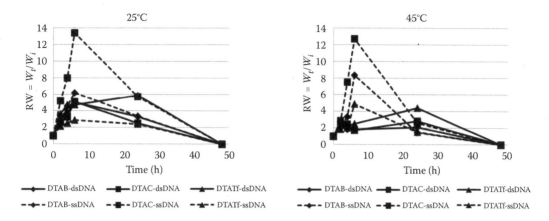

FIGURE 8.7 Time-dependent changes of the RW of DNA gel particles studied, where W_i stands for the initial weight of the particles and W_t for the weight of the particles at time t. (Adapted from Morán, M.C. et al., *Soft Matter* 8, 3200–3211, 2012.)

8.3.4 KINETICS OF DNA RELEASE

The observed accumulative DNA release has been determined, as shown in Figure 8.8. Generally, the release pattern resembles that observed in the swelling/dissolution profiles. Thus, particles prepared using the native nucleic acid had a faster release in the sequence DTATf > DTAB > DTAC as a consequence of the dissolution profile. In the case of particles formed with denatured DNA, slower kinetics were observed, which is congruent with that observed for the swelling/dissolution profiles (see Figure 8.7).

8.3.5 HEMOLYTIC ASSESSMENT

When used as DNA carriers, understanding the interactions of these DNA gel particles with blood components is crucial for improving their behavior *in vitro*. First of all, the hemolytic activity of this DTA-based surfactant was studied as a function of its concentration. The dependence of hemolysis on the concentration and the surfactant structure is shown in Figure 8.9. In these experiments,

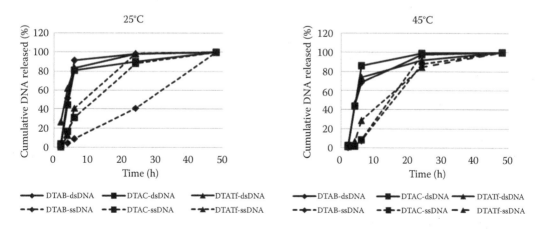

FIGURE 8.8 Time-dependent changes in DNA release profiles for the DNA gel particles studied. (Adapted from Morán, M.C. et al., *Soft Matter* 8, 3200–3211, 2012.)

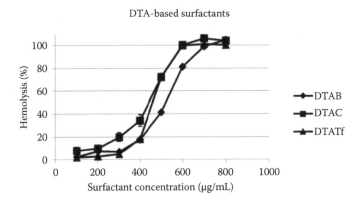

FIGURE 8.9 Dependence of rat erythrocyte hemolysis on DTA-based surfactant concentration. Erythrocytes were incubated for 10 min at room temperature at different surfactant concentrations, and the amount of hemoglobin released was determined. The data correspond to the average of three independent experiments ± standard deviation. (Adapted from Morán, M.C. et al., *Soft Matter* 8, 3200–3211, 2012.)

hemolysis was determined in the presence of a range of surfactant concentrations, which allows us to define the hemolytic potency (HC_{50}) of each surfactant.

One drawback of these surfactant–DNA gel particles, in toxicological terms, is the need for a cationic surfactant, which may cause some cellular damage. Our results indicate, however, that the effect of the surfactant can be modulated when administered in the DNA gel particles, unlike what happens in aqueous solutions. This modulation is due to the strong interaction between the surfactant and the biopolymer, which leads to a very slow release of the surfactant from the vehicle. Accordingly, although the HC_{50} values for these three surfactants are very close in aqueous solution, strong differences were found when the hemolysis kinetics of the corresponding surfactant–DNA gel particles was determined, as represented in Figure 8.10 for DTA–dsDNA particles. Because the hemolytic character of these surfactants in solution is almost identical, the differences found in the hemolysis responses induced by the different surfactants in the DNA particles can only be related with the capacity to form weaker or stronger surfactant–DNA complexes. It is then expected that for a higher degree of complexation, less surfactant, which could interact with the erythrocyte's membrane, would be released in solution.

Figure 8.10 also presents the hemolysis kinetics for particles prepared with ssDNA, in which significative differences from dsDNA particles can be seen. The differences found between particles

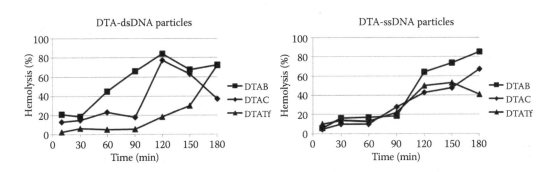

FIGURE 8.10 DTA–DNA particle-induced hemoglobin release from rat erythrocytes as a function of time. Erythrocytes were incubated at room temperature in the presence of individual DTA–DNA particles. (Adapted from Morán, M.C. et al., *Soft Matter* 8, 3200–3211, 2012.)

prepared with dsDNA and ssDNA can be attributed to differences between the two secondary structures. Previous studies of polyelectrolyte–surfactant systems, both experimental and theoretical, showed that the linear charge density of the polyelectrolyte, its flexibility and any amphiphilic character, play significant roles in the corresponding interactions [48–54]. We note that because the linear charge density of dsDNA (0.59 negative charges/Å) is considerably higher than for ssDNA (0.29 negative charges/Å), from a simple electrostatic mechanism, dsDNA should interact more strongly with oppositely charged polyelectrolytes. However, ssDNA is much more flexible than dsDNA, which is quite rigid and is characterized by a large persistence length (500 Å) [48,51]. In molecular simulations [48–50], the role of the flexibility of the polyelectrolyte has been documented in some detail, and it was found that a flexible chain tends to interact more strongly with an oppositely charged macro-ion than a rigid one. Both a higher flexibility of ssDNA and a higher hydrophobicity due to the exposed bases are found to play an important role in the interaction of DNA [55].

This trend in surfactant–DNA interaction reflects both the release of hemoglobin (degree of hemolysis) and the release of DNA into the medium, as a consequence of different dissolution kinetics of the polyelectrolyte–surfactant complexes. Under the experimental conditions in which the hemolysis studies were carried out, dsDNA–surfactant particles were fully dissolved by the end of the experiments. However, ssDNA–surfactant particles remained visible in the dispersion. Here, for the first time, both parameters were determined simultaneously, giving us information about the effectiveness of the two release processes. Figure 8.11 shows the relative kinetics of DNA and hemoglobin release.

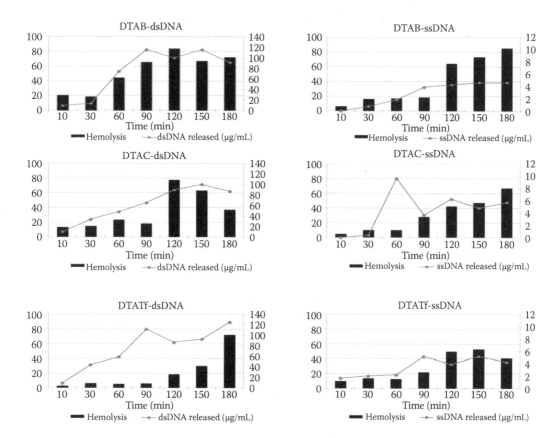

FIGURE 8.11 Relative kinetics of DTA–DNA particle-induced hemoglobin release from rat erythrocytes and DNA release. (Adapted from Morán, M.C. et al., *Soft Matter* 8, 3200–3211, 2012.)

The amount of DNA that is released and the hemolytic response are strongly dependent on both the structure of the counterion in the surfactant and the secondary structure of the DNA. In the case of particles prepared with native DNA, the amount of dsDNA that is released at the end of the experiment (180 min) reaches 100 μg/mL. However, with particles prepared with denatured DNA, only 10% of this amount is released into the medium. This behavior, which can be correlated with the degree of complexation, is higher in the case of ssDNA, thus decreasing the amount of noncomplexed DNA that could be detected in solution. These differences are also supported by visual inspection: surfactant–dsDNA particles are completely dissolved at the end of the experiment, whereas surfactant–ssDNA particles are still present after 180 min.

At this point, it is possible to establish which of these systems is the most hemocompatible. For this, the hemolysis values for a defined amount of released DNA are compared. In the case of the surfactant-dsDNA particles, for a concentration of dsDNA equal to 100 μg/mL, the degree of hemolysis is 30%, 60%, and 80% when DTATf, DTAC, and DTAB are used as cationic agents, respectively. In the case of surfactant–ssDNA particles, and for a concentration of ssDNA equal to 5 μg/mL, the degree of hemolysis is 20%, 50%, and 70% when DTATf, DTAC, and DTAB are used as surfactants, respectively. It is interesting to note that the hemolytic response follows the sequence DTATf <DTAC< DTAB, independently of the secondary structure of the DNA.

The surfactant content of the corresponding surfactant–DNA gel particles was calculated from the hemolysis responses (Figure 8.12). The surfactant content of these DNA gel particles mostly follows the sequence DTATf < DTAC < DTATB. As mentioned above, the formation of these DNA gel particles is based on associative phase separation, which is entropically driven, determined by the translational entropy of the counterions. Accordingly, the differences in surfactant contents found in these DNA gel particles can be correlated with differences in the apparent degrees of counterion dissociation in these surfactants from the corresponding micelles.

By control of the physicochemistry of the components on the DNA gel particles, a better assessment on the final properties of these particles can be achieved. In this context, the reduction in the amount of surfactant needed to form surfactant–DNA gel particles will most probably increase the potential of these systems in drug delivery. Recent studies point out why pDNA is much more efficiently transfected than linear DNA using cationic lipids as vectors in gene therapy. It has been shown that pDNA, in contrast with linear DNA, is compacted—retaining a significant number of counterions in its vicinity. This, in turn, drives a lower effective negative charge, and therefore a lower amount of cationic lipid is needed. For an effective DNA transfection, the lower the amount of the cationic lipid, the lower the cytotoxicity is. Current research is focused on characterizing the *in vitro* cytotoxicity of these surfactant–DNA particles.

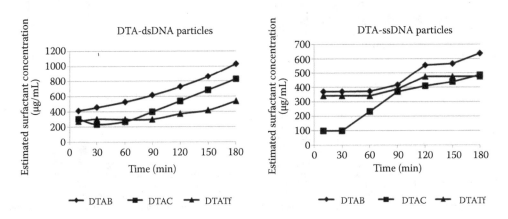

FIGURE 8.12 Surfactant released from DTA–DNA particles as a function of time. (Adapted from Morán, M.C. et al., *Soft Matter* 8, 3200–3211, 2012.)

8.4 SUMMARY

Chemical pDNA gels have been prepared by a cross-linking reaction. The double-stranded gels are porous, transparent, and have a three-dimensional structure as shown by scanning electron microscopy. These gels offer a novel opportunity for monitoring pDNA–cosolute interactions by simply following the change in gel volume. Cationic surfactants induced drastic volume transitions in pDNA gel volume starting abruptly at the critical aggregation concentration (cac); cac is found to decrease with increasing length of the hydrophobic tail. Moreover, the swelling/deswelling process is reversible. The deswelling of the pDNA gels upon the addition of different metal ions does not follow the pronounced surfactant-induced shrinking behavior, but seems to be of a more progressive type; it occurs at lower ion concentrations with increasing valency of the counterion. Concerning polycations, chitosan is the most efficient agent in collapsing pDNA gels. Larger concentrations of polyamines and lysozyme are required to produce similar gel deswelling. The photodisruption of pDNA gels was used as a strategy for the controlled and sustained release of pDNA. pDNA gels are nontoxic to cells, and so they have further merit as possible carriers for intracellular delivery of drugs and genes.

On the other hand, dodecyltrimethylammonium-based surfactants were used to prepare surfactant–DNA gel particles based on associative phase separation and interfacial diffusion. The formation of DNA gel particles by mixing DNA (either ssDNA or dsDNA) with different single-chain surfactants whose structures differ only in the corresponding counterion were evaluated. Analysis of the data indicates that the degree of counterion dissociation from the surfactant micelles and the polar/hydrophobic character of the counterion are important parameters in the final properties of the particles obtained. The stronger interaction of ssDNA than of dsDNA with amphiphiles suggests the important role of both flexibility and hydrophobic interactions. Although the hemolytic potency of the various surfactants in solution is very similar, strong differences were found when the hemolysis kinetics of the corresponding surfactant–DNA gel particles was determined. It was found that the stronger the surfactant–DNA interaction, the slower the hemolysis and DNA release kinetics. The surfactant content of the DNA gel particles was calculated from the hemolytic responses, following the degree of counterion dissociation from the micelle for the different surfactants. By control of the physicochemistry of the components on the DNA gel particles, a better assessment on the final properties of these particles can be achieved. In this context, the reduction in the amount of surfactant needed to form surfactant–DNA gel particles will most probably increase the potential of these systems in drug delivery.

8.5 PROSPECTS

Chemical pDNA gels were prepared by a cross-linking reaction with EGDE. These gels offer a novel opportunity for monitoring pDNA–cosolute interactions by simply following changes in the gel volume. Moreover, the swelling/deswelling is reversible. The features of gel swelling behavior along with the pDNA controlled release profile are crucial for the future rational design of pDNA-based carriers. The biocompatibility of pDNA gels ensures its biomedical use in practical applications. Future perspectives concerning pDNA-based vehicles include the encapsulation of the tumor suppressor p53 gene and the protein expression at cancer cells, with the aim of normal cell function reestablishment. Preliminary studies on a new p53 encoding pDNA microgel that is porous, biocompatible, and photodegradable show its suitability for the dual release of pDNA and anticancer drugs in a controlled and sustained manner. This bifunctionality may represent an enormous advance in comparison to individual chemotherapy treatments, and will certainly serve as a basis for improving cancer therapy.

As can be seen, a general understanding of DNA–surfactant interactions and, in particular, the phase behavior, gives a basis for developing novel DNA-based materials, including DNA gel particles. The systems can be used for the delivery of ssDNA or dsDNA. One drawback of the surfactant–DNA gel particles, in toxicological terms, is the need for a cationic surfactant, which may cause some cell damage. Our results indicate, however, that the effect of the surfactant can be modulated when administered in the DNA gel system, rather than in an aqueous solution. Although toxicity certainly applies for most classic surfactants, we are engaged in current work focusing on the hemocompatible and cytotoxic assessments of DNA gel particles prepared with cationic compounds with much improved intrinsic biocompatibility. These include surfactants with the cationic functionality based on an amino acid [28], polysaccharides [26], and proteins [27]. Special attention will be given to the decrease in size of these DNA gel particles [30], a prerequisite for cellular uptake and internalization, and subsequent DNA delivery and transfection.

ACKNOWLEDGMENTS

We are grateful for financial support from Fundação para a Ciência e a Tecnologia (FCT), (SFRH/BPD/47229/2008) and to PTDC/EBB-Bio/114320/2009, and the Project CTQ2009-14151-C02-02 from the Spanish Ministry of Science and Innovation. We thank Fani Sousa e Ângela Sousa for the 2.7-kpb plasmid pUC19 production and recovery. M. Carmen Morán acknowledges the support of the MICINN (Ramon y Cajal contract RyC 2009-04683).

REFERENCES

1. N. Somia and I.M. Verma. Gene therapy: Promises, problems and prospects. *Nature*, **389**, 239–242 (1997).
2. A.V. Gorelov, D.M. McLoughlin, J.C. Jacquier and K.A. Dawson. DNA complexes with cationic surfactant in mixed solutions. *Nuovo Cimento Soc. Ital. Fis. D-Conden.*, **20**, 2553–2557 (1998).
3. R. Dias, S. Mel`nikov, B. Lindman and M.G. Miguel. DNA phase behaviour in the presence of oppositely charged surfactants. *Langmuir*, **16**, 9577–9583 (2000).
4. I. Koltover, T. Salditt, J.O. Rädler and C.R. Safinya. An inverted hexagonal phase of cationic liposome–DNA complexes related with DNA release and delivery. *Science*, **281**, 78–81 (1998).
5. Y. Yamasaki and K. Yoshikawa. Higher order structure of DNA controlled by the Redox State of Fe2+/Fe3+. *J. Am. Chem. Soc.*, **119**, 10571–10578 (1997).
6. T. Iwataki, S. Kidoaki, T. Sakaue, K. Yoshikawa and S.S. Abramuchuk. Competion between compaction of single chains and bundling of multiple chains in giant DNA molecules. *J. Chem. Phys.*, **120**, 4004–4011 (2004).
7. B. Zhao and J.S. Moore. Fast pH and ionic strength responsive hydrogels in microchannels. *Langmuir*, **17**, 4758–4763 (2001).
8. J. Sjöström and L. Piculell. Simple gel swelling experiments distinguish between associating and non-associating polymer-surfactant pairs. *Langmuir*, **17**, 3836–3843 (2001).
9. C. Sayil and O. Okay. Swelling-shrinking hysteresis of poly(N-isopropyl)acrylamide) gels in sodium dodecylbenzenesulfonate solutions. *J. Appl. Polym. Sci.*, **83**, 1228–1232 (2002).
10. L.M. Bronstein, O.A. Platonova, A.N. Yakunin, I.M. Yanovskaya and P.M. Valetsky. Complexes of polyelectrolyte gels with oppositely charged surfactants: Interaction with metal ions and metal nanoparticles formation. *Langmuir*, **14**, 252–259 (1998).
11. O.E. Philippova, L.A. Chtcheglova, N.S. Karybiants and A.R. Khokhlov. Two mechanisms of gel/surfactant binding. *Polym. Gels Netw.*, **6**, 409–421 (1998).
12. S.G. Starodoubtsev, N.A. Churochkina and A.R. Khokhlov. Hydrogel composites of neutral and slightly charged poly(acrylamide) gels with incorporated bentonite. Interaction with salt and ionic surfactants. *Langmuir*, **16**, 1529–1534 (2000).
13. S.G. Starodoubtsev, A.T. Dembo and K.A. Dembo. Effect of polymer charge density and ionic strength on the formation of complexes between sodium arylamido-2-methyl-1-propane-sulfonate-co-acrylamide gels and cetylpyridinium chloride. *Langmuir*, **20**, 6599–6604 (2004).
14. P. Hansson. Surfactant self-assembly in polyelectrolyte gels: Aggregation numbers and their relation to the gel collapse and the appearance of ordered structures in the NaPA/C_{12}TAB System. *Langmuir*, **14**, 4059–4064 (1998).
15. P. Hansson, S. Schneider and B. Lindman. Phase separation in Polyelectrolyte gels interacting with Surfactants. *J. Phys. Chem. B*, **106**, 9777–9793 (2002).

16. N.A. Peppas. *Hydrogels in Medicine and Pharmacy*, Vols. I–III. CRC Press, Boca Raton. FL (1986).
17. D. Costa, A.J.M. Valente, A.A.C.C. Pais, M.G. Miguel and B. Lindman. Cross-linked DNA gels: Disruption and release properties. *Colloids Surf. A*, **354**, 28–33 (2010).
18. D. Costa, P. Hansson, S. Schneider, M.G. Miguel and B. Lindman. Interaction between covalent DNA gels and a cationic surfactant. *Biomacromolecules*, **7**, 1090–1095 (2006).
19. D. Costa, M.G. Miguel and B. Lindman. Effect of additives on swelling of covalent DNA gels. *J. Phys. Chem B.*, **111**, 8444–8452 (2007).
20. D. Costa, M.G. Miguel and B. Lindman. Responsive polymer gels: Double stranded DNA versus single stranded. *J. Phys. Chem B.*, **111**, 10886–10896 (2007).
21. D. Costa, M. Reischl, B. Kuzma, M. Brumen, J. Zerovnik, V. Ribitsch, M.G. Miguel and B. Lindman. Modelling the surfactant uptake in cross-linked DNA gels. *J. Dispersion Sci. Technol.*, **30**, 16–20 (2009).
22. D. Costa, A.J.M. Valente, M.G. Miguel and J. Queiroz. Gel network photodisruption: A new strategy for the co-delivery of plasmid DNA and drugs. *Langmuir*, **27**, 13780–13789 (2011).
23. M.C. Morán, M.G. Miguel and B. Lindman. DNA gel particles. *Soft Matter*, **6**, 3143–3156 (2010).
24. M.C. Morán, M.G. Miguel and B. Lindman. DNA gel particles: Particle preparation and release characteristics. *Langmuir*, **23**, 6478–6481 (2007).
25. M.C. Morán, M.G. Miguel and B. Lindman. Surfactant–DNA gel particles: Formation and release characteristics. *Biomacromolecules*, **8**, 3886–3892 (2007).
26. M.C. Morán, T. Laranjeira, A. Ribeiro, M.G. Miguel and B. Lindman. Chitosan–DNA particles for DNA delivery: Effect of chitosan molecular weight on formation and release characteristics. *J. Dispersion Sci. Technol.*, **30**, 1494–1499 (2009).
27. M.C. Morán, A. Ramalho, A.A.C.C. Pais, M.G. Miguel and B. Lindman. Mixed protein carriers for modulating DNA release. *Langmuir*, **25**, 10263–10270 (2009).
28. M.C. Morán, M.R. Infante, M.G. Miguel, B. Lindman and R. Pons. Novel biocompatible DNA gel particles. *Langmuir*, **26**, 10606–10613 (2010).
29. M.C. Morán, M.G. Miguel and B. Lindman. DNA gel particles from single and double-tail surfactants: Supramolecular assemblies and release characteristics. *Soft Matter*, **7**, 2001–2010 (2011).
30. M.C. Morán, F.R. Baptista, A. Ramalho, M.G. Miguel and B. Lindman. DNA gel nanoparticles: Preparation and controlling the size. *Soft Matter*, **5**, 2538–2542 (2009).
31. M.C. Morán, T. Alonso, F.S. Lima, M.P. Vinardell, M.G. Miguel and B. Lindman. Counter-ion effect on surfactant–DNA gel particles as controlled DNA delivery systems. *Soft Matter*, **8**, 3200–3211 (2012).
32. D. Costa, J. Queiroz, M.G. Miguel and B. Lindman. Swelling behavior of a new biocompatible plasmid DNA hydrogel. *Colloids Surf. B*, **92**, 106–112 (2012).
33. C. Staples, R. Boatman and M. Cano. Ethylene glycol ethers: An environmental risk assessment. *Chemosphere*, **36**, 1585–1613 (1998).
34. B. Jönsson, B. Lindman, K. Holmberg and B. Kronberg. *Surfactants and Polymers in Aqueous Solution*. John Wiley, Chichester (1998).
35. W.M. Gelbart, A. Ben-Shaul and D. Roux. *Micelles, Monolayers, and Biomembranes*. Springer-Verlag, New York (1992).
36. M.N. Jones and D. Chapman. *Micelles, Monolayers, and Biomembranes*. Wiley-Liss, New York (1995).
37. T.I. Smol'janinova, V.A. Zhidkov (deceased) and G.V. Sokolov. Analysis of difference spectra of protonated DNA: Determination of degree of protonation of nitrogen bases and the fractions of disordered nucleotide pairs. *Nucleic Acids Res.*, **10**, 2121–2134 (1982).
38. B.L. Bales, M. Benrraou and R. Zana. Characterization of micelles of quaternary ammonium surfactants as reaction media I: Dodeclytrimethylammonium bromide and chloride. *J. Phys. Chem. B.*, **106**, 1926–1939 (2002).
39. T.M. Perger and M. Bester-Rogac. Thermodynamics of micelle formation of alkyltrimethylammonium chlorides from high performance electric conductivity measurements. *J. Colloid Interface Sci.*, **313**, 288–295 (2007).
40. F.S. Lima, F.A. Maximiano, I.M. Cuccovia and H. Chaimovich. Surface activity of the triflate ion at the air/water interface and properties of *N, N, N*-trimethyl-*N*-dodecylammonium triflate aqueous solutions. *Langmuir*, **27**, 4319–4323 (2011).
41. S. Ichimura, M. Zama and H. Fujita. Quantitative determination of single/stranded sections in DNA using fluorescent probe acridine orange. *Biochim. Biophys. Acta*, **240**, 485–495 (1971).
42. A.R. Peacocke. The interaction of acridines with nucleic acids. In *Acridines*, R.M. Acheson (Ed.). Interscience Publishers, New York, pp. 723–754 (1973).
43. A. Cser, K. Nagy and L. Biczók. Fluorescence lifetime of Nile Red as a probe for the hydrogen bonding strength with its microenvironment. *Chem. Phys. Lett.*, **360**, 473–478 (2002).

44. G. Hungerford, E.M.S. Castanheira, M.E.C.D. Real Oliveira, M.G. Miguel and H.D. Burrows. Monitoring ternary systems of C12E5/water/tetradecane via the fluorescence of solvatochromic probes. *J. Phys. Chem. B*, **106**, 4061–4069 (2002).

45. P. Greenspan and S.D. Fowler. Spectrofluorometric studies of the lipid probe, Nile Red. *J. Lipid Res.*, **16**, 781–789 (1985).

46. R.D. Koynova, B.C. Tenchov and P.J. Quinn. Sugars favor formation of hexagonal phase at the expense of the lamellar liquid-crystalline phase in hydrated phosphatidylethanolamines. *Biochim. Biophys. Acta*, **980**, 377–380 (1989).

47. R.M. Epand and M. Bryszewska. Modulation of the bilayer to hexagonal phase transition and solvation of phosphatidylethanolamines in aqueous salt solutions. *Biochemistry*, **27**, 8776–8779 (1988).

48. T. Wallin and P. Linse. Monte Carlo simulations of polyelectrolytes at charged micelles. 1. Effects of chain flexibility. *Langmuir*, **12**, 305–314 (1996).

49. T. Wallin and P. Linse. Monte Carlo simulations of polyelectrolytes at charged micelles. 2. Effects of linear charge density. *J. Phys. Chem.*, **100**, 17873–17880 (1996).

50. T. Wallin and P. Linse. Monte Carlo simulations of polyelectrolytes at charged micelles. 3. Effects of surfactant tail length. *J. Phys. Chem. B*, **101**, 5506–5513 (1997).

51. J.C.T. Kwak. *Polymer-Surfactant Systems*. Marcel Dekker, New York (1998).

52. I. Lynch, J. Sjostrom and L. Piculell. Hydrophobicity and counterion effects on the binding of ionic surfactants to uncharged polymeric hydrogels. *J. Phys. Chem. B*, **109**, 4252–4257 (2005).

53. I. Lynch, J. Sjostrom and L. Piculell. Reswelling of polyelectrolyte hydrogels by oppositely charged surfactants. *J. Phys. Chem. B*, **109**, 4258–4262 (2005).

54. I. Lynch and L. Piculell. Presence or absence of counterion specificity in the interaction of alkylammonium surfactants with alkylacrylamide gels. *J. Phys. Chem. B*, **110**, 864–870 (2006).

55. B. Lindman, R.S. Dias, M.G. Miguel, M.C. Morán and D. Costa. Manipulation of DNA by surfactants. In *Highlights in Colloid Science*, D. Platikanov and D. Exerowa (Eds.). Wiley-VCH, Weinheim, pp. 179–202 (2009).

9 Advances in Poly(amino acid)s–Based Amphiphilic Graft Polymers and Their Biomedical Applications

Chan Woo Park, Hee-Man Yang,
Se Rim Yoon, and Jong-Duk Kim

CONTENTS

9.1 INTRODUCTION

Graft polymers comprise a general class of random copolymers that have a linear and long main backbone with randomly distributed side chains. Graft polymers have been used in polymer industrial applications for impact-resistant materials, compatibilizers, and emulsifiers. These comb-type graft polymers have adjustable polymer properties, which can be achieved by controlling the number and length of the branches, as well as through monomer selection.

In recent years, natural polymer-based graft copolymers using polysaccharide or poly(amino acid)s have been the object of growing scientific attention due to their degradability and biocompatibility, and have attracted considerable interest due to their widespread applications, including use as bioplastics, ecofriendly surfactants, and drug delivery carriers. Notably, poly(amino acid)s have remarkable diversity, obtained from the incorporation of various natural amino acids, and their properties are controllable. Moreover, amphiphilic poly(amino acid)s–based graft polymers consisting of a hydrophilic backbone with hydrophobic branches or a hydrophobic backbone with hydrophilic branches are surface-active and form diverse structures of self-aggregates such as spherical, cylindrical, tubular, and vesicular aggregates in aqueous solution. In addition, they are generally smaller than a few hundred nanometers and are water-soluble. Moreover, these poly(amino acid)s–based nanostructures are thermodynamically and kinetically more stable than conventional surfactant systems. These characteristics are known to be essential for biomedical applications, and thus amphiphilic poly(amino acid)s–based graft polymers are regarded as ideal materials for biomedical applications, including drug delivery, coatings for bioimplants, and bio-imaging agents.

Using drug delivery systems based on surfactants or polymer nanoparticles, the therapeutic properties of drugs can be improved with reduced side effects. Polymer nanocarriers can improve solubility, stability in blood, and cellular uptake of drugs, and the release of drugs can be controlled. In cancer treatments, in particular, nanoparticle carriers can accumulate at tumor sites due to leaky blood vessels in the tumor region, thereby enabling target tumor–specific delivery of anticancer drugs. Owing to successful achievements of drug delivery via nanocarriers, some of the liposome or drug-conjugated polymer systems have been approved for clinical use [1]. In addition, amphiphilic block copolymer micellar systems are in clinical trials, and poly(amino acid)s–based micellar systems have shown promising therapeutic effects in clinical trials [2,3]. With increasing interest in drug delivery systems, the global market for the nano-biomedical field is expected to be $70 to $160 billion by 2015 [4]. Although graft polymer systems have not been comprehensively studied in clinical trials thus far, graft polymer systems are expected to surpass block copolymer systems because they can achieve multifunctionality by multiple grafts.

This chapter describes the history of the development of amphiphilic poly(amino acid)s–based graft polymers, their synthesis strategies, and the behavior of self-assemblies in aqueous solution. Furthermore, various attempts to prepare functional amphiphilic poly(amino acid)s and promote their application in drug delivery and bioimaging are reviewed.

9.2   POLY(AMINO ACID)S–BASED AMPHIPHILIC GRAFT POLYMERS

9.2.1   POLY(AMINO ACID)S

Poly(amino acid)s, which are polymerized from natural or synthetic amino acids, have received a great deal of attention for biomedical applications due to their biocompatibility, biodegradability, nontoxicity, and *in vivo* stability [5]. Due to the diversity of amino acids, the properties of poly(amino acid)s including hydrophobicity, polarity, stimuli responsivity, and reactive side chains can be modulated by the selection of monomers. Furthermore, poly(amino acid)s can contribute to stable secondary structures such as α-helixes and β-sheets by hydrogen bonding.

Poly(amino acid)s can be synthesized by ring opening polymerization (ROP) of α-amino acid *N*-carboxyanhydride (NCA; Figure 9.1a). The polymerization of amino acid–NCA can be initiated by nucleophiles and aprotic bases. The NCA polymerization can lead to high molecular weight and relatively narrow polydispersity of poly(amino acid)s, making the process economic and fast [6].

Poly(succinimide) (PSI) has also been widely used as a reactive polymer platform to synthesize various poly(amino acid)s. PSI can be synthesized by acid-catalyzed polycondensation of L-aspartic acid. An outstanding advantage of PSI is that it can be converted into various hydrophilic poly(amino acid)s such as poly(L-aspartic acid) (PAsp), poly(aspartamide), and poly(L-asparagine) (PAsn) by simple hydrolysis or an aminolysis reaction in NaOH, ethanolamine, and NH_4OH solution, respectively (Figure 9.1b). One of the most extensively studied derivatives of PSI is poly-α,β-(*N*-2-hydroxyethyl-L-aspartamide) (PHEA). PHEA can be prepared by aminolysis of PSI using ethanolamine. The degradability of PHEA was tested *in vitro*, and amide bonds degraded in the presence

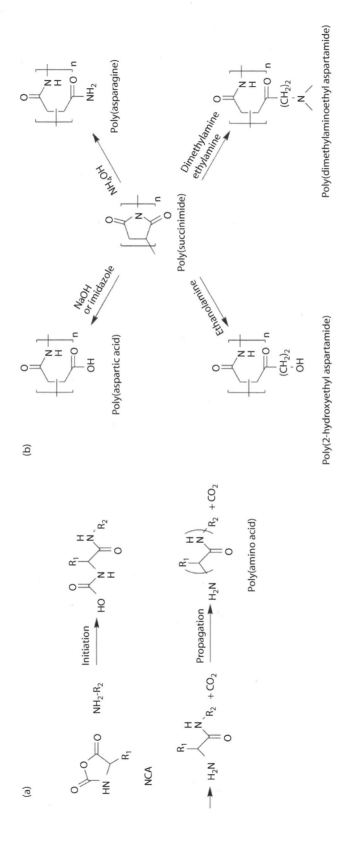

FIGURE 9.1 (a) Polymerization of amino acid–NCA with a primary amine initiator and (b) diverse poly(amino acid) derivatives prepared from PSI.

of protease and released corresponding amino acids or oligomers [7]. Furthermore, PHEA-coated particles showed significantly prolonged circulation in the bloodstream relative to PEG-coated particles due to the stealth property of PHEA, and they also showed a less pronounced accelerated blood clearance phenomenon than PEG-coated particles [8].

9.2.2 Synthesis of Poly(amino acid)s–Based Amphiphilic Graft Polymers

Amphiphilic poly(amino acid)s, which have hydrophilic and hydrophobic parts in a single polymer chain, exhibit surface-active properties and form self-assemblies in aqueous solution. From this point of view, amphiphilic poly(amino acid)s can be considered candidate materials to solubilize hydrophobic inorganic nanoparticles or hydrophobic molecules within hydrophobic cores of self-aggregates. For this reason, various types of amphiphilic poly(amino acid)s that can meet required properties for various applications have been proposed. Representative structures of amphiphilic poly(amino acid)s are block and graft polymers. Amphiphilic poly(amino acid) block copolymers can be synthesized from the ROP of amino acid–NCA using amino initiators such as α-methoxy-ω-aminopoly(ethylene glycol) [9,10].

On the other hand, graft polymers of poly(amino acid)s can be synthesized through various routes and with diverse materials. A major advantage of graft polymers is multifunctionality, which can be achieved by multiple grafts. The properties of graft polymers can be controlled by the chemical composition of the backbone and grafts. In addition, the graft density (degree of substitution; DS) and degree of polymerization (DP) of branches play important roles (Figure 9.2). In general, long polymers and small molecules (alkyls, lipids, cholesterol, etc.) have been used as side chains to prepare graft polymers [11–13]. However, a new type of amphiphilic poly(amino acid)s–based graft polymers that have nanoparticles as branches have recently been proposed [14]. In addition, supramolecular graft polymers that have side chains grafted by noncovalent interactions such as ionic interactions, π–π interactions, and hydrogen bonding instead of covalent bonding were also introduced [15].

Typically, synthesis routes for amphiphilic poly(amino acid)s–based graft polymers can be classified into either a "graft from" method or a "graft onto" method (Figure 9.2). In the next section, we introduce several representative synthesis methods for various amphiphilic poly(amino acid)s–based graft polymers, as summarized in Table 9.1.

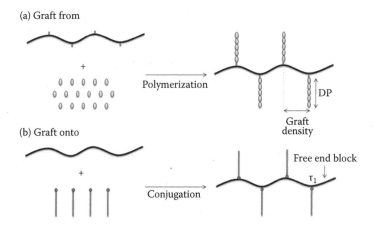

FIGURE 9.2 Synthesis of poly(amino acid) graft polymers using (a) the graft from and (b) the graft onto methods. (a) CAC of amphiphilic-based graft polymers of poly(amino acid)s as a function of DS and (b) plot of hydrodynamic diameter of PAsp-g-C$_{18}$ versus DS of octadecyl group. (a. From Jeong, J.H. et al., *Polymer* 44, 583–591, 2003. With permission. b. From Kang, H.S. et al., *Langmuir* 17, 7501–7506, 2001. With permission.)

TABLE 9.1

Various Amphiphilic Graft Polymers of Poly(amino acid)s and Their Synthesis Methods

Method	Polymer	Grafting Reaction	References
Graft from	PLA-*g*-PLL	ROP of amino acid–NCA	[16,17]
	PLA-*g*-PAsp		
	PLA-*g*-PAla		
	PCL-*g*-PLL	ROP of amino acid–NCA	[18]
	PHEA-*g*-PLeu	ROP of amino acid–NCA	[19]
	PHEA-*g*-PBMA	ATRP from PHEA	[20]
Graft onto	PHEA-*g*-C_n	Reaction of alkylamine with PSI	[11,21–24]
	PAsp-*g*-C_n		
	PAsn-*g*-C_n		
	PDMAEA-*g*-C_n		
	PAsp-*g*-PHS	Reaction of PHS with PSI	[13]
	PAsn-*g*-PCL	Reaction of PCL-NH_2 with PSI	[12]
	PBLG-*g*-PEG	Ester exchange reaction of PBLG	[26]
	PLL-*g*-PLGA	DCC/NHS chemistry	[27]
	PHEA-*g*-PLA	*N,N*,-carbonyldiimidazole coupling	[28]
	PHEA-*g*-Au nanoparticles	Coordinate bond	[14,29]
	PHEA-*g*-QD nanoparticles		
	PHEA-*g*-Fe_3O_4 nanoparticles		
	PLL-*g*-PLGA	Ionic interaction	[15]

9.2.2.1 The "Graft from" Method

The "graft from" reaction utilizes backbone polymers having randomly distributed initiating groups as macro-initiators, and the polymerization of branches is initiated from the polymer backbone. ROP of amino acid–NCA from the backbone polymer was attempted to synthesize poly(amino acid)–based graft polymers.

In earlier studies, biodegradable hydrophobic polyesters such as poly(L-lactide) (PLA) and poly(ε-caprolactone) (PCL), were used as macro-initiators. In the 1990s, Langer and coworkers [16] reported amphiphilic poly(amino acid)s–based graft polymers synthesized through the grafts from this method. Lysine (Lys) was randomly incorporated into the PLA backbone, and the resultant polyester–amino acid polymer was utilized as a macro-initiator. Cationic poly(L-lysine) (PLL) grafted PLA was obtained by ROP reaction of Lys-NCA from primary amines of a P(LA-co-LL) backbone. In addition, they demonstrated the versatility of the backbone platform by expansion of amino acid–NCA selection, and they synthesized anionic PAsp grafted PLA and poly(D,L-alanine) grafted PLA [17]. Similarly, Nottelet et al. [18] reported a synthesis method for PLL branches by ROP of Lys(z)-NCA from a PCL backbone. Notably, the polycarbanion derived from PCL initiated the polymerization reaction.

Poly(amino acid)s–based graft polymers consisting of poly(amino acid)s backbone and grafts were also developed [19]. Our group recently synthesized hydrophobic poly(L-leucine) or poly(L-valine) grafted PHEA using primary amine functionalized PHEA as a macro-initiator. The L-leucine-NCA or L-valine-NCA was polymerized by ROP initiated from the PHEA backbone. The DS was controlled by adjusting the number of amine groups in the PHEA backbone, and the DP of branches was modulated by controlling the feed moles of amino acid–NCA. NCA polymerization is a well-known method that is able to control the DP. However, it is not easy to confirm the molecular weight distribution of branches polymerized from the backbone.

Cavallaro et al. [20] meanwhile proposed a poly(amino acid)s–based macro-initiator for atomic transfer radical polymerization (ATRP). This transition metal-mediated living radical polymerization leads to uniform polymer chain length and narrow polydispersity. They conjugated 2-bromoisobutyryl bromide to PHEA to realize a macro-initiator, and hydrophobic butyl methacrylate (BMA) was polymerized from PHEA via ATRP.

9.2.2.2 The "Graft onto" Method

Amphiphilic poly(amino acid)s–based graft polymers can also be synthesized by the "graft onto" method, a simple process of attaching side chains onto the polymer backbone. The graft reaction has been achieved by aminolysis of reactive PSI or poly(γ-benzyl L-glutamate) (PBLG) or conjugation using reactive end groups of polymers by utilizing coupling agents. The graft onto method enables conjugation of not only well-defined polymers that have narrow molecular weight distribution but also grafting of various side chains such as small molecules, drugs, and nanoparticles. A major advantage of this method is that it provides the ability to simultaneously graft various branches onto a polymer backbone, and a great variety of graft polymers with adjustable properties and functionalities have been developed.

In 2000, our group proposed synthesis routes for alkyl-grafted poly(amino acid)s using reactive PSI [11]. Because PSI can be aminolyzed with primary amines by nucleophilic attack, alkylamines (C_n, n = 12, 16, 18) can be readily grafted to the PSI in dimethylformamide. After the grafting reaction, the PSI backbone can be converted into negatively charged hydrophilic poly(amino acid)s, PAsp, by hydrolysis in NaOH solution, and a series of amphiphilic PAsp-g-C_n were obtained [21]. Furthermore, the PSI backbone could be converted into various hydrophilic poly(amino acid)s including hydroxylated PHEA, neutral PAsn, and positively charged poly-α,β-(N-2-dimethylaminoethyl L-aspartamide) (PDMADA) [22–24]. On the other hand, Giammona and coworkers [25] conjugated poly(ethylene glycol) on PHEA-g-C_{12} to obtain high hydrophilicity with neutral charge and a stealth property. Furthermore, the feasibility of grafting various amine-containing molecules such as lipids and bulky polymers onto PSI has been demonstrated. Phytosphingosine (PHS), which is a primary amine-containing lipid, was reacted with PSI, and amphiphilic PHEA-g-PHS and PAsp-g-PHS polymers were synthesized [13]. In addition, a bulky and hydrophobic polymer, PCL, having a primary amine end, can also be readily grafted on a PSI backbone [12].

Poly(β-benzyl-L-aspartate) (PBLA) and PBLG have also received strong interest as reactive poly(amino acid)s backbones. PBLA and PBLG can be synthesized by ROP of BLA-NCA and BLG-NCA, and PEG or primary amine-containing molecules can be grafted by an ester exchange reaction or aminolysis reaction with the elimination of benzyl ester groups [26].

The grafting onto method using coupling agents has also been extensively studied. Unlike the use of a reactive polymer backbone platform, it requires coupling agents to conjugate polymers having functional groups such as hydroxyl, carboxylate, and amine groups. Jeong and Park [27] reported on an amphiphilic poly(L-lysine)-g-poly(D,L-lactic-co-glycolic acid) (PLL-g-PLGA) synthesized from the conjugation of primary amine groups in PLL with the carboxylic acid end group of PLGA using carbodiimide chemistry. On the other hand, Lee et al. [28] prepared PHEA-g-PLA by coupling hydroxyl end groups of PLA with amine-functionalized PHEA using a N,N,-carbonyldiimidazole coupling agent.

Recently, a new synthesis route for nanoparticle grafted poly(amino acid)s polymers (nanocrystallo-polymer) was realized. Functional groups such as thiols or carboxylates, which can form coordinate bonds with nanoparticles, were introduced to a hydrophilic poly(amino acid)s backbone. A ligand exchange reaction was then carried out with hydrophobic nanocrystals such as gold nanoparticles, quantum dots, and iron oxide nanoparticles. Consequently, various inorganic nanoparticle–grafted PHEA specimens were prepared and hydrophobic nanoparticle branches served as the hydrophobic component in the amphiphilic nanocrystallo polymer [14,29].

A supramolecular graft polymer of PLL-g-PLGA produced via noncovalent grafting was reported by Jo et al. [15] negatively charged carboxylic acid groups of PLGA and positively charged amine groups of PLL formed ionic bonds in a water/dimethyl sulfoxide mixture, and noncovalently grafted PLL-g-PLGA was obtained.

9.3 AQUEOUS SELF-AGGREGATES OF POLY(AMINO ACID)S–BASED AMPHIPHILIC GRAFT POLYMERS

9.3.1 Principle for Self-Aggregation of Amphiphilic Poly(amino acid)s

Amphiphilic poly(amino acid)s having hydrophilic and hydrophobic parts exhibit surface-active property, and form self-assemblies in an aqueous solution. The hydrophobic parts are not easily soluble in the aqueous solution, and they cluster, forming an oily core by hydrophobic interaction. On the other hand, hydrophilic poly(amino acid)s, such as PAsp, PHEA, and PLL, which have polar or ionic characteristics, positively interact with water molecules, thereby stabilizing self-aggregates in aqueous solution. Amphiphilic poly(amino acid)s usually form spherical aggregates with a hydrophilic shell and hydrophobic core due to the strong repulsive force of the hydrophilic parts.

Self-aggregation can be induced by direct dissolution or sonification of amphiphilic poly(amino acid)s in an aqueous solution if they have small hydrophobic parts such as alkyl groups. On the other hand, in the case of long hydrophobic parts with a small hydrophilic part system, self-aggregates can be prepared using common solvents for both the hydrophobic and hydrophilic parts. For self-aggregation, the polymer solution in the organic solvent is added to excess water (this is known as the precipitation-dialysis method) [12], or water drops are slowly added to the polymer solution [15]. The mixing condition of water and organic solvent is water-rich and dilute for the first method, whereas it is organic solvent–rich and concentrated for the latter. For this reason, these two different methods give different energy balances between entropic loss via restriction within aggregates and enthalpic gain by reduction of interface energy, and thereby show different self-aggregation behaviors.

As with low–molecular weight surfactants, amphiphilic poly(amino acid)s also start to form self-aggregates at a certain concentration, which is called the critical association concentration (CAC). Near the CAC, amphiphilic polymers form a loosely associated core, and thereby some water exists in the hydrophobic core region. As the concentration is increased, water molecules in the core are released, and aggregates are more tightly packed to attain the lowest energy state [30]. Typically, amphiphilic copolymers show much higher CAC values than low–molecular weight surfactants, and thus they are thermodynamically and kinetically more stable than surfactant systems. The most preferred method to measure CAC of amphiphilic poly(amino acid)s is to use fluorescent probes such as pyrene, which is sensitive to the environmental polarity [21]. The nature of hydrophobic groups and graft density govern the CAC values of poly(amino acid)s–based graft polymers. It has been reported that CAC of graft polymers consisting of hydrophilic backbone and hydrophobic branches decreased with increasing DS of hydrophobic side chains due to strong hydrophobic interaction. The PAsp-g-C$_n$, PAsn-g-PCL, and many other graft polymer systems also agreed with the following equation:

$$\log \text{CAC} = A - B\,[\text{DS}]$$

where A and B depend on the polymer nature and environment [21]. Moreover, as shown in Figure 9.3a, the PCL grafted PAsn showed much lower CAC than PAsp-g-C$_{18}$ and PAsp-g-dehydrochloric acid. Although it is not possible to compare PAsn and PAsp backbone systems directly, it is believed that crystallizable and long hydrophobic polymer branches have higher driving forces for aggregate formation compared with relatively short hydrophobic alkyl chains [12].

In general, amphiphilic polymers form spherical aggregates smaller than a few hundred nanometers, depending on the length of hydrophilic and hydrophobic parts and other factors. Notably, for the amphiphilic poly(amino acid)s–based graft polymers, the DS of hydrophobic branches is an important factor for the size of spherical aggregates. As shown in Figure 9.3b, the hydrodynamic diameter of self-aggregates of PAsp-g-C$_{18}$ decreased with DS. This is attributed to the stronger hydrophobic interaction of hydrophobic parts with increasing packing density of hydrophobic groups in self-aggregates [6,21].

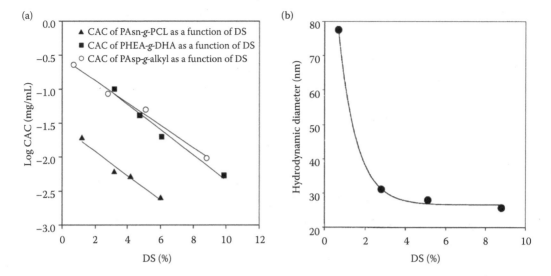

FIGURE 9.3 (a) CAC of poly(amino acid)s–based amphiphilic graft polymers as a function of DS and (b) plot of hydrodynamic diameter of PAsp-g-C$_{18}$ versus DS of octadecyl group. (a. From Jeong, J.H. et al., *Polymer* 44, 583–591, 2003. With permission. b. From Kang, H.S. et al., *Langmuir* 17, 7501–7506, 2001. With permission.)

9.3.2 DIVERSE MORPHOLOGIES OF AMPHIPHILIC POLY(AMINO ACID)S–BASED AMPHIPHILIC GRAFT POLYMERS

Amphiphilic polymers can form not only spherical aggregates but also cylindrical, tubular aggregates, vesicles, lamellar structures, and many more, by spontaneous self-assembly in a selective solvent. With the development of microscopic techniques such as transmission electron microscopy and scanning electron microscopy, the research on morphology transition of amphiphiles has attracted much interest. The favored morphology of self-assembled nanostructures depends mainly on the hydrophilic/hydrophobic ratio, molecular geometry, and chemical nature. In addition, the morphologies can be affected by external factors such as concentration, temperature, and solvent [31,32]. The concepts of packing parameter and curvature have been widely accepted to describe the effect of molecular geometry of amphiphiles on the final shape and structure of self-aggregates. The packing parameter (p) is defined as $p = v_0/a_0l_c$, where v_0 is the volume of hydrophobic unit, l_c is the extended length of the hydrophobic unit, and a_0 is the effective headgroup area [33]. Figure 9.4a illustrates the critical packing parameters and final structures of self-aggregates with three major morphologies (sphere, cylinder, and vesicles). Although the packing parameter was introduced to explain the aggregation of amphiphiles with low molecular weight, it can also be applied to an amphiphilic polymer system [34]. Graft polymers can be regarded as a connected surfactant system, and their aggregation behavior can also be explained by the packing parameter.

In a graft polymer system, the hydrophilic/hydrophobic ratio and number of branches are regarded as key parameters of the molecular geometry. Notably, Zhang et al. [35] theoretically demonstrated the importance of the graft point position (τ_i) using the self-consistent field theory. As shown in Figure 9.2, the length of the free end block can be decided according to the first graft point (τ_1), and its variance could induce a morphological transition when other parameters are fixed. In addition, they also conducted a simulation on the number of branches (m) dependent morphological transition. At large τ_1, in the case of an amphiphilic graft polymer with a hydrophilic backbone, with increasing m values, a morphological transition from sphere (C) → mixture of sphere and rod (CL) → mixture of sphere, rod, and vesicle (CLV) took place. In the small τ_1 region, on the other

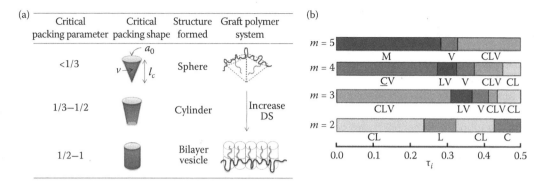

FIGURE 9.4   (a) Molecular packing parameters and the structure of aggregates, and schematic representation of graft polymers with different DS. (b) Aggregate morphology of graft polymers as a function of the first graft point (τ_i) with various numbers of branches (m). (From Zhang, L. et al., *J. Phys. Chem. B* 111, 9209–9217, 2007. With permission.)

hand, a morphological transition from CL → CLV → mixture of sphere and vesicle aggregates (CV) → large aggregates (M) was expected (Figure 9.4b).

The experimental observation of morphology transition for amphiphilic poly(amino acid)s–based graft polymers has been extensively studied by our group. An early study showed the morphology transition of PHEA-*g*-PLA as a function of DS of hydrophobic branches (Figure 9.5) [28]. In the low DS region, PHEA-*g*-PLA formed spherical aggregates, and particle size decreased with increasing DS. A further increase of DS results in starting of vesicle formation at DS = 4.37, and coexistence of sphere micelles and vesicles was observed. Finally, all aggregates showed vesicle structure at DS = 6.34. The root of morphology transition induced by changing DS is the increase of hydrophobic fraction in polymers with an increase in DS. However, it is not only related to the fraction of hydrophobic parts but also to the flexibility and bending of the backbone and molecular geometry. With an increase of graft density, the polymer becomes rigid and bulky, and the flexibility of the backbone is reduced. For this reason, the bilayer structure would be preferred over hydrophobic cores enclosed with hydrophilic backbone polymer. Recently, we observed the structural transition of PHEA-*g*-C$_{12}$ as a function of DS of hydrophobic alkyl chains. Therefore, unlike the PHEA-*g*-PLA system, aggregate morphologies changed from sphere to cylindrical micelle, then to tubular micelle, and finally to vesicles, as DS increased [36]. Graft polymers can be regarded as a connected surfactant system, and the packing parameters of each unit can be calculated from the surface area of hydrophilic backbone per hydrophobic chain measurement by the π-A isotherm. The

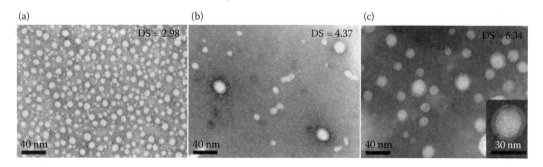

FIGURE 9.5   Morphological transition of PHEA-*g*-PLA aggregates from (a) sphere to (b) sphere + vesicles, and (c) vesicles observed by TEM with negative staining using phosphotungstic acid. (From Lee, H.J. et al., *Macromolecules* 39, 4938–4940, 2006. With permission.)

calculated packing parameters according to the DS and the structure of aggregates matched well with general packing parameter theory. On the other hand, Cai et al. [37] reported the effect of DS of PBLG-*g*-PEG on the morphology of aggregates. They observed vesicle structures at low DS, and the morphology changed from spindle micelles to spherical micelles with an increasing number of PEG chains.

More recently, diverse morphologies of self-assembled nanocrystallo-polymers were also observed by Jang et al. [14]. The gold nanoparticles–grafted PHEA forms micelle-like aggregates with a multiple gold nanoparticles core in aqueous solution. Dodecyl chains were grafted to the backbone as a cohydrophobic group to cause morphology transition, and DS was adjusted to control the hydrophilic/hydrophobic ratio. In the low DS region, the size of micelles decreased and the number of gold nanoparticles in the core was also reduced with an increasing number of dodecyl chains. Then the morphology changed to the core shell uni-gold nanoparticle micelles. A further increase of DS induced structure change to cylindrical aggregates with regularly embedded gold nanoparticles (Figure 9.6).

The effect of solvent has also been considered as a key parameter in the morphology transition. Jo et al. [15]. observed self-aggregate structures of PLGA-*g*-PLL supramolecular graft polymers prepared in DMSO/water. After micellization, a further increase of water content in the solution induced a morphological change from sphere to vesicles. Interestingly, recovery of the percentage of water in the solution by adding DMSO restored the spherical aggregates from the vesicle structure. The formation of vesicles is ascribed to the thermodynamically unfavorable increase in the spherical aggregates in the core chain stretching, which exceeds the driving force and reduces the interfacial area. Similarly, Cai et al. [37] observed that the vesicle structure of PBLG-*g*-PEG can be changed to spindles, and spindle aggregates are formed with the addition of *N,N*-dimethylformamide to the solvent; this is attributed to the capability of the PBLG backbone to interact well with dimethylformamide.

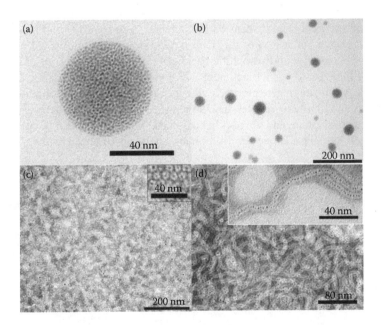

FIGURE 9.6 Various morphologies of self-assembled PHEA-*g*-Au nanoparticles observed by TEM, and (a) pristine PHEA-*g*-Au nanoparticles formed nanocrystallo-spherical aggregates. With the introduction of dodecyl chains, morphological transition is observed from (b) spherical aggregates to (c) core shell type aggregates, and (d) cylindrical aggregates. (From Jang, K.-S. et al., *Soft Matter* 4, 349–356, 2008. With permission.)

9.4 ADVANCES IN AMPHIPHILIC GRAFT POLYMERS IN BIOMEDICAL APPLICATIONS

9.4.1 Drug Delivery Using Poly(amino acid)s–Based Amphiphilic Graft Polymers

To deliver therapeutic agents to the target site efficiently and without any toxicity, polymeric micelles formed by amphiphilic copolymers are used to increase pharmacokinetic and pharmacodynamic profiles. Because current cancer treatments such as surgical intervention, chemotherapy, and radiation therapy often cause toxicity to the patient and kill healthy cells, efficient drug delivery carriers that either passively or actively target cancer cells are needed.

Passive targeting arises from tumor biology where tumor vessels have a defective architecture allowing nanoparticles to accumulate in the tumor selectively, by an enhanced permeability and retention (EPR) effect (Figure 9.7). Due to leaky blood vessels and poor lymphatic drainage of the tumor, nanosized carriers could escape into the tumor and accumulate, whereas free drugs just diffuse nonspecifically. Due to this unique characteristic of the tumor, this phenomenon was considered to be a landmark finding in tumor targeting [38].

Developing nanosized polymeric drug carriers is necessary to deliver chemotherapeutic drugs more efficiently, and amphiphilic copolymers have been developed as a promising delivery tool to form stable structures. Among the various amphiphilic copolymers, poly(amino acid)s have received considerable attention for their application as potential polymeric drugs based on various characteristics such as biocompatibility and biodegradability. Our group researched stable aggregates using poly(amino acid)s–based graft polymers as delivery carriers obtained by controlling the grafted hydrophobic moieties with long alkyl chains [12,21]. Some groups have grafted poly(L-lactide) (PLA) or poly(2,2-dimethyltrimethylene carbonate) (PDTC) sequences on a poly(amino acid)–based backbone [39,40]. The design of synthetic poly(amino acid)s–based graft polymers for drug delivery carriers should consider the feasibility of their synthesis, effective enzymatic degradability, and ease of conjugating or encapsulating therapeutic agents. Most therapeutic agents could be encapsulated physically in the inner core of self-assembled aggregates, or directly conjugated/grafted onto the main backbone. Amphiphilic poly(amino acid)s–based graft polymers that have both hydrophilic and hydrophobic parts in a single polymer chain would exhibit surface-active property, and form self-assemblies in aqueous solution.

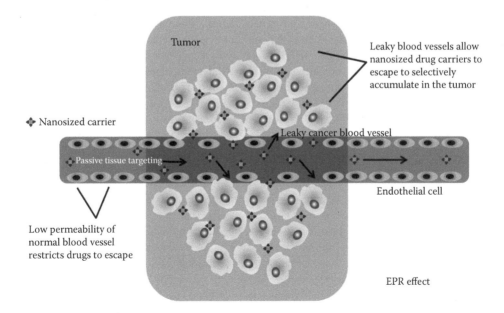

FIGURE 9.7 Schematic representation of EPR effect.

From this point of view, amphiphilic poly(amino acid)s–based graft polymers are candidate materials to solubilize hydrophobic molecules within the hydrophobic cores [41]. Release of the physically encapsulated drug can be controlled by the rate of drug diffusion from the micellar core or breakup of the micelles by external stimuli [49]. On the other hand, studies on grafting of therapeutic molecules directly onto the main backbone have been performed. Kang et al. incorporated methotrexate (MTX) in a poly(amino acid)–based graft polymer by both physical encapsulation and chemical conjugation. MTX is an antineoplastic agent used to treat acute lymphoblastic leukemia, psoriasis, and other trophoblastic tumors [22]. Jung et al. conjugated sphingolipid metabolite, phytosphingosine (PHS), to a PHEA backbone (PHEA-g-PHS). Amphiphilic graft polymers not only act as an efficient intracellular vehicle but also as a drug because sphingolipids and their metabolites are involved in essential biological processes such as cell growth, survival, and death in mammalian cells [13].

9.4.2 Surface Modification for Enhanced Cellular Uptake

Many studies have been conducted on anticancer drug delivery systems based on the EPR effect. However, this effect has shown some drawbacks and limitations because certain tumors do not show the EPR effect [38,42]. Therefore, surface modification for active targeting has become an important factor to enhance drug release and targeting accuracy. Active targeting could be achieved by modifying the particles' surface with targeting entities, such as attaching antibodies, peptides, or other specific adhesion molecules, or by conjugating specific functionalities to the polymer backbone. Our group previously conjugated Arg-Gly-Asp (RGD) peptide directly to the PHEA backbone for selective targeting of angiogenic tumors [19], and conducted further studies to attach HER2/neu antibody to PHEA for enhanced signal intensity for imaging breast cancer [43] (see Section 9.4.5 for further information). Yang et al. [44] conjugated oligoarginine (Arg_8), a cell-penetrating peptide, to PHEA via a thioether linkage, to be selectively adsorbed on the cell membrane and enhance cellular uptake. In addition to the electrostatic interaction between Arg_8 and the cell membrane, the specific interaction between the guanidinium group of Arg_8 and hydrogen bond acceptor moiety on the plasma membrane causes adsorptive endocytosis, and thus enhanced cellular uptake of the drug carrier. It was previously found that thiolated polymers have a mucoadhesive property arising from the specific interaction of thiols with mucus glycoproteins. Park and Kim [45] grafted L-cysteine, a mucoadhesive agent, to PHEA to synthesize a new type of thiolated polymer that could be applied in a mucoadhesive delivery system as well. PEGylation has been intensively applied in drug delivery systems because to its stealth character to increase circulating half-life, shield antigenic and immunogenic epitopes, avoid receptor-mediated uptake by the reticuloendothelial system (RES), and prevent recognition and degradation by proteolytic enzymes [46]. Some studies were conducted on grafting PEGs onto amphiphilic derivatives of PHEA. As previously reported, PEG containing PHEA-based copolymers not only showed enhanced solubility of polymeric carriers but also displayed a high stability profile. This ensures enhanced prevention of the drug from interaction with proteins, enzymes, or other circumstances in plasma [25,47].

9.4.3 Stimuli-Responsive Poly(amino acid)s–Based Graft Polymers

As drug delivery systems have been developed, delivering therapeutic drugs in a stable manner and selectively releasing them at desired sites have become increasingly important factors. Stimuli-responsive delivery carriers have been a major topic in drug delivery systems and a large amount of research has been conducted on the effects of pH, temperature, redox potential, magnetic field, light, ultrasound, and so on [48]. Stimuli-responsive polymers have the capability of producing conformational and chemical changes when an external signal is received, and thus the drug could be selectively released in a controlled manner where desired. Keeping in mind the significant differences between normal and pathological tissues, different designs were proposed for stimuli-responsive nanocarriers.

Among the stimuli mentioned previously, pH-responsive delivery carriers have been studied the most, and, as shown in Table 9.2, the pH of organs, tissues, and cellular compartments differ

TABLE 9.2
pH in Various Tissues and Cellular Compartments

Tissue/Cellular Compartment	pH
Blood	7.4
Human skin	5.5
Tumor, extracellular	6.8
Stomach	1.0–3.0
Duodenum	4.8–8.2
Colon	7.0–7.5
Early endosome	6.0–6.5
Late endosome	5.0–6.0
Lysosome	4.5–5.0

considerably [49]. For example, due to lactic acid accumulation during aerobic and anaerobic glycolysis, the tumor's extracellular environment is more acidic (pH 6.5) than that of the normal blood stream (pH 7.4). Moreover, the pH values of endosomes and lysosomes are even lower (pH 5.0–5.5) [50]. Therefore, biocompatible and biodegradable poly(amino acid) derivatives with sensitive response to slight pH changes could be used to design anticancer drug-delivery carriers to tumor sites. Our group synthesized various amphiphilic poly(amino acid)s grafted with various pH-sensitive moieties, as shown in Figure 9.8.

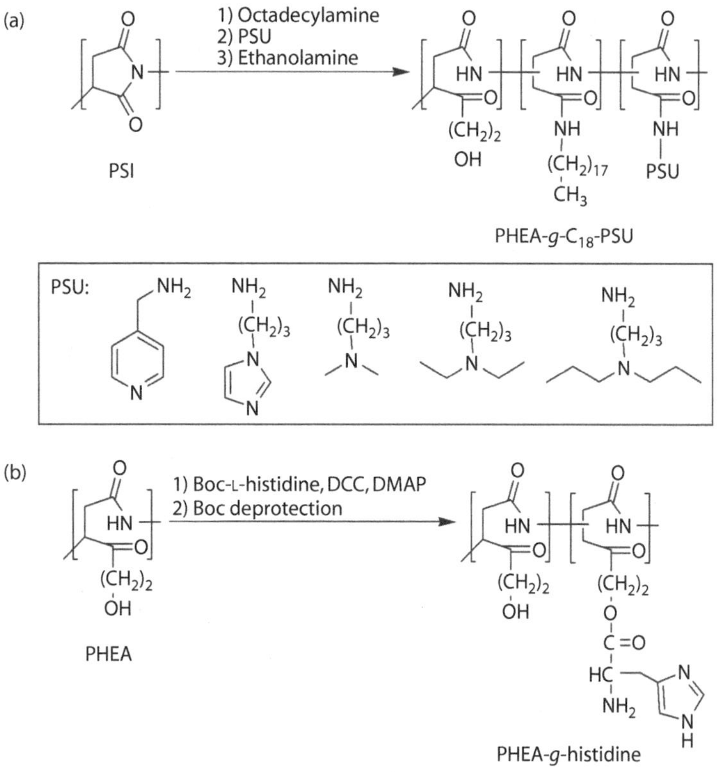

FIGURE 9.8 Various amphiphilic poly(amino acid)s grafted with various pH-sensitive moieties. (a) PHEA-g-C₁₈-PSU and (b) PHEA-g-C₁₈-His. (From Park, H.W. et al., *Colloid. Polym. Sci.*, 287, 919–926, 2009a; Yang, S.R. et al., *J. Control. Release* 114, 60–68, 2006. With permission.)

Park et al. [51] attached various cationic monomers such as 4-(aminomethyl) pyridine (PY), 1-(3 aminopropyl)imidazole (IM), and N-(3-aminopropyl)dibutylamine (BU) as pH-sensitive units onto a PHEA backbone. Because the molecules each have different pK_a values, different phase transitions at various pH ranges occur. This research therefore indicates that simple tuning of grafted groups onto amphiphilic copolymers could generate pH-sensitive polymeric aggregates that are insoluble in pH ranges of 4 to 6, 6 to 8, and 9 to 12, respectively [51]. Also, a pH-sensitive moiety, 1-(3-aminopropyl) imidazole (API), was attached to a PSI ring and reversible pH-dependent aggregation and deaggregation behavior was observed with high buffering capacity between pH 5.5 and 7 [52]. The buffering capacity of amphiphilic graft polymers indicates an effective endosome-disrupting behavior, followed by a proton sponge effect. The proton sponge effect is derived from an extensive influx of ions and water into the endosome when an acidic environment is formed. Inflow of water causes an osmotic imbalance, leading to the rupture of the endosome, which in turn, enables the drug carriers to escape to the cytoplasm and reach the nuclei in an effective manner [53]. Therefore, buffering capacity is an important parameter to decide whether the polymer itself has endosomolytic behavior when the endosomal pH is reached. Yang et al. [41]. successively grafted histidine, an endosomolytic synthetic molecule with buffering capacity (Figure 9.9a), to a PHEA backbone. Because of the tertiary amine from the imidazole ring (pK_a 6), the imidazole ring is partially protonated at physiological pH but serves as a proton sponge at endosomal pH (pH 5.0–5.5). Figure 9.9b represents enhanced nuclear access of doxorubicin and improved cell cytotoxicity when histidine is grafted onto a poly(amino acid)–based graft polymer [41].

Thermosensitive polymers based on biodegradable poly(amino acid) derivatives were also investigated by Tachibana et al. [54]. The reaction of PSI with a mixture of 5-aminopentanol and 6-aminohexanol was conducted, generating new thermoresponsive polymers that showed a clear lower critical solution temperature (LCST) in water. A novel dual-responsive polymer was synthesized by attaching N,N-diisopropylaminoethyl pendent groups to poly(aspartamide)s, and a crosslinked gel was prepared. By changing the pH and temperature, Moon and Kim [55] controlled the degree of swelling and morphology of a hydrogel structure, showing dual stimuli–responsive behavior. These novel stimuli-responsive nanocarriers hold promise for biomedical applications such as controlled drug delivery and tissue engineering and can be the subject of advanced study to design smart carriers for drug and gene delivery.

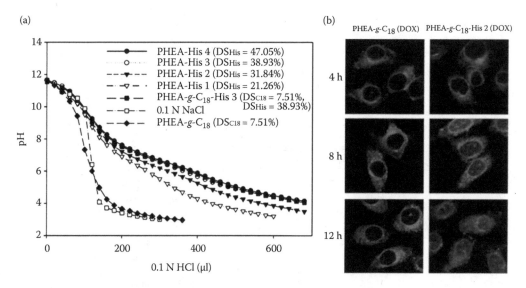

FIGURE 9.9 **(See color insert.)** (a) Acid–base titration profile of 0.1 N NaCl, PHEA-g-C$_{18}$, PHEA-His series, and PHEA-g-C$_{18}$-His 3; and (b) confocal laser scanning microscopy images of HeLa cells incubated with PHEA-g-C$_{18}$ (DOX) and PHEA-g-C$_{18}$-His 2 (DOX) with incubation time. (From Yang, S.R. et al., *J. Control. Release* 114, 60–68, 2006. With permission.)

9.4.4 POLY(AMINO ACID)S–BASED CARRIERS FOR GENE THERAPY

Gene therapy using short interfering RNA (siRNA) or plasmid DNA has attracted considerable interest due to the potential therapeutic ability to suppress the expression of selected genes related to aberrant gene expression such as in cancers and infectious diseases. After being delivered into the cell, they are incorporated into a RNA-induced silencing complex (RISC), and then silence gene expression of the target mRNA [56]. However, for successful gene therapy, siRNA and plasmid DNA need to be protected from degradation and delivered effectively into target cells. There are two strategies to deliver genes into cells [57]. The first involves a viral vector such as an RNA virus–based or a DNA virus vector, and the other a nonviral vector based on synthetic compounds such as polymers or liposomes. Due to the significant safety concerns of viral vectors, however, nonviral vectors have been widely applied as gene delivery carriers [58]. Polymeric biomaterials with cationic charges such as polyethylenimine (PEI), dendrimers, and chitosan were investigated as nonviral carriers to deliver genes into target cells. The cationic polymers can form ionic complexes with natively charged siRNA or DNA through electrostatic interactions between cationic amines of the polymer and anionic phosphates of the nucleotides [59].

Among the various poly(amino acid)s, PLL has been widely applied for cationic polymers to form an ionic complex with siRNA. Many groups have synthesized PEG-based block copolymers such as PEG-*b*-PLL to improve the colloidal stability of polymer complexes in physiological conditions and endow them with antifouling properties [60]. Although these systems have better biocompatibility than pure PLL, PEG-*b*-PLL shows low transfection efficiency compared with viral vectors or other cationic polymers. It is anticipated that modifications of the cationic polymer with small hydrophobic or lipid groups can address this drawback. Because the cell membranes are lipophilic in nature, the amphiphilic structure of the hydrophobic moiety grafted PLL can enhance the cell membrane compatibility, leading to improvement of siRNA delivery into the cell. Amphiphilic lipid-grafted PLLs (PLL-*g*-lipid) for plasmid DNA delivery have been previously reported [61–63]. A 25 kDa PLL was grafted with several lipids of variable chain lengths ranging from 8 to 18 carbon chains [63]. Gene delivery nanoparticles formed with myristic, palmitic, and stearic acid grafted PLL showed higher DNA delivery efficiency compared with nanoparticles formed with caprylic acid, oleic acid, and linoleic acid grafted and nongrafted PLL. Additionally, they reported that gene delivery efficiency depends on the extent of grafted lipids on PLL. Amphiphilic graft polymers composed of PLL as a hydrophilic part and biodegradable PLGA as a hydrophobic part were also prepared to mitigate the cytotoxicity problem and to improve transfection efficiency [27,64]. The Park group [27] reported that biodegradable and cationic PLL-*g*-PLGA shows low cytotoxicity and better gene transfer efficiency of plasmid DNA than pure PLL. The Saltzman group [64] also synthesized a PLGA-grafted PLL and fabricated an ionic complex with plasmid DNA using PLGA-*g*-PLL/PLGA with various weight ratios. The PLGA-*g*-PLL affected the encapsulation efficiency and the release profile of DNA. As the weight percentage of PLGA-*g*-PLL/PLGA was high, the loading amount of DNA in nanoparticles was remarkably increased by up to 90%. Moreover, the release profile was also controlled by the weight ratio of PLGA-*g*-PLL/PLGA. Increasing the weight percentage of PLGA-*g*-PLL/PLGA reduced the burst release on the first day.

Besides PLL, cationic polymers with poly(aspartamide) synthesized from PSI have also been used as gene delivery carriers after conjugation of ethylenediamine [65], glycidyltrimethyammonium chloride [66], or carboxypropyl-trimethylammonium chloride [67] onto a poly(aspartamide) backbone. Using these poly(aspartamide)-based cationic polymer systems, the Cavallaro group prepared spermine-grafted PHEA (PHEA-*g*-Spm) and amphiphilic spermine and butyramide–grafted PHEA(PHEA-*g*-Spm-C_4) to evaluate the transfection effect as a gene delivery carrier [68]. Although the PHEA-*g*-Spm-C_4–based complex showed smaller hydrodynamic diameter in aqueous solutions due to the hydrophobic interaction induced by grafted C_4, it did not produce any further improvement in transfection efficiency compared with that of the PHEA-*g*-Spm–based complex. However, the authors reported that both poly(aspartamide)-based complexes showed good biocompatibility and high transfection efficiency compared with a commercially available transfection agent. As mentioned previously, many studies have demonstrated that poly(amino acid)s–based amphiphilic graft polymers have better transfection efficiency than that of cationic polymers that are not modified with a hydrophobic group. We believe that poly(amino acid)s–based amphiphilic graft polymers have excellent potential as new gene delivery agents.

9.4.5 POLY(AMINO ACID)S–BASED NANOPARTICLES AS MAGNETIC RESONANCE IMAGING CONTRAST AGENTS

Magnetic resonance imaging (MRI) is an extremely useful diagnostic tool in medical science. MRI contrast agents have been developed to decrease relaxation times because faster relaxation leads to better contrast in MRI [69]. Commonly, MRI contrast agents can be classified into two groups, T1 contrast agents and T2 contrast agents. T1 contrast agents such as gadolinium diethylene triamine pentaacetic acid (Gd-DTPA) induce a positive contrast enhancement by a reduction of the T1 relaxation time, whereas T2 contrast agents, such as superparamagnetic nanoparticles, reduce the T2 relaxation time, which leads to a negative contrast enhancement in MRI [70].

Magnetite nanocrystals synthesized by aqueous phase processes, such as dextran-coated superparamagnetic iron oxide, have been widely used as commercial MRI contrast agents [71]. However, commercial products have limitations in terms of their applicability to specifically desired targeted imaging at the molecular level (i.e., molecular imaging), such as in targeted cancer cell detection [29,43].

PHEA-coated monolithic iron oxide nanoparticles were first reported by Yang et al. [29], who used an amphiphilic PHEA graft polymer (Figure 9.10). Hydrophobic alkyl chain and carboxylate group–grafted PHEA copolymers were synthesized from PSI. The ligand exchange reaction was carried out using a polymer with hydrophobic Fe_3O_4 nanocrystals (4, 6, 8, and 11 nm) in mixed organic solvents, and the carboxylate groups in the PHEA graft polymer formed coordinate bonds with Fe on the surfaces of Fe_3O_4 nanoparticles. Finally, hydrophobic Fe_3O_4 nanocrystal–grafted PHEA was transferred to an aqueous solution, and formed self-aggregates with core shell morphologies by hydrophobic interaction of hydrophobic Fe_3O_4 and the alkyl side chains of the PHEA graft polymer. The size of the core shell nanoparticles of the iron oxides was approximately 20 nm, and only negligibly changed in aqueous solutions of various pHs and concentrations of salt.

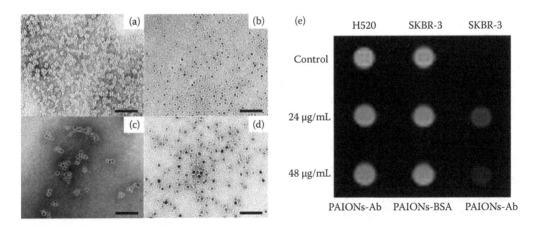

FIGURE 9.10 TEM images of PHEA-coated iron oxide nanoparticles that encapsulate (a) 4, (b) 6, (c) 8, and (d) 11 nm iron oxide particles (scale bar, 80 nm). (From Yang, H.-M. et al., *J. Mater. Chem.*, 19, 4566, 2009. With permission.) (e) T2-weighted MR images of [SKBR-3 breast cancer cells and H520 lung] cancer cells treated with BSA or HER2/neu antibody–conjugated iron oxide nanoparticles (particle concentration, 24 or 48 µg/mL; control, water). (From Yang, H.-M. et al., *Biomacromolecules* 11, 2866–2872, 2010. With permission.)

When iron oxide–based nanoparticles are smaller than 30 nm, they are classified as ultrasmall superparamagnetic iron oxide (USPIO) nanoparticles, which have longer half-life in the bloodstream. PHEA-coated USPIO showed high T2 relaxivity coefficients [r2 values, standardized T2 contrast-enhancement indicator, which is obtained as the slope of the plot of r2 = (T2 relaxation time^{-1}) against magnetic atom molarity] up to 190 L mmol^{-1} s^{-1}. In general, larger-sized Fe_3O_4 nanocrystals have higher r2 values. Although PHEA-coated USPIO has core shell type iron oxide nanoparticles, the r2 value of the particles is much higher than that of commercial iron oxide MR contrast agents such as Ferridex and Resovist, which contain many iron oxide nanoparticles in a capsule.

Poly(amino acid)s–coated iron oxide nanoparticles were also functionalized for further cancer targeting [43]. The HER2/neu antibody was chosen as a target ligand to bind the HER2/neu tyrosine kinase receptor, which is overexpressed on the surface of breast cancer cells. The results of Prussian blue staining showed that the HER2/neu antibody–conjugated iron oxide nanoparticles specifically bound with the target breast cancer cells. Furthermore, T2-weighted MR images of SKBR-3 breast cancer cells treated with bovine serum albumin (BSA)-conjugated iron oxide nanoparticles did not show any contrast differences compared with normal cells due to absence of interaction between BSA and breast cancer cells. On the other hand, SKBR-3 cells treated with HER2/neu antibody–conjugated iron oxide nanoparticles exhibited much enhanced signal intensities that were larger than those of BSA-conjugated nanoparticles and lung cancer, which has no HER2/neu (human epidermal growth factor receptor 2), treated with HER2/neu antibody–conjugated iron oxide nanoparticles (Figure 9.10).

For further enhancement of MR sensitivity, iron oxide nanoparticles encapsulated by poly(amino acid) nanoparticles with clustered magnetite nanoparticles as a core were reported by Lee et al. [23]. Amphiphilic PDMAEA-g-C$_{18}$ was prepared from PSI by aminolysis with octadecylamine and N,N-dimethylethylenediamine sequentially. Hydrophobic Fe_3O_4 nanoparticles were loaded within the core of PDMAEA-g-C$_{18}$ micelles by a simple emulsification method. Slightly cationic dimethylaminoethyl groups in the PDMAEA shell afford colloidal stability and prolonged circulation in the body. Furthermore, the surface can be changed to have strong positive charges in acidic environments, and this pH response can enhance the cellular uptake at acidic tumor sites. An *in vitro* MR experiment revealed that the T2 relaxivity coefficient of the PDMAEA-g-C$_{18}$–coated iron oxide nanoparticles was 333 L mmol^{-1} s^{-1} due to the aggregation effect of iron oxide nanoparticles; this value is much higher than that of single iron oxide nanoparticle encapsulated poly(amino acid) nanoparticles. The magnetic micelles were injected into CT26 tumor–implanted mouse, and T2-weighted MR images were observed. The PDMAEA-coated iron oxide nanoparticles accumulated in the tumor and T2 contrast enhancement was observed at 30 min postinjection. The relative signal decrease was in the range of 60% to 67%. Interestingly, a high signal decrease at the tumor was observed 72 h after postinjection (Figure 9.11a). The highly selective accumulation of nanoparticles in the tumor can be explained by the EPR effect.

Hydrophilic poly(amino acid)s grafted with a hydrophobic component are known as promising delivery carriers for poorly soluble drugs, and operate by entrapping the drugs in the hydrophobic domains. In this regard, amphiphilic poly(amino acid)s–based nanoparticles containing iron oxide nanocrystals and an anticancer drug were investigated for contrast enhancement of MRI and drug delivery to specific sites, respectively. Exploiting the hydrophobic interaction between the hydrophobic component and the hydrophobic part of octadecyl grafted PAsp, multifunctional PAsp nanoparticles containing iron oxide nanocrystals and doxorubicin were prepared for cancer diagnosis and therapy [72]. Iron oxide nanocrystals synthesized using the thermal decomposition method were incorporated into PAsp nanoparticles through an emulsion method. A hydrophobic drug, DOX, was also incorporated into the PAsp nanoparticles via a solvent diffusion method for cancer therapy application. The DOX and iron oxide coloaded PAsp nanoparticles show a twofold higher T2 relaxivity coefficient relative to that of a commercial product (Ferridex) due to the aggregation effect. In addition, DOX was successfully released from the nanoparticles and delivered into cancer cells, displaying anticancer activity.

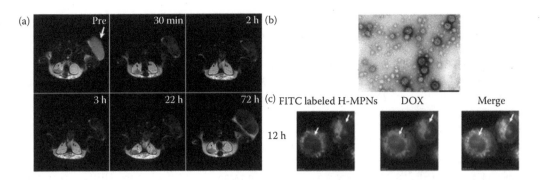

FIGURE 9.11 (See color insert.) (a) *In vivo* T2-weighted MR images taken postinjection of PDMAEA-coated magnetic nanoparticles into a mouse bearing a CT26 tumor on its back. (From Lee, H.J., *Chem. Commun.*, 46, 3559, 2010. With permission.) TEM image of (b) H-MPNs and (c) confocal laser scanning microscopic images of HeLa cells treated with FITC-labeled H-MPNs (DOX) with 12-h incubation time. Scale bar represents 200 nm. (From Yang, H.-M. et al., *Chem. Commun.* 47, 5322–5324, 2011. With permission.)

After cellular uptake, however, these nanoparticles should be able to escape from the endosome within the cell; otherwise, the endosome will capture the nanoparticles, leading to fusion with lysosomes and thereby degrading the drug or causing re-expulsion. Thus, escape of nanoparticles from the endosome must be considered for the successful delivery of anticancer drug. In this regard, histidine-conjugated poly(amino acid)s (PHEA-g-C$_{18}$-His) containing iron oxide nanocrystals and doxorubicin (DOX) was reported to enhance the contrast with high MR enhancement as well as to facilitate nuclear access for cancer therapy [73]. Histidine-conjugated magnetic poly(amino acid) nanoparticles (H-MPNs) were prepared using an emulsion method (Figure 9.11b). The loaded DOX in H-MPNs was released faster than that in magnetic poly(amino acid) nanoparticles (MPNs) that did not contain histidine moieties at pH 5.2. This is attributed to the swelling of H-MPNs induced by the ionization of histidine moieties on the H-MPNs surface. From confocal laser scanning microscopy images (Figure 9.11c), FITC-tagged H-MPNs successfully escaped the endosome through the proton sponge effect induced by ionization of the histidine moieties in the acidic environment of the endosomes. Moreover, DOX-loaded H-MPNs showed greater cytotoxicity than the DOX-loaded MPNs in a cytotoxicity test and exhibited excellent MR sensitivity with a high T2 relaxivity coefficient as well.

9.5   CONCLUSION AND FUTURE PROSPECTS

Over the last several decades, poly(amino acid)s–based amphiphilic graft polymers have been widely applied in a variety of biomedical fields, including drug delivery systems. Various types of amphiphilic graft polymers have been designed to carry cargo such as anticancer drugs, genes, and imaging agents for cancer treatment. However, there are a number of biological barriers to successfully delivering bioactive agents to the target site. The targeting ligands should be conjugated to recognize a specific cell. Cell-penetrating peptide (CPP) is also needed to facilitate cellular uptake. Physical (heat, sound, and light) or chemical (pH, temperature, and ionic strength) stimuli responsive groups might also enhance the cellular uptake or endosomal escape and control the release of the drug with a desired pattern. If the gene, imaging agent, and anticancer drug were coloaded or conjugated in the inner part of amphiphilic polymeric nanoparticles, multiple treatments with a single system could be realized. In this regard, multifunctional polymeric carriers should be developed as a new class of delivery carriers. Although various synthesis routes for block copolymers have already been established, graft polymers have advantages in the synthesis process because they can be easily tailored to provide a multivalent or multifunctional complex structure with multiple grafts. Thus, different kinds of moieties such as targeting ligands, stimuli-sensitive groups, and bioactive agents can be easily conjugated on the backbone of graft polymers. For example, poly(amino acid)s nanoparticles conjugated with endosomal pH-responsive histidine and containing an anticancer drug and magnetite in the hydrophobic core of nanoparticles were designed for simultaneous cancer diagnosis and therapy [73]. However, complicated chemistry for the synthesis of multifunctional polymers should be avoided, as it often becomes a hindrance for commercialization and compatibility with a wide range of drugs. Each drug has different solubility parameters, lipophilicity, and functional groups necessary for drug conjugation. Future amphiphilic graft polymers having multifunctional properties without complicated chemistries would make drug delivery systems more feasible.

Moreover, although amphiphilic polymeric nanostructures are thermodynamically and kinetically more stable than surfactant systems, they often have only temporary stability, and consequently become unstable at certain temperature, pH, and concentration conditions [74,75]. When polymeric micelles are used as drug delivery carriers, they face gradual disintegration by external stimuli and dilution in the bloodstream. To improve the structural stability of amphiphilic nanostructures, cross-linking can provide an alternative means of stabilizing polymeric carriers. Our group reported the synthesis of a new amphiphilic polymeric carrier with a cross-linked core using an amphiphilic block copolymer with an amino acid structure after cross-linking in the core part [31]. This new method can be applied to amphiphilic graft polymers to attain greater stability.

On the other hand, the effects of diverse morphologies prepared from poly(amino acid)s on drug delivery have not been reported, although polyester block copolymers have been reported. Discher and coworkers [76,77] demonstrated the prolonged blood circulation time and higher drug contents of cylindrical micelles compared with spherical micelles. Moreover, Muro et al. [78] reported the higher targeting efficiency of disc-shaped particles compared with spherical particles. The shape of the carriers can also be a crucial element to overcome challenges. For this reason, it would be worthwhile to investigate the diverse morphologies of poly(amino acid)s–based graft polymers, and its efficacy could be improved by a synergistic effect with the poly(amino acid)s–based multifunctional graft polymers.

In conclusion, the development of polymeric carriers with smart multifunction properties may enhance the desired activity of incorporated bioactive compounds. Also, simple chemistry for the synthesis of these carriers and enhanced structural stability in physiological conditions will facilitate their clinical application.

REFERENCES

1. T.M. Allen and P.R. Cullis. Drug delivery systems: Entering the mainstream. *Science*, **303**, 1818–1822 (2004).
2. T. Hamaguchi, K. Kato, H. Yasui, C. Morizane, M. Ikeda, H. Ueno, K. Muro, Y. Yamada, T. Okusaka, K. Shirao, Y. Shimada, H. Nakahama and Y. Matsumura. A phase I and pharmacokinetic study of NK105, a paclitaxel incorporating micellar nanoparticle formulation. *Br. J. Cancer*, **97**, 170–176 (2007).
3. Y. Matsumura. Poly (amino acid) micelle nanocarriers in preclinical and clinical studies. *Adv. Drug Deliv. Rev.*, **60**, 899–914 (2008).
4. J. Shi, A.R. Votruba, O.C. Farokhzad and R. Langer. Nanotechnology in drug delivery and tissue engineering: From discovery to applications. *Nano Lett.*, **10**, 3223–3230 (2010).
5. A. Chilkoti, M.R. Dreher and D.E. Meyer. Design of thermally responsive, recombinant polypeptide carriers for targeted drug delivery. *Adv. Drug Deliv. Rev.*, **54**, 1093–1111 (2002).
6. J.-D. Kim, S.R. Yang, Y.W. Cho and K. Park. Fast responsive nanoparticles of hydrophobically modified poly(amino acid)s and proteinoids. In *Reflexive Polymers and Hydrogels*, N. Yui, R. Mrsny and K. Park (Eds), Chap. 18. CRC Press, Boca Raton, FL, pp. 373–400 (2004).
7. B. Romberg, C. Oussoren, C. Snel, M.G. Carstens, W.E. Hennink and G. Storm. Pharmacokinetics of poly(hydroxyethyl-L-asparagine)-coated liposomes is superior over that of PEG-coated liposomes at low lipid dose and upon repeated administration. *Biochim. Biophys. Acta*, **1768**, 737–743 (2007).
8. B. Romberg, J.M. Metselaar, L. Baranyi, C.J. Snel, R. Bünger, W.E. Hennink, J. Szebeni and G. Storm. Poly(amino acid)s: Promising enzymatically degradable stealth coating for liposomes. *Int. J. Pharm.*, **331**, 186–189 (2007).

9. A. Lavasanifar, J. Samuel and G.S. Kwon. Poly(ethylene oxide)-*block*-poly(L-amino acid) micelles for drug delivery. *Adv. Drug Deliv. Rev.*, **54**, 169–190 (2002).

10. Y. Bae and K. Kataoka. Intelligent polymeric micelles from functional poly(ethylene glycol)-poly(amino acid) block copolymer. *Adv. Drug Deliv. Rev.*, **61**, 768–784 (2009).

11. H.S. Kang, M.S. Shin, J.-D. Kim and J.W. Yang. Self-aggregates of poly(aspartic acid) grafted with long alkyl chains. *Polym. Bull.*, **45**, 39–43 (2000).

12. J.H. Jeong, H.S. Kang, S.R. Yang and J.-D. Kim. Polymer micelle-like aggregates of novel amphiphilic biodegradable poly(asparagine) grafted with poly(caprolactone). *Polymer*, **44**, 583–591 (2003).

13. B.-K. Jung, C.Y. Baek, J.Y. Yang, J.H. Park and J.-D. Kim. Anticancer therapeutic self-aggregates of sphingolipid grafted poly (amino acid)-derivative and their enhanced intracellular delivery. *J. Ind. Eng. Chem.*, **16**, 1011–1018 (2010).

14. K.-S. Jang, H.J. Lee, H.-M. Yang, E.J. An, T.H. Kim, S.M. Choi and J.-D. Kim. Aqueous self-assembly of amphiphilic nanocrystallo-polymers and their surface-active properties. *Soft Matter*, **4**, 349–356 (2008).

15. Y.M. Jo, C.W. Park, B.-K. Jung, H.-M. Yang and J.-D. Kim. Size and morphology control of aggregates from supramolecular graft copolymers stabilized by ionic interaction. *Macromol. Chem. Phys.*, **211**, 2434–2442 (2010).

16. J.S. Hrkach, J. Ou, N. Lotan and R. Langer. Synthesis of poly(L-lactic acid-*co*-L-lysine) graft copolymers. *Macromolecules*, **28**, 4736–4739 (1995).

17. G. Caponetti, J.S. Hrkach, B. Kriwet, M. Poh, N. Lotan, P. Colombo and R. Langer. Microparticles of novel branched copolymers of lactic acid and amino acids: Preparation and characterization. *J. Pharm. Sci.*, **88**, 136–141 (1999).

18. B. Nottelet, A.E. Ghzaoui, J. Coudane and M. Vert. Novel amphiphilic poly(ε-caprolactone)-g-poly(L-lysine) degradable copolymers. *Biomacromolecules*, **8**, 2594–2601 (2007).

19. J.H. Jeong and J.-D. Kim. Self-assembled nanoparticle of poly(amino acid)s grafted with oligo(L-leucine) and the effect of substitution on self-aggregation, to be submitted.

20. G. Cavallaro, M. Licciardi, M.D. Stefano, G. Pitarresi and G. Giammona. New self-assembling polyaspartamide-based brush copolymers obtained by atom transfer radical polymerization. *Macromolecules*, **42**, 3247–3257 (2009).

21. H.S. Kang, S.R. Yang, J.-D. Kim, S.-H. Han and I.-S. Chang. Effects of grafted alkyl groups on aggregation behavior of amphiphilic poly(aspartic acid). *Langmuir*, **17**, 7501–7506 (2001).

22. H.S. Kang, J.-D. Kim, S.-H. Han and I.-S. Chang. Self-aggregates of poly(2-hydroxyethyl aspartamide) copolymers loaded with methotrexate by physical and chemical entrapments. *J. Controlled Release* **81**, 135–144 (2002).

23. H.J. Lee, K.-S. Jang, S. Jang, J.W. Kim, H.M. Yang, Y.Y. Jeong and J.-D. Kim. Poly(amino acid)s micelle-mediated assembly of magnetite nanoparticles for ultra-sensitive long-term MR imaging of tumors. *Chem. Commun.*, **46**, 3559–3561 (2010).

24. S.-I. Park, E.-O. Lee, B.-K. Jung and J.-D. Kim. The microfluidity and dissolution of hydrogenated PC liposome anchored with alkyl grafted poly(amino acid)s. *Colloids Surf. A*, **391**, 170–178 (2011).

25. P. Calicet, S.M. Quarta, F.M. Veronese, G. Cavallaro, E. Pedone and G. Giammona. Synthesis and biopharmaceutical characterisation of new poly(hydroxyethylaspartamide) copolymers as drug carriers. *Biochim. Biophys. Acta*, **1528**, 177–186 (2001).

26. H. Sugimoto, E. Nakanishi, T. Hanai, T. Yasumura and K. Inomata. Aggregate formation and release behaviour of hydrophobic drugs with graft copolypeptide-containing tryptophan. *Polym. Int.*, **53**, 972–983 (2004).

27. J.H. Jeong and T.G. Park. Poly(L-lysine)-g-poly(D,L-lactic-*co*-glycolic acid) micelles for low cytotoxic biodegradable gene delivery carriers. *J. Controlled Release*, **82**, 159–166 (2002).

28. H.J. Lee, S.R. Yang, E.J. An and J.-D. Kim. Biodegradable polymersomes from poly(2-hydroxyethyl aspartamide) grafted with lactic acid oligomers in aqueous solution. *Macromolecules*, **39**, 4938–4940 (2006).

29. H.-M. Yang, H.J. Lee, K.-S. Jang, C.W. Park, H.W. Yang, W.D. Heo and J.-D. Kim. Poly(amino acid)-coated iron oxide nanoparticles as ultra-small magnetic resonance probes. *J. Mater. Chem.*, **19**, 4566–4574 (2009).

30. Z. Gao and A. Eisenberg. A model of micellization for block copolymers in solutions. *Macromolecules*, **26**, 7353–7360 (1993).

31. C.W. Park, H.J. Lee, H.-M. Yang, M.-A. Woo, H.G. Park and J.-D. Kim. Size and morphology controllable core cross-linked self-aggregates from poly(ethylene glycol-*b*-succinimide) copolymers. *J. Polym. Sci. Part A*, **49**, 203–210 (2011).

32. L. Zhang and A. Eisenberg. Multiple morphologies of "crew-cut" aggregates of polystyrene-*b*-poly(acrylic acid) block copolymers. *Science*, **268**, 1728–1731 (1995).

33. J.N. Israelachvili. *Intermolecular and Surface Forces: With Applications to Colloidal and Biological Systems*, 2nd ed. Academic Press, New York (1992).

34. T. Smart, H. Homas, M. Massignani, M.V. Flores-Merino, L.R. Perez and G. Battaglia. Block copolymer nanostructures. *Nanotoday*, **3**, 38–46 (2008).

35. L. Zhang, J. Lin and S. Lin. Aggregate morphologies of amphiphilic graft copolymers in dilute solution studied by self-consistent field theory. *J. Phys. Chem. B*, **111**, 9209–9217 (2007).

36. E.J. An, H.J. Lee, K.-S. Jang and J.-D. Kim. Structural transition of self-assembled nanostructures from dodecyl chain grafted poly(2-hydroxyethyl aspartamide) in aqueous solution, to be submitted.

37. C. Cai, J. Lin, T. Chen and X. Tian. Aggregation behavior of graft copolymer with rigid backbone. *Langmuir*, **26**, 2791–2797 (2010).

38. J. Fang, H. Nakamura and H. Maeda. The EPR effect: Unique features of tumor blood vessels for drug delivery, factors involved and limitations and augmentation of the effect. *Adv. Drug Deliv. Rev.*, **63**, 136–151 (2011).

39. T. Peng, S.-X. Cheng and R.-X. Zhuo. Synthesis and characterization of novel biodegradable amphiphilic graft polymers based on aliphatic polycarbonate. *J. Polym. Sci. Part A*, **42**, 1356–1361 (2004).

40. Y.-X. Zhou, S.-L. Li, H.-L. Fu, S.-X. Cheng, X.-Z. Zhang and R.-X. Zhuo. Fabrication and in vitro drug release study of microsphere drug delivery systems based on amphiphilic poly-[*N*-(2-hydroxyethyl)-l-aspartamide]-*g*-poly (L-lactide) graft copolymers. *Colloids Surf. B*, **61**, 164–169 (2008).

41. S.R. Yang, H.J. Lee and J.-D. Kim. Histidine-conjugated poly(amino acid) derivatives for the novelendosomolytic delivery carrier of doxorubicin. *J. Controlled Release*, **114**, 60–68 (2006).

42. D. Peer, J.M. Karp, S. Hong, O.C. Farokhzad, R. Margalit and R. Langer. Nanocarriers as an emerging platform for cancer therapy. *Nat. Nanotechnol.*, **2**, 751–760 (2007).

43. H.-M. Yang, C.W. Park, M.-A. Woo, M.-I. Kim, Y.M. Jo, H.G. Park and J.-D. Kim. HER2/neu antibody conjugated poly(amino acid)-coated iron oxide nanoparticles for breast cancer MR imaging. *Biomacromolecules*, **11**, 2866–2872 (2010).

44. S.R. Yang, S.B. Kim, C.O. Joe and J.-D. Kim. Intracellular delivery enhancement of poly(amino acid) drug carriers by oligoarginine conjugation. *J. Biomed. Mater. Res. A*, **86**, 137–148 (2008).

45. H.W. Park and J.-D. Kim. Mucoadhesive interaction of cysteine grafted poly(2-hydroxyethyl aspartamide) with pig mucin layer of surface plasmon resonance biosensor. *J. Ind. Eng. Chem.*, **15**, 578–583 (2009).

46. M J. Roberts, M.D. Bentley and J.M. Harris. Chemistry for peptide and protein PEGylation. *Adv. Drug Deliv. Rev.*, **54**, 459–476 (2002).

47. G. Cavallaro, L. Maniscalco, M. Licciardi and G. Giammona. Tamoxifen-loaded polymeric micelles: Preparation, physico-chemical characterization and in vitro evaluation studies. *Macromol. Biosci.*, **4**, 1028–1038 (2004).

48. F. Meng, Z. Zhong and J. Feijen. Stimuli-responsive polymersomes for programmed drug delivery. *Biomacromolecules*, **10**, 197–209 (2009).

49. D. Schmaljohann. Thermo- and pH-responsive polymers in drug delivery. *Adv. Drug Deliv. Rev.*, **58**, 1655–1670 (2006).

50. J.-Z. Du, T.-M. Sun, W.-J. Song, J. Wu and J. Wang. A tumor-acidity-activated charge-conversional nanogel as an intelligent vehicle for promoted tumoral-cell uptake and drug delivery. *Angew. Chem. Int. Ed.*, **49**, 3621–3626 (2010).

51. H.W. Park, H.-S. Jin, S.Y. Yang and J.-D. Kim. Tunable phase transition behaviors of pH-sensitive polyaspartamides having various cationic pendant groups. *Colloid. Polym. Sci.*, **287**, 919–926 (2009).

52. K. Seo, J.-D. Kim and D. Kim. pH-dependent self-assembling behavior of imidazole-containing polyaspartamide derivatives. *J. Biomed. Mater. Res.*, **90**, 478–486 (2009).

53. A.K. Varkouhi, M. Scholte, G. Storm and H.J. Haisma. Endosomal escape pathways for delivery of biological. *J. Controlled Release*, **151**, 220–228 (2010).

54. Y. Tachibana, M. Kurisawa, H. Uyama, T. Kakuchic and S. Kobayashi. Biodegradable thermoresponsive poly(amino acid)s. *Chem. Commun.*, **1**, 106–107 (2003).

55. J.R. Moon and J.-H. Kim. Biodegradable stimuli-responsive hydrogels based on amphiphilic polyaspartamides with tertiary amine pendant groups. *Polym. Int.*, **59**, 630–636 (2010).

56. D. Ferber. Gene therapy: Safer and virus-free. *Science*, **294**, 1638–1642 (2001).

57. A. El-Aneed. An overview of current delivery systems in cancer gene therapy. *J. Controlled Release*, **94**, 1–14 (2004).

58. S. Lehrman. Virus treatment questioned after therapy death. *Nature*, **401**, 517–518 (1999).

59. T.G. Park, J.H. Jeong and S.W. Kim. Current status of polymeric gene delivery systems. *Adv. Drug Deliv. Rev.*, **58**, 467–486 (2006).

60. K. Osada, R.J. Christie and K. Kataoka. Polymeric micelles from poly(ethylene glycol)–poly(amino acid) block copolymer for drug and gene delivery. *J. R. Soc. Interface*, **6**, S325–S339 (2009).

61. M. Abbasi, H. Uludag, V. Incani, C. Olson, X. Lin, B.A. Clements, D. Rutkowski, A. Ghahary and M. Weinfeld. Palmitic acid-modified poly-L-lysine for non-viral delivery of plasmid DNA to skin fibroblasts. *Biomacromolecules*, **8**, 1059–1063 (2007).

62. M. Abbasi, H. Uludag, V. Incani, C. Yu, M. Hsu and A. Jeffery. Further investigation of lipid-substituted poly(L-lysine) polymers for transfection of human skin fibroblasts. *Biomacromolecules*, **9**, 1618–1630 (2008).

63. V. Incani, X. Lin, A. Lavasanifar and H. Uludag. Relationship between the extent of lipid substitution on poly(L-lysine) and the DNA delivery efficiency. *ACS. Appl. Mater. Interfaces*, **1**, 841–848 (2009).

64. J.S. Blum and W.M. Saltzman. High loading efficiency and tunable release of plasmid DNA encapsulated in submicron particles fabricated from PLGA conjugated with poly-L-lysine. *J. Controlled Release*, **129**, 66–72 (2008).

65. G. Cavallaro, M. Campisi, M. Licciardi, M. Ogris and G. Giammona. Reversibly stable thiopolyplexes for intracellular delivery of genes. *J. Controlled Release*, **115**, 322–334 (2006).

66. G. Giammona, G. Cavallaro, G. Pitarresi and E. Pedone. Cationic copolymers of α,β-poly-(*N*-2-hydroxyethyl)-D,L-aspartamide (PHEA) and α,β-polyasparthylhydrazide (PAHy): Synthesis and characterization. *Polym. Int.*, **49**, 93–98 (2000).

67. M. Licciardi, M. Campisi, G. Cavallaro, B. Carlisi and G. Giammona. Novel cationic polyaspartamide with covalently linked carboxypropyl-trimethyl ammonium chloride as a candidate vector for gene delivery. *Eur. Polym. J.*, **42**, 823–834 (2006).

68. G. Cavallaro, S. Scirè, M. Licciardi, M. Ogris, E. Wagner and G. Giammona. Polyhydroxyethylaspartamide-spermine copolymers: Efficient vectors for gene delivery. *J. Controlled Release*, **131**, 54–63 (2008).

69. H.B. Na, I.C. Song and T. Hyeon. Inorganic nanoparticles for MRI contrast agents. *Adv. Mater.*, **21**, 2133–2148 (2009).

70. J.W.M. Bulte and D.L. Kraitchman. Iron oxide MR contrast agents for molecular and cellular imaging. *NMR Biomed.*, **17**, 484–499 (2004).

71. C. Corot, P. Robert, J.-M. Idée and M. Port. Recent advances in iron oxide nanocrystal technology for medical imaging. *Adv. Drug Deliv. Rev.*, **58**, 1471–1504 (2006).

72. H.-M. Yang, B.C. Oh, J. Kim, T. Ahn, H.-S. Nam, C.W. Park and J.-D. Kim. Multifunctional poly(aspartic acid) nanoparticles containing iron oxide nanocrystals and doxorubicin for simultaneous cancer diagnosis and therapy. *Colloids Surf. A*, **391**, 208–215 (2011).

73. H.-M. Yang, H.J. Lee, C.W. Park, S. Lim, B.H. Jung and J.-D. Kim. Endosome-escapable magnetic poly(amino acid) nanoparticles for cancer diagnosis and therapy. *Chem. Commun.*, **47**, 5322–5324 (2011).

74. R.K. O'Reilly, C.J. Hawker and K.L. Wooley. Cross-linked block copolymer micelles: Functional nanostructures of great potential and versatility. *Chem. Soc. Rev.*, **35**, 1068–1083 (2006).

75. K.M. Huh, S.C. Lee, Y.W. Cho, J. Lee, J.H. Jeong and K. Park. Hydrotropic polymer micelle system for delivery of paclitaxel. *J. Controlled Release*, **101**, 59–68 (2005).

76. Y. Geng, P. Dalhaimer, S. Cai, R. Tsai, M. Tewari, T. Minko and D.E. Discher. Shape effects of filaments versus spherical particles in flow and drug delivery. *Nat. Nanotechnol.*, **2**, 249–255 (2007).

77. P. Dalhaimer, A. Engler, R. Parthasarthy and D.E. Discher. Targeted worm micelles. *Biomacromolecules*, **5**, 1714–1719 (2004).

78. S. Muro, C. Garnacho, J.A. Champion, J. Leferovich, C. Gajewski, E.H. Schuchman, S. Mitragotri and V.R. Muzykantov. Control of endothelial targeting and intracellular delivery of therapeutic enzymes by modulating the size and shape of ICAM-1-targeted carriers. *Mol. Ther.*, **16**, 1450–1458 (2008).

10 Polymeric Surfactants and Some of Their Applications

Tharwat Tadros

CONTENTS

10.1 INTRODUCTION

Polymeric surfactants are used for the preparation of many disperse systems and their stabilization, of which we mention dyestuffs, paper coatings, inks, agrochemicals, pharmaceuticals, personal care products, ceramics, and detergents [1]. One of the most important applications of polymeric surfactants is in the preparation of oil-in-water (O/W) emulsions as well as solid–liquid (S/L) dispersions [2,3]. In these cases, the hydrophobic portion of the surfactant molecule adsorbs "strongly" at the O/W interface or is dissolved in the oil phase, leaving the hydrophilic components in the aqueous medium, whereby they become strongly solvated by the water molecules. The extended hydrophilic chains provide effective steric stabilization, as will be discussed in the next section.

For the preparation of O/W emulsions, the polymeric surfactant is dissolved in the aqueous phase, and the oil is emulsified by the application of high-speed stirrers such as the Ultra Turrax (Germany). Because polymeric surfactants do not sufficiently reduce the interfacial tension at the oil–water interface, more energy is required to reduce the oil droplet size. This can be achieved by the use of high-pressure homogenizers. Alternatively, the interfacial tension can be reduced by the incorporation of a small proportion of small surfactant molecules such as sodium dodecyl sulfate or an alcohol ethoxylate. These surfactant–polymer mixtures may show synergy whereby the final interfacial tension will be lower than that produced by either component. The addition of the small surfactant molecules can also enhance the Gibbs elasticity, thus preventing the coalescence of the oil droplets during emulsification [1,4].

There are generally two methods for the preparation of suspensions, that is, condensation and dispersion. In the first case, start with molecular units and particle buildup by a process of nucleation and growth. A typical example is the preparation of polymer lattices. In this case, the monomer (such as styrene or methyl methacrylate [MMA]) is emulsified in water using an anionic or nonionic surfactant (such as sodium dodecyl sulfate or alcohol ethoxylate). An initiator such as potassium persulfate is added, and when the temperature of the system is increased, initiation occurs, resulting in the formation of the latex (polystyrene [PS] or poly[methyl methacrylate] [PMMA]). In the dispersion methods, preformed particles (usually powders) are dispersed in an aqueous solution containing a surfactant. The latter is essential for the adequate wetting of the powder (both external and internal surfaces of the powder aggregates and agglomerates must be wetted) [1]. This is followed by the dispersion of the powder using high-speed stirrers, and finally the dispersion is "milled" to reduce the particle size to the appropriate range.

For the stabilization of emulsions and suspensions against flocculation, coalescence, and Ostwald ripening, the following criteria must be satisfied: (i) a complete coverage of the droplets or particles by the surfactant (any bare patches may result in flocculation as a result of the van der Waals attraction or bridging); (ii) a strong adsorption (or "anchoring") of the surfactant molecule to the surface of droplet or particle; (iii) a strong solvation (hydration) of the stabilizing chain to provide effective steric stabilization; and (iv) a reasonably thick adsorbed layer to prevent weak flocculation [1].

Most of the previously mentioned criteria for stability are best served by using a polymeric surfactant. In particular, the molecules of the A-B, A-B-A blocks, and BA_n (or AB_n) grafts (see next section) are the most efficient for the stabilization of emulsions and suspensions. In this case, the B chain (called the anchoring chain) is chosen to be highly insoluble in the medium, with a high affinity to the surface in the case of suspensions, or soluble in the oil in the case of emulsions. The A chain is chosen to be highly soluble in the medium and strongly solvated by its molecules. These block and graft copolymers are ideal for the preparation of concentrated emulsions and suspensions, which are needed in many industrial applications.

This chapter starts with a section on the general classification of polymeric surfactants. The next section is devoted to the adsorption of polymeric surfactants at the S/L interface, and a summary is given for some of the methods that may be applied for studying polymeric surfactant adsorption and its conformation at the S/L interface. The same principles may be applied to the adsorption of polymeric surfactants at the liquid–liquid interface, although theoretical treatments of this problem are not as developed as those for the S/L interface. The next section is devoted to the principles involved in the stabilization of emulsions and suspensions using polymeric surfactants, that is, the general theory of steric stabilization. Two sections are devoted to describe the use of polymeric surfactants for the stabilization of suspensions and emulsions. The use of polymeric surfactants for the stabilization of nanoemulsions against Ostwald ripening is also described.

10.2   GENERAL CLASSIFICATION OF POLYMERIC SURFACTANTS

Perhaps the simplest type of a polymeric surfactant is a homopolymer that is formed from the same repeat units, such as poly(ethylene oxide) (PEO) or poly(vinyl pyrrolidone). These homopolymers have only slight surface activity at the O/W interface because the homopolymer segments (ethylene oxide or vinyl pyrrolidone) are highly water soluble and have little affinity to the interface. However, such homopolymers may adsorb significantly at the S/L interface. Even if the adsorption energy per monomer segment to the surface is small (the fraction of kT, where k is the Boltzmann constant and T is the absolute temperature), the total adsorption energy per molecule may be sufficient to overcome the unfavorable entropy loss of the molecule at the S/L interface.

Clearly, homopolymers are not the most suitable emulsifiers or dispersants. A small variant is to use polymers that contain specific groups that have high affinity to the surface. This is exemplified by partially hydrolyzed poly(vinyl acetate) (PVAc), technically called poly(vinyl alcohol) (PVA). The polymer is prepared by the partial hydrolysis of PVAc, leaving some residual vinyl acetate groups. Most commercially available PVA molecules contain 4% to 12% acetate groups. These acetate groups, which are hydrophobic, give the molecule its amphipathic character. On a hydrophobic surface such as PS, the polymer adsorbs with the preferential attachment of the acetate groups on the surface, leaving the more hydrophilic vinyl alcohol segments dangling in the aqueous medium. These partially hydrolyzed PVA molecules also exhibit surface activity at the O/W interface [5].

The most convenient polymeric surfactants are those of the block and graft copolymer types. A block copolymer is a linear arrangement of blocks of variable monomer composition. Poly-A-block-poly-B is the nomenclature for a diblock, and poly-A-block-poly-B-poly-A is the nomenclature for a triblock. One of the most widely used triblock polymeric surfactants is the Pluronic (BASF, Germany), which consists of two poly-A blocks of PEO and one block of poly(propylene oxide) (PPO). Several chain lengths of PEO and PPO are available.

The previously mentioned polymeric triblocks can be applied as emulsifiers or dispersants, whereby the assumption is made that the hydrophobic PPO chain resides at the hydrophobic surface, leaving the two PEO chains dangling in an aqueous solution and hence providing steric repulsion.

Several other diblock and triblock copolymers have been synthesized, although these are of limited commercial availability. Typical examples are diblocks of PS-block-PVA, triblocks of PMMA-block-PEO-block-PMMA, diblocks of PS-block-PEO, and triblocks of PEO-block-PS-PEO [5].

An alternative (and perhaps more efficient) polymeric surfactant is the amphipathic graft copolymer consisting of a polymeric backbone B (PS or PMMA) and several A chains (teeth) such as PEO [5]. This graft copolymer is sometimes called a "comb" stabilizer. This copolymer is usually prepared by grafting a macromonomer such as methoxy polyethylene oxide methacrylate with PMMA. The "grafting onto" technique has also been used to synthesize PS-PEO graft copolymers.

Recently, graft copolymers based on polysaccharides [6–8] have been developed for the stabilization of disperse systems. One of the most useful graft copolymers are those based on inulin, which is obtained from chicory roots. It is a linear polyfructose chain with a glucose end. When extracted from chicory roots, inulin has a wide range of chain lengths ranging from 2 to 65 fructose units. It is fractionated to obtain a molecule with narrow molecular weight distribution with a degree of polymerization lower than 23, and this is commercially available as Inutec N25. The latter molecule is used to prepare a series of graft copolymers by the random grafting of alkyl chains (using alky isocyanate) on the inulin backbone. The first molecule of this series is Inutec SP1 (ORAFTI, Belgium), which is obtained by the random grafting of C_{12} alkyl chains. It has an average molecular weight of approximately 5000 Da, and its structure is shown in Figure 10.1. The molecule is schematically illustrated in Figure 10.2, which shows the hydrophilic polyfructose chain (backbone) and the randomly attached alky chains.

The main advantages of using Inutec SP1 as a stabilizer for disperse systems are as follows: (i) strong adsorption to the particle or droplet by multipoint attachment with several alky chains, which ensures the lack of desorption and displacement of the molecule from the interface; and (ii) strong hydration of the linear polyfructose chains both in water and in the presence of high electrolyte concentrations and high temperatures, which ensures effective steric stabilization (see the next section).

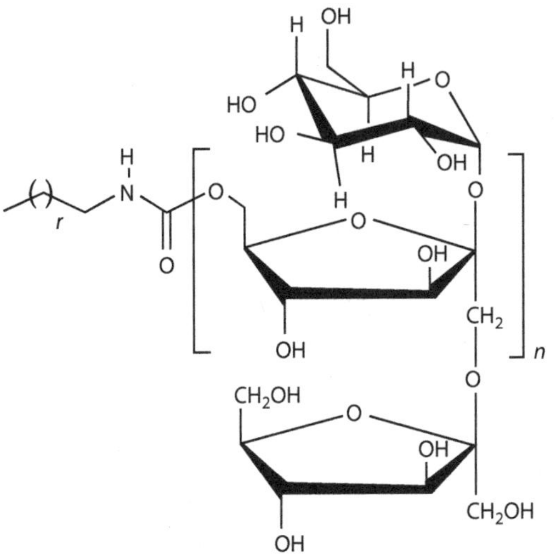

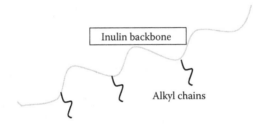

FIGURE 10.1 Structure of Inutec SP1; *r* is the number of alkyl groups attached and *n* is the number of fructose units in the inulin chain.

FIGURE 10.2 Schematic representation of Inutec SP1 polymeric surfactant. Gray, inulin backbone; black, the alkyl chains.

10.3 ADSORPTION AND CONFORMATION OF POLYMERIC SURFACTANTS AT INTERFACES

Understanding the adsorption and conformation of polymeric surfactants at interfaces is key to knowing how these molecules act as stabilizers. Most basic ideas on adsorption and conformation of polymers have been developed for the S/L interface [9,10].

The process of polymer adsorption is fairly complicated and involves the following: the polymer–surface interaction, the polymer–solvent interaction, the surface–solvent interaction, and the configuration (conformation) of the polymer at the S/L interface. The polymer–surface interaction is described in terms of adsorption energy per segment χ^s. The polymer–solvent interaction is described in terms of the Flory–Huggins interaction parameter χ. The polymer configuration is described by the following sequences: trains, segments in direct contact with the surface; loops, segments in between the trains that extend into solution; and tails, ends of the molecules that also extend into the solution. A schematic representation of the various polymer configurations is given in Figure 10.3.

For homopolymers, for example, PEO or poly(vinyl pyrrolidone) (PVP), a train–loop–tail configuration is the case. For adsorption to occur, a minimum energy of adsorption per segment χ^s is

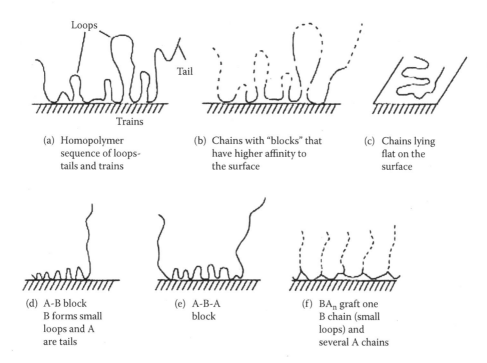

FIGURE 10.3 Various conformations of polymeric surfactant on a plane surface. (a) Homopolymer; (b) chains with "blocks" that have higher affinity to the surface; (c) chain lying flat on the surface; (d) A-B block; (e) A-B-A block; (f) BA_n graft.

required. When a polymer molecule adsorbs on a surface, it loses configurational entropy, and this must be compensated by an adsorption energy χ^s per segment. This is schematically shown in Figure 10.4, where the adsorbed amount Γ is plotted versus χ^s. The minimum value of χ^s can be very small (<0.1 kT) because a large number of segments per molecule are adsorbed. For a polymer with approximately 100 segments and 10% of these are in trains, the adsorption energy per molecule now reaches 1 kT (with $\chi^s = 0.1\ kT$). For 1000 segments, the adsorption energy per molecule is now 10 kT.

Homopolymers are not the most suitable for the stabilization of dispersions. For strong adsorption, one needs the molecule to be "insoluble" in the medium and to have a strong affinity (anchoring) to the surface. For stabilization, the molecule should be highly soluble in the medium and strongly solvated by its molecules. This requires a Flory–Huggins interaction parameter of less than 0.5. The previously mentioned opposing effects can be resolved by introducing "short" blocks in the molecule, which are insoluble in the medium and have a strong affinity to the surface, as illustrated

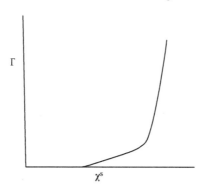

FIGURE 10.4 Variation of adsorption amount Γ with adsorption energy per segment χ^s.

in Figure 10.3b (e.g., partially hydrolyzed PVAc—88% hydrolyzed, i.e., with 12% acetate groups, usually referred to as PVA).

The previously mentioned requirements are better satisfied using A-B, A-B-A, and BA$_n$ graft copolymers. B is chosen to be highly insoluble in the medium, and it should have high affinity to the surface. This is essential to ensure strong anchoring to the surface (irreversible adsorption). A is chosen to be highly soluble in the medium and strongly solvated by its molecules. The Flory–Huggins interaction parameter χ can be applied in this case. For a polymer in a good solvent, χ has to be lower than 0.5; the smaller the χ value, the better the solvent for the polymer chains. Examples of B for hydrophobic particles in aqueous media are PS and PMMA. Examples of A in aqueous media are PEO, poly(acrylic acid), poly(vinyl pyrrolidone), and polysaccharides. For nonaqueous media such as hydrocarbons, the A chain(s) could be poly(12-hydroxystearic acid) (PHS).

For a full description of polymer adsorption, one needs to obtain information on the following [9,10]: (i) the amount of polymer adsorbed Γ (mg or mol) per unit area of the particles (It is essential to know the surface area of the particles in the suspension. Nitrogen adsorption on the powder surface may give such information, by the application of the BET equation, provided there will be no change in area upon dispersing the particles in the medium. For many practical systems, a change in surface area may occur upon dispersing the powder, in which case one has to use dye adsorption to measure the surface area; some assumptions have to be made in this case); (ii) the fraction of segments in direct contact with the surface, that is, the fraction of segments in trains p (p = number of segments in direct contact with the surface/total number); and (iii) the distribution of segments in loops and tails, $\rho(z)$, which extend to several layers from the surface. $\rho(z)$ is usually difficult to obtain experimentally, although recent application [10] of small-angle neutron scattering could provide such information. An alternative and useful parameter for assessing "steric stabilization" is the hydrodynamic thickness, δ_h (the thickness of the adsorbed or grafted polymer layer plus any contribution from the hydration layer). Several methods can be applied to measure δ_h, as will be discussed in the next section.

10.3.1 EXPERIMENTAL TECHNIQUES FOR STUDYING POLYMERIC SURFACTANT ADSORPTION

As mentioned previously, for a full characterization of polymeric surfactant adsorption, the following three parameters should be determined: (i) the adsorbed amount Γ (mg·m^{-2} or mol·m^{-2}) as a function of equilibrium concentration C_{eq}, that is, the adsorption isotherm; (ii) the fraction of segments in direct contact with the surface p (number of segments in trains relative to the total number of segments); and (iii) the segment density distribution $\rho(z)$ or the hydrodynamic adsorbed layer thickness δ_h.

It is important to obtain the adsorption parameters as a function of the important variables of the system: (i) the solvency of the medium for the chain, which can be affected by temperature, or the addition of salt or a nonsolvent (the Flory–Huggins interaction parameter χ could be separately measured); (ii) the molecular weight of the adsorbed polymer; (iii) the affinity of the polymer to the surface as measured by the value of χ^s, the segment–surface adsorption energy; and (iv) the structure of the polymer, which is particularly important for block and graft copolymers.

10.3.1.1 Measurement of the Adsorption Isotherm

This is by far the easiest to obtain. One measures the polymeric surfactant concentration before ($C_{initial}$, C_1) and after ($C_{equilibrium}$, C_2) to obtain the adsorbed amount Γ, calculated as follows:

$$\Gamma = \frac{(C_1 - C_2)V}{A} \tag{10.1}$$

where V is the total volume of the solution and A is the specific surface area ($m^2 \cdot g^{-1}$).

It is necessary in this case to separate the particles from the polymer solution after adsorption. This could be carried out by centrifugation or filtration. One should make sure that all particles are removed.

To obtain the previously mentioned isotherm, one must develop a sensitive analytical technique for the determination of the polymeric surfactant concentration in the parts per million range. It is essential to follow the adsorption as a function of time to determine the time required to reach equilibrium. For some polymer molecules such as PVA and PEO (or blocks containing PEO), analytical methods based on complexation with iodine/potassium iodide or iodine/boric acid–potassium iodide have been established. For some polymers with specific functional groups, spectroscopic methods may be applied, for example, ultraviolet, infrared, or fluorescence spectroscopy. A possible method is to measure the change in the refractive index of the polymer solution before and after adsorption. This requires very sensitive refractometers. High-resolution nuclear magnetic resonance (NMR) has been recently applied because the polymer molecules in the adsorbed state are in a different environment compared with those in the bulk. The chemical shifts of functional groups within the chain are different in these two environments [10]. This has the advantage of measuring the amount of adsorption without separating the particles.

10.3.1.2 Measurement of the Fraction of Segments p

The fraction of segments in direct contact with the surface can be directly measured using spectroscopic techniques: (i) infrared, if there is specific interaction between the segments in trains and the surface, for example, PEO on silica from nonaqueous solutions; (ii) electron spin resonance, which requires labeling of the molecule; and (iii) NMR, pulse gradient, or spin-echo NMR. This method is based on the fact that the segments in trains are "immobilized," and hence, they have lower mobility than those in loops and tails [10].

An indirect method of determining p is to measure the heat of adsorption ΔH using microcalorimetry. One should then determine the heat of adsorption of a monomer, H_m (or molecule representing the monomer, e.g., ethylene glycol for PEO); p is then given by the following equation:

$$p = \frac{\Delta H}{H_m n} \tag{10.2}$$

where n is the total number of segments in the molecule.

The previously mentioned indirect method is not very accurate and can only be used in a qualitative sense. It also requires very sensitive enthalpy measurements (e.g., using an LKB microcalorimeter).

10.3.1.3 Determination of the Segment Density Distribution $\rho(z)$ and Adsorbed Layer Thickness δ_h

The segment density distribution $\rho(z)$ is given by the number of segments parallel to the surface in the z-direction. The best technique to obtain $\rho(z)$ is to apply low angle neutron scattering using the so-called contrast matching procedure. By changing the isotopic composition of the particles and medium (using different ratios of deuterium to hydrogen atoms both in the particles and in the aqueous medium), one can contrast match the particle scattering length density with that of the medium.

An alternative and more practical method to obtain the layer extension is to measure the hydrodynamic thickness of the chain δ_h. This is defined as the distance of the plane of shear from the particle surface. δ_h is indeed the important parameter that determines the steric repulsion between the particles in a dispersion or emulsion droplets in an emulsion.

Several methods may be applied to obtain δ_h: (i) ultracentrifugation, (ii) dynamic light scattering (photon correlation spectroscopy [PCS]), and (iii) microelectrophoresis.

In all the techniques mentioned previously, one measures the hydrodynamic radius of the particle with and without the polymer layer and obtains δ_h by difference. For an accurate determination of δ_h, one should use small particles, and the layer thickness should be at least 10% of the particle radius (the accuracy of measuring the hydrodynamic radius is ±1%). In most cases, one has to use model particles for these measurements.

The dynamic light-scattering method is called quasi-elastic light scattering, intensity fluctuation spectroscopy, or most commonly, PCS. Laser light is used to measure the intensity fluctuation of scattered light by the particles as they undergo Brownian diffusion. The sample is diluted to an extent whereby the distances between the particles become comparable with the wavelength of the laser beam (i.e., coherent over the sample). In most modern instruments, this can be easily checked. The scattering intensity of the beam is measured as a function of time to obtain the fluctuations in intensity as a result of the diffusive motion of the particles.

These fluctuations decay exponentially with a time constant related to the diffusion coefficient D of the scatterer. From D, one can obtain the hydrodynamic radius R_h using the Stokes–Einstein equation, as follows:

$$D = \frac{kT}{6\pi\eta R_h} \tag{10.3}$$

where k is the Boltzmann constant, T is the absolute temperature, and η is the viscosity of the medium.

The microelectrophoresis technique is based on the measurement of the electrophoretic mobility u of the particles in the presence and absence of the polymer layer. From u, one can calculate the zeta potential ζ using the Hückel equation (which is applicable for small particles and extended double layers), that is, $\kappa R \ll 1$, where κ is the Debye–Hückel parameter that is related to the salt concentration, as follows:

$$u = \frac{2}{3}\frac{\varepsilon\varepsilon_0\zeta}{\eta} \tag{10.4}$$

where ε is the relative permittivity of the medium and ε_0 is the permittivity of free space.

By measuring ζ of the particles with and without the adsorbed polymer layer, one can obtain the hydrodynamic thickness δ_h. For accurate measurements, one should carry the measurements at various electrolyte concentrations and extrapolate the results in the plateau value. Several automatic instruments are available for the measurement of electrophoretic mobility (e.g., Malvern Zetasizer, Coulter Delsa sizer, and Brookhaven Instruments). All these instruments are easy to use, and the measurement can be carried out within a few minutes.

10.4 INTERACTION BETWEEN PARTICLES CONTAINING ADSORBED POLYMERIC SURFACTANT LAYERS (STERIC STABILIZATION)

When two particles or droplets, each with a radius R and containing an adsorbed polymer layer with a hydrodynamic thickness δ_h, approach each other to a surface–surface separation distance h that is

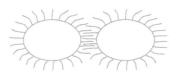

Interpenetration without   Compression without
compression   interpenetration

FIGURE 10.5 Schematic representation of the interaction between particles containing adsorbed polymer layers.

smaller than 2 δ_h, the polymer layers interact with one another, resulting in two main situations [11]: (i) the polymer chains may overlap with one another, and (ii) the polymer layer may undergo some compression. In both cases, there will be an increase in the local segment density of the polymer chains in the interaction region. This is schematically illustrated in Figure 10.5. The real situation is perhaps in between the two cases previously mentioned, that is, the polymer chains may undergo some interpenetration and some compression.

Provided the dangling chains (the A chains in A-B, A-B-A block, or BA$_n$ graft copolymers) are in a good solvent, this local increase in segment density in the interaction zone will result in a strong repulsion as a result of two main effects: (i) an increase in the osmotic pressure in the overlap region as a result of the unfavorable mixing of the polymer chains, when these are in good solvent conditions (this is referred to as osmotic repulsion or mixing interaction), and it is described by a free energy of interaction G_{mix}; and (ii) the reduction of the configurational entropy of the chains in the interaction zone; this entropy reduction results from the decrease in the volume available for the chains when these are either overlapped or compressed. This is referred to as volume restriction interaction, entropic or elastic interaction, and it is described by a free energy of interaction G_{el}.

The combination of G_{mix} and G_{el} is usually called the steric interaction free energy G_s, that is,

$$G_s = G_{mix} + G_{el} \tag{10.5}$$

The sign of G_{mix} depends on the solvency of the medium for the chains. If in a good solvent, that is, the Flory–Huggins interaction parameter χ is less than 0.5, then G_{mix} is positive and the mixing interaction leads to repulsion (see the next section). In contrast, if $\chi > 0.5$ (i.e., the chains are in a poor solvent condition), G_{mix} is negative, and the mixing interaction becomes attractive. G_{el} is always positive, and hence, in some cases, one can produce stable dispersions in a relatively poor solvent (enhanced steric stabilization).

10.4.1 Mixing Interaction G_{mix}

This results from the unfavorable mixing of the polymer chains, when these are in a good solvent condition. This is schematically shown in Figure 10.6.

Consider two spherical particles with the same radius and each containing an adsorbed polymer layer with thickness δ. Before the overlap, one can define in each polymer layer a chemical potential for the solvent μ_i^α and a volume fraction for the polymer in the layer ϕ_2.

In the overlap region (volume element dV), the chemical potential of the solvent is reduced to μ_i^β. This results from the increase in polymer segment concentration in this overlap region.

In the overlap region, the chemical potential of the polymer chains is now higher than that in the rest of the layer (with no overlap). This amounts to an increase in the osmotic pressure in the overlap region; as a result, solvent will diffuse from the bulk to the overlap region, thus separating the particles and hence a strong repulsive energy arises from this effect. The previously mentioned

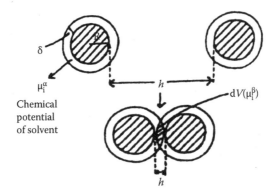

FIGURE 10.6 Schematic representation of polymer layers overlap.

repulsive energy can be calculated by considering the free energy of mixing of two polymer solutions, as for example treated by Flory and Krigbaum [11]. The free energy of mixing is given by two terms: (i) an entropy term that depends on the volume fraction of polymer and solvent and (ii) an energy term that is determined by the Flory–Huggins interaction parameter χ.

Using the previously mentioned theory, one can derive an expression for the free energy of mixing of two polymer layers (assuming a uniform segment density distribution in each layer) surrounding two spherical particles as a function of the separation distance h between the particles. The expression for G_{mix} is as follows:

$$\frac{G_{mix}}{kT} = \left(\frac{2V_2^2}{V_1}\right) v_2^2 \left(\frac{1}{2} - \chi\right)\left(\delta - \frac{h}{2}\right)^2 \left(3R + 2\delta + \frac{h}{2}\right) \qquad (10.6)$$

where k is the Boltzmann constant, T is the absolute temperature, V_2 is the molar volume of polymer, V_1 is the molar volume of solvent, and v_2 is the number of polymer chains per unit area.

The sign of G_{mix} depends on the value of the Flory–Huggins interaction parameter χ: if $\chi < 0.5$, G_{mix} is positive and the interaction is repulsive; if $\chi > 0.5$, G_{mix} is negative and the interaction is attractive; and if $\chi = 0.5$, $G_{mix} = 0$ and this defines the θ condition.

10.4.2 ELASTIC INTERACTION G_{el}

This arises from the loss in configurational entropy of the chains upon the approach of a second particle. As a result of this approach, the volume available for the chains becomes restricted, resulting in a reduction of the number of configurations. This can be illustrated by considering a simple molecule, represented by a rod that rotates freely in a hemisphere across a surface (Figure 10.7). When

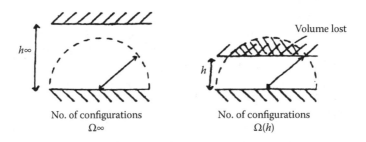

FIGURE 10.7 Schematic representation of configurational entropy loss upon the approach of a second particle.

the two surfaces are separated by an infinite distance ∞, the number of configurations of the rod is $\Omega(\infty)$, which is proportional to the volume of the hemisphere. When a second particle approaches to a distance h such that it cuts the hemisphere (losing some volume), the volume available to the chains is reduced and the number of configurations becomes $\Omega(h)$, which is less than $\Omega(\infty)$. For two flat plates, G_{el} is given by the following expression:

$$\frac{G_{el}}{kT} = 2v_2 \ln\left[\frac{\Omega(h)}{\Omega(\infty)}\right] = 2v_2 R_{el}(h) \tag{10.7}$$

where $R_{el}(h)$ is a geometric function whose form depends on the segment density distribution. It should be stressed that G_{el} is always positive and could play a major role in steric stabilization. It becomes very strong when the separation distance between the particles becomes comparable with the adsorbed layer thickness δ.

The combination of G_{mix} and G_{el} with the van der Waals attraction G_A gives the total energy of interaction G_T (assuming there is no contribution from any residual electrostatic interaction), that is,

$$G_T = G_{mix} + G_{el} + G_A \tag{10.8}$$

A schematic representation of the variation of G_{mix}, G_{el}, G_A, and G_T with surface–surface separation distance h is shown in Figure 10.8.

G_{mix} increases very sharply with the decrease in h, when $h < 2\delta$. G_{el} increases very sharply with the decrease in h, when $h < \delta$. G_T versus h shows a minimum, G_{min}, at separation distance comparable with 2δ. When $h < 2\delta$, G_T shows a rapid increase with the decrease in h.

The depth of the minimum depends on the Hamaker constant A, the particle radius R, and the adsorbed layer thickness δ. G_{min} increases with the increase in A and R. At a given A and R, G_{min} increases with the decrease in δ (i.e., with the decrease in the molecular weight, M_w, of the stabilizer). This is illustrated in Figure 10.9, which shows the energy–distance curves as a function of δ/R. The larger the value of δ/R, the smaller the value of G_{min}. In this case, the system may approach thermodynamic stability as is the case with nanodispersions.

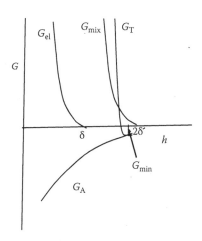

FIGURE 10.8 Energy–distance curves for sterically stabilized systems.

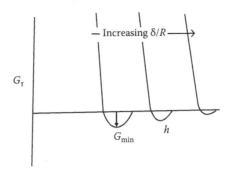

FIGURE 10.9 Variation of G_{min} with δ/R.

10.5 EMULSIONS STABILIZED BY POLYMERIC SURFACTANTS

The most effective method for emulsion stabilization is to use polymeric surfactants that strongly adsorb at the O/W or W/O interface and produce effective steric stabilization against strong flocculation, coalescence, and Ostwald ripening [12].

As mentioned previously, a graft copolymer of the AB_n type was synthesized by grafting several alkyl groups on an inulin (polyfructose) chain. The polymeric surfactant (Inutec SP1) consists of a linear polyfructose chain (the stabilizing A chain) and several alkyl groups (the B chains) that provide multianchor attachment to the oil droplets. This polymeric surfactant produces enhanced steric stabilization, both in water and in high electrolyte concentrations, as will be discussed in the next section.

The emulsions of isoparaffinic oil (Isopar M)/water and cyclomethicone/water were prepared using Inutec SP1. O/W emulsions (50:50 v/v) were prepared, and the emulsifier concentration was varied from 0.25 to 2 w/v% based on the oil phase. The emulsifier (0.5 w/v%) was sufficient for the stabilization of these 50:50 v/v emulsions [12].

The emulsions were stored at room temperature and 50°C, and optical micrographs were taken at intervals of time (for a year) to check the stability. Emulsions prepared in water were very stable, showing no change in droplet size distribution for more than a year, indicating the absence of coalescence. Any weak flocculation that occurred was reversible, and the emulsion could be redispersed by gentle shaking. Figure 10.10 shows optical micrographs for a dilute 50:50 v/v emulsion that was stored for 1.5 and 14 weeks at 50°C.

No change in droplet size was observed after storage for more than 1 year at 50°C, indicating the absence of coalescence. The same result was obtained when using different oils. Emulsions were also stable against coalescence in the presence of high electrolyte concentrations (up to 4 mol·dm^{-3} or ~25% NaCl).

(a)   (b)

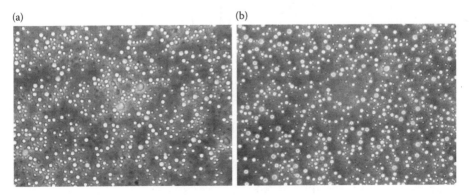

FIGURE 10.10 Optical micrographs of O/W emulsion stabilized with Inutec SP1 stored at 50°C for (a) 1.5 weeks and (b) 14 weeks.

The previously mentioned stability in high electrolyte concentrations is not observed with polymeric surfactants based on PEO. The high stability observed using Inutec SP1 is related to its strong hydration both in water and in electrolyte solutions. The hydration of inulin (the backbone of Inutec SP1) could be assessed using cloud point measurements. A comparison was also made with PEO with two molecular weights, namely, 4000 and 20,000.

Solutions of PEO 4000 and 20,000 showed a systematic decrease in cloud point, with an increase in NaCl or MgSO$_4$ concentration. In contrast, inulin showed no cloud point up to 4 mol·dm^{-3} NaCl and up to 1 mol·dm^{-3} MgSO$_4$.

The previously mentioned results explain the difference between PEO and inulin. With PEO, the chains show dehydration when the NaCl concentration is increased more than 2 or 0.5 mol·dm^{-3} MgSO$_4$. The inulin chains remain hydrated at much higher electrolyte concentrations. It seems that the linear polyfructose chains remain strongly hydrated at high temperature and high electrolyte concentrations.

The high emulsion stability obtained when using Inutec SP1 can be accounted for by the following factors: (i) the multipoint attachment of the polymer by several alkyl chains that are grafted on the backbone, (ii) the strong hydration of the polyfructose "loops" both in water and in high electrolyte concentrations (χ remains less than 0.5 under these conditions), (iii) the high-volume fraction (concentration) of the loops at the interface, and (iv) the enhanced steric stabilization—the case with multipoint attachment that produces strong elastic interaction.

Evidence for the high stability of the liquid film between emulsion droplets when using Inutec SP1 was obtained by Exerowa et al. [13] using disjoining pressure measurements. This is illustrated in Figure 10.11, which shows a plot of disjoining pressure versus separation distance between two emulsion droplets at various electrolyte concentrations. The results show that by increasing the capillary pressure, a stable Newton black film (NBF) is obtained at a film thickness of approximately 7 nm. The lack of rupture of the film at the highest pressure applied (4.5×10^4 Pa) indicates the high stability of the film in water and in high electrolyte concentrations (up to 2.0 mol·dm^{-3} NaCl).

The lack of rupture of the NBF up to the highest pressure applied of 4.5×10^4 Pa clearly indicates the high stability of the liquid film in the presence of high NaCl concentrations (up to 2 mol·dm^{-3}). This result is consistent with the high-emulsion stability obtained at high electrolyte concentrations and high temperature. The emulsions of Isopar M/water are very stable under such conditions, and this could be accounted for by the high stability of the NBF. The droplet size of 50:50 O/W emulsions prepared using 2% Inutec SP1 is in the region of 1 to 10 µm. This corresponds to a capillary

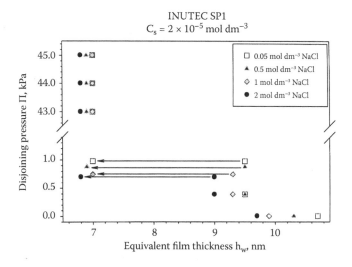

FIGURE 10.11 Variation of disjoining pressure π (in kPa, 10^3 Pa), with equivalent film thickness at various NaCl concentrations.

pressure of approximately 3×10^4 Pa for the 1 μm drops and approximately 3×10^3 Pa for the 10 μm drops. These capillary pressures are lower than those to which the NBF have been subjected, and this clearly indicates the high stability obtained against coalescence in these emulsions.

10.6   SUSPENSIONS STABILIZED USING POLYMERIC SURFACTANTS

As mentioned previously, there are generally two procedures for the preparation of S/L dispersions:

1. Condensation methods: the buildup of particles from molecular units, that is, nucleation and growth (a special procedure is the preparation of latexes by emulsion or dispersion polymerization).
2. Dispersion methods: in this case, one starts with preformed large particles or crystals, which are dispersed in the liquid by using a surfactant (wetting agent) with the subsequent breakup of the large particles by milling (comminution) to achieve the desirable particle size distribution. A dispersing agent (usually a polymeric surfactant) is used for the dispersion process and subsequent stabilization of the resulting suspension.

There are generally two procedures for the preparation of latexes:

1. Emulsion polymerization: the monomers that are essentially insoluble in the aqueous medium are emulsified using a surfactant, and an initiator is added while heating the system to produce the polymer particles that are stabilized electrostatically (when using ionic surfactants) or sterically (when using nonionic surfactants).
2. Dispersion polymerization: the monomers are dissolved in a solvent in which the resulting polymer particles are insoluble. A protective colloid (normally a block or graft copolymer) is added to prevent the flocculation of the resulting polymer particles that are produced in addition to an initiator. This method is usually applied for the production of nonaqueous latex dispersions and is sometimes called *nonaqueous dispersion* (NAD) polymerization.

Surfactants play a crucial role in the process of latex preparation because they determine the stabilizing efficiency, and the effectiveness of the surface-active agent ultimately determines the number of particles and their size. The effectiveness of any surface-active agent in stabilizing the particles is the dominant factor, and the number of micelles formed is relatively unimportant. In the NAD process, the monomer, normally an acrylic, is dissolved in a nonaqueous solvent, normally an aliphatic hydrocarbon, and an oil soluble initiator and a stabilizer (to protect the resulting particles from flocculation) are added to the reaction mixture. The most successful stabilizers used in NAD are block and graft copolymers. Preformed graft stabilizers based on PHS are simple to prepare and effective in NAD polymerization.

Dispersion methods are used for the preparation of the suspensions of preformed particles. The role of surfactants (or polymers) in the dispersion process can be analyzed in terms of the three processes involved: (i) the wetting of the powder by the liquid, (ii) the breakup of the aggregates and agglomerates, and (iii) the comminution of the resulting particles and their subsequent stabilization. All these processes are affected by surfactants or polymers, which adsorb on the powder surface, thus aiding the wetting of the powder, the breakup of the aggregates and agglomerates, and the subsequent reduction of particle size by wet milling.

10.6.1   POLYMERIC SURFACTANTS IN EMULSION POLYMERIZATION

Recently, the graft copolymer of hydrophobically modified inulin (Inutec SP1) has been used in emulsion polymerization of styrene, MMA, butyl acrylate, and several other monomers [14]. All lattices were prepared by emulsion polymerization using potassium persulfate as an initiator. The z-average particle size was determined by PCS, and electron micrographs were also taken.

The emulsion polymerization of styrene or MMA showed an optimum Inutec–monomer ratio of 0.0033 for PS and 0.001 for PMMA particles. The initiator–monomer ratio was kept constant at 0.00125. The monomer conversion was higher than 85% in all cases. Latex dispersions of PS reaching 50% and of PMMA reaching 40% could be obtained using such low concentration of Inutec SP1. Figure 10.12 shows the variation of particle diameter with monomer concentration.

The stability of the latexes was determined by determining the critical coagulation concentration (CCC) using $CaCl_2$. The CCC was low (0.0175–0.05 mol·dm^{-3}), but this was higher than that for the latex prepared without surfactant. The addition of Inutec SP1 resulted in a large increase in the CCC, as illustrated in Figure 10.13, which shows log W–log C curves (where W is the stability ratio) at various additions of Inutec SP1.

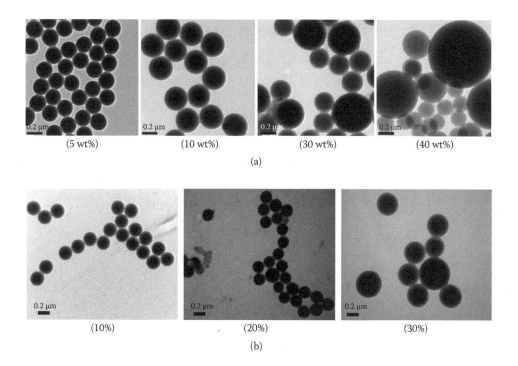

(5 wt%)   (10 wt%)   (30 wt%)   (40 wt%)

(a)

(10%)   (20%)   (30%)

(b)

FIGURE 10.12 Electron Micrographs of latexes prepared at different weight percent: (a) PS latexes: 5, 10, 20, 40 wt%; (b) PMMA latexes: 10, 20, 30 wt%.

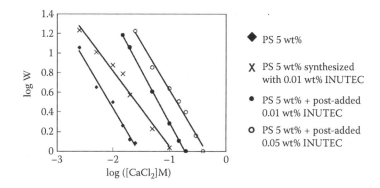

♦ PS 5 wt%

✗ PS 5 wt% synthesized with 0.01 wt% INUTEC

● PS 5 wt% + post-added 0.01 wt% INUTEC

○ PS 5 wt% + post-added 0.05 wt% INUTEC

FIGURE 10.13 Influence of the addition of Inutec SP1 on the latex stability.

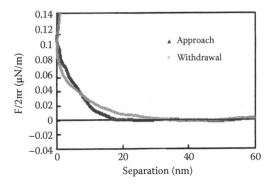

FIGURE 10.14 Force–distance curves between hydrophobized glass surfaces containing adsorbed Inutec SP1 in water.

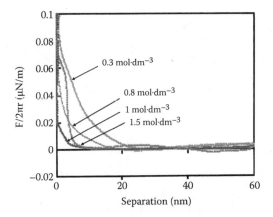

FIGURE 10.15 Force–distance curves for hydrophobized glass surfaces containing adsorbed Inutec SP1 at various Na_2SO_4 concentrations.

As with the emulsions, the high stability of the latex when using Inutec SP1 is due to the strong adsorption of the polymeric surfactant on the latex particles and formation of strongly hydrated loops and tails of polyfructose that provide effective steric stabilization. Evidence for the strong repulsion produced when using Inutec SP1 was obtained from atomic force microscopy investigations [15], whereby the force between hydrophobic glass spheres and hydrophobic glass plate, both containing an adsorbed layer of Inutec SP1, was measured as a function of the distance of separation both in water and in the presence of various Na_2SO_4 concentrations. The results are shown in Figures 10.14 and 10.15.

10.6.2 DISPERSION POLYMERIZATION

This method is usually applied for the preparation of nonaqueous latex dispersions; hence, it is called NAD. The method has also been adapted to prepare aqueous latex dispersions by using an alcohol–water mixture.

In the NAD process, the monomer, normally an acrylic, is dissolved in a nonaqueous solvent, normally an aliphatic hydrocarbon, and an oil-soluble initiator and a stabilizer (to protect the resulting particles from flocculation) are added to the reaction mixture. Graft stabilizers based on PHS are simple to prepare and are effective in NAD polymerization.

Commercial 12-hydroxystearic acid contains 8% to 15% palmitic and stearic acids, which limit the molecular weight during polymerization to an average of 1500 to 2000. This oligomer may be converted to a "macromonomer" by reacting the carboxylic group with glycidyl methacrylate. The macromonomer is then copolymerized with an equal weight of MMA or similar monomer to give a comb graft copolymer with an average molecular weight of 10,000 to 20,000. The graft copolymer contains, on average, 5 to 10 PHS chains pendent from a polymeric anchor backbone of PMMA. This graft copolymer can stabilize latex particles of various monomers. The major limitation of the monomer composition is that the polymer produced should be insoluble in the medium used.

NAD polymerization is carried out in two steps: (i) the seed stage: the diluent, the portion of the monomer, the portion of dispersant, and the initiator (azo or peroxy type) are heated to form an initial low-concentration fine dispersion; and (ii) the growth stage: the remaining monomer and more dispersant and initiator are then fed over the course of several hours to complete the growth of the particles. A small amount of transfer agent is usually added to control the molecular weight. Excellent control of particle size is achieved by the proper choice of the designed dispersant and the correct distribution of dispersant between the seed and the growth stages. NAD acrylic polymers are applied in automotive thermosetting polymers, and hydroxy monomers may be included in the monomer blend used.

10.6.3 Polymeric Surfactants for the Stabilization of Preformed Latex Dispersions

For this purpose, PS latexes were prepared using the surfactant-free emulsion polymerization. Two latexes with a z-average diameter of 427 and 867 (as measured using PCS) that are reasonably mono-disperse were prepared [16]. Two polymeric surfactants, namely, Hypermer CG-6 and Atlox 4913 (UNIQEMA, UK), were used. Both are graft (comb) type consisting of PMMA/poly(methacrylic acid) backbone with methoxy-capped PEO side chains ($M = 750$ Da). Hypermer CG-6 is the same graft copolymer as Atlox 4913, but it contains a higher proportion of methacrylic acid in the backbone. The average molecular weight of the polymer is approximately 5000 Da. Figure 10.16 shows a typical adsorption isotherm of Atlox 4913 on the two latexes.

Similar results were obtained for Hypermer CG-6, but the plateau adsorption was lower ($1.2\ \text{mg}\cdot\text{m}^{-2}$ compared with $1.5\ \text{mg}\cdot\text{m}^{-2}$ for Atlox 4913). It is likely that the backbone of Hypermer CG-6, which contains more poly(methacrylic acid), is more polar and hence is less strongly adsorbed. The amount of adsorption was independent of particle size.

The influence of temperature on adsorption is shown in Figure 10.17. The amount of adsorption increases with the increase in temperature. This is due to the poorer solvency of the medium for the PEO chains. The PEO chains become less hydrated at higher temperature, and the reduction in solubility of the polymer enhances adsorption.

The adsorbed layer thickness of the graft copolymer on the latexes was determined using rheological measurements. Steady-state (shear stress σ – shear rate γ) measurements were carried out,

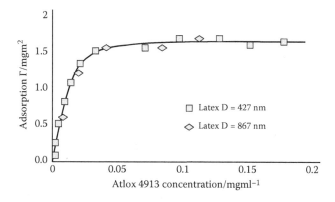

FIGURE 10.16 Adsorption isotherms of Atlox 4913 on the two latexes at 25°C.

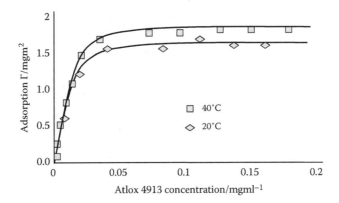

FIGURE 10.17   Effect of temperature on adsorption of Atlox 4913 on PS.

and the results were fitted to the Bingham equation to obtain the yield value σ_β and the high shear viscosity η of the suspension,

$$\sigma = \sigma_\beta + \eta\gamma \tag{10.9}$$

As an illustration, Figure 10.18 shows a plot of σ_β versus the volume fraction ϕ of the latex for Atlox 4913. Similar results were obtained for latexes stabilized using Hypermer CG-6.

At any given volume fraction, the smaller size latex has higher σ_β when compared with the larger latex. This is due to the higher ratio of adsorbed layer thickness to particle radius, Δ/R, for the smaller latex. The effective volume fraction of the latex ϕ_{eff} is related to the core volume fraction ϕ by the following equation:

$$\phi_{eff} = \phi\left[1 + \frac{\Delta}{R}\right]^3 \tag{10.10}$$

As discussed previously, ϕ_{eff} can be calculated from the relative viscosity η_r using the Dougherty–Krieger equation, as follows:

$$\eta_r = \left[1 - \left(\frac{\phi_{eff}}{\phi_p}\right)\right]^{-[\eta]\phi_p} \tag{10.11}$$

where ϕ_p is the maximum packing fraction.

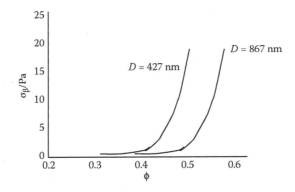

FIGURE 10.18   Variation of yield stress with latex volume fraction for Atlox 4913, at two latex diameters: $D = 427$ nm; $D = 867$ nm.

The maximum packing fraction ϕ_p can be calculated using the following empirical equation:

$$\frac{\left(\eta_r^{1/2} - 1\right)}{\phi} = \left(\frac{1}{\phi_p}\right)(\eta^{1/2} - 1) + 1.25 \tag{10.12}$$

The results showed a gradual decrease in the adsorbed layer thickness Δ with the increase in the volume fraction ϕ. For the latex with diameter D of 867 nm and Atlox 4913, Δ decreased from 17.5 nm at $\phi = 0.36$ to 6.5 at $\phi = 0.57$. For Hypermer CG-6 with the same latex Δ decreased from 11.8 nm at $\phi = 0.49$ to 6.5 at $\phi = 0.57$. The reduction in Δ with an increase in ϕ may be due to overlap or compression (or both) of the adsorbed layers as the particles come close to one another at a higher volume fraction of the latex.

The stability of the latexes was determined using viscoelastic measurements. For this purpose, dynamic (oscillatory) measurements were used to obtain the storage modulus G^*, the elastic modulus G', and the viscous modulus G'' as a function of strain amplitude γ_0 and frequency ω (rad·s^{-1}). The method relies on the application of a sinusoidal strain or stress, and the resulting stress or strain is measured simultaneously. For a viscoelastic system, the strain and stress sine waves oscillate with the same frequency, but out of phase. From the time shift Δt and ω, one can obtain the phase angle shift δ.

The ratio of the maximum stress σ_0 to the maximum strain γ_0 gives the complex modulus $|G^*|$, as follows:

$$|G^*| = \frac{\sigma_0}{\gamma_0} \tag{10.13}$$

$|G^*|$ can be resolved into two components: storage (elastic) modulus G', the real component of the complex modulus; and loss (viscous) modulus G'', the imaginary component of the complex modulus. The complex modulus can be resolved into G' and G'' using the vector analysis and the phase angle shift δ, as follows:

$$G' = |G^*| \cos \delta \tag{10.14}$$

$$G'' = |G^*| \sin \delta \tag{10.15}$$

G' is measured as a function of electrolyte concentration and temperature to assess the latex stability. As an illustration, Figure 10.19 shows the variation of G' with temperature for latex stabilized

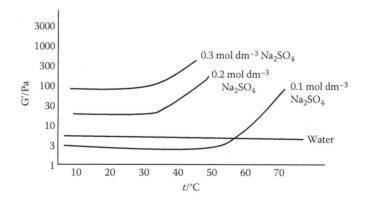

FIGURE 10.19 Variation of G' with temperature, $t°C$, in water and at various Na$_2$SO$_4$ concentrations.

with Atlox 4913 in the absence of any added electrolyte and in the presence of 0.1, 0.2, and 0.3 mol·dm^{-3} Na$_2$SO$_4$. In the absence of electrolyte, G' showed no change with temperature up to 65°C.

In the presence of 0.1 mol·dm^{-3} Na$_2$SO$_4$, G' remained constant up to 40°C, above which G' increased with further increase in temperature. This temperature is denoted as the critical flocculation temperature (CFT). The CFT decreases with the increase in electrolyte concentration, reaching approximately 30°C in 0.2 and 0.3 mol·dm^{-3} Na$_2$SO$_4$. This reduction in CFT with an increase in electrolyte concentration is due to the reduction in solvency of the PEO chains with an increase in electrolyte concentration. The latex stabilized with Hypermer CG-6 gave relatively higher CFT values when compared with that stabilized using Atlox 4913.

10.7 USE OF POLYMERIC SURFACTANTS FOR THE PREPARATION AND STABILIZATION OF NANOEMULSIONS

Nanoemulsions are systems that cover the size range 20 to 200 nm [17–19]. They can be transparent, translucent, or turbid, depending on the droplet radius and the refractive index difference between the droplets and the continuous phase. This can be understood by considering the dependence of light scattering (turbidity) on the two factors mentioned previously. For droplets with a radius that is less than 1/20 of the wavelength of the light, the turbidity τ is given by the following equation:

$$\tau = K N_o V^2 \tag{10.16}$$

where K is an optical constant that is related to the difference in refractive index between the droplets n_p and the medium n_o, and N_o is the number of droplets each with a volume V.

It is clear from Equation 10.16 that τ decreases with the decrease in K, that is, smaller ($n_p - n_o$), decrease in N_o, and decrease in V. Thus, to produce a transparent nanoemulsion, one has to decrease the difference between the refractive index of the droplets and the medium (i.e., try to match the two refractive indices). If such matching is not possible, then one has to reduce the droplet size (by high-pressure homogenization) to values less than 50 nm. It is also necessary to use a nanoemulsion with low oil volume fraction, in the region of 0.2.

Nanoemulsions are only kinetically stable. They have to be distinguished from microemulsions (which cover the size range 5–50 nm), which are mostly transparent and thermodynamically stable. The long-term physical stability of nanoemulsions (with no apparent flocculation or coalescence) make them unique, and they are sometimes referred to as *approaching thermodynamic stability*. The inherently high-colloid stability of nanoemulsions can be well understood by considering their steric stabilization (when using nonionic surfactants or polymers) and how this is affected by the ratio of the adsorbed layer thickness to droplet radius, as was discussed previously.

Unless adequately prepared (to control the droplet size distribution) and stabilized against Ostwald ripening (which occurs when the oil has some finite solubility in the continuous medium), nanoemulsions may show an increase in the droplet size, and an initially transparent system may become turbid upon storage.

The attraction of nanoemulsions for the application in personal care and cosmetics as well as in health care is due to the following advantages: (i) the very small droplet size causes a large reduction in the gravity force, and Brownian motion may be sufficient for overcoming gravity. This means that no creaming or sedimentation occurs on storage; (ii) The small droplet size also prevents any flocculation of the droplets. Weak flocculation is prevented, and this enables the system to remain dispersed with no separation; and (iii) The small droplets also prevent their coalescence because these droplets are nondeformable, and hence surface fluctuations are prevented. In addition, the significant surfactant film thickness (relative to droplet radius) prevents any thinning or disruption of the liquid film between the droplets.

The production of small droplets (submicrometer size) requires the application of high energy. The process of emulsification is generally inefficient. Simple calculations show that the mechanical energy required for emulsification exceeds the interfacial energy by several orders of magnitude. For example, to produce a nanoemulsion at $\phi = 0.1$ with an average radius R of 200 nm, using a surfactant that gives an interfacial tension $\gamma = 10$ m·Nm^{-1}, the net increase in interfacial free energy is $A\gamma = 3\phi\gamma/R = 1.5 \times 10^4$ J·m^{-3}. The mechanical energy required in a homogenizer is 1.5×10^7 J·m^{-3}, that is, an efficiency of 0.1%. The rest of the energy (99.9%) is dissipated as heat.

The intensity of the process or the effectiveness in making small droplets p is often governed by the net power density $(\varepsilon(t))$.

$$p = \varepsilon(t)dt \tag{10.17}$$

where t is the time during which emulsification occurs.

The breakup of droplets will only occur at high ε values, which means that the energy dissipated at low ε levels is wasted. Batch processes are generally less efficient than continuous processes. This shows why with a stirrer in a large vessel, most of the energy applied at low intensity is dissipated as heat. In a homogenizer, p is simply equal to the applied pressure in the homogenizer.

Several procedures may be applied to enhance the efficiency of emulsification when producing nanoemulsions: (i) One should optimize the efficiency of agitation by increasing ε and by decreasing dissipation time; (ii) The nanoemulsion is preferably prepared at high-volume faction of the disperse phase and diluted afterward. However, very high ϕ values may result in coalescence during emulsification; (iii) Adding more surfactant by creating a smaller γ_{eff} and possibly diminishing recoalescence; (iv) Use a surfactant mixture that shows higher reduction in γ of the individual components; (v) If possible, dissolve the surfactant in the disperse phase rather than in the continuous phase—this often leads to smaller droplets; and (vi) It may be useful to emulsify in steps of increasing intensity, particularly with nanoemulsions having highly viscous disperse phases.

The high kinetic stability of nanoemulsions can be explained from the consideration of the energy–distance curves for sterically stabilized dispersions shown in Figure 10.9. It can be seen from this figure that the depth of the minimum decreases with increasing δ/R. With nanoemulsions having a radius in the region of 50 nm and an adsorbed layer thickness of, say, 10 nm, the value of δ/R is 0.2. This high value (when compared with the situation with macroemulsions where δ/R is at least 10 times lower) results in a very shallow minimum (which could be less than kT). This situation results in very high stability with no flocculation (weak or strong). In addition, the very small size of the droplets and the dense adsorbed layers ensure the lack of deformation of the interface, the lack of thinning, and the disruption of the liquid film between the droplets and hence coalescence is also prevented.

One of the main problems with nanoemulsions is Ostwald ripening, which results from the difference in solubility between small and large droplets The difference in chemical potential of dispersed phase droplets between droplets with different sizes is given by Lord Kelvin [20],

$$s(r) = s(\infty)\exp\left(\frac{2\gamma V_m}{rRT}\right) \tag{10.18}$$

where $s(r)$ is the solubility surrounding a particle of radius r, $s(\infty)$ is the bulk phase solubility, and V_m is the molar volume of the dispersed phase.

The quantity $(2\gamma V_m/RT)$ is termed the characteristic length. It is of the order of approximately 1 nm or less, indicating that the difference in solubility of a 1 μm droplet is of the order of 0.1% or less. Theoretically, Ostwald ripening should lead to the condensation of all droplets into a single drop (i.e., phase separation). This does not occur in practice because the rate of growth decreases with the increase in droplet size.

For two droplets of radii r_1 and r_2 (where $r_1 < r_2$),

$$\frac{RT}{V_m} \ln\left[\frac{s(r_1)}{s(r_2)}\right] = 2\gamma\left(\frac{1}{r_1} - \frac{1}{r_2}\right) \tag{10.19}$$

Equation 10.19 shows that the larger the difference between r_1 and r_2, the higher the rate of Ostwald ripening.

Ostwald ripening can be quantitatively assessed from the plots of the cube of the radius versus time t [21–23], calculated as follows:

$$r^3 = \frac{8}{9}\left(\frac{s(\infty)\gamma V_m D}{\rho RT}\right) \tag{10.20}$$

where D is the diffusion coefficient of the disperse phase in the continuous phase and ρ is its density.

Ostwald ripening can be reduced by the incorporation of a second component, which is insoluble in the continuous phase (e.g., squalane). In this case, significant partitioning between different droplets occurs, with the component having low solubility in the continuous phase expected to be concentrated in the smaller droplets. During Ostwald ripening in the two-component disperse phase system, equilibrium is established when the difference in chemical potential between droplets with different sizes (which results from curvature effects) is balanced by the difference in chemical potential, resulting from the partition of the two components. If the secondary component has zero solubility in the continuous phase, the size distribution will not deviate from the initial one (the growth rate is equal to zero). In the case of limited solubility of the secondary component, the distribution is the same as that governed by Equation 10.20, that is, a mixture growth rate is obtained, which is still lower than that of the more soluble component.

Another method for reducing Ostwald ripening depends on the modification of the interfacial film at the O/W interface. According to Equation 10.20, the reduction in γ results in the reduction of Ostwald ripening. However, this alone is not sufficient because one has to reduce γ by several orders of magnitude. Walstra [24,25] suggested that by using surfactants that are strongly adsorbed at the O/W interface (i.e., polymeric surfactants) and that do not desorb during ripening, the rate could be significantly reduced. An increase in the surface dilational modulus and a decrease in γ would be observed for the shrinking drops. The difference in γ between the droplets would balance the difference in capillary pressure (i.e., curvature effects).

To achieve the previously mentioned effect, it is useful to use A-B-A block copolymers that are soluble in the oil phase and insoluble in the continuous phase. A strongly adsorbed polymeric surfactant that has limited solubility in the aqueous phase can also be used, for example, hydrophobically modified inulin, Inutec SP1. This is illustrated in Figure 10.20, which shows plots of the cube of the radius of R^3 versus time for 20 v/v% silicone O/W emulsions at the two concentrations of Inutec SP1 (1.6%, top curve and 2.4%, bottom curve) [26]. The concentration of Inutec SP1 is much lower than that required when using nonionic surfactants.

The rate of Ostwald ripening is 1.1×10^{-29} and 2.4×10^{-30} m^3·s^{-1} at 1.6% and 2.4% Inutec SP1, respectively. These rates are approximately three orders of magnitude lower than those obtained using a nonionic surfactant. The addition of 5% glycerol was found to decrease the rate of Ostwald ripening in some nanoemulsions.

Various nanoemulsions with the hydrocarbon oils of different solubilities were prepared using Inutec SP1. Figure 10.21 shows plots of r^3 versus t for the nanoemulsions of the hydrocarbon oils that were stored at 50°C. It can be seen that both paraffinum liquidum with low and high viscosities give almost a zero slope, indicating the absence of Ostwald ripening in this case. This is not

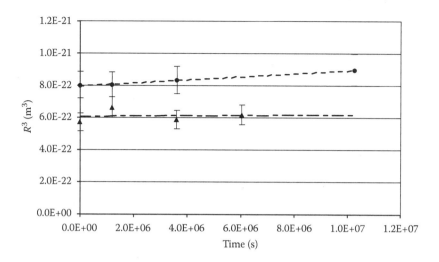

FIGURE 10.20 Silicone oil-in-water nanoemulsions stabilized with Inutec SP1. Top curve, 1.6% Inutec SP1; bottom curve, 2.4% Inutec SP1.

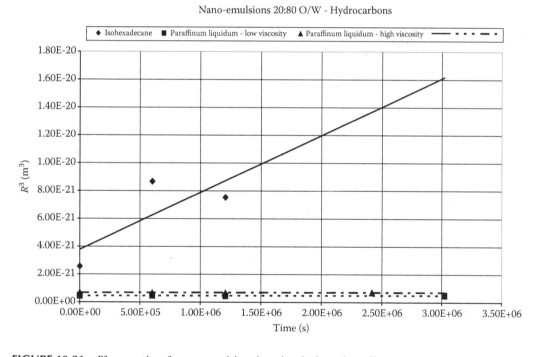

FIGURE 10.21 R^3 versus time for nanoemulsions based on hydrocarbon oils.

surprising because both oils have very low solubility, and the hydrophobically modified inulin, Inutec SP1, strongly adsorbs at the interface, giving high elasticity that reduces both Ostwald ripening and coalescence.

With the more soluble hydrocarbon oils, namely, isohexadecane, there is an increase in r^3 with time, giving a rate of Ostwald ripening of 4.1×10^{-27} m³·s⁻¹. The rate for this oil is almost three orders of magnitude lower than that obtained with a nonionic surfactant, namely, laureth-4 (C_{12}-alkyl chain with 4 mol ethylene oxide). This clearly shows the effectiveness of Inutec SP1 in reducing Ostwald ripening. This reduction can be attributed to the enhancement of the Gibbs dilational elasticity, which results from the multipoint attachment of the polymeric surfactant with several alkyl groups to the oil droplets. This results in a reduction of the molecular diffusion of the oil from the smaller to the larger droplets.

REFERENCES

1. Th. Tadros. *Applied Surfactants*. Wiley-VCH, Germany (2005).
2. Th. Tadros. In *Principles of Polymer Science and Technology in Cosmetics and Personal Care*, E.D. Goddard and J.V. Gruber (Eds.). Marcel Dekker, New York (1999).
3. Th. Tadros. In *Novel Surfactants*, K. Holmberg (Ed.). Marcel Dekker, New York (2003).
4. J.W. Gibbs. *Scientific Papers*. Longman Green, London (1906); M. Volmer. *Kinetik der Phase Building*, Steinkopf, Dresden (1939).
5. I. Piirma. *Polymeric Surfactants*. Surfactant Science Series, Vol. 42. Marcel Dekker, New York (1992).
6. C.V. Stevens, A. Meriggi, M. Peristerpoulou, P.P. Christov, K. Booten, B. Levecke, A. Vandamme, N. Pittevils and Th. Tadros. *Biomacromolecules*, **2**, 1256 (2001).
7. E.L. Hirst, D.I. McGilvary and E.G. Percival. *J. Chem. Soc.*, 1297 (1950).
8. M. Suzuki. In *Science and Technology of Fructans*, M. Suzuki and N.J. Chatterton (Eds.). CRC Press, Boca Raton, FL, p. 21 (1993).
9. Th. Tadros. In *Polymer Colloids*, R. Buscall, T. Corner and J.F. Stageman (Eds.). Elsevier Applied Sciences, London, p. 105 (1985).
10. G.J. Fleer, M.A. Cohen-Stuart, J.M.H.M. Scheutjens, T. Cosgrove and B. Vincent. *Polymers at Interfaces*. Chapman and Hall, London (1993).
11. D.H. Napper. *Polymeric Stabilization of Dispersions*. Academic Press, London (1983).
12. Th. Tadros, A. Vandamme, B. Levecke, K. Booten and C.V. Stevens. *Adv. Colloid Interface Sci.*, **108–109**, 207 (2004).
13. D. Exerowa, G. Gotchev, T. Kolarev, Khr. Khristov, B. Levecke and T. Tadros. *Langmuir*, **23**, 1711 (2007).
14. J. Nestor, J. Esquena, C. Solans, B. Levecke, K. Booten and Th.F. Tadros. *Langmuir*, **21**, 4837 (2005).
15. J. Nestor, J. Esquena, C. Solans, P.F. Luckham, B. Levecke and Th.F. Tadros. *J. Colloid Interface Sci.*, **311**, 430 (2007).
16. W. Liang, G. Bognolo and Th. Tadros, Th.F., *Langmuir*, **11**, 2899 (1995).
17. H. Nakajima, S. Tomomossa and M. Okabe. First Emulsion Conference, Paris (1993).
18. H. Nakajima. In *Industrial Applications of Microemulsions*, C. Solans and H. Kunieda (Eds.). Marcel Dekker, New York (1997).
19. Th. Tadros, P. Izquierdo, J. Esquena and C. Solans. *Adv. Colloid Interface Sci.*, **108–109**, 303 (2004).
20. W. Thompson (Lord Kelvin). *Phil. Mag.*, **42**, 448 (1871).
21. I.M. Lifshitz and V.V. Slesov. *Sov. Phys.*, **35**, 331 (1959).
22. C. Wagner. *Z. Electroche.*, **35**, 581 (1935).
23. A.S. Kabalnov and E.D. Schukin. *Adv. Colloid Interface Sci.*, **38**, 69 (1992).
24. P. Walstra. In *Encyclopedia of Emulsion Technology*, P. Becher (Ed.). Marcel Dekker, New York (1983).
25. P. Walstra and P.E.A. Smoulders. In *Modern Aspects of Emulsion Science*, B.P. Binks (Ed.). The Royal Society of Chemistry, Cambridge (1998).
26. Th. Tadros, E. Vandekerckhove, A. Vandamme, B. Levecke and K. Booten. *Cosmet Toilet*, **120**(2), 45 (2005).

Part IV

Biosurfactants

11 Biosurfactants*

Girma Biresaw

CONTENTS

11.1 INTRODUCTION

Biosurfactants are surfactants that have surface active properties (e.g., adsorb on surfaces, lower surface/interfacial tensions, emulsify water in oil or oil in water, etc.) and are also biodegradable [1]. Biodegradability and surface activity are common properties to all biodegradable surfactants. Biosurfactants can be designed and synthesized to have properties specific to a wide range of applications. However, regardless of the specific application, the biosurfactant is also designed to be biodegradable after use.

Biodegradability refers to the ability of materials to easily break down to harmless products (preferably water, carbon dioxide, and minerals) under natural conditions. This is an important property that distinguishes biosurfactants from nonbiodegradable surfactants. Nonbiodegradable surfactants

---

* Mention of trade names or commercial products in this chapter is solely for the purpose of providing specific information and does not imply recommendation or endorsement by the U.S. Department of Agriculture. The U.S. Department of Agriculture is an equal opportunity provider and employer.

account for almost two-thirds of the global surfactant market [2]. Biodegradability is important because most surfactants end up in waste streams after use, which eventually get discharged into waterways and landfills. The use of nonbiodegradable surfactants for such applications will require extra and sometimes costly treatment before discharge to ensure environmental compliance.

Biosurfactants derive their biodegradability from the structures of the ingredients used to synthesize them. These ingredients originate naturally in plants and animals and, hence, are themselves biodegradable. The three main ingredients used in biosurfactant synthesis are lipids, carbohydrates, and proteins. Lipids are triglycerides obtained from animals or plants. Lipids derived from triglycerides and used in biosurfactant synthesis include diglycerides and monoglycerides, fatty acids, fatty alcohols, and esters of fatty acids, including fatty acid methyl esters (FAMEs) widely used in biodiesel.

Lipids, carbohydrates, proteins, and their combinations are further functionalized with a variety of biobased or petroleum-based reagents by chemical, biological, and enzymatic methods or a combination of synthesis methods [1–14] to produce biosurfactants with nonionic, anionic, cationic, and zwitterionic structures. Some examples of nonionic biosurfactants are listed in Table 11.1. Examples of anionic and cationic biosurfactants are given in Table 11.2.

Biosurfactants are used in a wide range of products and applications, each requiring specific and unique properties in addition to biodegradability and surface activity. Table 11.3 lists some examples of the applications of biosurfactants. In examining Table 11.3, it is clear that biosurfactants used as ingredients in one application may not be suitable for a different application. For example, biosurfactants used as ingredients in foods or pharmaceuticals will not be suitable for use as biocides or disinfectants. Although biocompatibility, nontoxicity, and safety are important properties in food and pharmaceutical applications, they are most likely not as important for biocide and disinfectant applications.

In this chapter, the synthesis, properties, and applications of selected biosurfactants will be reviewed. The chapter is divided into two broad sections. The first part involves a review of biosurfactants synthesized from small molecules by chemical, enzymatic, and microbial methods or a combination of these methods. The small molecules include monosaccharides and disaccharides, amino acids, fatty acids, fatty acid esters, and various combinations of these. The second part involves a discussion of biosurfactants derived from agriculture-based high–molecular weight biopolymers. The starting materials for these biosurfactants include starches, plant proteins, and vegetable oils.

TABLE 11.1

Some Examples of Nonionic Biosurfactants

Alkyl ester sulfonates, MES
Alkyl glycosides
Erythritol and xylitol-based sugar-alcohol esters
Ethylene and propylene glycol esters
Fatty acid–polyol esters and their ethoxylates
Mannosylerythritol lipids
Mono-, di-, and triglycerides
Polyglycerol esters
Rhamnolipids
Saccharide fatty acid esters and amides
Saccharide fatty alcohol ethers
Saponins
Sophorolipids
Sorbitan derivatives—polysorbates (Span, Tween)
Starch anhydride esters
Sterols—ethoxylated or glycosylated
Trehalose lipids

TABLE 11.2
Some Examples of Anionic and Cationic Biosurfactants

Anionic

MES

Lignin sulfonates

Polysoaps of seed oils

Cationic

Amino acid-, peptide-, or protein-based surfactants, e.g., lipoproteins

Amphiphilic polypeptides, e.g., copolymers of amino acids

Esterquats

Ethoxylated fatty amines

Phospholipids and lung surfactants

Modified plant proteins

TABLE 11.3
Examples of Application/Properties of Biosurfactants

Application/Property	Reference
Agricultural crop protection	[15,18,28,35]
Cosmetic	[10,13–15,17–19,21,22,24,25,27,32,35–37,46,61]
Environmentally friendly	[10,13–16,18,32,34,36,39,40,42,43,45–48,53,58,60]
Environmental remediation	[10,15,17,18,25]
Food additive	[15,19,36,37]
Food processing	[13–15,17,18,22,27,32–37,46,61]
Medical	[24,27,35]
Medical—antibiotic, antibacterial, antifungal, antimicrobial	[12,15,18–22,24,26,27,32,34–37,46–48,53,54,57,58]
Medical—anticancer, antitumor	[22,26,27,34,35,59]
Medical—diagnostic	[23,55]
Medical—drugs, medicines, pharmaceuticals	[10,13–15,17,18,21,24,27,32,36,37,46,59,61]
Medical—personal body/skin care	[19,20,22,25,33–35,37,39,41,42]
Medical—therapeutic drug/gene delivery	[41,45,52,54–60,62]
Medical—therapy	[25,26]
Medical—treatment of wounds	[19]
Medical—treatment of cellulite	[19]
Petroleum—additives	[17]
Petroleum—enhanced/secondary oil recovery	[10,17,18,21,22,25]
Surfactant/detergent—dishwashing	[19,42]
Surfactant/detergent—emulsification	[19,22,26,27,33,36,42]
Surfactant/detergent—laundry	[10,15,19,22,25,32,33,35,37,39–44]
Surfactant/detergent—surface modification	[24,33,44,53,57]

11.2 BIOSURFACTANTS SYNTHESIZED FROM SMALL MOLECULES

11.2.1 BIOSURFACTANTS BASED ON LIPIDS AND MONOSACCHARIDES OR DISACCHARIDES

These are known by different names, most commonly as glycolipids [3,8–10,15–38]. As the name implies, these biosurfactants comprise glycans and lipids in their structures. The term glycan is a generic name for the carbohydrate portion of the biosurfactant. Glycolipids are synthesized by microorganisms, enzymatically, chemically, or combinations thereof, from lipids and carbohydrates. A wide range of glycolipid structures is possible because of the large number of lipids and saccharides available and the numerous possible ways they can be combined. Despite the wide structural variations, however, glycolipid biosurfactants have a common feature. They are all nonionic surfactants with a polyhydroxy (saccharide) hydrophilic head group linked to a lipophilic hydrocarbon chain of a lipid by an ester, ether, amide, or other functional group. The numerous glycolipid biosurfactants are categorized into several subgroups using a variety of criteria (e.g., linking group chemistry, method of synthesis, etc.). In this section, examples of glycolipid biosurfactants from two subcategories will be discussed. The subgroups are (a) naturally occurring glycolipid biosurfactants and (b) chemically synthesized glycolipid biosurfactants.

11.2.1.1 Naturally Occurring Glycolipid Biosurfactants

There are several naturally occurring glycolipid biosurfactants. They are naturally produced by enzymes (fermentation) or synthesized by microorganisms from saccharide and lipid components. Two of the well-known naturally occurring glycolipid biosurfactants are rhamnolipids and sophorolipids, whose structures comprise rhamnose and sophorose, respectively, as their saccharide components [8–10,15–26]. A wide range of lipids are used in the enzymatic synthesis of these two glycolipids.

The basic structure of rhamnolipids includes one (monorhamnose) or two (dirhamnose) sugar units linked to a single lipid chain with a terminal carboxylic group. The lipid chain may have one or more branches of four or more carbons. An example of rhamnolipid structure is given in Figure 11.1 [19].

Sophorolipids contain two sugar units (disaccharide and sophorose) linked to a lipid that also has a terminal carboxylic group. Unlike rhamnolipids, the terminal carboxylic group in sophorolipids can react with a free hydroxyl group in the saccharide to form a lactose. Thus, sophorolipids can have acidic or lactonic structures, as shown in Figure 11.2 [19].

Rhamnolipids and sophorolipids have been proposed or are being developed for potential applications in a wide range of products [8–10,15–26], including personal care/cosmetics (shampoos, body wash, detergents, skin care products, and skin moisturizing agents), food additives and processing, encapsulants for foods and pharmaceuticals, cleaning products (dishwashing and laundry detergents), degreasing agents, soil and bioremediation, wastewater treatment, pesticides, disinfectants, enhanced oil recovery, medicinal/clinical efficacy (antimicrobial, antiviral antitumor, and skin disease treatment), and agricultural crop protection (herbicides, pesticides, and insecticides).

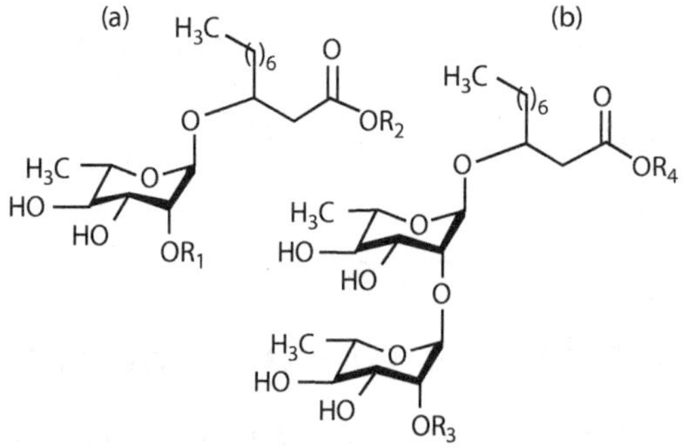

FIGURE 11.1 Generalized structure of monorhamnolipid (a) and dirhamnolipid (b). (From Kitamato, D., T. Morita, T. Fukuoka, and T. Imura. Self-assembling properties of glycolipid biosurfactants and their functional developments. In *Biobased Surfactants and Detergents Synthesis, Properties, and Applications*, edited by D.G. Hayes, D. Kitamoto, D.K.Y. Solaiman, and R.D. Ashby. AOCS Press, Urbana, IL. Chapter 9, 231–272, 2009.)

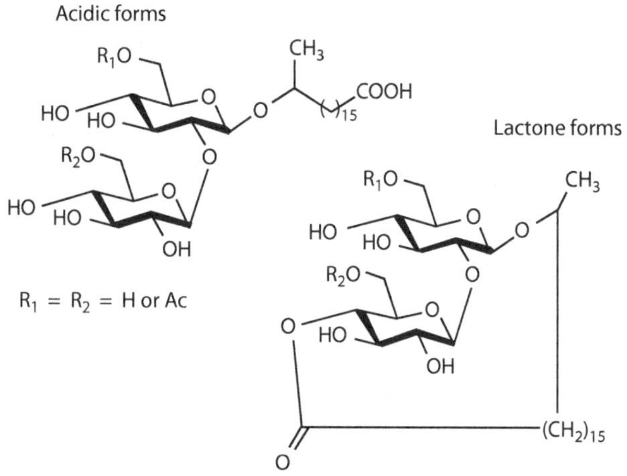

Acidic forms

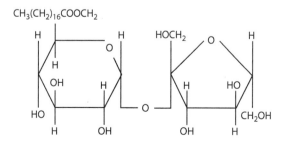

$R_1 = R_2 = H$ or Ac

Lactone forms

FIGURE 11.2 Structures of acidic (top) and lactone (bottom) sophorolipids. (From Kitamato, D., T. Morita, T. Fukuoka, and T. Imura. Self-assembling properties of glycolipid biosurfactants and their functional developments. In *Biobased Surfactants and Detergents Synthesis, Properties, and Applications*, edited by D.G. Hayes, D. Kitamoto, D.K.Y. Solaiman, and R.D. Ashby. AOCS Press, Urbana, IL. Chapter 9, 231–272, 2009.)

11.2.1.2 Synthetic Glycolipid Biosurfactants

The simplest of these are the saccharide fatty acid ester (SFE) biosurfactants [27–38]. They are commercially produced from lipids (fatty acids) and saccharides by chemical and enzymatic methods. SFE surfactants can be produced with the desired hydrophilic–lipophilic balance (HLB) by selecting the appropriate synthesis parameters, such as the structure of lipid (lauric, myristic, palmitic, stearic, oleic, etc.), the structure of the saccharide (glucose, sucrose, fructose, xylose, sorbitol, etc.), and the degree of substitution (DS). The DS refers to the degree of esterification reaction between the free hydroxyl groups in the saccharide and the fatty acid. For example, in the sucrose stearate of DS = 1, only one of the eight free hydroxyl groups of sucrose has reacted with stearic acid to give the structure shown in Figure 11.3 [29]. This molecule has high HLB because it has fewer lipophilic groups (which increases with increasing DS) and higher number of hydrophilic groups (which decreases with increasing DS). Accordingly, a mixture of sucrose esters rich in monoester (DS = 1) will have higher HLB than a mixture with multiple (DS > 1) ester substitutions. The effect of ester composition on the HLB of sucrose stearate is illustrated in Table 11.4 [29].

SFEs have a wide range of applications because they are safe, odorless, flavorless, nontoxic, non-irritant, and digestible materials [27–38]. As a result, they are widely used as ingredients in foods (emulsifiers, dispersants, and preservatives), detergent cleaners (for fruits, vegetables, baby bottles, etc.), cosmetics and personal care products (toothpastes, lotions, shampoos, and lipsticks), pharmaceutical (emulsifiers in liquids, plastic, semisolid and solid dosage forms, solubilizing agents,

$CH_3(CH_2)_{16}COOCH_2$

FIGURE 11.3 Structure of sucrose monostearate SFE. (From Mitsubishi Kagaku Foods Corp. Available at http://www.mfc.co.jp/english/whatsse.htm.)

TABLE 11.4

Effect of Composition of Sucrose Ester of Varying Degrees of Stearate Substitution on the HLB of Sucrose Stearate Ester Mixture

Percentage of Sucrose Ester with Stearate Substitution of								
Mono-	Di-	Tri-	Tetra-	Penta-	Hexa-	Hepta-	≥Octa-	HLB
2	2	8	12	28	22	18	8	1
16	32	30	14	8				3
36	46	16	2					7
58	30	12						11
73	22	5						16

Source: Mitsubishi Kagaku Foods Corp., http://www.mfc.co.jp/english/whatsse.htm.

skin-penetrating enhancers for ointments and anesthetics, and pore-forming agents), biobased and biocompatible emulsifiers, and bioactive ingredients (antitumor and antibiotic, and in agriculture as insecticides, herbicides, and plant growth inhibitors). They are also used in the processing of foods (e.g., viscosity control) and pharmaceuticals (e.g., lubricants in tablet production). Industrial uses, such as in textile and plastics, are being developed.

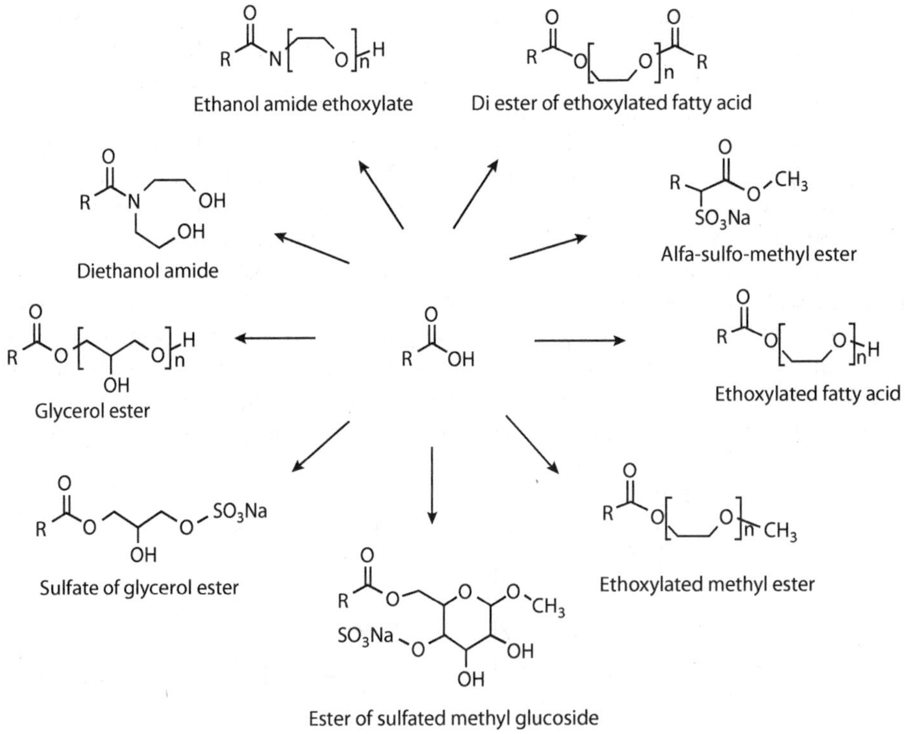

FIGURE 11.4 Examples of biosurfactants from transformation of fatty acids. (From Johansson, I., and M. Svensson, *Curr. Opin. Colloid Interface Sci.* 6, 178–188, 2001.)

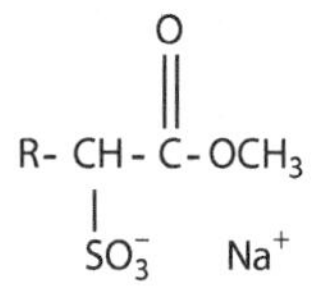

$$
\begin{array}{c}
\text{O} \\
\parallel \\
\text{R- CH- C- OCH}_3 \\
\mid \\
\text{SO}_3^- \quad \text{Na}^+
\end{array}
$$

FIGURE 11.5 Structure of methyl ester α-sulfonate obtained from vegetable oils.

11.2.2 BIOSURFACTANTS BASED ON LIPIDS

Lipids, particularly fatty acids and their methyl esters, can be chemically converted into a variety of biosurfactants. Some examples of biosurfactants from such transformation of fatty acids are shown in Figure 11.4 [3].

One important class of lipid-based biosurfactants is methyl ester sulfonate (MES), produced from the reaction of FAMEs with sulfur trioxide (SO_3). A generalized structure of vegetable oil based MES is given in Figure 11.5. This anionic biosurfactant is widely used in various cleaning products, most notably in laundry detergents, as potential replacement to the petroleum-based linear alkylbenzene sulfonates (LASs). LASs are the dominant anionic surfactants in laundry detergents. Recently, however, MES has become competitive with LAS in both cost and quality because of advances in manufacturing efficiency as well as increasing petroleum prices. MES biosurfactants have also become very competitive. MES has also been found to have superior tolerance to hard water compared with LAS and other nonionic surfactants used in laundry and dishwashing detergent formulations [39–44].

11.2.3 BIOSURFACTANTS BASED ON AMINO ACIDS

These are obtained by reacting amino acids with functionalized lipids [12–14,45–51]. The products are obtained by reaction on the acyl or amine group of the amino acid. The functionalized lipids include fatty acids, amines, and alcohols. The amino acids used include cystine, lysine, arginine, and so on. The biosurfactants of different structures (single chain, double chain, and gemini) and charge (cationic, anionic, nonionic, and amphoteric surfactants can be obtained depending on the

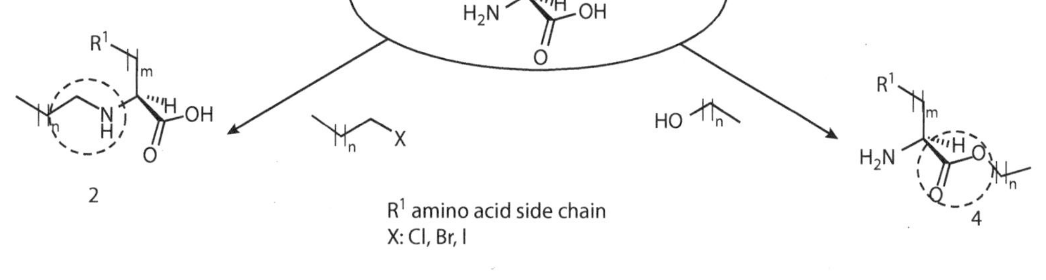

FIGURE 11.6 Generalized scheme for synthesis of amino acid–based biosurfactants. (From Infante, M.R. et al., *C.R. Chim.* 7, 583–592, 2004.)

free functional groups) can be produced. A generalized scheme for synthesis of biosurfactants from amino acids is given in Figure 11.6 [14].

Amino acid–based biosurfactants are generally safe, have low human and environmental toxicity, and have high surface activity and rate of biodegradation. As a result, these biosurfactants can be used as ingredients in foods, pharmaceuticals, and cosmetics [12–14,45,52–54]. Amino acid–based biosurfactants have been found to possess wide bactericidal activity, high biodegradability, and low toxicity profiles. Amino acid–based cationic biosurfactants have been found to be suitable for DNA compaction, which is an important requirement for application in gene and drug delivery therapeutics [11–14,45,52–62].

The surface properties of the amino acid–based biosurfactants [e.g., critical micelle concentrations (CMCs)] were found to be a function of their chain structures but not of the nature of the polar head (e.g., arginine or lysine) [12]. Thus, the CMC of the gemini biosurfactants was found to be three orders of magnitude lower than that of the linear biosurfactant of similar alkyl chain length [12]. In addition, the presence of a cationic charge in the amino acid provides antimicrobial activity to these compounds [12].

Another suggested application for amino acid–based biosurfactants is in dry cleaning as a replacement for perchloroethylene. In this process, the biobased surfactant in liquid carbon dioxide has been evaluated in dry-cleaning tests. It is reported that the amino acid–based biosurfactant in liquid carbon dioxide achieved the removal of 84% particulate soil and 98% nonparticulate soil relative to perchloroethylene, which is the industry standard [11].

11.2.4 Surface and Interfacial Properties of Biosurfactants Based on Small Molecules

Reported surface-active properties of biosurfactants include CMC, surface tension at CMC (γ_{CMC}) or minimum surface tension (γ_{min}), and minimum water–oil interfacial tension ($\gamma_{o/w}$). Hexadecane or kerosene oils were used in most of the interfacial tension investigations. Table 11.5 is a summary of some of the reported surface activity data for sophorose, rhamnose, amino acid, and FAME-based biosurfactants [13,15,18,40,63,64].

Abalos et al. [15] compared the CMC and γ_{CMC} of a single (pure) rhamnose biosurfactant structure with that of a mixture (blend) of six to seven rhamnose biosurfactant structures. The structure of the pure material consists of a dirhamnose hydrophilic unit (RR) attached to a C20 branched lipophilic unit. The structures of the rhamnose biosurfactants in the blend were a varying combination of monorhamnose or dirhamnose hydrophilic units attached to various structures (total carbons, unsaturations, and branching) of lipophilic units. Their data showed that the pure material had a much lower CMC than the blends. However, the γ_{CMC} of the pure material was slightly higher than that of the blends (Table 11.5).

Lang and Wullbrandt [18] investigated the surface properties of rhamnose biosurfactants whose structures were varied systematically. The structures of the biosurfactants were varied by varying the number of rhamnose units (R and RR) in the hydrophilic section and the number of isodecane units (i-C10, [i-C10]$_2$) in the lipophilic section of the molecule. Their CMC and γ_{CMC} results are summarized in Table 11.5. The observed CMC decreased with increasing lipophilic chain length, which correlates well with the decreasing water solubility of the biosurfactants as follows: RRi-C10 ≈ Ri-C10 > R[i-C10]$_2$ ≈ RR[i-C10]$_2$. All structures displayed similar γ_{CMC} values, which were in the range of 25 to 30 mN/m. Also, except the R[i-C10]$_2$ structure, the rhamnolipid biosurfactants displayed $\gamma_{o/w}$ of <1 mN/m.

Infante et al. [13] reported the CMC, γ_{CMC} and other surface properties of arginine-based biosurfactants with a varying combination of head group and lipophilic chain structures. The lipophilic group structures were linear alkyl chains of 10, 12, or 14 carbons. Three head group structures were obtained from synthesis with argentine: a monocationic amide (MA) from the reaction of the argentine amine group with a fatty acid, a dicationic amide (DA) from the reaction of the argentine carboxylic acid group with a fatty amine, and a dicationic ester (DE) from the reaction of the argentine

TABLE 11.5
Some Reported Surface and Interfacial Tensions of Biosurfactants Based on Small Molecules

Biosurfactant	Hydrophile	Lipophile	CMC (mg/L)	γ_{CMC} (mN/m)[a]	$\gamma_{o/w}$[b] (mN/m)
		Sophorolipids [63]			
	c	c	10–200	33–37	
		Rhamnolipids [15]			
Pure	RR[d]	[i-C10]$_2$[d]	106	28.8	
Mixed	e	e	150–230	26.8–27.3	
		Rhamnolipids [18]			
RRC10	RR[d]	i-C10[d]	200	30	<1
RC10	R[f]	i-C10[d]	200	25	<1
RRC10C10	RR[d]	[i-C10]$_2$[g]	10	27	<1
RC10C10	R[f]	[i-C10]$_2$[g]	20	26	4.0
		Arginine (Amino Acid)-Based Biosurfactants [13]			
MA	ArA$^+$ Cl$^-$[f]	C10, C12, C14	16, 5.8, 2	40, 32, 32	
DA	ArA^{2+} (Cl$^-$)$_2$[h]	C10, C12, C14	26, 1.8, 0.7	35, 37, 33	
DE	ArE^{2+} (Cl$^-$)$_2$[i]	C10, C12, C14	38, 13, 5	35, 34, 30	
		Alkyl Ester Sulfonates [45,64]			
MMS	Na$^+$SO$_3^-$	C14 (myristate)		39.9	10.9
MPS	Na$^+$SO$_3^-$	C16 (palmitate)		39.0	9.7
MSS	Na$^+$SO$_3^-$	C18 (stearate)		39.0	8.4
EMS	Na$^+$SO$_3^-$	C14 (myristate)		39.6	10.1
EPS	Na$^+$SO$_3^-$	C16 (palmitate)		36.4	8.0

Note: EMS, ethyl myristate sulfonate; EPS, ethyl palmitate sulfonate; MMS, methyl myristate sulfonate; MPS, methyl palmitate sulfonate; MSS, methyl stearate sulfonate; R, monorhamnose structure (see Figure 11.1); RR, dirhamnose structure (see Figure 11.1).

[a] Minimum surface tension or surface tension at CMC

[b] Minimum hexadecane–water interfacial tension.

[c] Various structures (see Figure 11.2).

[d] Branched C10 chain structure (see Figure 11.1).

[e] Mixture of six to seven biosurfactant structures with varying combinations of monorhamnose and dirhamnose hydrophiles and various chain structure lipophiles.

[f] Monocation amide.

[g] Two consecutive units of branched C10 chain structures (see Figure 11.1).

[h] Dication amide.

[i] Dication ester.

carboxylic acid group with a fatty alcohol. In all cases, the CMCs decreased with increasing lipophilic chain length. The DE biosurfactants displayed the highest CMCs at all lipophilic chain lengths. The γ_{CMC} values of DE and DA argentine biosurfactants of various chain lengths varied in a narrow range of 30 to 35 mN/m for DE and 33 to 37 mN/m for DA. The long-chain MA biosurfactants displayed an identical γ_{CMC} value of 32 mN/m, whereas the short-chain MA biosurfactant displayed the highest value of 40 mN/m.

Kapur et al. [40] compared the surface and interfacial tensions of various structures of alkyl ester α-sulfonates. These were methyl or ethyl esters of the following fatty acids: myristic (C10), palmitic (C12), and stearic (C18) acids. The authors did not specify the oil(s) used in the interfacial tension investigation. The surface tension results showed only a minor effect on γ_{CMC} value (overall range of 39.0–39.9 mN/m) due to changes in fatty acid chain length for the MES. On the other hand, for the ethyl ester sulfonates, a larger effect of fatty acid chain length on surface tension (overall range of γ_{CMC} 36.4–39.6 mN/m) was observed. The minimum interfacial tension ($\gamma_{o/w}$) of MES decreased with increasing fatty acid chain length within a narrow range of 8.4 to 10.9 mN/m. A similar effect of fatty acid chain length on the interfacial tension of ethyl ester sulfonates was observed (Table 11.5).

Stirton et al. [64] investigated the effect of alkyl chain length and univalent cation counterion (Li^+, Na^+, K^+) on the surface tension (γ_{CMC}) of α-sulfo-alkyl stearates. The surface tension decreased with increasing alkyl chain length in the following order: methyl > ethyl > n-propyl ~ i-propyl > n-butyl. The surface tension also decreased with the increasing size of the cation counterion in the following order: $Li^+ > Na^+ > K^+$ (Table 11.5).

11.3 BIOSURFACTANTS DERIVED FROM FARM-BASED BIOPOLYMERS

In this section, we discuss the biosurfactants obtained by the chemical modification of the major components of commodity crops such as corn, soybean, wheat, barley, and so on. Three components account for more than 95% of these grains: starches, 30% to 80%; proteins, 10% to 37%; and fats and oils, 1% to 20%. Table 11.6 compares the relative ratios of these components in selected commodity crops.

These three components have many common features. First, they have high molecular weights relative to those used for the synthesis of the biosurfactants discussed in the previous section. Starches and plant proteins are high–molecular weight biopolymers based on glucose and amino acids, respectively. The molecular weights of vegetable oils are not as high as those of starches and proteins but are still higher than the common lipids (fatty acids, FAME) used in biosurfactant synthesis discussed in the previous section. These three grain components are also insoluble in water. Thus, chemical modification is applied to make them soluble in water as well as to impart them surface-active properties. In this section, we will provide details on the chemical modification of each of these components as well as the investigation of their surface-active properties at the air–water and oil–water interfaces.

11.3.1 STARCH-BASED BIOSURFACTANTS

Starch is a biopolymer of 1,4-glucose units with linear or branched structures [65–68]. Starches comprising a linear chain of glucose units are called amylose, whereas those comprising branched structures are called amylopectin. Schematics of starch with amylose and amylopectin structures are given in Figure 11.7.

Starches from most plants comprise a mixture of these two structures, with a higher proportion of the branched amylopectin structure (75%–80%) than the linear amylose structure (20%–25%) [65,66]. Starches with such composition are called normal starches. On the other hand, starches from certain varieties of corn and rice comprise only a highly branched amylopectin structure and

TABLE 11.6
Major Components (%) of Seeds from Selected Commodity Crops

Cereal	Starch	Protein	Fat/Oil	β-Glucan
Corn	80.0	9.8	4.8	0.0
Barley	78.0	9.9	1.2	3.0
Oat	66.0	17.0	7.0	4.0
Wheat	52.0	23.0	9.7	0.0
Soy	30.0	37.0	20.0	0.0

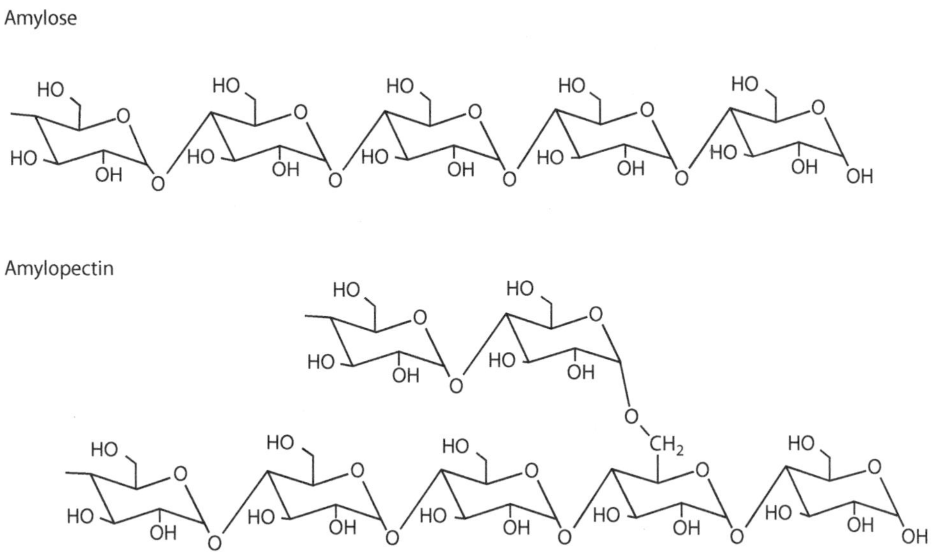

Amylose

Amylopectin

FIGURE 11.7   Structures of amylose and amylopectin starch.

are called waxy starches [67]. There are also corn varieties with a high proportion (>70%) of the linear starch structure amylose, which are referred to as high amylose starches [68].

In the investigations described here, normal, high amylose, and waxy starches obtained from commercial sources were used. The starches were chemically modified by reacting them with various anhydrides in a microwave reactor. The chemical reaction involves the esterification of free hydroxyl groups of the starch glucose units with fatty acids or anhydrides [69,70]. A schematic depicting such a reaction is shown in Figure 11.8.

A list of the starches used along with some pertinent data is given in Table 11.7. Esterifications were conducted using reagent grade acetic acid, acetic anhydrides (>99%), octenylsuccinic anhydrides

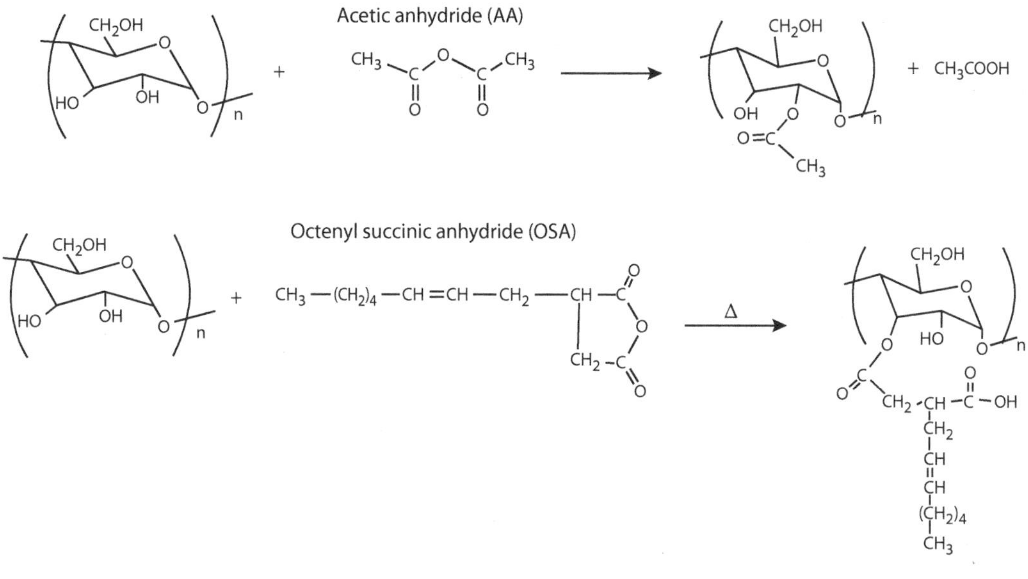

FIGURE 11.8   Schematic of chemical modification of starch by reaction with anhydrides.

TABLE 11.7
Amylose Composition of Starches

Starch	Amylose (%)
Normal corn	27
Maltodextrin	27
Waxy maize	1
High amylose corn	70

TABLE 11.8
Chemically Modified Starches Investigated

Starch	Anhydride		DS		Starch Ester
	Acetic	Succinic	Acetic	Succinic	
Maltodextrin[a]			0.00		
Normal corn[b]	Acetic		0.78		Normal acetates
Waxy maize[c]	Acetic		0.35		Waxy acetate 1
Waxy maize[c]	Acetic		0.70		Waxy acetate 2
High amylose corn[d]	Acetic		0.57		High amylose acetate
Waxy maize[c]	Acetic	Octenyl	0.36	0.046	Waxy acetate/octenylsuccinate 2
Waxy maize[c]	Acetic	Octenyl	0.80	0.030	Waxy acetate/octenylsuccinate 1
Waxy maize[c]	Acetic	Dodecenyl	0.31	0.022	Waxy acetate/dodecenyl succinate

[a] Star-dri, obtained from Tate & Lyle, Decatur, IL.
[b] Pure food grade (PFG), obtained from Tate & Lyle.
[c] 7350, obtained from Tate & Lyle.
[d] Hylon 7, obtained from National Starch, Bridgewater, NJ.

(97%), and dodecenyl succinic anhydrides (95%). The acetate was reacted with the starch by itself or in conjunction with one of the long-chain anhydrides. This allowed for the synthesis of starch acetate as well as starch with mixed ester substitution (acetate and octenylsuccinate or acetate and dodecenyl succinate). Starch esters with different degrees of ester substitution (DS) were prepared by varying the reaction conditions (time and temperature). A list of the synthesized starch esters is given in Table 11.8. A detailed description of the chemical modification of starch using a microwave reactor can be found elsewhere [69,70]. The chemically modified starches were readily soluble in water and used in surface and interfacial tension investigations.

11.3.2 Plant Protein–Based Biosurfactants

These biosurfactants were obtained by subjecting the protein enriched grain powders (also referred to as protein meals) through a multitude of processes, which can be divided into two broad categories of processing steps: (a) step 1, protein isolation, and (b) step 2, protein modification. In the protein isolation step 1, the protein meal is suspended in the appropriate buffer and maintained under specified conditions of temperature, pH, and time to induce dissolution of the proteins. The protein solution from such treatment (supernatant) is then separated by centrifugation and freeze-dried to produce the proteins. The product from step 1 is called protein isolate. In the protein modification step 2, the protein isolates are dissolved in the appropriate buffer and subjected to the required chemical modification. The chemically modified proteins are then isolated by freeze-drying

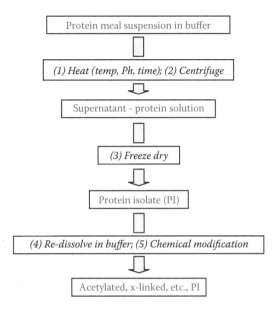

FIGURE 11.9 **(See color insert.)** Flowchart for the isolation and chemical modification of plant proteins.

and used in surface property investigations. A general flowchart for the isolation and chemical modification of plant proteins is given in Figure 11.9.

Plant proteins from barley, lupin, and oat were isolated, chemically modified, and their surface properties investigated. Chemical modification was conducted using one or more of the following methods: acetylation, succinylation, cross-linking, cross-linking and acetylation, and cross-linking and succinylation.

Lupin protein isolate (LPI) was obtained by treating milled lupin meal with phosphate buffers at different pH and temperatures [71]. Barley protein isolate (BPI) and oat protein isolate (OPI) were obtained by dissolving hexane defatted powders of barley and oat, respectively, in 0.015 N aqueous NaOH, followed by precipitation with 2 N aqueous HCl until pH 4.5 [72,73]. The precipitate protein isolates were recovered by centrifugation and were further chemically modified by one or more of the methods listed previously. A series of solutions of the protein isolates (LPI, BPI, and OPI) and their chemical modifications in phosphate buffer or 0.5 mol/L of aqueous NaCl were prepared and used in surface and interfacial tension investigations.

Detailed procedures for the isolation and chemical modification of the plant proteins from lupin, barley, and oat can be found elsewhere [71–73]. The protein isolates and their chemical modifications prepared and investigated are listed in Table 11.9.

11.3.3 Soybean Oil–Based Biosurfactants

Soybean oil makes up approximately 20% of the weight of soy grain, the rest being starch and proteins (Table 11.6). It is a triglyceride with more than 10 known fatty acid residues, of which only five are present in significant quantities [74,75]. A schematic of a triglyceride structure is given in Figure 11.10. Some structural features and composition of the fatty acid constituents of soybean oil are summarized in Table 11.10 [74].

Table 11.10 shows that the largest fatty acid component in soybean oil is linoleic acid (~54%), which has two *cis* double bonds in its structure. The second most abundant is oleic acid (~23%), which has a single *cis* double bond in its structure. Overall, unsaturated fatty acids account for up to 85% of the fatty acids in soybean oil.

TABLE 11.9

Isolated and Chemically Modified Plant Proteins

| Crop | Isolation Procedure | | Chemical Modification | Product |
	Temperature (°C)	pH		
Barley	Ambient	12	None	BPI
Barley	Ambient	12	Acetylation	BPI-A
Barley	Ambient	12	Cross-linking	BPI-X
Barley	Ambient	12	Acetylation and cross-linking	BPI-AX
Lupin	Ambient	4	None	LPI-A4
Lupin	Ambient	7	None	LPI-A7
Lupin	Ambient	8	None	LPI-A8
Lupin	100	4	None	LPI-B4
Lupin	100	7	None	LPI-B7
Lupin	100	8	None	LPI-B8
Oat	Ambient	12	None	OPI
Oat	Ambient	12	Acetylation	OPI-A
Oat	Ambient	12	Succinylation	OPI-S
Oat	Ambient	12	Cross-linking	OPI-X
Oat	Ambient	12	Acetylation and cross-linking	OPI-AX

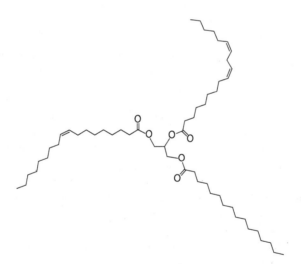

FIGURE 11.10   Schematic of a triglyceride structure such as that found in soybean oil.

Soybean oil is insoluble in water and can be adsorbed on polar surfaces [76] but has poor surface-active properties. Soybean oil can be used to produce water-soluble and surface-active materials by chemical modification. Such modification can be carried out at the two reactive sites of the triglyceride: the fatty acid double bonds and the ester groups that link the fatty acid with the glycerol (see Figure 11.10). Both reactive sites are used to synthesize water-soluble surface-active polysoaps using a three-step synthesis procedure as follows: step 1, conversion of the double bonds on the fatty acid chains to epoxides to produce epoxidized soybean oil (ESO); step 2, ring-opening polymerization of ESO using a cationic initiator to produce polymerized ESO (or PESO); step 3, hydrolysis and saponification of PESO into soybean-based polysoap (SOYP).

A schematic of reaction steps 2 and 3 is given in Figure 11.11. The catalyst used in step 2 (ring-opening PESO) was a purified and redistilled BF$_3$·diethyl etherate. The PESOs of varying molecular

TABLE 11.10
Fatty Acid Composition of Soybean Oil

Fatty Acid	Carbons	Double Bonds	% w/w
Myristic	14	0	0.1
Palmitic	16	0	10.6
Palmitoleic	16	1	0.1
Margaric	17	1	0.1
Stearic	18	0	4.0
Oleic	18	1	23.3
Linoleic	18	2	53.7
Linolenic	18	3	7.6
Arachidic	20	0	0.3
Behenic	22	0	0.3

Source: O'Brien, R.D. *Fats and Oils*, 9–10, 1998. Technomic Publishing Co., Inc., Lancaster, PA.

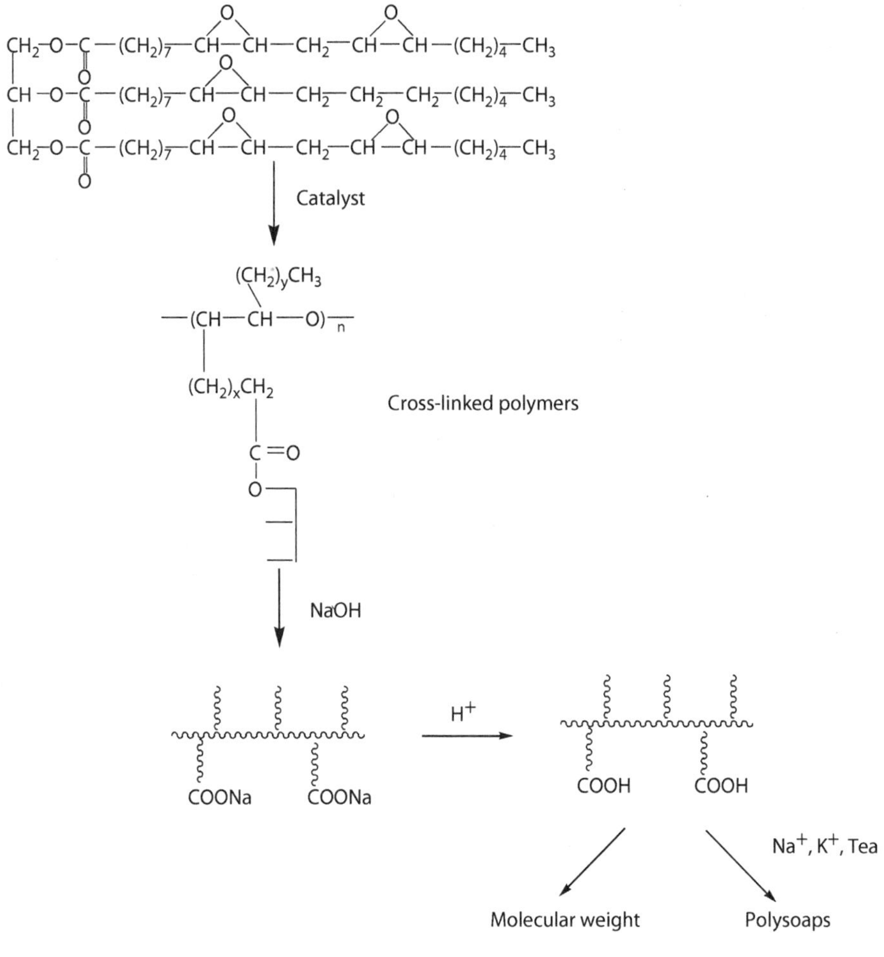

FIGURE 11.11   Ring-opening polymerization and hydrolysis of ESO.

TABLE 11.11

Molecular Weight of Soybean Oil–Based Polysoaps

Molecular Weight (g/mol)	Counterion
2439	TEA+, Na+, K+
2461	TEA+, Na+, K+
2615	TEA+, Na+, K+
3219	TEA+, Na+, K+

weights were obtained by varying the reaction temperature or the amount of catalyst (or both). The PESO products were positively identified using FTIR and NMR spectroscopy, and their molecular weights were determined using GPC.

In step 3, the ester groups in PESO were hydrolyzed by refluxing in 0.4 mol/L of aqueous NaOH and neutralized with 1.0 mol/L of HCl to precipitate the polycarboxylic acid product (see Figure 11.11). SOYPs with Na+ and K+ counterions were prepared by saponification of the polycarboxylic acid with one equivalent of aqueous NaOH and KOH, respectively. Polysoaps with triethanol ammonium counterion (TEA+) were prepared by saponification of the polycarboxylic acid with two equivalents of aqueous triethanol amine (TEA). The synthesized polysoap biosurfactants of varying molecular weights and counterions are listed in Table 11.11. More details about the synthesis, spectroscopic identification, and molecular weight determination of SOYP biosurfactant products are given elsewhere [77,78].

11.3.4 MEASUREMENT OF SURFACE AND INTERFACIAL TENSIONS

Deionized water, which was purified to a conductivity of 18.3 mΩ cm and then filtered on a 0.22 μL sterile disposable filter, was used to prepare aqueous solutions of the biosurfactants for surface and interfacial property investigations. The oils used in interfacial tension experiments were hexadecane (+99% anhydrous), canola, and soybean oil and were used without further purification.

Dynamic surface and interfacial tensions were measured using the axisymmetric drop shape analysis (ADSA) method [79]. In this method, surface or interfacial tensions are obtained by analyzing the change in the geometry of a pendant drop of a liquid suspended in air or in a second liquid, respectively, as a function of time. Initially, surface-active molecules diffuse from the bulk to the surface or interface of the pendant drop, which increases their concentration at the drop surface/interface and also changes the drop geometry as a function of time. An equilibrium surface/interface concentration and drop geometry is attained after a relatively long period. The geometry of a pendant drop is related to its surface or interfacial tension by the Bashforth–Adams equation, as follows [79–82]:

$$\gamma = (\Delta\rho g a^2)/H \tag{11.1}$$

where γ is the surface or interfacial tension, $\Delta\rho = (\rho_1 - \rho_2)$ is the difference in the densities of the drop and the medium, g is the gravitational acceleration, a is the maximum or equatorial diameter of drop, and H is the drop shape parameter. H is a function of the drop shape factor S and is obtained from drop geometry as follows:

$$S = b/a \tag{11.2}$$

where b is the diameter of the drop at height a from the bottom of the pendant drop.

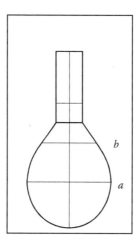

FIGURE 11.12 Schematic of a pendant drop geometry, where a is the maximum or equatorial drop diameter and b is the drop diameter at a height a from the bottom of the drop.

A schematic of a pendant drop depicting the dimensions a and b used in the Bashforth–Adams Equations 11.1 and 11.2 is illustrated in Figure 11.12.

Dynamic surface and interfacial tensions were measured on the FTA 200 automated goniometer (First Ten Angstroms, Portsmouth, VA). A schematic of the instrument configured for such measurement on a pendant drop is shown in Figure 11.13. The main features of the instrument are the automated pump that can be fitted with various syringe and needle sizes for the control of pendant drop formation; charge coupled device camera with a monitor for automated drop image viewing, adjusting, and capturing; and computer and software for setting experimental conditions (e.g., maximum drop volume, images per second, total run time, etc.), acquiring and storing droplet video images and numerical data, accurately measuring droplet dimensions (a and b in Figure 11.12) and using data to automatically calculate surface/interfacial tension, and displaying droplet images and measured surface/interfacial tension results as a function of time.

All surface and interfacial tension measurements were conducted at room temperature (23 ± 2°C). The instrument is calibrated with purified water and then checked by measuring the interfacial

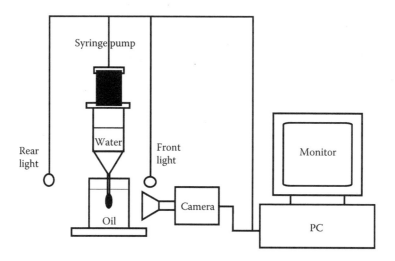

FIGURE 11.13 Schematic of an automated pendant drop goniometer.

tension between purified water and purified hexadecane (which has a value of 51.3 mN/m [82]) before measurements on biosurfactant solutions.

In a typical surface/interfacial tension experiment on the FTA200 instrument, a 10 mL disposable syringe equipped with a 17-gauge (1.499 mm OD) blunt disposable needle is used to generate a pendant drop of the solution in air (for surface tension) or oil (for interfacial tension) contained in a glass cuvette. The instrument is programmed to automatically deliver the specified volume of the aqueous solution at 1 µL/s to create the pendant drop, and also to automatically trigger image capture when the pump stops. The volume of aqueous solution to be automatically pumped is selected to generate the largest possible pendant drop that will not fall off before image acquisition is complete. All runs are programmed to acquire images at a rate of 0.067 s/image, with a predetermined trigger period multiplier to allow for a total of 35 images to be captured during the acquisition period. At the end of the acquisition period, each image is automatically analyzed and a plot of time versus surface or interfacial tension is automatically displayed. The data from each run were saved as both a spreadsheet and a movie. The spreadsheet contained the time and surface or interfacial tension for each image, and the movie contained each of the drop images as well as calibration information. Equilibrium surface or interfacial tension values are obtained by averaging the values at very long periods, where the surface or interfacial tension values showed little or no change with time. Repeat measurements (three to five) are conducted on each sample, and average values are used in data analysis.

11.3.5 SURFACE AND INTERFACIAL TENSIONS

11.3.5.1 Dynamic Surface Tension

A typical surface tension versus time plot using data from ADSA measurement is illustrated in Figure 11.14. The data in Figure 11.14 are for a 10% solution of normal corn starch acetate in purified water. Three distinct regions are observed: a short initial period of sharp surface tension decrease with time, an intermediate period of gradual surface tension decrease with time, and a final and a long period of more or less constant surface tension with time. These three regions correlate with the rate of biosurfactant diffusion from the bulk to the surface of the droplet. This diffusion rate is governed by the difference in the concentration of the biosurfactant at the surface versus

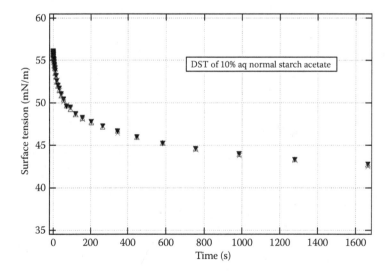

FIGURE 11.14   Multiple measurements of dynamic surface tension for 10% aqueous solution of normal corn starch acetate (DS = 0.33).

in the bulk. Initially, this difference is big because the surface concentration is very low, and a fast diffusion from the bulk to the surface occurs. With time, the surface concentration increases and the diffusion rate slows. At very long times, the surface concentration approaches its equilibrium value, and the diffusion rate approaches zero, that is, there will be no net diffusion of biosurfactant from the bulk to the surface. The equilibrium surface concentration corresponds to the maximum surface concentration for the given initial bulk concentration of this biosurfactant. Similar surface tension versus time profiles were obtained for all concentrations of aqueous solutions of starch, plant protein, and soybean-based biosurfactants investigated.

11.3.5.2 Equilibrium Surface Tension

The equilibrium surface tension is the minimum surface tension observed at very long times in surface tension versus time measurements such as that shown in Figure 11.14. The equilibrium surface tension value is estimated by averaging the last two or three data points from dynamic surface tension measurements. The equilibrium surface tension corresponds to the equilibrium or maximum biosurfactant concentration at the drop–air interface. Its value is dependent on the initial concentration and chemical structure of the biosurfactant.

Increasing the concentration of the biosurfactant in the droplet results in a decrease in the equilibrium surface tension, which levels off to a minimum value at very high concentrations. This decrease in surface tension occurs because of an increase in biosurfactant surface concentration, which is required to maintain the equilibrium between bulk and surface concentrations. The effect of concentration on equilibrium surface tension of selected starch ester, soy-based polysoap, and

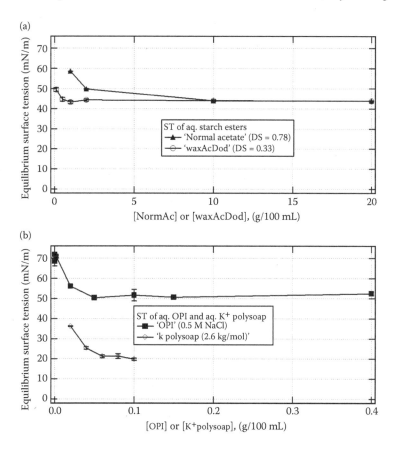

FIGURE 11.15 Effect of concentration on equilibrium surface tension of aqueous biosurfactant solutions: (a) aqueous starch esters; (b) aqueous OPI and aqueous K^+ polysoap.

plant protein–based biosurfactants is illustrated in Figure 11.15. As shown in the figure, increasing biosurfactant concentration initially causes a sharp reduction of the equilibrium surface tension, followed by a gradual decrease, and finally no effect. These observations are consistent with the effect of biosurfactant bulk concentration on biosurfactant equilibrium surface concentration. Thus, increasing biosurfactant concentration in bulk results in an increase of equilibrium surface concentration to reestablish the equilibrium between biosurfactants in bulk and at the surface. However, the increase of equilibrium surface concentration with increasing bulk concentration can occur only until full surface coverage is attained. Once the droplet surface is saturated with the biosurfactant, further increase of bulk concentration will cause no increase in surface concentration or decrease in surface tension, as shown in Figure 11.15.

Another factor that affects the equilibrium surface tension is the chemical structure of the biosurfactant. An examination of the effect of starch structure on equilibrium surface tension shows that dextrin, which is low molecular weight starch, was poorer at lowering the equilibrium surface tension than the starch acetates. Among the starches, the normal and waxy acetates were more effective than the high amylose acetates. A comparison of the effect of the ester structure on equilibrium surface tension of waxy starch esters shows that at low concentrations, the waxy acetates were less effective at lowering the equilibrium surface tension than the long-chain (octenyl and dodecenyl succinates) waxy esters. The data also show that DS had an effect on the low concentration equilibrium surface tension of long-chain waxy esters. However, at high concentrations, normal and waxy starch esters, regardless of the structure of the ester component, gave similar equilibrium surface tension.

An examination of the effect of molecular weight on the equilibrium surface tension of TEA^+ polysoaps shows that at low concentrations, the high–molecular weight TEA^+ polysoap was more effective at lowering the equilibrium surface tension. However, at high concentrations, the equilibrium surface tension was independent of molecular weight. The data also show that equilibrium surface tension of K^+ polysoaps was unaffected by molecular weight at any concentration. Thus, the equilibrium surface tensions of both TEA^+ and K^+ polysoaps at high concentrations were independent of molecular weight.

An analysis of the effect of protein source on the equilibrium surface tension of protein isolates from oat (OPI), barley (BPI), and lupin (LPI) shows that the equilibrium surface tension of the protein isolates was unaffected by the source of the protein.

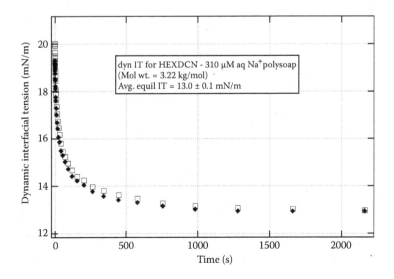

FIGURE 11.16 Multiple measurements of dynamic interfacial tension (IT) between hexadecane (hexdcn) and aqueous Na^+ polysoap (MW = 3.2 kg/mol, 310 μmol/L).

11.3.5.3 Dynamic Interfacial Tension

A typical interfacial tension versus time data between aqueous biosurfactant and hexadecane oil is illustrated in Figure 11.16. The data in Figure 11.16 are for the interface between 310 µmol/L aqueous Na$^+$ SOYP (molecular weight, 3.2 kg/mol) and hexadecane. As can be observed in Figure 11.16, the interfacial tension displays similar variation with time as was observed for surface tension of biosurfactants (see Figure 11.14). Three distinct regions are observed: a brief initial period of fast drop in interfacial tension, followed by a region of gradual decline in interfacial tension, and finally a region at very long periods where the interfacial tension is constant and does not change with time. As was the case with surface tension, the change in interfacial tension with time is related to the diffusion of the biosurfactant from the bulk of the aqueous droplet to the water/hexadecane interface. Initially, the concentration of biosurfactant at the interface is well below its equilibrium value because there is little or no biosurfactant at the interface when it is being formed. This condition triggers a fast diffusion of biosurfactant from bulk to the interface, which accounts for the fast drop in interfacial tension. As the biosurfactant interface concentration increases, the diffusion rate slows and so is the rate of decrease in interfacial tension. At very long time, the interface attains its equilibrium biosurfactant concentration, and no net diffusion of biosurfactant or reduction in interfacial tension is observed. Similar interfacial tension versus time profiles were observed for all biosurfactants investigated at all concentrations.

11.3.5.4 Equilibrium Interfacial Tension

The equilibrium interfacial tension is obtained from the dynamic interfacial tension data, such as shown in Figure 11.16. It is the minimum interfacial tension obtained by averaging two or more values at very long times, where the interfacial tension displays little or no change with time. The equilibrium interfacial tension is a function of the equilibrium concentration of the biosurfactant at the interface. Factors that affect the equilibrium interfacial concentration also affect the equilibrium interfacial tension. Two such factors are biosurfactant concentration and chemical structure.

Changing the concentration of biosurfactant in the aqueous phase also changes its concentration at the interface because the two concentrations are related by an equilibrium constant. Thus, increasing the concentration of biosurfactant in the aqueous phase increases the equilibrium interface concentration and, as a result, reduces the interfacial tension. This trend continues until the interface is

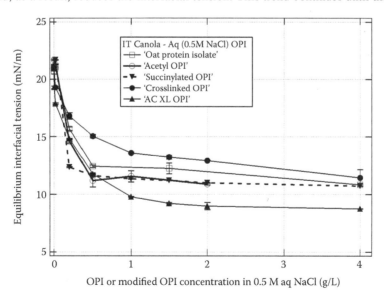

FIGURE 11.17 Effect of OPI and modified OPI concentration in 0.5 mol/L of aqueous NaCl on equilibrium interfacial tension with canola oil.

saturated (or full interface coverage is attained) with the biosurfactant. Once full interface coverage has been attained, further increase of biosurfactant concentration in the aqueous phase will not result in increased interfacial concentration of biosurfactant or decrease in interfacial tension.

The effect of concentration on equilibrium interfacial tension of modified OPI biosurfactants is illustrated in Figure 11.17. Similar results are observed for starch ester and SOYP biosurfactants. In all cases, the addition of biosurfactant in water results in a sharp reduction of the interfacial tension from the value for the pure water/oil values. Further increase of biosurfactant concentration results in gradual decrease in the equilibrium interfacial tension and reaches a constant value at very high concentrations.

The second factor that affects equilibrium interfacial tension is the structure of the biosurfactant. The structure of biosurfactants affects the way the biosurfactant molecules organize at the interface and, hence, the interface area per molecule. This will affect the biosurfactant concentration at the interface and, hence, its equilibrium interfacial tension. The analysis of equilibrium interfacial tension versus concentration data such as those shown in Figure 11.17 provides examples of the effect of biosurfactant structure on equilibrium interfacial tensions.

The data in Figure 11.17 compare the effect of OPI chemical structure on the equilibrium interfacial tension between water (with 0.5 mol/L NaCl) and canola oil. The data show that chemical modification had a mixed effect on the surface activity of OPI. Acetylation and succinylation resulted in lower equilibrium interfacial tensions, whereas cross-linking increased the equilibrium interfacial tensions relative to unmodified OPI. Moreover, a combined chemical modification involving both acetylation and cross-linking gave the best improvement of OPI surface properties, by giving the lowest equilibrium interfacial tension of the OPI biosurfactants.

An examination of the effect of acetate DS shows that at low concentrations (<10 g/100 mL), low DS waxy starch acetate biosurfactant was less effective than the other starch acetates in lowering the interfacial tension of water/hexadecane. However, at high concentrations, all starch acetates displayed similar equilibrium interfacial tensions.

Similarly, a comparison of the effect of ester chemical structure on the equilibrium interfacial tension between aqueous waxy starch ester biosurfactant and hexadecane indicates that the waxy starch with long-chain esters (octenyl and dodecenyl succinates) was more effective at lowering the interfacial tension than the short-chain acetate ester. The difference in equilibrium interfacial tension between these starch esters was more pronounced at low biosurfactant concentrations.

An examination of the effect of TEA$^+$ polysoap chemical structure on equilibrium interfacial tension shows that the equilibrium interfacial tension of TEA$^+$ polysoap is a function of the polysoap molecular weight. Thus, the lower the molecular weight of the TEA$^+$ polysoap, the more effective it is in lowering the equilibrium interfacial tension. The effect of molecular weight was more pronounced at low TEA$^+$ polysoap concentrations. A similar but less pronounced effect of molecular weight on equilibrium interfacial tension was observed for Na$^+$ polysoaps.

11.3.6 Surface Energy

As discussed earlier, the equilibrium biosurfactant surface tension at very high concentration is constant. It is also the minimum surface tension (γ_{min}) of the biosurfactant in the particular solvent. This is illustrated in the equilibrium surface tension versus concentration data of Figure 11.15.

This minimum surface tension is similar to the surface tension at the CMC (γ_{CMC}) or critical aggregate concentration (γ_{CAC}) reported for other biosurfactants (Table 11.5). At the CMC, the slope of the surface tension versus concentration curve abruptly changes. For most surfactants, the slope becomes zero, and as a result, γ_{CMC} is also γ_{min}. However, for some surfactants, the new slope is slightly less than zero, and the surface tension continues to decrease passed the CMC until it becomes constant at a much higher concentration. For such surfactants, (γ_{CMC}) > (γ_{min}).

We define γ_{min} as the surface tension at full surface coverage, that is, at a concentration when the air–solvent interface is completely replaced by the air–biosurfactant interface. As a result, γ_{min}

TABLE 11.12

Selected Measured Surface Energies of Soybean Oil, Starch, and Plant Protein–Based Biosurfactants

Biosurfactant	Surface Energy (mJ/m^2)
Plant Protein	
LPI-A4	40.2 ± 2.7
LPI-A7	38.8 ± 0.5
LPI-A8	42.0 ± 0.3
Starch Esters (DS)	
Waxy acetate 2 (0.7)	43.4 ± 1.1
Normal acetate (0.78)	43.0 ± 1.6
High amylose acetate (0.57)	48.4 ± 0.2
Waxy acetate/octenyl 1 (0.84)	41.4 ± 1.9
Waxy acetate/dodecyl (0.31)	44.0 ± 0.5
SOYPs (kg/mol, Counterion)	
2.44, TEA^+	26.4 ± 0.6
2.461, Na^+	26.4 ± 1.2
2.60, K^+	19.9 ± 0.6
3.20, TEA^+	23.9 ± 1.4

represents the surface energy of the biosurfactant. On the basis of this definition, the surface energies of the chemically modified starch, plant protein, and soybean oil biosurfactants were determined from surface tension versus concentration data, such as those shown in Figure 11.15. The representative surface energy values of the biosurfactants are given in Table 11.12.

A comparison of the surface energy data for modified starch biosurfactants showed that dextrin had the highest surface energy, which is expected based on the fact that its free hydroxyl groups are more readily accessible than those on starch. The other chemically modified starches, except high amylose starch acetate, displayed similar average surface energy values, within one standard deviation, regardless of the structure of the starch, ester group, or DS (Table 11.12). In addition, the surface energy values for these biosurfactants were within the 36 to 59 mJ/m^2 range reported for unmodified starch [83–86]. High amylose starch acetate gave the highest surface energy among the modified starch biosurfactants. This might be an indication that the predominantly linear structure in high amylose starch might favor orientation of more hydroxyl groups of the glucose units toward the interface than is the case in normal and waxy starches.

A comparison of the surface energies of modified protein-based biosurfactants shows that the surface energy of LPI varied, depending on the exact extraction procedure (pH and temperature). For modified BPI and OPI biosurfactants, the surface energy decreased with acetylation but was unchanged or slightly increased with cross-linking.

In general, the surface energies of soy-based polysoaps increased with decreasing molecular weight (Table 11.12).

11.3.7 INTERFACIAL ENERGY

As mentioned earlier, the equilibrium oil–water interfacial tension decreases with increasing biosurfactant concentration in water. At very high concentrations of biosurfactant, the equilibrium interfacial tension attains its minimum value and remains constant with further increase of biosurfactant concentration. This minimum interfacial tension corresponds to full coverage of the water–oil interface by the biosurfactant. This means that at these concentrations, the water–oil interface is completely replaced by the biosurfactant–oil interface. Thus, this minimum equilibrium interfacial

TABLE 11.13

Selected Values of Measured Versus Predicted[a] Biosurfactant–Oil Interfacial Energies

Biosurfactant—Oil	Measured	Interfacial Energy, mJ/m²			
		Calculated[a]			
		Ant	x_s^d	GM	HM
Plant Protein—Soybean[b]					
LPI-A4	6.6 ± 0.2	5.1	0.95	2.1	2.1
LPI-A7	4.4 ± 0.1	3.7	0.95	2.0	2.0
LPI-A8	3.8 ± 0.1	6.9	0.95	2.2	2.4
Starch Esters (DS)—Hexadecane[c]					
Waxy acetate/octenyl 1 (0.84)	10.8 ± 0.6	13.9	0.7	12.4	12.5
Normal acetate (0.78)	12 ± 0.6	15.5	0.7	13.0	13.0
Waxy acetate/dodecyl (0.31)	10.4 ± 0.8	16.5	0.7	13.3	13.4
Waxy acetate 2 (0.7)	13.5 ± 1.2	15.9	0.7	13.1	13.2
High amylose acetate (0.57)	14.2 ± 0.6	20.9	0.7	14.9	15.2
Soybean Polysoaps—Hexadecane (kg/mol, counterion)[c]					
2.44, K⁺	8.3 ± 0.6	0.8	0.6	12.2	13.7
2.46, TEA⁺	12.3 ± 0.1	6.3	0.6	11.3	13.9
2.60, Na⁺	13.0 ± 0.0	5.9	0.6	11.3	13.9
3.20, K⁺	12.7 ± 0.2	7.6	0.6	11.2	14.1

[a] Calculation methods: Ant, Antonoff's method (Equation 11.4); HM, harmonic mean method (Equation 11.10); GM, geometric mean method (Equation 11.11); x_s^d fraction of nonpolar component of biosurfactant surface energy (Equation 11.9).

[b] Surface energy of soybean oil, 35.1 mJ/m² [92].

[c] Surface energy of hexadecane, 27.5 mJ/m² [91].

tension corresponds to the interfacial energy between the biosurfactant and the oil. This interfacial energy, like the surface energy discussed previously, is obtained from equilibrium interfacial tension versus concentration data, such as those in Figure 11.17, by averaging the values at very high concentrations that display little or no change with concentration. This procedure was applied to determine the interfacial energies between the oils and the biosurfactants: hexadecane and starch esters, hexadecane and soy-based polysoaps, soybean or canola oils and protein isolates, soybean or canola oils, and modified protein isolates. Examples of the resulting interfacial energy data between biosurfactants and oils is summarized in Table 11.13.

11.3.8 POLARITY OF BIOSURFACTANTS

The surface energy of biosurfactants (γ_s) comprise polar $\left(\gamma_s^p\right)$ and dispersion $\left(\gamma_s^d\right)$ components related as follows [87,88]:

$$\gamma_s = \gamma_s^p + \gamma_s^d \tag{11.3}$$

The relative values of the polar and dispersion components are functions of the chemical structure of the biosurfactant. The degree of polarity has a profound effect on the performance of the biosurfactant in the various applications. Thus, the knowledge of the polar versus dispersion surface energy components provides insight into how chemical structural changes affect the performance of biosurfactants.

One method of estimating the polar and dispersion surface energy components is from the relationship between the surface energy of biosurfactant and the interfacial energy between the biosurfactant with another material, such as the various oils. There are several methods of relating surface and interfacial energies [87,88], and the simplest of these is Antonoff's method, given as follows [89,90]:

$$\gamma_{so} = |\gamma_s - \gamma_o| \tag{11.4}$$

where γ_s and γ_o are the surface energies of the surfactant and the oil and γ_{so} is the interfacial energy.

Unfortunately, Antonoff's method, although simple, does not involve polar and dispersion components in relating surface and interfacial energy and, hence, cannot be used to estimate the degree of polarity of biosurfactants.

However, there are two other methods that incorporate polar and dispersion surface energy components of the biosurfactant $\left(\gamma_s^p, \gamma_s^d\right)$ and the oil $\left(\gamma_o^p, \gamma_o^d\right)$ in relating surface and interfacial energies. These are called the geometric mean (GM) and the harmonic mean (HM) methods [87].

In the HM method, the relationship between surface and interfacial energy components is as follows:

$$\gamma_{so} = \gamma_s + \gamma_o - \frac{4\gamma_s^d\gamma_o^d}{\gamma_s^d + \gamma_o^d} - \frac{4\gamma_s^p\gamma_o^p}{\gamma_s^p + \gamma_o^p} \tag{11.5}$$

The corresponding relationship of the GM method is as follows:

$$\gamma_{so} = \gamma_s + \gamma_o - 2\left(\gamma_o^d\gamma_s^d\right)^{\frac{1}{2}} - 2\left(\gamma_o^p\gamma_s^p\right)^{\frac{1}{2}} \tag{11.6}$$

The application of the HM or the GM method to estimate the polar and dispersion surface energies of the biosurfactants will require, at the minimum, the values for the surface energies of the biosurfactant and oil, the interfacial energy between the two, and the polar and dispersion surface energy components for the oil. Even with these values known, there will be two unknowns $\left(\gamma_s^p \text{ and } \gamma_s^d\right)$ but only one equation, and as a result, the GM or HM equations cannot be solved.

One way around this problem is to generate interfacial energy data for each biosurfactant using two different oils with different polar and dispersion surface energy components. This will give, for each biosurfactant, two HM or two GM equations with two unknowns, and they can be solved to estimate the polar and dispersion surface energy components of the biosurfactants. The application of this procedure, however, will require selecting appropriate oils, conducting twice as many experiments, generating and processing twice as many data. This makes the procedure very cumbersome and undesirable.

In our work, we used a different approach to simplify the HM and GM equations to estimate the surface energy components, without resorting to conducting additional experiments. First, we selected oils (hexadecane, soybean, and canola) that have only dispersion surface energy component [88], that is, oils for which $\gamma_o^p = 0$ and $\gamma_o^d = \gamma_o$. This allowed for a great deal of simplification of the HM

$$\gamma_{so} = \gamma_s + \gamma_o - \frac{4\gamma_s^d\gamma_o}{\gamma_s^d + \gamma_o} \tag{11.7}$$

and GM equations,

$$\gamma_{so} = \gamma_s + \gamma_o - 2\left(\gamma_o \gamma_s^d\right)^{\frac{1}{2}} \tag{11.8}$$

Second, we defined γ_s^d and γ_s^p in Equations 11.5 and 11.6 in terms of fractional dispersity, x_s^d, as follows:

$$\gamma_S^d = x_S^d \gamma_S \text{ and } \gamma_S^p = \left(1 - x_S^d\right)\gamma_S \tag{11.9}$$

where $x_s^d = \gamma_s^d/\gamma_s$ is the fraction of nonpolar component of γ_s, and has values between 0 and 1 [87]. This led to simplified HM and GM equations for predicting interfacial energy as follows:

$$\gamma_{so} = \gamma_s + \gamma_o - \frac{4x_s^d \gamma_s \gamma_o}{x_s^d \gamma_s + \gamma_o} \tag{11.10}$$

$$\gamma_{so} = \gamma_s + \gamma_o - 2\left(x_s^d \gamma_s \gamma_o\right)^{\frac{1}{2}} \tag{11.11}$$

Equations 11.10 and 11.11 were used to predict the biosurfactant–oil interfacial energies from the analysis of measured biosurfactant surface energy (some of which is given in Table 11.12), literature or measured surface energy values for the oils (hexadecane, 27.0 mJ/m² [91]; soybean oil, used a measured value of 35.1 ± 0.3 mJ/m², which is within the 31.7–35.4 mJ/m² reported range [92,93]), and x_s^d as a fitting parameter with values in the range 0 to 1. The x_s^d values that predicted interfacial energies closest to measured values were used to estimate the polarities of the biosurfactants. The larger the x_s^d values, the less polar the biosurfactant.

Table 11.13 summarizes measured versus predicted biosurfactant–oil interfacial energies predicted using Equations 11.10 and 11.11, for x_s^d values that were varied in the range 0.6 to 0.95. Also included in Table 11.13 are interfacial energies predicted using Antonoff's method (Equation 11.4). The data in Table 11.13 show that Antonoff's method did well in predicting the interfacial energy between protein-based biosurfactants and soybean oil but underpredicted that between soy-based polysoaps and hexadecane, while overpredicting that between starch-based biosurfactants and hexadecane. The result also shows that for similar x_s^d values, the HM and the GM methods predicted similar values of interfacial energies. An examination of measured versus HM-predicted and GM-predicted interfacial energies indicates that the best fitted x_s^d values for the biosurfactants decrease in the following order: protein based > starch based > soy-based polysoaps. This means that the polarities of the biosurfactants decrease in the reverse order, that is, soy-based polysoaps > starch based > protein based.

11.4 PROSPECTS

The push to replace petroleum-based surfactants with biosurfactants will increase for two reasons. First is environmental, that is, to reduce the negative impact of surfactants on the environment. The common property of all biosurfactants is their ease of biodegradability. As a result, they will have much lower negative impact on the environment than petroleum-based surfactants. The second major reason is to reduce reliance on petroleum in favor of renewable farm-based raw materials for manufacture of surfactants. These two reasons will bring public and regulatory pressures on major manufacturers, wholesale suppliers, and retail chains to switch from petroleum-based surfactants to biosurfactants. This will result in the continued increase of market share of biosurfactants relative to petroleum-based surfactants. Currently, biosurfactants account for about one-third of the global surfactant market.

However, to increase market share, biosurfactant applications must expand into new areas. Currently, biosurfactants are mainly used in foods, cosmetics, cleaning/disinfecting, and pharmaceutical/medical markets. A major potential market area for biosurfactants is in materials and nanomaterials. Some, but very limited, work has been conducted to date at developing biosurfactants for materials applications [94–105]. This effort needs to be expanded and accelerated to have an impact on growing the biosurfactant market share.

Growing the biosurfactant market share will also require overcoming potential conflict related to the food versus nonfood application of renewable resources. Currently, such a debate is occurring about the food versus biofuel application of farm products. The various options being pursued to resolve the food versus biofuel debate will be of great interest to the food versus biosurfactant dilemma that may occur in the push to increases market share. This means, a great deal of effort will need to be exerted to develop biosurfactant raw materials from nonfood crops [106] or from by-products obtained from the processing of food crops. Some such effort has been reported [8,15,17,107,108], but more needs to be done.

11.5   SUMMARY AND CONCLUSIONS

Biosurfactants are surfactants that are primarily biodegradable, which is a major advantage over most nonbiodegradable surfactants in the market. Other than biodegradability, biosurfactants differ widely in their raw material base, method of manufacture, properties, and applications. In this chapter, biosurfactants are grouped into two broad categories based on the molecular weight of the starting materials. Biosurfactants synthesized from small molecules such as monosaccharides and disaccharides, lipids, amino acids, and their combinations are discussed first. Biosurfactants based on high–molecular weight farm-based biopolymers such as starches, plant proteins, and vegetable oils are grouped and discussed in the second half of the chapter.

Biosurfactants based on small molecules are obtained using chemical, enzymatic, and microbial methods or a combination of synthesis methods. Among the major categories are naturally occurring glycolipids, synthetic glycolipids, methyl ester α-sulfonates, and amino acid–based biosurfactants.

Because of their biological activity, excellent surface properties, and low toxicity, biosurfactants based on small molecules are used in a variety of industries. These include agriculture (crop protection), food (additives, processing, preserving, etc.), cosmetics/body care (skin, hair, oral, etc.), medical (antibiotic, antitumor, gene therapy, etc.), petroleum (enhanced oil recovery), cleaning (dishwashing and laundry detergents), and remediation (contaminated soil, water, beach, etc.).

Naturally occurring glycolipid biosurfactants are produced by fermentation processes. Two of the well-known such biosurfactants are rhamnolipids and sophorolipids. SFEs are glycolipids that are chemically produced from combinations of various structures of sugars and lipids. SFEs are of great commercial interest in a variety of industries, mostly in food, pharmaceutical, and medicinal products.

Methyl ester α-sulfonate biosurfactants are produced by chemical synthesis from FAMEs (obtained from vegetable oils in the biodiesel process) and are viewed as potential replacements to petroleum-based LASs in laundry and other cleaning applications. Amino acid–based biosurfactants are obtained by the reaction of amino acids and lipids with enzymatic, chemical, or combination methods. These are mostly cationic surfactants with a much lower toxicity than traditional cationic surfactants and have found a great deal of attention for application in drug and DNA delivery systems.

Biosurfactants based on starch, protein, and vegetable oils are of great interest because these three constitute the major components of commercial farm crops such as soybean, corn, wheat, barley, oat, and others. The successful development of value-added biosurfactants based on these components will have a positive effect on the farm economy.

Starch-based biosurfactants were produced by various degrees of chemical esterifications of normal, high amylose, and waxy corn starches with acetyl, dodecenyl succinyl, octenylsuccinyl, and mixed anhydrides. Plant protein–based biosurfactants were obtained by acetylation, cross-linking, succinylation, acetylation and cross-linking, and succinylation and cross-linking of protein isolates of oat, barley, and lupin. Soybean oil–based polysoaps were synthesized by ring-opening polymerization of ESO. Polysoaps varying in molecular weight and counterions were produced.

Aqueous solutions of biosurfactants derived from crop components were investigated for their surface and interfacial properties using the ADSA method. All of the farm-based biosurfactants were found to be effective at reducing the surface tension of water or aqueous salt or buffer solutions. They were also effective at reducing the interfacial tensions at water–hexadecane, water–soybean oil, and water–canola oil interfaces.

From the analysis of equilibrium surface tension versus concentration data, the surface energies of the biosurfactants were determined. The results showed that the surface energy of starch ester biosurfactants was within the range of reported values for unmodified starch. On the other hand, plant protein–based biosurfactants displayed a wide range of surface energy values that was dependent on the crop source of the protein isolate. Soybean oil–based polysoap biosurfactants displayed surface energy that was molecular weight–dependent.

The equilibrium interfacial tension between aqueous biosurfactants and the oils decreased with increasing biosurfactant concentration and leveled off at a constant minimum value at very high concentrations. From the analysis of equilibrium interfacial tension versus concentration data, the interfacial energies between the biosurfactants and the oils were determined.

The surface energy of the biosurfactants and the corresponding interfacial energies with the oils were analyzed using the GM and the HM methods. The analysis was used to estimate the relative polarities of these farm-based biosurfactants. The GM and the HM methods gave similar predictions. The result of the analysis showed that the polarity of the farm-based biosurfactants increased in the following order: modified plant protein isolates < modified starch esters < SOYPs. The predicted polarity trend is consistent with the expectation that the polyionic polysoap biosurfactants will be the most polar among the three categories of biosurfactants investigated.

Currently, biosurfactants account for about one-third of the global surfactant market. However, the prospect of expanding market share for biosurfactants at the expense of petroleum-based surfactants is rather good. There is a great deal of public and regulatory pressure on manufacturers, suppliers, and retail outlets to switch from petroleum based to biosurfactants. The driving force for these demands is the low negative impact of biosurfactants on the environment and also to reduce dependence on petroleum. However, meeting these demands and expanding market share will require the development of biosurfactants into new application areas, outside of the current food, cosmetics, pharmaceutical/medical, and cleaning/disinfecting markets.

The successful expansion of the market share of biosurfactants will also require solving the food versus nonfood conflict by developing raw materials from nonfood crops or by-products from the processing of food crops.

ACKNOWLEDGMENTS

The author expresses profound gratitude to colleagues who collaborated in the works cited in the second half of this chapter and to Megan Goers, Natalie Lafranzo, and Xiaozhou Cao for their help with the surface and interfacial tension measurements in the cited works.

REFERENCES

1. M. Patel. Surfactants based on renewable raw materials. Carbon dioxide reduction potential and policies and measures for the European Union. *J. Ind. Ecol.*, **7**(3–4), 47–62 (2004).

2. M.K. Patel, A. Theib and E. Worrell. Surfactant production and use in Germany: Resource requirements and CO2 emissions. *Resour. Conserv. Recyc.*, **25**, 61–78 (1999).

3. I. Johansson and M. Svensson. Surfactants based on fatty acids and other natural hydrophobes. *Curr. Opin. Colloid Interface Sci.*, **6**, 178–188 (2001).

4. D. Kitamato, T. Morita, T. Fukuoka and T. Imura. Self-assembling properties of glycolipid biosurfactants and their functional developments. In *Biobased Surfactants and Detergents Synthesis, Properties, and Applications*, D.G. Hayes, D. Kitamoto, D.K.Y. Solaiman and R.D. Ashby (Eds.), Chapter 9. AOCS Press, Urbana, IL, pp. 231–272 (2009).

5. A. Allam, J.-B. Behr, L. Dupont, V. Nardello-Rataj and R. Plantier-Royon. Synthesis, physico-chemical properties and complexing abilities of new amphiphilic ligands from D-galacturonic acid. *Carbohydr. Res.*, **345**, 731–739 (2010).

6. P. Laurent, H. Razafindralambo, B. Wathelet, C. Blecker, J.-P. Wathelet and M. Paquot. Synthesis and surface-active properties of uronic amide derivatives, surfactants from renewable organic raw materials. *J. Surfact. Deterg.*, **14**, 51–63 (2011).

7. T. Maugard, M. Remaud-Simeon, D. Petre and P. Monsan. Lipase-catalyzed synthesis of biosurfactants by transacylation of N-Methyl-Glucamine and fatty-acid methyl esters. *Tetrahedron*, **53**, 7629–7634 (1997).

8. M.V.P. Rocha, M.C.M. Souza, S.C.L. Benedicto, M.S. Bezerra, G.R. Macedo, G.A.S. Pinto and L.R.B. Goncalvez. Production of biosurfactant by *Pseudomonas aeruginosa* grown on cashew apple juice. *Appl. Biochem. Biotechnol.*, **136–140**, 185–194 (2007).

9. F.J.S. Oliveira, L. Vazquez, N.P. de Campos and F.P. de Franca. Production of rhamnolipids by a *Pseudomonas alcaligenes* strain. *Process Biochem.*, **44**, 383–389 (2009).

10. M.M. Müller, B. Hörmann, C. Syldatk and R. Hausmann. *Pseudomonas aeruginosa* PAO1 as a model for rhamnolipid production in bioreactor systems. *Appl. Microbiol. Biotechnol.*, **87**, 167–174 (2010).

11. M.J.E. van Roosmalen, G.F. Woerlee and G.J. Witkamp. Amino acid based surfactants for dry-cleaning with high-pressure carbon dioxide. *J. Supercrit. Fluids*, **32**, 243–254 (2004).

12. M.R. Infante, L. Pérez, M.C. Morán, R. Pons, M. Mitjans, M.P. Vinardell, M. T Garcia and A. Pinazo. Biocompatible surfactants from renewable hydrophiles. *Eur. J. Lipid Sci. Technol.*, **112**, 110–121 (2010).

13. M.R. Infante, L. Pérez, A Pinazo, P. Clapés, M.C. Morán, M. Angelet, M. T Garcia and M.P. Vinardell. Amino acid-based surfactants. *C. R. Chimie*, **7**, 583–592 (2004).

14. A. Pinazo, R. Pons, L. Pérez and M.R. Infante. Amino acids as raw material for biocompatible surfactants. *Ind. Eng. Chem. Res.*, **50**, 4805–4817 (2011).

15. A. Abalos, A. Pinazo, M.R. Infante, M. Casals, F. Garcia and A. Manresa. Physicochemical and antimicrobial properties of new rhamnolipids produced by *Pseudomonas aeruginosa* AT10 from soybean oil refinery wastes. *Langmuir*, **17**, 1367–1371 (2001).

16. A.M. Abdel-Mawgoud, F. Lépine and E. Déziel. Rhamnolipids: Diversity of structures, microbial origins and roles. *Appl. Microbiol. Biotechnol.*, **86**, 1323–1336 (2010).

17. C.J.B. De Lima, F.P. Franca, E.F.C. Servulo, M.M. Resende and V.L. Cardoso. Enhancement of rhamnolipid production in residual soybean oil by an isolated strain of *Pseudomonas aeruginosa*. *Appl. Biochem. Biotechnol.*, **136–140**, 463–470 (2007).

18. S. Lang and D. Wullbrandt. Rhamnose lipids—Biosynthesis, microbial production and application potential. *Appl. Microbiol. Biotechnol.*, **51**, 22–32 (1999).

19. N. Lourith and M. Kanlayavattanakul. Natural surfactants used in cosmetics: Glycolipids. *Int. J. Cosmet. Sci.*, **31**, 255–261 (2009).

20. R.D. Ashby, J.A. Zerkowski, D.K.Y. Solaiman and L.S. Liu. Biopolymer scaffolds for use in delivering antimicrobial sophorolipids to the acne-causing bacterium *Propionibacterium acnes*. *New Biotechnol.*, **28**(1), 24–30 (2011).

21. J.A. Casas, S.G. de Lara and F. Garcia-Ochoa. Optimization of a synthetic medium for *Candida bombicola* growth using factorial design of experiments. *Enzyme Microb. Technol.*, **21**, 221–229 (1997).

22. C. Jing, S. Xin, Z. Hui and Q. Yinbo. Production, structure elucidation and anticancer properties of sophorolipid from *Wickerhamiella domercqiae*. *Enzyme Microb. Technol.*, **39**, 501–506 (2006).

23. M. Kasture, S. Singh, P. Patel, P.A. Joy, A.A. Prabhune, C.V. Ramana and B.L.V. Prasad. Multiutility sophorolipids as nanoparticle capping agents: Synthesis of stable and water dispersible Co nanoparticles. *Langmuir*, **23**, 11409–11412 (2007).

24. I.N.A. Van Bogaert, J. Zhang and W. Soetaert. Microbial synthesis of sophorolipids. *Process Biochem.*, **46**, 821–833 (2011).

25. L. Zhang, P. Somasundaran, S.K. Singh, A.P. Felse and R. Gross. Synthesis and interfacial properties of sophorolipid derivatives. *Colloids Surf. A*, **240**, 75–82 (2004).

26. E. Zini, M. Gazzano, M. Scandola, S.R. Wallner and R.A. Gross. Glycolipid biomaterials: Solid-state properties of a poly(sophorolipid). *Macromolecules*, **41**, 7463–7468 (2008).

27. H. Seino, T. Uchibori, T. Nishitani and S. Inamasu. Enzymatic synthesis of carbohydrate esters of fatty acid (I) esterification of sucrose, glucose, fructose and sorbitol. *J. Am. Oil Chem. Soc.*, **61**, 1761–1765 (1984).

28. C.C. Akoh and B.G. Swanson. Synthesis and properties of alkyl glycoside and stachyose fatty acid polyesters. *J. Am. Oil Chem. Soc.*, **66**, 1295–1301 (1989).

29. Mitsubishi Kagaku Foods Corp. Available at http://www.mfc.co.jp/english/whatsse.htm. (2013).

30. K.S. Devulapalle, A.G. de Segura, M. Ferrer, M. Alcalde, G. Mooserz and F.J. Plou. Effect of carbohydrate fatty acid esters on *Streptococcus sobrinus* and glucosyltransferase activity. *Carbohydr. Res.*, **339**, 1029–1034 (2004).

31. S.W. Chang and J.F. Shaw. Biocatalysis for the production of carbohydrate esters. *New Biotechnol.*, **26**, 109–116 (2009).

32. Y. Shi, J. Lia and Y.-H. Chu. Enzyme-catalyzed regioselective synthesis of sucrose-based esters. *J. Chem. Technol. Biotechnol.*, **86**, 1457–1468 (2011).

33. H.-J. Altenbach, M. Berger, B. Jakob, R. Ihizane, K. Laumen, K. Lange, G. Machmuller, S. Muller and M.P. Schneider. Lipid modification of amino acids, carbohydrates and polyols. *Lipid Technol.*, **22**, 155–158 (2010).

34. M. Ferrer, F. Comelles, F.J. Plou, M.A. Cruces, G. Fuentes, J.L. Parra and A. Ballesteros. Comparative surface activities of di- and trisaccharide fatty acid esters. *Langmuir*, **18**, 667–673 (2002).

35. M. Ferrer, J. Soliveri, F.J. Plou, N. Lopez-Cortes, D. Reyes-Duarte, M. Christensen, J.L. Copa-Patino and A. Ballesteros. Synthesis of sugar esters in solvent mixtures by lipases from *Thermomyces lanuginosus* and *Candida antarctica* B, and their antimicrobial properties. *Enzyme Microb. Technol.*, **36**, 391–398 (2005).

36. R. Ye and D.G. Hayes. Lipase-catalyzed synthesis of saccharide-fatty acid esters utilizing solvent-free suspensions: Effect of acyl donors and acceptors, and enzyme activity retention. *J. Am. Oil Chem. Soc.*, **89**, 445–463 (2012).

37. R. Ye, S.-H. Pyo and D.G. Hayes. Lipase-catalyzed synthesis of saccharide–fatty acid esters using suspensions of saccharide crystals in solvent-free media. *J. Am. Oil Chem. Soc.*, **87**, 281–293 (2010).

38. S. Sabeder, M. Habulin and Z. Knez. Lipase-catalyzed synthesis of fatty acid fructose esters. *J. Food Eng.*, **77**, 880–886 (2006).

39. D. Martínez, G. Orozco, S. Rincón and I. Gil. Simulation and pre-feasibility analysis of the production process of α-methyl ester sulfonates (α-MES). *Bioresour. Technol.*, **101**, 8762–8771 (2010).

40. B.L. Kapur, J.M. Solomon and B.R. Bluestein. Summary of the technology for the manufacture of higher alpha-sulfo fatty acid esters. *J. Am. Oil Chem. Soc.*, **55**, 549–557 (1978).

41. L. Cohen, D.W. Roberts and C. Pratesi. Φ-Sulfo fatty methyl ester sulfonates (φ-MES): A novel anionic surfactant. *Chem. Eng. Trans.*, **21**, 1033–1038 (2010).

42. L. Cohen, F. Soto, A. Melgarejo and D.W. Roberts. Performance of φ-sulfo fatty methyl ester sulfonate versus linear alkylbenzene sulfonate, secondary alkane sulfonate and α-sulfo fatty methyl ester sulfonate. *J. Surfact. Deterg.*, **11**, 181–186 (2008).

43. S.N. Trivedi. Methyl ester sulfonates (MES)—An alternate surfactant from renewable natural resources. Chemithon Engineers Pvt. Ltd., Mumbai, December 20, 2005. Available at http://www.chemithon.com/Resources/pdfs/Technical_papers/Note%20On%20Methyl%20Esther%20Sulfonate.pdf.

44. W.B. Sheats and B.W. MacArthur. Methyl ester sulfonate products. Undated Technical Data, The Chemithon Corporation. Available at http://chemithon.com/Resources/pdfs/Technical_papers/Methyl%20Ester%20Sulfonate%20Products%205th%20Cesio%20v19,R1.pdf. (2009).

45. R.O. Britoa, E.F. Marques, S.G. Silva, M.L. do Vale, P. Gomes, M.J. Araujo, J.E. Rodriguez-Borges, M.R. Infante, M.T. Garcia, I. Ribosa, M.P. Vinardell and M. Mitjans. Physicochemical and toxicological properties of novel amino acid-based amphiphiles and their spontaneously formed catanionic vesicles. *Colloids Surf. B*, **72**, 80–87 (2009).

46. A. Pinazo, M. Angelet, R. Pons, M. Lozano, M.R. Infante and L. Perez. Lysine-bisglycidol conjugates as novel lysine cationic surfactants. *Langmuir*, **25**, 7803–7814 (2009).

47. C.M.C. Faustino, A.R.T. Calado and L. Garcia-Rio. New urea-based surfactants derived from α,ω-amino acids. *J. Phys. Chem. B*, **113**, 977–982 (2009).

48. C.M.C. Faustino, A.R.T. Calado and L. Garcia-Rio. Dimeric and monomeric surfactants derived from sulfur-containing amino acids. *J. Colloid Interface Sci.*, **351**, 472–477 (2010).

49. R. Bordes and K. Holmberg. Physical chemical characteristics of dicarboxylic amino acid-based surfactants. *Colloids Surf. A*, **391**, 32–41 (2011).

50. R. Mousli and A. Tazerouti. Synthesis and some surface properties of glycine-based surfactants. *J. Surfact. Deterg.*, **14**, 65–72 (2011).

51. S.G. Silva, J.E. Rodrıguez-Borges, E.F. Marques and M.L.C. do Vale. Towards novel efficient monomeric surfactants based on serine, tyrosine and 4-hydroxyproline: Synthesis and micellization properties. *Tetrahedron*, **65**, 4156–4164 (2009).

52. V. Jadhav, S. Maiti, A. Dasgupta, P.K. Das, R.S. Dias, M.G. Miguel and B. Lindman. Effect of the head-group geometry of amino acid-based cationic surfactants on interaction with plasmid DNA. *Biomacromolecules*, **9**, 1852–1859 (2008).

53. N. Lozano, A. Pinazo, R. Pons, L. Pérez and E.I. Franses. Surface tension and adsorption behavior of mixtures of diacyl glycerol arginine-based surfactants with DPPC and DMPC phospholipids. *Colloids Surf. B.*, **74**, 67–74 (2009).

54. N. Lozano, L. Perez, R. Pons and A. Pinazo. Diacyl glycerol arginine-based surfactants: Biological and physicochemical properties of catanionic formulations. *Amino Acids*, **40**, 721–729 (2011).

55. E.F. Marques, R.O. Brito, S.G. Silva, J.E. Rodrıguez-Borges, M.L. do Vale, P. Gomes, M.J. Araujo and O. Soderman. Spontaneous vesicle formation in catanionic mixtures of amino acid-based surfactants: Chain length symmetry effects. *Langmuir*, **24**, 11009–11017 (2008).

56. W. Mohammed-Saeid, J. Buse, I. Badea, R. Verrall and A. El-Anee. Mass spectrometric analysis of amino acid/di-peptide modified gemini surfactants used as gene delivery agents: Establishment of a universal mass spectrometric fingerprint. *Int. J. Mass Spectrom.*, **309**, 182–191 (2012).

57. D. Mondal, G.G. Zhanel and F. Schweizer. Synthesis and antibacterial properties of carbohydrate-templated lysine surfactants. *Carbohydr. Res.*, **346**, 588–594 (2011).

58. M.C. Moran, M.R. Infante, M.G. Miguel, B. Lindman and R. Pons. Novel biocompatible DNA gel particles. *Langmuir*, **26**, 10606–10613 (2010).

59. D.R. Nogueiraa, M. Mitjans, M.R. Infante and M.P. Vinardell. Comparative sensitivity of tumor and nontumor cell lines as a reliable approach for in vitro cytotoxicity screening of lysine-based surfactants with potential pharmaceutical applications. *Int. J. Pharm.*, **420**, 51–58 (2011).

60. L. Sanchez, V. Martınez, M.R. Infante, M. Mitjans and M.P. Vinardell. Hemolysis and antihemolysis induced by amino acid-based surfactants. *Toxicol. Lett.*, **169**, 177–184 (2007).

61. R.G. Shrestha, L.K. Shrestha, T. Matsunaga, M. Shibayama and K. Aramaki. Lipophilic tail architecture and molecular structure of neutralizing agent for the controlled rheology of viscoelastic fluid in amino acid-based anionic surfactant system. *Langmuir*, **27**, 2229–2236 (2011).

62. M.P. Vinardell, T. Benavides, M. Mitjans, M.R. Infante, P. Clapés and R. Clothier. Comparative evaluation of cytotoxicity and phototoxicity of mono and diacylglycerol amino acid-based surfactants. *Food Chem. Toxicol.*, **46**, 3837–3841 (2008).

63. N.M. Pinzon, Q. Zhang, S. Koganti and L.-K. Ju. Advances in bioprocess development of rhamnolipid and sophorolipid production. In *Biobased Surfactants and Detergents: Synthesis, Properties, and Applications*, D.G. Hayes, D. Kitamoto, D.K.Y. Solaiman and R.D. Ashby (Eds.), Chapter 4. AOCS Press, Urbana, IL, pp. 77–105 (2009).

64. A.J. Stirton, R.G. Bistline, J.K. Weil, W.C. Ault and E.W. Maurer. Sodium salts of alkyl esters of α-sulfo fatty acids. Wetting, lime soap dispersion, and related properties. *J. Am. Oil Chem. Soc.*, **39**, 128–131 (1962).

65. R.L. Whistler, J.N. BeMiller and E.F. Paschall. *Starch: Chemistry and Technology*. Academic Press, Orlando, FL, pp. 1–718 (1984).

66. T. Aberle, W. Burchard, G. Galinsky, R. Hanselmann, R.W. Klingler and E. Michel. Particularities in the structure of amylopectin, amylose and some of their derivatives in solution. *Macromol. Symp.*, **120**, 47–63 (1997).

67. J.A. Klavons, F.R. Dintzis and M.M. Millard. Hydrodynamic chromatography of waxy maize starch. *Cereal Chem.*, **74**, 832–836 (1997).

68. F.R. Senti. High amylose corn starch: Its production, properties and uses. In *Starch: Chemistry and Technology*, R.L. Whistler and E.F. Paschall (Eds.). Academic Press, New York, pp. 499–522 (1967).

69. R. Shogren and G. Biresaw. Surface properties of water soluble starch, starch acetates and starch acetates/alkenylsuccinates. *Colloids Surf. A*, **298**, 170–176 (2007).

70. G. Biresaw and R. Shogren. Friction properties of chemically modified starch. *J. Synth. Lubr.*, **25**, 17–30 (2008).

71. A.A. Mohamed, S.C. Peterson, M.P. Hojilla-Evangelista, D.J. Sessa, P. Rayas-Duarte and G. Biresaw. Effect of heat treatment and pH on the thermal, surface, and rheological properties of *Lupinus albus* protein. *J. Am. Oil Chem. Soc.*, **82**, 135–140 (2005).

72. A. Mohamed, M.P. Hojilla-Evangelista, S.C. Peterson and G. Biresaw. Barley protein isolate: Thermal, functional, rheological and surface properties. *J. Am. Oil Chem. Soc.*, **84**, 281–288 (2007).

73. A. Mohamed, G. Biresaw, J. Xu, M.P. Hojilla-Evangelista and P. Rayas-Duarte. Oats protein isolate: Thermal, rheological, surface and functional properties. *Food Res. Int.*, **42**, 107–114 (2009).

74. R.D. O'Brien. *Fats and Oils*. Technomic Publishing Co., Inc., Lancaster, PA, pp. 9–10 (1998).

75. N.O.V. Sonnatag. Composition and characteristics of individual fats and oils. In *Bailey's Industrial Fats and Oils*, 4th ed., Vol. 1, D. Swern (Ed.). John Wiley & Sons, New York, pp. 429–424 (1979).

76. G. Biresaw, A. Adhvaryu, S.Z. Erhan and C.J. Carriere. Friction and adsorption properties of normal and high oleic soybean oils. *J. Am. Oil Chem. Soc.*, **79**, 53–58 (2002).

77. G. Biresaw, Z.S. Liu and S.Z. Erhan. Investigation of the surface properties of polymeric soaps obtained by ring opening polymerization of epoxidized soybean oil. *J. Appl. Polym. Sci.*, **108**, 1976–1985 (2008).

78. Z. Liu and G. Biresaw. Synthesis of soybean oil based polymeric surfactants in supercritical carbon dioxide and investigation of its surface properties. *J. Agric. Food Chem.*, **59**, 1909–1917 (2011).

79. Y. Rotenberg, L. Boruvka and A.W. Neumann. Determination of surface tension and contact angle from the shapes of axisymmetric fluid interfaces. *J. Colloid Interface Sci.*, **93**, 169–183 (1983).

80. P.C. Hiemenz. *Principles of Colloid and Surface Chemistry*. Marcel Dekker, New York (1986).

81. A.W. Adamson and A.W. Gast. *Physical Chemistry of Surfaces*. Wiley, New York (1997).

82. C.J. van Oss, M.K. Chaudhury and R.J. Good. Monopolar surfaces. *Adv. Colloid Interface Sci.*, **28**, 35–64 (1987).

83. J.W. Lawton. Biodegradable coatings for thermoplastic starch. In *Cereals: Novel Uses and Processes*, G.M. Campbell, C. Webb and S.L. McKee (Eds.). Plenum Press, New York, pp. 43–47 (1997).

84. J.W. Lawton. Surface energy of extruded and jet cooked starch. *Starch*, **47**, 62–67 (1995).

85. I. Krycer, D.G. Pope and J.A. Hersey. An evaluation of tablet binding agents. Part I. Solution binders. *Powder Technol.*, **34**, 39–51 (1983).

86. G. Biresaw and C.J. Carriere. Correlation between mechanical adhesion and interfacial properties of starch/biodegradable polyester blends. *J. Polym. Sci., Polym. Phys.*, **39**, 920–923 (2001).

87. S. Wu. *Polymer Interface and Adhesion*. Marcel Dekker, New York (1982).

88. C.J. van Oss. *Interfacial Forces in Aqueous Media*. Marcel Dekker, New York (1994).

89. G. Antonoff. On the validity of Antonoff's rule. *J. Phys. Chem.*, **46**, 497–499 (1942).

90. G.N. Antonow. Sur la tension superficielle des solutins dans la zone critique. *J. Chim. Phys.*, **5**, 364 (1907).

91. J.J. Jasper. The surface tension of pure liquid compounds. *J. Phys. Chem. Ref. Data*, **1**, 841–1010 (1972).

92. R. Subramanian, S. Ichikawad, M. Nakajima, T. Kimurac and T. Maekawac. Characterization of phospholipid reverse micelles in relation to membrane processing of vegetable oils. *Eur. J. Lipid Sci. Technol.*, **103**, 93–97 (2001).

93. W.M. Formo. Physical properties of fats and oils. In *Bailey's Industrial Oil and Fat Products*, 4th ed., Vol. 1, D. Swern (Ed.). Wiley, New York, pp. 177–232 (1979).

94. R. Atluri, N. Hedin and A.E. Garcia-Bennett. Nonsurfactant supramolecular synthesis of ordered mesoporous silica. *J. Am. Chem. Soc.*, **131**, 3189–3191 (2009).

95. D. Chen, Y. Zang and S. Su. Effect of polymerically-modified clay structure on morphology and properties of UV-cured EA/Clay nanocomposites. *J. Appl. Polym. Sci.*, **116**, 1278–1283 (2010).

96. Y.A. Diaz-Fernandez, E. Mottini, L. Pasotti, E.F. Craparo, G. Giammona, G. Cavallaro and P. Pallavicini. Multicomponent polymeric micelles based on polyaspartamide as tunable fluorescent pH-window biosensors. *Biosens. Bioelectron.*, **26**, 29–35 (2010).

97. S.K. Filippov, L. Starovoytova, C. Konak, M. Hruby, H. Mackova, G. Karlsson and P. Stepanek. pH sensitive polymer nanoparticles: Effect of hydrophobicity on self-assembly. *Langmuir*, **26**, 14450–14457 (2010).

98. S.H. Kim, T.Y. Olson, J.H. Satcher and T.Y.-J. Han. Hierarchical ZnO structures templated with amino acid based surfactants. *Microporous Mesoporous Mater.*, **151**, 64–69 (2012).

99. H. Noritomi, N. Igari, K. Kagitani, Y. Umezawa, Y. Muratsubaki and S. Kato. Synthesis and size control of silver nanoparticles using reverse micelles of sucrose fatty acid esters. *Colloid Polym. Sci.*, **288**, 887–891 (2010).

100. X. Zhang and L.S. Loo. Morphology and mechanical properties of a novel amorphous polyamide/nanoclay nanocomposite. *J. Polym. Sci., Part B: Polym. Phys.*, **46**, 2605–2617 (2008).

101. C. Tano, S.-Y. Son, J. Furukawa, T. Furuike and N. Sakairi. Dodecyl thioglycopyranoside sulfates: Novel sugar-based surfactants for enantiomeric separations by micellar electrokinetic capillary chromatography. *Electrophoresis*, **29**, 2869–2875 (2008).

102. C. Tano, S.-Y. Son, J. Furukawa, T. Furuike and N. Sakairi. Enantiomeric separation by MEKC using dodecyl thioglycoside surfactants: Importance of an equatorially oriented hydroxy group at C-2 position in separation of dansylated amino acids. *Electrophoresis*, **30**, 2743–2746 (2009).

103. B. Thomas, N. Baccile, S. Masse, C. Rondel, I. Alric, R. Valentin, Z. Mouloungui, F. Babonneau and T. Coradin. Mesostructured silica from amino acid-based surfactant formulations and sodium silicate at neutral pH. *J. Sol-Gel Sci. Technol.*, **58**, 170–174 (2011).

104. Y. Wen, J. Xu, H. He, B. Lu, Y. Li and B. Dong. Electrochemical polymerization of 3,4-ethylene-dioxythiophene in aqueous micellar solution containing biocompatible amino acid-based surfactant. *J. Electroanal. Chem.*, **634**, 49–58 (2009).

105. P. Worakitsiri, O. Pornsunthorntawee, T. Thanpitcha, S. Chavadej, C. Weder and R. Rujiravanit. Synthesis of polyaniline nanofibers and nanotubes via rhamnolipid biosurfactant templating. *Synth. Met.*, **161**, 298–306 (2011).

106. S.R. Morcelle, C.S. Liggieri, M.A. Bruno, N. Priolo and P. Clapés. Screening of plant peptidases for the synthesis of arginine-based surfactants. *J. Mol. Catal. B.*, **57**, 177–182 (2009).

107. R.D. Ashby, D.K.Y. Solaiman and J.A. Zerkowski. Production and modification of sophorolipids from agricultural feedstocks. In *Biobased Surfactants and Detergents Synthesis, Properties, and Applications*, D.G. Hayes, D. Kitamoto, D.K.Y. Solaiman and R.D. Ashby (Eds.), Chapter 2. AOCS Press, Urbana, IL, pp. 29–49 (2009).

108. S.K. Karmee. Lipase catalyzed synthesis of ester-based surfactants from biomass derivatives. *Biofuels, Bioprod. Biorefin.*, **2**, 144–154 (2008).

12 Microbially Derived Biosurfactants

Sources, Design, and Structure-Property Relationships

Ponisseril Somasundaran, Partha Patra,
John D. Albino, and Indumathi M. Nambi

CONTENTS

12.1 INTRODUCTION

Surfactants are used for numerous applications in food industries, pharmaceuticals, personal and home care, and mineral processing and in many other sectors. Most of the surfactants used today are petroleum based, and their toxicity impact is widely debated. Besides the toxicity impact, the synthesis routes of surfactants can contribute to the increase in the carbon footprint and the emission of greenhouse gases. Thus, the development of environmentally benign reagents and processes requires surfactants that are biocompatible, biodegradable, nontoxic, and most importantly, produced by a greener route.

12.1.1 Criteria for "Greener" Surfactant Systems

Among the different surfactant systems, there are no currently established comparison criteria in terms of the degree of "greenness" and sustainable usage in the future. It has been recognized by various governments, industries, and the public that sustainability requires a balance in economy, ecology, and societal values. The idea of sustainability, or being green, has been characterized by several related but fundamentally different parameters, such as carbon footprint, biodegradability, toxicity, renewability, and so on. Certainly, these individual metrics cannot provide a full understanding of the impact of a surfactant system from "cradle to grave to cradle" approach. The life cycle impact analysis and the life cycle management developed recently by ecologists can yield a more comprehensive picture of the sustainability of a manufactured material. In this regard, one of the safest approaches is to start with a greener source and to follow a greener route of obtaining the final product.

12.1.2 Greener Surfactants

As mentioned earlier, no standard definition currently exists for "green surfactants" that is generally acceptable to both academic researchers and industrial practitioners. However, with regard to the source of surfactants and those which are not petroleum based, the scope of green surfactants includes two major classes of materials, as discussed.

12.1.2.1 Chemically Synthesized Surfactants from Renewable Sources

These surfactants are chemically synthesized based on renewable raw materials, that is, from non-petroleum natural sources, such as plants and animal fats.

12.1.2.2 Biosurfactants

Biosurfactants are a structurally diverse group of surface-active molecules synthesized by microorganisms, that is, bacteria, yeast, and fungi. They exist in either the cell membrane or the extracellular region. In recent years, much attention has been directed toward biosurfactants because of the broad range of their functional properties and the diverse synthesis capabilities of microbes. The structural analysis of biosurfactants has also given an insight into the possibilities for their chemical synthesis. Biosurfactants are likely to gain wide market acceptance because they are readily biodegradable and are often recognized as significantly less toxic than their chemically synthesized counterparts. The important effects of biosurfactants are as follows: increase in surface area available for reactions, bioavailability of hydrophobic water-insoluble substrates, heavy metal binding, bacterial pathogenesis, and quorum sensing and biofilm formation. Some biosurfactants have also proven to be effective at extreme temperature, pH, and salinity [1–8].

In this chapter, the structure–property relationships of a few bacterially derived surfactants will be addressed.

12.2 GREENER SURFACTANTS DERIVED FROM BACTERIA

Commonly used biosurfactants include hydroxylated and cross-linked fatty acids, glycolipids, polysaccharide–lipid complexes, lipoprotein–lipopeptide complexes, phospholipids, and even the complete cell surface itself. The following are some of the common biosurfactants [9].

12.2.1 Emulsan

Emulsan is an anionic polysaccharide produced by the gram-negative bacterium *Acinetobacter calcoaceticus* strain RAG-1 from a variety of hydrocarbon feed sources, including crude oil (Figure 12.1). Emulsan causes only a little reduction in surface tension at an oil–water interface (10 mM/m) but binds tightly to oil surfaces and protects oil droplets from coalescence [10,11].

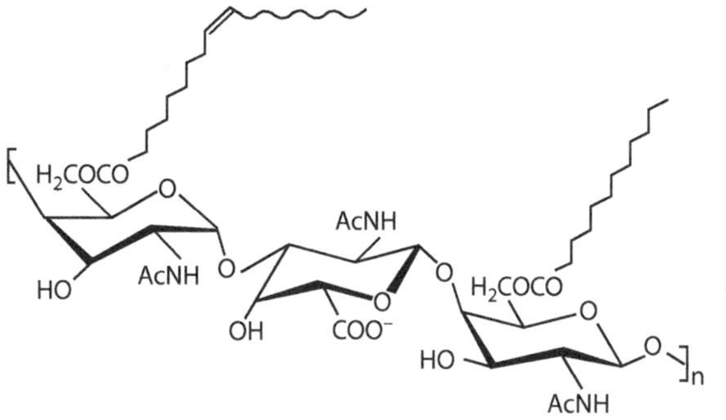

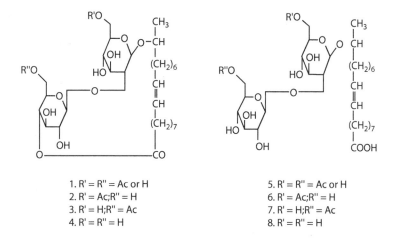

FIGURE 12.1 Structure of emulsan.

12.2.2 SOPHOROLIPIDS

Sophorolipids (Figure 12.2) are extracellular glycolipids produced mainly by the *Candida bombicola* species during their stationary phase. There has been considerable interest in the physiology of sophorolipid biosynthesis by the yeast strains of the *Candida* genus. It is very likely that the physiological role in sophorolipid production is related to the regulation of energy metabolism [12–15].

12.2.3 RHAMNOLIPIDS

Rhamnolipids, in which one or two molecules of rhamnose are linked to one or two molecules of β-hydroxydecanoic acid, are a family of microbial glycolipids that have thus far received the most attention. *Pseudomonas aeruginosa* produces rhamnolipids in two forms: L-rhamnosyl-L-rhamnosyl-β-hydroxydecanoyl-β-hydroxydecanoate and L-rhamnosyl-β-hydroxydecanoyl-β-hydroxydecanoate (Figure 12.3). The fermentative production of rhamnolipids having one β-hydroxydecanoic acid with one or two rhamnose lipids, methyl ester derivatives, and rhamnolipids with alternative fatty acid chains have also been reported. Several types of rhamnolipids produced by *P. aeruginosa* have been characterized [16,17].

1. R' = R" = Ac or H
2. R' = Ac;R" = H
3. R' = H;R" = Ac
4. R' = R" = H

5. R' = R" = Ac or H
6. R' = Ac;R" = H
7. R' = H;R" = Ac
8. R' = R" = H

FIGURE 12.2 Natural sophorolipids are composed of a mixture that contains the lactonic and ring-opened forms as well as various degrees of acetylation at the primary positions. (From Zhang, L. et al., *Colloids Surf. A* 240, 75–82, 2004. With permission.)

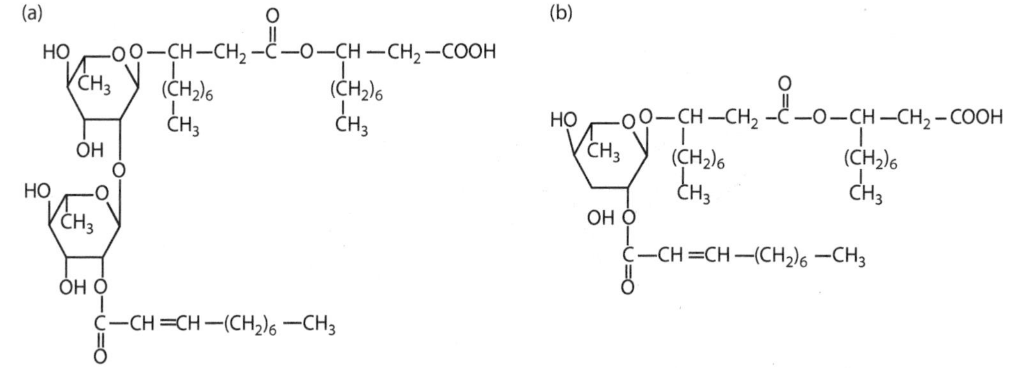

(a)

(b)

FIGURE 12.3 Structures of rhamnolipids from *P. aeruginosa*. (a) Two rhamnose head group and (b) one rhamnose head group. (From Zhang, L. et al., *Colloids Surf. A* 240, 75–82, 2004. With permission.)

12.2.4 LIPOPEPTIDES

Lipopeptides are extracellular secretions mainly from *Bacillus* spp., which contains numerous peptides linked to different fatty acid chains. More than 100 compounds belonging to this particular family of biosurfactant have been discovered since 1960. They are mainly classified into four main categories: the surfactins, the iturins, the fengycins or plipastatins, and the kurstakins [22]. Surfactin is highly water-soluble and has excellent biophysical properties that are important for its effective use [such as a low critical micelle concentration (cmc), 1.3 mM]. It is well established in the literature that surfactin is an effective oil dispersant [20]. In this chapter, the synthesis of surfactin by using molecular biology techniques and its surfactant characteristics are discussed.

Modular Genetics, Inc. (Woburn, MA) has developed an automated system for designing and generating engineered microorganisms. For example, Modular can rapidly produce *Bacillus subtilis* strains with specifically designed modifications of their genomic DNA. These modifications can include anything from changing a single base pair to deleting large regions of the chromosome or building large sets of "heterologous genes." This technology is used at Modular to produce engineered strains that synthesize novel lipopeptides. Each of these lipopeptides is a variant of the naturally occurring molecule surfactin, which is known to act as a powerful surfactant and oil dispersant [20]. Surfactin is synthesized by a peptide-synthetase enzyme, *srfA* operon. The three genes that encode these three synthetase subunits are shown in Figure 12.4.

Modular has automated the process of design and construction of engineered peptide synthetase enzymes and has the ability to rapidly create novel molecules using automated robotic equipment. The integration of computer design tools with robotic manipulation enables the use of cellular and molecular biology to produce new chemicals and materials.

The lipopeptide is highly water-soluble and has excellent biophysical properties that are important for its effective use (such as a low cmc, 1.3 mM).

Figure 12.5 shows a few of the novel lipopeptides that have been generated by Modular using automated, microbial strain engineering. In this particular case, all the variant molecules have amino acid substitutions at position 2 of the amino acid component of surfactin. By testing this set of compounds, along with sets of compounds with specific amino acid substitutions at other positions, researchers expect to identify amino acid positions (and particular amino acids at those positions) that are critical for it to function as a biodispersant and impart nontoxicity. On the other hand, it is well established in the literature that surfactin (Figure 12.6) is an effective oil dispersant [20].

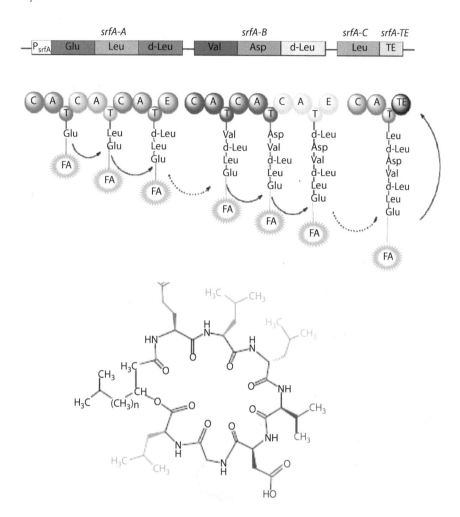

FIGURE 12.4 (**See color insert.**) Surfactin is a lipopeptide that is produced naturally by *B. subtilis*. It is composed of a fatty acid chain that is 12 to 17 carbons in length and has seven amino acid cyclic peptides.

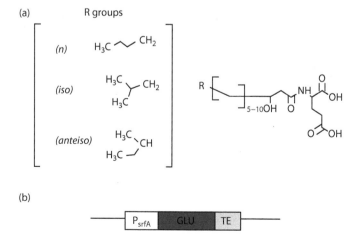

FIGURE 12.5 Modular's acyl amino acid. (a) Chemical structure of the acyl amino acid with glutamate attached to a lipid moiety. (b) Modular structure of the modified surfactin synthetase operon.

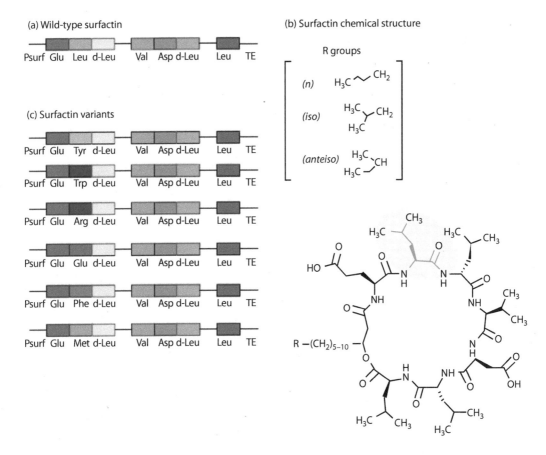

FIGURE 12.6 **(See color insert.)** (a) Wild-type surfactin synthase operon; (b) chemical structure of wild-type surfactin; (c) six of the surfactin analogues in Modular's collection have substitutions at position 2 of the cyclic amino acid.

12.3 STRUCTURE–PROPERTY RELATIONSHIPS OF BIOSURFACTANTS

12.3.1 INTERFACIAL PROPERTIES OF SOPHOROLIPIDS

Research with biosurfactants has led to a good understanding of the relationship between the alkyl ester chain length attached to the carboxyl of the sophorolipid lipid moiety and their interfacial properties (Figure 12.7). The *n*-alkyl chain length variation investigated includes C1, C2, C3, C4, and C6. From Figure 12.7, a relatively high activity of butyl and hexyl reagents was observed. cmc and minimum surface tension exhibited a normal inverse relationship with the alkyl ester chain length, and as usual, the cmc decreased to 1/2 per additional CH_2 group for the methyl, ethyl, and propyl series of chain lengths.

The surface tension results have been corroborated by fluorescence experiments. It can be seen from Figure 12.8 that all the esters are capable of forming hydrophobic domains (higher I_3/I_1), that is, an increase in polarity parameter—indicator of localization of polar styrene molecules in hydrophobic micellar domain—further indicating their potential to solubilize oleophilic materials (drugs, dyes, and oily dirt).

The adsorption of sophorolipid alkyl esters on hydrophilic solids was studied to determine the extent of lateral association in the adsorbed film. During the investigation of the surface activity of these compounds, the adsorption of these surfactants was surprisingly found to be higher on alumina than on silica (Figure 12.9). This selective adsorption behavior on hydrophilic solids is similar to that of sugar-based

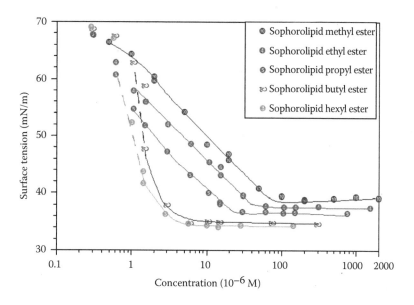

FIGURE 12.7 **(See color insert.)** Surface tension of esters of sophorolipids. (From Zhang, L. et al., *Colloids Surf. A* 240, 75–82, 2004. With permission.)

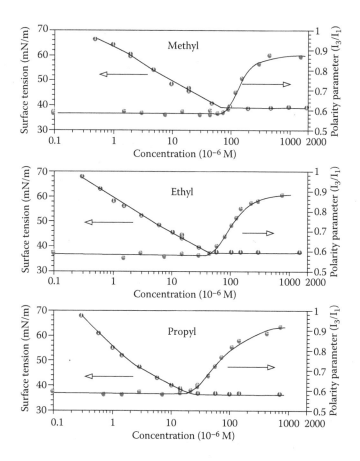

FIGURE 12.8 Surface tension and polarity parameters of sophorolipid methyl, ethyl, and propyl esters in solutions. (From Zhang, L. et al., *Colloids Surf. A* 240, 75–82, 2004. With permission.)

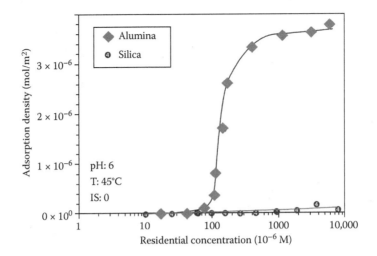

FIGURE 12.9 Adsorption isotherms of sophorolipid methyl ester on alumina and silica. IS indicates isoelectric point. (From Zhang, L. et al., *Colloids Surf. A* 240, 75–82, 2004. With permission.)

nonionic surfactants. Interestingly, it does not bear any similarity to that of nonionic ethoxylated surfactants. Although the mechanism of adsorption on solids is not understood, it has been proposed that hydrogen bonding is the primary driving force for the adsorption of the sophorolipids on alumina.

In addition, it was also observed from Figure 12.8 that an increase in the *n*-alkyl ester chain length of the sophorolipids makes it more surface active and causes a shift in the adsorption isotherms to lower concentrations, with the magnitude of the shift corresponding to the change in the cmc of these surfactants [18].

12.3.2 Role of Fatty Acid–Glutamate and Surfactin in Oil Dispersion

The cmc of both fatty acid–glutamate (FA-Glu) and surfactin was determined to obtain insight into the potential of these molecules for oil dispersion. It was observed that the cmc values of FA-Glu and surfactin were 0.01 and 0.01 g/L, respectively (Figure 12.10). These values are comparable with some of the commercially available surfactants and enzymes used for oil dispersion today.

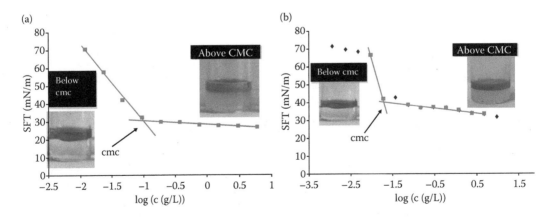

FIGURE 12.10 **(See color insert.)** cmc values of (a) FA-Glu and (b) surfactin. SFT on the *y* axis indicates surface tension, and "c" on the *x* axis indicates concentration. Pictures inserted in the two graphs indicate that below the cmc values (left pictures in the graphs), oil is not dispersed, whereas above the cmc values (right pictures in the graphs), oil is dispersed in the surfactant solution. SFT indicates surface tension (mN/m).

12.3.3 SYNERGISTIC/ANTAGONISTIC INTERACTIONS OF TRADITIONAL SURFACTANT AND BIOSURFACTANT MIXTURES

One important aspect that needs to be addressed before the introduction of biosurfactant to commercial use is the efficient way of replacement of conventional surfactant by the greener ones. In this regard, the synergistic behavior that exists between two surfactants in a mixture is discussed. As an example, the adsorption behavior of surfactant mixtures [nonionic sugar-based surfactant, *n*-dodecyl-β-D-maltoside (DM), in mixtures with anionic sodium dodecyl sulfate (SDS) and sodium dodecyl sulfonate] on alumina from pH 4 to 11 [21] is presented in this chapter. In this pH range, the adsorption behavior of both DM and anionic surfactant on alumina is pH dependent and exhibits opposite trends.

12.3.3.1 Adsorption of DM/SDS Mixtures at pH 6 (Active Nonionic/Active Anionic)

The isoelectric point of alumina was determined to be 8.9. Alumina is positively charged at pH 6, and therefore, anionic SDS can readily adsorb on alumina due to electrostatic interaction. Nonionic DM adsorbs on alumina through hydrogen bonding. The adsorption behavior of a mixture of both surfactants was studied at varying mixture ratios at pH 6. The adsorption isotherms of DM/SDS 3:1, 1:1, and 1:3 mixtures on alumina at pH 6 are shown in Figure 12.11, together with those of DM and SDS alone. Most notably, the amount of mixed surfactant adsorbed is higher than the individual component, which is indicated by a sharp rising part in the isotherm. This clearly establishes the presence of strong synergy between DM and SDS. This is the region where hydrophobic chain–chain interactions dominate the adsorption process because the surface is not yet saturated with the surfactants. At lower surfactant concentrations, SDS adsorbs more than DM because the electrostatic interactions are stronger than hydrogen bonding. In this region, the adsorption takes place mainly because of the electrostatic attraction between the negatively charged dodecyl sulfate and the positively charged alumina. However, some adsorption of the sugar-based surfactant is due to hydrogen bonding. At higher concentrations, the adsorbed SDS forms mixed aggregates with DM through hydrophobic chain–chain interactions and promotes the DM adsorption. The low critical micellar concentration of DM causes the aggregates to form

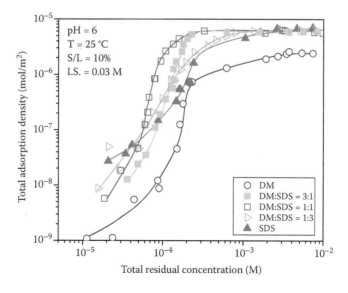

FIGURE 12.11 (See color insert.) Adsorption of DM, SDS, and DM/SDS mixtures on alumina. IS indicates ionic strength, and S/L indicates the percentage solids of alumina in suspension. (From Zhang, L., and P. Somasundaran, *Colloid Interface Sci.* 302, 20, 2006. With permission.)

at lower concentrations, promoting total adsorption as well. In the plateau region, the adsorption density of the mixture is slightly less than that of SDS. At this stage, the surface is saturated with surfactants. Because the sugar-based surfactant has a larger head group, the total adsorbed amount is less in molar terms.

12.4 FUTURE PROSPECTS

The correlation between toxicity and the origin of a material can sometimes be unexpected. Although a surfactant maybe produced by an organism, it may not necessarily be considered green. Very minor structural modifications, such as those due to metabolism, could result in very high toxicity. Furthermore, there is no absolute scale yet for toxicity, and there are no predictable rules relating molecular structure to toxicity. The complexity of toxicological properties of natural and man-made agents is only beginning to be understood. Biological toxicity may also result from physicochemical effects. For example, the colloidal properties of surfactant aggregates can determine the solubilization of oleophilic compounds.

Overall, it is crucial to build bridges between different scientific fields, especially materials sciences, bio/life sciences, and industrial ecology to understand the basic mechanisms involved to make further progress.

Future research should:

1. Define, characterize, and design greener model surfactant systems with desirable characteristics, particularly a lower carbon/chemical footprint
2. Provide a fundamental understanding of the colloidal and interfacial behavior of such greener surfactants in pure and mixed systems at interfaces and in solution
3. Establish a correlation between the microstructural and nanostructural properties and colloidal behavior for systems of academic and industrial interest
4. Study of synergetic systems

To develop greener, alternative model surfactant systems, we need to engage in the following efforts to bridge the knowledge gap:

1. Identify surfactant systems with significant effects on economy, ecology, and society
2. Work with selected ecologists or other relevant experts to define the right criteria, metrics, and knowledge framework for green surfactants

Currently, the basic knowledge of the interfacial and colloidal properties of biosurfactants is limited to the evaluation of surface tension and the critical micellar concentration of only a few compounds. Properties such as the aggregation behavior, the folding and unfolding of these surfactants, and the dynamics of these processes have not yet been studied.

In many applications in the nonconsumer product industry, solid surface modification to impart desired properties is of critical importance. More emphasis should be placed in these systems on designing reagents that impart a required property and on using the property to meet stringent specifications in subsequent processes and applications than on the solution and colloidal behavior. The development of these systems demands quite different surfactants from those typically used in consumer product industry. For example, chemisorbing surfactants with the additional capability of providing "anchoring" hydrogen bonds are more widely encountered in applications in the nonconsumer product industry than surfactants that function via electrostatic interactions. Despite their numerous advantages over the synthetic surfactants, the problem related with the large-scale and cheap production still exists and is a major hurdle in economic competitiveness.

ACKNOWLEDGMENTS

The authors acknowledge the support of the National Science Foundation, the researchers, and the industrial members of the IUCRC Center.

REFERENCES

1. M.I. Van Dyke, H. Lee and J.T. Trevors. Applications of microbial surfactants. *Biotechnol. Adv.*, **9**, 241–252 (1991).
2. K.-J. Hong, S. Tokunaga and T. Kajiuchi. Evaluation of remediation process with plant-derived biosurfactant for recovery of heavy metals from contaminated soils. *Chemosphere*, **49**, 379–387 (2002).
3. C. Schippers, K. Geßner, T. Muller and T.J. Scheper. Microbial degradation of phenanthrene by addition of a sophorolipid mixture. *J. Biotechnol.*, **83**, 189–198 (2000).
4. C.L. Royal, D.R. Preston, A.M. Sekelsky and G.S. Shreve. Reductive dechlorination of polychlorinated biphenyls in landfill leachate. *Int J Biodeterior Biodegradation*, **51**, 61–66 (2003).
5. N. Garti. What can nature offer from an emulsifier point of view: Trends and progress? *Colloids Surf. A*, **152**, 125–146 (1999).
6. Y. Uchida, S. Misava, T. Nakahara and T. Tabuchi. Factors affecting the production of succinoltrehalose lipids by *Rhodococcus erythropolis* SD-74 grown on *n*-alkanes. *Agric. Biol. Chem.*, **53**, 765–769 (1989).
7. D. Kitamoto, H. Yanagishita, T. Shinbo, T. Nakane, T. Nakahara and C. Kamisawa. Surface active properties and antimicrobial activities of mannnosylerythrytol lipids as biosurfactants produced by *Candida Antarctica*. *J. Biotechnol.*, **29**, 91–96 (1993).
8. H. Mager, R. Roetlisberger and F. Wagner. To Wella Aktiengesellschaft. Kosmetische mittel mit einem gehalt an einem sophoroselipid-lactone sowieseine verwendung. *European Patent*, 0 209 783 (1987).
9. M.G. Healy, C.M. Devine and R. Murphy. Microbial production of biosurfactants. *Resour Conserv Recy*, **18**, 41–57 (1996).
10. D.L. Gutnick, R. Allon, C. Levy, R. Petter and W. Minas. Applications of *Acinetobacter* as an industrial microorganism. In *The Biology of Acinetobacter*, K.J. Towner, E. Bergogne-Bérézin and C.A. Fewson (Eds.). Plenum Press, New York, pp. 411–441 (1991).
11. D.L. Gutnick and Y. Shabtai. Exopolysaccharide bioemulsifiers. In *Biosurfactants and Biotechnology*, N. Kosaric, W.L. Cairns and N.C.C. Gray (Eds.). Marcel Dekker, Inc., New York, pp. 211–246 (1987).
12. M. Davila, R. Marchal and J.P. Vandecasteele. Sophorose lipid fermentation with differentiated substrate supply for growth and production phases. *Appl. Microbiol. Biotechnol.*, **47**, 496–501 (1997).
13. K. Lottermoser, W.H. Schunck and O. Asperger. Cytochrome P450 of the sophorose lipid-producing yeast *Candida apicola*. *Yeast*, **12**, 565–575 (1996).
14. S. Ito and S. Inoue. Sophorolipids from *Torulopsis bombicola*: Possible relation to alkane uptake. *Appl. Environ. Microbiol.*, **43**, 1278–1283 (1982).
15. A. Albrecht, U. Rau and F. Wagner. Initial steps of sophorose lipid biosynthesis by *Candida bombicola* ATCC 22214 grown on glucose. *Appl. Microbiol. Biotechnol.*, **46**, 67–73 (1996).
16. J.R. Edward and J.A. Hayashi. Structure of a rhamnolipid from *Pseudomonas aeruginosa*. *Arch. Biochem. Biophys.*, **111**, 415–421 (1965).
17. K. Histasuka, T. Nakahara, N. Sano and K. Yamada. Formation of rhamnolipid by *Pseudomonas aeruginosa* and its function in hydrocarbon fermentation. *Agri. Biol. Chem.*, **35**, 686–692 (1971).
18. L. Zhang, P. Somasundaran, S. Singh, A. Felse and R. Gross. Synthesis and interfacial properties of sophorolipid derivatives. *Colloids Surf. A*, **240**, 75–82 (2004).
19. Scientists Brew 'Green' Dispersants in Gulf Spill's Wake, By PAUL VOOSEN of Greenwire, New York times, Published: April 20, 2011, http://www.nytimes.com/gwire/2011/04/20/20greenwire-scientists-brew-green-dispersants-in-gulf-spil-37018.html?pagewanted=all.
20. G.O. Reznik, P. Vishwanath, M.A. Pynn, J.M. Sitnik, J.J. Todd, J. Wu, Y. Jiang, B.G. Keenan, A.B. Castle, R.F. Haskell, T.F. Smith, P. Somasundaran and K.A. Jarrell. Use of sustainable chemistry to produce an acyl amino acid surfactant. *Appl. Microbiol. Biotechnol.*, **86**, 1387–1397 (2010).
21. L. Zhang and P. Somasundaran. Adsorption of mixtures of nonionic sugar-based surfactants with other surfactants at solid/liquid interfaces I. Adsorption of n-dodecyl-beta-D-maltoside with anionic sodium dodecyl sulfate on alumina. *Colloid Interface Sci.*, **302**, 20 (2006).
22. S.K. Satpute, S.S. Bhuyan, K.R. Pardesi, S.S. Mujumdar, P.K. Dhakephalkar, A.M. Shete and B.A. Chopade. Molecular genetics of biosurfactant synthesis in microorganisms. In *Biosurfactants*, R. Sen (Ed.). Springer and Landes Bioscience, New York, pp. 14–41 (2010).

Part V

Formulation and Application of Surfactant Aggregates

13 Triggered Drug Release Using Lyotropic Liquid Crystals as Delivery Vehicles

Dima Libster, Abraham Aserin, and Nissim Garti

CONTENTS

13.1 INTRODUCTION

Liquid crystals (LCs) are self-assembled organized mesophases with properties intermediate to those of crystalline solids and isotropic liquids [1]. In LC phases, long-range periodicity exists, although the molecules exhibit a dynamic disorder at atomic distances, as is the case in liquids. Accordingly, these materials can also be considered as ordered fluids [2]. Lyotropic LCs (LLCs) are materials that are composed from at least two molecules: an amphiphilic molecule and its solvent. A hydrophilic solvent, such as water, hydrates the polar moieties of the amphiphiles via hydrogen bonding, whereas the flexible aliphatic tails of the amphiphiles aggregate into fused hydrophobic regions based on van der Waals interactions. In addition to morphological dependence on the chemical composition, LLCs are also sensitive to external parameters, such as temperature and pressure [1–3]. As a function of the molecular shape of the surfactants, packing parameters, and interfacial curvature energy considerations, LLCs can be formed with aqueous domains ranging from planar bilayer lamellae to extended, cylindrical channels, to three-dimensional (3-D) interconnected channels and manifolds [4]. These mesophases are defined as lamellar (L_α), hexagonal (H), bicontinuous cubic (Q [or V]), and discontinuous cubic (I) phases, based on their symmetry [5]. In addition, most lyotropic mesophases exist as symmetrical pairs, a "normal" (type I) oil-in-water system, consisting of lipid aggregates in a continuous water matrix, and a topologically "inverted" (type II) water-in-oil version. The headgroups hydrated by water are arranged within a continuous nonpolar matrix, which is composed of the fluid hydrocarbon chains [5]. In addition to its biological significance, inverse lipid phases could be useful as host systems for the incorporation of food additives [6,7], the crystallization of membrane proteins for drug delivery [7,8], and for inorganic synthesis [9].

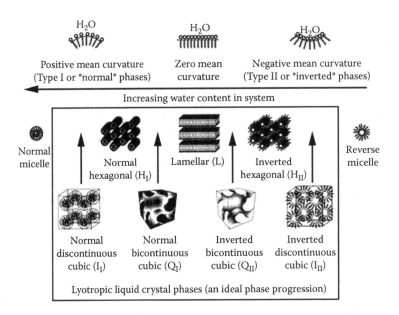

FIGURE 13.1 Schematic representations of common LLC phases formed by amphiphiles in water. (With kind permission from Springer Science+Business Media: *Struct. Bond.*, Functional lyotropic liquid crystal materials, 128, 2008, 181–222, D.L. Gin, C.S. Pecinovsky, J.E. Bara and L. Kerr.)

The lamellar structure does not possess any intrinsic curvature and is considered as the midpoint of an ideal, symmetrical LLC phase progression (Figure 13.1) [4,5]. The current review is focused mainly on the inverted (W/O) mesophases (cubic and hexagonal), representing an important class of nanostructures for potential applications.

13.1.1 Monoolein and Phytantriol: Main Building Blocks of Lipid Mesophases

Only a few synthetic amphiphiles can mimic the behavior of biological lipids and form inverted mesophases. The unsaturated monoglycerides (monoolein and monolinolein) belong to this category. Glycerol monooleate (GMO, monoolein) is a polar lipid that is commonly used as a food emulsifier. Presently, GMO is the preferred amphiphile for formulating LC phases for scientific research and drug delivery. This is a nontoxic, biocompatible, and biodegradable lipid, which possesses low water solubility, but swells and forms several LC phases in excess water [10].

Another lipid, phytantriol, was recently shown to form cubic LC. The advantage of this lipid, compared with GMO, is that it is not susceptible to esterase catalyzed hydrolysis and therefore provides additional stability to the mesophases. Phytantriol is commonly used as an ingredient in the cosmetics industry for improving moisture retention. Its phase behavior as a function of water concentration and temperature is very similar to that of GMO, although structurally they are very different (Figure 13.2) [11].

13.1.2 Cubic Phases

The bicontinuous cubic phase is the most complicated among all the LLCs [12]. Here, the lipids are located in a complex, optically isotropic 3-D lattice. Cubic structures can be either normal (type I, O/W) or inverse (type II, W/O) and their topology is differentiated as bicontinuous or micellar, resulting in seven cubic space groups. Only three space groups are both inverse and bicontinuous structures (Figure 13.3) [5]: the gyroid (G) type (*Ia3d*, denoted Q^{230}), the diamond (D) type (primitive lattice *Pn3m*, denoted Q^{229}), and the primitive (P) type (body-centered lattice *Im3m*, denoted

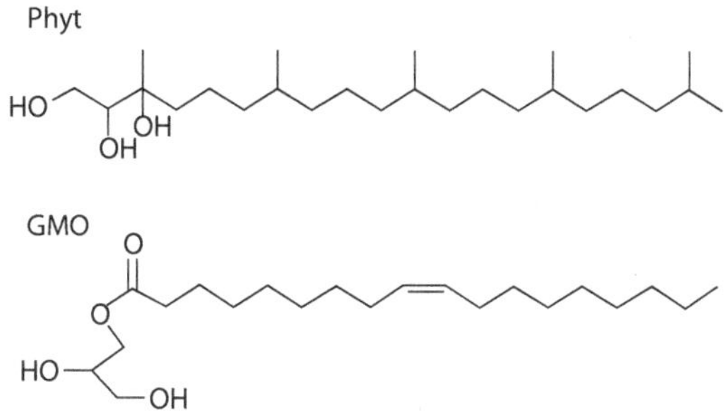

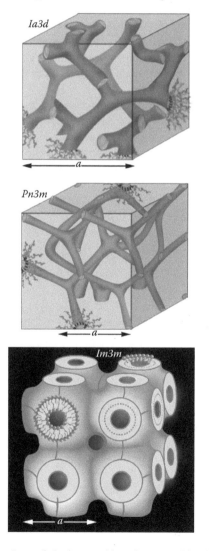

FIGURE 13.2 Chemical structures for phytantriol (Phyt) and glyceryl monooleate (GMO). (S.B. Rizwan, T. Hanley, B.J. Boyd, T. Rades and S. Hook: Liquid crystalline systems of phytantriol and glyceryl monooleate containing a hydrophilic protein: Characterisation, swelling and release kinetics. *J. Pharm. Sci.* 2009. 98. 4191–4204. Copyright Wiley-VCH Verlag GmbH & Co. KGaA. Reproduced with permission.)

FIGURE 13.3 Schematic illustrations of the inverse bicontinuous cubic phases *Ia3d* (gyroid type), *Pn3m* (diamond type), and *Im3m* (primitive type) (a = lattice parameter). (From J.M. Seddon, *Biochim. Biophys. Acta*, 1031, 169, 1990.)

Q^{224}) [13]. The aqueous channels of G-surface consist of two separate, left-handed and right-handed helical channels. The aqueous channels can extend through the matrix, but the centers of the water channels never intersect. Hence, they are connected and the structure adopts a helical arrangement. The structure D is characterized by a bilayer that separates two interpenetrating aqueous channel systems, forming a diamond lattice. In this configuration, four aqueous channels of the D-surface meet at a tetrahedral angle of 109.5°. Structure P contains two aqueous channel systems that are separated by a bilayer. The unit cell possesses three mutually perpendicular aqueous channels, connected to contiguous unit cells, taking the shape of a cubic array.

13.1.2.1 The Reverse Hexagonal Mesophase

The reverse hexagonal mesophase is of a primitive type (*P6mm*) and is characterized by one cylinder per unit cell corner. These densely packed, straight, water-filled cylinders exhibit two-dimensional ordering. Each cylinder is surrounded by a layer of surfactant molecules that are perpendicular to the cylinder surface such that their hydrophobic moieties point outward from the water rods (Figure 13.4). There is a growing indication that inverse hexagonal mesophases play structural and dynamic roles in biological systems [5,14,15]. These systems are assumed to be active as transient intermediates in biological phenomena that require topological rearrangements of lipid bilayers, such as membrane fusion/fission and the trans-bilayer transport of lipids and polar solutes [5,14,15]. H_{II} mesophases have recently been considered promising drug delivery vehicles, mainly owing to their unique structural features [9,16,17].

Another LC structure, based on a close packing of inverse micelles, is a 3-D hexagonal inverse micellar phase, of space group *P63/mmc*, recently discovered by Shearman et al. [18]. Ternary lipid mixtures comprising two lipids, dioleoylphosphatidylcholine (DOPC) and dioleoylglycerol (DOG), and cholesterol, in molar ratios of 1:2:1 and 1:2:2 in excess water induced the formation of 3-D hexagonal mesophase over a temperature range of 16°C to 52°C and a wide range of pressures (1–3000 bar). The new inverse micellar phase consists of an "hcp" packing of identical, spherical inverse micelles, with two lattice parameters, a and c, that were found to be $a = 71.5$ Å and $c = 116.5$ Å (Figure 13.5). This is the first new inverse LLC phase, which was reported two decades ago and it is the only known phase whose structure consists of a close packing of identical inverse micelles [18].

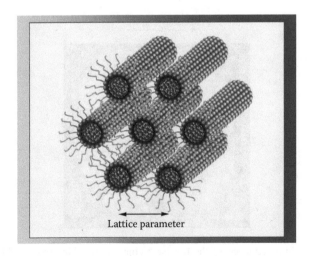

Lattice parameter

FIGURE 13.4 **(See color insert.)** Schematic presentation of H_{II} mesophase showing the packing of water-filled rods surrounded by lipid layers.

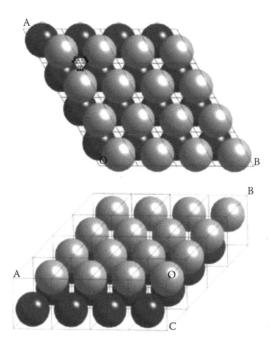

FIGURE 13.5 Plan (top) and perspective (bottom) views of the schematic structure of the 3-D hexagonal inverse micellar phase. The spheres represent the polar regions (water cores plus lipid headgroups), and the remaining fluid volume is filled by the hydrophobic regions of the lipid molecules. The different color shading of the two identical layers of spheres is purely for clarity. (From G.C. Shearman et al., *J. Am. Chem. Soc.*, 131, 1678–1679, 2009.)

13.2 PHASE BEHAVIOR

Most of the common surfactants form direct (type I) phases, where the interface bends away from the polar solvent toward the tail region. However, most biological amphiphiles (such as phospholipids) form "type II" LLC, where the interface curves toward the polar region. Only a few synthetic amphiphiles can mimic the behavior of biological lipids and form inverted mesophases. The unsaturated monoglycerides (monoolein and monolinolein) belong to this category [10]. GMO (monoolein) is a polar lipid that is commonly used as a food emulsifier. Presently, GMO is the preferred amphiphile for formulating LC phases for scientific research and drug delivery. This is a nontoxic, biocompatible, and biodegradable lipid, which possesses low water solubility, but in excess water it swells and forms several LC phases.

Monoolein/water is one of the most extensively investigated types of lipid-based mesophases [10]. The phase diagram of this system (Figure 13.6) revealed a complex structural behavior. At room temperature, the following phase sequence existed upon increasing hydration: lamellar crystalline phase (L_c) in coexistence with a L_2 phase, lamellar mesophase (L_α), and the inverted bicontinuous cubic mesophases-gyroid *Ia3d* and diamond *Pn3m*. Upon heating to approximately 85°C, the cubic phase is transformed into the H_{II} mesophase, followed by the micellar phase.

These concentration-dependent and temperature-dependent structural transitions can be qualitatively explained in terms of effective critical packing parameter (CPP), as developed by Israelachvili et al. [19]. According to this theory, amphiphiles possess geometric parameters characterized by the CPP (V_s/a_0l), where V_s is the hydrophobic chain volume, a_0 is the polar headgroup area, and l is the length of the chain in its molten state. The packing parameter is useful in predicting which phases can be preferentially formed by a given lipid because it connects the molecular shape and properties to the favored curvature of the polar–apolar interface and, therefore, the

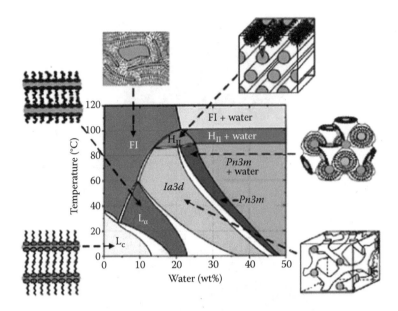

FIGURE 13.6 (See color insert.) Temperature-composition phase diagram of the monoolein/water system (up to 50 wt% water). A cartoon representation of the various phase states is included in which shaded zones represent water. The mesophases are as follows: L_c, crystalline lamellar; L_α, lamellar; *Ia3d*, gyroid inverted bicontinuous cubic; *Pn3m*, primitive inverted bicontinuous cubic; H_{II}, inverted hexagonal; and F1, reverse micelles fluid phases. (From V. Cherezov et al., *J. Mol. Biol.*, 357, 1605–1618, 2006.)

topology and shape of the aggregate. The main factors responsible for alterations of the mentioned parameters are the molecular shape of the surfactant, the chemical composition, and the temperature. According to the theory developed by Israelachvili and coworkers [19], inverse mesophases are formed by amphiphiles with CPP > 1, which adopt inversed cone-shaped geometry. Such lipids should possess a small headgroup area as compared with that of the tail region. In contrast, direct mesophases are preferred when the surfactant head cross-section is larger than that of the tail, resulting in CPP < 1. In the case of lamellar mesophase, these parameters are equal and hence result in CPP = 1. The size of the polar headgroup area is dictated by both molecular shape and the degree of hydration.

Within the boundaries of a given mesophase, the headgroup's area normally increases with increasing hydration and decreases upon temperature increase. The length and the volume of the lipophilic tails are also greatly affected by a temperature increase. Stronger thermal motion of the tails eventually decreases the length of the tails but increases their volume.

It was shown for monoglyceride-based LLC that H_{II} mesophases are formed from amphiphiles with CPP ~ 1.7, cubic phases with CPP ~ 1.3, and lamellar structures with CPP = 1 [2]. The lamellar phase is characterized by zero curvature because the cross sections of the polar heads and the lipophilic tails are similar. Upon increased hydration, cubic phases (*Ia3d* and *Pn3m*) with higher curvature and consequently higher CPP are preferred. The effective CPP theory can supply a reasonable explanation for the temperature-induced structural shifts from lamellar through cubic to reverse hexagonal phases, requiring higher curvature than in the lamellar phase. Increasing the thermal motion of both the hydrocarbon chains and the water molecules would increase the CPP values via expanding the volume of the lipophilic moiety, but decreasing the chain length and the headgroup area. This leads to an increase in curvature and therefore induces the formation of cubic and hexagonal mesophases.

It should be noticed that the CPP concept may only be used as a "rule of thumb" to predict transitions between various forms of LLCs. For example, it was shown that it fails to predict the appearance of the inverse bicontinuous phases and an intermediate phase [20,21]. This is the reason that a thermodynamic approach considering the total free energy of the LLC systems was applied by Seddon and coworkers [20,22,23]; however, this is beyond the scope of this chapter.

13.3 H_{II} MESOPHASE COMPOSED OF GMO/TRIGLYCERIDE/ WATER AS DRUG DELIVERY SYSTEMS

It was shown that in the monoolein-based system, the cubic phase is transformed into an H_{II} mesophase upon heating at approximately 85°C [10]. The CPP theory can supply a reasonable explanation for the temperature-induced structural shifts from lamellar through cubic to reverse hexagonal phases, requiring greater curvature than in the lamellar phase. Increasing the thermal motion of both the hydrocarbon chains and the water molecules would increase the CPP values via expanding the volume of the lipophilic moiety, but decreasing the chain length and the headgroup area. This leads to an increase in curvature and, therefore, induces the formation of cubic and hexagonal mesophases [1,19].

Bearing in mind the thermodynamic and structural considerations noted above, a systematic research was conducted to decrease the cubic to hexagonal transition temperature and stabilize the GMO-based H_{II} mesophase at room temperature [24]. To achieve this goal, experiments relative to the incorporation of triacylglycerides (TAGs) with various chain lengths into the binary GMO/water system were conducted. Amar-Yuli and Garti surmised that immobilization of a TAG between GMO tails would lead to a change in the geometry of monoolein molecules from cylindrical to wedge-shaped and thereby an increase in the CPP value of the system [24]. This should encourage a transition from lamellar or cubic phases to hexagonal structures. In addition, the immobilization of TAG in the system was expected to reduce the packing frustration, stabilizing the hexagonal LLC at room temperature. These experimental results showed that a critical and optimal chain length of the triglyceride is required to induce the formation of H_{II} at room temperature. Among the examined TAGs, tricaprylin (C_8) was the most successful, flexibly accommodating between the tails of the GMO.

The structural properties of ternary hexagonal mesophases composed of GMO, tricaprylin, and water were extensively and systematically studied, as was shown in numerous publications [25–29]. Several additives, including dermal penetration enhancers, were solubilized to control the physical properties of these carriers [27–29], such as viscosity and thermal stability. For instance, the synergistic solubilization of two major hydrophilic (vitamin C, ascorbic acid, AA) and lipophilic (vitamin E, D-α-tocopherol, VE) antioxidants within H_{II} mesophases was reported by Bitan-Cherbakovsky et al. [30,31]. This enabled expanding the conditions to obtain stable H_{II} mesophase at room temperature. In addition, it was shown that phosphatidylcholine (PC) can be embedded into the ternary GMO/TAG/water system [32–35]. PC was incorporated into the ternary mesophases because it is known that phospholipid-based structures possess relatively high thermal stability, and enhance transdermal drug permeation [36–39] and transmembrane transport across the digestive tract [40,41].

Incorporation of PC into the ternary system caused competition for water binding between the hydroxyl groups of GMO and the phosphate groups of PC, leading to dehydration of the GMO hydroxyls in favor of the phospholipid hydration [34]. On the macroscopic level, this was correlated with an improvement of elastic properties and thermal stability of the H_{II} mesophase [32]. Structural flexibility and stability is essential for a drug delivery system, especially for tuning its physical properties (such as viscosity) and composition.

Libster et al. [33] also explored and demonstrated correlations between the microstructural and mesostructural properties of the reverse hexagonal LLC, using environmental scanning electron microscope (ESEM) technique. It was shown that the mesoscopic organization of these systems is based on an alignment of polycrystalline domains. The topography of H_{II} mesophases, imaged directly in their hydrated state, as a function of aqueous phase concentration, was found to possess fractal characteristics, indicating a discontinuous and disordered alignment of the corresponding internal water rods on the mesoscale. Fractal analysis indicated that the mesoscale topography of the H_{II} phase was likely to be influenced by the microstructural parameters and the water content of the samples [33]. Garti and coworkers [32,34,42–45] also made considerable progress elucidating the solubilization of therapeutic peptides into the H_{II} mesophase.

13.4 CELL-PENETRATING PEPTIDES TRIGGERED DELIVERY

Lately, Cohen-Avrahami et al. [46] used for the first time cell-penetrating peptides (CPPs) as enhancers to overcome the stratum corneum barrier. Na-DFC is a common nonsteroid anti-inflammatory drug used to treat mild to moderate pain, particularly when inflammation is also present. This study utilized the advantages of the H_{II} mesophase as a transdermal vehicle, in addition to those of the CPPs as skin penetration enhancers, for the development of improved drug delivery vehicles in which the drug diffusion rate can be carefully controlled.

The selected CPPs, representing prominent members of these families, were chosen for their great efficiency and short length; their amino acid sequences are the following: "RALA," RALARALARALRALAR, is a 16-amino acid peptide belonging to a synthetic family of CPPs, based on GALA, amphipathic peptides named after their alanine-leucine-alanine repeats that exhibit improved membrane permeability. "PEN," penetratin (RQIKIWFQNRRMKWKK), is a 16-amino acid sequence based on the active penetrating peptide produced by the homeodomain of the antennapedia homeoprotein. "NONA," nona-arginine, RRRRRRRRR. Arginine was shown to play a key role in CPP attachment to membranes; thus, much research was done concerning the penetrating activity of polyarginine through membranes. The maximal solubilization load of Na-DFC, RALA, and PEN into the H_{II} structures were found to be 5 wt%, whereas NONA solubilization capacity was 6 wt%.

Small angle X-ray scattering (SAXS) data on the Na-DFC–loaded systems revealed that the drug caused a shrinkage of the hexagonal channel's diameter. The lattice parameter sharply decreases from 57.55 ± 0.5 Å in the blank system to 51.55 ± 0.5 Å in the 5 wt% loaded system (Figure 13.7), leading to denser packing of the system. The investigators assumed that such a "kosmotropic effect" of the Na-DFC might originate from the drug's location between the GMO molecules, causing an enhancement of the interactions between them. This assumption was supported by the slight increase in the domain lengths of the hexagonal clusters in the presence of Na-DFC from 392 to 493 ± 50 Å (Figure 13.7), hinting at a stabilizing effect. The main effect appeared in the low

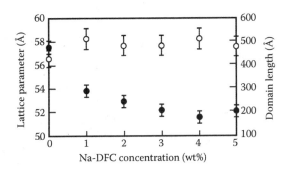

FIGURE 13.7 The lattice parameter (●) and the domain length (○) obtained from the SAXS measurements, versus solubilized Na-DFC concentration in the H_{II} mesophases. (Reprinted with permission from M. Cohen-Avrahami et al., D. Libster, A. Aserin and N. Garti. Sodium diclofenac and cell-penetrating peptides embedded in H_{II} mesophases: Physical characterization and delivery. *J. Phys. Chem. B*, 115, 10189–10197. Copyright 2011 American Chemical Society.)

concentration regime, whereas the additional Na-DFC quantities had a weaker influence. These effects were noticed until the system transformed into a micellar structure.

In addition, differential scanning calorimetry (DSC) results suggested enhanced water binding in the presence of the drug. Shrinkage of the aqueous columnar cylinders (as detected by SAXS) caused a change in the structure's curvature (curvature more concave toward the inner water phase) and an increase in the fraction of bound water at the expense of the free water that populated the inner channels. It also indicated that Na-DFC interacts with the surfactant tails, reducing the number of molecules that are free to participate in the melting process. Combined DSC measurements and ATR-FTIR analysis led the authors to conclude that the Na-DFC solubilization effect was primarily manifested in the interfacial region, when the drug intercalated between the GMO molecules. Na-DFC influenced the intermolecular bonding in this molecular region. It strengthened the interactions between GMO tails, loosened the repulsion between the carbon atoms at the headgroups, and enabled the formation of a tighter structure with higher curvature.

Each CPP was separately solubilized within the H_{II} for the identification of its specific location and chemical interactions within the mesophase. The SAXS measurements indicated two trends in the effect of the peptides. Up to 2 wt% of PEN solubilization, an initial shrinkage of the aqueous cylinders to a minimum of 53.4 ± 0.5 Å was detected. With increasing solubilization loads of PEN to 5 wt%, the shrinkage changed, and a swelling effect of the cylinders was detected until their diameter reached the original radius of 57.9 ± 0.5 Å (Figure 13.8).

NONA solubilization had a minor, yet consistent, swelling effect on the cylinders. The initial lattice parameter in the blank system was 57.9 ± 0.5 Å, and it slightly increased with NONA loading to a maximal value of 60.5 ± 0.5 Å at 6 wt% NONA. Its effect was found to be focused mostly on the water and the GMO carbonyls. The stretching mode of the water was gradually shifted from 3361 to 3346 cm^{-1} with increasing NONA concentration in the mesophase, indicating stronger hydrogen bonding of the water with the peptide.

It was established that structural investigation revealed that the solubilization sites of the different guest molecules depend on their molecular structure and differ significantly. Na-DFC populates the interfacial region, enhances the interactions between GMO tails, and shrinks and stabilizes the H_{II} mesophase (Figure 13.9c). PEN solubilization is concentration-dependent. The initial PEN loads populate the hydrophilic headgroup area, whereas the higher PEN loads pack closer to the GMO tails (Figure 13.9e, f). The hydrophilic NONA populated the inner channels region and swelled the mesophase (Figure 13.9d). RALA acted as a chaotropic agent in the H_{II} mesophase, interacting mostly with the water within the channels and enhancing the hydration of the GMO headgroups.

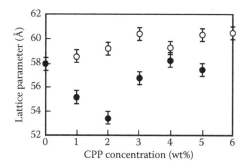

FIGURE 13.8   The lattice parameter (Å), calculated from the SAXS measurements, versus NONA (o) and PEN (•) concentration in the mesophase. (Reprinted with permission from M. Cohen-Avrahami et al., D. Libster, A. Aserin and N. Garti. Sodium diclofenac and cell-penetrating peptides embedded in H_{II} mesophases: Physical characterization and delivery. *J. Phys. Chem. B*, 115, 10189–10197. Copyright 2011 American Chemical Society.)

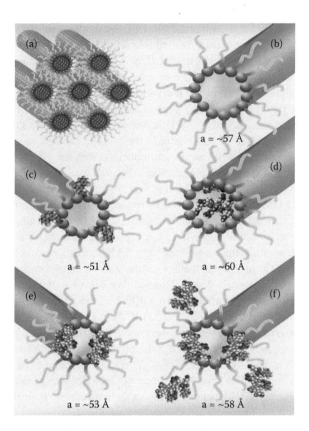

FIGURE 13.9   (See color insert.) (a) Schematic illustration of the H_{II} mesophase general structure. The water populates the inner region of the cylinders; the GMO tails as well as the TAG point outward. (b) A single blank, empty system of the H_{II} mesophase, (c) Na-DFC-loaded system. The drug populates the interfacial area and causes channel shrinkage. (d) NONA-loaded H_{II} mesophase. The peptide populates the aqueous channels and swells them. (e, f) PEN-loaded systems. At low PEN concentration (E), it is adsorbed on the GMO headgroups and causes channel shrinkage. With increasing PEN concentration (f), it populates an additional hosting region at the interface; a = lattice parameter. (Reprinted with permission from M. Cohen-Avrahami et al., D. Libster, A. Aserin and N. Garti. Sodium diclofenac and cell-penetrating peptides embedded in H_{II} mesophases: Physical characterization and delivery. *J. Phys. Chem. B*, 115, 10189–10197. Copyright 2011 American Chemical Society.)

Furthermore, the effect of the various peptides on skin permeation efficiency was determined by diffusion experiments on Na-DFC through porcine skin using Franz diffusion cells. All systems were composed of H_{II} mesophases loaded with 1 wt% Na-DFC and 1 wt% of each CPP (the control was a system with Na-DFC and no CPP). These experiments indicated that all three peptides significantly increased the diffusion of Na-DFC through the skin (Figure 13.10a).

NONA was found to be the most efficient CPP, significantly enhancing the transdermal penetration (a 2.2-fold increase in the total amount of Na-DFC that diffused through the skin). PEN and RALA caused 1.9-fold and 1.5-fold increases, respectively, compared with the control system. In all tested systems, there was a gradual and linear increase in the cumulative penetration of Na-DFC with time.

A calculation of Na-DFC percentage that permeated through the skin showed that in the blank systems, the total amount that diffused through the skin during the 24-h experiment was 0.9% from the initial applied dose. The amounts released in the NONA, PEN, and RALA systems were 1.9%, 1.8%, and 1.3%, respectively. The permeability coefficients (K_p, calculated as cm·h^{-1}), derived from the steady state flux of Na-DFC, revealed the same tendency (Figure 13.10b), a 1.5-fold increase in the presence of RALA and 2.3-fold and 2.2-fold increases with PEN and NONA, respectively.

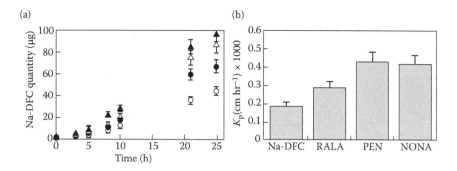

FIGURE 13.10 (a) Na-DFC cumulative skin permeation quantity versus time in the different systems: (i) control (Na-DFC only) (o), (ii) RALA-loaded (●), (iii) PEN-loaded (△), and (iv) NONA-loaded (▲). (b) The calculated permeability coefficient (K_p) based on the diffusion rate of 1.0 wt% Na-DFC *via* Franz diffusion cells as affected by the different peptides added to the mesophase. (Reprinted with permission from M. Cohen-Avrahami et al., D. Libster, A. Aserin and N. Garti. Sodium diclofenac and cell-penetrating peptides embedded in H_{II} mesophases: Physical characterization and delivery. *J. Phys. Chem. B*, 115, 10189–10197. Copyright 2011 American Chemical Society.)

Because the skin permeation studies revealed different profiles for each of the three peptides, a more detailed analysis using the well-established technique of "emptying" experiments was conducted by the authors. These experiments aimed to determine whether the rate-determining step in Na-DFC skin permeation from the mesophase was migration out of the mesophase or permeation through the skin. In addition, this analysis provides information on the rate-determining step in the CPP enhancement.

The "emptying" experiments revealed a surprising result. The control system, without CPP, released the largest quantities of Na-DFC with time. The release from the PEN systems was the second in order, whereas NONA and RALA systems released the lowest quantities of Na-DFC (Figure 13.11a).

The Higuchi diffusion equation [47] was applied to analyze Na-DFC release from the examined mesophases. Higuchi's model implies that drug release is primarily controlled by diffusion through the matrix and can be described by the following equation:

$$Q = [D_m (2A - C_d) C_d t]^{1/2} \tag{13.1}$$

Q is the mass of drug released at time t and is proportional to D_m, the apparent diffusion coefficient of the drug in the matrix; A, the initial amount of drug in the matrix; and C_d, the solubility of the drug in the matrix. The slope of a linear fit of the data from this plot is proportional to the "apparent diffusion coefficient" for the drug in the matrix and permits preliminary assessment of diffusion as the primary means of drug release from the correlation coefficient for the linear fit, and second, a means to compare the diffusion of a drug from the different matrices into the release medium. The linearity of the plots was found to be more than 0.95 for all CPPs, indicating the existence of a diffusion-controlled transport mechanism (Figure 13.11). The slope was the greatest for the blank system, 2.8 $h^{-1/2}$, and it decreased to 2.0, 1.3, and 1.2 in the PEN, NONA, and RALA systems, respectively.

The "blank" systems released the highest amounts of Na-DFC, and the release from the CPP-loaded systems was much lower. The order of Higuchi slopes was exactly the same: blank > PEN > NONA > RALA, and their values were 3.0, 0.9, 0.7, and 0.6 $h^{-1/2}$, respectively.

Considering the Na-DFC's amphiphilic nature and the H_{II} mesophase structure, in which the aqueous phase is tightly packed within the lipid structure, with the limited accessibility to the surrounding media, one can assume that the drug diffusion out of the mesophase occurs through the lipophilic oily regions. The CPPs incorporated within the mesophase slowed the drug diffusion

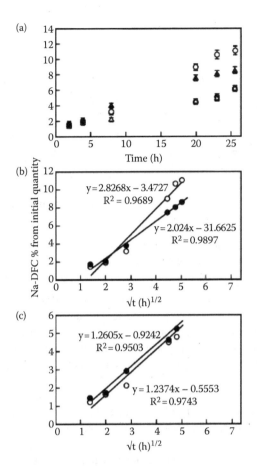

FIGURE 13.11 (a) Na-DFC release to water from the H_{II} mesophase containing: (i) no CPP (o), (ii) RALA (●), (iii) NONA (Δ), and (iv) PEN (▲). Na-DFC release versus the square root of time from (b) the blank (no CPP), (o) and PEN (●) systems, and (c) RALA (●) and NONA (o) systems. (Reprinted with permission from M. Cohen-Avrahami et al., D. Libster, A. Aserin and N. Garti. Sodium diclofenac and cell-penetrating peptides embedded in H_{II} mesophases: Physical characterization and delivery. *J. Phys. Chem. B*, 115, 10189–10197. Copyright 2011 American Chemical Society.)

rate by decreasing its mobility through the mesophase by any specific molecular interfacial interactions. Nevertheless, the emptying experiments showed that the main enhancing effect of the CPPs on skin penetration does not take place during drug migration out of the mesophase. It seems to be due to the enhanced permeation capability of the skin membrane and via any kinetic diffusion-controlled process. It should be stressed that RALA and NONA, which populate the inner aqueous channels, revealed a similar diffusion profile with the same slope, 1.2 ± 0.1 $h^{-1/2}$. PEN, which populates the outer interfacial region, is less disturbed in its diffusion, causing a release profile with a slope of 2 $h^{-1/2}$. Nevertheless, this interesting mechanism of the drug migration from the H_{II} mesophase in the presence of CPPs molecules cosolubilized within the mesophase should be further investigated.

13.5 LIPOLYSIS AND STRUCTURE-CONTROLLED DRUG RELEASE FROM REVERSED HEXAGONAL MESOPHASE

Garti et al. [48] investigated a system composed of a ternary reversed hexagonal mesophase (H_{II}) loaded with a lipase for modulating drug delivery capabilities of the system. *Thermomyces lanuginosa* lipase was solubilized into H_{II} mesophase for the benefit of continuing lipolysis of the lipids, consequently disordering and decomposing the hexagonal mesophase and thereby enhancing the diffusion of the encapsulated drug. The path of the lipolytic reaction of the examined GMO/ tricaprylin/water mesophase comprised gradual decomposition of the hexagonal structure, accompanied by the shrinking of the lattice parameter after 150 min of lipolysis, and finally, a structural shift to dispersion phase.

The effects of lipase concentration, water content in the H_{II} mesophase, and drug loading concentration on Na-DFC release from the H_{II} matrix were studied as a function of time. The effect of TLL's presence is depicted in Figure 13.12a. The Na-DFC release to a water continuous medium, in the absence of enzyme, is extremely slow even after approximately 450 min of reaction time. Close examination of the release profile in the presence of TLL (33 U/g) reveals much faster release with a two-stage process (Figure 13.12a) with two different rates. When the release is plotted against the square root of the time, we detected two lipolysis-induced release profiles that fit the Higuchi kinetics (data not shown).

The effect of water content within the H_{II} structure on the release of Na-DFC was also examined. At higher water phase contents (15–25 wt%), the Na-DFC release increased (by 1.8-fold; Figure 13.12b). The increased water contents hydrate the GMO polar headgroups, resulting in an enlarged area of the monoolein polar moieties and therefore increased water–lipid interfacial area.

It is well documented that the enzymatic activity of TLL is strongly dependent on its successful intercalation into the water–lipid interface [17,49–52]. Hence, the increased release of Na-DFC at high water content is attributed to the increased interfacial area, induced by additional hydration of the GMO headgroups.

The last parameter tested was the concentration of the loaded drug on its release. Surprisingly, the release from H_{II} mesophases loaded with lower Na-DFC contents (0.05–0.5 wt%) caused an increased drug release from the H_{II} mesophases (7%–53% after 7 h of TLL addition), as illustrated in Figure 13.12c. Na-DFC seems to be intercalated between the surfactant chains, mainly in the interfacial region. It enhanced the interactions between GMO molecules toward a denser packing of the nanostructure. Na-DFC solubilization resulted in a noticeable decrease in the lattice parameter and shrinkage of the system toward greater curvature and additional hydration of the GMO headgroups. This, in turn, led to the slower drug release with increasing initial drug concentration.

The lipolysis of the GMO and TC that constructs the H_{II} mesophase is, as expected, enzyme concentration–dependent. At enzyme concentrations higher than 20 U/g, the lipolysis shows a two-step hydrolysis. In the first one, the reaction is carried out at the interface of the inner water channels and is controlled by the H_{II} mesophase structure. In the second stage, the reaction is carried out in a partially disintegrated structure and its rate changes accordingly.

The lipolysis caused gradual two-stage structural changes in the H_{II} mesophase. In the first stage, the H_{II} structure is gradually hydrolyzed without noticeable disintegration and no change in the lattice parameter. The stability of the H_{II} is maintained due to the release of fatty acids that stabilize the H_{II} mesophase even with less GMO (gradually consumed). In the second stage (>150 min) a progressive disintegration of the H_{II} mesophase is observed (150–800 min), ending in phase separation (>800 min). In this reaction, a range of time shrinkage of the lattice parameter is observed due to a decrease in water content and enhanced water dehydration from the GMO headgroups, and by the formation of glycerol.

Garti et al. [48] demonstrated that sustained release from hexagonal mesophases can be tuned by three different approaches. *In vitro* release of Na-DFC from GMO matrices containing TLL enzyme was investigated. Release of Na-DFC was faster and the extent of release was greater using TLL because the decomposition of the mesophases is faster when the enzyme is present. The release profile in the presence of TLL (33 U/g) reveals higher drug release extent in the second stage of the reaction (>200 min), exhibiting a two-stage process (Figure 13.12a) with two different rates. These two lipolysis-induced release profiles that followed Higuchi kinetics were dictated by the described

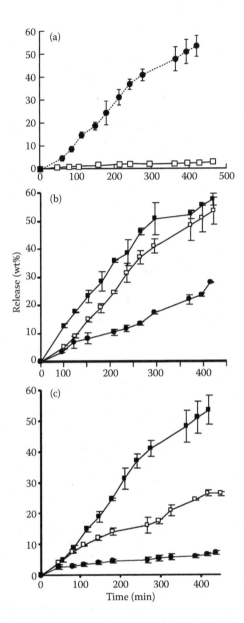

FIGURE 13.12 (a) Release of Na-DFC (0.05 wt%) from H_{II} mesophase as a function of time after the addition of increasing TLL contents (U/g): (□) 0, and (●) 33, demonstrating a strongly dependent release with increased lipase concentrations, attributed to the more rapid lipolysis of the H_{II} mesophase. (b) The release of Na-DFC (0.05 wt%) from H_{II} mesophase as a function of time after addition of 33 U/g TLL concentration with increasing water phase contents (wt%) of: (●) 15, (□) 20, (■) 25. (c) Release of Na-DFC after addition of 33 U/g TLL concentration from a H_{II} system as a function of time with increasing Na-DFC wt% contents of: (■) 0.05, (□) 0.1, and (●) 0.5. Lines are intended as a visual guide only. (From N. Garti et al., *Colloids Surf. B*, doi:10.1016/j.colsurfb.2012.01.013, 2012.)

structural changes of the carrier. Continuous hydrolysis of the liquid crystalline structure in the first reaction stage only slightly affected the lattice parameter and the microscopic structure of the mesophase. Hence, the release of diclofenac was relatively slow. In contrast, in the second reaction stage, a progressive collapse of the H_{II} mesophase was detected, which speeded up the release of the guest molecule.

Another approach that is able to regulate the release of the drug is to change the water concentration within the carriers. Release from the lipid systems was dependent on initial water content, resulting in greater extent of release and faster delivery with higher water contents (Figure 13.12b).

Such phenomena are attributed to greater enzyme activity in water-rich samples, due to the larger water–lipid interface of more hydrated monoolein polar headgroups. TLL was reported to be inactive in aqueous conditions, but upon binding to a lipid–water interface, its catalytic activity is greatly enhanced due to conformational changes of the enzyme leading to greater exposure of the active site. Therefore, the structure of the lipid–water interface dictates the degree of activation of the lipase [17,51–53]. Consequently, greater water concentrations imply better enzyme incorporation into the lipid–water interface and hence its greater activity.

In the particular case of Na-DFC, it was found that an additional approach, changing the drug concentration, was able to control the rate of its release (Figure 13.12c). In general, increasing the drug content in the formulation can increase the extent of release due to greater diffusion of the substance provided by greater concentration difference within and without the carrier. Examining Na-DFC release, we obtained the opposite trend—increasing the concentration of the drug in the carrier resulted in slower release. Normally in LCs, the drug release is diffusion controlled; however, in the presence of lipase, the drug release by diffusion is significantly controlled by the rate of lipolysis. As mentioned, DFC is incorporated in the water–lipid interface, and therefore, its increasing concentration interferes with the enzymatic reaction occurring at the interface, thereby slowing the reaction rate and the decomposition of the carrier. This, in turn, slows the diffusion rate of the drug from the carrier.

13.6 BIOLOGICALLY INSPIRED CARRIERS FOR BIO-MACROMOLECULES

Peptides and proteins, DNA and siRNA are increasingly considered for the development of new therapeutic compounds [54,55]. Peptides of various sizes are currently tested for a broad spectrum of diseases, including skin cancer, acne, psoriasis, hypertension, hepatitis, and rheumatism. These biomacromolecules are usually found to be very effective and, in most cases, low doses are needed for good medical treatment. Nevertheless, when delivered orally, these peptides and proteins are susceptible to cleavage by various enzymes, mainly the human digestive proteases, leading to practically poor bioavailability and poor pharmaceutical efficacy. As a result of these critical limitations, excessively large doses of the peptide-based drugs are usually required to obtain therapeutic effects *in vivo*, which in turn often cause a wide range of hazardous side effects [54,55]. This is the main reason that the medical application of these systems' otherwise high potential therapeutics is currently very limited, and even those few that are used nowadays are far from being efficient. In this respect, LLCs seem to be promising candidates as alternative means of delivery for various pharmaceuticals. Such vehicles can provide enhanced drug solubility, relative protection of the solubilized drugs, and controlled release of drugs while avoiding substantial side effects [56–58].

Although the cubic phase has been proposed and studied as a drug delivery vehicle, there is relatively little information about the interactions of peptides and proteins with the H_{II} mesophases and the therapeutic potential of these structures. Garti and coworkers [32,34,42,59] explored and controlled the physical properties of H_{II} mesophases to use these systems as drug delivery vehicles for biologically active peptides and proteins. The results of this structural research enabled significant expansion of the application spectrum of hexagonal LLCs, utilizing them for the solubilization of peptides and proteins [32,34,42–45,59], into this mesophase and its utilization as a sustained drug delivery vehicle. Two model cyclic peptides, cyclosporin A (11 amino acids) and desmopressin (9 amino acids), of similar molecular weight but with very different hydrophilic and lipophilic properties, were chosen to demonstrate the feasibility of using the H_{II} mesophase [58]. In addition, a larger peptide, RALA [43] (16 amino acids), was solubilized into the H_{II} structures as a model skin penetration enhancer. Finally, a larger macromolecule, LSZ protein (129 amino acids), was directly incorporated into a GMO-based H_{II} mesophase [44,45].

With the aim of designing a biologically inspired carrier in which the encapsulation and the delivery of DNA can be efficiently controlled, Amar-Yuli et al. [60] have designed two lipid-based columnar hexagonal LLCs that can accomplish two opposite roles while maintaining the same liquid crystalline symmetry. The first system was based on a nonionic lipid, such as monoolein, whereas the second system was modified by a low additional amount of the oleyl amine cationic surfactant. DNA was enzymatically treated to generate a broad distribution of contour lengths and diffusion characteristic times.

The effect of DNA confinement on these two columnar hexagonal structures was investigated by SAXS. Both the neat neutral and cationic LLC systems had a columnar hexagonal symmetry, the main difference being the reduced lattice parameter of 49.8 Å in the cationic system compared with the 55.5 Å of the neutral formulation. This difference was interpreted to be the result of the kosmotropic effect of the oleyl amine, which, due to its charged nature, dehydrates the surfactant polar heads and reduces the LLC lattice parameter.

A very different effect on the H_{II} lattice parameter was found when the DNA (1.4 wt% from the aqueous phase) was incorporated into the two systems. In the nonionic LLC system, the lattice parameter decreased from 55.5 to 50.8 Å (± 0.5 Å) in the presence of DNA. In contrast, in the cationic columnar phase, the lattice parameter increased from 49.8 to 59.2 Å. These effects can be rationalized by considering the negative charge of the DNA: when added to the neutral formulation, DNA leads to a negative effective charge that dehydrates the lipid polar heads and reduces the lattice parameter; on the other hand, when added to the cationic formulation, DNA partially neutralizes the overall positive charge, moderates the dehydration effect caused by the cationic surfactants, and induces, at least partially, swelling of the lattice.

The authors postulated two different possible arrangements for the DNA within the nonionic and cationic columnar hexagonal phases. Being strongly hydrophilic, the DNA double-strands must be either segregated outside the water channels, within the grain boundaries, or confined within the relatively narrow aqueous cylinders of the H_{II} structure. The strong effect of DNA on the lattice of the mesophases provides a robust indication that the DNA molecules were indeed confined within the water channels of the reverse hexagonal phases, although the different trends observed for the lattice evolution in the nonionic and cationic case suggest two different types of interactions among the lipids and the DNA [60].

This issue was directly addressed by performing ATR-FTIR analysis on the two formulations and indirectly by following the release of the DNA from each LLC system by UV spectrophotometry. In the nonionic-based H_{II} mesophase, the results provided evidence for the breakage of GMO–GMO and GMO–D_2O hydrogen bonds due to the incorporation of DNA molecules. Furthermore, carbonyl absorption revealed a slight modification in the area of their peaks when the DNA was present. In the absence of DNA, the number of hydrogen-bonded carbonyls was calculated to be 72% and it slightly increased to 78% in the presence of DNA. It was concluded that the DNA molecules interfere with the interfacial region of the cylinders, with tangible effects on the water (D_2O) and GMO molecules, including the hydroxyl and carbonyl groups and the ester moiety. The decrease in lattice parameter upon the addition of DNA, as detected by the SAXS measurements, is consistent with dehydration of the GMO hydroxyl groups, that is, a breakage of hydrogen bonds between GMO polar groups caused by the presence of the highly polar DNA. Additionally, the modifications within the vicinity of the GMO–water interface imply a less ordered state with increased content of gauche conformations, which is again coherent with the shrinkage of the H_{II} cylinders.

According to Amar-Yuli et al. [60], these results indicate that the DNA is confined within the aqueous domains of the H_{II} cylinders, stabilized by hydrogen bonds with the water or GMO headgroups, and leading to an observable rupture of the H-bonds in the H-bond donor part of the GMO polar heads (OH) and a decrease in the frequency of the H-bond acceptor bands of the GMO polar heads (CO–O). The role of DNA in this case is thus consistent with moving of bound water away from the interfacial region of the water channels.

Examination of the second H_{II} system, which contains the cationic surfactant oleyl amine shows very moderate changes upon incorporation of DNA in all the bands considered when compared with the nonionic mesophase. These data clearly suggested that in the presence of a cationic surfactant decorating the interface of the water channels, the interactions between the lipid polar heads and DNA change in nature, as the hydrogen bonding of D_2O and the GMO polar heads is mostly unaffected by the presence of DNA. It was inferred that the DNA is confined in the water channels also in the cationic H_{II} mesophases, and electrostatically bound to their surfaces.

The DNA cumulative release profile from the two mesophases is depicted in Figure 13.13, which illustrates the difference in release rates from the nonionic and cationic H_{II} host mesophases. Although practically no release of DNA molecules was found in the presence of oleyl amine even after 15 days, three main regions in the release profile can be identified when DNA is confined within the pure nonionic columnar system. During the first 3 days, no release is detected, and this can be attributed to the lag time needed for DNA molecules to diffuse out of the H_{II} cylinders. In the range of 4 to 9 days, the release of DNA from the LLC system into the excess water begins, progressing at a relatively high rate, as indicated by the sharp slope observed in Figure 13.13. After 9 days, the release of DNA slowed down significantly, and reached a plateau. At this point, a significant amount of DNA, approximately 65% of the initial content, had already been released by the nonionic H_{II} mesophase. These results are consistent with the conclusions based on SAXS and ATR-FTIR analyses stating that hydrogen bonds between DNA, water, and GMO headgroups are responsible for the DNA confinement within the water channels in the case of the nonionic hexagonal mesophase, which allows transport of DNA into the excess water. In contrast, the presence of cationic oleyl amine in the ionic hexagonal mesophase leads to such a strong electrostatic confinement of the DNA at the water–lipid interface that release into the excess water is prevented: this formulation is then more reliable for permanent encapsulation and protection of the DNA rather than a controlled release.

To gain insight into the correlations between the release loads and the molecular conformation of the DNA molecules released, single molecule AFM was used to provide the needed information on DNA contour length distributions (Figure 13.14). Figure 13.14a depicts the DNA contour length distribution before the release experiment. As can be seen, several DNA fragments are generated upon the enzymatic treatment and can be divided into three main lengths: relatively small fragments (≤100 nm contour length), medium-sized DNA strands (130–160 nm), and large macromolecules (approximately 190–200 nm). The contour length distribution of DNA after its

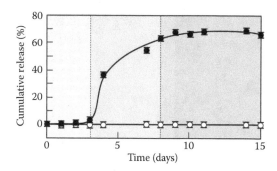

FIGURE 13.13 **(See color insert.)** The release profile of DNA from the two columnar H_{II} mesophases considered: Release from the nonionic system (filled circle, GMO/tricaprylin/water/ascorbic acid) and release from the cationic-based columnar hexagonal phase (open circle, GMO/oleyl amine/tricaprylin/water/ascorbic acid) at room temperature. (I. Amar-Yuli, J. Adamcik, S. Blau, A. Aserin and R. Mezzenga. Controlled embedment and release of DNA from lipidic reverse columnar hexagonal mesophases. *Soft Matter*, 7, 8162–8168, 2011. Reproduced by permission of The Royal Society of Chemistry.)

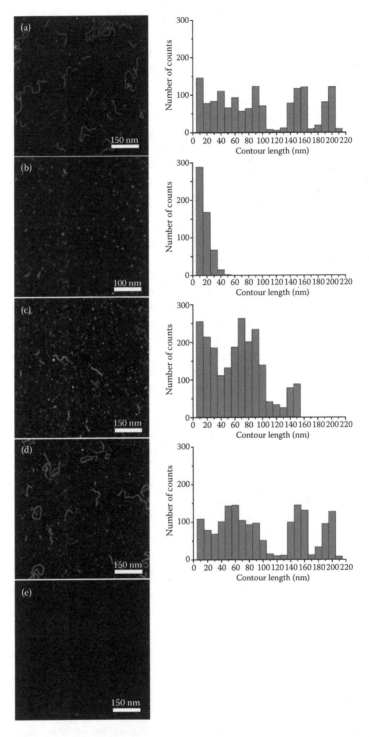

FIGURE 13.14 **(See color insert.)** AFM image and the corresponding contour length distribution of DNA fragments: (a) before incorporating DNA into the H_{II} nonionic LLC system; (b) after 3 days release in the excess water environment; (c) after 4 days release; and (d) after 15 days release at room temperature. (e) AFM image after 15 days of DNA release from the cation-based H_{II} LLC system. (I. Amar-Yuli, J. Adamcik, S. Blau, A. Aserin and R. Mezzenga. Controlled embedment and release of DNA from lipidic reverse columnar hexagonal mesophases. *Soft Matter*, 7, 8162–8168, 2011. Reproduced by permission of The Royal Society of Chemistry.)

release from the nonionic system was examined after 3, 4, and 15 days of release, corresponding to the lag time, the sharp slope, and plateau regions as identified in the release profile (Figure 13.14b through d).

During the lag time, on the third day of the release experiment, only very short length DNA fragments with contour lengths of less than 40 nm could be observed by AFM analysis (Figure 13.14b). These very short strands were too small and in such low concentrations to be detected by the spectrophotometer resulting in the flat absorption signal. The following day (day 4), according to the sharp increase in the diffusion as measured by the UV–vis absorption, the medium-sized DNA strands, mainly 40 to 100 nm in contour length, started to appear coexisting with the very short DNA fragments (Figure 13.14c). Finally, after 15 days, when the release profile had reached a plateau, the longest fragments, 190 to 200 nm in contour length, also appeared in the AFM analysis of the excess water (Figure 13.14d). Interestingly, after 15 days, and only after this time, the contour length distribution of the DNA released matches closely the initial distribution, indicating that release is nearly entirely completed (Figure 13.14d).

Quite in contrast, AFM examination of the excess water in the case of the cationic hexagonal phase reveals no traces of DNA including the smallest fragments detectable by high-resolution AFM (Figure 13.14e), confirming the strong binding of the DNA to the cationic lipids.

Figure 13.15 summarizes the main findings of this work. When confined within nonionic columnar hexagonal phases, DNA interacts with the polar heads of the lipids via hydrogen bonding (as illustrated in the figure), and this enables a controlled release of the DNA in excess water following three main release stages. When positive charges belonging to cationic lipids decorate the water

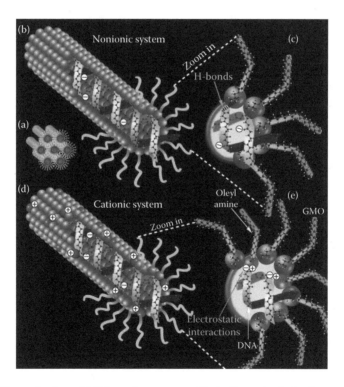

FIGURE 13.15 (See color insert.) Schematic summary illustrating a typical columnar hexagonal phase (a), the DNA entrapment and interactions with the surfactants forming the nonionic (b, c) and cationic (d, e) H_{II} systems. The molecular ratio between all the components is not to scale in the image. (I. Amar-Yuli, J. Adamcik, S. Blau, A. Aserin and R. Mezzenga. Controlled embedment and release of DNA from lipidic reverse columnar hexagonal mesophases. *Soft Matter*, 7, 8162–8168, 2011. Reproduced by permission of The Royal Society of Chemistry.)

channels, binding between the lipid and DNA becomes strong, and any release of DNA is completely suppressed at the charge ratio used. It is reasonable to anticipate that further adjustment of the positive/negative charge ratio or use of variable ionic strengths to partially screen electrostatic attraction can be used as additional means to fine-tune the DNA release kinetics.

13.7 "ON DEMAND" LLC-BASED DRUG DELIVERY SYSTEMS

Based on the principle that the H_{II} phases released model hydrophilic and hydrophobic drugs more slowly than the GMO cubic phase matrix, Fong et al. [57] designed phytantriol and GMO-based bicontinuous cubic (Q_2) and H_{II} nanostructures, which would allow changes to the nanostructure in response to an external change in temperature, with the intention of eventual control of drug release rates *in vivo*. Using glucose as a model hydrophilic drug, drug diffusion was shown to be reversible on switching between the H_{II} (very low release) and Q_2 nanostructures (high release), at temperatures above and below the physiological temperature, respectively (Figure 13.16).

However, the matrix used by Fong et al. [57] required the inclusion of a modifier (vitamin E acetate) to reduce the transition temperature to close to physiological temperatures for *in vivo* application, thereby enabling control over the structure by application of, for example, a heat pack to the skin surface after subcutaneous administration. This is a major limitation because there is no specificity in the heat source; hence, exposure to extremes of temperature may unintentionally induce drug release. Consequently, an alternative means was necessary to induce the phase transition that did not require direct heating, and did not require a reduction in the temperature at which the transition occurs, ideally occurring at the inherent transition temperature without additive (approximately 55°C for phytantriol), removing the potential for accidental activation.

In this context, toward the design of an advanced drug delivery system based on light-triggered phase transition of liquid crystalline phases, Fong et al. [61] reported the design of a novel liquid crystalline matrix–gold nanorod hybrid materials. Hydrophobized gold nanorods (GNRs) have been incorporated within the liquid crystalline matrix, composed of phytantriol and water, to provide remote heating and to trigger the phase transitions upon irradiation at close to their resonant wavelength. The surface of plasmonic metal nanoparticles delivers heat into the surrounding material upon exposure to an appropriate light source at the plasmon resonance. The application of near-infrared sensitive nanorods has significance in designing systems for ophthalmic, subcutaneous, or deeper tissue applications. These investigators used plasmonic nanoparticles to achieve reversible

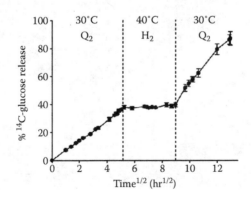

FIGURE 13.16 Dynamic release profiles for glucose into phosphate buffered saline from phytantriol +3% vitamin E acetate with changing temperature, plotted against square root of time. Temperature was switched from 30°C → 40°C → 30°C at the times indicated by the dashed lines. Note: $H_2 = H_{II}$. (I. Amar-Yuli, J. Adamcik, S. Blau, A. Aserin and R. Mezzenga. Controlled embedment and release of DNA from lipidic reverse columnar hexagonal mesophases. *Soft Matter*, 7, 8162–8168, 2011. Reproduced by permission of The Royal Society of Chemistry.)

phase transitions, hence offering a novel practical solution. The presence of nanorods at concentrations up to 3 nmol/L did not change the lattice parameter of cubic (v_2), hexagonal (H_{II}), or micellar (L_2) systems in the absence of laser irradiation.

Irradiation of the system (at 808 nm NIR diode laser), upon the inclusion of a low concentration of GNRs (0.3 nmol/L) in the matrix, did not induce a change in phase structure away from the *Pn3m* bicontinuous cubic phase. However, a small decrease in lattice dimension occurred, indicating that heating of the matrix had occurred. The structure relaxed back to the original position when the laser was off. Upon repeated application of the 5-s laser pulse, the system again displayed the heating effect and relaxed back to the starting position when irradiation was complete (Figure 13.17). This contraction and expansion of the lattice upon heating and cooling (the "breathing mode") was accompanied by concurrent expulsion and uptake of water from the matrix to satisfy the changes in lattice dimension.

In the case of higher concentrations of GNR, the 5 s on pulse for both the 1.5 and 3 nmol/L systems did induce a phase change to the H_2 and L_2 phase structures. At 3 nmol/L, complete transformation to the L_2 phase occurred by the end of the 5 s irradiation, whereas the lower 1.5 nmol/L concentration resulted in a mixed $H_2 + L_2$ phase, indicating a reduced heating effect (Figure 13.17). Again, when the laser was switched off, the system ultimately returned to the initial v_2 (*Pn3m*) phase structure. Interestingly, on conversion from the L_2 or $L_2 + H_2$ state back to the v_2 structure, the "gyroid" bicontinuous cubic phase with *Ia3d* space group was encountered. The gyroid phase coexisted with the H_2 phase initially and then with the v_2 (*Pn3m*) phase for approximately 5 to 6 s after the laser was switched off.

In addition, Figure 13.17 clearly demonstrated that in the case where no GNRs were added to the matrix, there was no significant change in apparent sample temperature (T_{app}) upon laser irradiation. The reversibility of the heating effect in the presence of the GNR is evident from the T_{app} profiles for the three samples containing the nanorods. The sample containing 0.3 nmol/L of GNR heated to approximately 50°C, just below the transition above which coexisting H_2 phase occurs. Increasing the nanorod concentration to 1.5 nmol/L induces heating to approximately 70°C, whereas 3 nmol/L of GNR heated the matrix to an apparent temperature of 75°C. The repeat

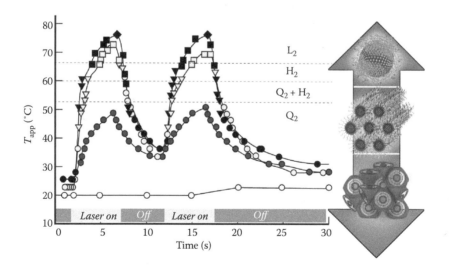

FIGURE 13.17 **(See color insert.)** Effect of laser irradiation on apparent temperature of the phytantriol + water matrix (T_{app}) with change in GNR concentration. GNR = 0 nmol/L are white symbols, 0.3 nmol/L gray symbols, 1.5 nmol/L striped symbols, and 3 nmol/L black symbols. Circles indicate v_2 phase, triangles indicate $v_2 + H_2$, squares indicate $H_2 + L_2$, and diamonds indicate L_2. The cartoon on the right indicates the type of phase structure present with increasing temperature. Note: $H_2 = H_{II}$. (Reprinted with permission from W.K. Fong, T. Hanley, B. Thierry, N. Kirby and B.J. Boyd. Plasmonic nanorods provide reversible control over nanostructure of self-assembled drug delivery materials. *Langmuir*, 26, 6136–6139. Copyright 2010 American Chemical Society.)

irradiation provided the same peak temperature within 1°C to 2°C and reproducible kinetics of heating and cooling. The heating effect observed in these experiments is clearly a function of nanorod concentration, although the relationship between concentration and maximum temperature at differing irradiation conditions requires further investigation and is likely complicated by concurrent cooling. The "breaking wave" shape of the profiles indicates nonlinear heating/cooling gradients in the material.

Fong et al. [61] concluded that GNRs embedded in liquid crystalline matrices produce localized plasmonic heating of the hybrid matrix, enabling fine control over the nanostructure. The phase transitions resulting from photothermal heating were fully reversible and specific to the GNR/laser wavelength combination. Localized plasmonic heating of the LC did not compromise the integrity of the lipid molecules in any of the mesophases. Undoubtedly, this research represented a significant advance toward effective, light-activated drug delivery systems with potential to solve unmet medical needs.

Phan et al. [62] extend the understanding of release to other self-assembled phases, the micellarcubic phase (I_2) and inverse micelles (L_2), in addition to reversed bicontinuous cubic (V_2) and inverse hexagonal (H_2) LCs. All these systems were prepared from GMO, which sequentially forms all four phases with increasing hexadecane (HD) content in excess water.

Figure 13.18 illustrates the *in vitro* release of ^{14}C-glucose from the four LC systems prepared from GMO/HD. There is a significant difference in the release rates of ^{14}C-glucose from the systems over 5 days. The release from the bicontinuous cubic phase was much more rapid than the other three mesophases. Approximately 100% was released over 50 h, compared with 3%, 2.5%, and 0.8% for L_2, H_2, and I_2, respectively. The profiles were approximately linear in all systems except for the bicontinuous cubic phase system, which was seen to plateau after 36 h, as the system started to run out of model drug to release from the matrix. Total recovery of glucose averaged 108% across the $n = 3$ samples for the bicontinuous cubic phase samples, influenced by variability in sampling for total starting concentration and individual release quantitation, but nevertheless close to the anticipated quantitative release based on similar past experiments [58,63,64]. The calculated diffusion coefficients for drug in the matrix over the first 36 h of drug release are presented in Table 13.1. In line with the release data, the magnitude of the diffusion coefficient ranks in the order $V_2 > L_2 > H_2 > I_2$. Diffusion of glucose in the V_2 matrix was shown to be significantly different from the other mesophases by one-way ANOVA test analysis ($p < 0.05$, $n = 3$).

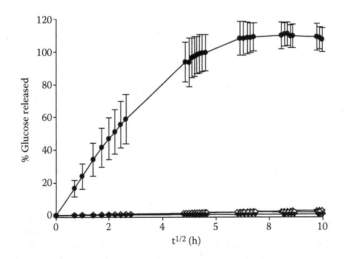

FIGURE 13.18 Model drug (glucose) released (%) from the four mesophases versus square root of time (h) where V_2 (●), L_2 (Δ), H_2 (◊) and I_2 (▽). Data shown as mean ± SD, $n = 3$. Note: $H_2 = H_{II}$. (Reprinted from *Int. J. Pharm.*, 421, S. Phan, W.-K. Fong, N. Kirby, T. Hanley and B.J. Boyd, Evaluating the link between self-assembled mesophase structure and drug release, 176–182, Copyright 2011, with permission from Elsevier.)

TABLE 13.1

Summary of Diffusion Coefficients (D) ^{14}C-Glucose from the Four Mesophases Calculated from Release Data Using Equation 13.1

Mesophase	Diffusion Coefficient (D) ($\times 10^{-8}$ cm^2 s^{-1})
V$_2$	36.3 ± 4.6[a]
L$_2$	2.95 ± 1.61
H$_2$	1.72 ± 0.44
I$_2$	0.12 ± 0.03

Note: mean ± SD, $n = 3$.

[a] Significantly different from the other diffusion coefficients ($p < 0.05$).

The investigators showed that the release kinetics provided a diffusion coefficient for glucose in the bicontinuous cubic phase of $D = 36 \times 10^{-8}$ cm^2 s^{-1}, much faster than the inverse hexagonal phase at $D = 1.7 \times 10^{-8}$ cm^2 s^{-1}. The "macro" viscosity of the stiff solid-like cubic phase is often mentioned as being responsible for the cubic phase's slow release characteristics. However, the closed rodlike micellar structure of the H$_2$ phase, despite its lower viscosity, retards drug release to a significantly greater extent than the V$_2$ phase. Glucose release was shown to be slow for the L$_2$ phase, and extremely slow for the I$_2$ phase. These phases generally have even lower viscosity than the H$_2$ phase, so there is no link between macroviscosity and drug release characteristics in these systems. The finding that the inverse micellar cubic phase has a larger lattice parameter than the bicontinuous cubic phase is not surprising as it has been previously reported by Mariani et al. [65] and by Seddon et al. [22] that due to the apparent mixed micellar structure of I$_2$, its lattice parameter, determined by the repeat distance of the unit cell, is approximately 1.6 times that of a comparable V$_2$ phase with *Pn3m* space group. The lattice parameter was slightly larger in the I$_2$ phase in the monolinolein/ tetradecane dispersions studied by Yaghmur et al. [66] compared with V$_2$, H$_2$, and L$_2$ phases. Hence, it is also apparent that the lattice parameter cannot be used directly to correlate with the drug release behavior. The authors concluded that there is a clear grouping of closed micellar type phase structures that promotes very slow release behavior despite their lower viscosity, whereas the bicontinuous structure offers less opportunity to maintain drug within the matrix reservoir over time [67].

It was suggested that liquid crystalline mesophases with slower release characteristics will have potential applications in sustained drug delivery systems. The results therefore indicate that bicontinuous V$_2$ cubic phase is actually a poor choice for a sustained release matrix material based on drug release alone and that the other mesophases, such as H$_2$, I$_2$, and L$_2$, may be better utilized as sustained-release drug depots in pharmaceutical therapies.

The most straightforward stimulus, which can be exploited for *in vivo* controlled release is certainly pH, because of the large pH changes occurring spontaneously within the mouth–stomach–intestinal tract. In this context, Negrini and Mezzenga [64] presented a pH-responsive lipid-based LLC, which has a number of significant advantages for real oral administration–controlled delivery. First, the system is based on a simple formulation of monolinolein and linoleic acid, which maintains it entirely food-grade. Second, it offers a general, tunable release and diffusion strategy that is not specific to the particular drug ingredients. Finally, the release can be controlled in a fully reversible way and makes it suitable for a targeted delivery to specific points (pH) of the gastrointestinal tract. The lipid (monolinolein), the pH-responsive molecule (linoleic acid), and the model hydrophilic polyphenol drug (phloroglucinol) were used in this study, whereas the main idea behind

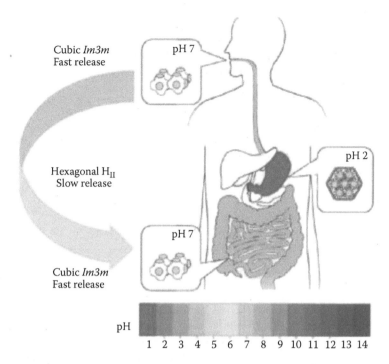

Cubic *Im3m*
Fast release

pH 7

Hexagonal H$_{II}$
Slow release

pH 2

pH 7

Cubic *Im3m*
Fast release

pH

1 2 3 4 5 6 7 8 9 10 11 12 13 14

FIGURE 13.19 **(See color insert.)** Schematics of the proposed pH-responsive drug delivery strategy across the gastrointestinal tract. At pH 7, the drug nanocarrier has a reverse bicontinuous cubic phase (*Im3m*) symmetry; at pH 2, in the stomach, the symmetry of the mesophase changes to reverse hexagonal phase, slowing down diffusion and preventing the premature release of the drug; and at neutral pH in the intestine, the symmetry reverts to bicontinuous cubic phase (*Im3m*), triggering the release of the active ingredients loaded in the mesophase. (Reprinted with permission from R. Negrini and R. Mezzenga. pH-responsive lyotropic liquid crystals for controlled drug delivery. *Langmuir*, 27, 5296–5303. Copyright 2011 American Chemical Society.)

the strategy proposed is highlighted in Figure 13.19. Because of the presence of linoleic acid, which can be in either the deprotonated or protonated state when changing the pH from neutral to acidic conditions, respectively, the LLC undergoes a structural change from reverse bicontinuous cubic phase to reverse columnar hexagonal phase. This is accompanied by a change in the release rate by a factor of 4, thus preventing the release in the stomach and making this system an ideal candidate for the targeted delivery of active ingredients in the basic environment typical of the intestinal tract.

This order–order transition can be explained by the presence of the ionizable carboxylic group of the linoleic acid (intrinsic p$K_a \approx$ 5). The linoleic acid is negatively charged at pH 7; the electrostatic repulsions between the negatively charged headgroups stabilize the *Im3m* bicontinuous cubic phase. When the pH decreases below the pK_a value, the carboxyl group reprotonates to a large extent, the surface charge density on the water channels at the water–lipid interface decreases, and the linoleic acid becomes highly hydrophobic, acting as a lipid molecule and stabilizing the hexagonal phase at 37°C.

This order–order transition is well-rationalized by the concept of the CPP expressed as *v/Al*, which is the ratio between the volume of the hydrophobic lipid tail, *v*, and the product of the cross-sectional lipid head area, *A*, and the lipid chain length, *l*. When linoleic acid is deprotonated (pH 7), the effective area *A* is large because of the electrostatic repulsive interactions among different lipid heads. When, however, the linoleic acid is mostly neutral (pH 2), *A* decreases and the CPP increases, promoting the transition from flat to reverse (water-in-oil) interfaces and inducing a bicontinuous cubic → reverse columnar hexagonal transition.

In vitro release studies were carried out first to establish the influence of the liquid crystalline symmetry on the release behavior. Figure 13.20a illustrates the drug release profiles from the liquid

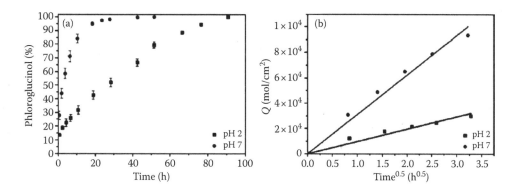

FIGURE 13.20 pH-induced changes in the release of phloroglucinol from the reverse bicontinuous cubic and hexagonal phases at 37°C. (a) Phloroglucinol released from the bicontinuous cubic phase at pH 7 (●) and reverse hexagonal phase at pH 2 (■) plotted against time. (b) Moles released per unit area plotted against the square root of time, illustrating the different diffusion-controlled behaviors in the two mesophases. (Reprinted with permission from R. Negrini and R. Mezzenga. pH-responsive lyotropic liquid crystals for controlled drug delivery. *Langmuir*, 27, 5296–5303. Copyright 2011 American Chemical Society.)

crystalline matrices, plotted as a percentage of the released drug against time. As can be observed, the release from *Im3m* is much more rapid and is nearly completed after 20 h; at this time, H_{II} has not yet released half of the initially loaded drug.

In Figure 13.20b, the profiles of the drugs released are plotted against the square root of time; the linear behavior confirms the Fickian diffusion release. Using the Higuchi equation (Equation 13.1), the diffusion coefficients are calculated to be $D_{pH 2} = 2.2 \times 10^{-8}$ cm^2/s and $D_{pH 7} = 30.6 \times 10^{-8}$ cm^2/s (slopes = 1.34×10^{-8} at pH 2 and 5×10^{-8} at pH 7), respectively. It is therefore possible to conclude that the bicontinuous cubic phase releases almost four times faster than the reverse hexagonal phase, consistent with previously reported work [64]. To demonstrate a pH-triggered on–off release behavior and tunable release profiles in dynamic conditions, the release study was also carried out by switching the pH between the values of 7 and 2 using the same setup, and the resulting release

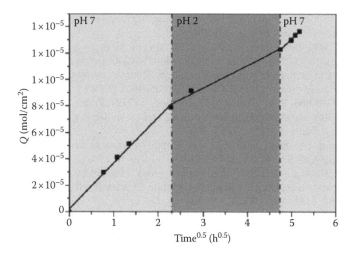

FIGURE 13.21 Release profiles of phloroglucinol plotted against the square root of time, upon sequential switch in the pH of the excess solution from pH 7 → pH 2 → pH 7. (Reprinted with permission from R. Negrini and R. Mezzenga. pH-responsive lyotropic liquid crystals for controlled drug delivery. *Langmuir*, 27, 5296–5303. Copyright 2011 American Chemical Society.)

data are given in Figure 13.21. The values of the slopes obtained in these conditions are 5.7×10^{-8} in the initial pH 7 conditions, 2.58×10^{-8} after switching the solution to pH 2, and 5.3×10^{-8} upon reverting the pH to 7 again. The values of the slopes at pH 7 agree well with those obtained by the static release experiment. Although samples measured at pH 2 show a slightly faster release; this results from the relatively long time needed to induce a complete change of *Im3m* into H_{II} in bulk mesophases, leading to a long-living coexistence of *Im3m* and H_{II} upon the switch of the pH from 7 to 2. It can be easily anticipated that, when applying these concepts to cubosomes and hexosomes of a few hundred nanometers in diameter, the order–order transitions will occur much faster, leading to a sharper pH-induced on–off–on release along the gastrointestinal tract.

13.8 SUMMARY

This chapter demonstrated that LC mesophases with complex architectures could be rationally designed and prepared for the solubilization and potential administration of bioactive molecules. It was shown that comprehensive understanding of the structural properties of the carriers is imperative for rational and successful tailoring of delivery vehicles. The principal strategy was to first characterize the different levels of organization of these special materials and then to explore "on demand" targeted release, based on controlling the polymorphism of lyotropic liquid crystalline mesophases. LLCs were shown to offer new approaches in the design technologies for DNA transfection and for gene delivery. Clearly, many additional experiments need to be carried out to clarify the detailed structure, the exact properties, and specific potential of these systems.

REFERENCES

1. D.L. Gin, C.S. Pecinovsky, J.E. Bara and L. Kerr. Functional lyotropic liquid crystal materials. *Struct. Bond.*, **128**, 181–222 (2008).
2. K. Larsson. Cubic lipid–water phases: Structures and biomembrane aspects. *J. Phys. Chem.*, **93**, 7304–7314 (1989).
3. I. Amar-Yuli. Hexagonal liquid crystals and hexosomes structural modifications and solubilization. Ph.D. dissertation, The Hebrew University of Jerusalem, Israel (2008).
4. S.M. Gruner. Stability of lyotropic phases with curved interfaces. *J. Phys. Chem.*, **93**, 7562–7570 (1989).
5. J.M. Seddon. Structure of the inverted hexagonal (H_{II}) phase, and non-lamellar phase-transitions of lipids. *Biochim. Biophys. Acta*, **1031**, 1–69 (1990).
6. L. Sagalowicz, R. Mezzenga and M.E. Leser. Investigating reversed liquid crystalline mesophases. *Curr. Opin. Colloid Interface Sci.*, **11**, 224–229 (2006).
7. L. Sagalowicz, M.E. Leser, H.J. Watzke and M. Michel. Monoglyceride self-assembly structures as delivery vehicles. *Trends Food Sci. Technol.*, **17**, 204–214 (2006).
8. V. Cherezov, J. Clogston, M.Z. Papiz and M. Caffrey. Room to move: Crystallizing membrane proteins in swollen lipidic mesophases. *J. Mol. Biol.*, **357**, 1605–1618 (2006).
9. B.J. Boyd, S.-M. Khoo, D.V. Whittaker, G. Davey and C.J.H. Porter. A lipid-based liquid crystalline matrix that provides sustained release and enhanced oral bioavailability for a model poorly water soluble drug in rats. *Int. J. Pharm.*, 340, 52–60 (2007).
10. C.J. Drummond and C. Fong. Surfactant self-assembly objects as novel drug delivery vehicles. *Curr. Opin. Colloid Interface Sci.*, **4**, 449–456 (2000).
11. S.B. Rizwan, T. Hanley, B.J. Boyd, T. Rades and S. Hook. Liquid crystalline systems of phytantriol and glyceryl monooleate containing a hydrophilic protein: Characterisation, swelling and release kinetics. *J. Pharm. Sci.*, **98**, 4191–4204 (2009).
12. R. Efrat, A. Aserin and N. Garti. On structural transitions in a discontinuous micellar cubic phase loaded with sodium diclofenac. *J. Colloid Interface Sci.*, **321**, 166–176 (2008).
13. S.M. Gruner, M.W. Tate, G.L. Kirk, P.T.C. So, D.C. Turner, D.T. Keane, C.P.S. Tilcock and P.R. Cullis. X-ray-diffraction study of the polymorphic behavior of *N*-methylated dioleoylphosphatidylethanolamine. *Biochemistry*, 27, 2853–2866 (1988).
14. M. Gradzielski. Investigations of the dynamics of morphological transitions in amphiphilic systems. *Curr. Opin. Colloid Interface Sci.*, **9**, 256–263 (2004).

15. L. Yang, L. Ding and H.W. Huang. New phases of phospholipids and implications to the membrane fusion problem. *Biochemistry*, **42**, 6631–6635 (2003).

16. L.B. Lopes, F.F.F. Speretta, M. Vitoria and L.B. Bentley. Enhancement of skin penetration of vitamin K using monoolein-based liquid crystalline systems. *Eur. J. Pharm. Sci.*, **32**, 209–215 (2007).

17. B.J. Boyd, D.V. Whittaker, S.-M. Khoo and G. Davey. Lyotropic liquid crystalline phases formed from glycerate surfactants as sustained release drug delivery systems. *Int. J. Pharm.*, **309**, 218–226 (2006).

18. G.C. Shearman, A.I.I. Tyler, N.J. Brooks, R.H. Templer, O. Ces, R.V. Law and J.M. Seddon. A 3-D hexagonal inverse micellar lyotropic phase. *J. Am. Chem. Soc.*, **131**, 1678–1679 (2009).

19. J.N. Israelachvili, D.J. Mitchell and B.W. Ninhan. Theory of self-assembly of hydrocarbon amphiphiles into micelles and bilayers. *J. Chem. Soc., Faraday Trans. 2*, **72**, 1525–1568 (1976).

20. G.C. Shearman, O. Ces, R.H. Templer and J.M. Seddon. Inversely otropic phases of lipids and membrane curvature. *J. Phys. Condens. Matter*, **18**, S1105–S1124 (2006).

21. M.H. Shah and A. Paradkar. Cubic liquid crystalline glyceryl monooleate matrices for oral delivery of enzyme. *Int. J. Pharm.*, **294**, 161–171 (2005).

22. J.M. Seddon, J. Robins, T. Gulik-Krzywicki and H. Delacroix. Inverse micellar phases of phospholipids and glycolipids. *Phys. Chem. Chem. Phys.*, **2**, 4485–4493 (2000).

23. J.M. Seddon, N. Zeb, R.H. Templer, R.N. McElhaney and D.A. Mannock. AnFd3m lyotropic cubic phase in a binary glycolipid/water system. *Langmuir*, **12**, 5250–5253 (1996).

24. I. Amar-Yuli and N. Garti. Transitions induced by solubilized fat into reverse hexagonal mesophases. *Colloids Surf. B*, **43**, 72–82 (2005).

25. I. Amar-Yuli, D. Libster, A. Aserin and N. Garti. Solubilization of food bioactives within lyotropic liquid crystalline mesophases. *Curr. Opin. Colloid Interface Sci.*, **14**, 21–32 (2009).

26. I. Amar-Yuli, E. Wachtel, E. Ben Shoshan, D. Danino, A. Aserin and N. Garti. Hexosome and hexagonal phases mediated by hydration and polymeric stabilizer. *Langmuir*, **23**, 3637–3645 (2007).

27. I. Amar-Yuli, E. Wachtel, D. Shalev, H. Moshe, A. Aserin and N. Garti. Thermally induced fluid reversed hexagonal (H-II) mesophase. *J. Phys. Chem. B*, **111**, 13544–13553 (2007).

28. I. Amar-Yuli, E. Wachtel, D. Shalev, A. Aserin and N. Garti. Low viscosity reversed hexagonal mesophases induced by hydrophilic additives. *J. Phys. Chem. B*, **112**, 3971–3982 (2008).

29. I. Amar-Yuli, A. Aserin and N. Garti. Solubilization of nutraceuticals into reverse hexagonal mesophases. *J. Phys. Chem. B*, **112**, 10171–10180 (2008).

30. L. Bitan-Cherbakovsky, I. Amar-Yuli, A. Aserin and N. Garti. Structural rearrangements and interaction within H_{II} mesophase induced by cosolubilization of vitamin E and ascorbic acid. *Langmuir*, **25**, 13106–13113 (2009).

31. L. Bitan-Cherbakovsky, I. Amar-Yuli, A. Aserin and N. Garti. Solubilization of vitamin E into H_{II} LLC mesophase in the presence and in the absence of vitamin C. *Langmuir*, **26**, 3648–3653 (2010).

32. D. Libster, A. Aserin, E. Wachtel, G. Shoham and N. Garti. An H_{II} liquid crystal-based delivery system for cyclosporin A: Physical characterization. *J. Colloid Interface Sci.*, **308**, 514–524 (2007).

33. D. Libster, P. Ben Ishai, A. Aserin, G. Shoham and N. Garti. From the microscopic to the mesoscopic properties of lyotropic reverse hexagonal liquid crystals. *Langmuir*, **24**, 2118–2127 (2008).

34. D. Libster, P. Ben Ishai, A. Aserin, G. Shoham and N. Garti. Molecular interactions in reverse hexagonal mesophase in the presence of cyclosporin A. *Int. J. Pharm.*, **367**, 115–126 (2009).

35. P. Ben Ishai, D. Libster, A. Aserin, N. Garti and Y. Feldman. Molecular interactions in lyotropic reverse hexagonal liquid crystals: A dielectric spectroscopy study. *J. Phys. Chem. B*, **113**, 12639–12647 (2009).

36. Y. Kurosaki, N. Nagahara, T. Tanizawa, H. Nishimura, T. Nakayama and T. Kimura. Use of lipid disperse systems in transdermal drug delivery: Comparative study of flufenamic acid permeation among rat abdominal skin, silicone rubber membrane and stratum corneum sheet isolated from hamster cheek pouch. *Int. J. Pharm.*, **67**, 1–9 (1991).

37. E. Touitou, H.E. Junginger, N.D. Weiner, T. Nagai and M. Mezei. Liposomes as carriers for topical and transdermal delivery, *J. Pharm. Sci.*, **83**, 1189–1203 (1994).

38. A. Spernath and A. Aserin. Microemulsions as carriers for drugs and nutraceuticals. *Adv. Colloid Interface Sci.*, **128–130**, 47–64 (2006).

39. A. Spernath, A. Aserin and N. Garti. Fully dilutable microemulsions embedded with phospholipids and stabilized by short-chain organic acids and polyols. *J. Colloid Interface Sci.*, **299**, 900–909 (2006).

40. H. Brondsted, H.M. Nielsen and L. Hovgaard. Drug-delivery studies in caco-2 monolayers. 3. Intestinal transport of various vasopressin analogs in the presence of lysophosphatidylcholine. *Int. J. Pharm.*, **114**, 151–157 (1995).

41. D.-Z. Liu, E.L. LeCluyse and D.R. Thakker. Dodecylphosphocholine-mediated enhancement of paracellular permeability and cytotoxicity in Caco-2 cell monolayers. *J. Pharm. Sci.*, **88**, 1161–1168 (1999).

42. D. Libster, A. Aserin, D. Yariv, G. Shoham and N. Garti. Concentration- and temperature-induced effects of incorporated desmopressin on the properties of reverse hexagonal mesophase. *J. Phys. Chem. B*, **113**, 6336–6346 (2009).

43. M. Cohen-Avrahami, A. Aserin and N. Garti. H_{II} mesophase and peptide cell-penetrating enhancers for improved transdermal delivery of sodium diclofenac. *Colloids Surf. B*, **77**, 131–138 (2010).

44. T. Mishraki, D. Libster, A. Aserin and N. Garti. Lysozyme entrapped within reverse hexagonal mesophases: Physical properties and structural behavior. *Colloids Surf. B*, **75**, 47–56 (2010).

45. T. Mishraki, D. Libster, A. Aserin and N. Garti. Temperature-dependent behavior of lysozyme within the reverse hexagonal mesophases (H_{II}). *Colloids Surf. B*, **75**, 391–397 (2010).

46. M. Cohen-Avrahami, D. Libster, A. Aserin and N. Garti. Sodium diclofenac and cell-penetrating peptides embedded in H_{II} mesophases: Physical characterization and delivery. *J. Phys. Chem. B*, **115**, 10189–10197 (2011).

47. T. Higuchi. Mechanism of sustained action medication. Theoretical analysis of rate of release of solid drugs dispersed in solid matrices. *J. Pharm. Sci.*, **52**, 5296–5303 (1963).

48. N. Garti, G. Hoshen and A. Aserin. Lipolysis and structure controlled drug release from reversed hexagonal mesophase. *Colloids Surf. B*, **94**, 36–43 (2012).

49. J. Borné, T. Nylander and A. Khan. Effect of lipase on different lipid liquid crystalline phases formed by oleic acid based acylglycerols in aqueous systems. *Langmuir*, **18**, 8972–8981 (2002).

50. J. Borné, T. Nylander and A. Khan. Effect of lipase on monoolein-based cubic phase dispersion (cubosomes) and vesicles. *J. Phys. Chem. B*, **106**, 10492–10500 (2002).

51. A. Misiūnas, Z. Talaikytė, G. Niaura, V. Razumas and T. Nylander. *Thermomyces lanuginosus* lipase in the liquid-crystalline phases of aqueous phytantriol: X-ray diffraction and vibrational spectroscopic studies. *Biophys. Chem.*, **134**, 144–156 (2008).

52. U. Derewenda, L. Swenson, Y.Y. Wei, R. Green, P.M. Kobos, R. Joerger, M.J. Haas and Z.S. Derewenda. Conformational lability of lipases observed in the absence of an oil–water interface. Crystallographic studies of enzymes from the fungi Humicola lanuginosa and Rhisopus delemar. *J. Lipid Res.*, **35**, 524–534 (1994).

53. A.M. Brzozowski, U. Derewenda, Z.S. Derewenda, G.G. Dodson, D.M. Lawson, J.P. Turkenburg, F. Bjorkling, B. Huge-Jensen, S.A. Patkar and L. Thim. A model for interfacial activation in lipases from the structure of a fungal lipase-inhibitor complex. *Nature*, **351**, 491–494 (1991).

54. C.R. Daas and P.F.M. Choong. Biophysical delivery of peptides: Applicability for cancer therapy. *Peptides*, **27**, 3479–3488 (2006).

55. T.R.S. Kumar, K. Soppimath and S.K. Nachaegari. Novel delivery technologies for protein and peptide therapeutics. *Curr. Pharm. Biotechnol.*, **7**, 261–276 (2006).

56. B. Ericsson, P.O. Eriksson, J.E. Löfroth and S. Engström. Cubic phases as delivery systems for peptide drugs. In *Polymeric Drugs and Drug Delivery System*, R.L. Dunn and R.M. Ottenbritte (Eds.), ACS Symp. Ser., Vol. 469. American Chemical Society, Washington, DC, pp. 251–265 (1991).

57. W.K. Fong, T. Hanley and B.J. Boyd. Stimuli responsive liquid crystals provide 'on-demand' drug delivery in vitro and in vivo. *J. Controlled Release*, **135**, 218–226 (2009).

58. J.C. Shah, Y. Sadhale and M.C. Dakshina. Cubic phase gels as drug delivery systems. *Adv. Drug Deliv. Rev.*, **47**, 229–250 (2001).

59. D. Libster, A. Aserin and N. Garti. Interactions of biomacromolecules with reverse hexagonal liquid crystals: Drug delivery and crystallization applications. *J. Colloid Interface Sci.*, **356**, 375–386 (2011).

60. I. Amar-Yuli, J. Adamcik, S. Blau, A. Aserin and R. Mezzenga. Controlled embedment and release of DNA from lipidic reverse columnar hexagonal mesophases. *Soft Matter*, **7**, 8162–8168 (2011).

61. W.K. Fong, T. Hanley, B. Thierry, N. Kirby and B.J. Boyd. Plasmonic nanorods provide reversible control over nanostructure of self-assembled drug delivery materials. *Langmuir*, **26**, 6136–6139 (2010).

62. S. Phan, W.-K. Fong, N. Kirby, T. Hanley and B.J. Boyd. Evaluating the link between self-assembled mesophase structure and drug release. *Int. J. Pharm.*, **421**, 176–182 (2011).

63. K.W.Y. Lee, T.-H. Nguyen, T. Hanley and B.J. Boyd. Nanostructure of liquid crystalline matrix determines in vitro sustained release and in vivo oral absorption kinetics for hydrophilic model drugs. *Int. J. Pharm.*, **365**, 190–199 (2009).

64. R. Negrini and R. Mezzenga. pH-responsive lyotropic liquid crystals for controlled drug delivery. *Langmuir*, **27**, 5296–5303 (2011).

65. P. Mariani, E. Rivas, V. Luzzati and H. Delacroix. Polymorphism of a lipid extract from pseudomonas-fluorescens—Structure-analysis of a hexagonal phase and of a novel cubic phase of ectinction symbol Fd. *Biochemistry*, **29**, 6799–6810 (1990).

66. A. Yaghmur, L. de Campo, S. Salentinig, L. Sagalowicz, M.E. Leser and O. Glatter. Oil-loaded mono-linolein-based particles with confined inverse discontinuous cubic structure (Fd3m). *Langmuir*, **22**, 517–521 (2006).
67. H. Ljusberg-Wahren, L. Nyberg and K. Larsson. Dispersion of the cubic liquid crystalline phase—Structure, preparation and functionality aspects. *Chim. Oggi*, **14**, 40–43 (1996).

14 Pharmaceutical Microemulsions and Drug Delivery

Maung Win, Paul Lang, Manu Vashishtha, and Dinesh O. Shah

CONTENTS

14.1 INTRODUCTION

A microemulsion can be defined as a thermodynamically stable, transparent dispersion of two immiscible liquids, stabilized by an interfacial film of an emulsifying agent (Figure 14.1) [1]. Due to the large interfacial area, isotropic nature, and thermodynamic stability of microemulsions, they show promise over macroemulsions in the solubilization and delivery of drugs [2–5]. A well-designed pharmaceutical microemulsion offers an extremely promising vehicle for drug delivery with very high interfacial area, small (10–100 nm) droplet size, infinite shelf-life, and tailored phase behavior to meet the requirements of drug distribution, bioavailability, and drug-release kinetics [6].

Although microemulsions represent a single liquid phase macroscopically, at a molecular level, they consist of three compartments, namely, oil, water, and the interfacial region (Figure 14.2). The most efficient surfactant + cosurfactant system for generating a microemulsion is the one that puts a maximum percentage of these molecules in the interfacial region or compartment as compared with that in oil or water. This in turn generates a large total interfacial area for the microemulsion system. The drug molecules can be partitioned in any of these three compartments preferentially. The total solubility of the drug in a microemulsion depends on the structure of the microemulsion as well as the partitioning of the drug in these three compartments (i.e., oil, water, and the interface). As shown schematically in Figure 14.3, by controlling the inherent stability of a microemulsion, one can modulate the flux or release rate of the drug from the microemulsion.

It should be emphasized that there is some confusion about the criterion used to call a system a microemulsion. Most researchers have used transparency as the only criterion to call a mixture of oil, water, surfactant, and cosurfactant a microemulsion. However, with the availability of dynamic

377

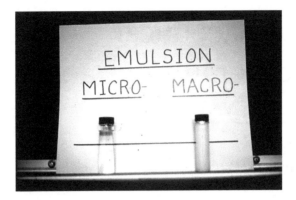

FIGURE 14.1 Samples of microemulsions versus macroemulsions. Extremely small size (<100 nm) of droplets makes the microemulsions transparent.

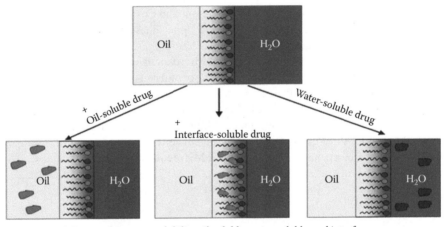

Microemulsions can solubilize oil-soluble, water-soluble, and interface-soluble drugs in a single phase liquid.

FIGURE 14.2 **(See color insert.)** Three invisible compartments of microemulsions: oil, water, and the interfacial region.

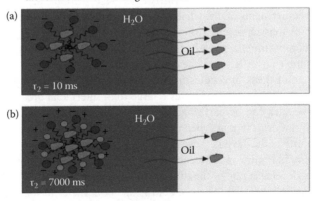

FIGURE 14.3 **(See color insert.)** Inherent stability of a microemulsion droplet and the flux of the drug from the droplet to the surrounding and the mixed micelles shown in the lower part of the figure have inherently greater stability. (a) Less stable micelle and (b) more stable micelle.

Components of Microemulsions

- Oil
- Water
- Surfactant (ionic or nonionic)
- Cosurfactant (short chain alcohols or nonionic surfactants)

- Oil – may be dispersed or continuous phase
- Water – may be dispersed or continuous phase
- Surfactant – lowers interfacial tension between oil and water phases
- Cosurfactant – usually short chain alcohols
  - Fluidizes the interface
  - Decreases interfacial viscosity
  - Destroys lamellar liquid crystalline structure
  - Reduces electrical repulsion between droplets
  - Induces appropriate curvature changes

FIGURE 14.4 Major components of microemulsions.

light scattering, small angle X-ray scattering (SAXS) and small angle neutron scattering (SANS), and pulsed gradient nuclear magnetic resonance (NMR), one can identify the different microstructures in the transparent region of the phase diagram of oil + water + surfactant + cosurfactant [7,8].

In general, a microemulsion can have four components, for example, oil, water, a surfactant, and a cosurfactant (Figure 14.4). However, there are reports of microemulsions formed by using a surfactant alone or with no cosurfactant [9,10]. A number of studies [11–13] have shown that the use of cosurfactants increases the area of one-phase microemulsion region [14–16] in the respective pseudoternary phase diagrams, which is important in controlling phase behavior upon injecting or ingesting a microemulsion into the body in relation to its *in vivo* stability. It must be noted that ultimately microemulsions will disintegrate through dilution by blood or body fluids, binding of the surfactants to other solutes, or adsorption of surfactants on the cellular walls. However, by controlling the structure of microemulsions, one can modulate the disintegration process of microemulsions. Therefore, depending on the structure of a microemulsion, it can either disintegrate very quickly or slowly upon administration into the body. Chain-length compatibility of surfactant with oil and alcohol has shown some very striking effects on the solubilization of water in microemulsions [17–21].

Surfactants and cosurfactants have to be nontoxic, biodegradable, and categorized as generally regarded as safe (GRAS) for the formulation of pharmaceutical microemulsions. Pharmaceutically acceptable cosurfactants are ethanol, medium chain monoglycerides and diglycerides, 1,2-alkanediols, sucrose–ethanol mixtures, alkyl monoglucosides, and geraniols [1,22]. Pharmaceutically acceptable surfactants are zwitterionic phospholipids, and nonionic surfactants such as Brijs, Tweens, Arlacel 186, sucrose esters, polyoxyethylene alkyl ethers, poly glycerol fatty acid esters, and sorbitan esters.

If alkyl chain length of oil is increased, the intermolecular interactions between the oil molecules increases with chain length and hence the polar molecules are pushed out to the interface or to the aqueous phase [18]. The empirical relation linking the lengths of oil, surfactant, and alcohol molecules with maximum solubilization was coined the Bansal, Shah, and O'Connell (BSO) equation by Garti and coworkers [19].

The primary role of surfactant is to lower the interfacial tension as well as to provide necessary total interfacial area to create the microemulsions. The role of cosurfactant is to lower the interfacial tension further into the milli-Newton range and make the spontaneous formation of microemulsions possible.

The cosurfactant also fluidizes the interface by disordering the alkyl chains of surfactant molecules (Figure 14.5). It should be noted that earlier studies on microemulsions were carried out

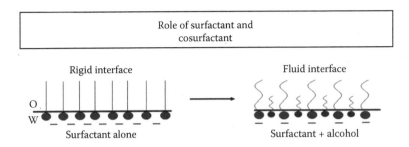

FIGURE 14.5 Role of surfactant and cosurfactant molecules at the oil/water interface.

using C16 or C18 soaps and short-chain alcohols like C5 or C6. Thus, the presence of short-chain alcohol molecules in between long fatty acid chains would not only disorder the long chains of soap molecules but would also facilitate the penetration of oil molecules into this mixed film and hence promote the wettability of the interface by oil. This can also influence the curvature of microemulsion droplets.

By titrating a water-in-oil microemulsion with hexanol, and oils of different chain length, one can draw the Schulman–Bowcott plot [23]. This is a simple mass–balance equation. However, when extrapolated to zero concentration of oil, the intercept on y axis gives the molar ratio of alcohol to surfactant at the interface. Thus, for most microemulsions, this ratio can be anywhere between 1 and 5. The more alcohol molecules per molecule of the surfactant at the interface, the larger the total interfacial area will be and hence the smaller the droplet size will be for the same volume of solubilized water in oil.

14.2 THE IMPORTANCE OF MOLECULAR SHAPE OF SURFACTANT AND COSURFACTANT MOLECULES

Figure 14.6 shows three samples of oil + water (50:50 v/v) emulsions using hexadecane, water, AOT, AOT + Arlacel-20 (1:1 w/w), or Arlacel-20 [9,10]. The AOT is a two-chain anionic surfactant, Arlacel-20 is a dodecyl sorbitol molecule. It is evident that the mixture creates the necessary condition to form transparent microemulsions. The interfacial tension as measured by drop-volume is very low when the surfactant and cosurfactant are mixed in this proportion (Figure 14.7). This also coincides with maximum solubilization of water-in-oil microemulsions. This can be interpreted as being due to the good fit of the intermolecular space left by AOT molecules with dodecyl sorbitol molecules. It was further shown [10] that these two surfactants and cosurfactants can be used as the main components and one

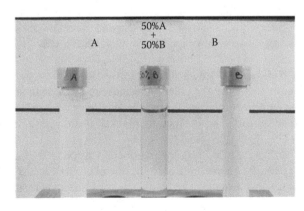

FIGURE 14.6 Three oil + water + emulsifier samples after shaking. Left, surfactant AOT; right, Arlacel-20; and middle, 1:1 w/w mixture of two surfactants (AOT and Arlacel-20 + oil + water).

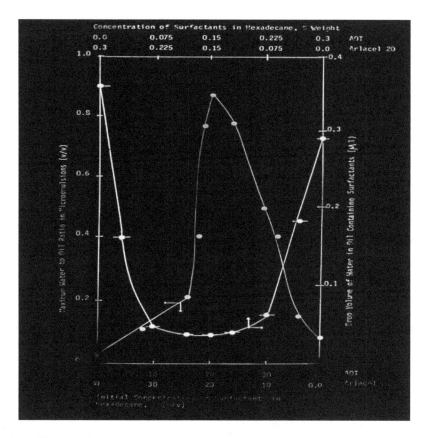

FIGURE 14.7 **(See color insert.)** Solubilization of water in hexadecane + AOT + Arlacel-20 mixture. The maximum solubilization of water occurs when the drop-volume is at a minimum or when the interfacial tension between oil and water is at a minimum in the presence of AOT and Arlacel-20.

can use any alkane to make water in oil microemulsions. For each oil, depending on its chain length, the maximum solubilization occurs at a specific molar ratio between these surfactants and cosurfactant.

14.3 FORMATION OF LOWER-PHASE, MIDDLE-PHASE, AND UPPER-PHASE MICROEMULSIONS BY VARIATION OF SEVERAL PARAMETERS

If one takes 50:50 oil and water and then adds approximately 5% surfactant and 3% alcohol, and shakes the mixture vigorously, then upon equilibration, it may form lower phase microemulsion in equilibrium with the upper oil phase. The reduction in the volume of oil phase taken initially indicates that the volume of oil solubilized in the water phase (Figure 14.8). If one repeats this experiment using increasing amounts of salt such as sodium chloride, one will find the situation shown in Figure 14.8.

Upon chemical analysis of each phase, one would find that increasing salt concentration induces a migration of surfactant plus some alcohol from the water phase to the middle phase to the upper phase microemulsions. This phase behavior is also accompanied by a change in the solubilization of oil or water (or both) in the microemulsion phase and in the interfacial tension between adjacent phases.

The salinity at which the middle phase microemulsion solubilizes equal volumes of oil and brine is called optimal salinity. The interfacial tension at both interfaces, microemulsion/water or microemulsion/oil, is very low in the milli-Newton range. It has been shown that at the optimal salinity, various phenomena involving oil + water + surfactant and alcohol show maxima or minima [24–26]. In summary, at the optimal salinity, the oil recovery from the porous media is maximum, the pressure-drop across the porous media is minimum, the surfactant loss in porous media is

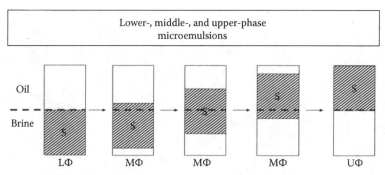

* S represents the surfactant rich phase

FIGURE 14.8 Effect of increasing salt concentration on the migration of surfactant + cosurfactant from lower to middle to upper phase.

minimum, the coalescence time or phase separation time for oil and water is minimum, and the interfacial tension is minimum [27–33].

Like salt concentration, one can change oil chain length, alcohol or surfactant chain length, or temperature to induce such phase behavior, that is, the formation of lower to middle to upper phase microemulsions. By mixing ethoxylated surfactants with petroleum sulfonates, one can increase the optimal salinity from 1% to 25% NaCl concentration. Thus, one can make the middle phase microemulsions in a wide range of salinity [27].

Each type of microemulsion has specific applications. The oil-in-water (lower phase) microemulsions are useful as metal-cutting fluids, or for agricultural sprays, the middle phase microemulsions are useful for enhanced oil recovery, whereas the upper phase microemulsions (i.e., water-in-oil) are relevant for preparing nanocrystals or nanoparticles of solid materials [28–39] using them as nanoreactors.

The requirements for pharmaceutical microemulsions are rather very strict. The components have to be nontoxic, nonirritating, blood compatible, and be of GRAS category. The following is a detailed description of the pharmaceutical microemulsions and their efficacy in drug delivery.

14.4 TRANSDERMAL APPLICATIONS OF PHARMACEUTICAL MICROEMULSIONS

Transdermal applications of microemulsions offer the ability to deliver drug locally, avoid the gastrointestinal tract, and deliver an alternative route for orally delivered drugs. The main barrier to transdermal application is the transport through the stratum corneum, the outermost 50 μm of skin. However, the large total interfacial area provided by microemulsion allows for greater contact area between the stratum corneum and microemulsion droplets, allowing for enhanced drug transfer. In addition, the ability to solubilize high concentrations of a drug in microemulsion droplets contributes to a high concentration gradient that aids drug transfer across the stratum corneum.

Several mechanisms are exploited to enhance transdermal drug delivery. The most popular being the use of permeation enhancers [40]. Transcutol, eucalyptus oil, and peppermint oil have been used as permeation enhancers with varying degrees of success (transcutol having the greatest effect) [41]. Menthol has also been used as a permeation enhancer [42]. The presence of a cosurfactant such as ethanol also increases transdermal drug flux compared with cosurfactant-free formulations [40]. Not only do permeation enhancers increase penetration, other components of microemulsion such as the surfactant and oil phase can also enter the skin. For example, lecithin has a high affinity for epidermal tissue and can mix with the skin lipid components. This can lead to a change in skin lipid fluidity and cause an enhanced absorption of drugs [43]. In general, drugs with lower molecular weights have higher permeation rates through skin. A predictive rule is that the flux of drug through the skin should decrease by a factor of 5 for an increase of 100 daltons in molecular weight of the drug [44].

Transdermal microemulsion formulations of nonsteroidal anti-inflammatory drugs (NSAIDs), antifungals, antivirals, hormones, local anesthetics, and other miscellaneous drugs have been reported (Table 14.1). Many transdermal microemulsion formulations must be thickened to have the proper viscosity for transdermal application. For example, a gelling agent, Carbopol ETD 220, was used to formulate antifungal gels of fluconazole and clotrimazole [45,46]. The formulations showed no irritation. In addition, the clotrimazole formulation showed greater antifungal activity compared with the commercial Candid-V due to greater bioadhesion and enhanced penetration of oil globules through fungal cell walls [46]. Another popular method to thicken oil-in-water microemulsion formulations is through the use of hydrogels. Gelling agents such as Carbomer 940, xanthan gum, and carrageenan have been used to increase the viscosity of microemulsion and form microemulsion-based hydrogels [47,48].

A new oil-in-water microemulsion-based gel containing 1% itraconazole (ITZ), which is nonirritating and produces no erythema or edema, has been developed for topical delivery [49]. Topical

TABLE 14.1
Overview of Transdermal Microemulsions and their *In vivo* Advantages

Drug	Composition	*In vivo* Advantages	Reference	Microemulsion Type
Aceclofenac	Oleic acid, linoleic acid, triacetin or Labrafac as oil phase, Transcutol, Labrasol	Greater reduction in pain compared with cream	Yang et al. [57]	o/w
Ketoprofen	Triglycerides, lecithin, *n*-butanol	Enhanced permeation through human skin with respect to conventional formulations	Paolino et al. [43]	o/w
Lidocaine	Glyceryl oleate, polyoxyl 40, tetraglycol, isopropyl palmitate	Shorter time to provide analgesia compared with commercial cream	Sintov and Shapiro [58]	o/w
Tetracaine hydrochloride	Lecithin, *n*-propanol, isopropyl myristate	Increase in local analgesic response compared with aqueous drug solution	Changez et al. [59]	w/o and o/w (both)
Theophylline	Oleic acid, Cremophor RH40, Labrasol	Higher AUC values compared with oral administration	Zhao et al. [60]	o/w
Cyclosporin A	Aerosol OT, Tween 84, isopropyl myristate	Higher local concentrations and minimal system distribution to other organs via the circulation	Liu et al. [61]	o/w
Fluconazole	Capryol 90, Cremophore EL, benzyl alcohol, chlorocresol, carbopol ETD 2020	Increased bioadhesion, faster onset of action	Bachhav and Patravale [45]	o/w
Acyclovir	Isopropyl myristate, Tween 20, Span 20, DMSO	Higher skin retention and same effectiveness as commercial cream	Shisu et al. [41]	o/w
Penciclovir	Oleic acid, Cremophor EL, ethanol	Increased permeation compared with commercial cream	Zhu et al. [48]	o/w

delivery of voriconazole by microemulsion has also been reported. Here, two permeation enhancers were incorporated, namely, sodium deoxycholate and oleic acid, which favored transdermal rather than dermal delivery [50]. Recently, investigation of real-time cutaneous absorption of the drugs via *in vivo* microdialysis and improvement of skin permeation of evodiamine and rutaecarpine for transdermal delivery has been reported [51]. Preparation of new microemulsions containing isopropyl myristate, water, cremophor RH40:PEG400, and terpenes has been reported for transdermal drug delivery of ketoprofen [52]. For a feasible strategy for the prevention of Alzheimer disease, a new microemulsion-based patch prepared using a lamination technique for transdermal delivery of huperzine A (HA) and ligustrazine phosphate (LP) has been reported. It is more beneficial for fighting against amnesia in comparison with monotherapy [53].

A microemulsion system with high water-soluble capacity using isopropyl myristate, Labrasol, and Cremophor EL as oil, surfactant, and cosurfactant, respectively, has been developed for the investigation of microemulsion microstructures, key formulation variables, and their relationship to drug transdermal permeation enhancement [54]. The occlusive versus nonocclusive application of microemulsion for the transdermal delivery of progesterone has been done [55], in which mechanisms of enhanced skin penetration were investigated. The assortment of components that will augment drug permeation across the skin using microemulsion systems has been reported [56].

14.5 INTRANASAL APPLICATIONS OF PHARMACEUTICAL MICROEMULSIONS

Intranasal applications are particularly novel in that they allow drugs targeted to the brain. Reports in the literature show that drug uptake into the brain from nasal mucosa occurs through three different pathways: one is the pathway by which drug is absorbed into the systemic circulation and reaches the brain by crossing the blood–brain barrier, the other two are via the olfactory and the trigeminal neural pathways where the drug is transported directly from the nasal cavity to the cerebrospinal fluid and brain tissue, respectively [62]. Studies conducted by Kaur and Kim show that diazepam molecules from the microemulsion reach the brain mainly by crossing the blood–brain barrier after intranasal administration with no significant direct nose-to-brain transport [63]. It is important to note that areas under the curve for brain and plasma (AUC_{brain}/AUC_{plasma}) after the delivery of clonazepam-loaded or diazepam-loaded microemulsions showed equivalent values for intravenous or intranasal delivery in rats, whereas the AUC_{brain}/AUC_{plasma} after delivery of diazepam in rabbit models show higher AUC_{brain}/AUC_{plasma} values after intravenous administration of diazepam microemulsion compared with intranasal diazepam microemulsion application [61,62]. Intranasal application is more effective in rat models compared with rabbit models. This indicates that the total surface area in the nasal cavity and the nose-to-brain distance for blood flow must be taken into account in interpreting the results from different species of animals versus humans. It can be inferred from nasal cavity anatomical characteristics that the rabbit model is much closer to a human model compared with rat models [64]. However, most of the *in vivo* work done via intranasal drug delivery has been conducted on rat models. The proper selection of animal models for studying intranasal delivery of pharmaceutical microemulsions is important to be able to properly extrapolate the findings to humans.

Much work has been done by Mishra and coworkers [62,65–68] in the treatment of insomnia, status epilepticus, Alzheimer disease, and migraines. In a study done by Porecha et al. [65], intranasal delivery of sleep-inducing drugs (diazepam, lorazepam, and alprazolam) by pharmaceutical microemulsions has been shown to be more effective than intranasal drug solutions and commercial oral tablets. More specifically, microemulsions containing Capmul MCM, Tween 80, Captex 200p, ethanol, and water were formulated and loaded with either diazepam, lorazepam, or alprazolam. Mucoadhesive microemulsions (MME) were formulated by the addition of 0.5% w/w polycarbophil to microemulsions. *In vitro* nasal toxicity studies on the effects of nasal mucosa treated with blank microemulsions and rinsed with phosphate buffer saline (PBS) after 1 h showed reversible contraction of the epithelial layer and no damage, suggesting that the effects of the microemulsion excipients on nasal mucosa are reversible. More importantly, the study showed the following order of time until onset of sleep in mice from longest time to shortest time: control (no treatment) > oral tablet > drug solution delivered intranasally > microemulsion delivered intranasally > MME delivered intranasally (i.e., MME had the shortest time or was most effective). This stresses the importance of a mucoadhe-sive agent in an intranasal formulation to increase residence time, and thus, the effect on drug delivery. In addition, these findings suggest that intranasal microemulsion enhances transport of the drug across the nasal mucosa compared with a drug solution. In other studies, pharmaceutical microemulsions of the following have been successfully formulated: zolmitriptan, clonazepam, sumatriptan, and tacrine [62,66–68]. *In vivo* studies on the MME formulation of each drug demonstrated rapid and larger extent of drug uptake compared with microemulsion and drug solutions [62,66–70]. In conclusion, intranasal application of pharmaceutical microemulsion allows for faster onset of action, ease of application, and greater residence time compared with oral tablets or commercial drug solutions.

Microemulsion systems composed of oleic acid, Tween 80, ethyl alcohol, and water for the rapid-onset delivery of sildenafil for the treatment of erectile dysfunction and for the delivery of udenafil for intranasal administration has been reported [69,70]. Topical intervention with the microemul-sions can be a useful option to reduce nasal mucosal exposure to allergens in perennial allergic rhi-nitis [71]. The efficacy of microemulsions to facilitating the bioavailability of puerarin (PUE) after oral and nasal administration has been investigated. The results showed that nasal administration might be a promising route to enhance the absorption of PUE in the form of microemulsions [72].

Investigation of the nasal absorption of insulin from a new microemulsion spray in rabbits has been done in which the bioavailability of insulin lispro via the nasal route using a water-in-oil microemulsion was found to reach 21.5% relative to subcutaneous administration, whereas the use of an inverse microemulsion as well as a plain solution yielded less than 1% bioavailability. The results show that the acceleration in the intramucosal transport process is the result of encapsulating insulin within the nano-droplet clusters of a water-in-oil microemulsion, whereas the microemul-sion ingredients seem to have no direct role [73]. Elshafeey et al. [74] studied the effect of nasal administration of ildenafil citrate microemulsions in comparison to the oral route.

14.6   OPHTHALMIC APPLICATIONS OF PHARMACEUTICAL MICROEMULSIONS

Approximately 90% of all ophthalmic drugs are delivered using aqueous eye drops. Although they are convenient, only about 5% of the drug applied via eye drops eventually reaches the target tissue, and the remaining 95% is drained into the nasal cavity or enters the systemic circulation through conjunctival uptake [75]. The use of microemulsions offers an alternative to suspensions and solu-tions typically used in conventional eye drops. The primary challenge hindering the use of micro-emulsion formulations is the high irritability of the cornea due to surfactants and cosurfactants of microemulsions. To solve the problem of rapid loss of drug due to drainage, several researchers have developed novel uses of microemulsion formulations. In one study, an electrolyte-triggered gelling system, Kelcogel, was added to a microemulsion of cyclosporine A [76]. The viscosity of the gel thickened upon contact with tear fluids to provide prolonged residence time in the anterior eye. Another method is to disperse nanoparticles containing drug into a contact lens matrix. Using this method, lidocaine, timolol, and cyclosporine A microemulsions have been successfully dispersed in p-HEMA gels [75,77,78]. These gels are capable of extended release of drug due to the high loading and controlled release provided by the microemulsion.

Preparation and characterization of solid lipid nanoparticles (SLNs) prepared with stearic acid (SLN-A) and a mixture of stearic acid and Compritol (SLN-B) as lipid matrix and Poloxamer-188 as surfactant using sodium taurocholate and ethanol as cosurfactant mixture has been reported. The results show that a mixture of stearic acid and compritol is a good ocular drug delivery system in compari-son with stearic acid alone [79]. Comparison of conductivity–water content curve (CWCC) method and visual method in determining the critical water content during the formation of oil-in-water type micro-emulsions has been reported. The results firmly confirmed the shortcomings of the visual method and the feasibility of the CWCC method in oil-in-water type microemulsions during formation [80]. A novel method to produce SLNs using *n*-butanol as an additional cosurfactant using an oil-in-water microemulsion quenching technique has been reported [81]. Eradication of superficial fungal infections by conventional and novel approaches has been reported by Kumar et al. [82].

14.7 ORAL APPLICATIONS OF PHARMACEUTICAL MICROEMULSIONS

Oral drug delivery is commonly used due to ease of administration. Approximately 40% of drugs are hydrophobic and exhibit low bioavailability due to poor dispersion and absorption in the gastrointestinal tract when taken orally. Alternative routes of administration, such as transdermal or parenteral, are good measures to overcome the low bioavailability of such drugs. However, the use of oral microemulsions remedy the problem of low bioavailability, as discussed below, and has added benefits. Also, self-microemulsifying drug delivery systems (SMEDDS) are of low cost and can form microemulsions spontaneously upon dilution with water or gastrointestinal fluids.

Several mechanisms are likely responsible for the increase in bioavailability of hydrophobic drugs when delivered by an oral microemulsion as follows:

1. The use of microemulsion provides very small droplets (10–100 nm) that can disperse in the intestine and achieve high interfacial areas for mass transfer.
2. Microemulsions increase intestinal epithelial permeability by disturbing the cell membrane [83]. Surfactant monomers can partition into the lipid bilayer of intestinal epithelial cells and thereby disrupt the lipid-packing arrangement.
3. Cell tight junction permeability also increases in the presence of microemulsion excipients. However, most hydrophobic drugs with low molecular weights are likely to be absorbed by the transcelullar route. They are less likely to be absorbed by the paracellular route as the tight junction is the absorption barrier [84].
4. The inhibition of P-glycoprotein (P-gp) by microemulsion excipients as intestinal expression of P-gp can bind to drugs making them less bioavailable [83,85].
5. The prevention of enzymatic degradation by microemulsions in the gastrointestinal tract as the drug molecules are solubilized in the oil phase and are not easily accessible to the enzymes in the aqueous phase [86,87].

Common permeation enhancers include oleic acid and transcutol [87,88]. *In vitro* studies have shown that oleic acid alters intercellular lipid fluidity and disrupts lipid bilayers [87]. Another method to increase bioavailability is to incorporate P-glycoprotein inhibitors in the microemulsion formulation. Cremophor EL and Cremophor EH, common excipients in oral microemulsion, which act as P-gp inhibitors by binding to the hydrophobic domain of the P-gp [88]. The use of tacrolimus and cyclosporine A as P-gp inhibitors has also been reported [85]. For example, the use of a microemulsion formulation of paclitaxel increased the bioavailability compared with the commercial macrodispersed formulation. A further increase in bioavailability was seen with the addition of P-gp inhibitors (cyclosporine A or tacrolimus) to the microemulsion formulation.

Novel methods have been developed to further simplify the administration of oral microemulsions. An overview of oral microemulsions and their *in vivo* advantages has been presented below (Table 14.2). A solid self-microemulsifying formulation of nimodipine was prepared by spray-drying the liquid SMEDDS by using dextran 400 as a solid carrier [89]. The same dose of nimodipine in the solid SMEDDS and in the liquid SMEDDS resulted in similar AUC values that were both higher than the AUC of the conventional tablet. Another study incorporated a microemulsion preconcentrate into a solid poly(ethylene glycol) (PEG 3350) matrix [90]. The study showed the solid microemulsion preconcentrate was a two-phase system consisting of the crystalline PEG 3350 clusters and the liquid microemulsion preconcentrate dispersed in between PEG 3350 crystals. The presence of the PEG 3350 matrix did not hinder self-microemulsification of the liquid preconcentrate in water.

TABLE 14.2

Overview of Oral Microemulsions and their *In vivo* Advantages

Drug	Composition	*In vivo* Advantage	Reference	Type of Microemulsion
Biphenyl dimethyl dicarboxylate	Tween 80, Neobee M-5, triacetin	Greater AUC and $(C_{max})^a$ compared with suspension	Kim et al. [93]	o/w
Paclitaxel	Vitamin E, DOC-Na, TPGS, Cremophor RH 40	Increase in bioavailability compared with commercial solution	Kang et al. [94]	o/w
Simvastatin	Carpryol 90, Cremophor EL, Carbitol	Higher release rate and increase in bioavailability compared with conventional tablet	Kang et al. [94]	o/w
Ibuprofen	MCT, DGMO-C, HCO-40, ethanol	Greater bioavailability compared with suspension	Araya et al. [95]	o/w
Acyclovir	Labrafac, Labrasol, Plurol Oleique	Increase in bioavailability compared with commercial tablet	Ghosh et al. [96]	o/w
Atorvastatin	Labrafil, propylene glycol, Cremophor RH40	Increase in bioavailability compared with conventional tablet	Shen and Zong [86]	o/w
Silymarin	Tween 80, ethanol, ethyl linoleate	Increase in relative bioavailability compared with solution and suspension	Wu et al. [97]	o/w
Fenofibrate	Labrafac CM10, Tween 80, PEG-400	Higher pharmacodynamic effect compared with plain drug	Patel and Vavia [84]	o/w
Nimodipine	Ethyl oleate, Cremophor RH 40, Labrasol, Dextran 40	Greater AUC and C_{max} compared with conventional tablet	Yi et al. [98]	o/w
Oridonin	Maisine, Labrafac CC, Cremophor EL, Transcutol P	Greater AUC and C_{max} compared with suspension	Zhang et al. [99]	o/w
Vinpocetine	Labrafac, oleic acid, Cremophor EL, Transcutol P	Increase in bioavailability compared with suspension	Chen et al. [83]	o/w
9-Nitrocamptothecin	Ethyl oleate, Tween 80 or Cremophor EL, PEG-400, ethanol	More tumor shrinkage compared with suspension	Lu et al. [100]	o/w
Curcumin	Emulsifier OP, Cremorphor EL, PEG 400, ethyl oleate	Greater bioavailability compared with suspension	Cui et al. [101]	o/w
Docetaxel	Capryol 90, Cremophor EL, Transcutol	Greater bioavailability compared with commercial product	Yin et al. [88]	o/w
Exemestane	Capryol 90, Transcutol P, Cremophor ELP	Increase in bioavailability compared with suspension	Singh et al. [102]	o/w

[a] C_{max} = mean maximum plasma level.

Oral insulin delivery is difficult due to enzymatic degradation and chemical degradation. The microemulsion vehicle alone does not provide protection for insulin. Several studies have successfully combined nanocapsulation and microemulsion technologies to form nanoparticles, and have yielded hypoglycemic effects *in vivo* [91,92].

The effects of silymarin, an inhibitor of the P-glycoprotein efflux pump, on oral bioavailability of paclitaxel in rats has been studied [103]. Comparison of the physicochemical properties of monoglycerides, diglycerides, and triglycerides of medium chain fatty acids for the development of oral pharmaceutical dosage forms of poorly water-soluble drugs using phase diagrams, drug solubility, and drug dispersion experiments were carried out. The results showed that the monoglyceride gave microemulsion (clear or translucent liquid) and emulsion phases, whereas diglycerides and triglycerides exhibited an additional gel phase, which clearly indicates that in all monoglycerides, diglycerides, and triglycerides, the oil-in-water microemulsion region was the largest for the diglyceride [104]. Transformation of lipid-based oral drug delivery systems into solid dosage forms has been done by Tan et al. [105]. Oral administration and their therapeutic applications using microemulsions have been reported [106]. Development of water-in-oil microemulsions with the potential of prolonged release for oral delivery of L-glutathione has been done by Wen et al. [107].

14.8 PARENTERAL APPLICATIONS OF PHARMACEUTICAL MICROEMULSIONS

The parenteral route for drug delivery is important considering that preclinical studies require the use of parenteral delivery to understand pharmacokinetics in the body. In addition, parenteral administration is preferred in cases in which rapid action and system circulation is needed. In some cases, parenteral administration is the only choice for poorly absorbed drugs via oral administration or drugs where gastrointestinal side effects are common. In addition, parenteral administration bypasses the first-pass hepatic metabolism. In considering a parenteral microemulsion formulation, the excipients are under more stringent regulation due to the lack of protection barrier compared with other routes, for example, oral and topical. The excipients should be biocompatible, sterilizable, available as nonpyrogenic grade, nonirritant to nerves, and nonhemolytic [108].

Several parenteral microemulsion formulations have been shown to be more effective than the commercial product [108]. In one case, a microemulsion of artemether, a poorly water-soluble antimalarial agent, showed greater survival rate in mice compared with the market formulation, larither [108]. Another advantage of microemulsion parenteral delivery includes the small droplet size of microemulsions. Small size droplet generate huge amount of surface area which provides high concentration gradient of drug and improved permeation [109]. Microemulsions also offer the advantage of low production costs and storage. Self-emulsifying formulations can be stored as anhydrous solutions that can be reconstituted with saline in minutes [110]. An overview of parenteral microemulsions and their *in vivo* advantages is presented in Table 14.3.

The application of parenteral propofol microemulsions has been extensively studied in multiple animal models including rats, cats, dogs, horses, and humans [111,112–115]. Induction time to anesthesia using propofol microemulsion and market macroemulsion formulations have shown no difference [112,116]. In addition, human trials comparing the market macroemulsion Diprivan and newly developed microemulsion Aquafol showed bioequivalency [113]. A longer shelf-life due to the thermodynamic stability of microemulsions, and a decreased propensity to support bacterial growth due to the absence of soybean oil used in the macroemulsion are probable benefits [112,116]. In fact, stability studies conducted on a microemulsion formulation containing 1% w/v propofol in a nonlipid, aqueous formulation that included 8% w/v Poloxamer 188, 3% w/v poly(ethylene glycol) 400, 1% w/v propylene glycol, 0.2% w/v citric acid, 0.18% w/v methylparaben, 0.02% propylparaben, and water as excipients showed stability at temperatures up to 40°C for periods of 3 years even after the first dose had been drawn [114]. In a recent study, the authors concluded that propofol loading (5% vs. 1%) in microemulsions did not affect pharmacodynamics and recovery [112] but free propofol concentration in aqueous phase presumably produces severe pain [115]. This problem

TABLE 14.3

Overview of Parenteral Microemulsions and their *In vivo* Advantages

Drug	Composition	*In vivo* Advantage	Reference	Type of Microemulsion
Paclitaxel	Lecithin, Poloxamer 188, Cremophor EL, Ethanol	Less hypersensitivity reaction, higher AUC value and prolonged circulation as compared with Taxol	He et al. [116]	o/w
	Cremophor EL, Glycofurol, Labrafil 1944 CS, Poly(lactic-co-glycolic acid)	Antitumor effects observed for prolonged time	Zhang et al. [118]	o/w
	Lecithin, Poloxamer 188, Ethanol, Tricaproin, Tributyrin	Higher AUC values and less toxicity and longer circulation time compared with Taxol	Nornoo et al. [119]	o/w
	Capmul MCM, Myvacet 9-45 Lecithin, Butanol	Higher AUC values and prolonged half-life and wide tissue distribution	Nornoo and Chow [120] and Zhang et al. [121]	o/w
Norcanthridine	Ethyl oleate, lecithin, ethanol	Higher concentrations in liver and AUC values as compared with commercial formulations. The drug is used for liver metastases. Hence, higher liver concentrations would be useful	Brime et al. [122]	o/w
	Isopropyl myristate, lecithin, Tween 80	Higher efficacy, survival rate and lesser nephrotoxicity as compared with Fungizone	Brime et al. [123] and Park and Kim [124]	o/w
Flurbiprofen	PEGylated phospholipid, ethanol, ethyl oleate, lecithin	Prolonged circulation and higher AUC values as compared with the solution	Nesamony et al. [125]	o/w
Clonixic acid	Tween 85 and 20, castor oil	Less painful as compared with marketed formulation	Lee et al. [126]	o/w
Vincristine	PEGylated phospholipid, vitamin E, cholesterol, oleic acid	Higher efficacy, survival rate and lesser side effects as compared with free drug	Junping et al. [127]	o/w
Ibuprofen eugenol ester	Solutol HS 15, ethanol, medium chain triglycerides	Prolonged circulation and higher AUC values as compared with the solution	Zhao et al. [128]	o/w
Quercetin	Tween 20, clove oil	Significant improvement in antileishmanial activity as compared with the free drug	Gupta et al. [129]	o/w
Artemether	Lecithin, labrasol, ethanol, Poloxamer 188, ethyl oleate	Significant improvement in antimalarial activity as compared with the conventional oily solution	Tayade [130]	
Bassic acid	Tween 20, clove oil	Significant improvement in antileishmanial activity as compared with the free drug	Lala et al. [131]	

(continued)

TABLE 14.3 (Continued)
Overview of Parenteral Microemulsions and their *In vivo* Advantages

Drug	Composition	*In vivo* Advantage	Reference	Type of Microemulsion
Itraconazole	POE-50-hydrogenated castor oil, benzyl alcohol, MCT, Ethanol	Higher AUC values as compared with the cyclodextrin-based formulation	Rhee et al. [132]	
Amphotericin B	Solutol HS 15, Peceol, Myrj-52	Higher LD_{50} value as compared with Fungizone	Darole et al. [133]	
Propofol	Solutol HS 15, Tween 80	Less painful as compared with marketed emulsion of propofol (Propovan)	Date and Nagarsenker [134]	

Source: Date, A.A., and M.S. Nagarsenker. Parenteral microemulsions: An overview. *Int. J. Pharm.* 355, 19–30, 2008.

can be alleviated by developing formulations that have a lower partition of propofol in the aqueous phase or it could be related to the rate of destabilization of microemulsions. It should be emphasized that a microemulsion can be tailored in such a way that it will not break or destabilize upon dilution by blood, plasma, or saline. Future research on more stable microemulsions against dilutions have the potential for providing a controlled drug delivery system. Preparation, characterization, sterility validation, and *in vitro* cell toxicity studies of microemulsions possessing potential parenteral applications has been done by Kang et al. [117].

14.9 OTHER APPLICATIONS OF PHARMACEUTICAL MICROEMULSIONS

Magnetic nanoparticles have important applications in the biomedical field. They have potential in hyperthermia for tumor therapy, contrast enhancement agents for magnetic resonance imaging, and magnetically guided drug delivery systems [135,136]. A precipitation of nanoparticles in reverse microemulsion is a well-established technique used for the preparation of magnetic nanoparticles [137]. A water-in-oil microemulsion containing an iron salt (e.g., $FeCl_2$, $FeCl_3$) is mixed with a microemulsion containing an inorganic base (e.g., $NaBH_4$, NH_4OH, $NaOH$) [135–137]. The use of microemulsion allows for particle sizes as small as 1 to 2 nm [136]. Particle size can be fine-tuned by changing the oil/water ratio as well as the amount of surfactant [137]. Increasing the aqueous region of the microemulsion allows for greater yield and smaller particle size. For example, the use of a bicontinous microemulsion, which allows greater aqueous phase dissolution compared with a water-in-oil microemulsion, increased product yield from 0.1 to 0.4 g of product per 100 g total mixture to 1.16 g of product per 100 g of total mixture [137]. Because iron oxide nanoparticles are highly reactive and have a high surface area, the particles must be coated to be inert to air and aqueous solution. The use of silane and silica coatings to yield stable particles prevents aggregation and reduces reactivity [135,136]. In addition, silica coatings can be surface-modified with phospholipids to yield biocompatibility or attached with proteins [136]. In conclusion, magnetic nanoparticles and the tailoring of their surfaces by functionalized molecules have a very promising future in biomedical applications.

Another application of pharmaceutical microemulsions is in drug detoxification. Due to the high interfacial area, injection of a biocompatible microemulsion in the blood of an overdosed person can absorb and solubilize drug molecules. A Pluronic F127–based oil-in-water microemulsion was shown to successfully sequester amitriptyline [138].

Controlling release from microstructure i.e. the lipidic cubic phase which act as a molecular sponge consisting of interpenetrating nanochannels filled with water and coated by lipid bi layers achieved by selective alkylation. In the case of derivatives with alkyl chains two and eight carbon atoms long, t20 values of 3 and 13 days, respectively, were observed and good control over the rate at which water soluble molecules diffuse from the lipidic cubic phase has been studied [139].

In the former study, it was found that a negative charge at the oil/water interface played a significant role in attracting positively charged amitriptyline, and the oil kept the drug molecules sequestered [138]. The latter study attributes greater drug extraction to a greater oil/water interface area, and reductions in particle size and coulombic interactions of the drug with anionic groups of fatty acids [2]. Another study showed the advantage of a nanoemulsion over the market macroemulsion, Intralipid, in extracting bupivacaine from PBS, human blood, and guinea pig heart [5].

Fluorocarbons are partially or fully fluorinated organic compounds [140]. They are studied as potential oxygen carriers or blood substitutes for their unique property of being able to dissolve large amounts of gasses (e.g., oxygen and carbon dioxide) [140,141]. Liquid fluorocarbons have the unique property of being biologically very inert as well [141]. A phase III clinical study of Oxygent, a perflurooctyl bromide emulsion, demonstrated reduction and avoidance of blood transfusion in surgery [142]. The advantage of fluorinated microemulsions compared with the macroemulsion would be the smaller droplet size, in the nanometer range, which would allow for the droplets to travel into the smallest capillaries. A fluorinated microemulsion can be formulated either by using fluorinated surfactants or by using a fluorocarbon liquid as the "oil" phase [140,141,143]. Despite the promising use of fluorinated microemulsions due to their bioinert and gas dissolution properties, there are no reports of *in vivo* studies.

14.10 FUTURE RESEARCH HORIZONS IN PHARMACEUTICAL MICROEMULSIONS

To date, pharmaceutical microemulsions have been shown to be able to protect labile drug, control drug release, increase drug solubility, increase bioavailability, and reduce patient variability. Furthermore, it has proven possible to formulate preparations suitable for most routes of administration such as topical, oral, or intravenous administration. Still, however, a considerable amount of fundamental work is required along the following directions:

1. Better control on the disintegration or destabilization of microemulsions once applied, ingested, or injected into the human body
2. Better control on the droplet size and size distribution of microemulsions
3. The selection of appropriate sensing or marker molecules for targeted deposition of microemulsions in a specific organ or tissue
4. The design of "smart microemulsions" that can combine the oil-soluble, water-soluble, and interface-soluble drugs into a single liquid to function efficiently in the body
5. Better design of organogels that combine the water-soluble polymers to form gels and trap microemulsions in it
6. Better design of microemulsions that can show residence times in the body that is much longer than a few hours

The field of pharmaceutical microemulsions is far from being fully utilized and researchers face many challenges. However, microemulsions have immense potential in many areas due to their extremely low interfacial tension, thermodynamic stability, large total interfacial area, and ability to dissolve immiscible liquids at the nano-level. With continued research and studies, pharmaceutical microemulsions could possibly play a much bigger role in our lives by affecting the ways we take drugs or nutraceutical molecules.

REFERENCES

1. M.J. Lawrence and G.D. Rees. Microemulsions-based media as novel drug delivery systems. *Adv. Drug Deliv. Rev.*, **45**, 89–121 (2000).

2. D.M. Dennis, J.A. Flint, T.E. Morey, B.M. Moudgil, C.N. Seubert, D.O. Shah and M. Varshney. Pluronic microemulsions as nanoreservoirs for extraction of bupivacaine from normal saline. *J. Am. Chem. Soc.*, **126**, 5108–5112 (2004).

3. S.R. Carino, A. Fitzpatrick, A. Jovanovic, D.O. Shah, R.S. Underhill and M. Varshney. Alkoxysilane microemulsions with skin layers for use in drug detoxification. *Polym. Mater. Sci. Eng.*, **84**, 979–980 (2001).

4. V. Aleksa, S.R. Carino, A.V. Jovanovic, D.O. Shah, R. Stephen, R.S. Underhill and M. Varshney. Oil-filled silica nanocapsules for lipophilic drug uptake: Implications for drug detoxification therapy. *Chem. Mater.*, **14**, 4919–4925 (2002).

5. J.A. Flint, T.E. Morey, S. Rajasekaran, D.O. Shah and M. Varshney. Treatment of local anesthetic-induced cardiotoxicty using drug scavenging nanoparticles. *Nano Lett.*, **4**, 757–759 (2004).

6. M. Kreilgaard. Influence of microemulsions on cutaneous drug delivery. *Adv. Drug Deliv. Rev.*, **54**, 77–98 (2002).

7. O. Regev, S. Ezrahi, A. Aserin, N. Garti, E. Wachtel, E.W. Kaler, A. Khan and Y. Talmon. A study of the microstructure of a four-component nonionic microemulsion by cryo-TEM, NMR, SAXS and SANS. *Langmuir*, **12**, 668–674 (1996).

8. M. Fanun. A study of the properties of mixed nonionic surfactants microemulsions by NMR, SAXS, viscosity and conductivity. *J. Mol. Liq.*, **142**, 103–110 (2008).

9. K.A. Johnson and D.O. Shah. Formulation and properties of an alcohol-free pharmaceutical microemulsion system. In *Surfactants in Solution*, Vol. 6, K.L. Mittal and P. Bothorel (Eds.). Plenum Press, New York, pp. 1441–1456 (1986).

10. P.D.T. Huibers and D.O. Shah. Effect of oil chain length and electrolytes on water solubilization in alcohol-free pharmaceutical microemulsions. *J. Colloid Interface Sci.*, **107**, 269–271 (1985).

11. R.G. Alany, S. Agatonovic-Kustrin, N.M. Davies, T. Rades and I.G. Tucker. Effects of alcohols and diols on the phase behavior of quaternary systems. *Int. J. Pharm.*, **196**, 141–145 (2000).

12. E. Kanaoka, K. Kawakami, K. Masuda, Y. Moroto, Y. Nishihara, K. Takahashi and T. Yoshikawa. Microemulsion formulation for enhanced absorption of poorly soluble drugs I. Prescription design. *J. Controlled Release*, **81**, 65–74 (2002).

13. T. Hayashi, K. Kawakami, K. Masuda, Y. Nishihara and T. Yoshikawa. Microemulsion formulation for enhanced absorption of poorly soluble drugs II. In vivo study. *J. Controlled Release*, **81**, 75–82 (2002).

14. M. Fanun. Conductivity, viscosity, NMR and diclofenac solubilization capacity studies of mixed nonionic surfactants microemulsions. *J. Mol. Liq.*, **135**, 5–13 (2007).

15. V. Clément, M. Fanun, N. Garti, O. Glatter, M.E. Leser, D. Orthaber, G. Scherf and A. Stradner. Sugar-ester nonionic microemulsion: Structural characterization. *J. Colloid Interface Sci.*, **241**, 215–225 (2001).

16. M. Fanun. Phase behavior, transport, diffusion and structural parameters of nonionic surfactants microemulsions. *J. Mol. Liq.*, **139**, 14–22 (2008).

17. M.J. Hou and D.O. Shah. Effects of the molecular structure of the interface and continuous phase on solubilization of water in water/oil microemulsions. *Langmuir*, **3**, 1086–1096 (1987).

18. V.K. Bansal, J.P. O'Connell and D.O. Shah. Influence of alkyl chain length compatibility on microemulsion structure and solubilization. *J. Colloid Interface Sci.*, **75**, 462–475 (1980).

19. A. Aserin, S. Ezrahi, N. Garti and E. Wachtel. Water solubilization and chain length compatibility in nonionic microemulsions. *J. Colloid Interface Sci.*, **169**, 428–436 (1995).

20. A. Aserin, S. Ezrahi, N. Garti and E. Tuval. The effect of structural variation of alcohols on water solubilization in nonionic microemulsions. 1. From linear to branched amphiphiles-General considerations. *J. Colloid Interface Sci.*, **291**, 263–272 (2005).

21. A. Aserin, S. Ezrahi, N. Garti and E. Tuval. The effect of structural variation of alcohols on water solubilization in nonionic microemulsions 2. Branched alcohols as solubilization modifiers: Results and interpretation. *J. Colloid Interface Sci.*, **291**, 273–281 (2005).

22. R.G. Alany, M.M. Maghraby, K.K. Goellner and A.X. Graf. Microemulsion systems and their potential as drug carriers. In *Microemulsions Properties and Applications*, M. Fanun (Ed.). CRC Press, Boca Raton, FL, pp. 247–291 (2000).

23. Y.K. Pithapurwala and D.O. Shah. Interfacial composition of microemulsions: Modified Schulman–Bowcott model. *Chem. Eng. Commun.*, **29**, 101–112 (1984).

24. S.I. Chou and D.O. Shah. The effect of counterions on coacervation and solubilization in oil-external and middle-phase microemulsions. *J. Colloid Interface Sci.*, **80**, 311–322 (1980).

25. K.S. Chan and D.O. Shah. The effect of surfactant partitioning on the phase behavior and phase inversion of the middle phase microemulsion. Paper presented at the Society of Petroleum Engineers of AIME Meeting (1979).

26. C. Ramachandran, D.O. Shah and S. Vijayan. Effect of salt on the structure of middle phase microemulsions using the spin-label technique. *J. Phys. Chem.*, **84**, 1561–1567 (1980).

27. V.K. Bansal and D.O. Shah. The effect of addition of ethoxylated sulfonate on salt tolerance, optimal salinity, and impedance characteristics of petroleum sulfonate solutions. *J. Colloid Interface Sci.*, **65**, 451–459 (1978).

28. V.K. Bansal and D.O. Shah. The effect of ethoxylated sulfonates on salt tolerance and optimal salinity of surfactant formulations for tertiary oil recovery. *Soc. Pet. Eng. J.*, **18**, 167–172 (1978).

29. M. Chiang, K.S. Chan and D.O. Shah. A laboratory study on the correlation of interfacial charge with various interfacial properties in relation to oil recovery efficiency during water flooding. *J. Colloid Interface Sci.*, **11**, 471 (1976).

30. V.K. Bansal and D.O. Shah. Micellar solutions for improved oil recovery. In *Micellization, Solubilization, and Microemulsions*, Vol. 1, K.L. Mittal (Ed.). Plenum Press, New York, pp. 87–113 (1977).

31. V.K. Bansal and D.O. Shah. Microemulsions and tertiary oil recovery. In *Microemulsions: Theory and Practice*, L.M. Prince (Ed.). Academic Press, New York, pp. 149–173 (1977).

32. W.C. Hsieh and D.O. Shah. The use of high resolution NMR spectroscopy for characterizing petroleum sulfonates. Paper presented at the Society of Petroleum Engineers of AIME, Meeting (1977).

33. W.C. Hsieh and D.O. Shah. The effect of chain length of oil and alcoholas well as surfactant to alcohol ratio on the solubilization, phase behavior and interfacial tension of oil/brine/surfactant/alcohol systems. Paper presented at the Society of Petroleum Engineers of AIME, June, pp. 46–50 (1977).

34. M.J. Hou and D.O. Shah. *Formation and Characterization of Ultrafine Silver Halide Particles*. Interfacial Phenomena in Biotechnology and Material Processing. Elsevier Science, Amsterdam, pp. 443–458 (1988).

35. P. Ayyub, A.N. Maitra and D.O. Shah. Formation of theoretical density microhomogeneous $YBa_2Cu_3O_7$-X, using a microemulsion-mediated process. *Physica C*, **168**, 571–579 (1990).

36. P. Kumar, V. Pillai and D.O. Shah. Preparation of Bi-Pb-Sr-Ca-Cu-O oxide superconductors by coprecipitation of nanosize oxalate presursor powders in the aqueous core of water-in-oil microemulsions. *Appl. Phys. Lett.*, **62**, 765–767 (1992).

37. M.S. Kumar, M.S. Multani, V. Pillai and D.O. Shah. Structure and magnetic properties of nanoparticles of barium ferrite synthesized using microemulsion processing. *Colloids Surf. A*, **80**, 69–75 (1993).

38. P. Kumar, V. Pillai and D.O. Shah. Preparation of $YBa_2Cu_4O_8$ superconductor from oxalate precursor under ambient pressure. *Solid State Commun.*, **85**, 373–376 (1993).

39. S. Hingorani, P. Kumar, M.S. Multani, V. Pillai and D.O. Shah. Microemulsion mediated synthesis of zinc oxide nanoparticles for varistor studies. *Mater. Res. Bull.*, **28**, 1303–1310 (1993).

40. G.M. El Maghraby. Transdermal delivery of hydrocortisone from eucalyptus oil microemulsion: Effects of cosurfactants. *Int. J. Pharm.*, **355**, 285–292 (2008).

41. Shishu, S. Rajan and Kamalpreet. Development of novel microemulson-based topical formulations of acyclovir for the treatment of cutaneous herpetic infections. *AAPS Pharm. Sci. Technol.*, **10**, 559–565 (2009).

42. X. Chang, H. Chen, D. Du, J. Liu, D. Mou, H. Xu, X. Yang and D. Zhu. Hydrogel-thickened microemulsion for topical administration of drug molecules at an extremely low concentration. *Int. J. Pharm.*, **341**, 78–84 (2007).

43. D. Paolino, C.A. Ventura, S. Nisticó, G. Puglisi and M. Fresta. Lecithin microemulsions for the topical administration of ketoprofen: Percutaneous absorption through human skin and in vivo human skin tolerability. *Int. J. Pharm.*, **244**, 21–31 (2002).

44. V.B. Junyaprasert, P. Boome, S. Songkro, K. Krauel and T. Rades. Transdermal delivery of hydrophobic and hydrophilic local anesthetics from o/w and w/o Brij 97-based microemulsions. *J. Pharm. Pharm. Sci.*, **10**, 288–298 (2007).

45. Y.G. Bachhav and V.B. Patravale. Microemulsion based vaginal gel of fluconazole: Formulation, *in vitro* and in vivo evaluation. *Int. J. Pharm.*, **365**, 175–179 (2009).

46. Y.G. Bachhav and V.B. Patravale. Microemulsion-based vaginal gel of clotrimazole: Formulation, *in vitro* evaluation, and stability studies. *AAPS Pharm. Sci. Techol.*, **10**, 476–481 (2009).

47. H. Chen, X. Chang, D. Du, J. Li, H. Xu and X. Yang. Microemulsion-based hydrogel formulation of ibuprofen for topical delivery. *Int. J. Pharm.*, **315**, 52–58 (2006).

48. W. Zhu, C. Guo, A. Yu, Y. Gao, F. Gao and G. Zhai. Microemulsion-based hydrogel formulation of penciclovir for topical delivery. *Int. J. Pharm.*, **13**, 152–158 (2009).

49. A. Chudasama, V. Patel, M. Nivsarkar, K. Vasu and C. Shishoo. Investigation of microemulsion system for transdermal delivery of itraconazole. *J. Adv. Pharm. Technol. Res.*, **2**, 30–38 (2011).

50. G.N. El-Hadidy, H.K. Ibrahim, M.I. Mohamed and M.F. El-Milligi. Microemulsions as vehicles for topical administration of voriconazole: Formulation and in vitro evaluation. *Drug Dev. Ind. Pharm.*, **38**, 64–72 (2012).

51. Y.T. Zhang, J.H. Zhao, S.J. Zhang, Y.Z. Zhong, Z. Wang, Y. Liu, F. Shi and N.P. Feng. Enhanced transdermal delivery of evodiamine and rutaecarpine using microemulsion. *Int. J. Nanomed.*, **6**, 2469–2482 (2011).

52. N. Worachun, P. Opanasopit, T. Rojanarata and T. Ngawhirunpat. Development of ketoprofen microemulsion for transdermal drug delivery. *Adv. Mater. Res.*, **506**, 441–444 (2012).

53. J. Shi, W. Cong, Y. Wang, Q. Liu and G. Luo. Microemulsion-based patch for transdermal delivery of huperzine A and ligustrazine phosphate in treatment of Alzheimer's disease. *Drug Dev. Ind. Pharm.*, **38**, 752–761 (2012).

54. J. Zhang and B. Michniak-Kohn. Investigation of microemulsion microstructure and its relationship to transdermalpermeation of model drugs: Ketoprofen, lidocaine and caffeine. *Int. J. Pharm.*, **421**, 34–44 (2011).

55. G.M. El Maghraby. Occlusive and non-occlusive application of microemulsion for transdermal delivery of progesterone: Mechanistic studies. *Sci. Pharm.*, **80**, 765–778 (2012).

56. Ritika, S.L. Harikumar and G. Aggarwal. Microemulsion system in role of expedient vehicle for dermal application. *J. Drug Deliv. Ther.*, **2**(4), 23–28 (2012).

57. J.H. Yang, Y.I. Kim and K.M. Kim. Preparation and evaluation of aceclofenac microemulsion for transdermal delivery system. *Arch. Pharm. Res.*, **29**, 534–540 (2002).

58. A.C. Sintov and L. Shapiro. New microemulsin vehicle facilitates percutaneous penetration *in vitro* and cutaneous drug bioavailability in vivo. *J. Controlled Release*, **95**, 173–183 (2006).

59. M. Changez, J. Chander and A.K. Dinda. Transdermal permeation of tetracaine hydrochloride by lecithin microemulsion: In vivo. *Colloid Surf. B*, **48**, 58–66 (2006).

60. X. Zhao, J.P. Liu, X. Zhang and Y. Li. Enhancement of transdermal delivery of theophylline using microemulsion vehicle. *Int. J. Pharm.*, **327**, 58–64 (2006).

61. H. Liu, Y. Wang, Y. Lang, H. Yao, Y. Dong and S. Li. Bicontinuous cyclosporin a loaded water-AOT/Tween 850 isopropylmyristate microemulsion: Structural characterization and dermal pharmacokinetics in vivo. *J. Pharm. Sci.*, **98**, 1167–1775 (2008).

62. T.K. Vyas, A.K. Barbar, R.K. Sharma and A. Misra. Intranasal mucoadhesive microemulsions of zolmitriptan: Preliminary studies on brain-targeting. *J. Drug Target.*, **13**, 317–324 (2005).

63. P. Kaur and K. Kim. Pharmacokinetics and brain uptake of diazepam after intravenous and intranasal administration in rats and rabbits. *Int. J. Pharm.*, **364**, 27–35 (2008).

64. L. Li, I. Nandi and K.H. Kim. Development of an ethyl laurate-based microemulsion for rapid-onset intranasal delivery of diazepam. *Int. J. Pharm.*, **237**, 77–85 (2002).

65. S. Porecha, T. Shah, V. Jogani, S. Naik and A. Misra. Microemulsion based intranasal delivery system for treatment of insomnia. *Drug Deliv.*, **16**, 128–134 (2009).

66. T.K. Vyas, A.K. Barbar, R.K. Sharma, S. Singh and A. Misra. Intranasal mucoadhesive microemulsions of clonazepam: Preliminary studies on brain targeting. *J. Pharm. Sci.*, **95**, 570–580 (2005).

67. T.K. Vyas, A.K. Barbar, R.K. Sharma, S. Singh and A. Misra. Preliminary brain-targeting studies on intranasal mucoadhesive microemulsions of sumatriptan. *AAPS Pharm. Sci. Technol.*, **7**, E8 (2006).

68. V.V. Jogani, P.J. Shah, P. Mishra, A.K. Mishra and A.R. Misra. Intranasal mucoadhesive microemulsion of tacrine to improve brain targeting. *Alzheimer Dis. Assoc. Disord.*, **22**, 116–124 (2008).

69. H.J. Cho, W.S. Ku, U. Termsarasab, I. Yoon, C.W. Chung, H.T. Moon and D.D. Kim. Development of udenafil-loaded microemulsions for intranasal delivery: In vitro and in vivo evaluations. *Int. J. Pharm.*, **423**, 153–160 (2012).

70. H.T. Lu, R.N. Chen, M.T. Sheu, C.C. Chang, P.Y. Chou and H.O. Ho. Rapid-onset sildenafil nasal spray carried by microemulsion systems: In vitro evaluation and in vivo pharmacokinetic studies in rabbits. *Xenobiotica*, **415**, 67–77 (2011).

71. M. Andersson, L. Greiff and P. Wollmer. Effects of a topical microemulsion in house dust mite allergic rhinitis. *Basic Clin. Pharmacol. Toxicol.*, **108**, 146–148 (2011).

72. A. Yu, H. Wang, J. Wang, F. Cao, Y. Gao, J. Cui and G. Zhai. Formulation optimization and bioavailability after oral and nasal administration in rabbits of puerarin-loaded microemulsion. *J. Pharm. Sci.*, **100**, 933–941 (2011).

73. A.C. Sintov, H.V. Levy and S. Botner. Systemic delivery of insulin via the nasal route using a new microemulsion system: In vitro and in vivo studies. *J. Controlled Release*, **148**, 168–176 (2010).

74. A.H. Elshafeey, E.R. Bendas and O.H. Mohamed. Intranasal microemulsion of sildenafil citrate: *In vitro* evaluation and *in vivo* pharmacokinetic study in rabbits. *AAPS Pharm. Sci. Tech.*, **10**, 361–367 (2009).

75. D. Gulsen and A. Chauhan. Dispersion of microemulsion drops in HEMA hydrogel: A potential ophthalmic drug delivery vehicle. *Int. J. Pharm.*, **292**, 95–117 (2005).

76. L. Gan, Y. Gan, X. Zhang, C. Zhu and J. Zhu. Novel microemulsion in situ electrolyte-triggered gelling system for ophthalmic delivery of lipophiliccyclosporine A: In vitro and in vivo results. *Int. J. Pharm.*, **365**, 143–149 (2009).

77. C.C. Li, M. Abrahamson, Y. Kapoor and A. Chauhan. Timolol transport from microemulsions trapped in HEMA gels. *J. Colloid Interface Sci.*, **315**, 297–306 (2007).

78. Y. Kapoor and A. Chauhan. Ophthalmic delivery of Cyclosporine A from Brij-97 microemulsion and surfactant-laden p-HEMA hydrogels. *Int. J. Pharm.*, **361**, 222–229 (2008).

79. M.A. Kalam, Y. Sultana, A. Ali, M. Aqil, A.K. Mishra and K. Chuttani. Preparation, characterization, and evaluation of gatifloxacin loaded solid lipid nanoparticles as colloidal ocular drug delivery system. *J Drug Target.*, **18**(3), 191–204 (2010).

80. D.W. Xiang, T.T. Tang, J.F. Peng, L.L. Li, X.B. Sun and D.X. Xiang. Comparison of conductivity-water content curve and visual methods for ascertaintation of the critical water content of O/W microemulsions formation. *Yao Xue Xue Bao*, **45**, 1052–1056 (2010).

81. M.M. Mojahedian, S. Daneshamouz, S.M. Samani and A. Zargaran. A novel method to produce solid lipid nanoparticles using n-butanol as an additional co-surfactant according to the o/w microemulsion quenching technique. *Chem. Phys. Lipids*, **174**, 32–38 (2013).

82. L. Kumar, S. Verma, A. Bhardwaj, S. Vaidya and B. Vaidya. Eradication of superficial fungal infections by conventional and novel approaches: A comprehensive review. *Artif. Cells Nanomed. Biotechnol.* 1–15, (2013).

83. Y. Chen, G. Li, X. Wu, Z. Chen, J. Hang, B. Qin, S. Chen and R. Wang. Self-microemulsifying drug delivery system (SMEDDS) of vinpocetine: Formulation development and in vivo assessment. *Biol. Pharm. Bull.*, **31**, 118–125 (2008).

84. A.R. Patel and P.R. Vavia. Preparation and in vivo evaluation of SMEDDS (self-microemulsifying drug delivery system) containing fenofibrate. *AAPS J.*, **9**, 334–352 (2007).

85. S. Yang, R.N. Gursoy, G. Lambert and S. Benita. Enhanced oral absorption of paclitaxel in a novel self-microemulsifying drug delivery system with or without concomitant use of P-glycoprotein inhibitors. *Pharm. Res.*, **21**, 261–270 (2004).

86. H.R. Shen and M.K. Zong. Preparation and evaluation of self-microemulsifying drug delivery systems (SMEDDS) containing atorvastatin. *J. Pharm. Pharmacol.*, **58**, 1183–1191 (2006).

87. J.Y. Zheng and M.Y. Fulu. Decrease of genital organ weights and plasma testosterone levels in rats following oral administration of leuprolide microemulsion. *Int. J. Pharm.*, **307**, 209–215 (2006).

88. Y.M. Yin, F.D. Cui, C.F. Mu, M.K. Choi, J.S. Kim, S.J. Chung, C.K. Shim and D.D. Kim. Docetaxel microemulsion for enhanced oral bioavailability: Preparation and in vitro and in vivo evaluation. *J. Controlled Release*, **140**, 86–94 (2009).

89. T. Yi, J. Wan, H. Xu and X. Yang. A new solid self-microemulsifying formulation prepared by spray-drying to improve oral bioavailability of poorly water soluble drugs. *Eur. J. Pharm. Biopharm.*, **70**, 439–444 (2008).

90. P. Li, S.R. Hynes, T.F. Haefele, M. Pudipeddi, A.E. Royce and A.T.M. Serajuddin. Development of clinical dosage forms for a poorly water-soluble drug II: Formulation and characterization of a novel solid microemulsion preconcentrate system for oral delivery of a poorly water-soluble drug. *J. Pharm. Sci.*, **98**, 1750–1764 (2008).

91. A. Graf, T. Rades and S.M. Hook. Oral insulin delivery using nanoparticles based on microemulsions with different structure-types: Optimization and in vivo evaluation. *Eur. J. Pharm. Sci.*, **37**, 53–61 (2009).

92. S. Watnasirichaikul, T. Rhades, I.G. Tucker and N.M. Davies. In-vitro release and oral bioactivity of insulin in diabetic rats using nanocapsules dispersed in biocompatible microemulsion. *J. Pharm. Pharmacol.*, **54**, 473–480 (2002).

93. C.K. Kim, Y.J. Cho and Z.G. Gao. Preparation and evaluation of biphenyl dimethyl dicarboxylate microemulsions for oral delivery. *J. Controlled Release*, **70**, 149–155 (2001).

94. B.K. Kang, J.S. Lee, S.K. Chon, S.Y. Jeong, S.H. Yuk, G. Khang, H.B. Lee and S.H. Cho. Development of self-microemulsifying drug delivery systems (SMEDDS) for oral bioavailability enhancement of simvastatin in beagle dogs. *Int. J. Pharm.*, **274**, 65–73 (2004).

95. H. Araya, M. Tomita and M. Hayashi. The novel formulation design of O/W microemulsion for improving the gastrointestinal absorption of poorly water soluble compounds. *Int. J. Pharm.*, **305**, 61–74 (2005).

96. P.K. Ghosh, R.J. Majithiya, M.L. Umrethia and R.S.R. Murthy. Design and development of microemulsion drug delivery system of acyclovir for improvement of oral bioavailability. *AAPS Pharm. Sci. Technol.*, **7**, 77 (2006).

97. W. Wu, Y. Wang and L. Que. Enhanced bioavailability of silymarin by self-microemulsifying drug delivery system. *Eur. J. Pharm. Biopharm.*, **63**, 288–294 (2006).

98. T. Yi, J. Wan, H. Xu and X. Yang. Controlled poorly soluble drug release from solid self-microemulsifying formulations with high viscosity hydroxypropylmethylcellulose. *Eur. J. Pharm. Sci.*, **34**, 274–280 (2008).

99. P. Zhang, Y. Liu, N. Feng and J. Xu. Preparation and evaluation of self-microemulsifying drug delivery system of oridonin. *Int. J. Pharm.*, **355**, 269–276 (2008).

100. J.L. Lu, J.C. Wang, S.X. Zhao, X.Y. Liu, H. Zhao, X. Zhang, S.F. Zhou and Q. Zhang. Self-microemulsifying drug delivery system (SMEDDS) improves anticancer effect of oral 9-nitrocamptothecin on human cancer xenografts in nude mice. *Eur. J. Pharm. Biopharm.*, **69**, 899–907 (2008).

101. J. Cui, H. Li, H. Lou, B. Yu, G. Zhai, Y.G. Zhao and W. Zhu. Enhancement of oral absorption of curcumin by self-microemulsifying drug delivery systems. *Int. J. Pharm.*, **371**, 148–155 (2009).

102. A.K. Singh, A. Chaurasiya, A. Awasthi, G. Mishra, D. Asati, R.K. Khar and R. Mukherjee. Oral bioavailability enhancement of exemestane from self-microemulsifying drug delivery system (SMEDDS). *APPS Pharm. Sci. Technol.*, **10**, 906–916 (2009).

103. J.H. Park, H.J. Hur, J.S. and H.J. Lee. Effects of silymarin and formulation on the oral bioavailability of paclitaxel in rats. *Eur. J. Pharm. Sci.*, **45**, 296–301 (2012).

104. H.N. Prajapati, D.M. Dalrymple and A.T. Serajuddin. A comparative evaluation of mono-, di- and tri-glyceride of medium chain fatty acids by lipid/surfactant/water phase diagram, solubility determination and dispersion testing for application in pharmaceutical dosage form development. *Pharm. Res.*, **29**, 285–305 (2012).

105. A. Tan, S. Rao and C.A. Prestidge. Transforming lipid-based oral drug delivery systems into solid dosage forms: An overview of solid carriers, physicochemical properties, and biopharmaceutical performance. *Pharm. Res.* **30**(12), 2993–3017 (2013).

106. S. Gibaud and D. Attivi. Microemulsions for oral administration and their therapeutic applications. *Expert Opin. Drug Deliv.*, **9**(8), 937–951 (2012).

107. J. Wen, Y. Du, D. Li and R. Alany. Development of water-in-oil microemulsions with the potential of prolonged release for oral delivery of L-glutathione. *Pharm. Dev. Technol.*, **18**(6), 1424–1429 (2013).

108. M. Joshi, S. Pathak, S. Sharma and V. Patravale. Design and in vivo pharmacodynamic evaluation of nanostructured lipid carriers for parenteral delivery of artemether: Nanoject. *Int. J. Pharm.*, **364**, 119–126 (2008).

109. S. Talegaonkar, A.A. Farhan, J. Ahmad, R.K. Khar, S.A. Pathan and Z.I. Khan. Microemulsions: A novel approach to enhanced drug delivery. *Recent Patents on Drug Delivery & Formulation*, **2**, 238–257 (2008).

110. P. Boscan, E.P. Steffey, T.B. Farver, K.R. Mama, N.J. Huang and S.B. Harris. Comparison of high (5%) and low (1%) concentrations of micellar microemulsion propofol formulations with a standard (1%) lipid emulsion in horses. *Am. J. Vet. Res.*, **67**, 1476–1483 (2006).

111. D.M. Dennis, T. Grand, N. Gravenstein, S.P. McGorray, J.H. Modell, T.E. Morey, D.O. Shah and D. Shekhawat. Preparation and anesthetic properties of propofol microemulsions in rats. *Anesthesiology*, **104**, 1184–1190 (2006).

112. T.M. Morey, J.H. Modell, D. Shekhawat, D.O. Shah, G.P. Thomas, F.A. Kero, M.M. Booth and D.M. Dennis. Anesthetic properties of a propofol microemulsion in dogs. *Anesthesiology*, **103**, 882–887 (2006).

113. K.M. Kim, B.M. Choi, S.W. Park, S.H. Lee, L.V. Christensen, J. Zhou, B.H. Yoo, H.W. Shin, K.S. Bae, S.E. Kern, S.H. Kang and G.J. Noh. Pharmacokinetics and pharmacodynamics of propofol microemulsion and lipid emulsion after an intravenous bolus and variable rate infusion. *Anesthesiology*, **106**, 924–934 (2007).

114. R.M. Cleale, W.W. Muir, A.C. Waselau, M.W. Lehmann, D.M. Amodie and P. Lerche. Pharmacokinetic and pharmacodynamic evaluation of propofol administered to cats in a novel, aqueous, nano-droplet formulation or as an oil-in-water macroemulsion. *J. Vet. Pharmacol. Ther.*, **32**, 436–445 (2009).

115. J.Y. Sim, S.H. Lee, D.Y. Park, J.A. Jung, K.H. Ki, D.H. Lee and G.J. Noh. Pain on injection with microemulsion propofol. *Br. J. Clin. Pharmacol.*, **67**, 316–325 (2008).

116. L. He, G. Wang and Q. Zhang. An alternative paclitaxel microemulsion formulation: Hypersensitivity evaluation and pharmacokinetic profile. *Int. J. Pharm.*, **250**, 45–50 (2003).

117. B.K. Kang, S.K. Chon, S.H. Kim, S.Y. Jeong, M.S. Kim, S.H. Cho, H.B. Lee and G. Khang. Controlled release of paclitaxel from microemulsion containing PLGA and evaluation of anti-tumor activity in vitro and in vivo. *Int. J. Pharm.*, **286**, 147–156 (2004).

118. X.N. Zhang, L.H. Tang, J.H. Gong, X.Y. Yan and Q. Zhang. An alternative paclitaxel self-emulsifying microemulsion formulation: Preparation, pharmacokinetic profile, and hypersensitivity evaluation. *J. Pharm. Sci. Technol.*, **60**, 89–94 (2006).

119. A.O. Nornoo, D.W. Osborne and D.L.S. Chow. Cremophor-free intravenous microemulsions for paclitaxel I: Formulation, cytotoxicity and hemolysis. *Int. J. Pharm.*, **349**, 108–116 (2008).

120. A.O. Nornoo and D.L.S. Chow. Cremophor-free intravenous microemulsions for paclitaxel II. Stability, in vitro release and pharmacokinetics. *Int. J. Pharm.*, **349**, 117–123 (2008).

121. L. Zhang, X. Sun and Z.R. Zhang. An investigation on liver-targeting microemulsions of norcantharidin. *Drug Deliv.*, **12**, 289–295 (2005).

122. B. Brime, P. Frutos, P. Bringas, A. Nieto, M.P. Ballesteros and G. Frutos. Comparative pharmacokinetics and safety of a novel lyophilized amphotericin B lecithin-based oil-water microemulsion and amphotericin B deoxycholate in animal models. *J. Antimicrob. Chemother.*, **1**, 103–109 (2003).

123. B. Brime, G. Molero, P. Frutos and G. Frutos. Comparative therapeutic efficacy of a novel lyophilized amphotericin B lecithin-based oil–water microemulsion and deoxycholate-amphotericin B in immunocompetent and neutropenic mice infected with *Candida albicans*. *Eur. J. Pharm. Sci.*, **22**, 451–458 (2004).

124. K.M. Park and C.K. Kim. Preparation and evaluation of flurbiprofen-loaded microemulsion for parenteral delivery. *Int. J. Pharm.*, **181**, 173–179 (1999).

125. J. Nesamony, C.L. Zachar, R. Jung, F.E. Williams and S. Nauli. Preparation, characterization, sterility validation, and in vitro cell toxicity studies of microemulsions possessing potential parenteral applications. *Drug Dev. Ind. Pharm.*, **39**(2), 240–251 (2013).

126. J.M. Lee, K.M. Park, S.J. Lim, M.K. Lee and C.K. Kim. Microemulsion formulation of clonixic acid: Solubility enhancement and pain reduction. *J. Pharm. Pharmacol.*, **54**, 43–49 (2002).

127. W. Junping, K. Takayama, T. Nagai and Y. Maitani. Pharmacokinetics and antitumor effects of vincristine carried by microemulsions composed of PEG-lipid, oleic acid, vitamin E and cholesterol. *Int. J. Pharm.*, **251**, 13–21 (2003).

128. X. Zhao, D. Chen, P. Gao, P. Ding and K. Li. Synthesis of ibuprofen eugenol ester and its microemulsion formulation for parenteral delivery. *Chem. Pharm. Bull.*, **53**, 1246–1250 (2005).

129. S. Gupta, S.P. Moulik, S. Lala, M.K. Basu, S.K. Sanyal and S. Datta. Designing and testing of an effective oil-in-water microemulsion drug delivery system for in vivo application. *Drug Deliv.*, **12**, 267–273 (2005).

130. N.G. Tayade. Studies on novel drug delivery systems for anti-malarial drugs. Ph.D thesis submitted to University of Mumbai, India (2006).

131. S. Lala, S. Gupta and N. Sahu. Critical evaluation of the therapeutic potential of bassic acid incorporated in oil-in-water microemulsions and poly-D,L-Lactide nanoparticles against experimental leishmaniasis. *J. Drug Target.*, **14**, 171–179 (2006).

132. Y.S. Rhee, C.W. Park, T.Y. Nam, Y.S. Shin, S.C. Chi and E.S. Park. Formulation of parenteral microemulsion containing itraconazole. *Arch. Pharm. Res.*, **30**, 114–123 (2007).

133. P. Darole, D. Hegde and H. Nair. Formulation and evaluation of microemulsion based delivery system for amphotericin B. *AAPS Pharm. Sci. Technol.*, **9**(1), 122–128 (2008).

134. A.A. Date and M.S. Nagarsenker. Design and evaluation of microemulsions for improved parenteral delivery of propofol. *AAPS Pharm. Sci. Technol.*, **9**(1), 138–145 (2008).

135. G. Zhang, Y. Liao and I. Baker. Surface engineering of core/shell/iron/iron oxide nanoparticles from microemulsions for hyperthermia. *Mater. Sci. Eng.*, **30**, 92–97 (2009).

136. S. Santra, R. Tapec, N. Theodoropoulou, J. Dobson, A. Hebard and W. Tan. Synthesis and characterization of silica-coated iron oxide nanoparticles in microemulsion: The effect of nonioninc surfactants. *Langmuir*, **17**, 2900–2906 (2001).

137. A.L. Loo, M.G. Pineda, H. Saade, M.E. Treviño and R.G. López. Synthesis of magnetic nanoparticles in bicontinuous microemulsions. Effect of surfactant concentration. *J. Mater. Sci.*, **43**, 3649–3654 (2008).

138. M.A. James-Smith, D. Shekhawat, B.M. Moudgil and D.O. Shah. Determination of drug and fatty acid binding capacity to Pluronic F127 in microemulsions. *Langmuir*, **23**, 1640–1644 (2007).

139. M. Caffrey, J. Clogston, G. Craciun and D.J. Hart. Controlling release from the lipidic cubic phase by selective alkylation. *J. Controlled Release*, **102**, 441–461 (2005).

140. P. LoNostro, S. Choi, C. Ku and S. Chen. Fluorinated microemulsions: A study of the phase behavior and structure. *J. Phys. Chem. B.*, **103**, 5347–5352 (1999).

141. H.M. Courrier, T.F. Vandamme and M.P. Krafft. Reverse water-in-fluorocarbon emulsions and microemulsions obtained with a fluorinated surfactant. *Colloids Surf. A*, **244**, 1441–1448 (2004).

142. M.P. Krafft, A. Chittofrati and J.G. Riess. Emulsions and microemulsions with a fluorocarbon phase. *Curr. Opin. Colloid Interface Sci.*, **8**, 251–258 (2003).

143. M. Stébé, G. Serratrice and J. Delpuech. Fluorocarbons as oxygen carriers. An NMR study of nonionic fluorinated microemulsions and of their oxygen solutions. *J. Phys. Chem.*, **89**, 2837–2843 (1985).

15 Hydrotropes
Structure and Function

Krister Holmberg

CONTENTS

15.1 INTRODUCTION

The term *hydrotrope* means "directed toward water" and is used for substances that increase the solubility of sparingly soluble organic molecules in water. The term, which was coined by Neuberg almost 100 years ago [1], is widely used in connection with liquid formulations but the meaning of the term differs depending on the application. In cleaning formulations, which volumewise may be the most important, the term hydrotrope has a very broad meaning, ranging from cosolvents, such as glycol ethers, to hydrophilic surfactants, such as medium chain alkyl phosphates or short alkyl glucosides. In a more strict sense, a hydrotrope is a compound that shows some amphiphilicity, yet does not exhibit the distinct break in the surface tension versus concentration plot that is indicative of a self-assembly process in solution. Because such a self-assembly process, or micellization, is often used as the definition of a surfactant, a true hydrotrope is not a surfactant. The most common traditional hydrotropes are short-chain alkylbenzene sulfonates. Two representative examples of such compounds are shown in Figure 15.1. Both sodium xylene sulfonate and sodium cumene sulfonate decrease the surface tension to approximately 40 mN/m, but they do not self-assemble into micelles; hence, they are not surfactants. When the alkyl chain of a sodium alkylbenzene sulfonate has four or more carbon atoms, a distinct self-association at a certain concentration can be detected by techniques such as light scattering or nuclear magnetic resonance (NMR) diffusometry. Such compounds should be regarded as surfactants.

Alkylnaphthalene sulfonates are alternatives to alkylbenzene sulfonates and produce approximately the same effect. Aromatic carboxylic acids are also used as hydrotropes. Sodium salicylate is a typical example of such a compound. An advantage with the sulfonates over the corresponding carboxylates is the lower pK_a values of the former acids. The carboxylic acid derivatives cannot be used for acidic formulations.

The term hydrotropy was introduced in a search for substances that increased the water solubility of lipophilic organic compounds [1]. Ways to transfer a poorly water-soluble substance into an aqueous formulation is one of the key issues in formulation science. There are three main approaches:

1. Formulation of a dispersed system, such as an oil-in-water emulsion or a solid suspension.
2. Formulation of a microemulsion, that is, a macroscopic one-phase system that, on the microscopic level, exhibits distinct hydrophilic and hydrophobic domains (a microemulsion

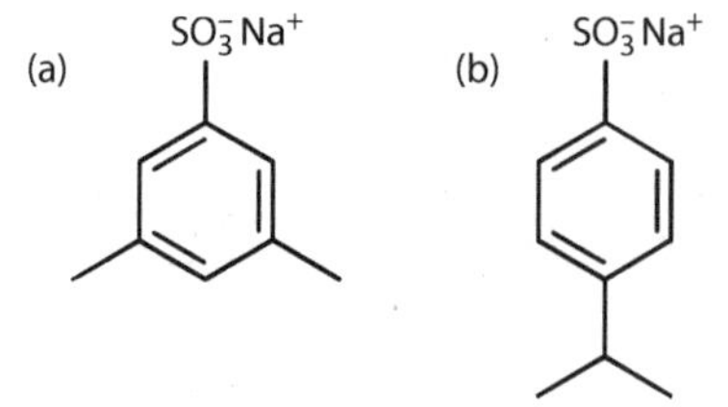

FIGURE 15.1 Sodium xylene sulfonate (a) and sodium cumene sulfonate (b).

may consist of oil droplets in water or water droplets in oil, or it may be bicontinuous, i.e., contain coexisting long channels of oil and water) [2].

3. Formulation of a system that is homogeneous on both the macroscopic and the microscopic levels. Use of a water-miscible organic solvent to improve the solubility of a lipophilic compound is the prime example.

Use of a hydrotrope may relate to either approach 2 or approach 3, depending on the definition of the term. A hydrophilic surfactant, often referred to as a cosurfactant, may be used to formulate a microemulsion, that is, a system of type 2 above. Use of a cosolvent such as a glycol ether falls within category 3. A classic hydrotrope such as xylene or cumene sulfonate is a borderline case and may be classified into either type 2 or type 3 of the formulation approaches listed previously. This will be further discussed in the next section. The term "coupling agent" is sometimes used synonymously with hydrotrope. This teminology can be confusing, however, because the term coupling agent is well established in materials science as a substance that imparts compatibility between otherwise incompatible solids, for instance, a hydrophobic polymer and a hydrophilic inorganic filler. It is then common practice to treat the surface of the filler with a coupling agent to make it hydrophobic.

Apart from the ways to improve the aqueous solubility mentioned previously, that is, use of a cosolvent, a classic hydrotrope, or a surfactant, one must be aware of yet another parameter of importance: the electrolyte. An electrolyte can be either salting-in or salting-out. Salting-in electrolytes help solubilize poorly soluble organic substances in water and the effect can be substantial. The order in which salts appear on this scale is given by the Hofmeister series, named after the inventor's famous article from the late 1800s in which he classified ions according to their ability to salt-out proteins [3]. The Hofmeister series for anions is:

$$SO_4^{2-} > HPO_4^{2-} > CH_3COO^- > Cl^- > NO_3^- > Br^- > ClO_3^- > I^- > ClO_4^- > SCN^-$$

For cations it is:

$$NH_4^+ > K^+ > Na^+ > Li^+ > Mg^{2+} > Ca^{2+} > guanidinium$$

This means that the addition of a salt such as guanidinium thiocyanate should enhance the solubility of an organic compound in water. This is indeed normally the case but the use of electrolytes to improve the aqueous solubility of poorly water-soluble substances lies outside the scope of this review and will not be further dealt with.

There are several previous reviews on hydrotropes in the literature [4–8]. They illustrate the concept from various angles and together they give a comprehensive treatment. The aim of this review is to give the current view on the role of hydrotropes in liquid formulations.

15.2   THE SERIES COSOLVENT—HYDROTROPE–SURFACTANT: A COMPARISON

Cosolvents are widely used in liquid formulations to improve the solvency of hydrophobic constituents. Due to low volatility, relatively low toxicity (there are exceptions), and their generally good performance, glycol ethers have been particularly popular in this respect. In general, the efficiency in terms of increasing solubility increases with the hydrophobicity of the glycol ether. Dipropylene glycol monopropyl ether and monopropylene glycol mono-*tert*-butyl ether, which are both on the borderline of being water soluble, belong to the most efficient solubilizers for organic dyes.

Hydrotropes are similar to cosolvents in that relatively large amounts are needed for the solubilization of hydrophobic organic substances. Typical hydrotrope concentrations are 1 mol/L and the cosolvents are used in even higher amounts. Surfactants, on the other hand, are effective as solubilizers at concentrations several orders of magnitude lower.

The mechanism of solubilization is clearly different for a cosolvent and for a surfactant. The cosolvent changes the solvency by affecting the character of the solvent. The change may be quantified by the use of solubility parameters. Replacing water with a water–glycol ether mixture leads to a reduction of the hydrogen bonding and the polar contributions of the solubility parameter, which in turn, means that the dispersion contribution gains in importance. This is favorable for the dissolution of most hydrophobic organic compounds. Table 15.1 gives the solubility parameters for two glycol ethers, one glycol and water [9].

The true hydrotropes lie between the cosolvents and the surfactants. As mentioned previously, a hydrotrope such as sodium xylene sulfonate or sodium cumene sulfonate does not seem to self-associate in water. There is no clear break in the surface tension versus log concentration curve. Also, the curve for the solubilization of a hydrophobic organic substance is different for a surfactant and for a hydrotrope of the alkylbenzene sulfonate type. The solubilization curve for a proper surfactant, that is, a molecule that gives a pronounced surface tension reduction and that self-assembles in solution at a certain concentration, is typically S-shaped, which is not the case for the hydrotrope. The curve for a hydrotrope resembles that for a glycol ether although the hydrotrope is usually more efficient, that is, it can be used in lower amounts. Figure 15.2 illustrates the difference in solubilization patterns.

The difference in solubilization capacity for sodium xylene sulfonate and ethylene glycol monopropyl ether shown in Figure 15.2 may seem small. However, the plot is made in a log-log diagram, which means that the hydrotrope is approximately 10 times more efficient. The difference in solubilization efficiencies of one hydrophobe, sodium xylene sulfonate, and a range of typical cosolvents is illustrated in Figure 15.3 in a nonlogarithmic diagram.

The efficiency in solubilization capacity shown in Figure 15.3 seems to match the ability of the hydrophobe/cosolvent to reduce the critical micelle concentration (CMC) of ionic surfactants. Bauduin and coworkers [10] investigated and compared a wide range of alcohols and glycol ethers

---

TABLE 15.1
Solubility Parameters of Selected Solvents

Solvent	δ_d	δ_p	δ_h
Diethylene glycol monoethyl ether	16.2	9.2	12.3
Ethylene glycol mono-*n*-butyl ether	16.0	5.1	12.3
Propylene glycol	16.8	9.4	23.4
Water	15.6	16.0	42.4

Note: δ_d, δ_p, and δ_h stand for the dispersion, polar bonding, and hydrogen bonding contributions to the solubility parameter, respectively, according to the relation $\delta^2 = \delta_d^2 + \delta_p^2 + \delta_h^2$, where δ^2 is the cohesive energy density.

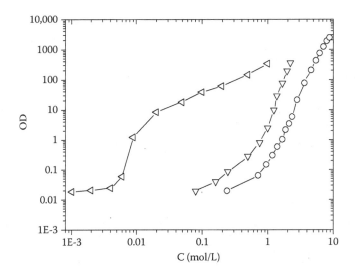

FIGURE 15.2 Solubilization of an organic dye (Disperse Red 13) in aqueous solution as a function of amount of added surfactant SDS (left), sodium xylene sulfonate (middle), and ethylene glycol monopropyl ether (right). The solubilization is quantified by the optical density (OD) given on the *y* axis. The higher the OD, the higher the extent of solubilization. (Redrawn from Bauduin, P. et al., *Langmuir*, 21, 6769–6775, 2005.)

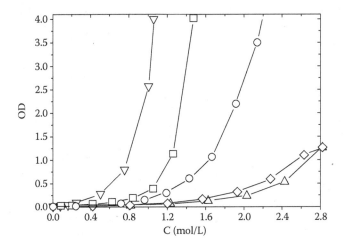

FIGURE 15.3 Solubilization of an organic dye (Disperse Red 13) in aqueous solution as a function of the amount of added sodium xylene sulfonate (extreme left curve) and (following from left to right) propylene glycol monopropyl ether, ethylene glycol monopropyl ether, acetone, and 1-propanol. The solubilization is quantified by the OD given on the *y* axis. (Redrawn from Bauduin, P. et al., *Langmuir*, 21, 6769–6775, 2005.)

for their ability to reduce the CMC of the anionic surfactant sodium dodecyl sulfate (SDS). The relative order of efficiency with respect to substances included in Figure 15.3 was propylene glycol monopropyl ether > ethylene glycol monopropyl ether > 1-propanol, that is, the same order as in Figure 15.3 [11]. One may note that this is not entirely in accordance with their relative hydrophobicities as given by their log *P* values, *P* being the partitioning beween octanol and water. The log *P* values for the the series propylene glycol monopropyl ether, ethylene glycol monopropyl ether, 1-propanol are 0.613, 0.265, and 0.344, respectively [10]. However, even if the relationship between hydrophobicity and hydrotrope efficiency is not entirely straightforward, it is obvious that the more hydrophobic glycol ethers usually perform best. Dipropylene glycol monopropyl ether and propylene glycol mono-*tert*-butyl ether both have borderline water solubility and exhibit excellent hydrotrope efficiency [10].

It is important to keep in mind that the performance of a traditional hydrotrope is influenced by the many components in a formulation. For instance, it has been demonstrated that the amount of hydrotrope required to transform a cloudy formulation into a clear solution, sometimes referred to as the minimum hydrotrope concentration (MHC), can be reduced by the presence of salts, lower alcohols, and urea [12].

15.3 THE STRUCTURE OF THE HYDROTROPE

It has long been known that the structure of the hydrotrope is decisive for its effect. However, there are not too many systematic investigations in the literature devoted to the structure–hydrophobe efficiency relation. In one of the few such studies, Burns [13] compared the efficiency of aromatic sulfonates as hydrotropes. The amount of hydrotrope required to clear cloudy formulations based on a hydrophobic fatty alcohol ethoxylate as surfactant was measured. It could be concluded from the study that alkyl substituted naphthalene sulfonates were approximately equal in performance to cumene sulfonate, which was the best of all alkylbenzene sulfonates. The efficiency was better with sodium than with potassium as counterion, which is in accordance with the salting-in power of cations, see previous section.

Hydrotropes also reduce the viscosity of liquid cleaning formulations. The smaller hydrotropes are usually the most efficient viscosity modifiers. In the abovementioned investigation, the toluene and xylene sulfonates performed best in this respect.

Figure 15.4 shows the effect of three alkylbenzene sulfonates on the cloud point of a nonionic surfactant, octylphenol ethoxylate, with eight oxyethylene units [14]. As can be seen, the most hydrophobic hydrotrope, sodium chlorobenzene sulfonate, is the most efficient. As the figure shows, a simple salt, such as sodium chloride, produces a cloud point depression instead. The degree of hydrophobicity of the anion is evidently decisive for the hydrotrope action.

Alkyl glucosides are known to function as hydrotropes in formulations based on alcohol ethoxylates. Matero [15] studied the effect of four alkyl glucosides with alkyl chains of 4, 8, 10, and 11

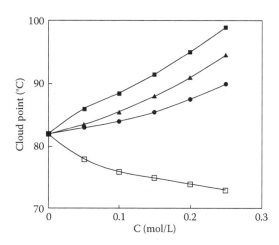

FIGURE 15.4 Cloud point as a function of concentration of added hydrotrope or NaCl in a 2% solution of the nonionic surfactant octylphenol-EO$_8$. The curves represent, from top to bottom: sodium chlorobenzene sulfonate, sodium *p*-toluene sulfonate, sodium xylene sulfonate, and sodium chloride. (Redrawn from Varade, D., and P. Bahadur, *J. Surfactants Deterg.*, 7, 257–261, 2004.)

carbons. The degree of polymerization of the sugar moiety was between 1 and 2 for all four com-pounds. A traditional hydrotrope, sodium toluene sulfonate, was also included in the study. The hydrotropic effect was investigated in formulations based on three different nononic surfactants, one very hydrophilic (C_6EO_2), one intermediate ($C_{11}EO_5$), and one hydrophobic ($C_{13}EO_6$). The tri-decyl chain of $C_{13}EO_6$ was highly branched, which renders extra hydrophobicity to the surfactant.

A hydrotrope is expected to increase the cloud point of a formulation based on a traditional nonionic surfactant, such as an alcohol ethoxylate. Table 15.2 shows the results obtained in the study. As can be seen, the hydrotropes with intermediate alkyl chain length performed well. For the most hydrophobic fatty alcohol ethoxylate, $C_{13}EO_6$, in particular, there is a very clear opti-mum in alkyl chain length of the hydrotrope. A glucoside with eight carbon atoms in the side chain gave the most pronounced effect in terms of cloud point increase. This is in accordance with several previous statements [6–8,15,16] that a good hydrotrope should be a surfactant of intermediate hydrophilicity (such surfactants are often referred to as cosurfactants). A cloud point increase of such formulations is practically important. Hydrophobic fatty alcohol ethoxylates, such as $C_{13}EO_6$, are efficient in removing oily soil but they obviously cannot be used if the cloud point of the formulations becomes too low. Some kind of hydrotrope is then needed to raise the cloud point of the system.

The results with the alkyl glucosides as hydrotropes are summarized in Figure 15.5. It is very clear from Figure 15.5 that there is an optimum in effect at an intermediate hydrophobicity of the hydrotrope.

TABLE 15.2

Cloud Point (°C) of Solutions of 1% Alcohol Ethoxylate with the Addition of 1% Hydrotrope

Fatty Alcohol Ethoxylate	No Hydrotrope	Butyl Glucoside	Octyl Glucoside	Decyl Glucoside	Undecyl Glucoside	Sodium Toluenesulfonate
C_6EO_2	46	55	68	70	40	56
$C_{11}EO_5$	27	28	59	52	35	36
$C_{13}EO_6$	<0	9	48	37	9	9

Source: Matero, A. et al., *J. Surfactants Deterg.*, 1, 485–489, 1998.

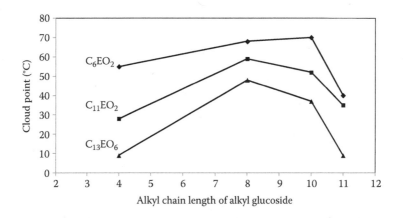

FIGURE 15.5 Cloud point as a function of chain length of a series of alkyl glucosides. The effect on the cloud point was investigated for the three different fatty alcohol ethoxylates shown in the figure. (Redrawn from Matero, A. et al., *J. Surfactants Deterg.*, 1, 485–489, 1998.)

TABLE 15.3
Amount of Hydrotropes Required to Clarify a Formulation

Hydrotrope	Required Amount (weight%)
Butyl glucoside	11.3
Octyl glucoside	2.9
Decyl glucoside	4.9
Undecyl glucoside	16.8
Sodium toluene sulfonate	3.1

Source: Matero, A. et al., *J. Surfactants Deterg.*, 1, 485–489, 1998.
Note: Formulations based on (in weight%) $C_{11}EO_5$ (5%), tetrapotassium pyro-
phosphate (5%), sodium metasilicate (5%), hydrotrope (X%), and water as
the balance. The pH of the formulation was 12.5.

Cleaning formulations are usually rich in electrolytes. The salts are needed to give the desired pH, to impart an anticorrosion effect (often exerted by silicates), and to sequester divalent and polyvalent ions. It is well-known that the addition of electrolytes decreases the solubility of a fatty alcohol ethoxylate. Most salts produce a pronounced cloud point depression. A hydrotrope acts to compensate for the salt addition by raising the cloud point. Table 15.3 shows the efficiency of the same series of hydrotropes that was used in the experiments of Table 15.2 in raising the cloud point of a formulation based on $C_{11}EO_5$ as the main surfactant.

Without added hydrotrope, a nonionic surfactant such as $C_{11}EO_5$ gives cloudy solutions in formulations with high electrolyte content. In such formulations, a hydrotrope is required. Table 15.3 shows that the alkyl glucosides with intermediate length alkyl chains were the most efficient, that is, these were needed at the lowest amounts. Also, the traditional hydrotrope, sodium toluene sulfonate, was very efficient.

Another role of hydrotropes is that they should expand the isotropic region of the formulation at the expense of the liquid crystalline regions. This was explored for the same series of alkyl glucosides with sodium toluene sulfonate used as a reference.

Figure 15.6 shows how the addition of the hydrotrope reduced the liquid crystalline region in the pseudoternary phase diagram based on the anionic surfactant SDS and pentanol. This is obviously useful for formulation work because the highly viscous liquid crystalline domains cause problems in mixing operations.

As can be seen, the hydrotrope can extend the isotropic microemulsion region to higher surfactant concentration and at the same time reduce the extension of the liquid crystalline phase. One may note that the shorter alkyl glucoside, butyl glucoside, is more eficient than its longer counterpart, octyl glucoside. One may also note that the traditional hydrotrope, sodium toluene sulfonate, is at least as efficient as the best alkyl glucoside in this respect. There is obviously no direct correlation between efficiency in increasing the cloud point of a formulation based on a hydrophobic fatty alcohol ethoxylate and in decreasing the extent of the liquid crystalline region in the SDS-pentanol-water system. This indicates that the two processes are based on different mechanisms. At the cloud point, there is a macroscopic separation into one more concentrated and one more dilute surfactant phase [17], and the surfactant is present in large elongated micelles with a critical packing parameter (CPP) close to 1. A medium chain amphiphile, such as octyl or decyl glycoside, will form mixed micelles with the fatty alcohol ethoxylate and reduce the CPP value to well below 1. This will lead to less elongated micelles, which do not favor phase separation. A higher temperature is then needed for the surfactant mixture to achieve the high CPP value associated with phase separation, that is, clouding. Shorter alkyl glucosides, as well as traditional hydrotropes of the alkylbenzene sulfonate type, will be too hydrophilic to form mixed micelles with the hydrophobic fatty alcohol ethoxylate.

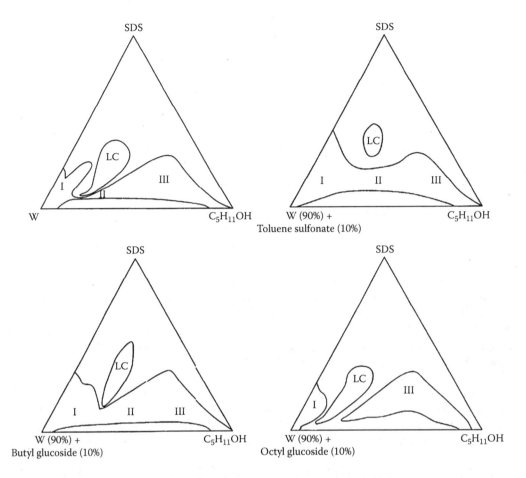

FIGURE 15.6 Effect of hydrotrope addition on the phase behavior of the system SDS, pentanol ($C_5H_{11}OH$), and water (W). I, II, and III indicate microemulsions of oil-in-water, bicontinuous, and water-in-oil types, respectively. LC stands for a liquid crystalline phase. (Redrawn from Matero, A. et al., *J. Surfactants Deterg.*, 1, 485–489, 1998.)

Longer alkyl glucosides, on the other hand, will act as hydrophobic surfactants themselves and not affect the CPP value much.

Disruption of a lamellar liquid crystalline phase is based on a different mechanism. The most effective breakers of such ordered structures are amphiphiles with short or highly branched alkyl chains or alkylaryl compounds. Xylene and cumene sulfonates with their planar structure are obviously effective in penetrating into the organized surfactant layer, disrupting the structure of the liquid crystal. The alkylbenzene sulfonates seem to be effective on most types of surfactant liquid crystals regardless of the packing geometry.

15.4 HYDROTROPES IN MICROEMULSION FORMULATIONS

Microemulsions are usually formulated with more than one amphiphile. In most cases, it is difficult to attain a broad microemulsion region with only a single surfactant. The main exception to this are microemulsions based on nonionic surfactants with polyoxyethylene chains as polar headgroup. With such surfactants as the only amphiphile, oil-in-water, water-in-oil, as well as bicontinuous microemulsions can be formulated although the amount of surfactant required is usually high. In the vast majority of cases, however, microemulsions are made from a mixture of one or more true surfactants and a small amphiphile, often a medium chain alcohol. The alcohol is usually referred to as "cosurfactant" (sometimes "cosolvent") but can easily be included in the hydrotrope category, by the broad definition of the term used here. Apart from medium chain alcohols, typically butanol, pentanol, or hexanol, glycol ethers of similar hydrophobicity can be used and are the preferred choice if volatile compounds are to be avoided, which is often the case in industrial formulations.

The traditional hydrotropes, that is, short-chain alkylbenzene sulfonates, usually do not offer advantages over the cheaper alcohols and glycol ethers and are not much used for microemulsion formulations. Friberg and Brancewicz [4], as well as Guo et al. [18], have, however, studied in detail the effect of sodium xylene sulfonate on the phase behavior of the SDS-pentanol-hydrocarbon-water system. They found that the hydrotrope extended the microemulsion region toward higher hydrocarbon content when the hydrocarbon was aliphatic. For white oil, which is a mixture of aliphatic and aromatic hydrocarbons, the effect was less pronounced.

Friberg [16] has stated that the main role of a hydrotrope in a microemulsion based on an ionic surfactant is to destabilize the lamellar liquid crystalline phase. The small and polar hydrotropes will disrupt the ordered packing of the main amphiphile. For all types of microemulsion formulations, it is essential to keep the liquid crystalline region as small as possible and the favorable action of hydrotropes in this respect is well documented.

As mentioned in the previous section (The Structure of the Hydrotrope), a hydrotrope, being a small and relatively hydrophilic molecule, will also change the curvature of an oil–water interface stabilized by a hydrophobic surfactant. This is another important role of hydrotropes in microemulsion formulations. In the so-called Winsor systems, a change in curvature of the oil–water interface from convex toward oil (the W/O situation) to a zero mean curvature (as for bicontinuous microemulsions) and on to convex toward water (the O/W situation) is responsible for the change from a Winsor II system (a W/O microemulsion in equilibrium with excess water), via a Winsor III system (a bicontinuous microemulsion in equilibrium with excess oil and excess water) to a Winsor I system (an O/W microemulsion in equilibrium with excess oil). Transition from one Winsor system to another can be made by a change in temperature for formulations based on nonionic surfactants and by a change in electrolyte concentration for formulations based on ionic surfactants. Use of a hydrotrope is a way to achieve such transitions for formulations based on all types of surfactants. The concept is illustrated in Figure 15.7.

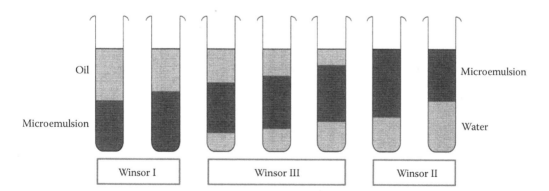

FIGURE 15.7 A mixture of oil, water, and surfactant may form a two-phase system consisting of an O/W microemulsion in equilibrium with excess oil (Winsor I), or a three-phase system comprising a bicontinuous microemulsion in equilibrium with both oil and water (Winsor III), ora two-phase system consisting of a W/O microemulsion in equilibrium with excess water (Winsor II). Increasing the temperature for a system based on a fatty alcohol ethoxylate or adding salt to a system based on an ionic surfactant is a way to induce a transition from left to right in the figure. The addition of a hydrotrope is a way to cause a transition from right to left.

15.5   SUMMARY

Hydrotropes play an important role in improving the solubilization of hydrophobic substances in water and in increasing the low-viscosity isotropic region in oil–water surfactant formulations. Although the mechanism by which hydrotropes exert their action is not entirely clear, and may vary from case to case, it seems clear that the disruption of well-ordered surfactant packing and change of the curvature of oil–water interfaces are key events. Ordered surfactant packing leads to liquid crystalline phases, which are usually highly unwanted in practical formulations. Hydrotropes are efficient in reducing the extent of liquid crystalline regions. The change in curvature of the oil–water interface that hydrotropes can induce is manifested as an increase of the cloud point of formulations based on nonionic surfactants and as a change from Winsor II, via Winsor III, to Winsor I systems for microemulsions.

REFERENCES

1. C. Neuberg. Hydrotropic phenomena. *J. Chem. Soc.*, **110**, 555 (1916).
2. K. Holmberg, B. Jönsson, B. Kronberg and B. Lindman. *Surfactants and Polymers in Aqueous Solution*, 2nd ed., Wiley, Chichester, pp. 139–156 (2003).
3. F. Hofmeister. Zur Lehre von der Wirkung der Saltze. *Arch. Expl. Pathol. Pharmacol.*, **24**, 247–260 (1888).
4. S. Friberg and C. Brancewicz. O/W microemulsions and hydrotropes: The coupling action of a hydrotrope. *Langmuir*, **10**, 2945–2949 (1994).
5. A. Matero, Å. Mattson and M. Svensson. Alkyl polyglucosides as hydrotropes. *J. Surfact. Deterg.*, **1**, 485–489 (1998).
6. B.K. Roy and S.P. Moulik. Effect of hydrotropes on solution behavior of amphiphiles. *Curr. Sci.*, **85**, 1148–1155 (2003).
7. T.K. Hodgdon and E.W. Kaler. Hydrotropic solutions. *Curr. Opin. Colloid Interface Sci.*, **12**, 121–128 (2007).
8. J. Eastoe, M. Hopkins Hatzopoulos and P.J. Dowding. Action of hydrotropes and alkyl-hydrotropes. *Soft Matter*, **7**, 5917–5925 (2011).
9. J.H. Hildebrand and R.L. Scott. *Regular Solutions*. Prentice-Hall, Englewoods Cliffs, NJ (1962).
10. P. Bauduin, A. Renoncourt, A. Kopf, D. Touraud and W. Kunz. Unified concept of solubilization in water by hydrotropes and cosolvents. *Langmuir*, **21**, 6769–6775 (2005).
11. P. Bauduin, A. Basse, D. Touraud and W. Kunz. Effect of short non-ionic amphiphiles derived from ethylene and propylene alkyl glycol ethers on the CMC of SDS. *Colloids Surf. A*, **8**, 270–271 (2005).
12. S. Kumar, N. Parveen and Kabir-ud-Din. Additive induced association in unconventional systems: A case of the hydrotrope. *J. Surfact. Deterg.*, **8**, 109–114 (2005).
13. R.L. Burns. Hydrotropic properties of some short-chain alkylbenzene- and alkylnaphthalene sulfonates. *J. Surfact. Deterg.*, **2**, 13–16 (1999).
14. D. Varade and P. Bahadur. Effects of hydrotropes on the aqueous solution behavior of surfactants. *J. Surfact. Deterg.*, **7**, 257–261 (2004).
15. A. Matero. Hydrotropes. In *Handbook of Applied Surface and Colloid Chemistry*, Vol. 1, K. Holmberg (Ed.). Wiley, Chichester, pp. 407–420 (2001).
16. S.E. Friberg. Hydrotropes. *Curr. Opin. Colloid Interface Sci.*, **2**, 490–494 (1997).
17. K. Holmberg, B. Jönsson, B. Kronberg and B. Lindman. *Surfactants and Polymers in Aqueous Solution*, 2nd ed., Wiley, Chichester, pp. 97–118 (2003).
18. R. Guo, M.E. Compo, S.E. Friberg and K. Morris. The coupling action of a hydrotrope and structure transition from lamellar liquid crystal to bicontinuous. *J. Disp. Sci. Technol.*, **17**, 493–500 (1996).

16 Surfactant Ionic Liquids
Potential Structured Reaction Media?

Paul Brown, Craig Butts, and Julian Eastoe

CONTENTS

16.1 INTRODUCTION

Ionic liquids (ILs) and surfactant micelles or mesophases are two interrelated fields of modern solution phase physical chemistry. The aim of this review is to describe the behavior of long-chain IL surfactants as compared with normal surfactants, and to explore potential applications of such structured media as phase tunable solvents for chemical reactions.

ILs are low–melting point organic salts resulting from extreme packing frustrations because of the sizes and shapes of the constituent cations and anions. As such, the common perception is that ILs are themselves weakly structured solvents owing to local electrostatic interactions. There is a vast matrix of possible IL structures, some comprising long alkyl chains coupled to much smaller counterions; these compounds are structural relatives of common surfactants (see next paragraphs) and have been coined surfactant ionic liquids (SAILs).

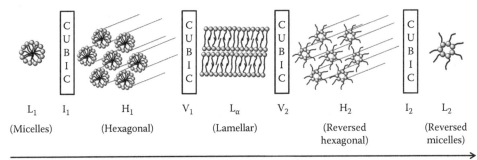

FIGURE 16.1 The "ideal" sequence of phases as a function of amphiphile concentration (subscripts 1 and 2 refer to "normal" and "reversed" phases, respectively). Lamellar phases can be found in different phase states, including lamellar crystalline (L_c), lamellar gel (L_β), and lamellar fluid (L_α). (From Drummond, C.J. and T. Kaasgaard. Ordered 2-D and 3-D nanostructured amphiphile self-assembly materials stable in excess solvent. *Chem. Phys. Chem.*, 8, 4957–4975, 2006. Reproduced by permission of The Royal Society of Chemistry.)

Surfactants are an important chemical class with multifarious applications and an industrial production exceeding 10 M tonnes per year. Common surfactants possess two moieties with dissimilar properties, linked together (either covalently or ionically) in the same molecule, and these have been termed "amphiphiles." It is traditional to classify these units as hydrophobic, normally long-chain hydrocarbons or fluorocarbons; or hydrophilic, being separate ions in the case of ionic surfactants; or zwitterionic and uncharged oligomers in the case of neutral ampholytic or nonionic surfactants. Charged ionic surfactants bear the closest structural relationship to ILs and SAILs. In aqueous solutions, common surfactant and SAIL molecules self-assemble to generate micelles above a critical micelle concentration (cmc), which is typically 10^{-6} to 10^{-2} mol dm^{-3} for most common compounds. This aggregation is driven by the hydrophobic effect, where the appropriate ions aggregate to minimize (hydrophobic ions) or maximize (hydrophilic ions) interactions with the polar water solvent. Micelles are dynamic aggregates showing rates of monomer uptake/exchange that are close to those of diffusion control [1]. The polar head groups residing at the micellar interface with the aqueous phase are highly hydrated and, in the case of ionic surfactants, form a Stern layer. This local region, typically of nanometric dimensions, has important consequences for reagent solubilization, and consequently "catalysis" because it influences the local concentrations and solvent environments of added reagents.

With increasing concentration, the amphiphile aggregate packing density increases and, in an attempt to minimize unfavorable interactions between surfactant aggregates, a range of lyotropic liquid crystalline (LC) mesophases may form: examples of mesophase structures can be seen in Figure 16.1.

For common surfactants, mesophase formation, structures, and properties have been extensively studied [3,4]. Interestingly, it is now appreciated that certain ILs and especially SAILs can also stabilize mesophases [5], as explored in more detail in the next section.

16.2 SOLUBILIZATION

Hydrophobic compounds solubilize more readily in aqueous surfactant solutions than in pure water because they can be accommodated into hydrophobic micellar cores [6]. The solvent penetrates to some extent into the micelles, so it can interact with both the hydrophilic head groups and the nonpolar alkyl chains, which is believed to be an important feature in "micellar catalysis." Duynstee and Grunwald [7] first demonstrated the catalytic effect of surfactant micelles, and since then much work has been done on investigating micellar effects on both rate and selectivity of different chemical reactions. There are in fact a number of comprehensive reviews [6,8] on the subject, so only key aspects will be covered here.

16.3 MICELLAR CATALYSIS KINETICS

There is indeed a vast literature in this field, and it is now appreciated that the major factor influencing rate enhancement in micellar catalysis is the increased local concentration of the reactants relative to the surrounding aqueous phases, owing to localized partitioning into the interface/micelles. The field is dominated by studies with certain common surfactants (e.g., anionic such as sodium dodecyl sulfate (SDS) or cationic such as cetylmethylammonium bromide [CTAB]). Substrates are solubilized into different regions of micellar structures through electrostatic and hydrophobic interactions [8]. This behavior has certain parallels with that typified by enzymes [9]; both enzymatic and micellar catalysis involve the "preassembly" of reagents owing to a balance of hydrophobic and hydrophilic interactions. Both also exhibit substrate specificity, with kinetic activity proportional to catalyst and substrate saturation. However, there are distinctions; rate and regioselectivity enhancements are generally weaker for micelles compared with enzymes.

The ways in which micellar media alter reaction pathways and kinetics have been accounted for by a pseudophase model [10]. This approach regards the micellar and the aqueous medium as discrete reaction pseudophases in which substrates exist in thermodynamic equilibrium. For unimolecular reactions, the catalytic effect is accounted for by considering the local properties of aqueous micelles. On the other hand, for bimolecular (or higher order) reactions, the local concentration of the reacting substrates is also important.

For a generic bimolecular reaction, the observed micellar kinetics can be accounted for with reference to Scheme 16.1, and a Michaelis–Menten type equation for the observed rate constant (Equation 16.1):

$$k_{obs} = \frac{k_2^w[B_w] + k_2^m K_s N_m [S_m]}{1 + K_s[S_m]}$$ (16.1)

k_2^w and k_2^m are rate constants for reactions in aqueous (w) and micellar (m) pseudophases, respectively. The other terms are K_s the substrate–micelle binding constant (see Equation 16.2), with S_m representing the concentration of micellized surfactant (Equation 16.3), and N_m the local molar

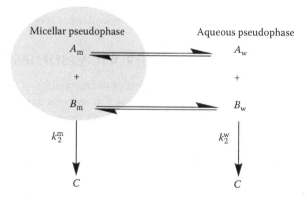

SCHEME 16.1 Kinetic model for a typical $A + B \rightarrow C$ bimolecular reaction in a micellar solution, where A and B are substrates located in the external water (w) or internal micellar (m) pseudophases, respectively.

concentration within the pseudophases (Equation 16.4). The equilibrium constant, K_s, describes the extent of substrate partitioning, via

$$K_s = \frac{[A_m]}{[A_w][S_m]} \tag{16.2}$$

and for hydrophobic substrates, it is found to increase with increasing surfactant hydrophobicity (alkyl chain length). Therefore, it can be appreciated that the chemical nature and structure of the amphiphile are of utmost importance for determining the solubility and penetration of solute and substrate molecules into the micellar pseudophase.

$$[S_m] = [S_t] - \text{cmc} \tag{16.3}$$

The total surfactant concentration is $[S_t]$ (Equation 16.3), so that when $[S_t] \gg \text{cmc}$, then $[S_m] \approx [S_t]$. The other terms in Equation 16.1 relate to the local molar concentration within the pseudophase, N_m

$$N_m = \frac{[B_m]}{[S_m]V_m} \tag{16.4}$$

where $[B_m]$ is the concentration of reactant ions in the micellar pseudophase, V_m is the micellar molar volume and $[S_m]V_m$ is the fractional volume in which the reaction occurs.

In the case of bimolecular reactions, competition of the incoming reactant for the surfactant counterion X must be taken into account, which is done by introducing a term for the ion-exchange equilibrium.

$$K_B^X = \frac{[B_w][X_m]}{[N_m][X_w]} \tag{16.5}$$

This modification is called the pseudo-ion exchange (PIE) model [11], requiring assumptions about B_m: first, a continual coverage of the micelle surface by counterions, expressed by the mole fraction, β, where $\beta = 1 - \alpha$, with α the fractional free charge at the micelle surface; second, it treats the micellar surface as a selective ion exchanger with competition between inert counterions X and reactive ions B.

For ionic bimolecular reactions, the second-order rate constant is generally quite similar to that for the pure aqueous phase reaction. This suggests a water-like medium for the majority of micelle catalyzed bimolecular reactions, and therefore indicates that any rate enhancement results mainly from an increase in local reactant concentrations in the micellar pseudophase [12].

16.4   REACTIONS IN MICELLES AND LYOTROPIC MESOPHASES

A good example of micellar enhanced reactivity is that of the metalation of picket fence porphyrins toward Cu^{2+} and Zn^{2+} [13]. A wide rate range is observed in aqueous anionic surfactant solutions, up to a factor of 3.5×10^4 greater compared with normal homogeneous solutions. Rate depends on alkyl chain length because the mechanism is a function of the location of the solubilized molecules within micelles [14].

An example demonstrating reaction control owing to electrostatic effects is the basic hydrolysis of nitrophenyl esters of carboxylic acids [15]. In the presence of CTAB, the substitution of a thiol for a nitro group was increased by a factor of 4×10^5, in addition, the concentration enhancement was affected by increased local effects on the acid–base relationship of the nucleophilic thiolation reaction as shown in Figure 16.2.

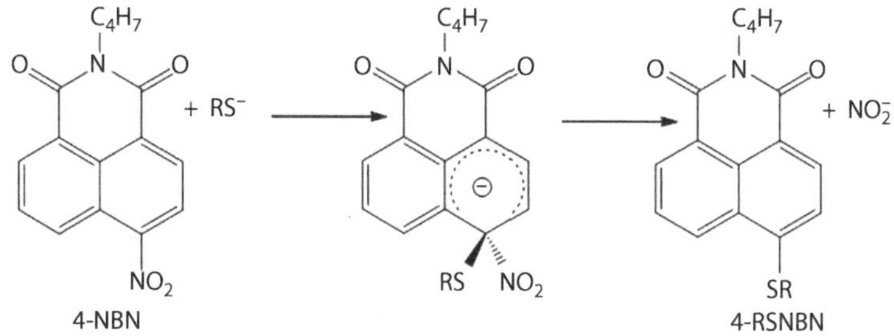

FIGURE 16.2 Thiol substitution for a nitro group where 4-NBN represents 4-nitro-*N*-*n*-butyl-1,8-naph-thalimide and 4-RSNBN represents 4-heptanothiolate-*N*-*n*-butyl-1,8-naphthalimide. (Cornils, B., and E.G. Kuntz: *Aqueous-Phase Organometallic Catalysis: Concepts and Applications*, 2nd ed. P. 351. 2004. Copyright Wiley-VCH Verlag GmbH & Co. KGaA. Reproduced with permission.)

Micellar systems can also act as phase mediators [8], bringing together insoluble or gaseous reagents such as in the case of hydroformylation. This process is normally carried out in aqueous systems [15], which present limitations because many higher olefin substrates suffer from low solubility. For example, Chen and coworkers [16] used a water-soluble [RhCl(CO)(tppts)$_2$] [tppts = tris(*m*-sulfonatophenyl)phosphine] complex to hydroformylate 1-dodecene in water, in the presence of surfactants such as SDS, resulting in enhancements in both yield and regioselectivity.

Lyotropic liquid crystal mesophases can also have a profound effect on chemical reactivity as they can exert strong control over the orientation of reactants. Reactions performed in surfactant mesophases are sensitive to phase type, with the more ordered mesophases less able to solubilize guest molecules, somewhat limiting the potential for higher ordering effects on reactivity [17]. At the time of writing, studies of kinetics and reaction outcomes in LCs were only very limited. This is surprising because the analogous dilute micellar systems, which are known to confer remarkable effects on catalysis and stereoselectivity, have been much studied. Ahmad and Friberg [18] reported the first reaction (*p*-nitrophenyl laurate hydrolysis) in a lyotropic LC phase (CTAB-water-hexanol). They reported that the reactions proceeded readily in isotropic, L_α and "middle" (hexagonal) phases. Interestingly, changes in the reaction rate caused by composition variation within each phase were also noted, for example, "an increase of factor of 3..." when the hexanol–CTAB ratio was increased from 0.2% to 0.7% w/w. However, they did not simultaneously analyze the mesophase structure, leaving doubts to the origin of this rate enhancement.

Bakeeva and coworkers [19] noted an up to tenfold increase in rate for an alkaline hydrolysis when carried out in a micellar phase as compared with water. The kinetics were faster with increasing surfactant concentrations. The same reaction in the hexagonal phase resulted in only a twofold rate enhancement but, interestingly, it was explained by a change in substrate orientation within the hexagonal cylinders rather than a change in mobility (diffusion) as seen within the micellar phase. Clearly, mesophases and related structures have interesting and potentially useful roles to play as media in the control of reaction outcomes, although systematic studies combining kinetics and structural effects are sorely lacking.

16.5 IONIC LIQUIDS

Ionic compounds are typically solids with high melting points. However, as early as in 1914, Waldon [20] described neutralization of ethylamine with concentrated nitric acid to create an IL (ethylammonium nitrate [EAN]) having an unusually low melting point of approximately 13°C. Today, ILs are classified as salts that form stable liquids below (arbitrarily) 100°C. These salts are poorly coordinated, where at least one ion has a delocalized charge preventing stable crystal lattice formation.

1,3-dialkylimidazolium cations

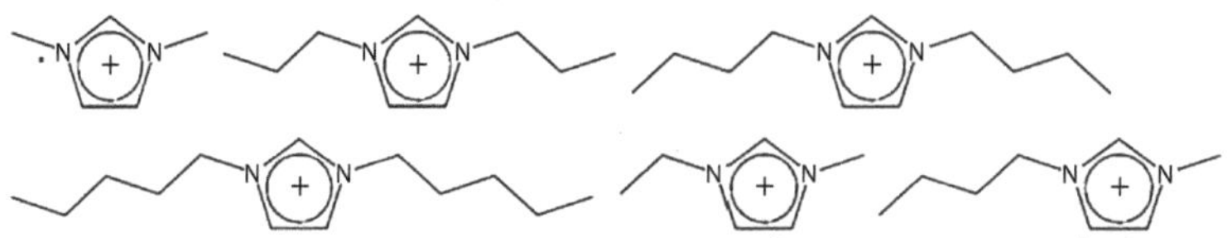

Anions

BF_4^-, NO_3^-, PF_6^-, ClO_4^-, $CF_4SO_3^-$, Cl^-, Br^-, TfO^-, $AlCl_3^-$

FIGURE 16.3   Various 1,3-dialkylimidazolium cation-anion combinations.

Barrer [21] synthesized the first imidazolium IL, on which most of today's ILs are still based (see Figure 16.3). However, it was not until much later when the first imidazolium halogenoaluminate salts were synthesized [22] that ILs really became of interest both academically and commercially. The field advanced in the early 1980s with the introduction of a homologous series of 1,3-dialkylimidazolium chlorides (Figure 16.3a–c), and studies of the properties of mixtures of these salts [23].

It was found that reacting $AlCl_3$ with 1,3-dialkylimidozalium halides gave rise to a series of equilibria to generate Cl^-, $AlCl_4^-$, $Al_2Cl_7^-$, $Al_3X_{10}^-$, etc. [24], and that by adjusting the mixing ratios of imidazolium halide to the halogenaluminate, properties such as viscosity, acidity, and refractive index were affected [25]. It was also shown that by changing the anions and cations, other physicochemical properties of ILs, such as solubility, density, and melting point could be controlled [26]. The chloroaluminate melt prepared from 1-methyl-3-ethylimidazolium chloride [emim]Cl, had the most favorable physical properties (large liquidus range and electrochemical window) and was easy to prepare [27]. These imidazolium salts (Figure 16.3) were used in organic synthesis [28] and proved to be an ideal starting point for the development of other ILs [24].

Inert anions such as toluenesulfonate [29], tetrafluoroborate [23], and hexafluorophosphates [30] were used with 1,3-dialkylimidozalium cations to create the first air and moisture stable room temperature ILs (RTILs). However, these compounds are quite expensive, are often difficult to prepare in high purity, and have unknown toxicity, and so fluorine-free (halogen free) and hydrophobic anions that are more environmentally friendly such as tetraalkylborates and tetraphenylborates are currently being reinvestigated [31,32].

The range of potential cations is also vast, with not only imidazolium but also pyridinium, ammonium, phosphonium, thiazolium, sulfonium, and pyrrolidinium salts (and more). These ILs are thermally stable [33] and have low saturated vapor pressures; in fact, for a long time, it was thought that distillation of these liquids was impossible until the work by Earle [34]. Because there is such a huge number of permutations ($\sim 10^9$) [35] ILs are highly tunable and many have been developed for specific synthetic purposes, replacing molecular solvents for greener chemistry and earning the designation "designer solvents" [36].

16.5.1   CUSTOMIZING ILs

By modifying the cations and anions of ILs, their properties can be dramatically altered: for example, [bmim]CH_3COO^- is miscible with water whereas [bmim][PF_6] is not [37]. The size, charge, and charge distribution and separation are acknowledged as major factors influencing the melting point [38,39] and viscosity [40]. In accordance with the Kapustinskii equation [41] (Equation 16.6), increasing the cation size or decreasing that of the anion (or both) decreases the lattice energy (U_L) of the salt (for example, NaCl has mp 803°C, whereas for 1-propyl-3-methylimidazolium chloride, mp ~60°C) [42]. Here, z^+ and z^- and r^+ and r^- represent the

$$U_L = -K \cdot \frac{\upsilon \cdot |z^+| \cdot |z^-|}{r^+ \cdot r^-} \cdot \left(1 - \frac{d}{r^+ + r^-}\right) \qquad (16.6)$$

elementary charges and the anion and cation radii, respectively, and υ is the number of ions in the empirical formula. The terms K and d are salt-specific constants.

Attractive Coulombic and van der Waals interactions decrease the energy of the salt, whereas repulsive Pauli forces increase it [43]. By minimizing this cohesive energy, the glass transition (T_g) is lowered; also, due to higher degrees of freedom, more asymmetric cations result in lower melting points by disrupting crystal packing [44], as shown in Figure 16.4 [45]. There is an optimum chain length asymmetry for lowering the melting point showing that organic chemical structure can have profound effects on the physicochemical properties of ILs.

Commonly used anions in ILs include hexafluorophosphates $\left(PF_6^-\right)$, tetrafluoroborates $\left(BF_4^-\right)$, tetrachloroaluminates $\left(AlCl_4^-\right)$, bistriflate imides (Tf_2N^-), and halides because they are symmetrical and pseudo-spherical. However, there are a whole host of possible anions, leading to a huge number of low–melting point salts.

There are also other interesting characteristics of ILs including density, which does not show much sensitivity to variations in temperature [46] but does depend greatly on structure. For example, increasing alkyl chain length decreases the density of 1,3-methylimidazolium hexafluorophosphates (Figure 16.5) [47,48].

Unsurprisingly, there are also strong effects on vapor pressure, often considered advantageous [38]. For many years, it was thought that ILs did not exhibit a measurable vapor pressure [49]; however, new generations of ILs can be reversibly vaporized at sufficiently high temperatures, and recondensed upon cooling [34].

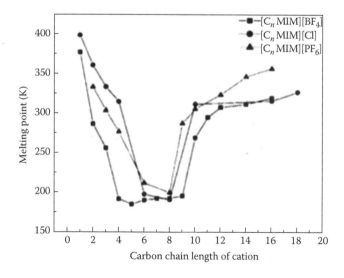

FIGURE 16.4 Observed melting points for a series of imidazolium-based ILs with increasing carbon chain length on the cation. (Reprinted with permission from Sun, N., X. He, X. Lu, and X. Zhang. Physical properties of ionic liquids: Database and evaluation. *J. Phys. Chem. Ref. Data*, 35, 1475–1517. Copyright 2006, American Institute of Physics.)

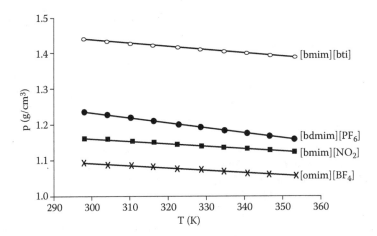

FIGURE 16.5 Apparent linear dependence of the density as a function of the temperature for ILs. [bmim] [bti]: 1-butyl-,3-methylimidazolium bis(trifluoromethylsulfonyl)imide; [bdmim][PF$_6$]: 1-butyl-2,3-dimethylimidazolium hexafluorophosphate; [omim][BF$_4$]: 1-octyl-,3-methylimidazolium tetrafluoroborate (modified and reproduced with permission from Elsevier). (From Valderrama, J.O., and K. Zarricueta, *Fluid Phase Equilibria*, 275, 145–151, 2009.)

16.6 REACTIONS USING ILs

In the mid 1980s, Fry and Pienta [50] and Boon et al. [51] realized that ILs could act as solvents for organic synthesis. ILs are often suitable as reaction media as they are thermally stable, having large liquidus ranges, for example, [emim]Cl/[emim]AlCl$_3$ mixtures (Figure 16.3b) are liquid between −90°C and 300°C compared with 0°C and 100°C for water or approximately 114°C and 78°C for ethanol [27,52]. In addition, ILs are able to dissolve a wide range of organic, inorganic, and organometallic compounds as well as gases such as H$_2$, CO, and O$_2$. All these features make ILs attractive solvents for catalytic hydrogenations, carbonylations, hydroformylations, and aerobic oxidations, and because they can be polar but noncoordinating, they have rate-enhancing effects on reactions involving cationic intermediates [53]. It has also been noted that they can facilitate reaction stereoselectivity, which is not possible in normal molecular solvents [54–56].

The unusual effects of ILs on reactions cannot be explained through traditional theories of polarity and Lewis acid–base properties, although they still exhibit hydrogen bonding, dispersion, electrostatic, dipolar, and hydrophobic interactions [24]. As with all solvents, altering IL functional groups affects solvent properties. Bartsch and Dzyuba [57] proposed that the polarity difference [based on the Dimroth–Reichardt E$_T$(30) parameter] between the alcohol and ether functionalized imidazolium salts of (Tf$_2$N$^-$) influences the *endo/exo* ratio of Diels–Alder products formed between cyclopentadiene and methyl methacrylate (Figure 16.6).

ILs are also useful for precious metal catalysis because the organic products can be removed without aqueous workup, the catalyst often remains in the IL, so that it can be directly reused [58]. There are different systems for catalysis in (or catalysis by) ILs in which the catalyst and substrate

| E$_T$(30) | 52.0 | 60.8 | 61.4 |
| *endo/exo* | 4.3 | 6.1 | 5.7 |

FIGURE 16.6 Diels–Alder product ratios for the reaction between cyclopentadiene and methyl methacrylate in IL media with different polarities. (From Bartsch, R.A., and S.V. Dzyuba, *Tetrahedron Lett.*, 43, 4657–4659, 2002.)

are dissolved in the IL, those in which the IL acts as both solvent and catalyst, and those in which the IL acts as a ligand for the catalyst [59,60], including biphasic [61,62] and triphasic [63] systems.

16.6.1 HYDROFORMYLATION

Hydroformylation is an important process in catalysis, and the first example with an IL was Pt-catalyzed hydroformylation of ethane [64]. Since then, much work has been done to improve product separation, catalyst recovery, stability [65], and selectivity. To increase alkene solubility and selectivity for the linear aldehyde, imidazolium triflate IL was used with a Rh catalyst in a biphasic system with organic solvents (Figure 16.7) [66], giving evidence that all catalytic activity occurred in the ionic phase.

16.6.2 CROSS-COUPLING REACTIONS

Heck couplings in [bmim][PF$_6$] were demonstrated by Carmichael et al. [60] using Pd(OAc)$_2$-Ph$_3$P as catalyst to couple 4-bromoanisole with ethylacrylate, which gave a yield of 98% ethyl 4-methoxycinnamate (Figure 16.8). Because the IL is virtually insoluble in water and alkenes, but dissolves the transition meal catalyst, simple extraction and recycling of the expensive palladium catalyst becomes possible.

Another example is Heck coupling of 1-bromonaphthalene to butyl vinyl ether: in regular solvents such as toluene or dimethylsulfoxide, a mixture of isomers is formed; however, Chen and coworkers [67] showed that in [bmim][PF$_6$] >99% selectivity to the α-arylation product occurs (Figure 16.9).

In Suzuki coupling reactions, [bmim][BF$_4$] afforded good yields [68,69] and a phosphonium IL allowed nanofiltration and recovery of the solvent and catalyst [70]. There were problems with Negeshi cross-coupling reactions though; by using an imidazolium salt with a phosphine ligand, the problems of cation deprotonation with an amine base could be avoided. Interestingly, where [bmim]-based ILs met with failure, [bdmim][BF$_4$]/toluene biphasic solvents gave high yields (~90%) [71]. This IL was also used for a copper-catalyzed, ligand-free Sonogashira reaction; however, ultrasonic irradiation had to be used to assist the reaction [72].

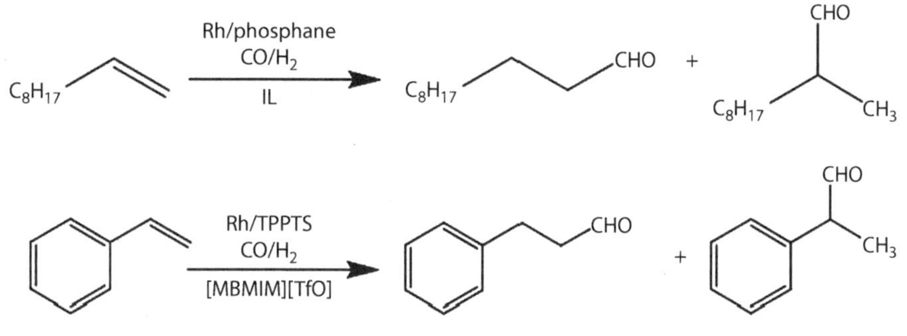

FIGURE 16.7 Hydroformylation of 1-decene and styrene in IL-biphasic systems. (From Leclercq, L. et al., *Chem. Commun.*, 311, 2008.)

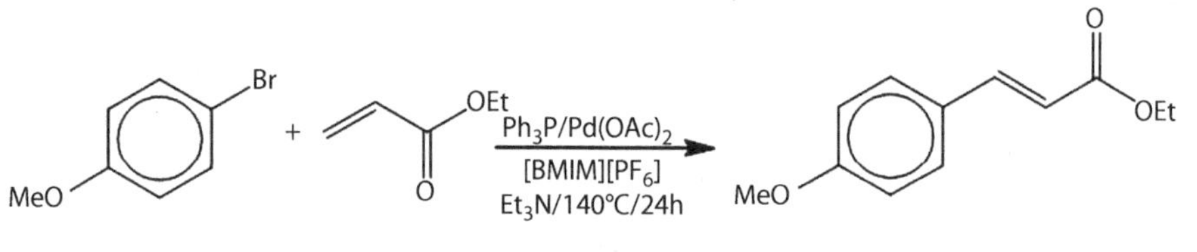

FIGURE 16.8 Heck reaction to couple 4-bromoanisole with ethylacrylate.

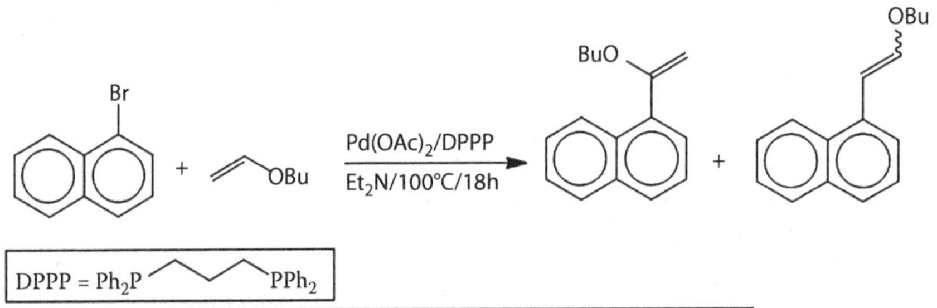

Solvent	Conv. %	α/β
[bmim][PF$_6$]	50	>99%
Toluene	23	46/54
Acetonitrile	36	69/31
DMF	100	60/40
DMSO	100	75/25

FIGURE 16.9 Coupling of 1-bromonaphthalene to butyl vinyl ether in different solvents. α/β is the isomer ratio. (From Chen, W. et al., *Org. Lett.*, 3, 295–297, 2001.)

Although it has been shown that ILs can enhance the activity and stability of catalysts in cross-coupling reactions as well as stereoselectivity, there are sometimes drawbacks to using ILs in synthesis. In particular, ILs may have high viscosities compared with conventional molecular solvents, which can slow heat and mass transfer, thereby lowering reaction rates. For example, Diels–Alder [73] reactions occur faster in water than in IL media due to the absence of hydrophobic interactions and the weaker hydrogen bonding, so it is important to realize that conventional solvents still offer the best media for certain reactions.

16.6.3 Other Reactions

The first reports of hydrogenation in ILs showed the rhodium catalyzed hydrogenation of pent-1-ene with reaction rates up to five times higher than in acetone solvent and Rh losses below detection [74]. Regioselective hydrogenation in ILs has also been reported [75]. Two other industrially important processes include aromatic substitution reactions, for which selectivity, reaction time, and yields can be improved by ILs [76]. For olefin dimerizations, Ni-catalyzed dimerization of propene in [bmim][AlCl$_4$] was the first to be reported [74]. Many catalyzed oxidations such as epoxidation, dehydroxylation, and the oxidation of alcohols [77], thiol compounds [78], oximes [79], and alkanes, among others, have made use of ILs. A good example of aromatic aldehyde oxidation to the corresponding acid was that studied by Howarth [80]. Nickel II acetylacetonate catalyst was employed in [bmim][PF6] at 60°C with oxygen, and the reaction was repeated three times using the recovered IL/Ni system without a decrease in yield.

Biocatalytic reactions in ILs have also been investigated because, unlike organic solvents of comparable polarity, they often do not deactivate enzymes, which simplifies reactions involving polar substrates such as sugars [81]. IL solvents permit faster rates and enantioselectivity, which can be advantageous [82]; work on amino acid/peptide chemistry shows not only that the polar nature of ILs is particularly suitable for synthesis but also that amino acids can be used as starting materials for the generation of chiral ILs [83–85]. ILs also offer new possibilities as buffers in nonaqueous systems, which is especially exciting for the optimization of acid–base conditions and enzymatic reactions [86], an area recently reviewed by Rantwijk and Sheldon [87].

16.7 IL APPLICATIONS AND INDUSTRIAL PROCESSES

Initial optimism about potential industrial utilization of ILs was perhaps misplaced; however, there are numerous real-life applications. There may still be an "ionic liquid revolution," especially because current as well as proposed applications of ILs are no longer limited to solvent replacement, but diverse applications ranging from catalysis, separations, data storage [88], photochemistry [89], CO_2 capture [90], hazardous gas storage [91], lubrication [92] and rocket propulsion [93,94] to name just a few. Commercial, as well as up-and-coming applications are reviewed in the next section.

16.7.1 THE BASIL PROCESS

The biphasic acid-scavenging using ionic liquids (BASIL) process was first introduced by BASF in 2002 and is used commercially in the preparation of alkoxyphenylphosphines (Scheme 16.2). The improved process uses a 1-methylimidazole to scavenge acid, the resulting salt is an IL, which forms a discrete phase, and is much more easily removed from the reaction mixture than traditional solid by-products [36].

Compared with conventional liquid processes, the space–time yield (mass of product formed per volume of the reactor and time, kg m^{-3} s^{-1}) is increased by a factor of more than 8.5×10^4, and yield increases from 50% to 98% have been possible. The IL produced is also recycled via base decomposition of the 1-H-3-imidazolium chloride [95].

16.7.2 CELLULOSE PROCESSING

Holbrey and coworkers [96] found that technically usable concentrations of cellulose could be produced by using ILs. This shows great potential as it prevents the disposal of huge quantities of waste water as well as various other compounds such as CS_2. Now their techniques are being commercially developed by BASF to make cellulose–polymer blends as novel plastics.

16.7.3 NUCLEAR FUEL WASTE DISPOSAL

ILs exhibit good stability against radiation and hence can be used in solvent extraction systems as alternate nonaqueous electrolyte media for high-temperature pyrochemical processing of spent nuclear fuel [97]. Recovery of uranium [98], and other useful fission products such as palladium

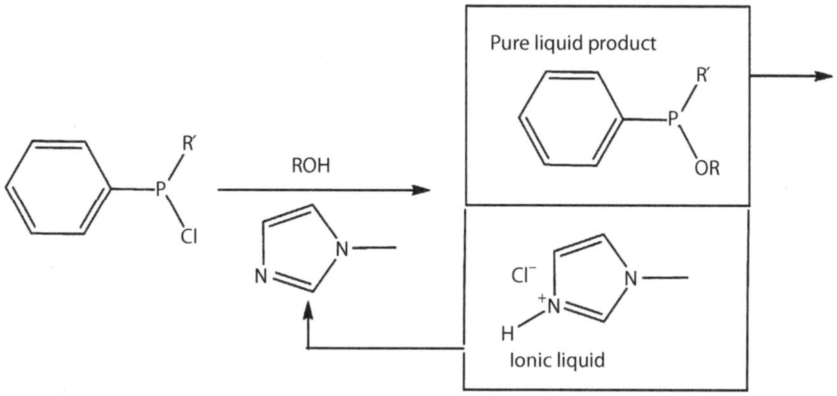

SCHEME 16.2 The BASIL process.

[99] using ILs have been reported. In a further development, uranyl ions in nitric acid have been extracted by using [bmim][Tf$_2$N] and transferred to supercritical CO$_2$ [100], because ILs do not dissolve in CO$_2$ and thus their recycling and repeated use should be possible.

16.7.4 PLASTIC RECYCLING

Although not yet an economic approach, recycling of thermosetting plastics has been demonstrated by depolymerizing nylon in ILs at 300°C to regenerate the cyclprolactam feedstock [101]. This is especially exciting because up until now, most thermosetting plastics have been simply unrecyclable.

16.7.5 IONIKYLATION

This process, developed by Petrochina (Beijing, China), employs aluminum chloride–based ILs in place of sulfuric acid catalysts for alkylation of isobutene, which is a key step in gasoline generation [102]. This process is employed on an industrial scale, with 65,000 tonnes produced per year being the largest commercialization of ILs to date. They are also employed as the transport medium for reactive gases at subambient pressures, and as paint additives to enhance the drying properties and finish of paint [103,104].

16.7.6 OTHER APPLICATIONS

ILs exhibit high thermal stability and thus show potential for use in solar energy thermal power plants [105]. Furthermore, owing to low volatility, they can be used for safer microwave synthesis because they reduce sudden pressure surges. The dipolar characteristics of ILs also facilitate rapid excitation by microwaves and, consequently, faster reactions [24].

16.8 MICELLIZATION AND LYOTROPIC MESOPHASES OF SAILS

In the early 1980s, Evans and coworkers [106,107] described micelle formation with the addition of surfactants in EAN, which was the first reported case of self-assembly in ILs. For typical cationic amphiphiles, the cmcs determined in EAN were 5 to 10 times larger than in water [108], and it was proposed that this was due to the enhanced solubility of the surfactant hydrocarbon chains in the IL [107].

Unlike water and other molecular solvents capable of supporting amphiphile self-assembly, ILs themselves consist of cations and anions that can obviously undergo ion exchange with surfactants [109]. ILs can also be mixed with other solvents, as either the major or minor constituent. The first systems studied were IL/H$_2$O systems because ILs are often very hygroscopic, and physicochemical properties and reactivity should be dependent on water content. Reverse micelles with IL polar cores and reverse microemulsions with nanosized IL droplets dispersed in cyclohexane (continuous phase) by nonionic Triton X-100 surfactant were first reported in 2004 by Gao and coworkers [110]. Since then, much work has continued concerning ILs both in the dispersed and continuous phases.

Some ILs can also act as amphiphiles, not just solvents, and interestingly, if the IL molecules comprise long alkyl chains, which are termed SAILs, then they have additional thermotropic properties, forming liquid crystals [111]. Examples of such IL amphiphiles include 1-alkylimidazolium nitrates or chlorides [112], which form LCs in water, acidic water, or tetrahydrofuran (THF). The lyotropic phases of SAILs based on 1-alkyl-3-methylimidazolium cations (Figure 16.3) have been investigated by Firestone and coworkers [113], and by Bowers and coworkers [114]. These systems showed aggregation in aqueous solution, yielding micellar-like structures just above the cmc. In 2007, Dong and coworkers [115] showed evidence of both lamellar (L$_\alpha$) and hexagonal (H$_1$) phases for the SAIL 1-dodecyl-3-methylimidazolium bromide with water in the range of 10% to 50% v/v (Figure 16.10). In fact, relatively short alkyl chain ILs (C$_8$) have been shown to form mesophases [114], which was further confirmed by ^{2}H-NMR and x-ray diffraction [116]. Recently, Galgano and

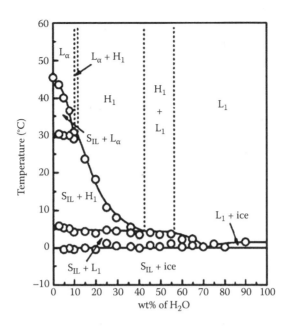

FIGURE 16.10 Temperature–concentration phase diagram for $C_{12}mimBr/H_2O$ mixtures. S_{IL} (solid phase of $C_{12}mimBr$), L_α (lamellar phase), H_1 (hexagonal phase), L_1 (micellar solution). (Reprinted from *J. Colloid Interface Sci.*, 307, Dong, B., T. Inoue, and L.Q. Zheng. Phase behavior of binary mixture of 1-dodecyl-3-methylimidazolium bromide and water revealed by differential scanning calorimetry and polarized optical microscopy, 578–581. Copyright 2007, with permission from Elsevier.)

El Seoud [117] and Brown and coworkers [118] clearly demonstrated that the IL nature of SAILS has no notable effect on aggregation properties. Nevertheless, the development of SAILs is advantageous because it presents the opportunity to combine the catalytic properties of surfactant systems with those unique to ILs, especially for separations and extractions. Furthermore, this also offers the possibility to switch between a "pure" IL solvent, with favorable solvating properties, and an IL mesophase, which may confer different selectivity and kinetics.

16.9 REACTIONS IN IL MESOPHASES

Many studies on 1-alkyl-3-methylimidazolium salts have been undertaken [119] showing self-organization [120] and a tendency to supercool to a glassy state. Bowlas and coworkers [121] produced imidazolium salts with long alkyl chain lengths (C_n = 12–18) that exhibited mesophases over large temperature ranges, showing promise as a solvent with catalytic ability [113]. Through using *N*-alkylimidazolium ILs (Figure 16.11), Huang and coworkers [112] demonstrated the stereoselectivity of a Diels–Alder reaction, a condensation reaction of cyclopentadiene with diethyl maleate. In ethanol, the *endo* product was favored 88:12, whereas in the L_α LC phase, the *endo* product was favored 46:54.

This thermotropic LC result is extremely promising and leads to intriguing questions about the possible combined effect of IL solvent properties and mesophase structure on organic reactions. Brinchi and coworkers [122] demonstrated, through an accelerated decarboxylation in an imidazolium-based SAIL, that microenvironment properties vary with SAIL concentration as well as modifications of SAIL structure, such as anion type or alkyl chain length. This presents tantalizing possibilities for control over reaction outcomes, owing to changes in the nature of the ILs and the solvent employed to induce the mesophases, as well as the composition of IL/solvent mixtures. Then, it would be possible to consider the entire mesophase as being the "supporting solvent" for any reaction under investigation.

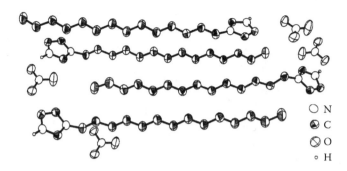

FIGURE 16.11 Bilayer lamellar structure of [C$_{14}$H$_{29}$-imH][NO$_3$] used as a solvent to enhance *endo* selectivity in a Diels–Alder reaction. (From Huang, H.W., C.K. Lee, and I. Lin. Simple amphiphilic liquid crystalline *N*-alkylimidazolium salts. A new solvent system providing a partially ordered environment. *Chem. Commun.*, 1911, 2000. Reproduced by permission of The Royal Society of Chemistry.)

16.10 CONCLUSIONS AND FUTURE PROSPECTS

This review has explored how the rate, stereoselectivity, and mechanism of reactions can be altered by performing them in structured reaction media comprising surfactants, mesophases, as well as less well-ordered ILs. It has also been shown that certain ILs self-assemble, opening up the possibility of newly ordered, tunable solvents. With a dearth of systematic studies in ordered ionic solvent systems, interesting questions can be posed.

- How does substrate partitioning, orientation, and reactivity in micelles compare with that in the pure structured IL solvent?
- Is there a balance between increased local concentrations and limited diffusion in more ordered mesophases?
- Where do reactions occur in a given mesophase? Could reaction pathways be switched during a reaction by modulating the solvent structural environment?

Reactions in ILs will continue to be studied due to the often facile extraction and separation properties, but by simultaneously investigating the control of the underlying solvent structure to manipulate reaction behavior, exciting new potential applications may be found.

ACKNOWLEDGMENT

Paul Brown thanks HEFCE and University of Bristol, School of Chemistry for a DTA PhD Scholarship.

REFERENCES

1. J. Rassing, P.J. Sams and E. Wyn-Jones. A new model describing the kinetics of micelle formation from chemical relaxation studies. *Chem. Phys. Lett.*, **13**, 233–236 (1972).
2. C.J. Drummond and T. Kaasgaard. Ordered 2-D and 3-D nanostructured amphiphile self-assembly materials stable in excess solvent. *Chem. Phys. Chem.*, **8**, 4957–4975 (2006).
3. K. Holmberg, B. Jönsson, B. Kronberg and B. Lindman. *Surfactants and Polymers in Aqueous Solution*, 2nd ed., Chap. 1–2. Wiley, Weinheim, pp. 39–95 (2003).
4. T.F. Tadros. *Applied Surfactants: Principles and Applications*, 1st ed., Chap. 2–3. Wiley, Weinheim, pp. 19–71 (2005).
5. T.L. Greaves and C.J. Drummond. Ionic liquids as amphiphile self-assembly media. *Chem. Soc. Rev.*, **37**, 1709–1726 (2008).

6. S. Tascioglu. Micellar solutions as reaction media. *Tetrahedron*, **34**, 11113–11152 (1996).

7. E.F.J. Duynstee and E. Grunwald. Organic reactions occurring in or on micelles. I. Reaction rate studies of the alkaline fading of triphenylmethane dyes and sulfonphthalein indicators in the presence of detergent salts. *J. Am. Chem. Soc.*, **81**, 4540 (1959).

8. T. Dwars, E. Paetzold and G. Oehme. Reactions in micellar systems. *Angew. Chem. Int. Ed.*, **44**, 7174–7199 (2005).

9. R.B. Dunlap and E.H. Cordes. Secondary valence force catalysis. Catalysis of hydrolysis of methyl orthobenzoate by sodium dodecyl sulfate. *J. Am. Chem. Soc.*, **90**, 4395 (1968).

10. G. Savelli, R. Germani and L. Brinchi. *Reactions and Synthesis in Surfactant Systems*, 1st ed. CRC Press, New York (2001).

11. M.N. Khan and E. Ismail. An apparent weakness of the pseudophase ion-exchange (PIE) model for micellar catalysis by cationic surfactants with nonreactive counterions. *J. Chem. Soc. Perkin*, **2**, 1346 (2001).

12. C.A. Bunton. Effects of submicellar aggregates on nucleophilic aromatic substitution and addition. *Catal. Rev. Sci. Eng.*, **20**, 1 (1979).

13. D.C. Barber, T.E. Woodhouse and D.G. Whitten. Diverse reactivities of picket fence porphyrin atropisomers in organized media: A probe of molecular location and orientation. *J. Phys. Chem.*, **96**, 5106–5114 (1992).

14. H. Al-Lohedan, C.A. Bunton and M.M. Mhala. Micellar effects upon spontaneous hydrolyses and their relation to mechanism. *J. Am. Chem. Soc.*, **104**, 6654–6660 (1982).

15. B. Cornils and E.G. Kuntz. *Aqueous-Phase Organometallic Catalysis: Concepts and Applications*, 2nd ed. Wiley, Weinheim, p. 351 (2004).

16. H. Chen, Y.E. He, M. Li, X.J. Li and Y.Z. Li. Studies on 1-dodecene hydroformylation in biphasic catalytic system containing mixed micelle. *J. Mol. Catal. A*, **194**, 13 (2003).

17. W.J. Leigh and M.S. Workentin. Liquid crystals as solvents for spectroscopic chemical reactions, and gas chromatographic applications. In *Handbook of Liquid Crystals*, D. Demus, J. Goodby, G.W. Gray, H.-W. Spiess and V. Vill (Eds.). Wiley, Weinheim, p. 849 (1998).

18. S.I. Ahmad and S. Friberg. Catalysis in micellar and liquid-crystalline phases. I. System water-hexadecyltrimethylammonium bromide-hexanol. *J. Am. Chem. Soc.*, **94**, 5196–5199 (1972).

19. R.F. Bakeeva, L.A. Kudryavtseva, G. Eme, E.M. Kosacheva, V.E. Bel'skii, D.B. Kudryavstev, R.R. Shagidullin and V.F. Sopin. Kinetics of alkaline hydrolysis of esters of phosphorus acids in micellar and hexagonal phases in the cetylmethylammonium bromide-NaOH-water system. *Russ. Chem. Bull.*, **47**, 1454–1459 (1998).

20. P. Waldon. Molecular weights and electrical conductivity of several fused salts. *Bull. Acad. Sci.* (St. Petersberg), **6**, 405–442 (1914).

21. R.M. Barrer. The viscosity of pure liquids. II. Polymerised ionic melts. *Trans. Faraday Soc.*, **39**, 59–67 (1943).

22. H.L. Chum, V.R. Koch, L.L. Miller and R.A. Osteryoung. Electrochemical scrutiny of organometallic iron complexes and hexamethylbenzene in a room temperature molten salt. *J. Am. Chem. Soc.*, **97**, 3264 (1975).

23. M.L. Druelinger, C.L. Hussey, J.A. Levisky and J.S. Wilkes. Dialkylimidazolium chloroaluminate molten salts. *Proc. Int. Symp. Molten Salts*, **81**, 245–255 (1980).

24. P. Keim and P. Wasserscheid. Ionic liquids—"New solutions" for transition metal catalysis. *Angew. Chem. Int. Ed.*, **39**, 3772–3789 (2000).

25. R.J. Gale and R.A. Osteryoung. Potentiometric investigation of dialuminum heptachloride formation in aluminum chloride-1-butylpyridinium chloride mixtures. *Inorg. Chem.*, **18**, 1603 (1979).

26. S. Chauhan, S.M.S. Chauhan, N. Jain and A. Kumar. Chemical and biochemical transformations in ionic liquids. *Tetrahedron*, **61**, 1015–1060 (2005).

27. C.L. Hussey, J.A. Levisky, J.S. Wilkes and R.A. Wilson. Dialkylimidazolium chloroaluminate melts: A new class of room-temperature ionic liquids for electrochemistry, spectroscopy and synthesis. *Inorg. Chem.*, **21**, 1263–1264 (1982).

28. H. Olivier. Recent developments in the use of non-aqueous ionic liquids for two-phase catalysis. *J. Mol. Catal. A*, **146**, 285–289 (1999).

29. E.I. Cooper and E.J.M. O'Sullivan. In *Molten Salts*, R.J. Gale, G. Blomgren and H. Kojima (Eds.). PV 92-16, The Electrochemical Society Proceedings Series, Pennington, NJ, p. 386 (1992).

30. C.M. Gordon, J.D. Holbrey, A.R. Kennedy and K.R. Seddon. Ionic liquid crystals: Hexafluorophosphate salts. *J. Mater. Chem.*, **8**, 2627–2636 (1998).

31. W.T. Ford, D.J. Hart and R.J. Hauri. Syntheses and properties of molten tetraalkylammonium tetraalkylborides. *J. Org. Chem.*, **38**, 3916–3918 (1973).

32. T. Kakiuchi, T. Kawakami, N. Nishi, F. Shigematsu and M. Yamamoto. Fluorine-free and hydrophobic room-temperature ionic liquids, tetraalkylammonium bis(2-ethylhexyl)sulfosuccinates, and their ionic liquid–water two-phase properties. *Green Chem.*, **8**, 349–355 (2006).

33. N.J. Audic. An ionic, liquid-supported ruthenium carbene complex: A robust and recyclable catalyst for ring-closing metathesis in ionic liquids. *J. Am. Chem. Soc.*, **125**, 9248–9249 (2003).

34. M. Earle. The distillation and volatility of ionic liquids. *Nature*, **439**, 831–834 (2006).

35. M.J. Earle and K.R. Seddon. Ionic liquids. Green solvents for the future. *Pure Appl. Chem.*, **72**, 1391–1398 (2000).

36. N.V. Plechkova and K.R. Seddon. Applications of ionic liquids in the chemical industry. *Chem. Soc. Rev.*, **37**, 123–150 (2008).

37. M. Koel. Physical-chemical properties of ionic liquids based on dialkyl-imidazolium cation. *Proc. Est. Acad. Sci. Chem.*, **49**, 145–155 (2000).

38. P. Wasserscheid and T. Welton. *Ionic Liquids in Synthesis*. Wiley, Weinheim, p. 364 (2002).

39. E. Burello, P.N. Davey, I. Lopez-Martin, G. Rothenberg and K.R. Seddon. Anion and cation effects on imidazolium salt melting points: A descriptor modelling study. *J. Phys. Chem.*, **8**, 690–695 (2000).

40. A. Bagno, C.P. Butts, C. Chiappe, F. D'Amico, J.C.D. Lord, D. Pieraccini and F. Rastrelli. The effect of the anion on the physical properties of trihalide-based N, N-dialkylimidazolium ionic liquids. *Org. Biomol. Chem.*, **3**, 1624–1630 (2005).

41. A.F. Kapustinskii. Lattice energy of ionic crystals. *Q. Rev. Chem. Soc.*, **10**, 283 (1956).

42. K.R. Seddon. Room-temperature ionic liquids-neoteric solvents for clean catalysis. *Kinet. Catal.*, **37**, 743–748 (1996).

43. C.A. Angell, E.I. Cooper and W. Xu. Ionic liquids: Ion mobilities, glass temperatures and fragilities. *J. Phys. Chem. B.*, **107**, 6170–6178 (2003).

44. A.K. Abdul-Sada, A.M. Greenway, P.B. Hitchcock, J.M. Thamer, K.R. Seddon and J.A. Zora. Upon the structure of room temperature halogenoaluminate ionic liquids. *J. Chem. Soc. Chem. Commun.*, 1753–1754 (1986).

45. N. Sun, X. He, X. Lu and X. Zhang. Physical properties of ionic liquids: Database and evaluation. *J. Phys. Chem. Ref. Data*, **35**, 1475–1517 (2006).

46. R.A. Mantz and P.C. Truelove. Viscosity and density of ionic liquids. In *Ionic Liquids in Synthesis*, P. Wasserscheid and T. Welton (Eds.). Wiley, Weinheim, pp. 56–68 (2002).

47. J.A. Boxall, R. Lichtenhalter and K.N. Marsh. Room temperature ionic liquids and their mixtures—A review. *Fluid Phase Equilib.*, **219**, 93–98 (2004).

48. J.O. Valderrama and K. Zarricueta. A simple and generalized model for predicting the density of ionic liquids. *Fluid Phase Equilib.*, **275**, 145–151 (2009).

49. A.V. Blokhin, M. Frenkel, G.J. Kabo, J.W. Magee, Y.U. Paulechka and O.A. Vydrov. Thermodynamic properties of 1-butyl-3-methylimidazolium hexafluorophosphate in the ideal gas state. *J. Chem. Eng. Data*, **48**, 457 (2003).

50. S.E. Fry and N.J. Pienta. Effects of molten salts on reactions. Nucleophilic aromatic substitution by halide ions in molten dodecyltributylphosphonium salts. *J. Am. Chem. Soc.*, **107**, 6399–6400 (1986).

51. J.A. Boon, J.A. Levisky, J.L. Plug and J.S. Wilkes. Friedel-Crafts reactions in ambient-temperature molten salts. *J. Org. Chem.*, **51**, 480–483 (1986).

52. A.A. Fannin, D.A. Floreani, L.A. King, J.S. Landers, B.J. Piersma, D.J. Stech, R.L. Vaughn, J.S. Wilkes and J.L. Williams. Properties of 1, 3-dialkylimidazolium chloride-aluminum chloride ionic liquids. 2. Phase transitions, densities, electrical conductivities, and viscosities. *J. Phys. Chem.*, **88**, 2614 (1984).

53. R. Sheldon. Catalytic reactions in ionic liquids. *Chem. Commun.*, 2399–2407 (2001).

54. M.S. Rao, B.V.S. Reddy, P.N. Reddy and J.S. Yadav. Bi(OTf)3-[Bmim]PF6: A novel and reusable catalytic system for the synthesis of cis-aziridine carboxylates. *Synthesis*, **9**, 1387 (2003).

55. Z.C. Chen, Y. Hi, Z.G. Le and Q.G. Zheng. Organic reactions in ionic liquids: A simple and highly regioselective N-substitution of pyrrole. *Synthesis*, **12**, 1951–1954 (2004).

56. Z. Hou, Y. Hu and B. Feng. The functionalized ionic liquid-stabilized palladium nanoparticles catalyzed selective hydrogenation in ionic liquid. *Catal. Commun.*, **10**, 1903–1907 (2009).

57. R.A. Bartsch and S.V. Dzyuba. Expanding the polarity range of ionic liquids. *Tetrahedron Lett.*, **43**, 4657–4659 (2002).

58. R. Bernini, A. Coratti, G. Fabrizi and A. Goggiamani. CH3ReO3/H2O2 in room temperature ionic liquids: A homogeneous recyclable catalytic system for the Baeyer–Villiger reaction. *Tetrahedron Lett.*, **44**, 8991–8994 (2003).

59. A. Ahosseini, W. Ren and A.M. Scurto. Understanding biphasic ionic liquid/CO2 systems for homogeneous catalysis: Hydroformylation. *Ind. Eng. Chem. Res.*, **48**, 4254–4265 (2009).

60. A.J. Carmichael, M.J. Earle, D.J. Holbrey, P.B. McCormac and K.R. Seddon. The Heck reaction in ionic liquids: A multiphasic catalyst system. *Org. Lett.*, **1**, 997–1000 (1999).

61. G.W. Parshall. Catalysis in molten salt media. *J. Am. Chem. Soc.*, **94**, 8716–8719 (1972).

62. C.M. Gordon. New developments in catalysis using ionic liquids. *Appl. Catal. A*, **222**, 101–117 (2001).

63. J. Dupont, R.F. de Souza and P.A.Z. Suarez. Ionic liquid (molten salt) phase organometallic catalysis. *Chem. Rev.*, **102**, 3667–3692 (2002).

64. R. Fehrmann, M. Haumann, A. Riisager and P. Wassersheid. *Catalytic SILP Materials. Topics in Organometallic Chemistry*, Vol. 23. Springer, Heidelberg, pp. 149–161 (2008).

65. R. Fehrmann, M. Haumann, A. Riisager and P. Wassersheid. Stability and kinetic studies of supported ionic liquid phase catalysts for hydroformylation of propene. *Ind. Eng. Chem. Res.*, **44**, 9853–9859 (2005).

66. L. Leclercq, I. Suisse and F. Agnossu-Niedercorn. Biphasic hydroformylation in ionic liquids: Interaction between phosphane ligands and imidazolium triflate, toward an asymmetric process. *Chem. Commun.*, 311 (2008).

67. W. Chen, J. Ross, J. Xiao and L. Xu. Palladium-catalyzed regioselective arylation of an electron-rich olefin by aryl halides in ionic liquids. *Org. Lett.*, **3**, 295–297 (2001).

68. C.J. Mathews, P.J. Smith, T. Welton, A.J.P. White and D.J. Williams. In situ formation of mixed phosphine-imidazolylidene palladium complexes in room-temperature ionic liquids. *Organometallics*, **20**, 3848–3850 (2001).

69. C.J. Mathews, F. McLachlan, P.J. Smith and T. Welton. Palladium-catalyzed Suzuki cross-coupling reactions in ambient temperature ionic liquids: Evidence for the importance of palladium imidazolylidene complexes. *Organometallics*, **22**, 5350–5357 (2003).

70. F.C. Ferreria, G. Livingston, J.P. Pink and H. Wong. Recovery and reuse of ionic liquids and palladium catalyst for Suzuki reactions using organic solvent nanofiltration. *Green Chem.*, **8**, 373–379 (2006).

71. B. Betzemeier, P. Knochel, M. Ossberger and J. Sirieix. Palladium catalyzed cross-couplings of organozincs in ionic liquids. *Synlett*, 1613–1615 (2000).

72. A.R. Gholap, K. Venkatesan, R. Pasricha, T. Daniel, R.J. Lahoti and K.V. Srinivasan. Copper- and ligand-free Sonogashira reaction catalyzed by Pd(0) nanoparticles at ambient conditions under ultrasound irradiation. *J. Org. Chem.*, **70**, 4869–4872 (2005).

73. S. Tiwari and A. Kumar. Diels–Alder reactions are faster in water than ionic liquids at room temperature. *Angew. Chem. Int. Ed.*, **118**, 4824–4825 (2006).

74. Y. Chauvin, B. Gilbert and I. Guibard. Catalytic dimerization of alkenes by nickel complexes in organochloroaluminate molten salts. *J. Chem. Soc. Chem. Commun.*, 1715–1757 (1990).

75. J. Arras, P. Claus, Y. Shayeghi and M. Steffan. The promoting effect of a dicyanamide based ionic liquid in the selective hydrogenation of citral. *Chem. Commun.*, 2348–2349 (2008).

76. N. Llewellyn Lancaster and V. Llopis-Mestre. Aromatic nitrations in ionic liquids: The importance of cation choice. *Chem. Commun.*, 2812–2813 (2003).

77. N. Jiang and J. Ragauskas. Selective aerobic oxidation of activated alcohols into acids or aldehydes in ionic liquids. *J. Org. Chem.*, **72**, 7030–7033 (2007).

78. S.M.S. Chauhan, A. Kumar and K.A. Srinivas. Oxidation of thiols with molecular oxygen catalyzed by cobalt(II) phthalocyanines in ionic liquid. *Chem. Commun.*, 2348 (2003).

79. S.M.S. Chauan, A. Jain and A. Kumar. Metalloporphyrin and heteropoly acid catalyzed oxidation of C=NOH bonds in ionic liquids: Biomimetic models of Nitric Oxide Synthase. *Tetrahedron Lett.*, **46**, 2599–2602 (2005).

80. J. Howarth. Oxidation of aromatic aldehydes in the ionic liquid [bmim]PF6. *Tetrahedron Lett.*, **41**, 6627–6629 (2000).

81. S. Park and R.J. Kazlauskas. Biocatalysis in ionic liquids—Advantages beyond green technology. *Curr. Opin. Biotechnol.*, **14**, 432–437 (2003).

82. N. Kaftzik, U. Kragl, S.H. Schoefer and P. Wassersheid. Enzyme catalysis in ionic liquids: Lipase catalysed kinetic resolution of 1-phenylethanol with improved enantioselectivity. *Chem. Commun.*, 425 (2001).

83. A.D. Headley and B. Ni. Chiral imidazolium ionic liquids: Their synthesis and influence on the outcome of organic reactions. *Aldrichim. Acta*, **40**, 107–117 (2007).

84. A. Gaumont, F. Guillan, C. Malhaic, J. Levillain and J. Plaquevent. Ionic liquids: New targets and media for a-amino acid and peptide chemistry. *Chem. Rev.*, **108**, 5035–5060 (2008).

85. C. Chiappe, G. Imperato and B. Koenig. Ionic green solvents from renewable resources. *Eur. J. Org. Chem.*, 1049–1058 (2007).

86. G. Ou, J. She, Y. Yuan and H. Zhou. Ionic liquid buffers: A new class of chemicals with potential for controlling pH in non-aqueous media. *Chem. Commun.*, 4626–4628 (2006).

87. F.R. Rantwijk and R.A. Sheldon. Biocatalysis in ionic liquids. *Chem. Rev.*, **107**, 2757–2785 (2007).

88. P. Licence, F.J.M. Rutten and H. Tadesse. Rewritable imaging on the surface of frozen ionic liquids. *Angew. Chem.*, **46**, 4163–4165 (2007).

89. C.M. Gordon and A.J. McClean. Photoelectron transfer from excited-state ruthenium(II) tris(bipyridyl) to methylviologen in an ionic liquid. *Chem. Commun.*, 1395–1396 (2000).

90. E.D. Bates, J.H. Davis Jr, R.D. Mayton and I. Ntai. CO_2 capture by a task-specific ionic liquid. *J. Am. Chem. Soc.*, **124**, 926 (2002).

91. J.R. Brozozowski, H. Cheng, P.B. Henderson, R.M. Pearlstein and D.J. Tempel. High gas storage capacities for ionic liquids through chemical complexation. *J. Am. Chem. Soc.*, **130**, 400–401 (2008).

92. Y. Chen, W. Liu, C. Ye and L. Yu. Room-temperature ionic liquids: A novel versatile lubricant. *Chem. Commun.*, 2244 (2001).

93. G. Gamero-Catano and V. Hruby. Electrospray as a source of nanoparticles for efficient colloid thrusters. *J. Prop. Power*, **17**, 977 (2001).

94. C.M. Jin, C. Ye, B.S. Phillips, J.S. Zabinski, X. Liu, W. Liu and J.M. Shreeve. Polyethylene glycol functionalized dicationic ionic liquids with alkyl or polyfluoroalkyl substituents as high temperature lubricants. *J. Mater. Chem.*, **16**, 1529–1535 (2006).

95. M. Maase, K. Massonne, K. Halbritter, R. Noe, M. Bartsch, W. Siegel, V. Stegmann, M. Flores, O. Huttenloch and M. Becker. Method for the separation of acids from chemical reaction mixtures by means of ionic fluids. Wo. Pat. 062171 (2003).

96. J.D. Holbrey, R.D. Rogers, S.K. Spear and R.P. Swatloski. Dissolution of cellose with ionic liquids. *J. Am. Chem. Soc.*, **124**, 4974–4975 (2002).

97. S. Mekki, C.M. Wai, I. Billard, G. Moutiers, J. Burt, B. Yoon, J.S. Wang, C. Gaillard, A. Ouadi and P. Hesemann. Extraction of lanthanides from aqueous solution by using room-temperature ionic liquid and supercritical carbon dioxide in conjunction. *Chem. Eur. J.*, **12**, 1760–1766 (2006).

98. P. Giridhar, T.G. Srinivasan, P.R. Vasudeva and K.A. Venkatesan. Electrochemical scrutiny of organometallic iron complexes and hexamethylbenzene in a room temperature molten salt. *Electrochim. Acta*, **52**, 3006–3012 (2007).

99. M. Jayakumar, T.G. Srinivasav and K.A. Venkatesan. Electrochemical behavior of fission palladium in 1-butyl-3-methylimidazolium chloride. *Electrochim. Acta*, **52**, 7121–7127 (2007).

100. J.S. Wang, C.N. Sheaff, B. Yoon, R.S. Addleman and C.M. Wai. Extraction of uranium from aqueous solutions by using ionic liquid and supercritical carbon dioxide in conjunction. *Chem. Eur. J.*, **15**, 4458–154463 (2009).

101. A. Kamimura and S. Yamamoto. An efficient method to depolymerize polyamide plastics: A new use of ionic liquids. *Org. Lett.*, **9**, 2533–2535 (2007).

102. Z.C. Liu, R.G. Xia, C.M. Xu and R. Zhang. Ionic liquid alkylation process produces high-quality gasoline. *Oil Gas J.*, **104**, 52–56 (2006).

103. A. Hoff, C. Jost, A. Prodi-Schwab, F.G. Schmidt and B. Weyershausen. Ionic liquids: New designer compounds for more efficient chemistry. *Elements: Degussa Science Newsletter*, **9**, 10–15 (2004).

104. K. Lehmann and B. Weyerhausen. Industrial application of ionic liquids as performance additives. *Green Chem.*, **7**, 15–19 (2005).

105. B. Wu, R.G. Reddy and R.D. Rogers. *Solar Energy: The Power to Choose*. Proc. Solar Forum, Washington, DC, pp. 445–451 (2001).

106. D.F. Evans, S.H. Chen, G.W. Schriver and E.M. Arnett. Thermodynamics of solution of nonpolar gases in a fused salt. Hydrophobic bonding behavior in a nonaqueous system. *J. Am. Chem. Soc.*, **103**, 481–482 (1981).

107. W.J. Benton, D.F. Evans and E.W. Kaler. Liquid crystals in a fused salt: Beta, gamma-distearoylphosphatidylcholine in N-ethylammonium nitrate. *J. Phys. Chem.*, **87**, 533 (1983).

108. E.Z. Casassa, D.F. Evans, R. Roman and A. Yamauchi. Micelle formation in ethylammonium nitrate, a low-melting fused salt. *J. Colloid Interface Sci.*, **88**, 89 (1982).

109. C.J. Drummond and T.L. Greaves. Protic ionic liquids: Properties and applications. *Chem. Rev.*, **108**, 206 (2008).

110. H.X. Gao, J.C. Li, B.X. Han, W.N. Chen, J.L. Zhang, R. Zhang and D.D. Yan. Microemulsions with ionic liquid polar domains. *Phys. Chem. Chem. Phys.*, **6**, 2914–2916 (2004).

111. K. Binnemans. Ionic liquid crystals. *Chem. Rev.*, **105**, 4148–4204 (2005).

112. H.W. Huang, C.K. Lee and I. Lin. Simple amphiphilic liquid crystalline *N*-alkylimidazolium salts. A new solvent system providing a partially ordered environment. *Chem. Commun.*, 1911 (2000).

113. M.A. Firestone, J.A. Dzielawa, P. Zapol, L.A. Curtiss, S. Seifert and M.L. Dietz. Lyotropic liquid-crystalline gel formation in a room-temperature ionic liquid. *Langmuir*, **18**, 7258–7260 (2002).

114. J. Bowers, C.P. Butts, P.J. Martin and M.C. Vergara-Gutierrez. Aggregation behavior of aqueous solutions of ionic liquids. *Langmuir*, **20**, 2191–2198 (2004).

115. B. Dong, T. Inoue and L.Q. Zheng. Phase behavior of binary mixture of 1-dodecyl-3-methylimidazolium bromide and water revealed by differential scanning calorimetry and polarized optical microscopy. *J. Colloid Interface Sci.*, **307**, 578–581 (2007).

116. I. Goodchild, L. Collier, S.L. Millar, I. Prokes, J.C.D. Lord, C.P. Butts, J. Bowers, R.P. Webster and R.K. Heenan. Structural studies of the phase, aggregation and surface behaviour of 1-alkyl-3-methylimidazolium halide + water mixtures. *J. Colloid Interface Sci.*, **307**, 455–468 (2007).

117. P.D. Galgano and O.A. El Seoud. Surface active ionic liquids: Comparison of the micellar properties of 1-hexadecyl-3-methylimidazolium chloride with other cationic surfactants. *J. Colloid Interface Sci.*, **345**, 1–11 (2010).

118. P. Brown, C.P. Butts, R. Dyer, J. Eastoe, I. Grillo, F. Guittard, S. Rogers and R.K. Heenan. Anionic surfactants and surfactant ionic liquids with quaternary ammonium counterions. *Langmuir*, **27**, 4563–4571 (2011).

119. C.M. Paleos and D. Tsiourvas. Supramolecular hydrogen-bonded liquid crystals. *Liq. Cryst.*, **28**, 1127 (2001).

120. K.M. Lee, Y.T. Lee and I.J.B. Lin. Supramolecular liquid crystals of amide functionalized imidazolium salts. *J. Mater. Chem.*, **13**, 1079–1084 (2003).

121. C.J. Bowlas, C.J. Bruce and K.R. Seddon. Liquid-crystalline ionic liquids. *Chem. Commun.*, 1625–1626 (1996).

122. L. Brinchi, R. Germani, E. Braccalenti, N. Spreti, M. Tiecco and G. Savelli. Accelerated decarboxylation of 6-nitrobenzisoxazole-3-carboxylate in imidazolium-based ionic liquids and surfactant ionic liquids. *J. Colloid Interface Sci.*, **348**, 137–145 (2010).

17 Stimuli-Responsive Surfactants
History and Applications

John Texter

CONTENTS

17.1 INTRODUCTION

Interest in stimuli-responsive surfactants is steadily growing, although the stimuli-responsiveness of surfactants is in fact an essential property of surfactants that exhibit self-assembly. We will see in the sequel that the development and use of various kinds of stimuli-responsive surfactants in producing novel stimuli-responsive polymers and advanced materials has become a very active research area [1]. However, it is fair to say that the self-assembly exhibited by surfactants in micelle formation, reversible vesicle formation, monolayer and thin-film formation, vesicle formation, and the formation of various types of lyotropic liquid crystalline phases are all examples of surfactant stimuli-responsiveness. This response is simply an association or aggregation in response to thermodynamic driving forces emanating from chemical potential (concentration) effects and the effects of other intensive or field variables such as temperature and electromagnetic fields.

Our understanding of the micellization process was greatly aided by Israelachvili's monograph [2]. Subsequent advances in understanding the various exotic complexes formed by surfactants in solution as well as the formation of lyotropic mesophases have come mostly from Monte Carlo and molecular dynamics simulation studies. Monolayer and interfacial surfactant film structures have also succumbed to experimental scattering studies. An early comprehensive focus on interfacial structure was held in Chicago at the 210th National Meeting of the American Chemical Society in 1995 and a 1996 Gordon Research Conference on Chemistry of Interfaces. Important articles were collected in a review volume [3]. Computer (simulation) experiments have since become reasonable alternatives to many association and phase transition processes.

Understanding the intermolecular forces that govern association and phase transitions (including precipitation from solution) among surfactants and between surfactants and other materials remains fundamentally important. Such processes are central to microemulsion formulation and to how we use surfactants in mixtures with polymers and dispersions of all types. The extensive literature covering these last topics will not be further developed here, but intermolecular associations are among the most important types of stimuli-responsiveness exhibited by almost all surfactants.

Important earlier reviews of stimuli-responsive surfactants systems include that of Hubbard and Abbott [4] on giant micellar systems responsive to temperature, electrochemical potential, light, and pH. Some of these same topics were discussed by Xie and Feng [5], and stimuli coming from electrolyte and organic molecules in addition to a variety of applications were reviewed. We will see that many of the new stimuli-responsive surfactants are Gemini surfactants. An excellent review covering the formation and early use of Gemini surfactants has been provided by In [6].

17.1.1 REACTIVE SURFACTANTS

The next most important type of stimuli-responsiveness exhibited by *some* surfactants is their ability to form chemical bonds. An important subclass of reactive surfactants has long been known as surfmers, surfactants that are reactive monomers suitable for participating in polymerization reactions. The development of surfmers overcame an important problem in emulsion polymerization, wherein latex particles formed thereby no longer carried surfactants into applications in which surfactants could leach away and cause untoward problems. The stabilizing surfactants used were covalently incorporated as comonomers into the latex particles. This topic is treated in greater detail later. However, a treatise [7] on reactions and synthesis in surfactant systems presented reviews of the aforementioned association processes as well as the use of surfmers in various polymerization processes, in addition to focusing on various reaction pathways incorporated into the use of diverse surfactants. These pathways include synthetic transformations of certain types of surfactants into others [8], cleavable surfactants [9], Gemini and oligomeric surfactants [6], biologically active glycolipids [10], acid and oxidatively labile surfactants [11], and electrochemically active surfactants [12].

Stimuli-responsive block copolymer surfactants received a relatively early review [13]. It focused on copolymeric vinyl ethers and poly(ethylene oxide) (PEO) blocks and accompanying thermoreversibility effects with examples exhibiting lower critical solution temperatures (LCST). The effects of added compounds and pH were also examined. Another early review examined the stimuli-responsiveness arising from nonionic surfactant complexation with polyelectrolytes [14]. These associations produced micelles decorating various portions of the polyelectrolyte, according to the bead and necklace model popularized by Cabane and Duplessix [15]. The effects of six types of stimuli—pH, electrolyte, molecular binding, temperature, light, and electrochemical have been recently reviewed [5].

17.1.2 EARLY EXAMPLES

Stimuli-responsive surfactants that are responsive to concentrations of an agent other than the surfactant itself are exemplified by the simple soaps comprising alkyl carboxylates [16]. Such materials have long been important for their detergency and personal care applications, and are also famous for their sensitivity to transition metals and "hard water" effects. Zwitterionic or amphoteric surfactants that are pH-sensitive are also among the oldest classes of stimuli-responsive surfactants. Some examples have been reviewed in various contexts [17].

17.1.3 COVERAGE AND SCOPE

In the following, we address two types of topics. In Sections 17.2 through 17.9, we discuss various types of stimuli-responsive surfactants. Subsections about thermoreversibility, pH sensitivity, etc., are not exclusive because many or most such amphiphiles exhibit multi-stimuli–responsiveness. Rather, these subsections serve to focus on the diversity of stimuli that have been studied. Neither the number of types of stimuli-responsive surfactants nor the coverage in each category are exhaustive. The example articles selected were chosen to illustrate the rich diversity that exists. Our choices are not entirely restricted to molecular surfactants, and they include polymers and polymeric nanoparticles that are surface-active agents. Thereafter, in Section 17.10, we discuss a variety of applications of stimuli-responsive surfactants. We end with a short summary of our personal outlook for the future of stimuli-responsive surfactants in colloid and materials sciences.

17.2 SURFMERS AND INISURFS

An excellent historical perspective of the use of surfmers in emulsion polymerization has been presented [18], and addresses the critical issues surrounding where and how surfmers are incorporated into the resulting latex particles. The great majority of surfmers studied are simple vinyl derivatives, whether charged or nonionic. Surfactant initiators are also described [18], and key examples are amphiphilic azo derivatives. These surfmers are stimuli-responsive to free radicals and essentially irreversibly undergo addition in chain polymerization. Furthermore, stimuli-responsive surfmers are also described in later sections (see discussions of ionic liquid surfactant acrylates and methacrylates and resulting materials).

17.3 BIOSURFACTANTS AND PEPTIDES

Efficient surfactants designed using natural protein motifs have been found to be switchable in application to emulsion and foam stability in response to mild chemical stimuli [19]. These surfactants are nontoxic and effective in various pharmaceutical and personal care applications. A 21-peptide AM1 (Ac-MKQLADSLHQLARQVSRLEHA-CONH$_2$) was demonstrated to form a robust film at fluid–fluid (emulsion) and fluid–air (foam) interfaces when divalent zinc was available [20–22]. Under acidic conditions or in the absence of Zn^{2+}, the AM1 is mobile and the interface is stable but much more flexible. AM1 was derived from the amphipathic peptide Lac21 (Ac-MKQLADSLMQLARQVS-RLESA-CONH$_2$).

An interesting family of peptide surfactants was designed for siRNA delivery [23]. Delivery carriers for siRNA have ranged from cationic lipid/surfactant complexes to cationic nanoparticles. These peptide surfactants have a high degree of biodegradability. The amine groups, when protonated, provide means for binding the siRNA, and the lipid tail groups provide a driving force to form condensed nanoparticles of the peptide/siRNA complexes. Several examples were demonstrated as having promising performance in siRNA delivery to U87-luc cells. These transfection peptides are illustrated in Figure 17.1. In essence, this family may be viewed as a family of double-tailed surfactants with fatty tails on each end and short to long strands of ehtylenediamines that can be protonated and that bind to the siRNA or other DNA matter.

A recombinant stimuli-responsive surfactant has been reported [24] that is synthesized in *Escherichia coli*. The resulting polypeptide, GAM1 (GMKQLADS LHQLARQVSRLEHA), was also synthesized using a reverse-phase method. Both the recombinant and reverse-phase synthesized

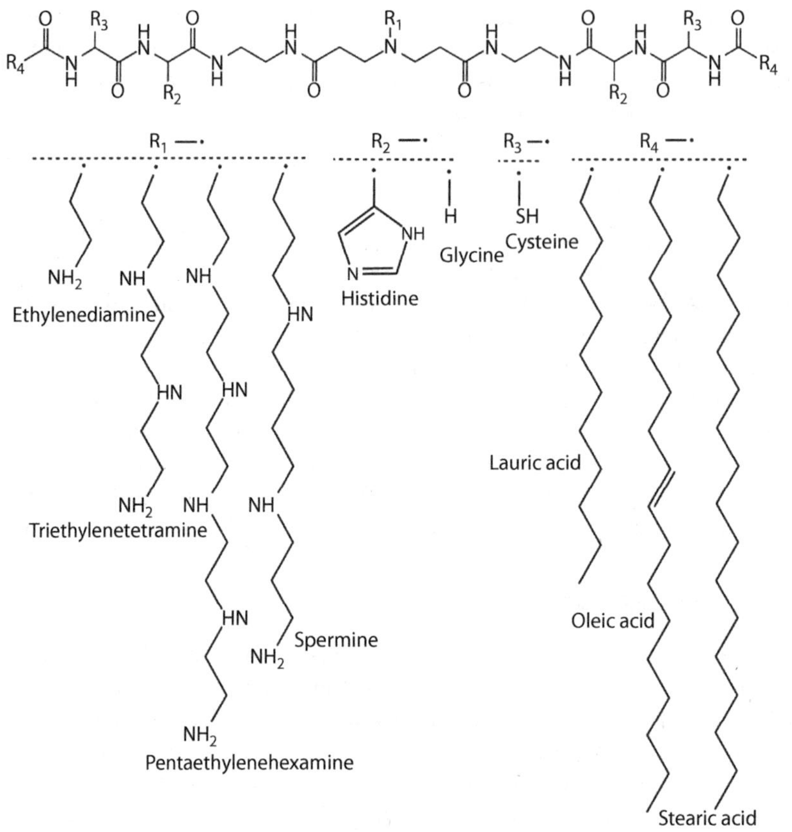

FIGURE 17.1 Peptide surfactants developed for transfection applications and siRNA delivery by Wang et al. [23]. Top, the symmetric peptide structure; bottom, the range of R groups investigated.

GAM1 stabilized a foam column. Addition of aqueous EDTA to the top of the foam column resulted in rapid collapse of the column.

17.4 REDOX SURFACTANTS

Interest in the behavior of redox surfactants seems to have emanated from Dan Buttry's laboratory in Laramie [25,26]. It was noted that the influence of a variety of surface-active species on polarographic maxima suppression motivated interest in how surfactants interacted with electrodes and with each other. Decyl, dodecyl, and tetradecyl bromides were quaternized with (dimethylamino) methyl ferrocene. The oxidized ferricinium surfactants were found to be less strongly adsorbed on gold electrodes than the reduced ferrocene forms. In this early work, it was found that the formal electrochemical potentials were different for monomeric surfactant than for surfactants aggregated in micelles, and this realization led to the development of a method for calculating the free energy changes accompanying micellization.

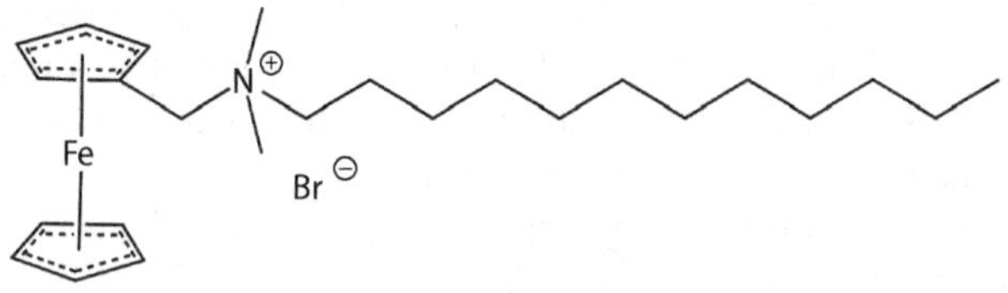

DDMFB

These introductory studies were followed by examining the detailed surface chemistry of disulfide-modified electrodes by using dodecyldimethyl (methylferrocene) ammonium bromide (DDMFB) as an electrochemical probe molecule [27]. The hexadecyl and octadecyl homologues of DDMFB above as well as FcADS, bis(aminophenyl-4-yl ferrocenylformamide) disulfide were examined in self-assembled monolayers and in the context of mutual ionic interactions [28].

The packing of N-[(cholesteryloxy)carbonyl]-N'-ethylviologen (ChEV) and N-ethyl-N'-octadecylviologen (EODV; see below) on gold electrodes was found to be very close at 3.7×10^{-10} mol/cm^2 and 4.2×10^{-10} mol/cm^2, respectively [29]. Although there is a significant disparity between the cross-sectional area of the cholesteryl group in comparison with the octadecyl derivative, the ethylviologen seems to determine the packing. Relative blocking effects for oxidation of ruthenium hexamine suggested that the cholesteryl moieties were less organized in their monolayer in comparison to the octadecyl derivative.

ChEV

EODV

The synthesis of ferrocenyl bipyridine surfactants has been described with a view to coupling transition metal complexes with redox-active ferrocenyl-based surfactants [30]. Friedel–Crafts acylation with ω-Br-alkyl acid chlorides of ferrocenyl bipyridines produces active complexes that can be used to couple with oligomers of ethylene oxide and to alkylate tertiary amines to make cationic redox-active surfactants.

A ferrocenylundecyltrimethylammonium bromide surfactant undergoing reversible one-electron oxidation was found to enable reversible dynamic surface tension control in aqueous solutions [31]. Redox conversions of this surfactant were used to create surfactant concentration gradients and accompanying Marangoni phenomena [32]. This reduction to practice of redox-controlled surface fluidics suggested the plausibility of utilizing electrode arrays for fluid-based sorting, mixing, and reaction control. The ω-ferrocenylundecyldimethylammonium bromide (FUDMAB) [33] moiety was demonstrated to reversibly control surface tension by oxidizing the ferrocene to ferricinium, as illustrated in Figure 17.2 at the air/water interface, in which the reduced form is transformed into a bolaform surfactant by the one-electron oxidation. The lower areal density packing exhibited by the one-electron oxidation product results in a higher interfacial tension. The interfacial gas to expanded liquid phase transition was studied by fluorescence microscopy.

The effect of an anionic redox-active surfactant, sodium ω-sulfoundecyl ferrocene (NaSUF; see below), was examined [34,35]. One electron oxidation produces a mesoionic species of lower surface activity consistent with greater water solubility. Surprisingly, however, although the sulfo and trimethylammonium moieties form micellar structure of approximately 6 nm diameter as shown by dynamic light-scattering measurements, the mesoionic sulfo species forms aggregates of approximately 70 nm diameter upon one-electron oxidation to the ferricinium form. Electrolyte effects and extensive comparative discussions of vesicle formation by mixtures of cationic and anionic surfactants [36,37] were used to conclude that these aggregates were vesicle-like, although transmission

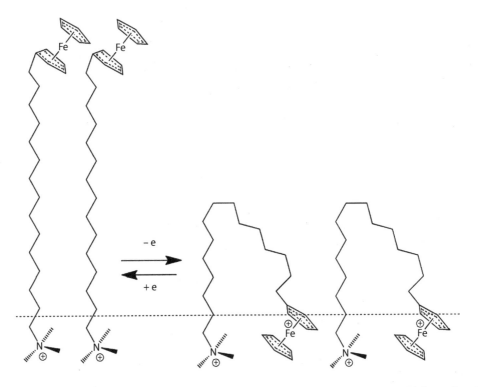

FIGURE 17.2 Illustration of how electrochemical redox control of surface tension at the air/water interface can be achieved using ferrocene-based surfactants. The FUDMA example illustrated here becomes an unsymmetrical bolaform surfactant upon one-electron oxidation.

electron microscopy (TEM) evidence was not offered. If vesicles were formed reversibly and spontaneously, cryo-TEM would be needed to image such structures.

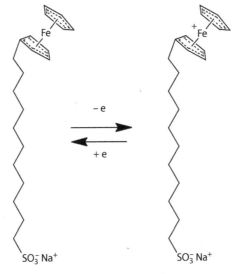

NaSUF and NaSUF⁺

Two tempo-amides of the C16 and C18 carboxylates (4-alkaneamide derivatives of 2,2,6,6-tetramethyl-1-piperidynyloxy radical), both insoluble, were examined for their mobility at the air/water interface [38]. The hexadecyl derivative was too water-soluble to get reproducible measurements. Packing densities corresponding to 50 to 450 Å²/molecule were examined. At 52 Å²/molecule, the C18 derivative (see below) was found to orient with only the amide group in the water and with both the piperidynyloxy radical and the alkane chain sticking into the air. A maximal self-diffusion coefficient of 1.5×10^{-5} cm²/s was obtained at the lowest packing density of 450 Å²/molecule.

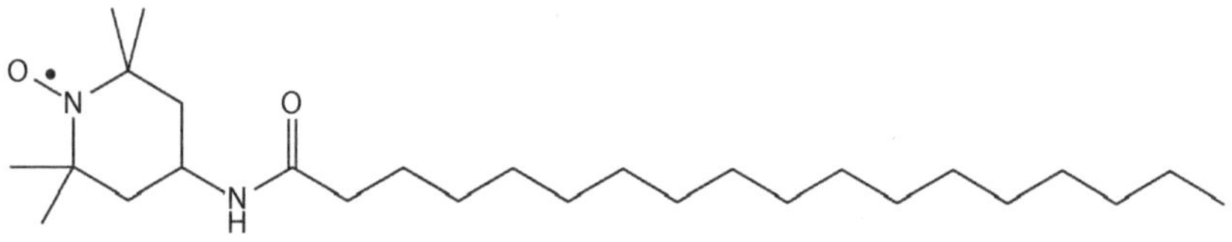

$TC_{18}A$

The tunable disassembly of micelles using a redox trigger was also demonstrated by Ghosh et al. [39]. In this case, a redox-active and cleavable surfactant was designed using a disulfide linking group to connect the surfactant head and tail groups. Two examples were illustrated. Both were simple sodium carboxylates having tetradecyl chains with disulfide inserted in one case at the beta (third) carbon (NaC_3SSC_{11}) and in the other case at the eighth carbon (NaC_8SSC_6; see below). Dithiothreitol (DTT) was used as the reducing agent, and tunability was obtained by forming mixed micelles with sodium laurate. The reaction products obtained upon cleavage using DTT were (1) ring closure in the DTT by formation therein of a disulfide bond; (2) an alkane thiol; and (3) an ω-thiolalkane carboxylate. The addition of sodium laurate increased the (mixed micelle) cmc and also lengthened the release (disassembly) time of a hydrophobic fluorescent probe, Nile red. The use of a cosurfactant yielded a design feature for formulating tunably releasing surfactant structures, wherein one considers the solubility of any guest-releasing material and the solubility of surfactant fragmentation products in micelles of the co-surfactant.

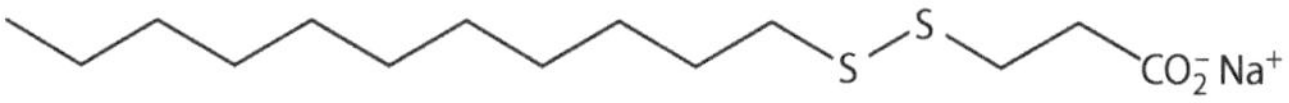

NaC_3SSC_{11}

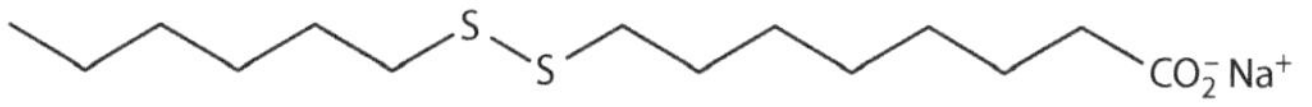

NaC_8SSC_6

17.5 PHOTOCHROMIC SURFACTANTS

A significant and very early study of photoactive surfactants was that of Kunitake et al. [40], in which a bilayer forming and circular dichroic (CD) double-tailed surfactant L-bisC$_{13}$ (see below) was used to prepare temperature-responsive bilayers and a series of azobenzene quaternary bromides, $C_{13}C_nAz$ ($n = 2, 4$, and 10) that undergo *cis-trans* isomerization. The membrane melting transition at 31°C to 32°C was studied using the CD spectra observed for the L-bisC$_{13}$ moieties, above and below this transition temperature range. These L-bisC$_{13}$ synthetic membranes were doped with the azobenzenes $C_{13}C_nAz$ (see below) in expectation that the bilayer integrity would be decreased upon *trans-cis* isomerization and that the resulting synthetic organelles would simulate photoreceptor cell membranes [40]. Electron microscopy showed that inclusion of the azobenzenes $C_{13}C_nAz$ in bilayer aggregates of these synthetic membranes induced morphological transitions upon light-induced *cis-trans* isomerization.

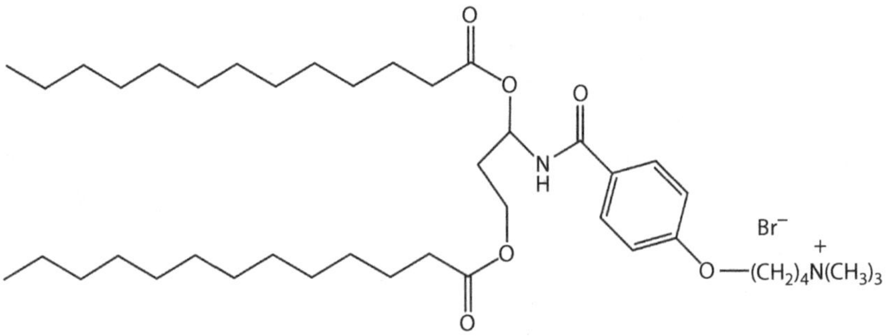

L-bisC$_{13}$

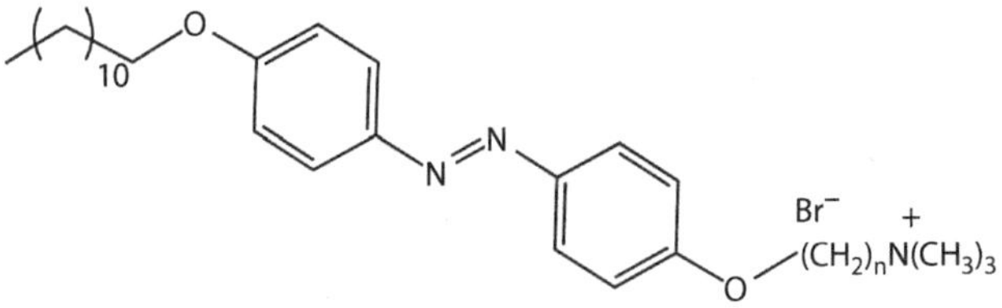

C$_{13}$C$_n$Az

Mixed surfactants comprising sodium dodecylsulfate (SDS) and the photoactive 4,4'-bis(trimethyl ammoniumhexyloxy)azobenzene dibromide (BTHA; see below) at the air/water interface exhibit photoactive changes in surface tension [41]. It was shown that *cis* to *trans* photoconversion yielded up to 25 mN/m changes in dynamic surface tension, although equilibrium changes of only 2 mN/m could be realized. An interesting application of this dynamic effect was demonstrated by photoejecting pendent drops from a capillary.

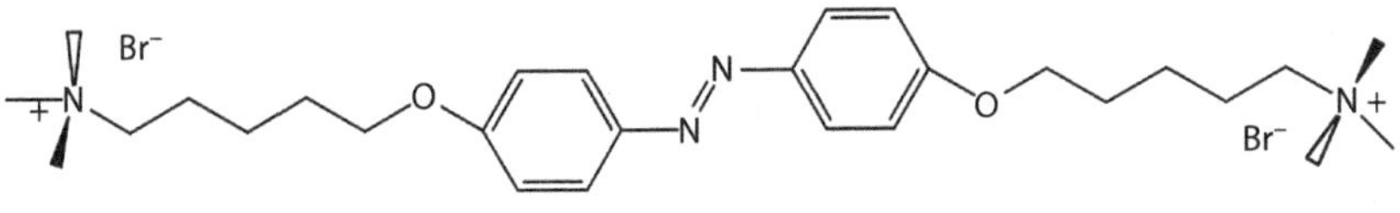

BTHA

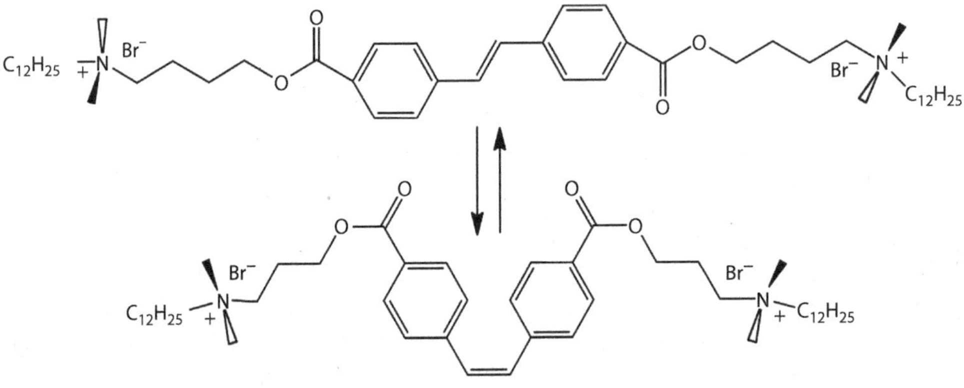

E-SGP ⇆ Z-SGP

The family of photoactive surfactants with stilbene groups linking head groups and exemplified by the *trans* (E-SGP) and *cis* (Z-EGP) homologues were reported to undergo photoactive isomerization [42]. It was also reported that the vesicles formed by E-SGP were transformed into spherical micelles upon phototransformation to the *cis* Z-SGP form as well as photodimerization into the cyclobutyl dimers (thermally disallowed but photochemically allowed) ZEZ-DiSGP and EEE-DiSGP (see below).

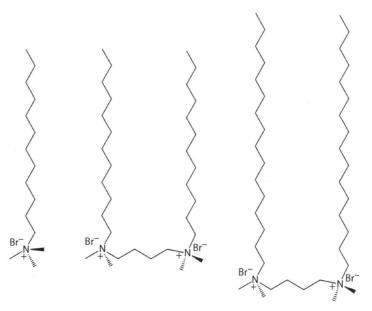

ZEZ-DiSGP (left) and EEE-DiSGP (right)

In a subsequent study [43], these stilbene-connected Geminis were used to formulate vesicles by mixing in photoinert dodecyltrimethyl ammonium bromide (DTAB) and the 12-4-12 (butanediyl-1,4-bis(dodecyldimethyl ammonium) and 16-4-16 (butanediyl-1,4-bis(hexadecyldimethyl ammonium bromide) Geminis (see below). UV irradiation of the 2 to 4 nm diameter vesicles resulted in the disruption of the vesicles and transformation into small micelles, along with photoactivated dimerization to form cyclobutyl derivatives. A key advance demonstrated in this extension was that the optical switches could be diluted significantly by the inert components while maintaining photoswitchability of the vesicle to micellar transformation.

DTAB (left), 12-4-12 (middle), 16-4-16 (right)

Cationic azobenzene surfactants (4,4′-phenylazophenyl)diammonium ions and (4-phenyl-azophenyl)ammonium ions have been studied as cationic exchange photochromic intercalcated aluminosilicate (montmorillonite) materials using molecular dynamics simulations [44]. *Trans-cis* isomerization resulted in interplanar spacing increases of up to 14% in these simulations. The bulkier cationic azobenzenes dimethyl-2-hydroxy-ethoxy-yl ammonium-4-azobenzene (DHEAAz), trimethyl-methyleneyl ammonium-4-azobenzene (TMAAz), and bis(trimethyl-methylenyl ammonium)-4,4′-azobenzene, bis(DMA)Az (see below), were used experimentally. The bis(DMA)Az example, a diamino, has tunable cation behavior because the tertiary amines are pH sensitive. DHEAAz is the most flexible example, and although *trans-cis* isomerization was achieved experimentally, changes in basal plane spacing were not observed [45,46].

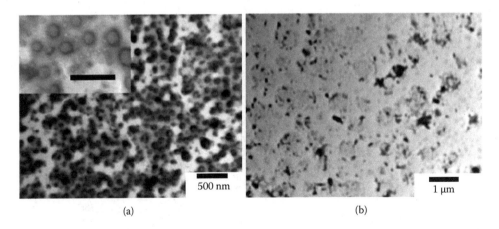

TMAAz (left), DHEAAz (lower), bis(DMA)Az (right)

An interesting, optically switchable nonionic surfactant based on malachite green was reported by Jiang et al. [47]. The diphenyl ketone carbonyl of malachite green is reacted with an anionic polymerization activated initiator such as 2-(4-lithiumphenoxy)ethyl lithium, quenched with water, reacted with KCN to convert the tertiary OH to nitrile, and finally extending the polyethyleneglycol (PEG) tail by reacting with 4-methoxypolyethoxysulfotoluene to yield PEG-MG-CN (see structure below). In water, this nonionic amphiphile organizes into vesicles as shown in the TEM of Figure 17.3. Under UV irradiation from a high-pressure mercury arc lamp at 900 mW/cm^2, these vesicles disassemble into triphenylmethyl (TPM) cations and cyanide anions. The destruction of vesicles

500 nm

1 μm

(a) (b)

FIGURE 17.3 TEM illustrating (a) the vesicle-like aggregates formed by PEG-MG-CN in buffer solution before UV irradiation and (b) condensed aggregates formed after UV irradiation and the photoinduced disassembly of the vesicles. (Reprinted with permission from Jiang, Y., Y. Wang, N. Ma, Z. Wang, M. Smet, and X. Zhang. Reversible self-organization of a UV-responsive PEG-terminated malachite green derivative: Vesicle formation and photoinduced disassembly. *Langmuir* 23:4029–4034. Copyright 2007 American Chemical Society.)

after 500 s of irradiation to form TPM-stabilized PEG-MG-CN aggregates is illustrated in Figure 17.3. This TEM shows the destruction of vesicles by transformation to other aggregates. Reformation of the PEG-MG-CN from TPM cations and cyanide occurs thermally in the dark, and this alternative form of optical switches is an interesting addition to the stimuli-responsive design toolbox.

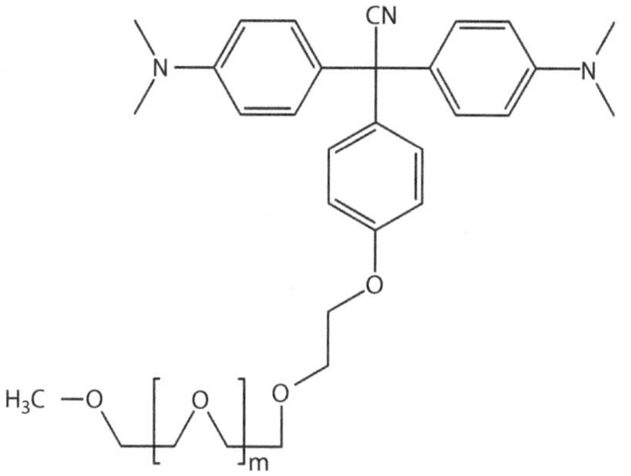

PEG-MG-CN

17.6   THERMOREVERSIBLE SURFACTANTS

Peptides (alanine, glycine, and serine monopeptides, dipeptides, and tripeptides) conjugated to N-isopropyl acrylamide-co-acrylic acid oligomers are thermoreversibly responsive to the LCST imparted by the N-isopropylacrylamide (NIPAM) components of the conjugate [48]. At pH 4, the nonconjugated oligomer exhibited an LCST at 37.7°C. At this pH, the conjugated monopeptides cause a decrease in the LCST, but the dipeptides and tripeptides caused increases of 38.4°C to 43.3°C and 42.6°C to 50.8°C, respectively. At pH 7.4, the nonconjugated oligomer did not seem to exhibit an LCST and the tripeptide conjugates (except for the gly-ala-ser tripeptide) similarly did not exhibit an LCST. Conjugates of the monopeptides and dipeptides exhibited LCST in the range of 41.6°C to 43.9°C and 46.2°C to 60.2°C, respectively, at pH 7.4.

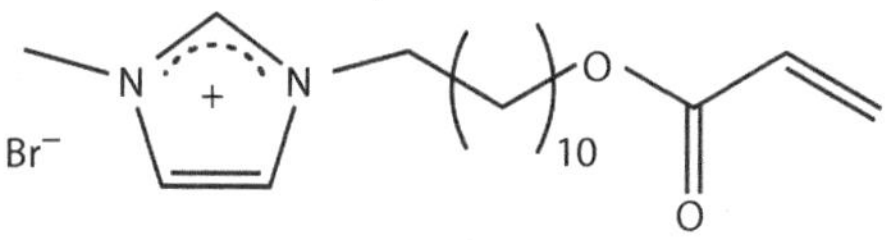

ILBr

A very interesting class of diblock copolymers was reported that was composed of a NIPAM block and a poly(ILBr) block (see ILBr monomer above) [49]. These materials exhibited reversible nanoparticle formation, as illustrated in Figure 17.4 for an aqueous solution of 1% (w/w) in diblock. The highly water-soluble diblock, when heated above the LCST, condensed to form nanoparticles having a core composed of poly(NIPAM) blocks and a stabilizing corona of pol(ILBr) blocks. This nanoparticle condensation and stabilization process is completely reversible.

Dual stimuli-responsive mushroom-like Janus polymer particles were derived from poly(methyl methacrylate)/poly(styrene-2-(2-bromoisobutyryloxy ethyl methacrylate)-*graft*-poly(2-(dimethylamino) ethyl methacrylate) [50]. The poly(2-(dimethylamino)ethyl methacrylate) or PDM chains exhibit temperature-responsive and pH-responsive volume phase transitions. A LCST is exhibited at 34°C. At neutral pH (7.2) and at 25°C, these Janus particles stabilize 1-octanol in water emulsions. At 60°C, the emulsion was destabilized as the particulates became solubilized entirely in the oil phase. Similarly, at 25°C, the emulsion was destabilized by increasing the pH to more than 7.7.

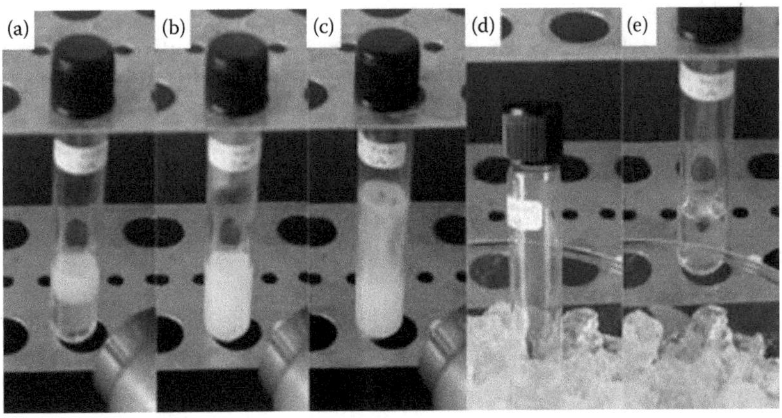

FIGURE 17.4 Heating–cooling cycle showing the thermoreversible transformation of aqueous poly(ILBr-*b*-NIPAM) solution to a suspension of poly(NIPAM) core particles, very well stabilized by corona-forming poly(ILBr) blocks: (a) initial condensation as solution warms; (b) the whole solution has warmed; (c) the suspension is actively boiling; (d) the suspension is cooled in ice; (e) clear solution after re-dissolution of poly(NIPAM) cores after cooling in (d). (Reprinted from Tauer, K., N. Weber, and J. Texter. Core-shell particle interconversion with di-stimuli-responsive diblock copolymers. *Chem. Commun.* 6065–6067, 2009. Reproduced by permission from the Royal Chemical Society.)

Recently, a new class of triblock copolymers has been reported comprising a center block of poly(propylene oxide) and symmetric end blocks of poly(ILr) [51–53]. The end blocks were added to the Br-PPO-Br macroinitiator by atom transfer radical polymerization (ATRP), and the approximate composition can be represented as $ILBr_{6.5}$-PO_{61}-$ILBr_{6.5}$ [52]. Over the composition range of 0.34 to 0.65 weight fraction in water, a line of LCST exists that decreases, respectively, from 45°C to 8°C. Below this line, the two-phase solution comprises micelles and monomers of this triblock in addition to water. As this line is traversed by warming, the solutions transform thermoreversibly into a lyotropic liquid crystalline gel phase. These triblocks have also been demonstrated to be excellent waterborne dispersion stabilizers for carbon nanotubes and nano-hydrothermal carbons [54,55].

17.7 pH-SENSTIVE SURFACTANTS

A molecular–thermodynamic theory of micellization of multicomponent surfactant mixtures [56] has also been extended to include pH-sensitive surfactants [57]. The types of pH sensitivity considered in the examples illustrated were limited to surfactants of the amine oxide and betaine type, such as dodecyldimethylamine oxide (C_{12}DAO), tetradecyldimethylamine oxide (C_{14}DAO), and hexadecyldimethyl betaine (C_{16}Bet).

Two lipidic β-amino acids that contain fatty acids, 2-(aminomethyl)-2-pentadecylheptadecanoic (AMPDHD) were synthesized for composing pH-sensitive vesicles [58]. The single-chain β-amino acid rac-2-(aminomethyl)heptadecanoic acid (AMHDA) was also synthesized as a comparative control. Using the double-tailed moiety at 10% (molar) with 1-palmitoyl-2-oleoyl-*sn*-glycero-3-phosphocholine (POPC), the vesicles obtained exhibited pH-sensitive surface charge.

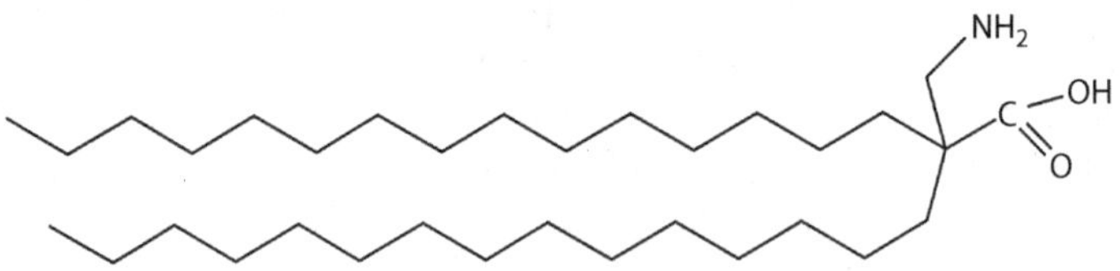

AMPDHD

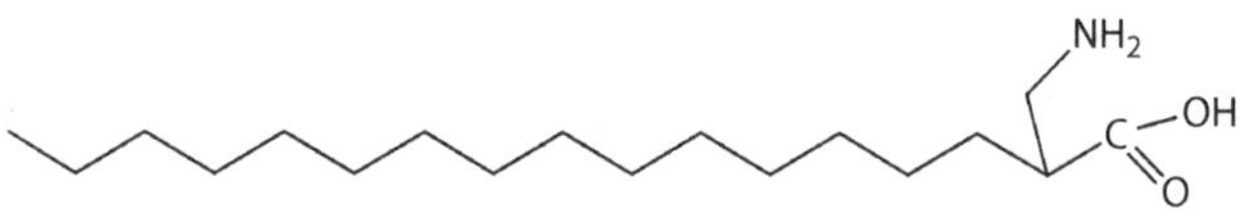

AMHDA

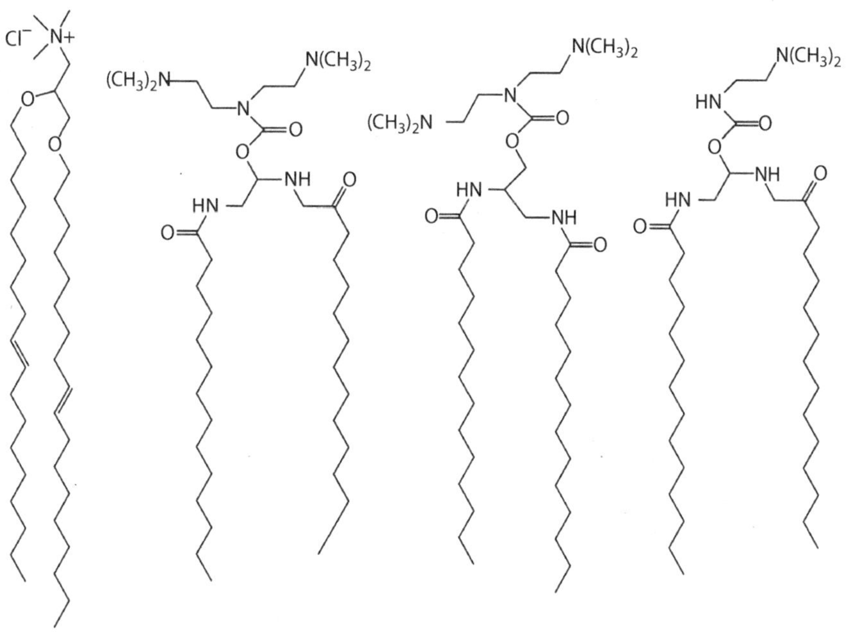

(left to right) DOTMA, 1,31b2, 1,21b2, 1,31mt2

The cationic double-tailed DOTMA is a long-reputed carrier for gene therapy transfection studies [59], and three myristoyl derivatives were created as possible competitors [60]. Two of these have several tertiary amines that can be protonated and one has a single tertiary amine. The 1,21b2 moiety was found to exhibit higher transfection activity.

A pH-sensitive dye, CI disperse red 60, was modified by turning it into a surfactant by linking poly(propylene oxide-b-ethylene oxide) oligomers to it through a urea linkage [61]. This form of surfactant design, CI-Red-PEO (see below), represents a novel but as yet not widely practiced form of molecular recognition in surfactant design (RL100: x/y = 19/3, 1314 Da; RL200: x/y = 42/3, 2260 Da; RL207: x/y = 33/10, 1610 Da; RL300: x/y = 58/8, 3217 Da). Such surfactants have very high stickiness because they almost perfectly match the underlying molecular crystalline material substrate. The dispersion stability obtained most likely would have been improved if small media milling had been used instead of sonication.

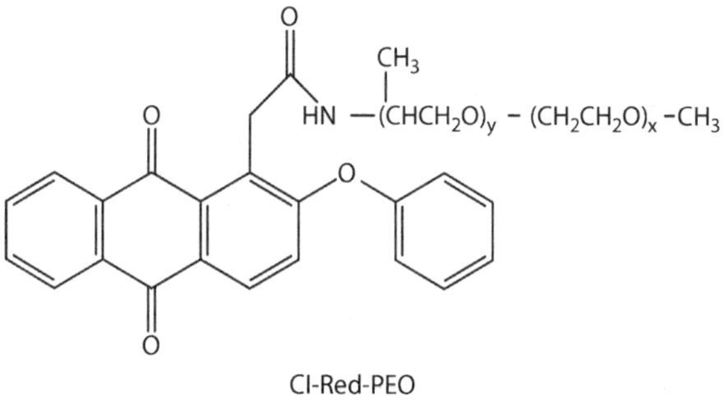

CI-Red-PEO

A very interesting approach to forming thermodynamically stable polymeric nanoparticles has been described wherein pH-sensitive hydrophilic diblocks are complexed with pH-sensitive oligomers [62]. A starting point is a double hydrophilic diblock copolymer of PEO and poly(methacrylic acid) (PMAA). A diblock PMAA-*b*-PEO having 2100 methacrylic acid units and approximately 5000 ethylene oxide units was used; its overall pK_a was 5.5. Poly-L-lysine (PLL) with molecular weight in the range of 15 to 30 kDa and chitosan oligomers less than 5 kDa with more than 90% deacetylation and $pK_a = 6.6$ were used as counter-polyelectrolytes. Solutions at concentrations of approximately 1 mg/mL were combined at equal molar ratios of methacrylic acid and amine for each PLL and oligomeric chitosan at a pH of 7.5 and 6.5, respectively. Driven by charge neutralization, coacervation produces PMAA-PLL (28 nm) and PMAA-oligochitosan (60 nm) nanoparticles stabilized by a PEO corona (shell). Rather than experiencing macroscopic phase separation, as is the usual case in coacervation, the phase separation is only nanoscopic because of the stabilization provided by the PEO block. This phenomenon is analogous to the reversible nanoparticle condensation described previously for PNIPAM-*b*-PILBr subjected to heating [49]. In this case, the nanoparticles can also be reversibly disassembled by lowering the pH to 5. Making this pH change protonates the PMAA block carboxyl groups and switches off the coulombic attraction between anionic carboxyl groups and cationic amino groups. We carefully avoid calling these nanoparticles "micelles" because these particles are not in dynamic equilibrium with the component "monomers." The formation of these core-shell nanoparticles is triggered by a first-order phase change, albeit on a nanoscale of less than 100 nm, called coacervation!

17.8   IONIC LIQUID SURFACTANTS

Ionic liquid (IL) surfactants are of wide interest because of the wide range of applications being developed for them. Furthermore, because they often are liquid at room temperature or close to room temperature, they represent a new solvent class as replacements for hydrocarbon solvents as well as aqueous solvents. Some such examples, such as ethylammonium nitrate (EAN), are intrinsically pH sensitive. The imidazolium-based IL surfactants, in aqueous systems, are known to be very sensitive to anions, and this anion sensitivity provides a significant form of stimuli-responsiveness. The behavior of three imidazolium bromide surfactants, $[C_n\text{bmim}]\text{Br}$ ($n = 8$, 10, and 12) in EAN has been described [63], and the effects of counterion basicity and chain length are investigated. The synthesis of $[C_{14}\text{bmim}]\text{Br}$ has been reported [64,65]. The effects of chloride, bromide, nitrate, and iodide on the cmc and other aggregation-related effects of aqueous 1-decyl-3-methyl imidazolium chloride were found to follow a Hofmeister series in the order $\text{Cl}^- < \text{NO}_3^- < \text{Br}^- < \text{I}^-$ with respect to lowering the cmc and lowering micellar ionization [66]. Other important stimuli-responsive anions in modifying imidazolium surfactants include BF_4^-, PF_6^-, CF_3SO_3^-, and N(CN)_2^- [67,68]. These imidazolium–anion based interactions have been examined in the context of various supramolecular complexes [69], ion exchange resins [70], and approaches to quantitative analyses [71] that emanate from imidazolium–anion pairing [68].

The synthesis and micellar properties of 1-alkyl-3-methylimidazolium chlorides have been reported for $[C_n\text{bmim}]\text{Cl}$ with $n = 10$, 12, 14, and 16 [72]. The cmc ranged from approximately 1 to 40 mmol/L as chain length decreased.

Bara et al. [73] recently reported the synthesis and characterization of 20 Gemini imidazolium-based surfactants. Bromide and tetrafluoroborate salts were prepared, and the resulting liquid crystalline phases were characterized by x-ray scattering and polarized light microscopy. Mono and Gemini imidazolium-based aromatic ether cationic surfactants have been designed and prepared using 4-hydroxy benzoic acid [74]. The benzoic acid is converted to the methyl ester followed by *O*-alkylation with long-chain alcohols, followed by reduction and bromination to yield 4-alkoxy benzyl bromides. *N*-Methylimidazole is then quaternized with this benzyl bromide to obtain the monosurfactants. Geminis are obtained by reacting the benzyl bromides with imidazole followed by quaternization with dibromoalkanes. The Gemini 1,4-bis(3-tetradecylimidazolium-1-yl) butane dibromide has been reported [65], and homologous decyl, dodecyl, and tetradecyl dibromides have been prepared and compared with the corresponding [C$_n$bmim]Br for n = 10, 12, and 14 [75]. The cmc values for the Geminis were much lower than the cmc of the mono imidazolium surfactants. Gemini imidazolium-based surfactants and pyridinium-based surfactants with thioether spacers have very recently been described and characterized [76,77].

A class of imidazolium POSS surfactants has been described [78], and two examples have been synthesized. The starting materials, chlorobenzyl-hepta(isobutyl)-POSS and chlorobenzylethyl-hepta(isobutyl)-POSS were quaternized with 2,3-dimethyl imidazole to make the respective imidazolium chlorides. The more rigid example provided thermally more robust intercalated clay (montmorillonite) composites.

Polymerizable IL surfactants are a new type of surfmer and these compounds are leading to the development of many new stimuli-responsive polymers and materials [1,79]. These new materials include elastic conducting thin films, open cell porating monoliths and thin films [80,81], and stimuli-responsive block copolymers, including a new class of thermoreversible (LCST) gelating lytotropic liquid crystalline mesophases [52].

Blocks of polymerized reactive surfactants may be thought of as polymer brushes when the surfactant monomer has a hydrophobic tail group. Homopolymers of ILBr synthesized by a two-phase ATRP process that produced very broad molecular weight distributions (by essentially imparting continuous initiation throughout the reaction period) have been found to exhibit interesting properties [82,83]. Such homopolymers can be precipitated into nanoparticles by the addition of a suitably stimulus-inducing anions, such as PF$_6^-$ [82]. Although it was expected that thin films of such homopolymers might exhibit poration in response to bromide ion exchange with hexafluorophosphate, a dewetting phenomenon was observed instead [82,83]. Extensive examination of the conditions and available surface energy data yielded the conclusion that this dewetting was not driven by surface energy decreases but rather by mechanical stress relaxation. This is an exciting new application area, and there is considerable interest in new polymeric materials that change dimension upon exposure to suitable stimuli.

Yu et al. [84] have followed the IL-stabilized microemulsion polymerizations of Yan and Texter [80,81] by synthesizing the methacrylate analogue of ILBr, MAUMBr. This new surfmer was used to stabilize microemulsions of styrene and [bmim]BF$_4$. A microemulsion of these components comprising 30% (w/w) [bmim] BF$_4$, 47% styrene, 23% MAUMBr, and approximately 3% (relative to styrene and MAUMBr) divinyl benzene (DVB) as a cross-linker was polymerized at 60°C using 2,2'-azoisobutyronitrile (AIBN) as thermal initiator. The resulting transparent and flexible membrane (~80 µm thick) was physically characterized. Its intrinsic electrical conductivity exceeded 10 mS/cm at 140°C.

MAUMBr

Polymerizable vinyl imidazolium surfactants, N-vinyl-N'-alkyl imidazolium bromides (dodecyl, hexadecyl, and octadecyl alkyl groups) were synthesized and used to ion exchange and intercalate montmorillonite to form new polystyrene/clay nanocomposites [85]. Very similar compositions were reported using 1-methyl-3-(4-vinylbenzyl)imidazolium chloride, 1-hexyl-3-(4-vinylbenzyl) imidazolium chloride, and 1-dodecyl-3-(4-vinylbenzyl)imidazolium chloride [86]. Similar vinyl imidazolium bromides (butyl, octyl, dodecyl, octadecyl, and docosyl) were evaluated as corrosion inhibitors for carbon steel in 1 mol/L H$_2$SO$_4$ [87]. The octadecyl homologue performed best in a series of tests including weight loss and polarization measurements.

Yuan and Antonietti [88,89] have synthesized a variety of such vinyl imidazolium bromide surfactants, including a bis-cross-linking agent (two vinyl imidazoles linked by quaternizing with 1,4-dibromobutane), and have demonstrated micellar (dispersion) polymerization of these monomers. Even-numbered carbon chain lengths of 8 to 18 were investigated. Latexes formulated in this way after ion exchange (of bromide) with bis(trifluoromethylsulfonyl)imide ($CF_3SO_2)N^-$ were stable in nonpolar solvents such as toluene. This stimulus response transforms the latexes from being compatible with polar solvents to being compatible with nonpolar solvents. Further work on this system demonstrated that lengthy nanoworms of polymer were produced by micellar polymerization. The resulting polymers condensed into nanoparticles having an apparent vesicular structure, but after forming spheroids of 50 nm diameter, an adjacent spheroid was condensed, and so on, yielding "nanoworms" [89]. Under other conditions, unilamellar and multilamellar vesicles were formed and it was found that the shapes ranged from spherical to polyhedral with flat facets, depending on the alkyl chain length of the composing vinylimidazolium monomer.

17.9 FLUORINATED SURFACTANTS

Perfluoronated surfactants are mainly known for their chemical inertness and high surface activity. Nonionic perfluoronated surfactants have found wide use in formulating release layers in multilayer coatings, but their use has been largely discontinued because of untoward environmental effects. Perfluoronated carboxylates obviously are pH sensitive and corresponding sulfonates are less so. Block copolymers having perfluoronated blocks can exhibit interesting switching phenomena, as discussed below, depending on the solvent that contacts such block copolymers.

The synergistic cmc for SDS in the presence of 1-methyl-3-(pentafluoro phenyl)imidazolium chloride or the H-phenyl homologue (each at 30 mmol/L) was 0.04 mmol/L for the perfluoro compound and 18 mmol/L for the H-phenyl moiety, a difference of 450-fold [90]. Perfluorinated end-caps for PEG surface brushes showed dramatic interfacial energy switching [91]. Contact with hexadecane drew perfluorinated end-caps to the interface, whereas contact with water resulted in the burying of these end-caps and facile wetting. No hysteresis in solvent responsiveness was observed.

Perfluorolauric acid (PFLA) and its sodium (PFL-Na) and lithium (PFL-Li) salts were used to make pH-sensitive vesicles [92]. These vesicles, when formed, have increased stability relative to their hydrocarbon counterparts because of the increased interchain attractive forces. The mechanism for vesicle formation illustrated in Figure 17.5 suggests the importance of singly protonated PFL^- dimers as building units of the vesicles. These dimers then self-assemble to form a bilayer.

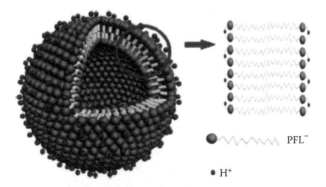

FIGURE 17.5 Schematic of the self-assembly of PFL^- and H^+ into vesicles through hydrogen bonds. (Reprinted with permission from Long, P.F., A.X. Song, D. Wang, R.H. Dong, and J.C. Hao. pH-sensitive vesicles and rheological properties of PFLA/NaOH/H_2O and PFLA/LiOH/H_2O systems. *J. Phys. Chem. B* 115:9070–9076. Copyright 2011 American Chemical Society.)

17.10 MAGNETIC SURFACTANTS

Several of the redox-active surfactants discussed previously would be expected to be paramagnetic and to be collectively organized in an applied magnetic field. However, the individual and collective behavior of oxidized ferrocenyl-based surfactants in magnetic fields have not yet been reported. Eastoe and coworkers, following recent work on making paramagnetic ILs [93,94], have fabricated paramagnetic surfactants by converting halide counterions, to paramagnetic lanthanide halide anionic counterions.

The Eastoe group [95] first reported the conversion of 1-methyl-3-decyl imidazolium chloride, dodecyltrimethylammonium bromide, and didodecyldimethylammonium bromide to the corresponding $FeCl_4^-$, $FeCl_3Br^-$, and $FeCl_3Br^-$ salts, respectively. These larger and paramagnetic anions caused pronounced decreases in the melting points of the corresponding solid materials. The cmc values increased slightly for the imidazolium and didecyldimethyl ammonium moieties and decreased slightly for the dodecyltrimethyl-ammonium upon conversion to the paramagnetic form. An applied magnetic field influences the surface tension at the air/water interface for all species, whether only diamagnetic or paramagnetic. The imidazolium chloride moiety increased its surface tension by approximately 5% in an applied field, whereas the imidasolium $FeCl_4^-$ moiety decreased its surface tension by approximately 9% in the same magnetic field [95]. Similar results were reported for $GdCl_3Br^-$, $HoCl_3Br^-$, and $CeCl_3Br^-$ salts of dodecyltrimethylammonium and for $GdCl_4^-$, $HoCl_4^-$, and $CeCl_4^-$, salts of 1-methyl-3-decyl imidazolium [96]. Polarized microscopy was used to show effects on mesophase behavior and SANS was used to show detailed effects on micellization [96]. An interesting dicationic surfactant based on the IL dimer cation, $[C_4mim_2]$ (1,2-ethyl bis[1'-yl,3'-butyl imidazolium]), the AOT anion, and $FeCl_3Br^-$ exhibited melting below $-25°C$ has also been reported [97]. This kind of exotic room temperature IL surfactant is multireponsive to magnetic fields, to anions that bind strongly to the imidazolium ring, and to solvent shifting.

The $GdCl_3Br^-$ and $HoCl_3Br^-$ salts of dodecyltrimethylammonium mentioned previously were used to demonstrate how such stimuli-responsive surfactants could be used to easily harvest or concentrate DNA in an aqueous environment [98]. DNA and many proteins readily bind cationic surfactants and anionic surfactants, and the Eastoe group has shown very clearly that when these cationic surfactants bring along a paramagnetic anion, low-intensity magnetic fields can be used to induce their aggregates to migrate to a magnet placed in an aggregate suspension. This phenomenon was demonstrated for herring sperm DNA and for myoglobin [98].

17.11 OTHER APPLICATIONS

17.11.1 ELECTRODE COATINGS

The physical surface electrochemistry of redox-active ferrocenyl ammonium surfactants of varying alkyl chain lengths have been examined in detail [99,100]. Adsorption–desorption kinetics and equilibria were quantified, and it was found that hydrophobicity or chain length drives adsorption free energy much more so than electrochemical potential. Interactions between redox active surfactants and between redox and nonredox (e.g., CTAB) surfactants were also studied. Coulombic repulsive interactions seemed stronger than hydrophobic attractive interactions.

17.11.2 EMULSIONS

A long-chain amidoamine surfactant, octadecyl-di(2-aminoethyl propan-amide-3-yl)amine ($C_{18}AA$), was found to stabilize toluene in water emulsions and to impart thermoreversible gelation behavior [101]. An LCST was tunable in 0.12 mol/L HCl over $10°C$ to $55°C$ by varying $C_{18}AA$ concentration from 1.6% to 0.6% (w/w). The LCST could also be tuned from $5°C$ to $55°C$ by pH over the 9.3 to 10.8 range.

17.11.3 VESICLES

Many key new materials have been derived by polymerizing surfactants in interesting mesophases and lyotropic liquid crystalline phases. The polymerization of various monomers partitioned into vesicular bilayers has been much studied. Much more significant results have been obtained by using polymerizable surfactants in formulating the vesicular bilayers. A key review of such work was given in which most of the important features of the processes and applications were well articulated [102].

Sodium cholate (SC; see below) is an interesting cosurfactant and has an obviously pH-dependent carboxyl group. When dodecyl triethylammonium bromide (DEAB) and SDS are mixed at a molar ratio of 3:1 in water at a total concentration of 10 mmol/L, vesicles are formed that are unresponsive to pH and temperature variations. However, when SC is added, the vesicles are driven across a transition boundary to form small micelles, cosolubilizing the SC [103]. SC has a polar surface (see hydroxys and carboxyl) as well as a hydrophobic surface and is intrinsically amphiphilic.

SC (sodium cholate)

In a subsequent study focusing on SC-induced temperature and pH effects on vesicles [104], DEAB/SDS/SC molar ratios of 75:25:4.5 (10.45 mmol/L total surfactant concentration) were used to study the effects of SC. Increasing temperature transformed the micellar solution into a vesicle solution. It seems that vesicle formation equilibria becomes evident at approximately 40°C and continues up to 80°C. Vesicle formation was also reversibly triggered by lowering pH to 3.3. These same phenomena were demonstrated for other catanionic micelle–vesicle transition systems.

An intriguing dynamic vesicle system has recently been designed and constructed wherein the single-tailed and double-tailed surfactants are synthesized *in situ* by a reversible linking chemistry [105]. These surfactants then reversibly and dynamically self-assemble to form vesicles. The cationic bisaldehyde A reacts with amine B to form the single-tailed and double-tailed species AB_1 and AB_2; the reaction sequence is shown in Figure 17.6. The dynamic nature of this system was verified by confocal laser microscopy and concentration studies that demonstrated assembly and disassembly.

17.11.4 INTERFACIAL TRAPPING OF NUCLEOPHILES

The diazotization of functional aryl groups has become a facile approach to making highly stable aqueous dispersions of diverse nanoparticulates such as carbon black and nanocolorants. The Romsted laboratory has capitalized on such chemistries to create a variety of cationic surfactants useful for probing the local concentrations of nucleophiles (water, halides, etc.) at interfaces such as in the palisades layer of micelles and microemulsions [106].

A recent comprehensive review [105] of these trapping results summarizes many of these studies. The basic stimuli-responsive surfactants in these studies use the 1-diazonium-2,6-dimethyl-4-hexadecyl benzene cation that thermally loses nitrogen at fixed temperature according to a first-order rate law to yield a highly reactive arene radical cation that reacts with diverse nucleophiles at

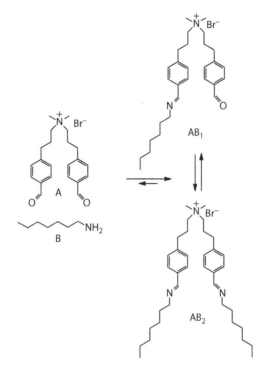

FIGURE 17.6 *In situ* dynamic formation of imine vesicles and imine surfactants AB$_1$ and AB$_2$ in water from the reactive quaternary salt A and the amine B.

diffusion-limited rates. This cationic surfactant can be doped into various cationic micellar and microemulsion structures without significantly perturbing such structures. The resulting chemical trapping reactions allow one to determine the relative concentrations of various nucleophiles in the interfacial regions of micelles and microemulsions. Various early studies established the relative concentrations of halides and water in these interfacial regions that could not be conveniently probed by any other technique. The reaction products obtained from micellar and microemulsion interface experiments (Figure 17.7a) were compared with those obtained in simple solutions of similar ratios of nucleophiles (Figure 17.7b) using a 1-diazonium-2,4,6-trimethyl benzene to generate control results. These reaction products are typically analyzed using high-performance liquid chromatography methods.

Comprehensive component concentrations, for example, have been determined for reverse microemulsion droplets including water pool and interfacial water concentrations, water pool and interfacial bromide concentrations, and various other component concentrations. For example, in CTAB, hexanol, iso-octane reverse microemulsions at a water to CTAB molar ratio of 12, the water pool concentration is 43.8 mol/L and the interfacial water concentration is 29 mol/L. The water pool bromide was 1.91 mol/L and the interfacial bromide concentration was 3.51 mol/L. The interfacial hexanol concentration was 7.82 mol/L. Trapping can also be done in anionic surfactant structures, and in an AOT/2,2,4-trimethylpentane reverse microemulsions at a water to AOT ratio of 12; the interfacial water concentration was 27.9 mol/L and the AOT concentration was 2.75 mol/L.

Similar information has been obtained for aqueous and reverse micelles, oil-in-water to bicontinuous to water-in-oil microemulsions, anionic micelles and polyelectrolytes, zwitterionic micelles and vesicles, nonionic micelles, and emulsions. These results provide quantitative insight into how interfacial concentrations of various components correlate with various surfactant aggregate morphologies. A cumulative result of these studies is that the "minimization of the hydrocarbon–water contact drives hydrated amphiphile chains together, creating the hydrocarbon core of the aggregate

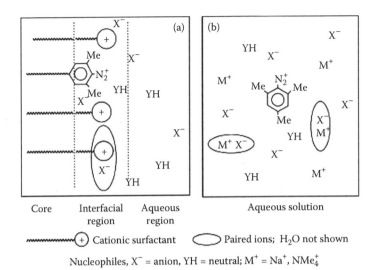

FIGURE 17.7 (a) Schematic illustrating interfacial cross section populated by cationic surfactants, a probe diazonium reporter surfactant, and various other species such as nucleophiles X⁻, YH, and water (data not shown). (b) An aqueous control volume cross section illustrating the same nucleophiles as in (a), a more water-soluble and nonsurfactant diazonium reporter probe, and nonsurfactant electrolytes having the same nucleophilic counterions as the surfactants in (a). (Reprinted with permission from L.S. Romsted. Do amphiphile aggregate morphologies and interfacial compositions depend primarily on interfacial hydration and ion-specific interations? The evidence from chemical trapping. *Langmuir*, **23**, 414–424 (2007). Copyright 2007 American Chemical Society.)

and the water of hydration is released into the surrounding bulk water. The increase in entropy from release of water is believed to be a major contributor to spontaneous aggregation at ambient temperatures" [107].

17.11.5   Separations

Microscale separations based on solubilization in and deposition from redox-active surfactant micelles was demonstrated [108] using the redox-active surfactant FTMA (11-ferrocenylundecyltrimethylammonium bromide) and example mixtures of two model drug-like compounds, *o*-tolueneazo-*â*-naphthol (**I**) and 1-phenylazo-2-naphthylamine (**II**). An equimolar mixture of **I** and **II** was subjected to six cycles of selective solubilization and deposition achieving a mixture 98.4% in **I**. This study set the stage for further development and application to microfluidic-based separations.

The same redox active surfactant, FTMA, was also used to generate surfactant concentration gradients across microfluidic channels for new separation processes [109]. These steady-state gradients in monomeric surfactant and micelle concentrations are believed to enable new approaches to separating diverse organic and biomolecules molecules known to interact with surfactants and micelles. These principles were further developed in showing how different types of molecules [110] interact primarily with monomeric surfactant, 2-(4,4-difluoro-5,7-dimethyl-4-bora-3a,4a-diaza-*s*-indacene-3-pentanoyl)-1-hexadecanoyl-*sn*-glycero-3-phosphocholine (BODIPY C5-HPC), or primarily with micelles of a hydrophobic dye, 1-phenylazo-2-naphthylamine.

17.11.6   Latex Coalescence

The effects of electrolyte, temperature, and enzymatic stimuli on latex film formation and dispersion stability have been examined by Urban and coworkers. Films formed from methylmethacrylate/

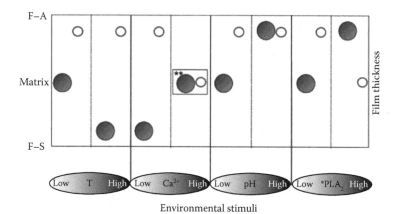

FIGURE 17.8 Schematic depicting mobility of AOT (open circles) and AOT/DLPC complexes (larger dark circles) diffusing through PLA_2-stabilized MMA/BA latex films toward the film–air (F–A) and film–substrate (F–S) interfaces in response to various external stimuli: T, Ca^{2+}, pH, and PLA_2. The *PLA_2 signifies PLA2 at low and high pH. The double asterisk indicates extensive coagulation. (Reprinted with permission from D.J. Lestage, M. Yu and M.W. Urban. Stimuli-responsive surfactant/phospholipid stabilized colloidal dispersions and their film formation. *Biomacromolecules*, **6**, 1561–1572 (2005). Copyright 2005 American Chemical Society.)

butylacrylate (MMA/BA) latexes stabilized by AOT (sodium bis[2-ethylhexyl]sulfosuccinate) and 1-myristoyl-2-hydroxy-*sn*-glycero-phosphocholine (MHPC) [111], by AOT and 1,2-dilauroyl-*sn*-glycero-3-phosphocholine [112] (DLPC), and by AOT and SDS [113] were exposed to thermal and high electrolyte external stimuli. An ionic cluster model was invoked to explain the formation of phospholipid rafts on the film surface at the film–air interface.

Although both amphiphiles exhibited a blooming phenomenon, the added thermal and electrolyte stimuli resulted in well-defined phospholipid rafts revealed by AFM measurements. The combination of AOT and DLPC in similarly formed MMA/BA films was found to exhibit diverse responses to various stimuli (Figure 17.8). Similar studies of combinations of AOT and SDS in response to a variety of stimuli have been reviewed [111].

Stable and stimuli-responsive nanolatexes based on copolymers of reactive and stimuli-responsive IL surfactants following Yan and Texter [80] were prepared and found to form highly robust thin films [68]. The imidazolium bromide–based IL surfactant monomers induce an ion exchange capacity in the films. These transparent films are transformed into opaque and porous films and membranes as a stimulus response to anion exchange of bromide by hexafluorophosphate (see next section).

17.11.7 DISPERSION STABILIZATION/DESTABILIZATION

Micelles of azobenzene-based surfactants have been shown to be effective in solubilizing a variety of organic pigments to create stable pigment dispersions. When such dispersions are exposed to certain base metals with opportune reduction potentials relative to the azo (dye) surfactants, the azo group is partially to fully reduced and dramatically increased in water solubility [114]. The micelle-solubilized pigments stabilized are released, and they precipitate onto the base metal and form a highly colored pigment film.

A CO_2 switchable surfactant, N'-dodecyl-N,N-dimethylacetamidinium bicarbonate, was shown to exhibit switchable dispersion stabilization [115]. Stable and coagulated dispersions are obtained, respectively, by purging with CO_2 and an inert gas (N_2, Ar, or air).

The stability of the nanolatexes discussed in the previous section [68], based on imidazolium bromide surfmers in aqueous dispersion, can also be tuned by varying anion and anion concentrations. Although the bromide imidazolium ion pair is highly hygroscopic, more "hydrophobic" anions destabilize the nanolatex dispersions. The concentrations needed for destabilization by Na_2S, NaBr, $NaBF_4$, and KPF_6 decrease in this order and are, respectively, 1.8 mol/L, 0.24 mol/L, 9.4 mmol/L, and 0.4 mmol/L.

Charged surfactants are popular dispersants for waterborne dispersions and charge stabilization is very effective. Colloidal stability in such dispersions is sensitive, however, to indifferent electrolyte concentration due to Debye–Hückel screening effects that decrease interparticle repulsion potentials. It has been known for some time that osmotic brushes of polymeric electrolytes produce softly repulsive electrosteric stabilization that is insensitive to indifferent electrolytes [116]. Coating particles with such brushes, however, is technically feasible but too expensive for many applications. Osmotic nanospheres that exhibit high adsorption affinity to charged and uncharged particle surfaces and are easily applied in processing that relies on adsorption from solution have been developed [68]. Stimuli-responsive surfactant monomers and conventional monomers were used to formulate such nanospheres as lightly cross-linked nanogels. Turbidity studies show that the stability of such nanogels can be maintained at ionic strengths as high as 9 mol/L and switched off at ionic strengths of 0.35 mmol/L using appropriate stimulus-inducing ions. The efficacy of such nanogels is demonstrated by producing concentrated dispersions of hydrophobic carbon including nanotubes, graphene, and hydrothermal carbon as well as model colloids and latexes [54,116,117]. The colloidal stability of such dispersions in high electrolyte concentrations has demonstrated immunity to Debye–Hückel charge screening. Stimulus-inducing ions are used to produce controlled dispersion destabilization to produce advanced material coating and deposition [118]. These nanogels have also been demonstrated to self-assemble into transparent and mechanically robust thin films. The films are shown to retain the stimuli responsiveness of the constituent monomers, and to undergo spinodal decomposition to form nano, meso, and microporous open cell porous coatings when exposed to appropriate stimulus-responsive ions [68,119].

17.11.8 SHAPE MEMORY POLYMERS

An exciting new class of polymers based on pH and thermal responsive single-tail and double-tail phosphate amphiphiles has been disclosed by Iijima et al. [120]. This class is represented by $VPEOPO_4$ (see below) wherein R_1 and R_2 can be H, a decyl chain, or a dodecyl chain. Nano TiO_2 particles surface-treated with $VPEOPO_4$ enabled the dispersion of these nanoparticles in a diverse range of solvents, including DMSO, CH_3CN, NMP, MEK, methanol, ethanol, isopropanol, acetic acid, THF, ethylacetate, toluene, and MMA. The phosphate group imparts strong binding affinity to the titania surfaces, and similar binding would be expected for any metal oxide. In addition, such modified nanoparticles could also be transparently dispersed in a variety of polymers, including epoxy resin and PMMA. It was discovered that these TiO_2/epoxy nanocomposites possess reversible shape memory properties. These shape memory properties are illustrated in Figure 17.9.

VPEOPO$_4$

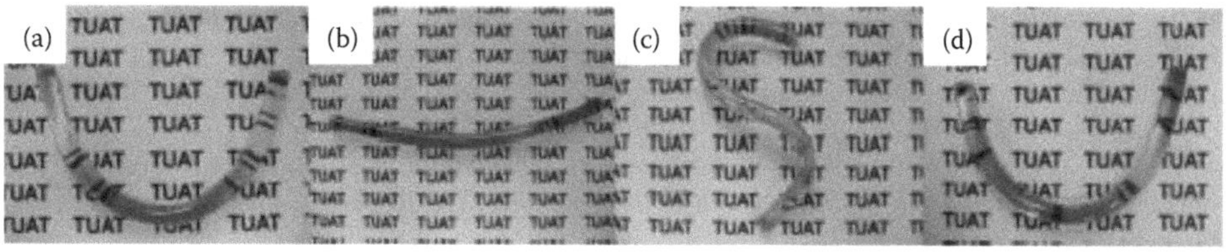

FIGURE 17.9 TiO$_2$ nanoparticles surface-modified with VPEOPO$_4$ and homogeneously dispersed at 18% (w/w) in an epoxy resin to form a shape memory nanocomposite. (a) Original "hard" structure and shape; (b) and (c), shapes produced by bending after heating to 60°C and cooling to room temperature; (d) after reheating to 60°C until shape recovery is achieved. (Reprinted with permission from M. Iijima, M. Kobayakawa, M. Yamazaki, Y. Ohta and H. Kamiya. Anionic surfactant with hydrophobic and hydrophilic chains for nanoparticle dispersion and shape memory polymer nanocomposites. *J. Am. Chem. Soc.*, **131**, 16342–16343 (2009). Copyright 2009 American Chemical Society.)

17.12 SUMMARY AND OUTLOOK

Surfactants have been found to exhibit irreversible and reversible responses to a variety of field variables such as chemical potential (concentration), temperature, and light. In almost every case, these responses affect the assembly or aggregation of the surfactants. We have seen that redox control of specialized surfactants has advanced from dynamic control of surface tension at the air/water interface to the development of advanced microfluidic separation processes. We also have seen a variety of assembly–disassembly processes. From a thermodynamic control perspective for manufacturing, we have illustrated two examples of reversible nanoparticle suspension formation, one emanating from thermoreversibility and another responding to pH. These phenomena emanate mainly from diblock-based phenomena that show the possibility of producing thermodynamically stable nanoparticle suspensions driven by free energy changes rather than by high energy and high shear processing. We have also seen that stimuli-responsive surfactants can provide new control aspects to more or less conventional dispersions. Dispersions can now be destabilized by many different stimuli, and this capability provides new processing control features and more features can be imagined as we perfect and finesse the reversibility of such destabilization and restabilization processes. Finally, we have seen that the introduction of imidazolium-based surfactants and stabilizers provides nearly ideal stabilization and exemplifies affordable osmotic brushes and spheres that render dispersions immune to Debye–Hückel charge screening destabilization and eliminate sensitivity of dispersions to electrolyte variations.

The development of reactive stimuli-responsive surfactants is demonstrating a myriad of new polymeric materials that are also stimuli-responsive. These new materials include nanoparticle surfactant stabilizers illustrated herein by nanolatexes based on ILBr, reversibly porating gels produced by microemulsion polymerization, and stimuli-responsive vesicles and star polymers.

REFERENCES

1. F. Yan, J. Liu and J. Texter. Advanced applications of ionic liquids in polymer science. *Prog. Polym. Sci.*, **34**, 431–448 (2009).
2. J. Israelachvili. *Intermolecular and Surface Forces*, 2nd ed. Academic Press, Boston, MA (1992).
3. J. Texter (Ed.). *Amphiphiles at Interfaces*. Steinkopff Verlag, Darmstadt (1977); *Prog. Colloid Polym. Sci.*, **103**, 1–328 (1997).
4. F.P. Hubbard, Jr. and N.L. Abbott. Stimuli-responsive giant micellar systems. *Surfactant Sci. Ser.*, **140**, 375–395 (2007).
5. Z.-F. Xie and Y.-J. Feng. Environmentally stimuli-responsive surfactants. *Huaxue Jinzhan (Prog. Chem.)*, **21**, 1164–1170 (2009).

6. M. In. Gemini surfactants and surfactant oligomers. In *Reactions and Synthesis in Surfactant Systems*, J. Texter (Ed.). Surfactant Sci. Ser., Vol. 100. Marcel Dekker, New York, pp. 59–110 (2001).

7. J. Texter (Ed.). *Reactions and Synthesis in Surfactant Systems*. Surfactant Sci. Ser., Vol. 100. Marcel Dekker, New York (2001).

8. A. Behler, M. Biermann, K. Hill, H.-C. Raths, M.-E. Saint Victor and G. Uphues. Industrial surfactant synthesis. In *Reactions and Synthesis in Surfactant Systems*, J. Texter (Ed.). Surfactant Sci. Ser., Vol. 100. Marcel Dekker, New York, pp. 1–44 (2001).

9. K. Holmberg. Cleaveable surfactants. In *Reactions and Synthesis in Surfactant Systems*, J. Texter (Ed.). Surfactant Sci. Ser., Vol. 100. Marcel Dekker, New York, pp. 45–58 (2001).

10. A. Lattes, I. Rico-Lattes, E. Perez and M. Blanzat. New glycolipids having biological activities: Key role of their organization. In *Reactions and Synthesis in Surfactant Systems*, J. Texter (Ed.). Surfactant Sci. Ser., Vol. 100. Marcel Dekker, New York, pp. 111–128 (2001).

11. J.-M. Kim and D.H. Thompson. Acid and oxidatively labile vinyl ether surfactants: Synthesis and drug delivery applications. In *Reactions and Synthesis in Surfactant Systems*, J. Texter (Ed.). Surfactant Sci. Ser., Vol. 100. Marcel Dekker, New York, pp. 145–154 (2001).

12. J.Y. Shin, L.I. Jong, N. Aydogan and N.L. Abbott. Three principles for active control of interfacial properties of surfactant solutions. In *Reactions and Synthesis in Surfactant Systems*, J. Texter (Ed.). Surfactant Sci. Ser., Vol. 100. Marcel Dekker, New York, pp. 154–173 (2001).

13. S. Aoshima and S. Sugihara. Stimuli-responsive polymer surfactants: Self-assembly of amphiphilic block copolymers. *Nihon Yukagakkaishi*, **49**, 1061–1069 (2000).

14. A. Hashidzume, T. Noda and Y. Morishima. Stimuli-responsive associative behavior of polyelectrolyte-bound nonionic surfactant moieties in aqueous media. *ACS Symp. Ser.*, **780**, 14–37 (2001).

15. B. Cabane and R. Duplessix. Neutron scattering study of water-soluble polymers adsorbed on surfactant micelles. *Colloids Surf.*, **13**, 19–33 (1985).

16. J. Texter. Characterization of surfactants. In *Surfactants—A Practical Handbook*, K.R. Lange (Ed.). Carl Hanser Verlag, Munich, pp. 1–68 (1999).

17. M.F. Cox. Surfactants. In *Detergents and Cleaners*, R.K. Lang (Ed.). Carl Hanser Verlag, Munich, pp. 43–90 (1994).

18. A. Guyot and K. Tauer. Polymerizable and polymeric surfactants. In *Reactions and Synthesis in Surfactant Systems*, J. Texter (Ed.). Surfactant Sci. Ser., Vol. 100. Marcel Dekker, New York, pp. 547–575 (2001).

19. Y.S. Dexter, K. Hu and A.P.J. Middelberg. Designed peptides as non-toxic switchable surfactants. *House. Person. Care Today*, **34**(3), 30–37 (2008).

20. A.S. Malcolm, A.F. Dexter and A.P.J. Middelberg. Reversible active switching of the mechanical properties of a peptide film at a fluid–fluid interface. *Nature Mat.*, **5**, 502–506 (2006).

21. A.S. Malcolm, A.F. Dexter and A.P.J. Middelberg. Foaming properties of a peptide designed to form stimuli-responsive interfacial films. *Soft Matter*, **2**, 1057–1066 (2006).

22. A.S. Malcolm, A.F. Dexter and A.P.J. Middelberg. Peptide surfactants (Pepfactants) for switchable foams and emulsions. *Asia-Pac. J. Chem. Eng.*, **2**, 362–367 (2007).

23. X.-L. Wang, S. Ramusovic, T. Nguyen and Z.-R. Lu. Novel polymerizable surfactants with pH-sensitive amphiphilicity and cell membrane disruption for efficient siRNA delivery. *Bioconjugate Chem.*, **18**, 2169–2177 (2007).

24. W. Kaar, B.M. Hartmann, Y. Fan, L.H.L. Lua, A.F. Dexter, R.J. Falconer and A.P.J. Middelberg. *Biotech. Bioeng.*, **102**, 176–187 (2008).

25. J.J. Donohue and D.A. Buttry. Adsorption and micellization influence the electrochemistry of redox surfactants derived from ferrocene. *Langmuir*, **5**, 671–678 (1989).

26. D.A. Buttry, L. Nordyke and J. Donohue. In situ measurements of monolayer mass changes at the solid/liquid interface. In *Chemically Modified Surfaces in Science and Industry*, D. Leyden and W. Collins (Eds.). Gordon and Breach, New York, pp. 377–392 (1988).

27. L.L. Nordyke and D.A. Buttry. Redox surfactants are chemical probes of electrode surface functionalization derived from disulfide immobilization on gold. *Langmuir*, **7**, 380–388 (1991).

28. H.C. De Long, J.J. Donohue and D.A. Buttry. Ionic interactions in electroactive self-assembled monolayers of ferrocene species. *Langmuir*, **7**, 2196–2202 (1991).

29. J. Li and A.E. Kaifer. Surfactant monolayers on electrode surfaces: Self-assembly of a viologen derivative having a cholesteryl hydrophobic residue. *Langmuir*, **9**, 591–596 (1993).

30. L.J. Hobson and I.R. Butler. Modified metallocenes for material science. *Synth. Met.*, **71**, 2199–2200 (1995).

31. B.S. Gallardo, M.J. Hwa and N.L. Abbott. In situ and reversible control of the surface activity of ferrocenyl surfactants in aqueous solution. *Langmuir*, **11**, 4209–4212 (1995).

32. D.E. Bennett, B.S. Gallardo and N.L. Abbott. Dispensing surfactants from electrodes: Marangoni phenomenon at the surface of aqueous solutions of (11-ferrocenylindecyl)trimethylammonium bromide. *J. Am. Chem. Soc.*, **118**, 6499–6505 (1996).

33. S.S. Datwani, V.N. Truskett, C.A. Rosslee, N.L. Abbott and K.J. Stebe. Redox-dependent surface tension and surface phase transitions of a ferrocenyl surfactant: Equilibrium and dynamic analyses with fluorescence images. *Langmuir*, **19**, 8292–8301 (2003).

34. N. Aydogan and N.L. Abbott. Comparison of the surface activity and bulk aggregation of ferrocenyl surfactants with cationic and anionic headgroups. *Langmuir*, **17**, 5703–5706 (2001).

35. N. Aydogan and N.L. Abbott. Effect of electrolyte concentration on interfacial and bulk solution properties of ferrocenyl surfactants with anionic headgroups. *Langmuir*, **18**, 7826–7830 (2002).

36. E.W. Kaler, A.K. Murthy, B.E. Rodruguez and J.A.N. Zasadzinski. Spontaneous vesicle formation in aqueous mixtures of single-tailed surfactants. *Science*, **245**, 1371–1374 (1989).

37. E.W. Kaler, K.L. Herrington, A.K. Murthy and J.A.N. Zasadzinski. Phase-behavior and structures of mixtures of anionic and cationic surfactants. *J. Phys. Chem.*, **96**, 6698–6707 (1992).

38. D.G. Wu, A.D. Malec, J. Majewski and M. Majda. Orientation and lateral mobility of insoluble Tempo amphiphiles at the air/water interface. *Electrochim. Acta*, **51**, 2237–2246 (2006).

39. S. Ghosh, K. Irvin and S. Thayumanavan. Tunable disassembly of micelles using a redox trigger. *Langmuir*, **23**, 7916:9 (2007).

40. T. Kunitake, N. Nakashima, M. Shimomura, Y. Okahata, K. Kano and T. Ogawa. Unique properties of chromophore-containing bilayer aggregates: Enhanced chirality and photochemically induced morphological change. *J. Am. Chem. Soc.*, **102**, 6642–6644 (1980).

41. J.Y. Shin and N.L. Abbott. Using light to control dynamic surface tensions of aqueous solutions of water soluble surfactants. *Langmuir*, **15**, 4404–4410 (1999).

42. J. Eastoe, M. Sanchez Dominguez, P. Wyatt, A. Beeby and R.K. Heenan. Properties of a stilbene-containing Gemini photosurfactant: Light-triggered changes in surface tension and aggregation. *Langmuir*, **18**, 7837–7844 (2002).

43. J. Eastoe, M. Sanchez Dominguez, P. Wyatt and A.J. Orr-Ewing. UV causes dramatic changes in aggregation with mixtures of photoactive and inert surfactants. *Langmuir*, **20**, 6120–6126 (2004).

44. H. Heinz, R.A. Vaia, H. Koerner and B.L. Farmer. Photoisomerization of azobenzene grafted to layered silicates: Simulation and experimental challenges. *Chem. Mater.*, **20**, 6444–6456 (2008).

45. M. Ogawa, T. Ishii, N. Miyamoto and K. Kuroda. Intercalation of a cationic azobenzene into montmorillonite. *Appl. Clay Sci.*, **22**, 179–185 (2003).

46. T. Okada, Y. Watanabe and M. Ogawa. Photoregulation of the intercalation behavior of phenol for azobenzene-clay intercalation compounds. *J. Mater. Chem.*, **15**, 987–992 (2005).

47. Y. Jiang, Y. Wang, N. Ma, Z. Wang, M. Smet and X. Zhang. Reversible self-organization of a UV-responsive PEG-terminated malachite green derivative: Vesicle formation and photoinduced disassembly. *Langmuir*, **23**, 4029–4034 (2007).

48. V. Bulmus, S. Patir, S. Ali Tuncel and E. Piskin. Stimuli-responsive properties of conjugates of *N*-isopropylacrylamide-co-acrylic acid oligomers with alanine, glycine and serine mono-, di- and tripeptides. *J. Controlled Release*, **76**, 265–274 (2001).

49. K. Tauer, N. Weber and J. Texter. Core-shell particle interconversion with di-stimuli-responsive diblock copolymers. *Chem. Commun.*, **45**, 6065–6067 (2009).

50. T. Tanaka, M. Okayama, H. Minami and M. Okubo. Dual stimuli-responsive "mushroom-like" Janus polymer particles as particulate surfactants. *Langmuir*, **26**, 11732–11736 (2010).

51. J. Texter and K. Bian. $(IL)_m(PO)_n(IL)_m$—Ionic liquid-based triblock copolymers. *Polym. Prepr.*, **51**, 105–106 (2010).

52. J. Texter, V. Arjunan Vasantha, R. Maniglia, L. Slater and T. Mourey. Triblock copolymer based on poly(propylene oxide) and poly(1-[11-acryloylundecyl]-3-methyl-imidazolium bromide). *Macromol. Rapid Commun.*, **33**, 69–74 (2012).

53. J. Texter, V. Arjunan Vasantha, R. Maniglia, L. Slater and T. Mourey. Ionic liquid-based anion and temperature responsive triblock copolymers. *Polym. Mater. Sci. Eng.*, **106**, 786–787 (2012).

54. J. Texter. Nanoparticle dispersions with ionic liquid-based stabilizers. US Patent Application Publication, US 2011/0233458 A1, September 29 (2011).

55. J. Texter, V. Arjunan Vasantha, K. Bian, X. Ma, L. Slater, T. Mourey and G. Slater. Stimuli-responsive triblock copolymers—Synthesis, characterization, and application. In *Non-Conventional Functional Block Copolymers*, B. Coughlin, P. Theato and A. Kilbinger (Eds.), Chap. 9. ACS Symp. Ser., Vol. 1066. American Chemical Society, Washington, DC, pp. 117–130 (2011).

56. A. Goldsipe and D. Blankschtein. Molecular-thermodynamic theory of micellization of multicomponent surfactant mixtures: 1. Conventional (pH-insensitive) surfactants. *Langmuir*, **23**, 5942–5952 (2007).

57. A. Goldsipe and D. Blankschtein. Molecular-thermodynamic theory of micellization of multicomponent surfactant mixtures: 2. pH-Sensitive surfactants. *Langmuir*, **23**, 5953–5962 (2007).

58. S. Capone, P. Walde, D. Seebach, T. Ishikawa and R. Caputo. pH-Sensitive vesicles containing a lipidic β-amino acid with two hydrophobic chains. *Chem. Biodivers.*, **5**, 16–30 (2008).

59. P.L. Felgner, T.R. Gadek, M. Holm, R. Roman, H.W. Chan, M. Wenz, J.P. Northrop, G.M. Ringold and M. Danielsen. Lipofection: A highly efficient, lipid mediated DNA-transfection procedure. *Proc. Natl. Acad. Sci. U. S. A.*, **84**, 7413–7417 (1987).

60. M. Spelios, S. Nedd, N. Matsunaga and M. Savva. Effect of spacer attachment sites and pH-sensitive headgroup expansion on cationic lipid-mediated gene delivery of three novel myristoyl derivatives. *Biophys. Chem.*, **129**, 137–147 (2007).

61. X. Dong, Z. Zheng and J. He. pH-Sensitive dye-polyether derivatives as dispersants for its parent dye. Part 2: Dispersion stability and dyeing performance. *J. Disp. Sci. Technol.*, **31**, 1188–1194 (2010).

62. A. Boudier, A. Aubert-Pouëssel, C. Gérardin, J.-M. Devoisselle and S. Bégu. pH-sensitive double-hydrophilic block copolymer micelles for biological applications. *Int. J. Pharm.*, **379**, 212–217 (2009).

63. L.J. Shi and L.Q. Zheng. Aggregation behavior of surface active imidazolium ionic liquids in ethylammonium nitrate: Effect of alkyl chain length, cations, and counter-ions. *J. Phys. Chem. B*, **116**, 2162–2172 (2012).

64. A.F. Ding, M. Zha, J. Zhang and S.S. Wang. Synthesis of a kind of geminal imidazolium ionic liquid with long aliphatic chains. *Chin. Chem. Lett.*, **18**, 48–50 (2007).

65. A.F. Ding, M. Zha, J. Zhang and S.S. Wang. Synthesis, characterization and properties of geminal imidazolium ionic liquids. *Colloids Surf. A Physicochem. Eng. Aspects*, **298**, 201–205 (2007).

66. J. Łuczak, M. Markiewicz, J. Thöming, J. Hupka and C. Jungnickel. Influence of the Hofmeister anions on self-organization of 1-decyl-3-methylimidazolium chloride in aqueous solutions. *J. Colloid Interface Sci.*, **362**, 415–422 (2011).

67. D. Mecerreyes. Polymeric ionic liquids: Broadening the properties and applications of polyelectrolytes. *Prog. Polym. Sci.*, **36**, 1629–1648 (2011).

68. D. England, N. Tambe and J. Texter. Stimuli-responsive nanolatexes—Porating films. *ACS Macro Lett.*, **1**, 310–314 (2012).

69. E. Alcade, I. Dinares and N. Mesquida. Imidazolium-based receptors. *Top. Heterocycl. Chem.*, **24**, 267–300 (2010).

70. E. Alcade, N. Mesquida, A. Ibañez and I. Dinares. A halide-for-anion swap using an anion-exchange resin (A(-) form) method: Revisiting imidazolium-based anion receptors and sensors. *Eur. J. Org. Chem.*, 298–304 (2012).

71. J. Zhao, F. Yan, Z.-Z. Chen, H.B. Diao, F.Q. Chu, A.M. Yu and J.M. Lu. Microemulsion polymerization of cationic pyrroles bearing an imidazolum-ionic liquid moiety. *J. Polym. Sci.: A Polym. Chem.*, **47**, 746–753 (2009).

72. O.A. El Seoud, P.A.R. Pires, T. Abdel-Moghny and E.L. Bastos. Synthesis and micellar properties of surface-active ionic liquids: 1-Alkyl-3-methylimidazolium chlorides. *J. Colloid Interface Sci.*, **313**, 296–304 (2007).

73. J.E. Bara, E.S. Hatakeyama, B.R. Wiesenauer, X. Zeng, R.D. Noble and D.L. Gin. Thermotropic liquid crystal behaviour of Gemini imidazolium-based ionic amphiphiles. *Liq. Cryst.*, **37**, 1587–1599 (2010).

74. S. Sunitha, P.S. Reddy, R.B.N. Prasad and S. Kanjital. Synthesis and evaluation of new imidazolium-based aromatic ether functionalized cationic mono and Gemini surfactants. *Eur. J. Lipid Sci. Technol.*, **113**, 756–762 (2011).

75. M.Q. Ao, G.Y. Xu, Y.Y. Zhu and Y. Bai. Synthesis and properties of ionic liquid-type Gemini imidazolium surfactants. *J. Colloid Interface Sci.*, **326**, 490–495 (2008).

76. A. Bhadani and S. Singh. Synthesis and properties of thioether spacer containing Gemini imidazolium surfactants. *Langmuir*, **27**, 14033–14044 (2011).

77. A. Bhadani, H. Kataria and S. Singh. Synthesis, characterization and comparative evaluation of phenoxy ring containing long chain Gemini imidazolium and pyridinium amphiphiles. *J. Colloid Interface Sci.*, **361**, 33–41 (2011).

78. C.L. Toh, L.F. Xi, S.K. Lau, K.P. Pramoda, Y.C. Chua and X.H. Lu. Packing behaviors of structurally different polyhedral oligomeric silsesquioxane-imidazolium surfactants in clay. *J. Phys. Chem. B*, **114**, 207–214 (2010).

79. J. Yuan and M. Antonietti. Poly(ionic liquid)s: Polymers expanding classical property profiles. *Polymer*, **52**, 1469–1482 (2011).

80. F. Yan and J. Texter. Surfactant ionic liquid-based microemulsions for polymerization. *Chem. Commun.*, **42**, 2696–2698 (2006).

81. F. Yan and J. Texter. Solvent-reversible poration in ionic liquid copolymers *Angew. Chem.*, **119**, 2492–2495 (2007).

82. X. Ma, Md. Ashaduzzaman, M. Kunitake, R. Crombez, J. Texter, L. Slater and T. Mourey. Stimuli-responsive poly(1-[11-acryloylundecyl]-3-methyl-imidazolium bromide)—Dewetting and nanoparticle condensation phenomena. *Langmuir*, **27**, 7148–7157 (2011).

83. X. Ma, R. Crombez, Md. Ashaduzzaman, M. Kunitake, L. Slater, T. Mourey and J. Texter. Polymer dewetting via stimuli-responsive structural relaxation—Contact angle analysis. *Chem. Commun.*, **47**, 10356–10358 (2011).

84. S.M. Yu, F. Yan, X.W.S. Zhang, J.B. You, P.Y. Wu, J.M. Lu, Q.F. Xu, X.W. Xia and G.L. Ma. Polymerization of ionic liquid-based microemulsions: A versatile method for the synthesis of polymer electrolytes. *Macromolecules*, **41**, 3389–3392 (2008).

85. F.A. Bottino, E. Fabbri, I.L. Fragala, G. Malandrino, A. Orestano, F. Pilati and A. Pollicino. Polystyrene-clay nanocomposites prepared with polymerizable imidazolium surfactants. *Macromol. Rapid Commun.*, **24**, 1079–1084 (2003).

86. A. Pucci, V. Liuzzo, B. Melai, C.S. Pomelli and C. Chiappe. Polymerizable ionic liquids for the preparation of polystyrene/clay composites. *Polym. Int.*, **61**, 426–433 (2012).

87. D. Guzmán-Lucero, O. Olivares-Xometl, R. Martínez-Palou, N.V. Likhanova, M.A. Domínguez-Aguilar and V. Garibay-Febles. Synthesis of selected vinylimidazolium ionic liquids and their effectiveness as corrosion inhibitors for carbon steel in aqueous sulfuric acid. *Ind. Eng. Chem. Res.*, **50**, 7129–7140 (2011).

88. J. Yuan and M. Antonietti. Poly(ionic liquid) latexes prepared by dispersion polymerization of ionic liquid monomers. *Macromolecules*, **44**, 744–750 (2011).

89. J.Y. Yuan, S. Sebastian, M. Dreschler, A.H.E. Müller and M. Antonietti. Self-assembly of poly(ionic liquid)s: Polymerization, mesostructure formation, and directional alignment in one step. *J. Am. Chem. Soc.*, **133**, 17556–17559 (2011).

90. A. Beyaz, W.S. Oh and V. Prakash Reddy. Synthesis and CMC studies of 1-methyl-3-(pentafluorophenyl) imidazolium quaternary salts. *Colloids Surf.*, **36**, 71–74 (2004).

91. J.A. Howarter and J.P. Youngblood. Fluorinated surfactants as stimuli-responsive polymers and brushes. Abstracts of Papers, 230th ACS National Meeting, Washington, DC, August 28–September 1, 2005, POLY-018 (2005).

92. P.F. Long, A.X. Song, D. Wang, R.H. Dong and J.C. Hao. pH-Sensitive vesicles and rheological properties of PFLA/NaOH/H_2O and PFLA/LiOH/H_2O systems. *J. Phys. Chem. B*, **115**, 9070–9076 (2011).

93. R.E. Del Sesto, T.M. McCleskey, A.K. Burrell, G.A. Baker, J.D. Thompson, B.L. Scott, J.S. Wilkes and P. Williams. Structure and magnetic behavior of transition metal based ionic liquids. *Chem. Commun.*, **44**, 447–449 (2008).

94. M. Li, S.L. De Rooy, D.K. Bwambok, B. El-Zahab, J.F. DiTusa and I.M. Warner. Magnetic chiral ionic liquids derived from amino acids. *Chem. Commun.*, **45**, 6922–6924 (2009).

95. P. Brown, A. Bushmelev, C.P. Butts, J. Cheng, J. Eastoe, I. Grillo, R.K. Heenan and A.M. Schmidt. Magnetic control over liquid surface properties with responsive surfactants. *Angew. Chem. Int. Ed.*, **51**, 2414–2416 (2012).

96. P. Brown, A. Bushmelev, C.P. Butts, J.-C. Eloi, I. Grillo, P.J. Baker, A.M. Schmidt and J. Eastoe. Properties of new magnetic surfactants. *Langmuir*, **29**, 3246–3251 (2013).

97. P. Brown, C.P. Butts, J. Eastoe, E.P. Hernndez, F.L. de Araujo Machadob and R.J. de Oliveira. Dication magnetic ionic liquids with tuneable heteroanions. *Chem. Commun.*, **49**, 2765–2767 (2013).

98. P. Brown, A.M. Khan, J.P.K. Armstrong, A.W. Perriman, C.P. Butts and J. Eastoe. Magnetizing DNA and proteins using responsive surfactants. *Adv. Mater.*, **24**, 6244–6247 (2012).

99. W. Peng, D.L. Zhou and J.F. Rusling. Adsorption-desorption dynamics of amphiphilic ferrocenes on electrodes studied by flow votammetry. *J. Phys. Chem.*, **99**, 6986–6993 (1995).

100. W. Peng and J.F. Rusling. Intermolecular interactions between surfactants coadsorbed on electrodes. *J. Phys. Chem.*, **99**, 16436–16441 (1995).

101. C. Morita, H. Sugimoto, Y. Imura and T. Kawai. Double-stimuli-responsive O/W emulsion gel based on a novel amidoamine surfactant. *J. Oleo Sci.*, **60**, 557–562 (2011).

102. J. Hotz and W. Meier. Vesicular polymerization. In *Reactions and Synthesis in Surfactant Systems*, J. Texter (Ed.). Surfactant Sci. Ser., Vol. 100. Marcel Dekker, New York, pp. 501–514 (2001).

103. L. Jiang, K. Wang, M.L. Deng, Y.L. Wang and J.B. Huang. Bile salt-induced vesicle-to-micelle transition in catanionic surfactant systems: Steric and electrostatic interactions. *Langmuir*, **24**, 4600–4606 (2008).

104. L.X. Jiang, K. Wang, F.Y. Ke, D.H. Liang and J.B. Huang. Endowing catanionic surfactant vesicles with dual responsive abilities via a noncovalent strategy: Introduction of a responser, sodium cholate. *Soft Matter*, **5**, 599–605 (2009).

105. C.B. Minkenberg, F. Li, P. van Rijn, L. Florusse, J. Boekhoven, M.A. Stuart-Cohen, G.J.M. Koper, R. Eelkema and J.H. van Esch. Responsive vesicles from dynamic covalent surfactants. *Angew. Chem. Int. Ed.*, **50**, 3421–3424 (2011).

106. L.S. Romsted. Interfacial compositions of surfactant assemblies by chemical trapping with arenediazonium ions: Method and applications. In *Reactions and Synthesis in Surfactant Systems*, J. Texter (Ed.). Surfactant Sci. Ser., Vol. 100. Marcel Dekker, New York, pp. 265–294 (2001).

107. L.S. Romsted. Do amphiphile aggregate morphologies and interfacial compositions depend primarily on interfacial hydration and ion-specific interactions? The evidence from chemical trapping. *Langmuir*, **23**, 414–424 (2007).

108. C.A. Rosslee and N.L. Abbott. Principles for microscale separations based on redox-active surfactants and electrochemical methods. *Anal. Chem.*, **73**, 4808–4814 (2001).

109. X.-Y. Liu and N.L. Abbott. Electrochemical generation of gradients in surfactant concentration across microfluidic channels. *Anal. Chem.*, **81**, 772–781 (2009).

110. X.-Y. Liu and N.L. Abbott. Lateral transport of solutes in microfluidic channels using electrochemically generated gradients in redox-active surfactants. *Anal. Chem.*, **83**, 3033–3041 (2011).

111. D.J. Lestage and M.W. Urban. Release and formation of surface-localized ionic clusters (SLICs) into phospholipid rafts from colloidal solutions during coalescence. *Langmuir*, **21**, 2150–2157 (2005).

112. D.J. Lestage, M. Yu and M.W. Urban. Stimuli-responsive surfactant/phospholipid stabilized colloidal dispersions and their film formation. *Biomacromolecules*, **6**, 1561–1572 (2005).

113. D.J. Lestage, W.R. Dreher and M.W. Urban. Stimuli-responsive behavior of sodium dodecyl sulfate and sodium dioctyl sulfosuccinate during coalescence of colloidal particles. *ACS Symp. Ser.*, **912**, 122–136 (2005).

114. T. Saji. Electroless plating of organic pigment thin films using surfactants with an azobenzene group. In *Reactions and Synthesis in Surfactant Systems*, J. Texter (Ed.). Surfactant Sci. Ser., Vol. 100. Marcel Dekker, New York, pp. 407–412 (2001).

115. M. Mihara, C. Fowler, C. O'Neill, P.G. Jessop and M.F. Cunningham. Stimuli-responsive polymer colloids using carbon dioxide switchable surfactants: Synthesis and reversible stabilization/aggregation. Abstracts of Papers, 242nd ACS National Meeting & Exposition, Denver, CO, August 28–September 1, 2011, IEC-19 (2011).

116. P. Pincus. Colloid stabilization with grafted polyelectrolytes. *Macromolecules*, **24**, 2912–2919 (1991).

117. J. Texter, R. Crombez, X. Ma, L. Zhao, F. Perez-Caballero, M.-M. Titirici and M. Antonietti. Waterborne nanocarbon dispersions for electronic and fuel applications. *Prepr. Symp.-Am. Chem. Soc., Div. Fuel Chem.*, **56**, 388–389 (2011).

118. J. Texter, D. Ager, V. Arjunan Vasantha, R. Crombez, D. England, X. Ma, R. Maniglia and N. Tambe. Advanced nanocarbon materials facilitated by novel stimuli-responsive stabilizers. *Chem. Lett.*, **41**, 1377–1379 (2012).

119. J. Texter, N. Tambe, R. Crombez, M. Antonietti and C. Giordano. Stimuli-responsive coatings of carbon nanotubes and nanoparticles using ionic liquid based nanolatexes. *Polym. Mater. Sci. Eng.*, **102**, 401–402 (2010).

120. M. Iijima, M. Kobayakawa, M. Yamazaki, Y. Ohta and H. Kamiya. Anionic surfactant with hydrophobic and hydrophilic chains for nanoparticle dispersion and shape memory polymer nanocomposites. *J. Am. Chem. Soc.*, **131**, 16342–16343 (2009).

Part VI

---

*Formulation and Application
of Emulsions*

18 Progress in Over a Century of Designing Emulsion Properties

Emerging Phenomenological Guidelines from Generalized Formulation and Prospects to Transmute the Knowledge into Know-How

Jean-Louis Salager, Ana Forgiarini, and Johnny Bullón

CONTENTS

18.1 INTRODUCTION: MANY CASES OF EMULSION AND A LARGE QUANTITY OF AVAILABLE INFORMATION

Emulsions are found in nature, such as in milk or plant saps, but most emulsions are available as man-made vehicles for a large variety of products, often prepared according to some recipe and production method empirically developed a long time ago for a specific application. Typical emulsions include the following: a viscoplastic oil-in-water (O/W) mayonnaise dressing with a high content of triglyceride oils, a viscous cold cream of the water-in-oil (W/O) type with very small water droplets in paraffin or polar oil, a bimodal drop size distribution O/W emulsion to transport heavy crude oil over long distances, a water-in-oil-in-water (W/O/W) multiple emulsion for controlled drug release in a medicinal patch, a pseudoplastic acrylic paint that can be applied with a brush, or a nanoemulsion in diesel fuel producing reduced CO and NO_x exhaust [1].

Emulsions are involved in many different applications and purposes such as for dispersing a liquid phase into another, gathering immiscible phases in a persistent system, attaining a particular rheological behavior, reaching a specific taste, spontaneously producing a mixture or separating two liquids quickly or slowly, or changing a system property according to a designed process, etc. Emulsions are made or handled according to distinct approaches and various techniques dealing with different phenomena taking place in the emulsification and concerning the attained properties: type or morphology, drop size average and distribution, stability or persistence, and rheological behavior. The general understanding of these properties has been developed in the past century since the Bancroft's rule in the early 1900s. Approximately 40 years ago, it was limited to only qualitative trends, which are not sufficient today as far as accuracy and prediction are concerned in many practical cases, particularly when strict property requirements are demanded. These general (qualitative) trends, which started to develop a century ago, are indicated in the extreme left part in Figure 18.1. The emulsion know-how (indicated in the middle zone of Figure 18.1) was first organized in Becher's pioneering book [2] followed by an edited encyclopedia, which started approximately 30 years ago [3], and was complemented by other treatises to keep track of the developments [4–8].

The advances in emulsion science and technology were carried out by dealing with specifics in mechanics, physics, or chemistry—most of the time with some specific application in mind. Because of the variety of phenomena and practical cases, there are a huge number of studies and publications with good scientific basis, but most of them have very well-defined circumstances as well as a limited area of application. This is because in essentially all cases, the quantitative aspects deal with only a few variables among the very large number involved. As a consequence, a large part of the knowledge coming from specifics cannot be widely generalized. This means that a new particular research study often has to be carried out for each new practical problem. It also means that there is no warranty that even a very good research would produce the best of all results that cannot be improved by a clever competitor. This is, of course, not an ideal situation in science and technology development. The many specific contributions to knowledge are indicated as squares and circles with a relatively small overlapping on the right side of Figure 18.1. These correspond to hundreds of particular cases dealing with electrical phenomena at interfaces such as the

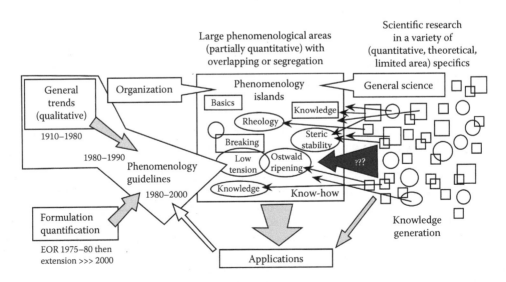

FIGURE 18.1 Organization of the available information in emulsion science and technology.

Derjaguin, Landau, Vervey, and Overbeek (DLVO) theory; the Lifshitz, Slezov, and Wagner theory; equilibrium and dynamics of adsorption; polymer/surfactant interactions, etc.

Because some studies in different instances could be organized together or were extended to some wider range of cases, the resulting combination could be applied in real applications, sometimes as a mixture of theory and empirical experience. This organization of results created phenomenology islands in which some general understanding is attained at some partially quantitative level, which allows us to solve problems. However, the generality applies only to specific circumstances and conditions, that is, particular systems involving only a limited number of variables or phenomena. For instance, emulsion rheology has been accurately related to emulsion characteristics such as phase proportion and viscosity, drop size average, and distribution [9], although the systematic attainment of these characteristics, particularly through formulation or stirring, was not considered. The so-called Ostwald ripening evolution [10,11], was well organized as a special phenomenology, but it only applies to some cases of very small droplets such as nanoemulsions.

Another island was created by extending the DLVO theory to emulsion stability dealing with the attractive and repulsive forces' balance when two drops approach [12–14]. Nevertheless, in this case, the variety of repulsive forces kept this island phenomenology at a qualitative level, that is, as a guideline rather than a quantitative knowledge with accurate predictions.

These islands are roughly independent because each deals with different situations and variables. However, a common variable effect could produce the overlapping of two islands over two or more cases, thus enlarging the phenomenology or deepening it, as for instance, in the association of ultralow tension with emulsion breaking. Conversely, two conflicting effects produced by the same variable could result in segregation, that is, an area in which the two phenomenologies are not compatible, for instance, Ostwald ripening phenomenology with low interfacial tension formulation. These phenomenology islands, some of which are indicated in the central part of Figure 18.1, are not numerous (possibly 20 to 30). However, they are quite important in practice because they result in know-how with a partial level of quantification that allows us to solve practical problems. Many areas of science and technology are involved when dealing with emulsions, as indicated in the first available reviews [3–15].

This is why most application cases deal with very complex phenomena both in surfactant/polymer phase behavior and emulsion morphology inversion or hysteresis, which have been detected and even studied, but are still far from being mastered. In practice, a huge number of variables and procedures, say 10 or 20, have to be considered to solve problems, and cannot be scrutinized in a reasonable amount of time. Because this knowledge has not yet been transmuted into a ready-to-use know-how, trial and error methods are often used without a guarantee of success in research and development.

Building up expertise currently takes decades of experience, and it may be said that despite an exponentially increasing number of studies, there are still a lot of unchecked possibilities that will generate "magic" patents in the future.

This situation indicates that there is a critical need for improvement in emulsion science and technology, particularly as far as the currently available information is processed, organized, and transferred to the potential users. Before recommending a proper evolution for the near future, it is worth analyzing first what has been done up to now.

Figure 18.1 indicates that in the last 25 years of the 20th century, an extension of the general trends emerged from the published and applied knowledge to contribute to the practical know-how. It may be said that general guidelines are now available as far as the essential phenomenologies are concerned, with some semiquantitative predictions regarding distinguishable cases and special applications.

Progress in the creation and organization of phenomenologically accurate guidelines has been significantly linked with the scientific treatment, generalization, and considerable simplification of the physicochemical formulation of surfactant–oil–water (SOW) systems, the issue to be considered first.

18.2   THE KEY DRIVING FORCE FOR PROGRESS: CONCEPTUALIZATION OF PHYSICOCHEMICAL FORMULATION

Emulsions are dispersed systems containing two immiscible phases called water and oil, that is, a polar one, which is most of the time an aqueous solution, and a nonpolar or slightly polar one, which may range from petroleum hydrocarbons to natural triglycerides. The dispersion of the two immiscible phases results in the formation of drops of one phase in the other. Because of the free energy driving force, drops tend to coalesce and the system tends to rapidly return to separated phases, unless the coalescence is inhibited, and then the drop dispersion persists for some time, forming what is generally called an emulsion. The persistence of the dispersed morphology implies that approaching drops do not coalesce immediately on contact, that is, the interdrop film does not drain readily. This is generally due to some repulsive effect produced by the presence of simple or polymeric surfactants, or particles at the interface (Figure 18.2). The last instance, so-called Pickering emulsions [16,17], are beyond the scope of the present study, and in what follows, only the surfactant (simple or polymeric) case will be treated.

Because of its polar/apolar duality, the surfactant molecule spontaneously tends to adsorb at the oil/water interface and to self-associate when there is an excess of it in solution. The actual surfactant effect depends on its relative interactions with the water and oil, referred to as hydrophilicity and lipophilicity, respectively. This obviously depends on the surfactant structure and the oil and water nature, as well as temperature, and even pressure in some cases. The first relationship between the surfactant structure and the emulsion type was illustrated almost a century ago by Langmuir's wedge theory (Figure 18.3). The adsorbed surfactant molecule "grabs" a higher number of oil or water molecules and thus results in a wedge, with the wider part on the side with more interactions. The association of wedged elements results in the curvature of the interface, hence, a type of emulsion. According to this model, a surfactant with a higher interaction with water will produce an O/W emulsion and vice versa. This geometric interpretation is quite easily understandable and may be used as a first approach, even if it is probable that it is correct only for extremely small domains like micelles, rather than for micrometric drops. Consequently, it is usually mentioned as a historical milestone rather than a scientific breakthrough [18,19].

In any case, it indicates that the emulsion type depends on what happens on the two sides of the interface, hence it is likely to vary with both the surfactant and oil/water characteristics.

This wedge theory idea, developed by Harkins et al. [20] and Langmuir [21], was a molecular mechanism explaining an experimental trend, the so-called Bancroft's rule [22,23]. According to this, the phase in which the surfactant is predominantly solubilized tends to be the continuous or external phase of the emulsion. For instance sodium oleate, a water-soluble salt, tends to stabilize O/W emulsions, whereas calcium dioleate, an oil-soluble surfactant, stabilizes the W/O morphology.

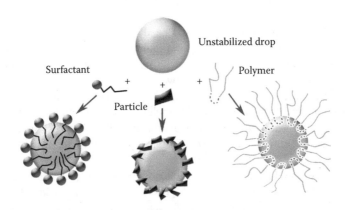

FIGURE 18.2 Emulsion drop stabilization with the interfacial presence of surfactants, polymers, or particles.

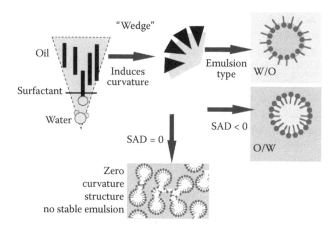

FIGURE 18.3 How "wedge" theory, which depends on surfactant interactions with water and oil, dictates the curvature and the emulsion type. The SAD is a formulation concept explained later. SAD > 0 (respectively, SAD < 0) refers to a surfactant with more interaction with oil (respectively, water).

Bancroft's rule was essentially correct as a general trend in most cases and was well summarized a few years later [24]. Other studies on the effect of the nature of oils and electrolytes reported by Seifriz [25] indicated that a fairly acceptable know-how had been reached by the 1920s.

It may be said that the situation did not progress much for two decades until Griffin [26] proposed the first quantification of the formulation by postulating a characteristic parameter for the surfactant. It was the so-called hydrophilic–lipophilic balance (HLB), an empirical numerical scale first ranging from 1 (for oleic acid) to 20 (for potassium oleate soap). Surfactants with low HLB value (~4–6) tend to stabilize W/O emulsions, whereas those with a high HLB value (~15–20) would stabilize O/W emulsions. The HLB of a surfactant was experimentally determined from a technique based on the attainment of the maximum stability of emulsions, which was a long procedure. It was mentioned that surfactants with intermediate HLB (7–14 range) were usually ineffective to stabilize any of these emulsion types and, moreover, that the attained type could depend on other variables. A few years later, Griffin introduced a new way to calculate the HLB number from the chemical structure of nonionic surfactants of the ethoxylated akylphenol type [27].

In 1957, Davies [28] proposed that because the HLB was linked with the "size" of the parts of the nonionic surfactants, a simple group contribution method could be used as a generalized way of calculating the HLB. An analysis of these data indicated that the idea was correct because it explicitly linked HLB to energetic contributions, but that there were very serious discrepancies in the numbers, particularly for ionic surfactants. What should be remembered from this is that the balanced surfactants with equal hydrophilic and lipophilic tendencies had an HLB of 10 according to Griffin, and 7 according to Davies. This is perfectly in agreement with the imprecision exhibited when comparing different species of surfactants such as alkyl sulfate and alkyl phenol ethoxylates. It also has to do with the fact that the surfactant relative affinity for oil and aqueous phases also depends on the nature of these phases, as well as temperature, and sometimes pressure [18–25].

Nevertheless, it was found that in a surfactant group and for given oil, water, and temperature conditions, the HLB number gives a significant indication of the relative tendencies, for instance, in the sorbitan esters and sorbitan ethoxylated esters [3–29]. This is the reason why HLB is still used in practice for some nonionic surfactant series.

In 1954, Winsor [30] proposed a fundamental approach to explain the phase behavior of SOW ternary systems. The formulation state was described as R, the ratio of the interactions of the surfactant with the oil phase to its interactions with the water phase. However, the interactions cannot be calculated with accuracy and only trends may be used. Consequently, the Winsor R ratio is a way to make guesses and to convey trends, but it cannot be used to carry out accurate calculations. This theoretical approach clearly indicated that the hydrophilic–lipophilic tendency of a surfactant with respect to the oil and water phases depends on the variables describing the nature of all the substances, as well as temperature and pressure. The experimental evidence indicated that the variation of the phase behavior could be very sensitive to small variations in any of the formulation variables. Of course, Winsor phase diagrams were established in the case of relatively pure systems and were probably not sophisticated enough to describe any system, in particular those containing commercial surfactant blends or mixtures of surfactants and cosurfactants. However, the three-phase behavior occurrence was quite characteristic of the experimental evidence for a perfect balance between hydrophilic and lipophilic affinities of the surfactant for the oil and water phases.

In the 1960s, a new experimental approach was presented by Shinoda and Arai [31] to attain a more accurate formulation expression, particularly with the inclusion of the nature of oil and brine phases. It was based on the fact that polyethoxylated surfactants exhibited a large variation in their relative affinity for the oil and water as the temperature was changed. This is essentially the same effect as the so-called cloud point, but with a second (oil) phase that can alter the phase separation process.

This phase inversion temperature (PIT) phenomenon is based on the dehydration of the polyethoxylated head group as the temperature increases. At low temperatures (below PIT), the surfactant is mostly in the aqueous phase and conversely [32,33]. The PIT is the temperature at which the surfactant migrates from water to oil, that is, the equivalent of the Winsor R = 1 situation. In many cases, a system in this condition exhibits a three-phase behavior, which is easy to detect. On the other hand, the actual emulsion type is found to switch at the PIT in most cases, and thus could be used for the detection of the PIT. Experimental evidence that is attained with good accuracy indicates that the PIT varies not only with the surfactant but also with the nature of the oil and water phases. Hence, it means that PIT is quite superior to HLB as an accurate measurement of the formulation state, involving all the parameters likely to produce a variation.

However, the PIT technique is based on the phase behavior of SOW systems containing a temperature-sensitive surfactant, in most cases, polyethoxylated ones, and furthermore, the useful temperature range is the liquid water zone, that is, 0°C to 100°C. This corresponds to the PIT of the system octane/water containing ethoxylated nonylphenol with the number of ethylene oxide groups (EON = 3–10), which is a limited range in practice.

Then, in 1974, enhanced oil recovery research studies seeking the occurrence of an ultralow crude–brine interfacial tension found it in very particular situations in which the surfactant hydrophilic/lipophilic tendencies were exactly balanced. At this time, the Winsor R ratio was rediscovered and the situation sought was described as corresponding exactly to R = 1 [34,35].

However, because the tension variation is considerable in the zone about the minimum whatever the scan variable (see Figure 18.4), the numerical handling of the so-called optimum formulation was a matter of extreme accuracy, for example, R = 1.00 with two decimal digits which was incompatible with the previously proposed formulation concepts. Consequently, a new concept had to be created to substitute the HLB, R, or PIT approaches in a more general and more accurate way.

The occurrence of a minimum in interfacial tension and of three-phase behavior was studied in a multidimensional space with formulation variables such as brine salinity, nature of the oil, surfactant characteristics, cosurfactant type and contribution (generally an alcohol needed to avoid the precipitation of ionic surfactants), temperature and even pressure. For the sake of simplicity, the first research was carried out using a simplified SOW system, that is, a NaCl aqueous solution, n-alkane, and a relatively pure surfactant.

Correlations for the attainment of an optimum formulation with minimum interfacial tension or three-phase behavior were found to be expressed by very simple equations such as

$$\mathrm{Ln}\,S - K\,\mathrm{ACN} + \sigma - f(A) - a_\mathrm{T}\,(T - T_\mathrm{ref}) = 0 \qquad (18.1)$$

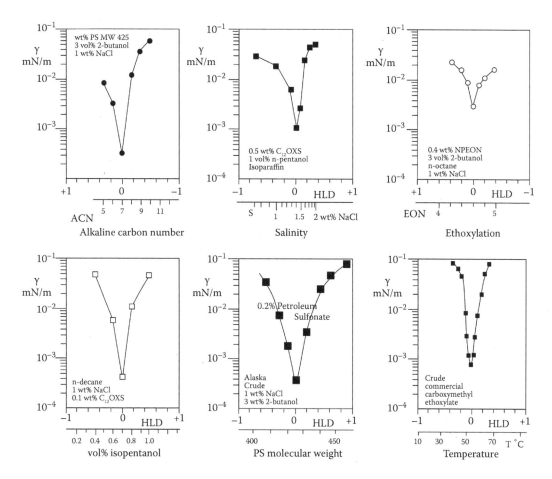

FIGURE 18.4 Interfacial tension versus different formulation variables. Data from Arandia et al. [36] and Strey [37] adapted to the same formulation scale according to HLD equations (Equations 18.4 and 18.5). PS, petroleum sulfonate; $C_{12}OXS$, dodecyl orthoxylene sulfonate; NPEON, nonylphenol ethoxylate with N ethylene oxide groups; $C_{12}EO_5$, dodecylalcohol ethoxylate with five ethylene oxide groups.

for anionic surfactants [38]

$$\alpha - EON - K\, ACN + b\, S - \phi(A) + c_T\,(T - T_{ref}) = 0 \qquad (18.2)$$

for ethoxylated nonionic surfactants [39]

In these equations, S is the salinity in wt% NaCl in water, ACN the alkane carbon number, and σ the characteristic parameter of an anionic surfactant. For an ethoxylated nonionic surfactant, EON is the average number of ethylene oxide groups and α is a parameter depending on the hydrophobic part. Both σ and α parameters increase linearly with the number of carbon atoms in the hydrophobic part of the surfactant. Functions $f(A)$ and $\phi(A)$ represent the contribution of the alcohol cosurfactant, and may be expressed as $(m_A\, C_A)$, that is, as a function of the alcohol type (m_A) and essentially proportional to the alcohol concentration C_A. T_{ref} is a reference temperature, for example, 25°C, and a, b, and c are positive constants. It is worth noting that the temperature effect has a different sign, that is, an increase in temperature renders an anionic surfactant more hydrophilic, and a polyethoxylated one more hydrophobic.

With the exception of the surfactant parameters, the terms correspond to variables that are easy to measure with a good accuracy. Hence, the precision of these equations is much better than what could be expected from HLB or R equivalents. It allows an improved and accurate handling of the generalized formulation, which can produce quick variations in the interfacial tension. In these equations, an increasing positive term results in a tendency to increase the interactions of the surfactant with the oil phase, or to decrease the interaction with the water phase, whereas the negative terms produce the opposite. This left part of the relationship was related to the standard chemical potential change when a surfactant molecule passes from water to oil, that is, $\Delta\mu^*_{O\to W}$, and is related to the surfactant partition coefficient between the phases [40,41].

This research effort involved a huge amount of experimental work, which after expressing the optimum formulation occurrence using Equations 18.1 and 18.2 in very simple systems, studied the way to attain it with realistic systems (salt mixture, oil mixtures, or surfactant mixtures) [35,42–48].

Problems that arise from complex systems such as fractionation of mixture species, and interfacial formulation changes with surfactant concentration and water-to-oil ratio (WOR) were also studied and understood, and partially solved [49–51].

In the 1980s, crude oil prices went down and enhanced oil recovery was no longer of commercial interest. As a result, only very few research and development institutions pursued this topic during the next two decades.

Some studies were, however, carried out in universities to clarify some questions about the significance of the correlations; it was found that they were a numerical measurement of the free energy change when a surfactant molecule passed from one phase to the other. This was called *surfactant affinity difference* (SAD), whose dimensionless expression was named *hydrophilic–lipophilic deviation* (HLD) [52,53].

At optimum formulation, HLD was exactly zero. Away, it was measuring a driving force or a difference of bending energies from one side to the other, which was related to the curvature, and hence to emulsion properties.

$$\Delta\mu^*_{O\to W} = \mu^*_W - \mu^*_O = \text{SAD} \tag{18.3}$$

$$\text{HLD} = \text{SAD/RT} = \text{Ln}S - K\,\text{ACN} + \sigma - f(A) - a_T\,(T - T_{\text{ref}}) = 0 \tag{18.4}$$

for ionic surfactants

$$\text{HLD} = \text{SAD/RT} = b\,S - K\,\text{ACN} + \beta - \phi(A) + c_T\,(T - T_{\text{ref}}) = 0 \tag{18.5}$$

for ethoxylated nonionic surfactants

This concept clearly indicates the same as the previous ones, that is, HLB, PIT, or Winsor R. However, with these relationships, it is very easy to make a calculation of the contributions of the different variables, and thus to calculate the value with good accuracy. In Figure 18.4, the effect of each variable on the interfacial tension can be compared with the others because the formulation scale is the same, as expressed in HLD units. The extreme similarity of all curves when the same HLD scale is used is worth noting. It means that the actual variation from one case to the other is only the depth of the interfacial tension minimum. This minimum interfacial tension is hence the significant yardstick to estimate the formulation quality, just as the solubilization parameter at optimum [54], which is the inverse according to the Huh [55] relationship.

In HLD equations, the terms are energies and can thus be added or subtracted from the total energy balance, without giving special importance to a particular variable. Studies dedicated to the interfacial curvature showed that it is directly related to HLD [56–59]. This is beyond the scope of this article, but it is worth noting that optimum formulation corresponds to the attainment of zero curvature. The first advantage of this formulation expression is its simplicity, for example, the effect of many variables could be expressed through a single variable, called the generalized formulation SAD or HLD. The second advantage is its generality, that is, it is valid for any SOW system and describes the current physicochemical situation in terms of energies that could be added or subtracted. As a consequence, it is easy to modify the formulation in practice and to know how to do it.

This was quite important for emulsion applications because many studies indicated that the formulation seen as a single (although multivariable) HLD value was linked with the emulsion properties, that is, there is some phenomenology that relates the phase behavior and the emulsion properties.

18.3 GENERAL PHENOMENOLOGICAL GUIDELINES CONCERNING EMULSION PROPERTIES

18.3.1 EMULSION TYPE OR MORPHOLOGY

Some general trends have been known for a long time, starting with Bancroft's rule, and then the value of the surfactant HLB. However, even such a simple rule exhibits some exceptions, essentially for two reasons. The first one occurs when the SOW system is out of equilibrium when it is stirred. A few studies have reported evidence of exception when all the surfactant was in the wrong phase [60–63]. Others dealt with more complex situations in which some partial equilibration had taken place [64,65]. The deviation from Bancroft's rule seems to be linked with the actual formulation departure from HLD = 0. However, it may be said that it is advisable to avoid nonequilibrium systems in practice if an exact prediction is wanted.

The second reason, that is, the use of systems containing a water/oil ratio very different from unity, cannot be avoided in practice. The fact is that when one of the phases is in large excess, it becomes the external phase upon stirring, no matter what the formulation. Because this was an actual case that could not be avoided in practice, it was extensively studied in the 1980s in relation to the inversion frontier in a bidimensional map in which the first variable was the generalized formulation as HLD (including all formulation variables) and the second one, the water/oil composition as WOR [66,67].

A general phenomenology emerged and allowed us to find not only trends but also very general guidelines for all systems. Figure 18.5 shows such a map in which the bold limit, indicated as an inversion line, is the frontier between the two simplest cases of oil external (W/O) and water external (O/W) emulsions.

In the central part of the plot, when the oil and water proportions are not too different, the inversion frontier (between A+ and A− zones) is horizontal at HLD = 0, no matter what the specific values of all the formulation variables. This is a considerable simplification in practice. Additionally,

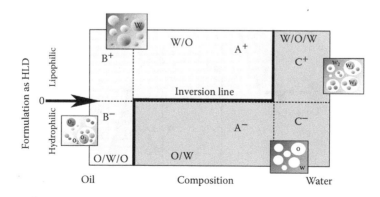

FIGURE 18.5 Emulsion type phenomenology on the two-dimensional formulation–composition map. A is the central zone with similar amounts of oil and water, whereas B (respectively, C) is the region containing an excess of oil (respectively, water). The associated sign corresponds to HLD.

the accuracy of the HLD results in an extremely general and precise phenomenology. If HLD > 0 (respectively, HLD < 0) in this central zone, the emulsion type is W/O (respectively, O/W), which is essentially equivalent to Bancroft's rule.

When one of the phases is in an excessive amount, that is, on the right and left extremes in Figure 18.5, it becomes the external phase and the inversion line appears in the map as an almost vertical segment between a so-called normal emulsion ($A^+ B^+$ and $A^- C^-$), in which the type and interface curvature follow Bancroft's rule, and abnormal ones (C^+ or B^-), in which the external phase is the one with the larger amount. As a matter of fact, these abnormal zones are very likely to exhibit multiple or double emulsion morphologies in which the inner one (droplets in drops) is normal and the outer one (drops in continuous phase) is abnormal. These emulsions are beyond the scope of this report but have been reviewed in the literature in relation to their formation and applications [68,69].

The morphology of these abnormal emulsions is not easy to determine by current methods such as conductimetry [70,71]. The actual location of the vertical segments has been found to depend on many variables among which the following have been reported: surfactant nature, surfactant concentration, oil and water viscosity, stirring energy, etc., which shift the vertical branches one way or the other [72]. An available summary of the current consistency in the know-how [52,73] and its straightforward utilization in practical applications has placed this phenomenology as an island in the center part of Figure 18.1. This know-how could be used to solve problems once the inversion frontier is experimentally determined.

The practical importance of this know-how motivated studies to use it in nonequilibrated systems. The most studied cases were the ones in which the formulation or composition was changed as it is done using a variation of temperature or dilution with water. In such cases, the emulsion inversion takes place when the frontier is crossed. In such cases, the inversion is not delayed, for instance, when the central horizontal segment in Figure 18.6 is crossed. This happens with nonionic systems when temperature is changed to alter the formulation [74]. In other cases, particularly when a vertical branch is crossed, the process exhibits a delay in both directions, thus producing a hysteresis zone (gray area in Figure 18.6). Almost a hundred articles have been published on this subject [75], which is complex but well documented, and has been interpreted by catastrophe theory with the corroboration that phase behavior and emulsion type are linked through the same model [76].

The determination of an inversion map including nonequilibrium processes takes more time than the standard inversion map. However, it is often worth the investment because it can bring new ways to solve problems [77], like the spontaneous formation of nanoemulsion by dilution, without spending considerable energy [78,79].

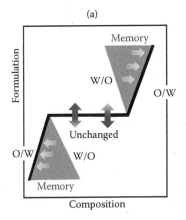
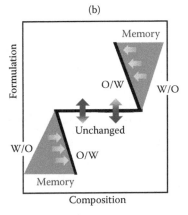

FIGURE 18.6 Dynamic inversion frontiers depending on the direction of change (arrows). In the case of the change in composition (horizontal arrows), inversion from a normal emulsion type to an abnormal one (a) and conversely (b). The hysteresis zones are indicated in gray.

18.3.2 EMULSION STABILITY/PERSISTENCE

Biphasic emulsions are nonequilibrium systems and thus are not stable, and will end up sooner or later as separated phases. However, the time scale for the coalescence to take place may be very long with respect to the time of use, and the persistence may be measured according to some kinetic stability criterion, such as flocculation, creaming, or partial phase separation.

Emulsion rupture depends on many different mechanisms that are involved in three steps. The first step involves the approach of two drops at a long distance, that is, to an interdrop distance of a micrometer or a fraction of it. This approaching step is driven by gravity or Brownian motion, and slowed down by Stokes' law or other viscosity and size effects. The result of this step is generally an increased gathering of drops, thus resulting in a higher internal phase ratio (HIPR) in a more concentrated emulsion, sometimes called a cream. If the process is not associated with drop coalescence, the emulsion stability is not affected and a slow stirring will disperse it again. This step is typically slowed down by a high external phase viscosity and a small drop size, and ends up as segregated drops in the direction of the gravity field. Unless a Bingham plastic type rheology with a yield stress is exhibited in the external phase, this step will not stop the coalescence process but will just slow it down, although sometimes significantly.

The second step deals with the drainage of the interdrop film and also involves the different mechanisms resulting from the approach of the two interfaces in which surfactant or polymer molecules are adsorbed. Solid particles located close to the interface, and more or less solvated by one of the fluids, may produce similar effects in a so-called Pickering emulsion; however, these topics are beyond the scope of this article.

When the two interfaces approach at a short distance, the van der Waals attraction forces between drops are opposed by different kinds of repulsions illustrated in Figure 18.7, such as (1) electrostatic potential due to the presence of an electrical double layer close to the interface; this happens in particular with simple or polymeric ionic surfactants; (2) steric repulsion due to the interactions of bulky portions of adsorbed molecules; other factors opposing drainage such as (3) a viscous effect due to the interdrop fluid motion, including a reduction of the flow as the thickness decreases; (4) an interfacial tension gradient that results in the well-known Gibbs–Marangoni film elasticity in foams; (5) a steaming potential resulting from the separation of adsorbed ions and displaced ones; and (6) interfacial rheology phenomena and the formation of gels close to the interface [80]. These last two are illustrated in Figure 18.7. Some other effects favor drainage by capillary effects due to drop flattening and depletion flow.

The competition between electrostatic repulsion and van de Waals attraction was described by the so-called DLVO theory for colloidal particles and its extension to emulsions. A similar reasoning may be carried out with other repulsion forces with essentially the same kind of result; however, only a qualitative phenomenology is available [8] and specific studies have to be carried out in particular cases.

This second step is most often the crucial one because the last stage, that is, coalescence when the film has ruptured, is essentially instantaneous. It is worth noting that the only thing that is common to all possibly participating phenomena in this second step is the presence of some adsorbed

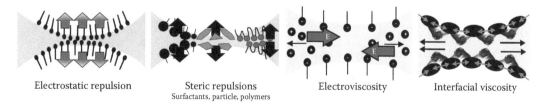

Electrostatic repulsion   Steric repulsions   Electroviscosity   Interfacial viscosity
Surfactants, particle, polymers

FIGURE 18.7 Different static or dynamic mechanisms stabilizing emulsions during the interdrop film thinning.

material at the interface. As a consequence, it may be anticipated that this presence or absence is what is likely to result in some general effect. This is what happens when using the generalized formulation as the variable.

As far as we know, the first indication linking the formulation with emulsion coalescence rate over a complete range was published in 1972 [81]. The experiments were carried out with a mixture of two surfactants, each previously placed in the phase where it would be at equilibrium. Hence, the interfacial mixture was probably at equilibrium very quickly, thus avoiding any deviation from out-of-equilibrium circumstances. The full formulation range from very hydrophilic to very lipophilic surfactant was covered thanks to a large mixture variation. Figure 18.8 (left plot) shows the relative emulsion stability for some experiments with pre-equilibrated systems. The observed variation exhibits (1) the presence of a maximum persistence of W/O (respectively, O/W) emulsion for a slightly lipophilic (respectively, hydrophilic) surfactant mixture, (2) with a high instability at the intermediate formulation that results in emulsion inversion, and (3) an increase in coalescence for systems with very hydrophilic and very lipophilic HLB surfactants. Similar relative stability variations from early studies [82–85] exhibit the same trends. This early evidence also showed that the HLB scale was not very accurate because the variation, although a consistent trend with all systems, was not corroborated by actual HLB value with a good precision as clearly apparent in Figure 18.8 (right plot).

It seems that the pioneering articles on rate coalescence by Boyd et al. [81] was overlooked at that time, and the phenomenology was rediscovered 10 years later, pointing out that emulsions were extremely unstable exactly at optimum formulation. Of course, it may be noticed that in the 1972 article, there was no data at optimum formulation (as in Figure 18.8, left plot). The use of a generalized formulation scale HLD clearly shows the systematic phenomenology of an extremely unstable emulsion exactly at HLD = 0. Several explanations were given with a variable degree of sophistication, from the wedge theory to advanced ones [19,86,87], which are worth reading. However, the simplest, and maybe more understandable, explanation involves a very simple experiment. The separation times of emulsions made by stirring WIII three-phase systems with different amounts of middle phase, thus different amounts of surfactant were compared [88], and it was shown that the separation times were essentially the same whatever the amount of surfactant (provided it was above the cmc or cμc).

This evidence plainly indicates that in WIII optimum systems, the surfactant does not adsorb at the oil–water interface which is formed by the stirring process and consequently does not produce any repulsion effect that could delay coalescence. In other words, it may be said that the surfactant is trapped in the microemulsion's third phase and is not available to adsorb at the oil–water interface newly created by the stirring procedure. This is likely to be a simplistic, although clever, reasoning

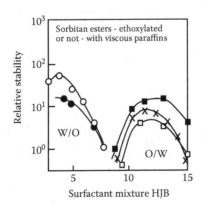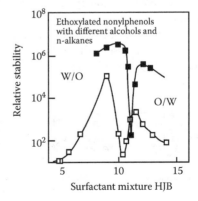

FIGURE 18.8 Aspects of the variation of emulsion stability with the HLB of the surfactant mixture. Data replotted from early reports [81–83].

that if the surfactant does not adsorb at the interface, none of the repulsion mechanisms (which all depend on the surfactant's presence) is going to work and a surfactant-free situation would be mimicked. This is why the observed phenomenology versus generalized formulation is very general [73–89]. The emulsions are extremely unstable at optimum formulation (HLD = 0), whatever way it is attained, and the very short persistence is essentially the same as in the absence of surfactant, that is, it is due to fluid separation by gravity. As shown in Figure 18.9a, on each side of the optimum formulation, the stability increases rapidly as HLD departs from zero, then stays at a high value over some formulation range, typically from 2 to 6 (in absolute values); then it decreases when the surfactant becomes very hydrophilic or very lipophilic and tends to fractionate in one of the phases rather than to adsorb at the interface.

This general trend applies in the central A^+/A^- region of the map in Figure 18.5, with an extension of the properties of A^+ to B^+ and A^- to C^-. In the conflicting zones of the map (C^+ and B^-), the results from Figure 18.9 apply only to the inner emulsion (droplets in drops) of a multiple emulsion, whereas the external emulsion (drops in continuous phase) is extremely unstable. The actual values and range over which the phenomenology indicated in Figure 18.9a takes place are typically $-0.2 < HLD < +0.2$ for the very low stability central zone, about $2 < |HLD| < 4$ for the high stability ranges and then $|HLD| > 6$ for a decreasing stability.

However, the actual values for the ranges and stability maxima depend on the numerous factors producing so-called second-order effects that are able to alter the different driving forces and kinetics in various mechanisms taking place in the different steps. Figure 18.9b and c indicates the possible phenomenology variations (arrows) attainable by changing physical factors like the drop size or the external phase rheology, or physicochemical features such as the surfactant structure, its concentration, its interfacial adsorption density, the interactions between surfactant and polymers, or the increase in interfacial rheology, etc.

There are dozens of possibilities found in hundreds of specific publications (Figure 18.1, right), and although some weak trends have been perceived [3,7], most are based on common sense such as the outcome of an increase in repulsion by using a bulkier surfactant steric group or its electrostatic charge. However, it may be said that there is no a priori systematic solution for a given case, and that several are always possible and will have to be tried by experimentation in practical cases. Some phenomenological reviews are available to understand the extreme complexity of the issue and its many possibilities, [13,14,90], but there is no straightforward guideline for the inexpert formulator,

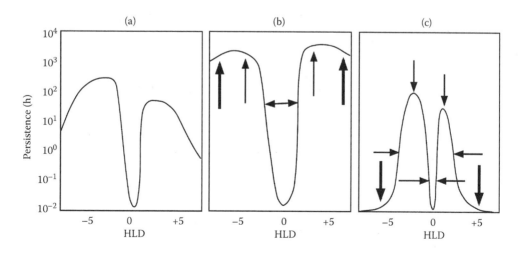

FIGURE 18.9 Variation of emulsion stability versus generalized formulation HLD. General phenomenology (a); actual variations in the exact location and magnitude of the high-stability zones are due to second order effects (b and c).

who is probably going to select an alternative found in the literature by chance, and might not apply it properly.

This, of course, is not the best way to reach a solution for an application. The point is that there are semiquantitative phenomenological guidelines for the general trends, but there is no island of know-how (Figure 18.1, center) for the second-order effects. The necessary and worthwhile information is available only in the specifics (Figure 18.1, right), provided that a strictly adaptable case has been studied.

In practice, the typical approach is to select by expert analysis a preferred mechanism that would dominate the other factors and surely provide what is required. This is probably the quickest way to solve a problem, although probably not the most elegant or clever. With top-notch formulation experts, a more astute procedure would be found by putting together various mechanisms, up to the point where some synergy appears and results in a much better performance. This is the way new discoveries emerge.

18.3.3 Emulsion Drop Size

The drop size average and distribution is an important characteristic because it has a great influence on the emulsion properties. Some basic trends have been well known for a long time, like the effect of the interfacial tension and of the stirring intensity and duration, which directly influence the rupture of drops [91–93]. Less obvious effects are related to simultaneous coalescence during the emulsification process, which equilibrates the rupture to produce the available drop size [94–97].

The coalescence probability depends on second-order factors that are the same as those influencing emulsion stability. It has been seen that they are numerous and competitive, and that their intricate combination is not yet mastered. Hence, it can be said that the related influence of the surfactant is not yet completely understood because it is extremely complex, and even more elusive because it deals with the first instant of the emulsification.

After some pioneering work [63] with a partial formulation scan with HLB variations, a few studies [98–99] have recently shown that variation of the generalized formulation around HLD = 0 has a systematic phenomenology. As the formulation approaches optimum from both sides, the interfacial tension decreases, sometimes considerably, hence the drop rupture is easier and there is a tendency for a drop size decrease. This happens at first when the absolute departure from the optimum formulation is relatively large (e.g., three units), at least large enough for the coalescence rate to be moderate (see later sections). When the formulation is very close to HLD = 0 (e.g., −0.5 < HLD < 0.5), the coalescence is very quick and the emulsion drops coalesce upon contact. Consequently, the available average drop size tends to increase considerably unless a second-order effect blocks the coalescence such as the formation of liquid crystals [99].

This complex combination of two phenomena produces a drop size variation with two minima located on both sides at some distance (typically one to three HLD units, two in Figure 18.10, map) from optimum formulation (Figure 18.10b). This very general phenomenology is quite important in practice because a very small variation in the generalized formulation can produce a large drop size variation (Figure 18.10b). The attainment of very small droplets at a certain formulation distance from HLD = 0 is of crucial importance in application for cost and other issues, like avoiding overheating in producing nanoemulsions with foodstuff or pharmaceuticals [100–102].

The actual formulation distance at which the minimum has been found to occur depends on second-order effects such as surfactant nature, interfacial tension value, and stirring energy. Although no systematic information is available to make any forecast, it is easy to determine the proper formulation with a few experiments.

Figure 18.10 drop size map indicates that the oil/water composition variation also produces a very important effect with a considerable decrease in attained drop size when the representative point of the emulsion on the map approaches the vertical (B$^-$/A$^-$) branch of the inversion line on the

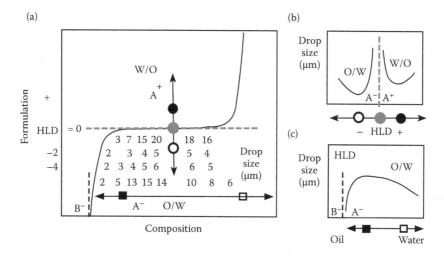

FIGURE 18.10 Drop size variation in the formulation–composition map adapted from data by Pérez et al. [98]. The numbers in (a) indicate the average drop size in micrometers. (b and c) Average drop size variation along the arrows indicated in (a).

A^- side. In this zone, the emulsion has a HIPR and exhibits a considerable viscosity. This occurrence is also a very general phenomenology, which was proposed a long time ago [103,104] and explained as a particularly efficient use of the stirring energy in a viscoplastic medium [98]. Even if the drop size values actually attained cannot be predicted everywhere in the map, the generality of these effects warranties a very useful guideline for application purposes, with an easy way to experimentally optimize the conditions.

18.3.4 Emulsion Viscosity and Rheology

When the drops of a stable emulsion are in contact, that is, when the internal phase constitutes more than 60% to 70% for 5 to 10 μm drop size emulsions, the rheology becomes non-Newtonian, in general, pseudoplastic or viscoelastic. This happens at an even lower internal phase content for nanoemulsions. In many emulsion applications such as cosmetics, foods, or paints, special rheological properties have been of primary importance in the past 50 years. This is why many studies have developed an extensive know-how of this subject, which may be considered as an independent phenomenology island in Figure 18.1. The emulsion viscosity has been shown to essentially depend on the external phase viscosity, the internal phase ratio, and the drop size average and distribution. Some chapters or journal reviews dealing with these principal effects (with the exception of physicochemical formulation) are available to understand most of the generalities at a very good pedagogical level [9,105–109]. Many publications are available on emulsion rheology, most of the time, with a sound physical basis and general applications [110–120].

Formulation and composition have a contribution to emulsion rheology, although not as dominant as in relation to other properties such as emulsion type or stability. The main general trend is a low emulsion viscosity at optimum formulation [121], probably due to the ultralow tension that allows an easy elongation of the drops. This is quite a favorable occurrence for enhanced oil recovery applications because the displacement of any emulsion produced in a porous medium is made easier. However, optimum formulation is also associated with instability, hence this low-viscosity occurrence is not useful in practice in most other applications where some emulsion persistence is desirable.

When a low viscosity is required, even in the presence of HIPR, the usual solution is to generate a bimodal distribution, generally a mixture of two emulsions with quite different drop sizes [122,123].

This feature has been used in large-scale applications like the emulsified transport of heavy crude oil over long distances [124].

When a solid-like behavior is required, an increase in the internal phase content of more than 80% to 90% has been a commonly used method, which only requires an effective stabilization against coalescence, often with polymeric amphiphiles that provide some kind of encapsulation such as in the case of proteins. However, the attainment of such systems often requires special procedures because the mechanical agitation of an almost solid substance is quite inefficient to produce small size droplets, and no conventional high-energy stirring can be used. The foam-like behavior found in the extreme case of very concentrated emulsions have to be produced by special techniques such as phase inversion [125,126].

The gelification of the external emulsion phase is another way to reach extreme viscoelastic properties. Increases in phase viscosity have previously been achieved by introducing polymeric molecules that are able to generate interactions with the solvent and produce agglomerates. The necessary presence of surfactants in emulsions has brought new possibilities, as discussed in the next section.

18.3.5 Formation of Structural Assemblies in the Emulsion External Phase

Emulsion rheology can be significantly altered by a special behavior of the external phase not only by using polymers as viscosifiers but also by generating surfactant arrangements, sometimes well beyond the classic spherical micelles. The shape of the micelles is quite important, not only for its influence on the rheology of phases but also because of its influence on the water–oil cosolubilization and the resulting alteration of the boundaries in the phase behavior map. Israelachvili et al. [127] have developed a geometric concept, the so-called packing parameter, which is based on the shape of the surfactant, and roughly determines the type of aggregate likely to occur.

This is, for instance, the case with the generation of worm micelles, which can produce three-dimensional networks with amazing rheological properties when they merge in different ways (linear, branched, or cross-linked) as seen in Figure 18.11 [128–134].

Other structures with surfactant bilayers like vesicles [135–136], liposomes, niosomes, etc., have resulted in rheological properties that have been used to generate a transport vehicle for personal care products or for drug delivery in pharmaceuticals. By the way, vesicles, particularly giant vesicles [137], are just like water-in-water emulsions, with possibly different aqueous phases inside and outside the vesicles.

If an oil phase is added, the surfactant assemblies can become extremely complex in a variety of mesophases like microemulsions and liquid crystals, both of which are misleading terms. Microemulsions are not emulsions, that is, they are single-phase systems in which the oil and water are made compatible due

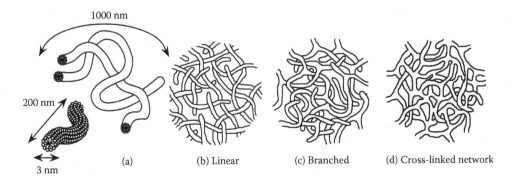

FIGURE 18.11 (a) Worm micelle structure and self-association mechanisms by (b) entanglement, (c) partial, or (d) complete fusion and the resulting rheology characteristics.

to a more or less random bicontinuous arrangement with surfactants [138]. Although they are beyond the scope of this article, it should be said that the occurrence of bicontinuous microemulsions takes place close to the optimum formulation, and is thus related to emulsion properties. As a matter of fact, microemulsions are the oil/water preferred arrangement at HLD = 0, and are actually permanently blended systems when there is no way to attain a persistent drop dispersion with zero curvature interface [139,140]. Microemulsion studies are the easier way to quantify formulation issues such as extending the simple HLD relations in Equations 18.4 and 18.5 to real problems for emulsion cases [141]. For instance, this is the case in the determination of the oil's characteristics, that is, the equivalent alkane carbon number (EACN) when the oil is not an alkane [142–143]. Up-to-date reviews on microemulsions and their applications are available in the recent literature [144–146].

Liquid crystals are neither liquids nor crystals, but semiorganized arrangements [147]. They are also found in macroemulsion systems as an additional phase, sometimes in equilibrium with a microemulsion [148–149]. It is worth noting that a layer of liquid crystal at the interface can considerably increase the emulsion's persistence [150].

By the way, this occurrence could also drastically change the assumed general properties and produce exceptions. It is known, for instance, that the presence of liquid crystals can inhibit the occurrence of a strong emulsion instability at HLD = 0. This results in a relatively persistent nanoemulsion formation exactly at optimum formulation, which is a key feature in the formation of nanoemulsions by dilution [78,151,152].

The two-drop size minima occurring on both sides at some distance from the optimum formulation (Figure 18.10b) disappears in the presence of liquid crystals. As an exception to the general trend, a single and even smaller drop size is found at optimum because the nanoemulsion produced is encapsulated by the liquid crystal layers and the drops do not coalesce instantly [99].

Different surfactant assemblies and liquid crystal types may be formed depending on the surfactant's nature and the SOW composition; for instance, the ones illustrated in the hypothetical diagram in Figure 18.12. Liquid crystals may also be formed with polymeric amphiphiles like polyethyleneoxide–polypropyleneoxide–polyethyleneoxide triblocks [153].

In very simple systems, as those described by Winsor in the 1950s, there is an easily perceived relationship between phase behavior and formulation. This is no longer the general case, and the complexity beyond the oil–water–microemulsion–liquid crystal phase behavior is often too intricate

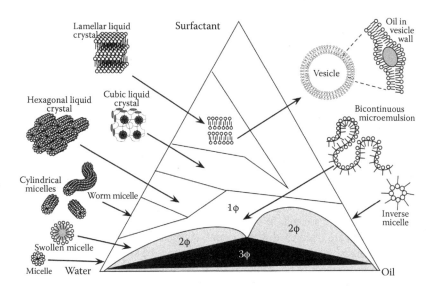

FIGURE 18.12 Hypothetical diagram showing the typical structures found in such map for a system close to optimum formulation.

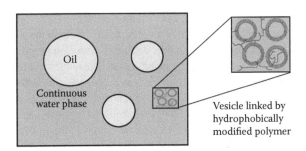

FIGURE 18.13 Method to attain a high viscosity with an O/W emulsion containing a low content of oil and a very low amount of amphiphile substances. The water external phase is made extremely viscous by vesicles connected by hydrophobically modified polymer links.

to handle and even more difficult to predict. Sometimes, the proper desired behavior just happens in extremely strict circumstances. The famous neat–nigre–brine slim triangle industrially used to recover neat soap is an example of a very elusive case of strong practical interest, which was probably discovered by chance [154]. Another example with low forecasting probability is spontaneous emulsification by dilution [78], a quite complex procedure depending on a fine-tuned phase diagram structure that could occur only in a narrow temperature range of a few degrees [155].

A quarter century of studies following the pioneering work of Goddard [156–158] has produced a phenomenological island on complex synergies, both for rheological and stability effects, through the interactions between surfactants and polymers.

This involves many formulation issues as seen in available reviews [159], and it is a research area still in development, with published cases being more complex and often more specific [160–163].

If some intervesicle links are established, as with a very small amount of hydrophobically modified polymer molecules (see Figure 18.13), an almost solid water gel can be produced. The nice situation is that the water gelification may be generated after the emulsion has been formed in the usual way [164]. This procedure may produce a gelified emulsion with a very small internal phase ratio, a useful trick for light mayonnaise as an alternative to a multiple emulsion.

Of course, the presence of structures, particularly in the external phase, can modify not only the emulsion rheology but also its stability, and this is why the two properties are sometimes linked [165]. With some clever use of the current know-how, both requirements can be matched.

18.4 CURRENT SITUATION

In recent years, information organization has evolved considerably (Figure 18.1). The very approximate trends available 25 years ago, quite separated from application and scientific research, have evolved into phenomenological guidelines. This was the result of a large effort of systematization of the formulation concepts and tools, and the finding of general relationships between emulsion properties, formulations, and other factors, which started with enhanced oil recovery being the driving force that demanded better accuracy.

The actual general guidelines clearly teach the existence of emulsion properties in semiquantitatively described circumstances. This allows the emulsion formulator to make some logical reasonings and to generate some predictions. As a result, researchers can considerably avoid the inefficient random trial and error methods, which are often used when a weak understanding of the properties being studied is cumulated with an excessive number of variables to test. It may be said that these guidelines have penetrated the know-how archipelago of phenomenological islands, as indicated by the wedge in Figure 18.1, to produce a better generality in many islands and a better overlapping between them [73]. At the same time, new phenomenological islands have been generated by specialists from the academic and industrial worlds who have concentrated their efforts in organizing the science and technological results from case by case studies into consistent phenomenologies with some degree of generality. This has resulted into an available know-how, so-called formulation engineering, which has been divided into unit operations [166], to efficiently solve problems and develop new applications [53,167–168].

Unfortunately this know-how is not yet appropriately acquired and assimilated, particularly by people involved in industrial research and development activities. An extremely simple way to confirm this diagnostic is to notice that a large majority of people formulating with surfactants in many different applications are still using the 60-year-old HLB as the only classification for their products, when a more performing approach has been readily available for at least the past two decades, according to what has been said by famous lecturers in congresses [169–170], and has boosted the research and development performance of people who use it.

There are several reasons for the lack of systematic utilization of the current available know-how. The first one is that surfactant science and technology, particularly on emulsions, has not significantly infiltrated the undergraduate programs in chemistry and related careers. Only a very few courses are taught at the graduate level in formulation or in fashionable topics like nanoemulsions. This situation seems quite inconsistent with the use of emulsions in numerous applications, but it is probably due to the impossibility to introduce new topics. Academic and professional associations have been created in the past decade in Europe and America, and are trying to develop a general area about the concept of formulation and its applications. This might one day generate a new specialty or career in formulation with the proper course load, but it is not the case today. This means that any prospective researcher in this area has to start from scratch after his or her university degree, and such an initiation is not generally in emulsion science, but in the application area, that is, it is generally not a way to progressively learn from general concepts to specific cases.

As was discussed previously, there are a huge number of phenomena and many different applications as far as emulsions are concerned, hence a huge amount of information is available most of the time in very specialized and quite disconnected areas. This dispersion and segregation are drawbacks with unfortunate consequences. People working in cosmetics and personal care products are obviously not going to look for useful information in food, paint or petroleum engineering books or journals. This is regrettable, because according to what is known, the actual solution for a drilling fluid problem might have been used for years in food conditioning or vice versa. A typical example is the emulsification of thousands of barrels a day of heavy crude oil by using the HIPR method developed for pounds of cosmetics 10 years before. This means that a large part of the divulged practical information is not likely to be available to people working in another branch as far as applications are concerned. It is also worth noting that "magic" solutions, which are sometimes called high-performance results in a very complex multidimensional domain, are not published in a pedagogic way. They are rather patented with a completely different purpose, and sometimes hidden from a laboratory to the next one in the same building.

University research also suffers from this scattering of information between many domains. Academic research activities have been increasing in some exponential way, but the topics in emulsion science (as in many other areas) are becoming more and more specialized for several reasons. The first one is that it is easier to produce an original work on a very specific case that nobody else has considered previously. The second one is that the wider the topic is, the more published the information is and thus the longer and more difficult the screening, learning, and assimilation of the basics. On the other hand, trying to generate a very general knowledge, in particular if it is related to a practical case, is always more complex and would require a much wider expertise than the one acquired by a graduate student or a postdoctoral student. If 20 years of experience is required to work in a wider and more general domain, then fewer people become available.

Another reason is that the financing resources of a research activity are generally limited in time and money. People that are involved in research and development are faced with two conflicting activities: spending time in acquiring information and assimilating knowledge, or spending time on using their knowledge to realize their work and achieve experience. If they only spend time to learn, they will never produce effective research, and if they realize research without proper understanding of what they are doing, they will exhibit poor performance. This situation exists both in academic and in industrial research centers, and even if the priorities and actual splitting of the time are quite different, in both cases, the problem is that there is definitely too much information nowadays.

A simple analysis indicates that this is mainly because more studies are published, both because of the increasing number of researchers involved, and because of the driving force to publish more as a proof of performance, as in the "publish or perish" policy or the "impact factor" criterion. This is also because computerized search methods, which have been helping for some time to retrieve useful information, are becoming less efficient. Any general computerized screening of the literature with one or two keywords would collect thousand of publications, but many of them are useless because they were picked through a misleading criterion. On the contrary, a very specified screening with several cumulative keywords will leave out fundamental publications. The overwhelming presence of search by the Internet is not solving the problem either, but rather worsening it. For instance, a search trial with the word "emulsion" gives 53 million results, "stable emulsion" detects only 7 million results, and "stable emulsion formulation" goes down to 2 million results. A quite specific search like "stable emulsion formulation for face cosmetics" returns 20,000 results, a number that cannot be consulted in a decent amount of time anyway. Hence, the problem is which pertinent criterion selects the 20 first results likely to be read.

Some people have been saying that in a few years it would be easier to reinvent something than to find out the original information in the scientific and technical literature. This may happen faster than expected because of the rate of the changes. Thirty years ago, a new academic researcher could be up-to-date with the emulsion literature and pretend to be an expert in 5 to 10 years. Today, even a specialist with already 30 years of experience has trouble finding the time to keep up with new things and to stay up-to-date.

This huge amount of information is generated in the specifics, that is, in the spots located in the right part in Figure 18.1, where the individual advances are produced in some quantitative but limited way. To be useful in practice, this information must be transferred to the phenomenological islands archipelago (black arrow with question marks in Figure 18.1); this is the main challenge in the future and will be discussed next. Such transfer implies an increase in generality without losing understanding. This implies gathering cases, and simplifying interpretations and conclusions to the point where they can be used in real cases. This implies the intervention of experience and expertise to organize the knowledge and transmute it into know-how organized in an island with predictive capability, even with some loss of specificity. Up to now, the creation and development of new islands has been done by a few experts or groups of experts either working together in parallel or in succession in the same group. However, the current rising number of domains and increasing complexity indicates the need for a radical reorganization. This will be the subject of the last section.

18.5   PROSPECTS FOR THE NEAR FUTURE

To master in one or two decades an adequate know-how in emulsion science and technology, two aspects have to be dealt with. The first one is the deep study of some topics, which have been lagging, and the generation of some new understanding of complex effects resulting from many variables. The second has to do with a better collaboration among the people working in emulsion science and technology. This includes several improvement procedures tending to generate better information transfer (printed material, courses, computerized services to share), and an extensive review work from experts to connect the know-how islands and determine the missing links to be studied and to be incorporated in a computerized system.

18.5.1   Prospective Topics

Some topics, which are relatively well organized in an island, might require more connections between the different aspects, in particular, the multidimensional effects when many variables are changed at the same time. This involves first the equilibrium data needed to be able to handle emulsions, that is, the phase behavior diagrams versus formulation with novel surfactants like gemini, extended, short polymeric, dendritic, multihead, and multitail surfactants, especially the "green" ones.

The influence of the combination of surfactant and the nature of the oil on phase behavior is not well known and is still puzzling. Also, the only cosurfactants studied have essentially been the alcohols. Other usual substances like natural or synthetic esters, sugar, or terpenic derivative components as well as hydrophilic/lipophilic linkers should be tested. The effect of the presence of "green" solvents of the ether or polyol type is not known. The use of double tails, such as the Gerbet hydrophobe, often results in a better performance in solubilization, and this should be related to the phase behavior.

The effect of surfactant mixtures has been focused up to now on the critical micelle concentration, to estimate the deviation from ideality. Data on the phase behavior variations produced by mixing species is required, particularly when the self-association results in worm micelles or other fancy structures.

Formulations with surfactants and polymers produce many different structures that are able to modify emulsion stability and rheology through various mechanisms. The current knowledge is far from exhaustive, and cases involving more than two species have not been studied in detail, although there is a considerable novelty prospect.

Some mechanisms of stabilization are not really well understood, like the occurrence of high interfacial viscosity with an adsorbed gel. The development of an interfacial rheometer that is able to make measurements close to the optimum formulation is necessary to understand the breaking mechanism in processes like water treatment or crude oil dehydration, which are currently solved in a quite empirical fashion.

The many different mechanisms involved in emulsion inversion are not well understood, in particular, if a multiple emulsion is produced as an intermediate. Very slight variations in the formulation or in the procedure have been found to result in drastic changes in practice, but essentially with no satisfactory understanding. These phenomena are of great interest to develop optimized techniques, and consequently have to be studied exhaustively. This is critical with very viscous phases because low-energy emulsification is probably going to be the preferred method and may be an unique method in the future.

The programming of formulation, composition, and other variables during the emulsification process should be studied using systems exhibiting some degree of nonequilibrium, with features like hysteresis or apparent equilibration delay. The current evidence indicates that some magic results can be attained the day the emulsion properties can be altered on demand, with some mastering of nonequilibrium procedures such as spontaneously triggered instabilities. Spontaneous emulsification from a microemulsion is known in principle, and has been used in some applications, but many studies are still required to be able to use the phenomenon to optimize performance in applications like detergency, oil recovery, foodstuff, cosmetics, etc.

Nanoemulsions are a vehicle or an intermediate in many applications, ranging from the synthesis of materials, the encapsulation of toxic materials, or the controlled release of drugs. Attaining some specific drop size distribution is still not a straightforward issue. Some low-energy effects based on formulation may be favorably associated with efficient shear phenomena in outstanding performance membrane or microchannel devices.

Emulsions can be produced by techniques that are quite efficient compared with high-energy stirring. This is the case with ultrasonic techniques, which are not well mastered as a function of formulation. It is known that the stirring effectiveness of standard devices such as a static mixer or an impeller in tanks is considerably altered by a slight change in formulation. However, no systematic study has been done to optimize the combination of two completely different aspects generally handled by different people.

Special cases like multiple emulsions and Pickering emulsions still require many additional studies to reach the level of the conventional simple emulsions stabilized by surfactants and polymers, particularly concerning the way they are produced.

The current drop size analysis methods involve drop-by-drop measurements with light-scattering techniques, which require a separation of the drops, that is, emulsion dilution. Nondestructive global techniques like backscattering, acoustic propagation, or nuclear magnetic resonance methods have been proposed but they are not yet satisfactory, and should be improved.

18.5.2 PROSPECTIVE ORGANIZATION AND COMPUTERIZED HELP TO THE FORMULATOR

An effective research and development in emulsion science and technology requires a very efficient use of the time of the formulator. This implies organizing the knowledge and the know-how so that it is better divulged and assimilated by the users.

The huge amount of information has to be presented in two ways. First, in a progression of quantity and quality, as well as difficulty, for people that could study it in a systematic way for a long time. Second, and mainly in islands, an information transfer should be provided for people who do not have much time to study, but want to acquire a well-organized phenomenology in a very limited area of self-consistent knowledge and know-how.

The current selected divulgence through publications such as current contents, advance assessments, and review chapters or books is useful, but it has to be enhanced and complemented by targeted short training courses given by experts who are able to transfer knowledge, know-how, and experience. This will probably require fair play cooperation between academe and industry worldwide, with a self-prohibition of hiding important features. This would be difficult to bear, and contrary to the usual business research strategy. However, it is worth remembering that the impressive progress in enhanced oil recovery and the birth of high-tech formulations from 1975 to 1980 was only possible because of an open cooperation between many petroleum companies and academic institutes. The reason for such an uncommon behavior at that time was that it was understood that the problems were too complex and that none of the companies, not even the number one company at that time (Exxon), would be able to solve their problems alone. This is probably the case in emulsion science and technology today, and evidence of it is that the delay between an actual discovery and its application is tending to increase.

Only the group of the best 50 to 100 experts in the world are able to create education and training instruments that will organize the know-how, make the complex issues easy to assimilate, and create a computerized system that will help the user pick the useful items in a jungle of specifics, and to dedicate his or her remaining small amount of time to some experimental work. This computerized advisor system will have to be built and improved upon as soon as possible before everybody, even the best world experts, would be overwhelmed by the flooding of the specifics, and would fail to stay up-to-date. The existence of such a computerized service is necessary to allow the formulator to use only the feasible and worthwhile information to tackle the problems and to create new programming methods, that is, to generate a new solution in each new case. This sounds quite complex, but it seems to be an extrapolation of what current experts are doing to invent a new magic trick by scratching their head for a few days, after armies of regular researchers have spent months in intricate trial-and-error wanderings without approaching a solution.

Fortunately, the world's best experts have often reunited in a few congresses, some of them strictly dealing with emulsions, such as the "World Congress on Emulsion," which has been held for the past 20 years. Other more general scientific meetings are even older, such as "Surfactants in Solution," a top serial congress that has kept going thanks to the initiative of an exceptionally clever individual, Kash Mittal, to whom this present contribution is dedicated.

18.6 SUMMARY

Emulsion-making evolved over the past century from an art to almost a science. A milestone in this progression was a research effort carried out in the mid-1970s to understand the properties of SOW systems at equilibrium and the attainment of low interfacial tension between oil and water. The concept of physicochemical formulation was quantified, and was then available as a framework to study emulsified systems in a systematic way in the late 1990s. This progress is still expanding in many areas, from macroemulsions to miniemulsions or nanoemulsions, but the knowledge is still scattered and more organized work is required to improve the current know-how, particularly with more cooperation between academia and industry.

REFERENCES

1. M. Chappat. Some applications of emulsions. *Colloids Surf. A*, **91**, 57–77 (1994).
2. P. Becher. *Emulsions: Theory and Practice*. Van Nostrand Reinhold Co., New York (1957).
3. P. Becher (Ed.). *Encyclopedia of Emulsion Technology*, 4 vols. (vol. 1 [1983], vol. 2 [1985], vol. 3 [1988], and vol. 4 [1996]). Marcel Dekker, New York (1983–1996).
4. K.J. Lissant (Ed.). *Emulsions and Emulsion Technology*, Vol. 2. Marcel Dekker, New York (1974).
5. F. Nielloud and G. Marti-Mestres (Eds.). *Pharmaceutical Emulsions and Suspensions*. Marcel Dekker, New York (2000).
6. J. Sjöblom. *Emulsions and Emulsion Technology*. Marcel Dekker, New York, 1st ed. (1996) and 2nd ed. (2001).
7. Th.F. Tadros. *Applied Surfactants*. Wiley VCH-Verlag, Weinheim (2005).
8. Th.F. Tadros (Ed.). *Emulsion Science and Technology*. Wiley VCH-Verlag, Weinheim (2009).
9. T.G. Mason. New fundamental concepts in emulsion rheology. *Curr. Opin. Colloid Interface Sci.*, **4**, 231–238 (1999).
10. P. Taylor. Ostwald ripening in emulsions. *Colloids Surf. A*, **99**, 175–185 (1995).
11. J. Weiss, N. Herrmann and D.J. McClements. Ostwald ripening of hydrocarbon emulsion droplets in surfactant solutions. *Langmuir*, **15**, 6652–6657 (1999).
12. P.M. Claesson. Experimental evidence for repulsive and attractive forces not accounted by the conventional DLVO theory. *Prog. Colloid Polym. Sci.*, **74**, 48–54 (1987).
13. D. Melik and H.S. Fogler. Fundamentals of colloidal stability in quiescent media. In *Encyclopedia of Emulsion Technology*, Vol. 3, P. Becher (Ed.). Marcel Dekker, New York, pp. 3–78 (1988).
14. P. Walstra. Emulsion stability. In *Encyclopedia of Emulsion Technology*, Vol. 4, P. Becher (Ed.). Marcel Dekker, New York, pp. 1–62 (1996).
15. E. Dickinson. *Emulsions. Annual Reports C*. The Royal Society of Chemistry, London, pp. 31–58 (1986).
16. B. Binks. Particles as surfactants—Similarities and differences. *Curr. Opin. Colloid Interface Sci.*, **7**, 21–41 (2002).
17. S. Tcholakova, N.D. Denkov and A. Lips. Comparison of solid particles, globular proteins and surfactants as emulsifiers. *Phys. Chem. Chem. Phys.*, **10**, 1608–1627 (2008).
18. J.H. Hildebrand. Emulsion type. *J. Phys. Chem.*, **45**, 1303–1305 (1941).
19. A. Kabalnov and H. Wennerström. Macroemulsions stability: The oriented wedge theory revisited. *Langmuir*, **12**, 276–292 (1996).
20. W.D. Harkins, E.C.H. Davies and G.L. Clark. The orientation of molecules in the surfaces of liquids, the energy relations at surfaces, solubility, adsorption, emulsification, molecular association, and the effect of acids and bases on interfacial tension (Surface Energy VI). *J. Am. Chem. Soc.*, **39**, 541–596 (1917).
21. I. Langmuir. The constitution and fundamental properties of solids and liquids. *J. Am. Chem. Soc.*, **39**, 1848–1906 (1917).
22. W.D. Bancroft. The theory of emulsification—Part V. *J. Phys. Chem.*, **17**, 501–519 (1913).
23. W.D. Bancroft. The theory of emulsification—Part VI. *J. Phys. Chem.*, **19**, 275–309 (1915).
24. P. Finkle, H.D. Draper and J.H. Hildebrand. The theory of emulsification. *J. Am. Chem. Soc.*, **45**, 2780–2788 (1923).
25. W. Seifriz. Studies in emulsions. Parts I & II. *J. Phys. Chem.*, **29**, 587–600 (1925).
26. W.C. Griffin. Classification of surface active agents by HLB. *J. Soc. Cosmet. Chem.*, **1**, 311–326 (1949).
27. W.C. Griffin. Calculation of HLB values of nonionic surfactants. *J. Soc. Cosmet. Chem.*, **5**, 249–256 (1954).
28. J.T. Davies. A quantitative kinetic theory of emulsion type. I. Physical chemistry of emulsifying agent. In Gas/Liquid and Liquid/Liquid Interfaces. *Proc. 2nd Int. Congr. Surface Activity*, London, pp. 426–438 (1957).
29. ICI Americas. *The HLB System—A Time Saving Guide to Emulsifier Selection*. Chemunique, ICI Americas, Wilmington, DE (1976).
30. P. Winsor. *Solvent Properties of Amphiphilic Compounds*. Butterworths, London (1954).
31. K. Shinoda and H. Arai. The correlation between phase inversion temperature in emulsion and cloud point in solution of nonionic cmulsifier. *J. Phys. Chem.*, **68**, 3485–3490 (1964).

32. K. Shinoda and H. Saito. The effect of temperature on the phase equilibria and the type of dispersions of the ternary system composed of water, cyclohexane and nonionic surfactant. *J. Colloid Interface Sci.*, **26**, 70–74 (1968).

33. K. Shinoda. The comparison between PIT system and the HLB-value system to emulsifier selection. In *Proc. 5th Int. Congr. Surface Activity*, Vol. 2, Barcelona, Spain, pp. 275–283 (1969).

34. D.O. Shah and R.S. Schechter (Eds.). *Improved Oil Recovery by Surfactant and Polymer Flooding.* Academic Press, New York (1977).

35. M. Bourrel and R.S. Schechter. *Microemulsions and Related Systems.* Marcel Dekker, New York (1988).

36. M.A. Arandia, A.M. Forgiarini and J.L. Salager. Resolving an enhanced oil recovery challenge: Optimum formulation of a surfactant-oil-water system made insensitive to dilution. *J. Surfact. Deterg.*, **13**, 119–126 (2011).

37. R. Strey. Microemulsion microstructure and interfacial curvature. *Colloid Polym. Sci.*, **272**, 1005–1019 (1994).

38. J.L. Salager, E. Vasquez, J. Morgan, R.S. Schechter and W.H. Wade. Optimum formulation of surfactant-water-oil systems for minimum interfacial tension and phase behavior. *Soc. Pet. Eng. J.*, **19**, 107–115 (1979).

39. M. Bourrel, J.L. Salager, R.S. Schechter and W.H. Wade. A correlation for phase behavior of nonionic surfactants. *J. Colloid Interface Sci.*, **75**, 451–461 (1980).

40. W.H. Wade, J. Morgan, R.S. Schechter, J.K. Jacobson and J.L. Salager. Interfacial tension and phase behavior of surfactant systems. *Soc. Pet. Eng. J.*, **18**, 242–252 (1978).

41. W.H. Wade, E. Vasquez, J.L. Salager, M. El-Emary, Ch. Koukounis and R.S. Schechter. Interfacial tension and phase behavior of pure surfactant systems. In *Solution Chemistry of Surfactants*, Vol. 2, K.L. Mittal (Ed.). Plenum Press, New York, pp. 801–817 (1979).

42. P.H. Doe, W.H. Wade and R.S. Schechter. Alkyl benzene sulfonates for producing low interfacial tensions between hydrocarbons and water. *J. Colloid Interface Sci.*, **59**, 525–531 (1977).

43. V.K. Bansal and D.O. Shah. The effect of ethoxylated sulfonates on salt tolerance and optimum salinity of surfactant formulation for tertiary oil recovery. *Soc. Pet. Eng. J.*, **18**, 167–172 (1978).

44. J.L. Salager, M. Bourrel, R.S. Schechter and W.H. Wade. Mixing rules for optimum phase behavior formulations of surfactant-oil-water systems. *Soc. Pet. Eng. J.*, **19**, 271–278 (1979).

45. M.C. Puerto and R.L. Reed. A three-parameter representation of surfactant/oil/brine interaction. *Soc. Pet. Eng. J.*, **23**, 669–683 (1983).

46. P. Fotland and A. Skauge. Ultralow tension as a function of pressure. *J. Dispersion Sci. Technol.*, **7**, 563–579 (1986).

47. A. Skauge and P. Fotland. Effect of pressure and temperature on the phase behavior of microemulsions. *Soc. Pet. Eng. Reserv. Eng.*, **5**, 601–608 (1990).

48. J.L. Salager, R.E. Antón, J.M. Andérez and J.M. Aubry. Formulation des micro-émulsions par la méthode HLD. In *Techniques de l'Ingénieur*, Vol. Génie des Procédés J2, Chap. 157, Paris, France, pp. 1–20 (2001).

49. A. Graciaa, J. Lachaise, J.G. Sayous, P. Grenier, S. Yiv, R.S. Schechter and W.H. Wade. The partitioning of complex surfactant mixtures between oil-water-microemulsion phases at high surfactant concentration. *J. Colloid Interface Sci.*, **93**, 474–486 (1983).

50. A. Graciaa, J. Lachaise, M. Bourrel, I. Osborne-Lee, R.S. Schechter and W.H. Wade. Partitioning of nonionic and anionic surfactant mixtures between oil/microemulsion/water phases. *Soc. Pet. Eng. Reserv. Eng.*, **2**, 305–314 (1987).

51. A. Graciaa, J.M. Andérez, C. Bracho, J. Lachaise, J.L. Salager, L. Tolosa and F. Ysambertt. The selective partitioning of the oligomers of polyethoxylated surfactant mixtures between interface, and oil and water bulk phases. *Adv. Colloid Interface Sci.*, **123–126**, 67–73 (2006).

52. J.L. Salager, N. Márquez, A. Graciaa and J. Lachaise. Partitioning of ethoxylated octylphenol surfactants in microemulsion-oil-water systems. Influence of temperature and relation between partitioning coefficient and physicochemical formulation. *Langmuir*, **16**, 5534–5539 (2000).

53. J.L. Salager, M.I. Briceño and C.L. Bracho. Heavy hydrocarbon emulsions—Making use of the state of the art in formulation engineering. In *Encyclopedic Handbook of Emulsion Technology*, J. Sjöblom, (Ed.). Marcel Dekker, New York, pp. 455–495 (2001).

54. J.L. Salager, R.E. Antón, D.A. Sabatini, J.H. Harwell, E.J. Acosta and L.I. Tolosa. Enhancing solubilization in microemusions—State of the art and current trends. *J. Surfact. Deterg.*, **8**, 3–21 (2005).

55. C. Huh. Interfacial tension and solubilizing ability of a microemulsion phase that coexists with oil and brine. *J. Colloid Interface Sci.*, **71**, 408–426 (1979).

56. E.J. Acosta, E. Szekeres, D.A. Sabatini and J.H. Harwell. Net-average curvature model for solubilization and supersolubilization in surfactant microemulsion. *Langmuir*, **19**, 186–195 (2003).

57. E.J. Acosta, J.S. Yuan and A.S. Bhakta. The characteristic curvature of ionic surfactants. *J. Surfact. Deterg.*, **11**, 145–158 (2008).

58. E.J. Acosta. The HLD-NAC equation of state for microemulsions formulated with nonionic alcohol ethoxylate and alkylphenol ethoxylate surfactants. *Colloids Surf. A.*, **320**, 193–204 (2008).

59. W. Kunz, F. Testard and T. Zemb. Correspondence between curvature, packing parameter, and hydrophilic-lipophilic deviation scales around the phase-inversion temperature. *Langmuir*, **25**, 112–115 (2009).

60. P. Becher. The effect of the nature of the emulsifying agent on emulsion inversion. *J. Soc. Cosmet. Chem.*, **9**, 141–148 (1958).

61. T.J. Lin. Effect of initial surfactant location on the viscosity of emulsions. *J. Soc. Cosmet. Chem.* **19**, 683–697 (1968).

62. T.J. Lin and J.C. Lambrechts. Effect of initial surfactant location on emulsion phase inversion. *J. Soc. Cosmet. Chem.*, **20**, 185–198 (1969).

63. T.J. Lin. Surfactant location and required HLB. *J. Soc. Cosmet. Chem.*, **21**, 365–375 (1970).

64. J.L. Salager, N. Moreno, R.E. Antón and S. Marfisi. Apparent equilibration time required for a surfactant-oil-water system to emulsify into the morphology imposed by formulation. *Langmuir*, **18**, 607–611 (2002).

65. G. Alvarez, R.E. Antón, S. Marfisi, L. Marquez and J.L. Salager. Apparent equilibration time required for a surfactant-oil-water system to emulsify into the morphology imposed by formulation. Part 2: Effect of sec-butanol concentration and initial location. *Langmuir*, **20**, 5179–5181 (2004).

66. J.L. Salager, M. Miñana-Perez, M. Perez-Sanchez, M. Ramirez-Gouveia and C I. Rojas. Surfactant-oil-water systems near the affinity inversion—Part III: The two kinds of emulsion inversion. *J. Dispersion Sci. Technol.*, **4**, 313–329 (1983).

67. J.L. Salager. Guidelines for the formulation, composition and stirring to attain desired emulsion properties (type, droplet size, viscosity and stability). In *Surfactants in Solution*, A. Chattopadhyay and K.L. Mittal (Eds.), Chap. 16. Surfactant Science Series, Vol. 64. Marcel Dekker, New York, pp. 261–295 (1996).

68. N. Garti. Double emulsions—Scope, limitations and new achievements. *Colloids Surf. A*, **123–124**, 233–246 (1997).

69. N. Garti and A. Benichou. Double emulsions for controlled release applications—Progress and trends. In *Encyclopedic Handbook of Emulsion Technology*, J. Sjöblom (Ed.). Marcel Dekker, New York, pp. 337–407 (2001).

70. D.H. Smith and K.H. Lim. Morphology and inversion of two fluids in systems of three and four thermodynamic dimensions. *J. Phys. Chem.*, **94**, 3746–3752 (1990).

71. J.M. Lee, K.H. Lim and D.H. Smith. Formation of two-phase multiple emulsions by inclusion of continuous phase into dispersed phase. *Langmuir*, **18**, 7334–7340 (2002).

72. J.L. Salager, L. Márquez, A.A. Peña, M.J. Rondón, F. Silva and E. Tyrode. Current phenomenological know-how and modeling of emulsion inversion. *Ind. Eng. Chem. Res.*, **39**, 2665–2676 (2000).

73. J.L. Salager, J. Bullón, A. Pizzino, M. Rondón-González and L. Tolosa. Emulsion formulation engineering for the practitioner. In *Encyclopedia of Surface and Colloid Science*, Vol. 1:1, P. Somasundaran (Ed.). Taylor & Francis, London, UK, pp. 1–6 (2010).

74. P. Izquierdo, J. Esquena, Th.F. Tadros, J.C. Dederen, J. Feng, M.J. Garcia-Celma, N. Azemar and C. Solans. Phase behavior and nanoemulsion formation by the phase inversion temperature method. *Langmuir*, **20**, 6594–6598 (2004).

75. J.L. Salager. Emulsion phase inversion phenomena. In *Emulsions and Emulsion Stability*, 2nd ed., J. Sjoblöm (Ed.), Chap 4. Taylor & Francis, London (2006).

76. J.L. Salager. Phase transformation and emulsion inversion on the basis of catastrophe theory. In *Encyclopedia of Emulsion Technology*, Vol. 3, P. Becher (Ed.), Chap. 2. Marcel Dekker, New York (1988).

77. J.L. Salager, A. Forgiarini, L. Marquez, A. Peña, A. Pizzino, M.P. Rodriguez and M. Rondón-Gonzalez. Using emulsion inversion in industrial processes. *Adv. Colloid Interface Sci.*, **108–109**, 259–272 (2004).

78. A. Forgiarini, J. Esquena, C. Gonzalez and C. Solans. Formation of nanoemulsions by low-energy emulsification methods at constant témperature. *Langmuir*, **17**, 2076–2083 (2001).

79. M. Porras, C. Solans, C. González, A. Martínez, A. Guinart and J.M. Gutiérrez. Studies of formation of W/O nanoemulsions. *Colloids Surf. A*, **249**, 115–118 (2004).

80. D. Langevin. Influence of interfacial rheology on foam and emulsions properties. *Adv. Colloid Interface Sci.*, **88**, 209–222 (2000).

81. J. Boyd, C. Parkinson and P. Sherman. Factors affecting emulsion stability and the HLB concept. *J. Colloid Interface Sci.*, **41**, 359–370 (1972).

82. M. Bourrel, A. Graciaa, R.S. Schechter and W.H. Wade. The relation of emulsion stability to phase behavior and interfacial tension of surfactant systems. *J. Colloid Interface Sci.*, **72**, 161–163 (1979).

83. J.L. Salager, L. Quintero, E. Ramos and J. Andérez. Properties of surfactant-oil-water emulsified systems in the neighborhood of three-phase transition. *J. Colloid Interface Sci.*, **77**, 288–289 (1980).

84. J.E. Vinatieri. Correlation of emulsion stability with phase behavior in surfactant systems for tertiary oil recovery. *Soc. Pet. Eng. J.*, **20**, 402–406 (1980).

85. F.S. Milos and D.T. Wasan. Emulsion stability of surfactant systems near the three-phase region. *Colloids Surf.*, **4**, 91–96 (1982).

86. R.D. Hazlett and R.S. Schechter. Stability of macroemulsions. *Colloids Surf.*, **29**, 53–69 (1988).

87. A. Kabalnov and J. Weers. Macroemulsions stability within the Winsor III region: Theory versus experiments. *Langmuir*, **12**, 1931–1937 (1996).

88. R.E. Antón and J.L. Salager. Emulsion Instability in the three-phase behavior region of surfactant-alcohol-oil-brine systems. *J. Colloid Interface Sci.*, **111**, 54–59 (1986).

89. J.L. Salager. Emulsion properties and related know-how to attain them. In *Pharmaceutical Emulsions and Suspensions*, F. Nielloud and G. Marti-Mestres (Eds.). Marcel Dekker, New York, pp. 73–125 (2000).

90. T. Tadros and B. Vincent. Emulsion stability. In *Encyclopedia of Emulsion Technology*, Vol. 1, P. Becher (Ed.). Marcel Dekker, New York, pp. 129–285 (1983).

91. H.P. Grace. Dispersion phenomena in high viscosity immiscible fluid systems and application of static mixers as dispersion devices in such systems. *Chem. Eng. Commun.*, **14**, 225–277 (1982).

92. P. Walstra. Formation of emulsion. In *Encyclopedia of Emulsion Technology*, Vol. 1, P. Becher (Ed.). Marcel Dekker, New York, pp. 57–127 (1983).

93. P. Brochette. Emulsification—Elaboration et étude des émulsions. In *Techniques de l'Ingénieur*, Vol. Génie des Procédés J2, Chap. 150, Paris, France, pp. 1–18 (2000).

94. L. Lobo, A. Svereika and M. Nair. Coalescence during emulsification—1. Method development. *J. Colloid Interface Sci.*, **253**, 409–418 (2002).

95. L. Lobo and A. Svereika. Coalescence during emulsification 2. Role of small molecule surfactants. *J. Colloid Interface Sci.*, **261**, 498–507 (2003).

96. J. Floury, J. Legrand and A. Desrumaux. Analysis of a new type of high pressure homogenizer. Part B. Study of droplet break-up and recoalescence phenomena. *Chem. Eng. Sci.*, **59**, 1285–1294 (2004).

97. S.M. Jafari, E. Assadpoor, Y. He and B. Bhandari. Recoalescence of emulsion droplets during high-energy emulsificaction. *Food Hydrocolloid*, **22**, 1191–1202 (2008).

98. M. Pérez, N. Zambrano, M. Ramirez, E. Tyrode and J.L. Salager. Surfactant-oil-water systems near the affinity inversion. Part XII: Emulsion drop size versus formulation and composition. *J. Dispersion Sci. Technol.*, **23**, 55–63 (2002).

99. L.I. Tolosa, A. Forgiarini, P. Moreno and J.L. Salager. Combined effects of formulation and stirring on emulsion drop size in the vicinity of three-phase behavior of surfactant-oil-water system. *Ind. Eng. Chem. Res.*, **45**, 3810–3814 (2006).

100. C. Solans, J. Esquena, A.M. Forgiarini, N. Uson, D. Morales, P. Izquierdo, N. Azemar and M.J. Garcia-Celma. Nanoemulsions: Formation, properties and applications. In *Adsorption and Aggregation of Surfactants in Solution*, K.L. Mittal and D.O. Shah (Eds.). Marcel Dekker, New York, pp. 525–554 (2003).

101. C. Solans, P. Izquierdo, J. Nolla, N. Azemar and M.J. Garcia-Celma. Nanoemulsions. *Cur. Opin. Colloid Interface Sci.*, **10**, 102–110 (2005).

102. T.G. Mason, J.N. Wilking, K. Meleson, C.B. Chang and S.M. Graves. Nanoemulsions: Formation, structure and physical properties. *J. Phys. Condens. Matter*, **18**, R635–R666 (2006).

103. T.J. Lin, T. Akabori, S. Tanaka and K. Shimura. Low energy emulsification. Part V: Mechanism of enhanced emulsification. *Cosmet. Toilet.*, **98**, 67–73 (1983).

104. T.J. Lin and Y.F. Shen. Low energy emulsification. Part VI: Applications in high-internal phase emulsions. *J. Soc. Cosmet. Chem.*, **35**, 357–368 (1984).

105. P. Sherman. Rheological properties of emulsions. In *Encyclopedia of Emulsion Technology*, Vol. 1, P. Becher (Ed.). Marcel Dekker, New York, pp. 405–437 (1983).

106. T.F. Tadros. Fundamental principles of emulsion rheology and their applications. *Colloids Surf. A*, **91**, 39–54 (1994).

107. H.A. Barnes. Rheology of emulsions—A review. *Colloids Surf. A*, **91**, 89–95 (1994).

108. L. Bécu, P. Grondin, A. Colin and S. Manneville. How does a concentrated emulsion flow? Yielding, local rheology, and wall slip. *Colloids Surf. A*, **263**, 146–152 (2004).

109. I. Masalova and A.Y. Malkin. Master curves for elastic and plastic properties of highly concentrated emulsions. *Colloid J.*, **70**, 327–336 (2008).

110. R. Pal, S.N. Bhattacharya and E. Rhodes. Flow behavior of oil-in-water emulsions. *Can. J. Chem. Eng.*, **64**, 3–10 (1986).

111. R. Pal. Yield stress and viscoelastic properties of high internal phase ratio emulsions. *Colloid Polym. Sci.*, **277**, 583–588 (1999).

112. R. Pal. Slippage during the flow of emulsions in rheometers. *Colloids Surf. A.*, **162**, 55–66 (2000).

113. G.A. Nuñez, M.I. Briceño, C. Mata and H. Rivas. Flow characteristics of concentrated emulsions of very viscous oil in water. *J. Rheol.*, **40**, 405–423 (1996).

114. N. Aomari, R. Gaudu, F. Cabioch and A. Omari. Rheology of water-in-crude oil emulsions. *Colloids Surf. A*, **139**, 13–20 (1998).

115. N. Jager-Lézer, J.F. Tranchant, V. Alard, C. Vu, P.C. Tchoreloff and J.L. Grosssiord. Rheological analysis of highly concentrated w/o emulsions. *Rheol. Acta*, **37**, 129–138 (1998).

116. I. Vinckier, M. Minale, J. Mewis and P. Moldenaers. Rheology of semi-dilute emulsions: Viscoelastic effects caused by interfacial tension. *Colloids Surf. A*, **150**, 217–228 (1999).

117. C. Mabille, V. Schmitt, P. Gorria, F. Leal Calderon, V. Faye, B. Deminiere and J. Bibette. Rheological and shearing conditions for the preparation of monodispersed emulsions. *Langmuir*, **16**, 422–429 (2000).

118. A.Y. Malkin, I. Masalova, P. Slatter and K. Wilson. Effect of droplet size on the rheological properties of highly concentrated w/o emulsions. *Rheol. Acta*, **43**, 584–591 (2004).

119. M.A. Farah, R.C. Oliveira, J.N. Caldas and K. Rajagopal. Viscosity of water-in-oil emulsions: Variation with temperature and water volume fraction. *J. Pet. Sci. Eng.*, **48**, 169–184 (2005).

120. D. Dan and G. Jing. Apparent viscosity prediction of non-newtonian water-in-crude oil emulsions. *J. Pet. Sci. Eng.*, **53**, 113–122 (2006).

121. J.L. Salager, M. Miñana-Perez, J. Andérez, J. Grosso, C. Rojas and I. Layrisse. Surfactant-oil-water systems near the affinity inversion—II: Viscosity of emulsified systems. *J. Dispersion Sci. Technol.*, **4**, 161–173 (1983).

122. R. Pal. Viscosity and storage/loss moduli for mixtures of fine and coarse emulsions. *Chem. Eng. J.*, **67**, 37–44 (1997).

123. M. Ramirez, J. Bullón, J. Andérez, I. Mira and J.L. Salager. Drop size distribution bimodality and its effect on O/W emulsion viscosity. *J. Dispersion Sci. Technol.*, **23**, 309–321 (2002).

124. G.A. Nuñez, G. Sanchez, X. Gutierrez, F. Silva, C. Dalas and H. Rivas. Rheological behavior of concentrated bitumen in water emulsions. *Langmuir*, **16**, 6497–6502 (2000).

125. J. Esquena, G.R.S. Ravi-Sankar and C. Solans highly concentrated W/O emulsions prepared by the PIT method as template for solid foam. *Langmuir*, **19**, 2983–2988 (2003).

126. J. Kizling, B. Kronberg and J.C. Erikson. On the formation and stability of high internal phase O/W emulsions. *Adv. Colloid Interface Sci.*, **123–126**, 295–302 (2006).

127. J.N. Israelachvili, D.J. Mitchell and B.W. Ninham. Theory of self assembly of hydrocarbon amphiphiles into micelles and bilayers. *J. Chem. Soc. Faraday Trans. II*, **72**, 1525–1568 (1976).

128. S.J. Candau, A. Khatory, F. Lequeux and F. Kern. Rheological behavior of wormlike micelles: Effect of salt content. *J. Physique IV* (Colloque C1, Suplem. J. Physique II) **3**, 197–209 (1993).

129. S.J. Candau and R. Oda. Linear viscoelasticity of salt-free wormlike micellar solutions. *Colloids Surf. A*, **183–185**, 5–14 (2001).

130. D.P. Acharya and H. Kunieda. Wormlike micelles in mixed surfactant solutions. *Adv. Colloid Interface Sci.*, **123–126**, 401–413 (2006).

131. S. Ezrahi, E. Tuval and A. Aserin. Properties, main applications and perspectives of worm micelles. *Adv. Colloid Interface Sci.*, **128–130**, 77–102 (2006).

132. S. Parathakkatt, J. George, M.S. Sajev and L. Sreejith. A rheological approach to viscoelasticity wormlike micelles of tunable properties. *J. Surfact. Deterg.*, **12**, 219–224 (2009).

133. B. Song, Y. Hu, Y. Song and J. Zhao. Alkyl chain length-dependent viscoelastic properties in aqueous wormlike micellar solution of anionic gemini surfactants with an azobenzene spacer. *J. Colloid Interface Sci*, **341**, 94–100 (2010).

134. M.R. Rojas, A. Müller and A.E. Saez. Effect of ionic environment on the rheology of wormlike micelle solutions of mixtures of surfactants with opposite charge. *J. Colloid Interface Sci.*, **342**, 103–109 (2010).

135. S. Segota and D. Tezak. Spontaneous formation of vesicles. *Adv. Colloid Interface Sci.*, **121**, 51–75 (2006).

136. G. Danker, C. Verdier and C. Misbah. Rheology and dynamics of vesicle suspension in comparison with droplet emulsion. *J. Non-Newtonian Fluid Mech.*, **152**, 156–167 (2008).

137. E.A. Kubatta and H. Rehage. Characterization of giant vesicles formed by phase transfer processes. *Colloid Polym. Sci.*, **287**, 1117–1122 (2009).

138. C. Stubenrauch (Ed.). *Microemulsions—Background, New Concepts, Applications, Perspectives.* Blackwell Publishing, UK (2009).

139. J.L. Salager. Microemulsions. In *Handbook of Detergents—Part A: Properties*, G. Broze (Ed.). Marcel Dekker, New York, pp. 253–302 (1999).

140. J.L. Salager and R.E. Antón. Ionic microemulsions. In *Handbook of Microemulsions Science and Technology*, P. Kumar and K.L. Mittal (Eds.). Marcel Dekker, New York, pp. 247–280 (1999).

141. J.L. Salager, R.E. Antón, A. Forgiarini and L. Marquez. Formulation of microemulsions. In *Microemulsions—Background, New Concepts, Applications, Perspectives*, C. Stubenrauch, (Ed.). Blackwell Publishing, UK, pp. 84–121 (2009).

142. S. Queste, J.L. Salager, R. Strey and, J.M. Aubry. The EACN Scale for oil classification revisited thanks to the fish diagram. *J. Colloid Interface Sci.*, **312**, 98–107 (2007).

143. F. Bouton, M. Durand, V. Nardello-Rataj, M. Serry and J.M. Aubry. Classification of terpene oils using the fish diagrams and the Equivalent Alkane Carbon Number (EACN) Scale. *Colloids Surf. A.*, **338**, 142–147 (2009).

144. V.B. Patravale and A.A. Date. Microemulsions: Pharmaceutical applications. In *Microemulsions—Background, New Concepts, Applications, Perspectives*, C. Stubenrauch, (Ed.). Blackwell Publishing, UK, pp. 259–301 (2009).

145. W. Von Rybinski, M. Hloucha and I. Johansson. Microemulsions in cosmetics and detergents. In *Microemulsions—Background, New Concepts, Applications, Perspectives*, C. Stubenrauch (Ed.). Blackwell Publishing, UK, pp. 230–258 (2009).

146. F.H. Haegel, J.C. Lopez, J.L. Salager and S. Engelskirchen, Microemulsions in large-scale applications. In *Microemulsions—Background, New Concepts, Applications, Perspectives*, C. Stubenrauch, (Ed.). Blackwell Publishing, UK, pp. 302–344 (2009).

147. S. Singh. Phase transition in liquid crystals. *Phys. Rep.*, **324**, 107–269 (2000).

148. D. Langevin. Microemulsions and liquid crystals. *Mol. Cryst. Liq. Cryst.*, **138**, 259–305 (1986).

149. A. Al-Bawab and S.E. Friberg. Some pertinent factors in skin care emulsion. *Adv. Colloid Interface Sci.*, **123–126**, 313–322 (2006).

150. J. Rouviere. Stabilisation d'émulsions huile-dans-eau par formation de multicouches de tensioactifs: Cas des huiles de silicone. *Inf. Chim.*, **325**, 158–164 (1991).

151. N. Uson, M.J. Garcia and C. Solans. Formation of water-in-oil (W/O) nanoemulsion in a water/mixed nonionic surfactant/oil systens prepared by low energy emulsification method. *Colloids Surf. A*, **250**, 415–421 (2004).

152. N. Sadurni, C. Solans, N. Azemar and M. Garcia. Studies on the formation of O/W nanoemulsions by low energy emulsification methods, suitable for pharmaceutical applications. *Eur. J. Pharm. Sci.*, **26**, 438–445 (2005).

153. P. Alexandridis and R.J. Spontak. Solvent regulated ordering in block polymers. *Curr. Opin. Colloid Interface Sci.*, **4**, 130–139 (1999).

154. L. Spitz. *Soap Manufacturing Technology*. AOCS Press, Urbana, IL, pp. 130 & 207 (2009).

155. H. Kunieda and K. Shinoda. Phase behavior in systems of nonionic surfactant-water-oil around the hydrophile-lipolile balance temperature (HLB-Temperature). *J. Dispersion Sci. Technol.*, **3**, 233–244 (1982).

156. E.D. Goddard, T.S. Phillips and R.B. Hannan. Water soluble polymer-surfactant interaction—Part I. *J. Soc. Cosmet. Chem.*, **26**, 461–475 (1975).

157. E.D. Goddard. Polymer/surfactant interaction. *J. Soc. Cosmet. Chem.*, **41**, 23–49 (1990).

158. E.D. Goddard, P.S. Leung and K.P.A. Padmanabhan. Novel gelling structures based on polymer/surfactant systems. *J. Soc. Cosmet. Chem.*, **42**, 19–34 (1991).

159. K. Holmberg, B. Jönsson, B. Kronberg and B. Lindman. *Surfactants and Polymers in Aqueous Solution*, 2nd ed. J. Wiley, UK (2002).

160. P. Deo, S. Jockush, M.F. Ottaviani, A. Moscatelli, N.J. Turro and P. Somasundaran. Interactions of hydrophobically modified polyelectrolytes with surfactants of the same charge. *Langmuir*, **19**, 10747–10752 (2003).

161. P. Deo, N. Deo, P. Somasundaran, A. Moscatelli, S. Jockusch, N.J. Turro, K.P. Ananthapadmanabhan and M.F. Ottaviani. Interactions of a hydrophobically modified polymer with oppositely charged surfactants. *Langmuir*, **23**, 5906–5913 (2007).

162. F.E. Antunes, B. Lindman and M.G. Miguel. Mixed systems of hydrophobically modified polyelectrolytes: Controlling rheology by charge and hydrophobe stoichiometry and interaction strength. *Langmuir*, **21**, 10188–10196 (2005).

163. C. Ishizuka, T. Ahmed, S. Arima and K. Aramaki. Viscosity boosting effect of added ionic surfactant in nonionic wormlike micellar aqueous solutions. *J. Colloid Interface Sci.*, **339**, 511–516 (2009).

164. B.B. Niraula, T.N. Seng and M. Misran. Vesicles in fatty acid salt-fatty acid stabilized O/W emulsion–emulsion structure and rheology. *Colloids Surf. A*, **236**, 7–22 (2004).

165. T.M. Dreher, J. Glass, A.J. O'Connor and G.W. Stevens. Effect of rheology on coalescence rates and emulsion stability. *AIChE J.*, **45**, 1182–1190 (1999).

166. J.L. Salager, L. Marquez, I. Mira, A. Peña, E. Tyrode and N. Zambrano. Principles of emulsion formulation engineering. In *Adsorption and Aggregation of Surfactants in Solution*, K.L. Mittal and D.O. Shah (Eds.). Marcel Dekker, New York, pp. 501–524 (2003).

167. S.E. Salager, E. Tyrode, M.T. Celis, J.L. Salager. Influence of the stirrer initial position on emulsion morphology. Making use of the local water-to-oil ratio concept for formulation engineering purpose. *Ind. Eng. Chem. Res.*, **40**, 4808–4814 (2000).

168. I. Cuéllar, J. Bullón, A. Forgiarini, A. Cardenas and M.I. Briceño. More efficient preparation of parenteral emulsion or how to improve a pharmaceutical recipe by formulation engineering. *Chem. Eng. Sci.*, **60**, 2127–2134 (2005).

169. H.T. Davis. Factors determining the emulsion type: Hydrophile-lipophile balance and beyond. *Colloids Surf. A*, **91**, 9–24 (1994).

170. I. Johansson. About characterization of surfactants outside the HLB-System. In Paper 065, CD *Proceedings 6th World Surfactant Congress CESIO*, Berlin, Germany (2004).

19 An Overview of Surfactants in Enhanced Oil Recovery

Paulina M. Mwangi and Dandina N. Rao

CONTENTS

19.1 INTRODUCTION

Surfactants are extremely important in the oil and gas industry [1]. Primary depletion and secondary recovery processes applied in oil reservoirs typically recover about a third of the original oil in place (OOIP); thus, nearly two trillion barrels of conventional oil and five trillion barrels of heavy oil remain in reservoirs worldwide after these methods have been exhausted [2]. The low oil recoveries from conventional recovery methods are the result of inefficient macroscopic sweep efficiencies due to the lack of mobility control and poor microscopic displacement efficiencies caused by the capillary trapping of oil that is attributed mainly to wettability and interfacial forces. The enhanced oil recovery (EOR) method of choice and the expected recovery depend on many considerations such as technology, reservoir geology, reservoir fluids, and overall economics that include the price of oil. It has long been an objective of the industry to develop processes to improve the overall recovery. However, the low oil prices from the mid-1980s until recently provided little incentive for research on EOR methods, especially surfactant processes that require significant initial costs

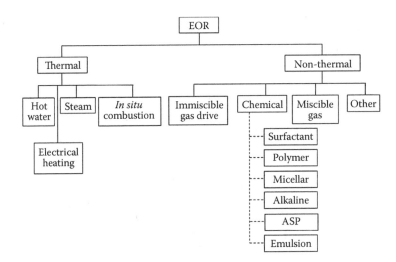

FIGURE 19.1 Classification of EOR processes. ASP indicates alkaline–surfactant–polymer flooding. (Adapted from S. Thomas, *Oil Gas Sci. Technol., Rev. IFP*, 63, 9–19, 2008.)

of the chemicals. In light of the current higher prices and the end of easy oil, the use of surfactants and other chemicals to improve oil recovery is becoming more attractive because of the growing energy demand, diminishing reserves, advancements in technologies, and better understanding of failed projects. Therefore, much interest and research is ongoing to develop effective and economical methods to increase oil recovery.

Before discussing the role of surfactants in the oil and gas industry, we need to distinguish the terms EOR and improved oil recovery (IOR). These terms are sometimes used loosely and interchangeably. However, IOR implies improving oil recovery by any means, for example, water and gas injection for pressure maintenance, operational strategies such as infill drilling and use of horizontal wells, improving vertical and areal sweep, and so on. On the other hand, EOR is more specific in concept and can be considered as a subset of IOR. EOR processes involve the injection of fluids to supplement the natural energy in the reservoir to displace oil to the producing well. More importantly, the injected fluids interact with the reservoir rock-fluid system to create conditions favorable for increasing oil recovery. These favorable interactions include oil swelling, interfacial tension (IFT) reduction, rock wettability modification, and phase behavior effects. Surfactants are typically used to decrease IFT and favorably change the wettability of the rock to increase oil recovery. As illustrated by Figure 19.1, this method of recovery is a subset of chemical flooding. This chapter will cover the history of surfactants in the petroleum industry, the types of surfactants used in EOR, the mechanism by which they increase oil recovery, their effect on IFT and wettability alteration, and their limitations and challenges. Basic theories of surfactant flooding will also be briefly discussed.

19.2 HISTORY OF SURFACTANTS IN THE PETROLEUM INDUSTRY

The concept of adding surfactant to injected water in order to reduce oil–water IFT and alter wettability has been around since the 1920s. In 1927, Uren and Fahmy [3] concluded that an inverse relationship exists between oil–water IFT and the fraction of oil recovered by water injection. In the same year, a patent was issued to Atkinson [4], who proposed the use of surfactants and other materials to decrease the IFT between oil and water, thereby increasing oil recovery. Early field trials resulted in small increases in oil recovery, although the field data reported were found to be inconclusive for evaluating the mechanism responsible for the increase in recovery. The results from these trials were not sufficiently promising to encourage the use of surfactants on a larger scale. In addition, the relative importance of the mechanisms responsible for increasing oil recovery was not well understood. During the next 25 years, a major part of the reported research on the use of surfactants in improving oil recovery was carried out by a group at Penn State University. They recognized that IFT, wettability, and surfactant adsorption were important factors in the use of surfactants in EOR [5].

In the 1960s, surfactants were made either by direct sulfonation of aromatic groups in the refinery streams of crude oils or by organic synthesis of alkyl/aryl sulfonates, which allowed the surfactants to be tailored to the reservoir of interest. These two different approaches encouraged significant advances in surfactant EOR processes. The advantages of these surfactants were their lower cost, wide range of properties, and availability of raw materials in somewhat large quantities [6]. In 1977, Gogarty and Tosch [7] discussed the principles of micellar surfactant systems and distinguished between water-continuous and oil-continuous systems. They were also successful in developing a miscible oil recovery process using an oil-continuous surfactant system. In the same year, Holm [8] advocated a similar process using soluble oil, which was an oil-continuous microemulsion that was more successful than other solvents in increasing oil recovery. However, it was not initially recognized that the process success also depended on maintaining an ultralow IFT at the rear of the slug where it was displaced by an aqueous polymer solution [9].

By the end of the 1980s, surfactants used in EOR underwent extensive research and development, including a significant amount of field pilot tests. In the 1970s, many surfactant flooding pilot tests did not adequately recognize the importance of reservoir characterization and mobility control. By the mid-1980s, surfactants had been developed with high salinity tolerance and other characteristics far superior to the earlier products. By the mid-1990s, surfactant structure could be tailored to the specific conditions needed [10]. In recent years, many advances are being made to create surfactants that are cost-effective and can withstand high temperatures and salinities and other parameters encountered in harsh and extreme reservoir environments [6,11–13].

19.3 TYPES OF SURFACTANTS USED IN EOR

Surfactants are surface-active agents that adsorb or concentrate at the fluid–fluid or fluid–solid interface, thus altering the IFT and the wettability of the system, respectively. This characteristic of surfactant makes them useful in the oil and gas industry, especially in EOR. Surfactant molecules consist of a lipophilic (oil soluble) tail and a hydrophilic (water soluble) head group. A balance between the hydrophilic and lipophilic parts of the surfactant molecule, characterized by the HLB number, gives it the characteristics of a surface-active agent. There are several ways of grouping surfactants; however, the most common is by the ionic nature of their head group, described as follows [1,6,14–16]:

1. Anionic surfactants have a surface-active portion that bears a negative charge. This type of surfactants are widely used in chemical EOR processes because they are relatively resistant to adsorption, are stable, and can be made relatively cheaply. Anionics are more resistant to adsorption because of the repulsion between the negative charges of the clays interweaved in the sandstone rock matrix and that found on the anionics. The opposite is true in carbonate rocks because of the positive charge that attracts the negatively charged anionic surfactants, thus leading to adsorption. Anionics such as sulfates and sulfonates are typically used for low and high temperature applications, respectively.
2. Cationic surfactants have a surface-active portion that bears a positive charge. This group of surfactants are rarely used because they are highly adsorbed by the negatively charged surface of sandstone rocks. However, they can be used in positively charged carbonate rocks to alter wettability from oil wet to water wet.
3. Nonionic surfactants have been mostly used as cosurfactants but are increasingly used as a primary surfactant. These surfactants do not form ionic bonds, but when dissolved in

aqueous solutions, they exhibit surfactant properties due to the electronegativity contrast between their constituents. Nonionics are much more tolerant of high salinities than anionics; however, their ability to reduce IFT is not as high as anionic surfactants. A mixture of anionic and nonionic is used to increase the tolerance to salinity.

4. Amphoteric surfactants, also known as zwitterionic surfactants, have a surface-active portion that contains both positive and negative charges. This type of surfactant can be nonionic–anionic, nonionic–cationic, or anionic–cationic [18]. Such surfactants are temperature and s~alinity tolerant but they are very expensive.

19.4   RECOVERY OF RESIDUAL OIL

The mobilization of residual oil that is left behind in the reservoirs after primary and secondary recovery processes is influenced by two factors: the mobility ratio (M) and the capillary number (N_c). The mobility ratio is defined as $M = \lambda_{displacing}/\lambda_{displaced}$, where $\lambda = k / \mu$, $\lambda_{displacing}$ is the mobility of the displacing fluid (e.g., water and gas), $\lambda_{displaced}$ is the mobility of the displaced fluid (e.g., oil), μ is the viscosity of the fluid, and k is the effective permeability. To achieve a favorable mobility ratio ($M \leq 1$), a polymer can be added to increase the viscosity of the displacing fluid. The capillary number is defined as the ratio of viscous to capillary forces and is calculated as $N_c = v\mu/(\sigma \cos\theta)$, where v is the fluid velocity, μ is the displacing fluid viscosity, σ is the IFT between the fluid phases, and θ is the contact angle between the fluid–fluid interface and the solid surface. The most effective way of increasing the capillary number is by reducing the IFT and favorably altering the contact angle. However, the majority of the published correlations of residual oil saturation with capillary number have disregarded the contact angle term in the capillary number equation by assuming $\cos\theta = 1$. Klins [20] showed several correlations in his work, where this assumption has been made. In addition, he summarized the effect of capillary number on residual oil saturation and showed that an increase of four to six orders of magnitude in capillary number is required for significant improvements in oil recovery. Surfactants are capable of increasing the capillary number either by reducing oil–water IFT or favorably modifying wettability and, thus, changing the contact angle. Significant improvements in oil recovery through capillary number increase by IFT reduction mechanism require several orders of magnitude reduction in IFT, down to ultralow IFTs. Surfactants capable of generating such huge reductions in oil–water IFT are expensive and are required in large quantities because of their adsorption on the rock surface. This has resulted in high costs, which have contributed to the continuous decline of conventional surfactant floods operating in the field.

19.5   EFFECT OF SURFACTANTS ON IFT

To achieve low residual oil saturations (neglecting wettability alteration by surfactants), the IFT has to be reduced from oil–brine values of approximately 20 to 30 mN/m to 0.001 to 0.01 mN/m [1]. Research groups have found that an ultralow IFT in the required range could be achieved by using petroleum sulfonate or alcohol surfactants [6]. In addition, they found that the IFT of an oil–brine–surfactant system is a function of salinity, oil composition, surfactant type and concentration, cosurfactant, electrolytes, temperature, and the phase behavior of the system [11].

19.5.1   INFLUENCE OF SALINITY ON IFT

Winsor [21] identified three types of phase equilibria in microemulsion phase behavior as type I, type II, and type III. In 1974, Healy and Reed [22] explained how the Winsor-type behavior describes the changes in phase behavior, solubilization of oil and water, and IFT as a function of salinity for anionic surfactants. The surfactant–brine–oil phase behavior is strongly affected by the salinity of the brine. This phase behavior is represented by a ternary diagram, such as those shown in Figure 19.2. For low brine salinities, the surfactant flood will exhibit good aqueous phase solubility

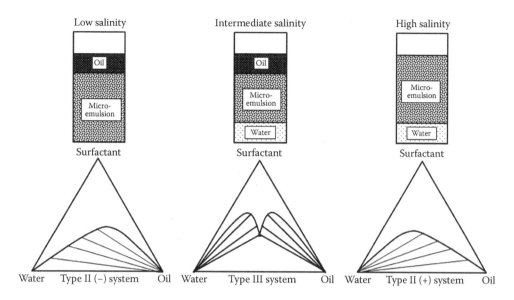

FIGURE 19.2 Three types of microemulsions and the effect of salinity on phase behavior. (Adapted from J.J. Sheng. Surfactant flooding. Gulf Professional Publishing, Kidlington, Oxford, pp. 239–335, 2011.)

and poor oleic phase solubility, thus forming a type I phase behavior. In a type I system, an oil-in-water microemulsion is formed, and the surfactant stays in the aqueous phase [1]. This system is also referred to as the lower phase microemulsion system or type II (−) system, where II means no more than two phases can form and (−) means the tie-lines have a negative slope. This behavior is not favorable to achieve ultralow IFT. In contrast, a type II system is a water-in-oil microemulsion with an excess oil phase also defined as the upper phase microemulsion or as type II (+). This phase behavior is formed in high salinity brines where the surfactant solubility is decreased in the aqueous phase by electrostatic forces. This behavior leads to surfactant retention in the oil phase and is unfavorable for EOR. In a type III microemulsion, the surfactant forms a microemulsion in a separate phase between the oil and the aqueous phases. This phase is a continuous layer containing surfactant, water, and dissolved hydrocarbons. Usually, type III provides low IFTs (~0.001 mN/m), especially where equal volumes of water and oil are solubilized in the microemulsion. This condition is defined as optimal salinity, which exhibits the lowest IFT between the brine and the oil phase. In addition, optimal salinity can be expressed as the midpoint salinity where IFT between microemulsion and water is the same or nearly close to the IFT between microemulsion and oil. This phase behavior system is desirable for EOR processes [18].

19.5.2 Influence of Surfactant Structure on IFT

Extensive research on surfactants has established a clear relationship between surfactant structure and fluid properties related to EOR performance [24]. The structure of a surfactant also determines its solubility in either brine or oil. Increasing the influence of the nonpolar end of the surfactant will increase oil solubility. This can be accomplished by increasing the nonpolar molecular weight, decreasing the tail branching, decreasing the number of polar groups, and decreasing the strength of the polar part of the surfactant [11]. The best surfactants used in EOR applications typically have a branched hydrophobe. Hydrophobe branching is a desirable trait for EOR surfactants. Linear surfactants have a tendency to form highly viscous gels [17] that could be detrimental to oil recovery. In addition, the chain length of the hydrophobe has a good correlation with the equivalent alkane carbon number of the crude of interest [18]. Wellington and Richardson [25] showed that branched alkyl chains with propylene oxide (PO) and ethylene oxide (EO) groups could yield ultralow IFT and high oil recovery at very low concentrations. Wu et al. [26] studied the effect of PO and EO groups in sulfonate surfactants for EOR. Levitt [17] investigated branched alcohol propoxy sulfates with hydrophobes ranging from C_{12} to C_{24} and with three to seven PO groups with a Texas crude oil and concluded they were promising EOR surfactants for reservoirs with low temperatures. Jayanti et al. [27] reported that branched alcohol propoxylated sulfates were excellent surfactants for removing organic liquid contaminants from soil.

19.5.3 INFLUENCE OF OIL PROPERTIES ON IFT

In addition, oil properties do affect the surfactant solubility in oil. High specific gravity crude oils tend to be rich in organic acids; thus, the surfactant–oil solubility is lower in high gravity oils. Some correlations have been found for the tendency of a surfactant to dissolve in oil as the temperature increases [11]. For most anionics, higher temperatures mean better solubility in brine. This trend is reversed for nonionics. Lastly, cosurfactants can be used to modify solubility so that the transition from type II (−) system to type II (+) system can occur at different salinities.

19.6 EFFECTS OF SURFACTANT ON WETTABILITY

Wettability is the ability of one fluid to spread or adhere on a (rock) surface in the presence of another immiscible fluid. This parameter is governed by three intermolecular surface forces: the London–van der Waals force, the electrostatic force, and the structural forces [28]. Consequently, this parameter has a profound effect on multiphase rock–fluid interactions. In porous media, wettability affects the efficiency of oil recovery methods, electrical properties, capillary pressure, relative permeability, and saturation profiles and determines the distribution of fluids in a reservoir [29,30]. The spread of a liquid on a solid surface depends on the solid surface properties as well as the liquid properties. Therefore, by manipulating the properties of the rock or liquid, one can optimize the performance to achieve the desired wetting condition. Surfactants can modify wettability by adsorbing the liquid–rock interface. Surfactant flooding schemes for recovering residual oil have been less satisfactory because of the loss of surfactant by adsorption on reservoir rocks and precipitation. Adsorption and wettability changes are determined mainly by the surfactant structure, rock mineralogy, oil chemistry, brine chemistry, and temperature [1,6]. The mineralogical composition of reservoir rocks and the reservoir fluids properties play an important role in determining surfactant interaction at their interfaces [30].

Wettability has been stated to be the most important factor in secondary recovery methods after geology [10]. However, most of the previous research conducted in the area of surfactants has focused on their ability to lower IFT while largely disregarding the wettability effects. Significant enhancements in oil recovery require several orders of magnitude reduction in IFT. The amount of surfactant required to generate this large IFT reduction would be large and thus expensive. However, wettability alteration can be induced by low-cost surfactants at moderate concentrations. Therefore, combining the effects of IFT reduction and favorable wetting conditions would make the use of surfactants more effective at lower concentrations. Most importantly, the effect of surfactants on wettability depends not only on how much is adsorbed but also on how they are adsorbed on the rock. A water-wet rock surface that is beneficial for displacement of oil can be obtained by manipulating the orientation of the adsorbed layers [30]. It is being recognized that the reservoir wettability is strongly influenced by the compositional effects of rock and fluids existing at the reservoir conditions. Therefore, when performing experiments, it is crucial to simulate reservoir conditions and use live reservoir fluids to understand the impact of true in situ surfactant-induced interfacial interactions in crude oil reservoirs.

Spinler et al. [31] conducted spontaneous imbibition tests on North Sea reservoir chalk plugs that were moderately water wet at both ambient and reservoir temperatures using dilute concentrations of surfactants. Low concentrations (in the range of 100–500 ppm) of surfactants were used to improve oil recovery in both spontaneous and forced water imbibition mechanisms. These improvements in oil recoveries were attributed to wettability alteration rather than reduction in IFT. Chen et al. [32] performed dilute surfactant imbibition tests on the vertically oriented dolomite cores of the Yates field at ambient conditions and found that additional oil was produced when compared with normal brine imbibition. They attributed these additional oil recoveries to both oil–water IFT reduction and wettability alteration. The contact angle measurements showed that the dilute surfactants shifted the wetting characteristics of the Yates dolomite rock from strongly oil wet toward less oil wet. Najurieta et al. [33] studied the effect of temperature on spontaneous imbibition and forced displacement mechanisms in water-wet porous media with and without surfactants and found that spontaneous imbibition was an important oil recovery mechanism in water-wet rocks and that surfactants would accelerate the imbibition mechanism. Hirasaki and Zhang [34] conducted spontaneous imbibition measurements at 80°C and reported improvements in oil recovery from oil-wet dolomite cores in the presence of an alkaline-anionic surfactant solution. Babadagli [35] compared the oil recovery of four different rock types (sandstone, limestone, dolomite, and chalk), a wide variety of oils (light and heavy crude, kerosene, and engine oil), and different types (nonionic and anionic) and concentrations of surfactants by conducting capillary imbibition experiments at room temperature. Except for light oils such as kerosene and light crude oil in sandstones, the nonionic surfactant solution yielded higher ultimate oil recoveries at faster rates in all the other systems. This was attributed to wettability alterations induced by the surfactant. In chalks, higher anionic surfactant concentrations yielded higher recoveries, but lower surfactant concentrations resulted in even lower oil recoveries than that observed with normal brine. Standnes and Austad [36] investigated the wettability alteration mechanism in oil-wet carbonate reservoir cores using surfactants. The surfactants were able to improve spontaneous imbibition at 40°C and 70°C into the matrix blocks because of wettability alterations to water wet and hence increased the oil recovery. Rao et al. [37] studied the impact of a surfactant-induced wettability alteration process and reported the development of mixed wettability using a nonionic surfactant. At this mixed-wet state, they recovered approximately 94% of the OOIP. Rao et al. [39–41] has also been on the forefront of characterizing wettability through contact angle measurements at reservoir conditions using the dual-drop-dual-crystal (DDDC) technique. This sophisticated technique is quick, repeatable, and accurately measures the advancing contact angle. Most researchers have typically relied on the sessile drop technique to measure contact angles; however, this technique is static and thus only measures the receding contact angle, which is not a true representative of the reservoir wettability. The DDDC technique measures both the advancing and receding contact angles. The advancing contact angle properly characterizes the reservoir wettability because it accounts for the reservoir fluid dynamics and its effect on wettability.

Ayirala et al. [38] studied the beneficial effects of wettability-altering surfactants in oil-wet fractured reservoirs by conducting contact angle measurements using the DDDC technique at ambient conditions. They proposed a sequential process of "diffusion and imbibition" in which the surfactant present in the fractures first diffuses into the rock matrix and alters wettability, enabling the imbibition of even more surfactants into the matrix for significant improvements in oil recovery. Seethepalli et al. [42] reported that the surfactant-induced wettability alterations of calcite to an intermediate/water-wet condition due to anionic surfactants were found to be better than a cationic surfactant with a West Texas crude oil in the same study. Kumar et al. [43] investigated the mechanism of wettability alteration due to surfactants using contact angle measurements at ambient conditions and atomic force microscopy. They reported that greater wettability alterations were possible with anionic rather that cationic surfactants and that water imbibition rates did not increase monotonically with increase in surfactant concentration.

19.7   VARIATIONS OF SURFACTANT FLOODING

Surfactant flooding has appeared in the literature under many names, such as detergent flooding, low-tension flooding, soluble oil flooding, microemulsion flooding, surfactant-assisted waterflooding, chemical flooding, and micellar–polymer flooding, to name a few. When a surfactant solution is injected into an oil–water system, it mobilizes the oil until the surfactant is diluted or otherwise lost due to adsorption on the rock surface. Because of the unsuccessful nature of some of the early implemented surfactant floods, alkali and polymers were added to combat the challenges faced when using surfactants only. Processes such as alkaline–surfactant (AS), polymer–surfactant (PS), and alkaline–surfactant–polymer (ASP) floods were created. The addition of alkali increases the pH, which lowers the surfactant adsorption so that low surfactant concentrations can be used, which reduces cost. Carbonate formations are typically positively charged at neutral pH, which favors adsorption of anionic surfactants. However, when Na_2CO_3 (alkali) is present, carbonate surfaces become negatively charged, and adsorption decreases severalfold. Another additional benefit of this chemical is that it reacts with the acid in oil to form soap; however, not all crude oils are reactive with alkaline chemicals [54]. The generation of soap allows the surfactant to be injected at lower concentrations than if used alone, which further reduces adsorption and facilitates the incorporation of polymer in the surfactant slug. In addition, alkali can alter reservoir rock wettability to reach either a more water-wet or a more oil-wet state [6]. On the other hand, polymers are usually added to the process for mobility control. The favorable mobility ratio aids in flood-front control and thus improves the vertical and areal sweep/recovery efficiencies. Therefore, the creation of ASP flooding holds high potential for substantial recovery at a reasonable cost because of the utilization of all three chemicals in one process. This process combines the macroscopic volumetric sweep efficiency improvement from the polymer due to the reduction in water-oil mobility ratio with the ability of surfactants to enhance microscopic sweep efficiency [6]. The major mechanisms of this process are IFT reduction, wettability modification, and improvement in mobility ratio [2].

19.7.1 SURFACTANT SELECTION

Surfactant selection is a crucial step that affects the success of this EOR process. Before the implementation of the process, extensive laboratory studies are needed to ensure the surfactant chosen is right for the reservoir of interest. Also, parameters such as optimum concentration, injection rate, and surfactant behavior at reservoir conditions have to be tested and determined. This grants the operator knowledge of the surfactant's strengths and weaknesses with respect to the reservoir of interest, which helps in predicting the oil recovery. Some of the experiments that can be used in selecting a surfactant are oil solubilization test, effect of brine concentration, microemulsion density test, surfactant and microemulsion viscosity test, coalescence time test, identification of the optimum surfactant–cosolvent formulation, and identification of optimal formulation for core flooding experiments [11]. Initial screening of surfactant effectiveness commonly involves examining microemulsion phase behavior of the chemical, with a specific crude oil and reservoir brine at the reservoir temperature of interest to determine where low IFTs can be obtained [23]. Some of the key surfactant selection criteria are as follows: high solubilization; favorable wettability alteration; little to no retention on reservoir rock; economics; ease with which it can be tailored to specific crude oil, salinity, and temperature; amount of branching needed to form low viscosity micelles and microemulsions; and propensity to form liquid crystals, gels, and macroemulsions.

19.7.2 APPLICATION OF SURFACTANT-BASED PROCESSES IN SANDSTONE RESERVOIRS

Almost all surfactant-based EOR processes have been in sandstone reservoirs, except a few well-stimulation/clean-up projects in carbonate reservoirs. Favorable reservoir characteristics in a sandstone reservoir are as follows: high permeability, high porosity, and good geological continuity. Low clay content is crucial to have low surfactant and polymer retention [18]. Most of the current chemicals are more effective at temperatures less that 150°C for surfactant flooding and less than 100°C when polymer is added [10]. It is also preferable to have the remaining oil in place more than 25% [19] and with a viscosity of less than 50 cP. In addition, a majority of implemented projects

TABLE 19.1
ASP Flood Field Cases

Field Name	Country	Reference	Field Name	Country	Reference
Cressford	Canada	[10]	Driscoll Creek	United States	[10]
David	Canada	[6,10]	Enigma	United States	[10]
Etzikom	Canada	[10]	White Castle	United States	[6,10]
Sa-Zhong-Xi	China	[6,18]	Tanner	United States	[46]
Xing-5	China	[18]	West Kiehl	United States	[10]
Xing-2-Xi	China	[18,44]	W. Moorcroft	United States	[10]
Bei-1-Xi	China	[44,45]	Beverly Hills	United States	[10]
Bei-2-Dong	China	[44,45]	Sho-Vel-Tum	United States	[10,47]
Xing-2-Zhong	China	[44,45]	Big Sinking	United States	[48]
Bei-2-Dong	China	[18]	Isenhaur	United States	[6,10]
Gudong	China	[44,45]	Mellot Ranch	United States	[10]
Minas I	Indonesia	[10]	Adena	United States	[10]
Minas II	Indonesia	[10]	Cambridge	United States	[10]
Angsi	Malaysia	[49]	Lagomar	Venezuela	[6,10]

have been in onshore reservoirs because of the effect of salinity on surfactants and the need for a reliable source of high quality water.

Table 19.1 lists ASP field projects that have been reported since the 1980s. As is evident, these projects are scattered around the world; however, the majority are in the United States and China. Projects in the United States were implemented in the 1980s and have since tapered off without addition of any new projects. However, China has been on the forefront of developing and implementing chemical flooding processes. Reported additional recoveries by ASP flooding from the listed fields are in the range of 10% to 28% of the OOIP.

19.7.3 APPLICATION OF SURFACTANT-BASED PROCESSES IN CARBONATE RESERVOIRS

One of the reasons that surfactant EOR is not as common in carbonate reservoirs is that anionic surfactants are highly adsorbed because of the positive charge on the rock surface. In addition, anhydrite is often present in carbonates, and it causes precipitation and high alkaline consumption. Because of this, ASP has only been tested in the laboratory for carbonate reservoirs [50–53]. In fractured reservoirs, spontaneous water imbibition can occur from the rock matrix into fractures. Subsequently, this mechanism leads to oil drainage from the matrix toward the fracture network, making surfactants more attractive to improve oil recovery in oil-wet carbonate reservoirs by changing rock wettability to a mixed or water-wet state, thus promoting the imbibition process [53]. Because a majority of the world's oil reserves are contained in carbonate reservoirs [54,55], the application of surfactant-based methods in carbonate reservoirs has recently become an active area of research as a strategy to increase oil recovery.

19.7.4 APPLICATION OF SURFACTANT-BASED PROCESSES IN GAS CONDENSATE RESERVOIRS

Gas condensate reservoirs suffer a rapid decline in productivity as the flowing bottom-hole pressure falls below the dew point. This reduction is caused by a condensate accumulation near the wellbore. Altering the wettability of rocks in the near-wellbore region from strongly liquid wet to intermediate gas wet provides a long-term strategy for the restoration of gas well productivity [56]. Li and Firoozabadi [57] pioneered the wettability alteration from strongly liquid wet to intermediate gas wet using polymers resulting in enhanced liquid mobility and gas productivity. Their work has been extended to high temperatures with similar results obtained [58,59]. However, the limitations of these studies are that reservoir fluids were not used to perform the experiments. Zheng and Rao [60,61] experimentally investigated potential remedies using low-cost surfactants to modify the fluid–fluid spreading coefficient and wettability in condensate buildup regions to enhance condensate recovery and gas productivity. Their results showed that the wettability of the representative gas condensate reservoir characterized through contact angles using the DDDC technique [39] was altered from initially oil wet to intermediate wet by using a low-cost anionic surfactant under both ambient and reservoir conditions. A core flooding study demonstrated 82% improvement in gas relative permeability because of the anionic surfactant-induced wettability alteration [62].

19.8 OPTIMIZATION OF SURFACTANT-BASED PROCESSES

This section will give a brief summary of the information provided in literature on the optimization of chemical flooding. There are many parameters that could affect a chemical flooding process; however, only a few are typically investigated in an optimization study. Zerpa et al. [66] considered the slug size and concentration in their optimization of the ASP flood, whereas Anderson et al. [67] examined the slug size, adsorption, permeability, and polymer mass. Delshad et al. [68] considered chemical concentrations and slug sizes as optimization parameters using an experimental design approach. Mwangi [69] investigated the effects of slug size, soaking time, and surfactant slug placement in an enhanced waterflooding project. In this study, a novel EOR method was proposed, named as surfactant-slug enhanced waterflood process technique, in which the underlying motivation was to achieve high oil recoveries without using large quantities of surfactants. As illustrated in Figure 19.3, this was achieved by soaking the area around the injection or production well with an ideally concentrated surfactant slug before conducting a waterflood. Two variations of this novel process were investigated: (1) the area around the production well was soaked, before conducting a waterflood, with either a 0.2 pore volume (PV) or a 0.3 PV surfactant slug, resulting in improving waterflood recovery from 47% to 55% and 62%, respectively; and (2) the area around the injection well was soaked with a 0.2 PV surfactant slug, thereby improving waterflood recovery from 47% to 67%. Table 19.2 shows the results of varying surfactant slug size and slug placement before waterflooding or surfactant flooding. Comparing results with a conventional waterflood or surfactant flood shows that when a surfactant is used efficiently, a dilute solution is sufficient to significantly improve recovery.

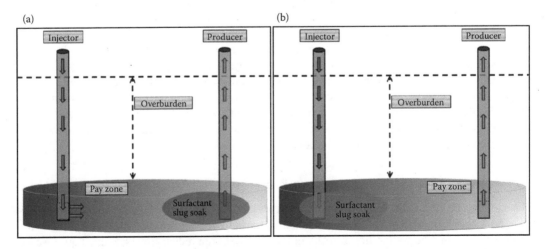

FIGURE 19.3 (See color insert.) Surfactant-slug enhanced waterflooding process: (a) surfactant-soaked production zone; (b) surfactant-soaked injection zone. (Adapted from P.M. Mwangi. An experimental study of surfactant enhanced waterflooding. Masters' Thesis, Louisiana State University, 2010.)

TABLE 19.2

Experimental Results for Process Optimization by Varying Surfactant Slug Size and Surfactant Placement

EOR Process	Surfactant Slug		Fluid Injected After Surfactant Slug Soak		Total Recovery (% OOIP)
	Slug Size	Slug Placement	Brine	Surfactant	
Waterflood	None	None	2 PV	None	47
Improved waterflood	0.2 PV	Producer	2 PV	None	55
Improved waterflood	0.3 PV	Producer	2 PV	None	62
Low concentrated (LC) surfactant flood	None	None	None	2 PV	52
Improved LC surfactant flood	0.2 PV	Producer	None	2 PV	58
Improved waterflood	0.2 PV	Injector	2 PV	None	67
Ideal surfactant flood	None	None	None	2 PV	94
Experimental parameters	Ideal soaking time: 24 h				
	Low concentrated surfactant flood: 1500 ppm				
	Ideal surfactant flood: 3000 ppm				
	Crude oil type: Yates oil				
	Brine type: Yates brine				

Note: The oil recovery efficiency of each process was evaluated by the OOIP recovered.

A significant amount of work has been done to determine the most effective placement of injected slugs in the PS, AS, and ASP processes. Yang and Me [63] investigated the effect of polymer placement in the ASP process, and they found that the incremental recovery was higher when polymer was injected separately from the alkaline and surfactant slug. Li [64] concluded from his experimental results that it was better to place polymer in the preflush slug than the post flush slug. On the other hand, Sheng [19] varied the location of the polymer slug with respect to the surfactant slug in his simulations, by either having a mixed surfactant–polymer slug, a polymer only slug, or a combination of both. On the basis of his reservoir simulations, he found that the incremental oil recovered was the same in all cases and, therefore, indicated that the placement of the polymer slug did not make a significant difference. These results may be caused by using a one-dimensional homogeneous simulation model as opposed to the heterogeneous cores used by Li [64]. Also, Sheng's [19] simulations were of a PS process, whereas Li [64] and Yang and Me [63] carried out an ASP flood. Gogarty's [65] analysis explained that field results indicated that line-drive patterns were superior to five-spot patterns for PS floods. Two injection schemes were proposed for a given polymer amount: (1) a small slug with a high concentration and (2) a large slug with a low concentration. Scheme 1 was reported [68] to be better because a high concentration polymer slug had a higher mobility ratio; therefore, the sweep efficiency was better.

19.9 LIMITATIONS AND CHALLENGES

The two major limitations of surfactant usage in EOR are typically related to the formation of high viscosity emulsions and high retention. This section will give a brief overview of these two challenges.

19.9.1 SURFACTANT RETENTION

The control of surfactant retention in the reservoir is one of the most important factors in determining the success or failure of a surfactant flooding project. About half or more of the total cost of the project typically goes to the chemical cost [18]. Surfactant retention is defined as the surfactant left behind in the reservoir due to various mechanisms, including adsorption onto the rock and phase trapping. Factors affecting the surfactant retention in a reservoir include temperature, pH, oxidative potential (Eh), salinity, microemulsion viscosity, surfactant type, activity of crude oil, cosolvent, and rock mineralogy. Usually, the factors that can be manipulated for EOR purposes are the surfactant type, the microemulsion viscosity, and the pH, whereas the rest are governed by reservoir conditions.

In the surfactant selection (disregarding wettability modification), anionic surfactants are preferred because they have low adsorption at neutral to high pH on both sandstones and carbonates. For sandstone rocks, surfactant adsorption depends more on the clay surfaces than on the quartz surfaces. Silica is negatively charged at reservoir conditions and exhibits a negligible adsorption of anionic surfactants at high pH [6]. However, the clays interbedded in the rock matrix have a negative charge on the faces and a positive charge on their edges, which are pH dependent. Several research groups [6,30,70] have observed this behavior and concluded that surfactant adsorption results primarily from the presence of clays in sandstones. In addition, Wang [70] reported that the Eh of a system has a significant effect on the adsorption of the surfactant. Wang observed a significant difference in retention values between the laboratory core floods and the field data. Many oil reservoirs exist in anaerobic and reducing environments; however, laboratory core floods are typically conducted in uncontrolled, aerobic, oxidizing conditions. Wang [70] reported higher retention values in core floods conducted in aerobic environment than those conducted in reduced environment.

The presence of multivalent cations also influences the amount of surfactant adsorbed due to ion bridging. Ion bridging occurs when an ion (especially cations such as Ca^{2+} and Mg^{2+}) acts as a bridge between the surfactant molecule and the rock surface, which results in the adsorption of the surfactant to the rock surface. Increased surfactant adsorption with an increase in multivalent cation concentration was observed by several researchers [71,72]. Newer surfactants are showing relatively higher tolerance to multivalent cations [73].

Glover et al. [72] suggested that phase trapping due to immobile oil/microemulsion phase significantly contributes to surfactant retention. They observed that surfactant retention increased linearly with salinity until the point where it deviated abnormally from linearity and almost all the surfactant injected was retained at that salinity. Novosad [71] extended the study and devised a method to quantify the surfactant retained due to adsorption and entrapment of immobile oil phase and surfactant precipitation due to divalent cations. He concluded that better-performing processes are usually accompanied by lower surfactant retention but not vice versa. In addition, microemulsion viscosity also plays a significant role in surfactant retention. This parameter is difficult to interpolate its behavior for it is a strong function of composition and temperature. Viscous microemulsions enhanced phase trapping due to entrapment, thus increasing surfactant retention. One way to reduce microemulsion viscosity is by adding alcohol [6] or by using highly branched surfactants [17] as branching in surfactant reduces the tendency to form ordered structures.

All in all, many factors affect surfactant retention, which makes the study of this parameter and its optimization difficult for researchers to master. However, the economics of surfactant flooding is significantly impacted by this parameter, and thus it is of utmost importance to reduce surfactant retention.

19.9.2 PRODUCTION OF HIGH VISCOSITY EMULSIONS

Although the surfactant-based processes have been effective in increasing oil production, they also contribute to the operation and production challenges due to the formation of emulsions [75]. The formation of emulsions during oil production is a costly problem both in terms of chemicals used and production lost [70,75]. In an ASP process, the polymer in the injection fluid is produced with emulsion. Its physical and chemical characteristics are altered in the process due to adsorption in the reservoir, shearing, and chemical reactions [75,76]. This leads to tight (small and closely distributed oil and water droplets) emulsions, increased water viscosity, stronger chemical interactions, limitations on the processing methods and process conditions, reduced efficiency on production equipment, and excessive equipment failures. In addition, the problem is further exacerbated by small oil and water droplets in the production stream due to the low IFT between oil and water produced by surfactants [75]. Therefore, the complexity and uncertainty of the nature of produced fluids when ASP chemicals are used adds to the processing difficulties. Emulsions are difficult to treat and cause several operational problems, such as failure of separation equipment in gas/oil separation plants, production of off-specification crude oil, and creating high pressure drops in flow lines [70]. Emulsions have to be treated to remove the dispersed water and salts to meet crude oil specifications for transportation and storage and to reduce corrosion and catalyst poisoning in downstream processing facilities. The demulsification process is used to break the emulsions using thermal, chemical (surfactant), electrical, and mechanical methods. Kokal [74] gives an elaborate review of production and operational problems related to crude oil emulsions. Zheng et al. [76] reported that the use of polymers increased failures in heating equipment, including heat exchangers and in-vessel heating elements such as fire tubes, and thus caused a significant increase in produced fluid processing difficulties and cost.

19.9.3 OVERVIEW

In summary, it is crucial to select surfactants that do not exhibit the previously mentioned problems. Once an appropriate surfactant is selected, surfactant modeling is relatively straightforward, with a few well-designed experiments needed to provide the most important simulation parameters. The key to reducing both viscosity and surfactant retention is the same: surfactant structure with special emphasis on branched hydrophobes [10]. The high salinity encountered in some reservoir brines can affect the effectiveness of the surfactants. New generations of surfactants are being designed to tolerate high salinity and hardness so that there is no practical limit for high salinity oil reservoirs. In addition, there are surfactants available for both low and high temperature reservoirs, such as sulfonates that are stable to very high temperatures but sensitive to divalent ions. Good reservoir characterization, reservoir engineering, and process optimization are also crucial to have a successful surfactant flood.

19.10 ECONOMICS

Currently, high-performance surfactants cost less than $2 per pound of pure surfactant [18]. In addition, the use of alkaline chemicals such as sodium carbonate with surfactants resulted in the reduction of chemical costs from $10 per incremental barrel of oil to less than $5 per incremental barrel of oil in certain cases. This has been applied commercially on a limited scale and is dependent on the reservoir conditions.

Technical feasibility of surfactant flooding has already been established; however, the economic feasibility depends on complex factors such as oil price, surfactant consumption, and surfactant cost. The implementation environment (onshore or offshore) also plays an important role in the economics of the project. Typically, surfactant-based EOR projects acquire their cost from the large volumes of injection chemicals required, as well as demulsifiers to break produced water/oil emulsions and inhibitors to control scale [75]. If the project under consideration is in an offshore environment, the cost significantly increases because of the large shipment and storage cost resulting from the large volumes of chemicals and additional equipment needed. Generally, the cost of the surfactant itself is the single most expensive item in the total cost of a chemical flood [1]. These costs include both the initial investment in purchasing the surfactant as well as the cost of replacing surfactant lost due to adsorption. It is frequently found that the amount of surfactant adsorbed accounts for most of the cost of the surfactant.

Because oilfield surfactants are synthesized from petroleum sources, their cost will rise at least as fast as that of oil. Hence, simply waiting for oil prices to increase will not necessarily make surfactant flooding economically feasible. The revenue from the oil produced by surfactant flooding must at least pay for the cost of surfactant, additional engineering services, equipment, and operating costs for the duration of the flood to provide a reasonable return on investment. Producing more barrels of oil for each pound of surfactant injected into the reservoir is a technological problem that has direct bearing on the economics of this EOR process. Therefore, low-cost dilute surfactants aimed at well-designed alteration of wettability to yield high oil recoveries seem to be the way forward in surfactant EOR.

19.11 SUMMARY AND PROSPECTS

Surfactants have a long history in EOR. The concept of lowering oil–water IFT by adding surfactants to injected water in a waterflood to recover more of the oil trapped by capillary forces is not only easy to understand but also easy to implement in the field. However, the cost of these chemical surfactants and their loss in the reservoir due to adsorption on the rock surface have made this an expensive EOR process that resulted in a prolonged decay in the number of field projects. The recent findings about the role of simple and inexpensive surfactants in altering reservoir wettability in favor of increased oil recoveries, as demonstrated in several laboratory studies, seems to have rekindled the interest in this area of research and development. The ability now to confidently and reproducibly characterize wettability through dynamic contact angle measurements at realistic pressures and temperatures using actual reservoir fluids has enabled us to quantify the wettability-altering effects of surfactants as well as the resulting oil recovery enhancements. Thus, the future of surfactants in EOR seems to be bright, despite some of the challenges involving emulsions that invariably form at low IFTs and the difficulty of their subsequent separation.

ACKNOWLEDGMENTS

The authors thank Wagirin (Ruiz) Paidin and Yu Zheng of LSU and Dr. Subhash Ayirala of Shell for their review and valuable suggestions.

REFERENCES

1. L.L. Schramm and D.G. Marangoni. Surfactants and their solutions: Basic principles. In *Surfactants: Fundamentals and Applications in the Petroleum Industry*, L.L. Schramm (Ed.), Chap. 1. Cambridge University Press, Cambridge, pp. 3–50 (2000).
2. S. Thomas. Enhanced oil recovery—An overview. *Oil Gas Sci. Technol., Rev. IFP*, **63**, 9–19 (2008).
3. L.C. Uren and E.H. Fahmy. Factors influencing the recovery of petroleum from unconsolidated sands by waterflooding. *Trans. AIME*, **77**, 318–335 (1927).
4. H. Atkinson. Recovery of petroleum from oil bearing sands. US Patent No. 1,651,311 (1927).
5. H.J. Hill, J. Reisberg and G.J. Stegemeier. Aqueous surfactant systems for oil recovery. *J. Pet. Technol.*, **25**, 186–194 (1973).
6. G.J. Hirasaki, C.A. Miller and M. Puerto. Recent advances in surfactant EOR. *SPE J.*, **16**, 889–907 (2011).
7. W.B. Gogarty and W.C. Tosch. Miscible type water flooding: Oil recovery with micellar solutions. *J. Pet. Technol.*, **20**, 1407–1414 (1968).
8. L.W. Holm. Soluble oils for improved oil recovery. In *Improved Oil Recovery by Surfactant and Polymer Flooding*, D.O. Shah and R.S. Schechter (Eds.). Academic Press, New York, pp. 453–495 (1977).
9. G.J. Hirasaki. Application of the theory of multicomponent, multiphase displacement to three-component, two phase surfactant flooding. *SPE J.*, **21**, 191–204 (1981).
10. G.A. Pope. Overview of chemical EOR. Paper presented at the Casper EOR Workshop, Casper, Wyoming (2007).
11. S. Adkins, G. Pinnawala Arachchilage, S. Solairaj, J. Lu, U. Weerasooriya and G.A. Pope. Development of thermally and chemically stable large-hydrophobe alkoxy carboxylate surfactants. Presented at the SPE/DOE Symposium on Improved Oil Recovery, Tulsa, OK (2012).

12. P.J. Liyanage, S. Solairaj, G. Pinnawala Arachchilage, H.C. Linnemeyer, D.H. Kim, U. Weerasooriya and G.A. Pope. Alkaline surfactant polymer flooding using a novel class of hydrophobe surfactants. Presented at the SPE/DOE Symposium on Improved Oil Recovery, Tulsa, OK (2012).

13. P. Chen and K.K. Mohanty. Surfactant-mediated spontaneous imbibition in carbonate rocks at harsh reservoir conditions. Presented at the SPE/DOE Symposium on Improved Oil Recovery, Tulsa, OK (2012).

14. D. Green and P.G. Willhite. *Enhanced Oil Recovery*. SPE Textbook Series, Society of Petroleum Engineers, Vol. 6. Richardson, TX (1998).

15. R. Nelson, J.B. Lawson, D.R. Thigpen and G.L. Stegemeier. Co-surfactant enhanced alkaline flooding. SPE/DOE 12672, Presented at the SPE/DOE fourth symposium on Enhanced Oil Recovery, Tulsa, OK (1984).

16. M.J. Rosen, H. Wang, P. Shen and Y. Zhu. Ultralow interfacial tension for enhanced oil recovery at very low surfactant concentrations. *Langmuir*, **21**, 3749–3756 (2005).

17. D.B. Levitt. Experimental evaluation of high performance EOR surfactants for a dolomite oil reservoir. MS. Thesis, University of Texas, Austin, TX (2006).

18. M. Aoudia, W.H. Wade and V. Weerasooriya. Optimized microemulsions formulated with propoxylated guerbet alcohol and propoxylated tridecyl alcohol sodium sulfates. *J. Dispersion Sci. Technol.*, **16**, 135–155 (1995).

19. J.J. Sheng. Surfactant flooding. In *Modern Chemical Enhanced Oil Recovery: Theory and Practice*, Chap. 7. Gulf Professional Publishing, Kidlington, Oxford, pp. 239–335 (2011).

20. M.A. Klins. *Carbon Dioxide Flooding: Basic Mechanism and Project Design*. International Human Resources Development Corporation, Boston, MA, 267–275 (1984).

21. P.A. Winsor. *Solvent Properties of Amphiphilic Compounds*. Butterworth Scientific Publications, London (1954).

22. R.N. Healy and R.L. Reed. Physicochemical aspects of microemulsion flooding. *SPE J.*, **14**, 491–501 (1974).

23. C. Huh. Interfacial tensions and solubilizing ability of a microemulsion phase that coexists with oil and brine. *Colloid Interface Sci.*, **21**, 408–426 (1979).

24. M. Bourrel and R.S. Schechter. *Microemulsions and Related Systems, Formulation, Solvency, and Physical Properties*. Marcel Dekker, New York (1988).

25. S.L. Wellington and E.A. Richardson. Low surfactant concentration enhanced waterflood. SPE 30748, Presented at the Annual Technical Conference and Exhibition of the Society of Petroleum Engineers, Dallas, TX (1997).

26. Y.S. Wu, P. Blanco, Y. Tang and W.A. Goddard. A study of branched alcohol propoxylate sulfate surfactants for improved oil recovery. SPE 95404, Presented at the SPE Annual Technical Conference and Exhibition, Dallas, TX (2005).

27. S. Jayanti, L.N. Britton, V. Dwarakanath and G.A. Pope. Laboratory evaluation of custom designed surfactants to remediate NAPL source zones. *Environ. Sci. Technol.*, **36**, 5491–5497 (2002).

28. G.J. Hirasaki. Wettability: Fundamentals and surface forces. *SPE Formation Eval.*, **6**, 217–226 (1991).

29. W.G. Anderson. Wettability literature survey—Part 2: Wettability measurement. *J. Pet. Technol.*, **38**, 1246–1262 (1986).

30. P. Somasundaran and L. Zhang. Adsorption of surfactant on minerals for wettability control in improved oil recovery processes. *J. Colloid Interface Sci.*, **191**, 202–208 (1997).

31. E.A. Spinler, D.R. Zornes, D.P. Tobola and A. Moradi-Araghi. Enhancement of oil recovery using a low concentration of surfactant to improve spontaneous and forced imbibition in chalk. SPE 59290, Presented at the SPE/DOE Improved Oil Recovery Symposium, Tulsa, OK (2000).

32. H.L. Chen, L.R. Lucas, L.D. Nogaret, H.D. Yang and D.E. Kenyan. Laboratory monitoring of surfactant imbibition with computerized tomography. *SPE Reserv. Eval. Eng.*, **4**, 16–25 (2001).

33. H.L. Najurieta, N. Galacho, M.E. Chimienti and N. Silvia. Effects of temperature and interfacial tension in different production mechanisms. SPE 69398, Presented at the SPE Latin American and Caribbean Petroleum Engineering Conference, Buenos Aires (2001).

34. G. Hirasaki and D.L. Zhang. Surface chemistry of oil recovery from fractured, oil-wet, carbonate formations. *SPE J.*, **9**, 151–162 (2004).

35. T. Babadagli. Analysis of oil recovery by spontaneous imbibition of surfactant solution. SPE 84866, Presented at the SPE International Improved Oil Recovery Conference in Asia Pacific, Kuala Lumpur (2003).

36. D.C. Standnes and T. Austad. Wettability alteration in carbonates by low cost ammonium surfactants based on bio-derivatives from the coconut palm as active chemicals to change wettability from oil-wet to water-wet conditions. *Colloids Surf.*, **218**, 161–173 (2003).

37. D.N. Rao, S.C. Ayirala, A.A. Abe and W. Xu. Impact of low-cost dilute surfactants on wettability and relative permeability. SPE 99609-MS, Presented at the SPE/DOE Symposium on Improved Oil Recovery, Tulsa, OK (2006).

38. S.C. Ayirala, C.S. Vijapurapu and D.N. Rao. Beneficial effects of wettability altering surfactants in oil-wet fractured reservoirs. *J. Pet. Sci. Eng.*, **52**, 261–274 (2006).

39. D.N. Rao and M.G. Girard. A new technique for reservoir wettability characterization. *J. Can. Pet. Technol.*, **35**, 31–39 (1996).

40. D.N. Rao. The concept, characterization, concerns and consequences of contact angles in solid-liquid-liquid systems. In *Contact Angle, Wettability and Adhesion*, Vol. 3, K.L. Mittal (Ed.). VSP International Science Publishers, Utrecht, the Netherlands, pp. 191–210 (2003).

41. C.S. Vijapurapu and D.N. Rao. The effect of rock surface characteristics on reservoir wettability. In *Contact Angle, Wettability and Adhesion*, Vol. 3, K.L. Mittal (Ed.). VSP International Science Publishers, Utrecht, the Netherlands, pp. 407–426 (2003).

42. A. Seethepalli, B. Adibhatla and K.K. Mohanty. Physicochemical interactions during surfactant flooding of fractured carbonate reservoirs. *SPE J.*, **9**, 411–418 (2004).

43. K. Kumar, E.K. Dao and K.K. Mohanty. Atomic force microscopy study of wettability alteration by surfactants. *SPE J.*, **13**, 137–145 (2008).

44. W. Demin, C. Jiecheng, W. Junzheng, Y. Zhenyu, Y. Yuming and L. Hongfu. Summary of ASP pilots in Daqing Oil Field. SPE 57288, Presented at the SPE Asia Pacific Improved Oil Recovery Conference, Kuala Lumpur, Malaysia (1999).

45. W. Demin, Z. Zhenhua, C. Jiecheng, Y. Jingchun, G. Shutang and L. Lin. Pilot tests of alkaline/surfactant/polymer flooding in Daqing oil field. *SPE Reserv. Eng.*, **12**, 229–233 (1997).

46. M.J. Pitts, P. Dowling, K. Wyatt and H. Surkalo. Alkaline-surfactant-polymer flood of the tanner field. Presented at the SPE/DOE Symposium on Improved Oil Recovery, Tulsa, Oklahoma (2006).

47. B.J. Felber. Selected U.S. Department of Energy's EOR technology applications. Presented at the SPE International Improved Oil Recovery Conference in Asia Pacific, Kuala Lumpur, Malaysia (2003).

48. B.J. Miller, M.J. Pitts, P. Dowling and D. Wilson. Single well alkaline-surfactant injectivity improvement test in the Big Sinking field. Presented at the SPE/DOE Symposium on Improved Oil Recovery, Tulsa, OK (2004).

49. M. Othman, M.O. Chong, R.M. Sai, S. Zainal, M.S. Zakaria and A.A. Yaacob. Meeting the challenges in alkaline surfactant pilot project implementation at Angsi field, Offshore Malaysia. Presented at the Offshore Europe, Aberdeen (2007).

50. N.R. Morrow. Wettability and its effect on oil recovery. *J. Pet. Technol.*, **42**, 1476–1484 (1990).

51. H.S. Al-Hashim, V. Obiora, H.Y. Al-Yousef, F. Fernandez and W. Nofal. Alkaline surfactant polymer formulation for saudi arabian carbonate reservoirs. SPE 35353, Presented at the SPE/DOE Improved Oil Recovery Symposium, Tulsa, OK (1996).

52. V. Bortalotti, P. Macini and F. Srisuriyachai. Laboratory evaluation of alkali and alkali–surfactant–polymer flooding combined with intermittent flow in carbonatic rocks. SPE 122499, Presented at the Asia Pacific Oil and Gas Conference and Exhibition, Jakarta, Indonesia (2009).

53. E.J. Manrique, V.E. Muci and M.E. Gurfinkel. EOR field experiences in carbonate reservoirs in the United States. *SPE Reserv. Eval. Eng.*, **10**, 667–686 (2007).

54. E.J. Manrique, C. Thomas, R. Ravikiran, M. Izadi, M. Lantz, J. Romero and V. Alvarado. EOR: Current status and opportunities. Presented at the SPE/DOE Improved Oil Recovery Symposium, Tulsa, OK (2010).

55. L.E. Treibel, D.L. Archer and W.W. Owen. A laboratory evaluation of the wettability of fifty oil-producing reservoirs. *SPE J.*, **12**, 531–540 (1972).

56. C.E. Johnson, Jr. Status of caustic and emulsion methods. *J. Pet. Technol.*, **28**, 85–92 (1976).

57. K. Li and A. Firoozabadi. Experimental study of wettability alteration to preferential gas-wetness in porous media and its effect. *SPE Reserv. Eng.*, **3**, 139–149 (2000).

58. G. Tang and A. Firoozabadi. Relative permeability modification in gas/liquid systems through wettability alteration to intermediate gas wetting. *SPE Reserv. Eng.*, **5**, 427–436 (2002).

59. M. Fahes and A. Firoozabadi. Wettability alteration to intermediate gas-wetting in gas-condensate reservoirs at high temperatures. SPE 96184, Presented at the SPE Annual Technical Conference and Exhibition, Dallas, TX (2005).

60. Y. Zheng and D.N. Rao. Surfactant-induced spreading and wettability effects in condensate reservoirs. SPE 129668, Presented at SPE Improved Oil Recovery Symposium, Tulsa, OK (2010).

61. Y. Zheng and D.N. Rao. Experimental study of spreading and wettability effects by surfactants in condensate reservoirs at reservoir conditions. SPE 141016, Presented at the SPE International Symposium on Oilfield Chemistry, The Woodlands, TX (2011).

62. Y. Zheng, B. Saikia and D.N. Rao. Correlation of surfactant-induced flow behavior modification in gas condensate reservoirs with dynamic contact angles and spreading. SPE 146786, Presented at the SPE Annual Technical Conference and Exhibition, Denver, CO (2011).

63. L. Yang and S.C. Me. *Research and Testing of Combinations of Chemical Enhanced Oil Recovery, Research and Development of Enhanced Oil Recovery in Daqing.* Petroleum Industry Press, pp. 326–351 (2006).

64. H.B. Li. *Advances in Alkaline-Surfactant-Polymer Flooding and Pilot Tests.* Science Press (2007).

65. W.B. Gogarty. Enhanced oil recovery through the use of chemicals—Part 1. *J. Pet. Technol.*, **35**, 1581–1590 (1983).

66. L.E. Zerpa, N.V. Queipoa, T.S. Pintosa and J.L. Salagerb. An optimization methodology of alkaline-surfactant-polymer flooding processes using field scale numerical simulation and multiple surrogates. *J. Pet. Sci. Eng.*, **47**, 197–208 (2005).

67. G.A. Anderson, M. Delshad, C.B. King, H. Mohammadi and G.A. Pope. Optimization of chemical flooding in a mixed-wet dolomite reservoir. SPE 100082, Presented at the SPE/DOE Symposium on Improved Oil Recovery, Tulsa, OK (2006).

68. M. Delshad, G.A. Pope and K. Sepehrnoori. A framework to design and optimize chemical flooding processes. DOE Report 835937 (2004).

69. P.M. Mwangi. An experimental study of surfactant enhanced waterflooding. Masters' Thesis, Louisiana State University (2010).

70. F.H. Wang. Effects of reservoir anaerobic, reducing conditions on surfactant retention in chemical flooding. SPE 22648, *SPE Reserv. Eng.*, **8**, 108–116 (1993).

71. J. Novosad. Surfactant retention in Berea sandstone—Effects of phase behavior and temperature. SPE 10064, *SPE J.*, **22**, 962–970 (1982).

72. C.J. Glover, M.C. Puerto, J.M. Maerker and E.L. Sandvik. Surfactant phase behavior and retention in porous media. SPE 7053, *SPE J.*, **19**, 183–193 (1979).

73. S. Solairaj. New method of predicting optimal surfactant structure for EOR. MS. Thesis, University of Texas, Austin, TX (2011).

74. S. Kokal. Crude-oil emulsions: A state-of-the-art review. *SPE Prod. Facil.*, **20**, 5–13 (2005).

75. K. Raney, S. Ayirala, R. Chin and P. Verbeek. Surface and subsurface requirements for successful implementation of offshore chemical enhanced oil recovery. OTC 21188, Presented at the Offshore Technology Conference, Houston, TX (2005).

76. F. Zheng, P. Quiroga and G.W. Sams. Challenges in processing produced emulsion from chemical enhanced oil recovery—Polymer flood using polyacrylamide. SPE 144322, Presented at the SPE Enhanced Oil Recovery Conference, Kuala Lumpur, Malaysia (2011).

20 Soil Removal by Surfactants during Cleaning Processes

Clarence A. Miller

CONTENTS

20.1 INTRODUCTION

The single largest market for surfactants is that for cleaning materials, such as household and industrial laundry detergents, dishwashing products, hand and body soaps, shampoos, and cleaners for hard surfaces such as floors, tile, and glass. Because of their importance, such products have been and continue to be the focus of research and development to improve performance, to reduce cost, and to minimize adverse environmental impact. In recent years, the last of these objectives has included a shift toward the greater use of sustainable ingredients, that is, those derived from renewable raw materials.

The emphasis in this article is on the mechanisms by which surfactants remove "soils" during cleaning processes carried out with aqueous phases. Other components of detergent formulations, such as "builders" to prevent the precipitation of anionic surfactants by calcium, magnesium, and other multivalent cations, polymers to prevent the redeposition of soils once removed, bleaches, fabric softeners, enzymes, and so on, are beyond the scope of the article. Much of the work cited is from articles on cleaning of fabrics during laundry processes because of the extensive literature on this topic.

The discussion below is divided into two main parts. Section 20.2 deals with the removal of oily liquid soils, and Section 20.3 deals with the removal of solid organic soils. Particulate soils, although important in many cases, are mentioned only in passing in Section 20.3 to keep the length of the article reasonable. Of course, mixtures of liquid and solid soils are frequently encountered, and formulations may need to be optimized to achieve the best overall cleaning performance.

Surfactant Science and Technology

20.2 OILY LIQUID SOILS

In laundry and other cleaning processes, it is often necessary to remove liquid phase "oily" soils such as dirty motor oil, liquid triglycerides (cooking oils), liquid fatty acids or alcohols, or combinations of these soils from fabrics or other solid surfaces. The chief mechanisms of removal are widely considered to be "roll-up" or "roll-back" and some combination of solubilization and emulsification. Roll-up involves using surfactant solutions to produce a large contact angle (measured through the soil, a convention followed throughout this article) at the periphery of soil layers on the solid (see Figure 20.1). As a result, the contact area between the individual soil drops and the solid decreases, and drops project farther into the surfactant solution, where they can be more readily broken off by agitation. The breakoff step is facilitated by the low interfacial tension (IFT) between soil and washing bath and is a type of emulsification, although the emulsification mechanisms discussed in the literature are usually limited to situations where contact angle effects are not central for soil removal. If the contact angle reaches 180° and water spreads spontaneously, the entire soil drop can detach. Roll-up is the chief mechanism of oily soil removal from hydrophilic surfaces such as cotton using conventional anionic surfactants in laundry products such as linear alkylbenzene sulfonates.

Solubilization and emulsification mechanisms, although long considered as removal mechanisms for solid organic soils, have received increased attention for liquid oily soils since the 1980s. One reason is that the use of synthetic fabrics has increased greatly in recent decades. These materials are more hydrophobic than cotton, and roll-up is less easily achieved. Another is that microemulsion behavior has become much better understood, which has enabled researchers to develop and exploit its possible application in detergency. The formation of microemulsions by solubilization can produce very low IFTs. Figure 20.2 illustrates a simple case where reduction in IFT with little change in contact angle leads to emulsification caused by gravity. Of course, roll-up and solubilization/ emulsification are not mutually exclusive because the changes in contact angle and IFT and the formation of microemulsions often occur simultaneously. Systems where roll-up dominates at some conditions and solubilization/emulsification at others for the same soil–surfactant system are considered in the discussion below.

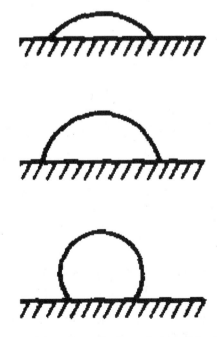

FIGURE 20.1 Schematic diagram of the roll-up of a drop of oily soil due to increase in contact angle measured through the soil.

508

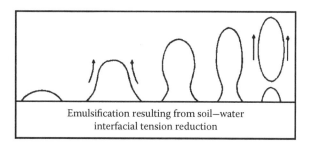

Emulsification resulting from soil–water
interfacial tension reduction

FIGURE 20.2 Schematic diagram of emulsification produced by gravity for a drop of oily soil with low IFT. There is a small change in the contact angle as the drop elongates until it becomes unstable and pinches off.

20.2.1 ROLL-UP

Young's equation gives the relationship between the contact angle and the various interfacial tensions for an oil drop immersed in an aqueous phase and resting on a solid surface (see Figure 20.3) as follows:

$$\gamma_{sw} = \gamma_{so} + \gamma_{ow} \cos \theta \tag{20.1}$$

Here, γ_{sw}, γ_{ow}, and γ_{so} are interfacial tensions at the solid–water, oil–water, and solid–oil interfaces, respectively, and θ is the equilibrium contact angle measured through the oil. The work of adhesion between solid and oil, that is, the energy required to replace a unit area of solid–oil interface by unit areas of solid–water and oil–water interfaces, is given by

$$W_{so} = \gamma_{sw} + \gamma_{ow} - \gamma_{so} = \gamma_{ow} (1 + \cos \theta) \tag{20.2}$$

where Equation 20.1 has been invoked to obtain Equation 20.2. W_{so} is useful in assessing the strength of the adhesion of oil to solid but is not equal to the energy required to detach an oil drop of a given volume initially immersed in water but in contact with the solid with contact angle θ and disperse it as a spherical drop in the aqueous phase. The energy requirements based on the thermodynamics for this process and for partial detachment of a drop are given by Carroll [1].

An interesting phenomenon affecting roll-up is that two different basic configurations are possible for an oil drop on a fiber of uniform, circular cross section. For strongly oil-wet fibers and large oil lenses, stable equilibrium exists for the axisymmetric configuration illustrated in Figure 20.4a. As indicated there, a large lens in this case means a large value of n, the ratio of the maximum radius of the lens to the fiber radius. As contact angle increases, especially above 60°, the axisymmetric configuration is stable only for very large drops. Smaller drops move to one side of the fiber, assuming the "clamshell" configuration of Figure 20.4b, which is more readily detached during washing. The condition for transition between the two configurations is given by Carroll [1,2].

It is beyond the scope of this article to discuss in detail the behavior near moving contact lines, which continues to be an area of active research. However, a few recent articles particularly relevant to roll-up are of interest. Kondiparty et al. [3–5] present theoretical and experimental evidence that,

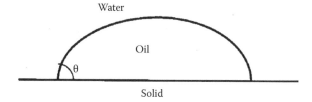

FIGURE 20.3 A drop of oil at equilibrium on a horizontal solid surface with contact angle θ.

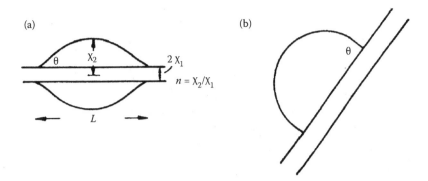

FIGURE 20.4 Possible configurations of a drop on a fiber: (a) axisymmetric about fiber axis; (b) "clamshell." (Reprinted with permission from Carroll, B.J. Equilibrium conformations of liquid drops on thin cylinders under forces of capillarity. A theory for the roll-up process. *Langmuir* 2, 248–250. Copyright 1986 American Chemical Society.)

for an oil drop on a solid surface immersed in an aqueous solution containing spherical nanoparticles or surfactant micelles, the layering of the spheres in the wedge-shaped film near the contact line can lead to spreading of the aqueous phase even when the equilibrium contact angle in the absence of layering is small but positive. Figure 20.5 illustrates this situation schematically.

Kolev et al. [6] focused on the dynamics of the roll-up process. On the basis of their experiments with aqueous solutions of C_{14-16} alpha olefin sulfonate surrounding hexadecane drops on horizontal glass microscope slides, they developed a phenomenological theory in which a dynamic drag force opposing motion of the contact line was added to Equation 20.1:

$$\gamma_{sw} - \gamma_{ow} \cos \theta - \beta \, (dr_c/dt) = \gamma_{so} \tag{20.3}$$

where β is a line drag coefficient and r_c is the radius of the contact line. They were able to calculate β using their measured values of θ, r_c, and γ_{ow} as a function of time during roll-up obtained from

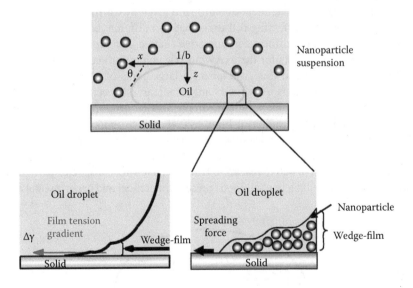

FIGURE 20.5 **(See color insert.)** Schematic diagram of nanoparticle structuring in wedge-shaped thin aqueous film spreading under an oil drop during roll-up. (Reprinted with permission from Kondiparty, K., A. Nikolov, S. Wu, and D.T. Wasan. Wetting and spreading of nanofluids on solid surfaces driven by the structural disjoining pressure: Statics analysis and experiments. *Langmuir* 27, 3324–3335. Copyright 2011 American Chemical Society.)

videomicroscopy experiments. γ_{ow} was found using drop shape analysis as for a sessile drop. Figure 20.6 shows the variation with time of r_c, and $\alpha = (\pi - \theta)$ for one of their experiments. Both r_c and α initially decrease rapidly, that is, θ increases rapidly, in stage 1 until reaching nearly constant values. Roll-up then proceeds in stage 2 until the drop becomes gravitationally unstable, which starts the detachment process of stage 3.

Subsequently, Kralchevsky et al. [7] added to the theory by providing an analysis of the possible behavior near the contact line. Basically, they supposed that water interacted with the surface layer of the glass slide to form a thin gel layer through which water could diffuse laterally beneath the oil drop from its periphery toward its axis. By fitting roll-up data similar to that of Figure 20.6, they were able to obtain values of the parameter β defined above and a penetration time t_p characteristic of the gel formation process.

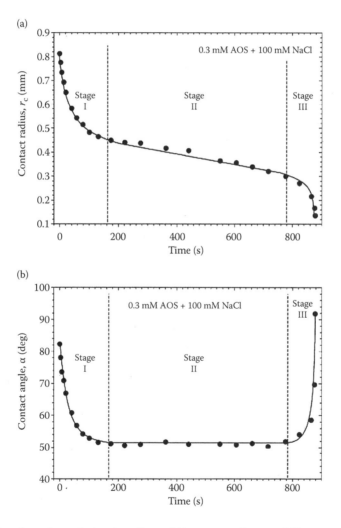

FIGURE 20.6 Time dependence during the roll-up of (a) contact radius r_c and (b) contact angle α measured through water = $(\pi - \theta)$. Stage 1 represents the initial rapid decrease in r_c and α, stage 2 represents the roll-up at nearly constant α, and stage 3 represents the detachment after gravitational instability sets in. The oil is 1-hexadecane, and the surfactant is C_{14-16} alpha-olefin sulfonate (AOS). (From V.L. Kolev et al., *J. Colloid Interface Sci.* 257, 357–363, 2003.)

20.2.2 SOLUBILIZATION

Solubilization rates of oily soils into micelles in surfactant solutions above the critical micelle concentration (CMC) can be measured by, for instance, determining n as a function of time for an axisymmetric oil drop on a fiber immersed in a surfactant solution (Figure 20.4a; [8]) or using videomicroscopy to determine the variation with time of the diameter of a spherical oil drop in a surfactant solution [9]. However, these rates are typically too low to be of practical interest for the low surfactant concentrations and short cleaning times used in household laundry and most other cleaning processes. Thus, solubilization alone is not a major mechanism of oily soil removal. However, such solubilization of oily soils as well as solubilization of water and surfactant by such soils can lead to the formation of microemulsions, which can have low IFTs and low viscosities, thereby facilitating emulsification. Hence, solubilization and emulsification can combine to cause oily soil removal. For example, when the soil initially contains polar components such as acids or alcohols, it can often be spontaneously emulsified during mass transfer processes, which include solubilization. Even when no polar components are present initially, some surfactant may be transferred to the soil during the washing step, and spontaneous emulsification may remove the soil during a subsequent rinse. These phenomena are discussed in the next section.

20.2.3 EMULSIFICATION AND SOLUBILIZATION/EMULSIFICATION

An oil drop resting on a horizontal solid surface in a pool of aqueous surfactant solution and having a nonzero contact angle provides an example of emulsification. The drop is drawn upward by gravity to form a pendant drop, as shown in Figure 20.2 (the 180° contact angle situation where water spreads spontaneously on the solid is excluded here because the drop would be removed by roll-up, as discussed previously). If the drop is sufficiently large or the IFT sufficiently low, the drop ultimately breaks off and rises owing to the effect of gravity, leaving a smaller drop attached to the solid. This behavior is that of stage 3 for the special case of roll-up on a horizontal surface (Section 20.2.1 and Figure 20.6).

Breakoff does not occur in the absence of external flow for drops small enough to be stabilized by interfacial tension. Although a detailed analysis including the contact angle and conditions for capillary instability is required for a precise value of the maximum stable drop volume V_{max}, its order of magnitude is the cube of the capillary length:

$$V_{max} = [\gamma/\Delta\rho g]^{3/2} \qquad (20.4)$$

where γ is the oil–water IFT, $\Delta\rho$ is the density difference between phases, and g is the gravitational acceleration. For a density difference of 200 kg/m³, this equation yields estimates of V_{max} of 11, 0.011, and 1.1×10^{-5} μL for IFTs of 10, 0.1, and 0.001 mN/m, respectively. That is, a nearly complete removal of oil occurs in the last case when IFT is ultralow, but removal is substantial even when IFT is reduced only to 0.1 mN/m. Removal by this mechanism does depend, however, on having suitable orientation of the soiled surface.

The emulsification of a drop can also be produced by flow parallel to the solid surface. Mahe et al. [10] showed that when the equilibrium contact angle has a large value close to 180° and when contact angle hysteresis exists, the drop can detach and roll along the surface. They presented an equation for the critical shear stress required for detachment when the flow is laminar. Basu et al. [11] developed a model, indicating that for smaller contact angles, hydrodynamic lift, enhanced or decreased by gravity depending on surface orientation, can lead to detachment.

In addition to these mechanical effects, emulsification can occur with the complete removal of soil from the solid, whatever the orientation, when the phase behavior of the system combined with heat or mass transport leads to spontaneous emulsification caused by local supersaturation. Basic knowledge of phase behavior in oil–water–surfactant systems, as summarized next, is important in understanding this mechanism of emulsification.

20.2.3.1 Microemulsion Phase Behavior

Consider first a simple case when comparable amounts of a pure alkane and water are present with a surfactant at a concentration of order 1% by weight. When conditions are such that the surfactant is preferentially soluble in the aqueous phase, for example, at low salinities for ionic surfactants, but present at sufficiently high concentration to form aggregates, the latter will solubilize some oil, and a thermodynamically stable oil-in-water (O/W) microemulsion is said to coexist with excess oil. This behavior is shown in the schematic diagram labeled "low salinity" and in the samples on the left side of the photograph in Figure 20.7. In the literature, this behavior is often called *Winsor I*, after an early observer of phase behavior in microemulsion systems [12]. For nonionic surfactants of the alcohol ethoxylate type, it is typically seen at low temperatures. When oil solubilization is substantial, the situation of most interest here, the oil droplets are nearly spherical and have diameters of order 10 nm.

In contrast, when the surfactant is preferentially soluble in the oil phase, a water-in-oil (W/O) microemulsion forms and coexists with excess aqueous phase, as seen in the "high salinity" diagram and in samples on the right side of the photograph in Figure 20.7. This behavior is called *Winsor II*. It occurs at high temperatures for nonionic alcohol ethoxylates. Here, the water or brine drops are also nearly spherical with typical diameters of order 10 nm.

Figure 20.7 shows that between the Winsor I and the Winsor II regions is a third region where a microemulsion coexists with both excess oil and excess aqueous phase. In this Winsor III region, the microemulsion is continuous in both oil and water, and the spontaneous curvature of surfactant films within the microemulsions is near zero. In contrast, the spontaneous curvature of droplets in the Winsor I and Winsor II regions has substantial magnitude, although of opposite sign in the two regions. For ionic surfactants, the "optimal salinity" occurs when a Winsor III microemulsion solubilizes equal volumes of oil and water. Likewise, for nonionic surfactants, the phase inversion

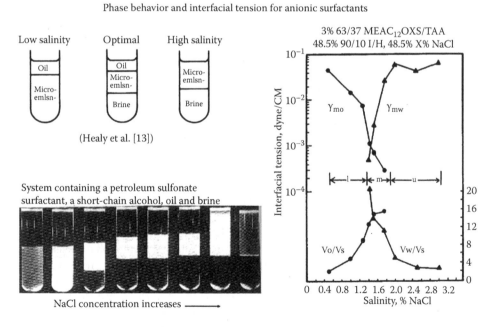

Phase behavior and interfacial tension for anionic surfactants

FIGURE 20.7 Variation with the salinity of microemulsion phase behavior for anionic surfactant. The diagram at the upper left shows schematically the phases present in the photograph, and the plots at the right show interfacial tensions (left scale) and corresponding solubilization parameters (ratios of oil and water volumes to surfactant volume in the microemulsion phase, right scale) for a system similar to that in the photograph. (From R.N. Healy et al., *Soc. Petrol. Eng. J.* 16, 147–160, 1976.)

temperature (PIT) or the hydrophilic/lipophilic balance temperature is that where a Winsor III microemulsion solubilizes equal volumes of oil and water.

The plots on the right side of Figure 20.7 show solubilization ratios, expressed as ratios of volumes of oil or water solubilized to surfactant volume in the microemulsions, and the corresponding IFTs between microemulsions and coexisting oil and water phases for a salinity scan of Winsor I, Winsor III, and Winsor II in an anionic surfactant system [13]. High solubilization ratios for both oil and brine and ultralow IFTs near 0.001 mN/m with both excess phases are seen near optimal salinity in the Winsor III region. Oil solubilization increases and water solubilization decreases with increasing salinity. In general, the high solubilization of oil (brine) correlates with the low microemulsion/oil (microemulsion/brine) IFT.

The Winsor I, Winsor III, and Winsor II sequence seen in nonionic surfactant systems with increasing temperature was studied by Shinoda et al. ([14], and references cited therein). The same sequence is also seen in more complex oil–water–surfactant systems, such as those with oils and surfactants that are mixtures. It remains true that any changes in temperature or composition that cause spontaneous curvature to change from favoring the O/W configuration to near zero to favoring W/O cause a shift in phase behavior from Winsor I to Winsor III to Winsor II and vice versa. In particular, if two surfactants are present, increasing the percentage of the more lipophilic species causes a shift in phase behavior in a direction away from Winsor I and toward Winsor II.

It is worth remarking, in view of the above discussion, that the phase behavior of nonionic alcohol ethoxylates does shift toward Winsor II with increasing salinity although variation is much less than with temperature. Similarly, the phase behavior of ionic surfactants varies with increasing temperature although the direction of the shift depends on the surfactant. The effect of salinity is usually much stronger than that of temperature, except for ionic surfactants having ethylene oxide (EO) or propylene oxide (PO) chains, for which, like nonionics, phase behavior shifts strongly toward Winsor II with increasing temperature.

20.2.3.2 Spontaneous Emulsification and Oily Soil Removal

Rang and Miller [15] used phase behavior determination and videomicroscopy to study spontaneous emulsification at 30°C in a system containing water, 1-hexadecane, 1-octanol, and pure ethoxylated alcohol, $C_{12}E_6$. As the nomenclature indicates, this surfactant has a 12-carbon chain connected to a six-EO chain. They conducted two sets of videomicroscopy experiments, one where drops containing hexadecane and octanol were injected into dilute aqueous solutions of $C_{12}E_6$ and another where drops containing hexadecane, octanol, and $C_{12}E_6$ were injected into water. In both cases they found that when the drop contained enough octanol, there was an initial period of diffusion followed by a spontaneous emulsification of oil drops. Figure 20.8 illustrates this behavior for one of the experiments where a hexadecane/octanol/$C_{12}E_6$ drop was injected into water. In this case, water initially diffused into the drop, converting it to a W/O microemulsion, whereas some octanol diffused into the aqueous phase. Under these conditions, the solubility of octanol in water is much greater than that of hexadecane and $C_{12}E_6$, with the result that little mass transfer of the latter components occurred. As a result, the ratio of octanol to $C_{12}E_6$ in the surfactant films of the microemulsion droplets decreased over time, shifting phase behavior away from Winsor II toward Winsor III. At the same time, the solubilization capacity for water increased and that for oil decreased. At some point, the microemulsion became supersaturated in oil, causing macroscopic oil droplets to nucleate at the microemulsion/water interface, as depicted in Figure 20.8d. Discrete droplets or lenses formed instead of a continuous oil film because oil did not spread spontaneously at the interface. At this point, the microemulsion drop was bicontinuous, but further mass transfer eventually caused it to become water continuous and miscible with the surrounding water phase. The oil droplets, which had been at the interface, became dispersed in the Winsor I microemulsion. Had the initial oil drop been attached to a solid surface as in a cleaning process, the oil droplets or lenses nucleated at the microemulsion/water interface during this inversion process would never have been in contact with the solid and, hence, would have been dispersed in the washing bath at the end of the inversion

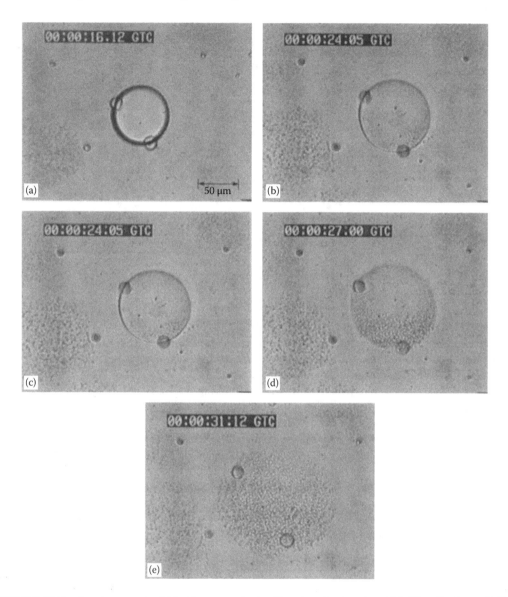

FIGURE 20.8 Spontaneous emulsification on contact with water of a drop in which 20 wt% of pure $C_{12}E_6$ was added to a 90:10 mixture by weight of 1-hexadecane/1-octanol. Drop diameter increases as it takes up water and becomes a microemulsion in frames (a to c), the spontaneous emulsification of oil on the drop surface appears in frame (d), and the microemulsion becomes miscible with water after inversion in frame (e). The two large drops around the periphery of the main drop played no role in the process. (From M.J. Rang and C.A. Miller, *Progr. Colloid Polym. Sci.* 109, 101–117, 1998.)

process and removed at the end of the washing cycle. Figure 20.9 shows a schematic representation of this spontaneous emulsification process.

As mentioned previously, a similar behavior was observed when hexadecane/octanol drops with sufficient octanol content were injected into dilute solutions of $C_{12}E_6$. In this case, the drop initially took up both water and $C_{12}E_6$, whereas octanol diffused into the aqueous phase as before. Both octanol and $C_{12}E_6$ transport lowered the ratio of octanol to $C_{12}E_6$ in the surfactant films of the microemulsion droplets, and spontaneous emulsification occurred by the same inversion mechanism as before.

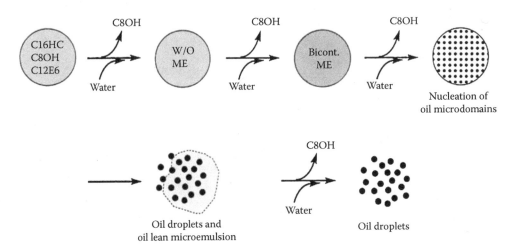

FIGURE 20.9 Schematic behavior showing spontaneous emulsification process for drops of 1-hexadecane/1-octanol/$C_{12}E_6$ contacting water as shown in the video frames of Figure 20.8. See text for detailed explanation. (From M.J. Rang and C.A. Miller, *Progr. Colloid Polym. Sci.* 109, 101–117, 1998.)

In the sections below, it is suggested that this inversion mechanism can play an important role not only in the removal of mixed soils initially containing some polar species but also in the removal of nonpolar soils during the rinsing process after washing in the Winsor II or Winsor III region. In the Winsor II case, surfactant partitions into the soil during the washing process, forming a W/O microemulsion. For anionic surfactants, it is proposed that rinsing causes salt to transfer from the soil drop into the lower salinity rinse water, thereby causing the microemulsion to invert from Winsor II to Winsor I with accompanying spontaneous emulsification. For nonionic surfactants, the rinse water must be below the PIT, so that the cooling of the drop causes similar inversion and spontaneous emulsification. As indicated below, there is recent evidence [16] that sometimes drops of bicontinuous microemulsions formed while washing in the Winsor III region become trapped in fabrics. These microemulsions contain surfactant and considerable soil and presumably could be removed by spontaneous emulsification, which develops as a result of inversion to Winsor I microemulsions during the rinse process.

20.2.3.3 Solubilization/Emulsification: Literature Review

Miller and Raney [17] reviewed the then-current knowledge of solubilization/emulsification mechanisms of detergency. Their main contribution compared with earlier reviews and textbooks was including results obtained in the years immediately preceding their article, which showed that important aspects of oily soil detergency could be understood in terms of phase behavior and interfacial tension of microemulsion systems. The following subsections summarize the relevant mechanisms and review more recent developments.

20.2.3.3.1 Hydrocarbon Soils

Raney et al. [18] showed using small laboratory-scale washing machines that the removal of pure hydrocarbon soils from polyester/cotton fabric by solutions of pure nonionic surfactants (alcohol ethoxylates) reached a maximum value near the PIT. They used radioactively tagged soils to measure the soil content of the washing bath at the end of the experiments and thus determined soil removal. Schambil and Schwuger [19] independently found similar results. At low temperatures in the Winsor I region, detergency increased monotonically with increasing temperature as solubilization increased and IFT decreased, facilitating emulsification (see Figure 20.10). Above the PIT, detergency decreased monotonically with increasing temperature. For these conditions, surfactant and water partitioned into the oil, more of which remained on the fabric as temperature and IFT increased. Near the PIT (approximately 30°C and 50°C, respectively, for the pure nonionic surfactants $C_{12}E_4$ and $C_{12}E_5$ with 1-hexadecane), a bicontinuous microemulsion formed with IFT low enough that a large portion of the oil was emulsified into the washing bath by agitation. Maximum 1-hexadecane removal by pure $C_{12}E_5$ was near 70%.

Using polyester fabric, Thompson [20] confirmed that when salinity is varied for anionic surfactants, a similar maximum in detergency occurs near the optimal salinity where spontaneous curvature is near zero, as phase behavior shifts from Winsor I at lower salinities to Winsor II at higher salinities. He also noted that a second maximum in detergency occurred in the Winsor I region for some systems. The monotonic increase in detergency with decreasing IFT was interrupted in these systems by an abrupt increase in contact angle, which hindered roll-up, the chief mechanism of soil removal for conditions well below the PIT. The increase in contact angle was caused by the formation of a surfactant-rich liquid intermediate phase following contact of surfactant solution and oil. At higher temperatures or salinities near the PIT or optimal salinity, the ultralow IFT caused detergency to rise again to a maximum value for the reasons explained in the preceding paragraph.

The previously mentioned results were for washing at fixed conditions. Solans and Azemar [21] reported that although hydrocarbon detergency for a constant wash temperature is indeed poor above the PIT, it is greatly improved by rinsing with water at a lower temperature. That is, if rinse temperature is below the PIT, the soil–water interface must pass through the PIT as it cools from the washing temperature. As a result, IFT decreases, spontaneous curvature reverses, and much of the remaining soil is removed, perhaps by the spontaneous emulsification mechanism discussed above.

In recent years, Scamehorn et al. [16,22–24] have carried out a careful and extensive study of oily soil removal as a function of salinity using mixtures of surfactants, some of which were anionic. One part of the study involved the removal of 1-hexadecane or a commercial motor oil from polyester/cotton fabric at 30°C by blends of anionic surfactants sodium dioctyl sulfosuccinate (Aerosol OT or AOT) and alkyldiphenyloxide disulfonate (Dowfax 8390) with nonionic sorbitan monooleate

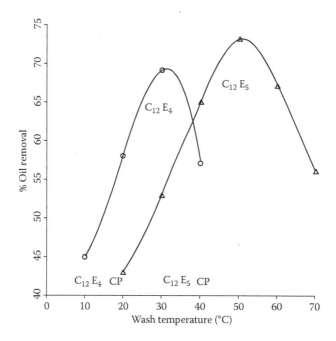

FIGURE 20.10 Removal of 1-hexadecane from 65:35 polyester–cotton fabric using 0.05 wt% aqueous solutions of pure $C_{12}E_4$ and $C_{12}E_5$. The arrows on the horizontal axis show the cloud point temperatures of the solutions. (From K.H. Raney et al., *J. Colloid Interface Sci.* 117, 282–290, 1987.)

(Span 80). The procedure involved washing in a laboratory-scale washing machine using water with a selected concentration of dissolved NaCl followed by rinsing with freshwater.

The authors reported excellent soil removal near optimal salinity as well as, in many cases, the Winsor I region very near the transition to Winsor III. The detergency studies in this part of the study did not include the Winsor II region. For washing near optimal salinity, a substantial portion of the oil removal occurred during the rinse step. Moreover, IFT between the washing bath and a fresh oil drop was more than 10 times higher at the end of the wash cycle, and anionic surfactant concentration in the washing bath was more than 10 times lower than corresponding values at the beginning of the cycle. The authors attributed these last observations to considerable surfactant-containing middle-phase microemulsion remaining on or within the fabric instead of being dispersed in the washing bath as suggested by earlier studies (see previous paragraphs). This microemulsion would then be removed during the freshwater rinse step when salinity decreased to values shifting phase behavior well into the Winsor I region and thereby causing surfactant to transfer to the aqueous phase, the same concept as for soil removal in the Winsor II region during low-temperature rinse with nonionic surfactants discussed earlier. They further suggested that because contact angle is smaller in the Winsor I than in the Winsor III region, roll-up may be the dominant removal mechanism during rinsing.

The roll-up mechanism is a possibility, although it should be recalled that fractional oil removal by washing at low salinities in the Winsor I region, where roll-up is important, is well below the excellent values achieved near optimal salinity after the rinsing step. Spontaneous emulsification produced by a mechanism similar to that described above is another possibility, not only near optimal conditions but also in the Winsor II region. Suppose that a microemulsion with substantial surfactant content remains on the fabric after washing. When exposed to freshwater of the rinse, the microemulsion takes up water and loses salt to the rinse water by diffusion. Eventually, the resulting inversion of the microemulsion due to a decrease in salinity results in the spontaneous emulsification of oil and miscibility between an O/W microemulsion and the rinse water as described in Section 20.2.3.2 above, leaving the oil drops dispersed in the aqueous phase. It is noteworthy that in this case, the hydrocarbon soil need not initially contain surfactant or other polar components as in the examples of Section 20.2.3.2. In a sense, the wash becomes a pretreatment process in which the surfactant enters the oil to form a microemulsion, which remains on the fabric, and the rinse becomes the main cleaning process. A similar inversion process leading to spontaneous emulsification can be envisioned when the surfactant is nonionic and the rinse is at a temperature below the PIT. As the microemulsion remaining on the fabric after the wash step cools, it inverts and reaches a condition where local supersaturation of oil occurs, leading to spontaneous emulsification and an O/W microemulsion that becomes miscible with the aqueous phase.

Comparing the results of this study with the earlier work cited at the beginning of this section, one may conclude that if IFT during the actual washing process is sufficiently low near the PIT or optimal salinity, high soil removal is obtained. A higher contact angle measured through the soil can help by allowing drops of microemulsion or soil attached to the solid to reach a configuration where gravity or agitation in the washing bath can more easily cause most of the drop to break off into the bath (see a further example in Section 20.2.3.3.2). However, if IFT is not sufficiently low near the PIT or optimal salinity, soil removal during the wash step is lower, and rinsing at a lower temperature or salinity allows inversion to occur and greater soil removal to be achieved.

As indicated previously, a mixture of three surfactants was used in this study. Dowfax 8390 is quite hydrophilic, Span 80 is quite hydrophobic, and AOT is a reasonably balanced intermediate. This choice was based on earlier work by the same group discussing advantages of the use of a balanced surfactant with added hydrophilic and lipophilic "linkers" [25]. Phase behavior presented along with the detergency results suggests that the use of such mixtures broadens the salinity range of the Winsor III region. As salinity increases, both anionic surfactants become less hydrophilic. However, this effect is reduced in magnitude compared with that for a single anionic surfactant because the salinity increase also causes more of the Dowfax 8390 to partition from the aqueous phase to oil–water interfaces and some of the AOT to partition from the interfaces into the oil phase. That is, the interfacial surfactant composition becomes more weighted toward the hydrophilic and less weighted toward the hydrophobic anionic surfactant. Thus, once the system enters the Winsor III region, more salt must be added to reach the transformation to Winsor II. This effect is augmented by the increased partitioning of Span 80 into the oil at high salinities. It is noteworthy that the use of the very hydrophilic and very hydrophobic species implies that a significant amount of surfactant is not located at the interfaces but instead in the bulk oil and aqueous phases as well as in the oil and brine solubilized in the microemulsion. Accordingly, more total surfactant may be required than for a surfactant or surfactant blend whose molecules are located mainly at the interfaces. However, it should be noted that this particular three-component mixture performed well in detergency studies at a total surfactant concentration of 0.11 wt%.

20.2.3.3.2 Triglyceride Soils

Mori et al. [26] reported that the pure $C_{12}E_5$/triolein/water system showed Winsor III behavior with a PIT near 65°C. The removal of triolein from polyester/cotton fabric using solutions of this surfactant exhibited a maximum of approximately 75% in the same temperature range. That is, behavior was similar to that discussed in the preceding section for pure hydrocarbons (Figure 20.10). Thompson [20] observed a maximum in triolein detergency of nearly 90% for the same system, although at a lower temperature near 50°C. The difference is likely due to small differences in the amounts of oleic acid present as an impurity in the triolein used in the two studies because the PIT is sensitive to polar impurities. Both studies found a second maximum with lower soil removal at a lower temperature similar to that mentioned above for hydrocarbon soils. Thompson [20] also showed that a maximum in detergency occurs near the optimal salinity for an anionic surfactant with triolein.

More recently, Tongcompou et al. [27] studied the removal of technical-grade (65%) triolein, which presumably contained substantial amounts of the more polar oleic acid formed by triolein hydrolysis. A mixture of three surfactants was used: Dowfax 8390 as the most hydrophilic surfactant, the same as in the hydrocarbon detergency studies discussed in the preceding section; secondary alcohol ethoxylate Tergitol 15-S-5 instead of Span 80 as the uncharged lipophilic surfactant; and Aerosol MA (AMA, sodium dihexyl sulfosuccinate) instead of AOT, its dioctyl counterpart.

For the three blends of these surfactants that were studied, plots of detergency as a function of salinity exhibited two maxima, one well below the salinity for transformation from Winsor I to Winsor II and the second at a salinity higher by 10 times than the first. All three blends exhibited 70% to 80% soil removal at the second maximum, which was located near the optimal salinity where IFT at the beginning of the washing cycle reached its lowest value. However, the minimum IFT values exceeded 0.1 mN/m, greater than those seen for the hydrocarbons. Behavior was similar to that of hydrocarbon soils near maximum detergency in that IFT was significantly higher at the end than at the beginning of the washing cycle and a significant amount of soil removal occurred during the freshwater rinse step. Hence, the previous discussion of behavior near the optimal salinity and in the Winsor II region for hydrocarbon soils, including the possibility of spontaneous emulsification, is relevant for this technical-grade triglyceride.

The removal of other liquid triglyceride soils (vegetable oils) with propoxylated sulfate surfactants was later investigated by the same group, using canola oil as the soil and methyl-branched $C_{14,15}$ 8PO sulfate as the only surfactant [28]. The combination of a long hydrocarbon chain and a PO chain facilitates solubilization and microemulsion formation with the large triglyceride molecules [29], thereby obviating the need for a lipophilic linker. Branching of long hydrocarbon chains is desirable because it promotes the formation of microemulsions instead of viscous liquid crystals near optimal conditions. Similar surfactants have been used in recent years for designing surfactant processes for enhanced recovery of crude oils, which typically have some components with high molecular weights [30,31].

Here, two maxima in soil removal as a function of salinity were also observed, one in the Winsor I region, where virtually all soil removal for the entire process occurred during the wash step, and the other in the Winsor II region, where most of the soil removal occurred during the rinse step. Soil removal was near 80% and 90% for the first and second maxima, respectively.

Another study [32] used palm oil and the same propoxylated sulfate but blended (25%) with a secondary alcohol ethoxylate similar to Tergitol 15-S-5 (75%). Soil removal as a function of salinity exhibited a single maximum showing more than 80% soil removal at the same salinity as the minimum IFT between soil and washing bath. However, soil removal remained nearly as great for higher salinities, including the Winsor II region, presumably owing to the freshwater rinse.

Microscopic observations of a single pendant soil drop on a horizontal fused polyester/cotton fabric surface at the maximum detergency condition showed that contact angle increased with time until reaching a value of approximately 125°. At that time, gravitational instability set in, leading to necking and subsequent detachment of most of its initial volume. These observations are similar to the experiments of Kolev et al. [6] described in Section 20.2.1. In both cases, IFT was not low enough to cause extensive emulsification with the formation of many small drops, and the drop was removed by roll-up followed by the breakoff of a single large drop. However, spontaneous emulsification would likely have been observed had the initial surfactant solution been replaced by lower salinity brine once surfactant had partitioned into the drop at the initial salinity. Similar behavior of individual drops was observed for salinities in the Winsor II region. Although quantitative comparisons of soil removal during washing and rinsing steps were not given, the authors indicate that both steps were important.

20.2.3.3.3 Mixed Soils

In practice, oily soils are often mixtures of different types of nonpolar and polar soils. Mori et al. [26] studied the phase behavior of the system pure $C_{12}E_4$/1-hexadecane/pure triolein/water as a function of temperature. For a 50:50 hexadecane/triolein blend, they found a Winsor III region near 40°C with minimal triolein solubilization below this temperature. The PIT was 32°C for hexadecane alone, whereas only minimal oil solubilization was seen below approximately 55°C in the pure triolein system. Detergency experiments were performed for the 50:50 blend as a function of temperature with no rinse. Different radiotracers were used for the two oils, which allowed the removal of both to be determined. As shown in Figure 20.11, soil removal—approximately 80% for hexadecane (cetane) and 75% for triolein—was nearly constant between 30°C and 40°C but decreased at both lower and higher temperatures. That is, behavior was generally similar to that for hexadecane and triolein alone. Figure 20.11 shows that pure $C_{12}E_5$ was also effective but in a higher temperature range.

The addition of polar components such as long-chain alcohols or fatty acids to nonpolar soils has a large effect on phase behavior. The polar species are present not only in excess and solubilized oil but also at oil–water interfaces, where they have a large effect on spontaneous curvature and hence microemulsion microstructure. Typically, they lower the PIT for a nonionic surfactant or the optimal salinity for an ionic surfactant. Lim and Miller [33] used videomicroscopy to observe behavior when drops of 1-hexadecane/oleyl alcohol mixtures were injected into dilute solutions of nonionic surfactants without stirring. If the temperature was below the PIT of the system, no significant activity was observed, although presumably oil was being slowly solubilized into the surfactant solution and a small amount of surfactant dissolved in the oil. However, at temperatures above the PIT, the drop began to take up surfactant and water, forming a W/O microemulsion. The continuing diffusion of surfactant into the drop caused the interface to become more hydrophilic and take up water. Thus began an inversion process similar to the spontaneous emulsification described earlier. However, because the long-chain alcohol was present, the first new phase that nucleated was not the oil phase as in the previous examples but instead the lamellar liquid crystal. It appeared as rather fluid myelinic figures (Figure 20.12) growing outward into the surfactant solution. When the oil was rich in hydrocarbon and not far above the PIT, as in Figure 20.12, there was some breakup into drops, but considerably more breakup and dispersion of lamellar phase drops would be expected in an agitated washing bath.

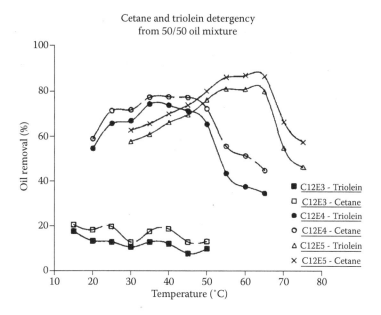

FIGURE 20.11 Removal of triolein and 1-hexadecane (cetane) from 65:35 polyester–cotton fabric by washing with pure nonionic surfactants. Initial soil contained 50 wt% triolein. (From F. Mori et al., *Colloids Surf.* 40, 323–345, 1989.)

The detergency of 1-hexadecane/oleyl alcohol soils was investigated by Raney and Benson [34]. Figure 20.13 shows that with the pure alcohol ethoxylate $C_{12}E_7$ and 1-hexadecane (cetane), soil removal increases throughout the temperature range shown. The system is below its PIT, which is 82°C. For a 90:10 blend of hexadecane/oleyl alcohol, the system is below its PIT at the lowest temperatures shown, and soil removal is relatively low, although greater than for hexadecane alone. Near the PIT, which is approximately 40°C, according to separate phase behavior experiments, soil removal approaches a maximum and remains high until the temperature exceeds approximately 60°C. For a 75:25 blend, the system is already above its PIT at the lowest temperature shown. Soil removal is high and does not drop significantly until the temperature exceeds approximately 40°C. The decrease at high temperatures occurs for a constant washing time as in Figure 20.13 because more time is needed until the lamellar phase begins to form and because the lamellar phase has a higher oleyl alcohol content and, hence, is more viscous and harder to disperse.

Srivastava et al. [35] used reflectivity to determine the removal of synthetic sebum (body soils) from cotton and cotton/polyester fabrics at 28°C. Instead of an aqueous surfactant solution, they used an O/W nanoemulsion formed spontaneously on dilution with soft freshwater of a W/O

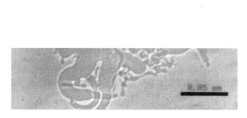
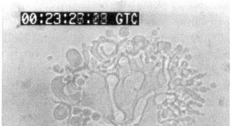

FIGURE 20.12 Video frame showing myelinic figures of lamellar phase several minutes after contact of a drop containing 1-hexadecane/oleyl alcohol (5.67:1 by weight) with a 0.05 wt% aqueous solution of pure $C_{12}E_8$ at 50°C. (From J.C. Lim and C.A. Miller, *Langmuir* 7, 2021–2027, 1991.)

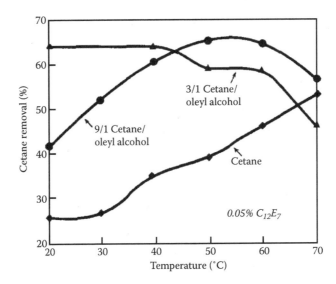

FIGURE 20.13 Effect of oleyl alcohol content on 1-hexadecane (cetane) removal. (From K.H. Raney and H.L. Benson, *J. Am. Oil Chem. Soc.* 67, 722–729, 1990.)

microemulsion containing 17% by weight of a mixture of commercial nonionic surfactants and equal weights of concentrated NaCl brine and an oil consisting of 95% paraffin oil and 5% octanoic acid. The mechanism of the formation of the nanoemulsion, which had oil drops of order 50 nm, was inversion similar to the spontaneous emulsification mechanism discussed previously, except that no solid was present. Its continuous phase was an O/W microemulsion.

Although the precise mechanism for sebum removal was not directly observed, it seems likely, as suggested by the authors, that behavior was similar to the spontaneous emulsification they saw using optical microscopy when the nanoemulsion contacted a sessile drop of the above model oil on a cellulose-coated glass slide. Probably surfactant and water diffused into the model oil, forming a W/O microemulsion, which then inverted as it continued to take up surfactant and water, thereby increasing the ratio of surfactant to octanoic acid in its droplets' monolayers and making the monolayers more hydrophilic. That is, soil removal probably occurred to a significant degree by spontaneous emulsification along the lines discussed above. Spreading of the model oil on the glass slide was also observed. Although the contact angle θ no doubt decreased for sebum, it is not clear whether it reached zero resulting in spreading, as found for the model oil.

20.3 SOLID ORGANIC SOILS

Another type of cleaning process involves the removal of solid organic soils such as waxes, fats, or greases from fabrics or "hard surfaces" such as dishes, metals, tiles, counters, kitchen and bathroom fixtures, and so on. One approach is simply to carry out the cleaning process at a temperature above the melting point of the relevant solid, in which case methods for removing oily liquid soils discussed previously are applicable. However, this approach may not be feasible or economically attractive when the melting point of the soil is high. Instead, suitable aqueous surfactant solutions or organic solvents may be able to remove the soil at temperatures below the melting point. A limitation of many solvents, although they may be effective, is that their flammability, volatility, or toxicity make their use undesirable. Hence, if solvent is required for adequate cleaning, a less hazardous combination of surfactant and solvent in the form of an emulsion of solvent in water stabilized by surfactant can be used. The case of aqueous solvent-free surfactant solutions is emphasized here because it is dealt with more extensively in the journal literature.

20.3.1 DISSOLUTION AND SOLUBILIZATION

Dissolution in the cleaning fluid, including solubilization in surfactant solutions, is one method for removing solid organic soils. Most soils of interest have such low solubility in water that dissolution is far too slow to be of interest for cleaning. Both equilibrium solubility and rate of dissolution can be enhanced by the use of surfactants above their CMCs—usually well above because solubilization effects increase, typically proportionally, with micelle concentration. Even so, solubilization alone in the absence of phase changes or chemical reactions is often too slow for most cleaning applications. For example, Shaeiwitz et al. [36] reported that the solubilization rate of a film of palmitic acid in a 4 wt% sodium dodecyl sulfate (SDS) solution at 30°C ranged from 0.25 to 0.38 µm/min, depending on the angular velocity during the experiment of the rotating disk in their apparatus. They used a radioactive tracer technique and measured the amount of the acid in the aqueous surfactant solution as a function of time. However, solubilization, especially of surfactant and water by the soil, often occurs concurrently with other mechanisms, somewhat analogously to the situation for oily liquid soils. Examples are given in Section 20.3.2.

Gotoh [37] used the quartz crystal microbalance (QCM) technique to study the removal of various soils, including carbon black and soot/olive oil mixtures from cotton and polyester fabrics by stirred surfactant-free mixtures of water and organic solvents such as alcohols. The frequency of crystal oscillation is highly sensitive to its mass and hence changes as soil deposited on its surface is removed. Little carbon black was removed by any solvent–water mixture studied. As expected, oily soils were best removed by dissolution in washing baths with low dielectric constants and better removed from cotton than from more hydrophobic polyester fabrics to which they adhered more strongly. In contrast, water-soluble soils were more readily removed from polyester than cotton and from both cotton and polyester by dissolution in water rather than organic solvents. Related studies were performed on multifiber fabrics [38] and of carbon black and stearic acid removal from polymer substrates [39,40].

20.3.2 SWELLING AND PENETRATION

The melting point of palmitic acid is 64°C, much higher than 30°C at which the solubilization experiments described in Section 20.3.1 were conducted. In further experiments with aqueous SDS solutions, Shaeiwitz et al. [36] found that the palmitic acid removal rate increased dramatically when the temperature was raised higher than 37°C. For the highest speed of rotation studied, it was more than three times greater at 40°C and eight times greater at 50°C than the higher values given in the preceding section. The reason is that instead of simply adsorbing on the solid surface and forming small surface admicelles with palmitic acid that subsequently desorb, the mechanism proposed by the authors to explain their results at 30°C, enough SDS can diffuse into the solid soil at higher temperatures to change the microstructure in such a way that water can enter in significant amounts. The results are increases in soil weight and volume and an accompanying decrease in shear modulus or shear viscosity. If the effect is large enough, applied shear from the flow can break off parts of the swollen soil and disperse them into the cleaning bath. The lowest temperature at which this behavior occurs in various systems has been called the penetration temperature T_p by Lawrence [41], who determined its value for a range of fatty acids with some common surfactants. As might be expected, both melting point and T_p increase with the increasing hydrocarbon chain length of the acid for a given surfactant, and T_p decreases with increasing surfactant concentration.

For palmitic acid, the lamellar liquid crystalline phase develops in the form of outwardly growing cylindrical myelinic figures at the solid–liquid interface at temperatures higher than 37°C in SDS solutions, as observed by optical microscopy [36]. Similar microscopy observations had been made earlier in the same and other systems by Stevenson [42] and Lawrence ([41], and earlier articles cited there), who proposed this behavior as a mechanism of detergency but did not report quantitative rates of soil removal. The swelling and growth of the lamellar phase as myelinic figures for solid soils is similar to that described in Section 20.2.3.3.3 as an important mechanism of detergency in systems where slightly polar liquid oily soils, for example, hexadecane containing oleyl alcohol, are contacted with aqueous solutions of hydrophilic nonionic surfactants above the PIT of the system. On the basis of optical microscopy observations and DSC results, Robb and Stevenson [43] also suggested that mesophases formed as a result of penetration during their measurements of trilaurin solubilization in nonionic surfactant systems.

Solid soil detergency has been studied by placing soiled samples of substrate in a cleaning bath and weighing them as a function of time [44,45]. When penetration occurs, soil weight initially increases as surfactant and water enter the soil faster than soil is solubilized by the bath. As indicated earlier, this process promotes the liquefaction of the soil. The penetration of commercial nonionic surfactants into slightly polar soils was found to be favored by surfactants with short, straight hydrocarbon chains and short EO chains. In the absence of agitation or other mechanical action, soil weight often continues to increase with time. With agitation, soil weight more typically reaches a maximum and then declines as shear forces become capable of removing liquefied soil.

Another technique that has been used to study solid soil removal is FT-IR [46,47]. Scheuing and Hsieh [46], for example, tracked decrease in eicosane content on a solid surface in solutions of nonionic surfactants. They found that the penetration of surfactant and water caused "disordering" of the eicosane layer. Possibly, a liquid crystalline phase was formed at the interface between soil and surfactant solution.

Bäckström et al. [48] demonstrated that ellipsometry can be used to determine the amount of soil removal from solid surfaces as a function of time. In this and subsequent articles [49,50], the same group presented examples of trilaurin and tripalmitin removal at 25°C by aqueous solutions of nonionic and anionic surfactants.

Using the same method, Malmsten and Lindman [51] found that a maximum in soil removal occurred in several solid triglyceride or solid triglyceride/fatty acid systems as the temperature or the ratio of two surfactants in a mixture was varied. That is, the extent of the removal of these solid soils apparently depended on achieving an optimal hydrophilic/lipophilic balance in the system, similar to the behavior discussed above for oily soil removal by solubilization/emulsification. Because the optimal temperature for experiments with pure nonionic surfactants occurred near the surfactant cloud point temperature, the authors suggested that optimal soil removal conditions corresponded to the packing of surfactant into planar layers. However, the situation seems complex and may depend on optimizing a tradeoff between the surfactant's ability to solubilize the soil and its ability to allow water entry to fluidize the soil, as discussed in the following paragraphs. The effect of soil composition on optimal removal conditions also needs clarification. Further studies are needed.

More recently, the QCM technique has been used to monitor sample weight during cleaning and similarly obtain information on effectiveness of this detergency mechanism. Weerawardena et al. [52,53] used QCM to track the removal of thin layers of tripalmitin and dotriacontane ($C_{32}H_{66}$) coated on the quartz crystal. They immersed the coated crystal at 21°C in unstirred solutions of the pure nonionic surfactant $C_{12}E_8$ at a concentration approximately 10 times the CMC. For tripalmitin, they observed that penetration produced a maximum in soil layer mass some 3 to 4 min after initial contact with the surfactant solution for layers ranging from approximately 0.05 to 0.5 μm in initial thickness. More than 90% of the soil was removed within 5 to 12 min after contact, longer times being required for the thicker layers. Qualitatively similar behavior was seen for dotriacontane, except that the entire process for this nonpolar soil was slower and only 70% to 80% of the soil was ultimately removed after 40 to 50 min. When a tripalmitin layer of similar thickness and the less hydrophilic surfactant $C_{12}E_5$ were used, the maximum increase in mass in the initial minutes of the experiment was less. Indeed, soil mass oscillated around its initial value for the first 7 to 8 min before decreasing monotonically until 75% of the initial mass had been removed after approximately 20 min. The existence of oscillations suggests that some of the swollen soil may have broken off early in the $C_{12}E_5$ experiment. However, more of the tripalmitin was removed in a shorter time for the more hydrophilic surfactant $C_{12}E_8$. As no optical microscopy was performed, it is uncertain whether myelinic figures of the lamellar liquid crystalline phase were formed during these experiments although it seems a strong possibility.

The same group used the same technique to investigate the removal of tristearin layers with solutions of alkyl glucose-based nonionic surfactants at 50°C [54,55]. In the former article, it was shown that soil removal was poor unless surfactant concentration exceeded the CMC. Above their CMCs, alkyl glucosides and alkyl maltosides with straight alkyl chain lengths of 8, 10, and 12 all removed 70% to 80% of the tristearin, although the process was faster for the more hydrophilic maltosides with their larger head groups and for the hydrophobe with 10 carbon atoms. The latter article is interesting in that it relates detergency results to the surfactant tail structure for the six possible locations of the glucoside head group along the alkyl chain of dodecyl β-D-glucoside. The authors found that the more highly branched isomers with the head group near the center of the hydrocarbon chain were unable to remove appreciable amounts of soil, apparently because their more bulky hydrophobes hindered penetration into the tristearin crystals.

Favrat et al. [56] used QCM to investigate the removal of stearic acid from quartz surfaces by solutions of two commercial formulations. Stearic acid is an important component of polishes and waxes. Results were qualitatively similar to those of the preceding two paragraphs, that is, initial swelling followed by a decrease in soil mass to a final constant value.

Cui et al. [57] found using QCM that the removal of tristearin was improved in some cases by exposing the soil to a dispersion of functionalized nanodiamond particles before placing it in contact with a suitable surfactant solution. The mechanism is not clear, but after reaching the soil–liquid interface, the nanoparticles may adsorb some tristearin, perhaps causing defects in the soil layer that facilitate penetration or solubilization.

Gotoh and Tagawa [58] and Gotoh [59] used QCM to investigate the removal of arachidic acid multilayers deposited by the Langmuir–Blodgett technique on gold surfaces and gold surfaces with polymer coatings. The cleaning solutions, which were subjected to ultrasonic waves, consisted of dilute aqueous NaCl and NaCl/NaOH solutions with added SDS, alkyl polyglucoside, or ethanol at 25°C. Surfactant concentrations were varied but were maintained below their CMCs in the initial cleaning solutions, precluding solubilization into micelles in the bulk liquid. However, the formation of surfactant/acid or surfactant/soap (in NaOH solutions) aggregates at the film surface might still have occurred with the subsequent desorption of the aggregates into the cleaning solution. The experiments showed that NaOH, ethanol, and both surfactants improved cleaning performance. Because information on variation of mass as a function of time was not presented, the swelling behavior of the films cannot be determined from the article. The authors considered that the primary mechanism of cleaning was a complete detachment of portions of the Langmuir–Blodgett film from the substrate, a hypothesis supported by information showing that low adhesion energies correlated with high soil removal.

Beaudoin et al. [60,61] and Kabin et al. [62,63] studied the removal of a 75% solution of abietic acid (AA) in isopropanol, which was deposited as layers a few micrometers thick on rotating disks. The soil in this case is not a solid but a highly viscous liquid. AA is a primary component of solder flux, whose removal from printed circuit boards is required during processing. AA has very low solubility in water and is not removed in significant amounts by water alone. The rotating disk experiments used solutions of pure nonionic surfactants of the form C_mE_n at ambient temperature, with an ultraviolet detector to measure the AA content of the washing bath as a function of time. The authors identified three stages of the cleaning process when the nonionic surfactant was $C_{12}E_5$ (Figure 20.14). During the first stage, surfactant and water penetrate the soil layer, making it more fluid. Relatively small amounts of AA (10%–20%) were removed by a micellar solubilization process, according to the authors, in which both interfacial resistance to mass transfer and mass transfer resistance in the bulk solution were significant for the conditions studied. The amount of AA removed increased linearly with time. Optical microscopy of the disk surface showed that the soil layer consisted of small particles and aggregates of particles both initially and throughout

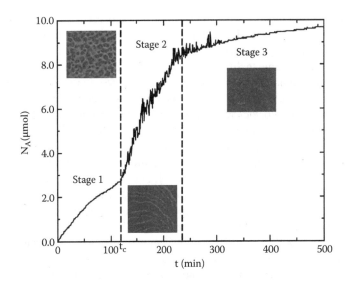

FIGURE 20.14 Typical cleaning curve of an AA film from a rotating disk, showing the number of moles N_A removed as a function of time t. The cleaning solution is 6.0×10^{-5} M pure $C_{12}E_5$, and the rotational speed of the disk is 750 rpm. Representative photographs show that in stage 1, the initially uniform film has broken up into discrete particles; in stage 2, starting at time t_c, the solid has become liquefied and flows outward in a spiral pattern of curved rivulets; in stage 3, drops remaining after breakup of the rivulets are removed. (From J.A. Kabin et al., *J. Colloid Interface Sci.* 206, 102–111, 1998.)

this stage, as shown by the first photograph in Figure 20.14. The authors' model of solubilization in stage 1 has many similarities to that of Shaeiwitz et al. [36], in particular the use of the known rotating disk analysis for mass transfer in the bulk solution, but it is not identical in its treatment of interfacial transport.

A relatively sharp transition was observed from stage 1 to stage 2, during which most of the AA was removed at a much faster rate than observed during stage 1. The time duration of stage 1 decreased with the increasing surfactant concentration and speed of disk rotation. The plot of AA concentration in the solution as a function of time in stage 2 exhibited spikes (Figure 20.14), suggesting that solubilization at the interface was replaced as the dominant mechanism by removal from the surface of small but macroscopic amounts of the liquefied film, which were subsequently solubilized in the bulk aqueous phase. The liquefied material flowed outward along the disk surface in a spiral pattern of curved rivulets, as shown by microscopy, eventually reaching the edge of the disk and entering the aqueous solution. Because the solubilization of the ejected material was not instantaneous, AA concentration in the solution became nonuniform, leading to the spikes in the measured time dependence of concentration. In stage 3, the rivulets broke up into discrete drops, which were individually detached from the surface by roll-up and shear.

The time required for the removal of most of the AA decreased when the hydrocarbon chain length of the surfactant was decreased below 12 for a constant EO chain length of 5. That is, more hydrophilic surfactants gave better detergency. The main reason was that less time until transition to stage 2 with its higher removal rates was needed for the surfactants with shorter hydrocarbon chains.

When the surfactant was changed from $C_{12}E_5$ to $C_{12}E_8$ or $C_{16}E_8$, stage 2 was no longer observed. That is, there was no abrupt increase in AA removal rate as seen in Figure 20.14. Moreover, the microscopy of the disk surface showed no rivulets as found for stage 2 for $C_{12}E_5$. However, the solubilization rate in stage 1 was greater for the more hydrophilic $C_{12}E_8$ than for $C_{12}E_5$. As previously suggested for the QCM experiments, it may be that the more lipophilic C_mE_5 surfactants penetrate deeper into the AA film and provide a greater degree of liquefaction, allowing the rivulets of stage 2 to develop, whereas the C_mE_8 surfactants provide a faster removal of the soil near its surface with the surfactant solution. Both types of soil removal may have to be considered to optimize performance in a given system.

Later, Kabin et al. [64] studied the removal of AA layers containing small amounts of benzoic acid (BA) using a similar technique. With $C_{12}E_5$ as the surfactant, AA was removed in three stages as before with only 10% to 20% removed in stage 1. However, more than 60% of the more soluble BA was removed during stage 1. When $C_{16}E_8$ was used instead, no stage 2 was observed, the same as in the absence of BA with this surfactant, and the proportions of AA and BA removed at a given time during stage 1 were the same. The authors suggested that this difference in behavior occurred because BA dissolves in the water, which penetrates the soil layer, and because water penetrates deeper into the AA layer for $C_{12}E_5$ than for $C_{12}E_8$, as discussed earlier. As a result, surfactant micelles leaving the interfacial region have a higher content of solubilized BA when the surfactant is $C_{12}E_5$.

20.3.3 EMULSIONS OF SOLVENTS IN SURFACTANT SOLUTIONS

In some applications, for example, the removal of asphalt and other viscous organic residues from metal surfaces such as the interior surfaces of storage tanks, the use of suitable solvents facilitates cleaning. However, cleaning with solvents alone is frequently undesirable owing to their flammability, volatility, or toxicity. These adverse effects can be mitigated by using an emulsion in which drops of the solvent are dispersed in an aqueous surfactant solution.

Kabin et al. [65] investigated the use of such emulsion cleaning at ambient temperature to remove layers of phenanthrene approximately 115 μm in thickness deposited on the surface of a stainless steel disk, which was part of a rotating disk apparatus. The solvent was d-limonene, and the emulsion was stabilized by the nonionic surfactant Tween 20 (ethoxylated sorbitol monolaurate). Solvent and surfactant contents up to approximately 30% by weight were studied, although water content was never less than 50% for any experiment. A small amount of radioactive phenanthrene was included in the film, which enabled phenanthrene concentration in the emulsion to be measured as a function of time.

It was found that the initial constant removal rate increased severalfold when emulsion was used compared with surfactant solution alone. The authors considered that the penetration of the film resulted in the removal of small but macroscopic particles of phenanthrene from the disk surface in both cases. However, the main advantage of using solvent was that it was able to reach the phenanthrene–stainless steel interface quickly with the result that the entire film detached after a few minutes. Some 90% of the initial phenanthrene present was removed by detachment. In contrast, solubilization with surfactant alone continued at nearly a constant rate for much longer times before detachment eventually occurred. For a given emulsion, the initial solubilization rate increased and the detachment time decreased with the increasing speed of rotation.

Chew et al. [66] studied the removal of polystyrene copolymer films from solid surfaces. They investigated ways of replacing an organic solvent (methyl-ethyl ketone), which caused the swelling and removal of the film but had undesirable health and environmental effects. The use of alkali (aqueous sodium hydroxide or sodium metasilicate solutions) alone produced swelling of the films but not their removal. However, an aqueous sodium metasilicate solution containing a "commercial blend of surfactants and dispersants" did remove the film by the swelling and continual breakoff of small amounts of material for some copolymers and by the detachment of the film from the solid surface for others.

Hunek and Cussler [67] conducted experiments on the removal of films of the phenolic resins present in photoresist. They provided equations for various special cases of a general dissolution model, which included interfacial chemical reaction, for example, with alkali, as well as resistance to mass transfer both at the interface and in the bulk cleaning solution.

20.4 SUMMARY

This article has summarized advances in knowledge during the past two decades of the mechanisms by which surfactants remove soils during cleaning processes. For oily (liquid) soils, it is well known that the most important mechanisms are roll-up and solubilization/emulsification. For solid soils, the two chief mechanisms are solubilization and liquefaction/detachment caused by the penetration of surfactant and water into the soil, which leads to subsequent soil removal.

Recent work has provided information on how contact angle, drop radius, and interfacial tension vary during the roll-up of oily soils in model systems as well as insight into fundamental behavior near the contact line during the process. Further research on the latter topic is desirable.

Experiments with a wider range of surfactants and oils than studied previously generally confirmed the previously known relationship between the detergency and the equilibrium phase behavior of oil–water–surfactant systems, which is central to soil removal by solubilization/emulsification. Surfactant types not investigated previously included propoxylated sulfates and blends consisting of a principal surfactant supplemented by hydrophilic and lipophilic linkers. Oils studied included not only pure hydrocarbons and triolein as in earlier work but also motor oil and vegetable oils such as canola and palm oils. The recent experiments indicated that lower than expected soil removal found in some cases near salinities previously believed to be near optimal for detergency was caused by trapped bicontinuous microemulsion in the fabric. That is, factors such as insufficient IFT or unfavorable wettability prevented the microemulsion with its solubilized soil from being dispersed into the washing bath, as evidently occurred in previous research. Detergency in systems with such trapped microemulsions, including those with high salinities in which a W/O microemulsion is trapped, could be increased greatly by rinsing at lower salinities. Possible mechanisms for this improvement are the roll-up and, as suggested here, inversion of the microemulsion to an O/W type with accompanying spontaneous emulsification.

The mechanisms involved in the liquefaction of solid soils and their subsequent removal from solid substrates have been made clearer by experiments using rotating disk and QCM techniques. More work with a wider range of systems and with the use of aqueous emulsions of solvents stabilized by surfactants is needed. Of particular interest is obtaining a better understanding of conditions when surfactants and solvents can quickly reach the soil–substrate interface and cause detachment of the soil layer, thus avoiding the need to solubilize or liquefy all the soil, which requires more time and more surfactant.

ACKNOWLEDGMENT

The author is grateful to Gautam Kini for reviewing the article and helping with the figures.

REFERENCES

1. B.J. Carroll. Physical aspects of detergency. *Colloids Surf. A*, **74**, 131–167 (1993).
2. B.J. Carroll. Equilibrium conformations of liquid drops on thin cylinders under forces of capillarity. A theory for the roll-up process. *Langmuir*, **2**, 248–250 (1986).
3. K. Kondiparty, A. Nikolov, S. Wu and D.T. Wasan. Wetting and spreading of nanofluids on solid surfaces driven by the structural disjoining pressure: Statics analysis and experiments. *Langmuir*, **27**, 3324–3335 (2011).
4. K. Kondiparty, A. Nikolov, D.T. Wasan and K.L. Liu. Dynamic spreading of nanofluids on solids. Part I: Experimental. *Langmuir*, **28**, 14618–14623 (2012).
5. K. Kondiparty, A. Nikolov, D.T. Wasan and K.L. Liu. Dynamic spreading of nanofluids on solids. Part II: Modeling. *Langmuir*, **28**, 16274–16284 (2012).
6. V.L. Kolev, I.I. Kochijashky, K.D. Danov, P.A. Kralchevsky, G. Broze and A. Mehreteab. Spontaneous detachment of oil drops from solid substrates: Governing factors. *J. Colloid Interface Sci.*, **257**, 357–363 (2003).

7. P.A. Kralchevsky, K.D. Danov, V.L. Kolev, T.D. Gurkov, M.I. Temelska and G. Brenn. Detachment of oil drops from solid surfaces in surfactant solutions: Molecular mechanisms at a moving contact line. *Ind. Eng. Chem. Res.*, **44**, 1309–1321 (2005).

8. B.J. Carroll. The kinetics of solubilization of nonpolar oils by nonionic surfactant solutions. *J. Colloid Interface Sci.*, **79**, 126–135 (1981).

9. A.A. Peña and C.A. Miller. Solubilization rates of oils in surfactant solutions and their relationship to mass transport in emulsions. *Adv. Colloid Interface Sci.*, **123–126**, 241–257 (2006).

10. M. Mahe, M. Vignes-Adler, A. Rousseau, C.G. Jacquin and P.M. Adler. Adhesion of droplets on a solid wall and detachment by a shear flow. 1. Pure systems. *J. Colloid Interface Sci.*, **126**, 314–328 (1988).

11. S. Basu, K. Nandakumar and J.H. Masliyah. A model for detachment of a partially wetting drop from a solid surface by shear flow. *J. Colloid Interface Sci.*, **190**, 253–257 (1997).

12. P.A. Winsor. *Solvent Properties of Amphiphilic Compounds*. Butterworth, London (1954).

13. R.N. Healy, R.L. Reed and D.G. Stenmark. Multiphase microemulsion systems. *Soc. Petrol. Eng. J.*, **16**, 147–160 (1976).

14. K. Shinoda. *Principles of Solution and Solubility*. Marcel Dekker, New York (1978).

15. M.J. Rang and C.A. Miller. Spontaneous emulsification of oil drops containing surfactants and medium-chain alcohols. *Progr. Colloid Polym. Sci.*, **109**, 101–117 (1998).

16. C. Tongcompou, E.J. Acosta, L.J. Quencer, A.F. Joseph, J.F. Scamehorn, D.A. Sabatini, N. Yanumet and S. Chavadej. Microemulsion formation and detergency with oily soils: III. Performance and mechanisms. *J. Surfact. Deterg.*, **8**, 147–156 (2005).

17. C.A. Miller and K.H. Raney. Solubilization-emulsification mechanisms of detergency. *Colloids Surf. A*, **74**, 169–215 (1993).

18. K.H. Raney, W.J. Benton and C.A. Miller. Optimum detergency conditions with nonionic surfactants: 1. Ternary water-surfactant-hydrocarbon systems. *J. Colloid Interface Sci.*, **117**, 282–290 (1987).

19. F. Schambil and M.J. Schwuger. Correlation between the phase behavior of ternary systems and removal of oil in the washing process. *Colloid Polym. Sci.*, **265**, 1009–1017 (1987).

20. L. Thompson. The role of oil detachment mechanisms in determining optimum detergency conditions. *J. Colloid Interface Sci.*, **163**, 61–73 (1994).

21. C. Solans and N. Azemar. Detergency and the HLB temperature. In *Organized Solutions*, S. Friberg and B. Lindman (Eds.), Chap. 19. Marcel Dekker, New York, pp. 273–288 (1992).

22. C. Tongcompou, E.J. Acosta, L.J. Quencer, A.F. Joseph, J.F. Scamehorn, D.A. Sabatini, S. Chavadej and N. Yanumet. Microemulsion formation and detergency with oily soils: I. Phase behavior and interfacial tension. *J. Surfact. Deterg.*, **6**, 191–203 (2003).

23. C. Tongcompou, E.J. Acosta, L.J. Quencer, A.F. Joseph, J.F. Scamehorn, D.A. Sabatini, S. Chavadej and SN. Yanumet. Microemulsion formation and detergency with oily soils: II. Detergency formulation and performance. *J. Surfact. Deterg.*, **6**, 205–214 (2003).

24. P. Tanthakit, S. Chavadej, J.F. Scamehorn, D.A. Sabatini and C. Tongcumpou. Microemulsion formation and detergency with oily soils: IV. Effect of rinse cycle design. *J. Surfact. Deterg.*, **11**, 117–128 (2008).

25. D.A. Sabatini, E. Acosta and J.H. Harwell. Linker molecules in surfactant mixtures. *Curr. Opin. Colloid Interface Sci.*, **8**, 316–326 (2003).

26. F. Mori, J.C. Lim, O.G. Raney, C.M. Elsik and C.A. Miller. Phase behavior, dynamic contacting and detergency in systems containing triolein and nonionic surfactants. *Colloids Surf.*, **40**, 323–345 (1989).

27. C. Tongcompou, E.J. Acosta, J.F. Scamehorn, D.A. Sabatini, N. Yanumet and S. Chavadej. Enhanced triolein removal using microemulsions formulated with mixed surfactants. *J. Surfact. Deterg.*, **9**, 181–189 (2006).

28. T.T. Phan, A. Witthayapanyanon, J.H. Harwell and D.A. Sabatini. Microemulsion-based vegetable oil detergency using an extended surfactant. *J. Surfact. Deterg.*, **13**, 313–319 (2010).

29. M. Miñana-Perez, A. Graciaa, J. Lachaise and J.L. Salager. Solubilization of polar oils with extended surfactants. *Colloids Surf. A*, **100**, 217–224 (1995).

30. S. Liu, D.L. Zhang, W. Yan, M. Puerto, G.J. Hirasaki and C.A. Miller. Favorable attributes of alkaline-surfactant-polymer flooding. *SPE J.*, **13**, 5–16 (2009).

31. D.B. Levitt, A.C. Jackson, C. Heinson, L.N. Britton, T. Malik, D. Varadarajan and G.A. Pope. Identification and evaluation of high-performance EOR surfactants. *SPE Res. Eval. Eng.*, **12**, 243–253 (2009).

32. P. Tanthakit, P. Ratchatawetchakul, S. Chavadej, J.F. Scamehorn, D.A. Sabatini and C. Tongcumpou. Palm oil removal from fabric using microemulsion-based formulations. *J. Surfact. Deterg.*, **13**, 485–495 (2010).

33. J.C. Lim and C.A. Miller. Dynamic behavior and detergency in systems containing nonionic surfactants and mixtures of polar and nonpolar oils. *Langmuir*, **7**, 2021–2027 (1991).

34. K.H. Raney and H.L. Benson. The effect of polar soil components on the phase inversion temperature and optimum detergency conditions. *J. Am. Oil Chem. Soc.*, **67**, 722–729 (1990).

35. V.K. Srivastava, G. Kini and D. Rout. Detergency in spontaneously formed emulsions. *J. Colloid Interface Sci.*, **304**, 214–221 (2006).

36. J.A. Shaeiwitz, A.F.-C. Chan, E.L. Cussler and D.F. Evans. The mechanism of solubilization in detergent solutions. *J. Colloid Interface Sci.*, **84**, 47–56 (1981).

37. K. Gotoh. Investigation of optimum liquid for textile washing using artificially soiled fabrics. *Textile Res. J.*, **80**, 548–556 (2010).

38. K. Gotoh. Evaluation of detergency using artificially soiled multifiber fabrics. *J. Oleo Sci.*, **59**, 477–482 (2010).

39. Y. Tagawa and K. Gotoh. Removal of carbon black particles from polymer substrates in water/ethanol mixtures. *J. Oleo Sci.*, **59**, 109–112 (2010).

40. K. Gotoh, N. Yu and Y. Tagawa. Evaluation of removal of model particulate and oily soils from poly(ethylene terephthalate) films by microscopic image analysis. *J. Oleo Sci.*, **62**, 73–79 (2013).

41. A.S.C. Lawrence. Polar interactions in detergency. In *Surface Activity and Detergency*, K. Durham (Ed.), Chap. 7. Macmillan, London, pp. 158–192 (1961).

42. D.G. Stevenson. Ancillary effects in detergent action. In *Surface Activity and Detergency*, K. Durham (Ed.), Chap. 6. Macmillan, London, pp. 146–157 (1961).

43. I.D. Robb and P.S. Stevenson. Solubilization of trilaurin in surfactant solutions. *Langmuir*, **16**, 7939–7945 (2000).

44. M.F. Cox. Surfactants for hard-surface cleaning: Mechanisms of solid soil removal. *J. Am. Oil Chem. Soc.*, **63**, 559–565 (1986).

45. M.F. Cox, D.L. Smith and G.L. Russell. Surface chemical processes for removal of solid sebum soil. *J. Am. Oil Chem. Soc.*, **64**, 273–276 (1987).

46. D.R. Scheuing and J.C.L. Hsieh. Detergency of nonionic surfactants toward a solid hydrocarbon soil studied by FT-IR. *Langmuir*, **4**, 1277–1283 (1988).

47. T. Hilgers, T. Bluhm and H. Krussmann. Penetration as physico-chemical factor in detergent efficiency. *Tenside Surf. Det.*, **33**(1), 37–41 (1996).

48. K. Bäckström, S. Engström, B. Lindman, T. Arnebrant, T. Nylander and K. Larsson. Cleaning of polymer and metal surfaces studied by ellipsometry. *J. Colloid Interface Sci.*, **99**, 549–552 (1984).

49. K. Backstrom and S. Engström. Removal of triglycerides from hard surfaces by surfactants: An ellipsometry study. *J. Am. Oil Chem. Soc.*, **65**, 412–420 (1988).

50. K. Backstrom, B. Lindman and S. Engström. Removal of triglycerides from polymer surfaces in relation to surfactant packing. Ellipsometry studies. *Langmuir*, **4**, 872–878 (1988).

51. M. Malmsten and B. Lindman. Ellipsometry studies of cleaning of hard surfaces. Relation to spontaneous curvature of the surfactant monolayer. *Langmuir*, **5**, 1105–1111 (1989).

52. A. Weerawardena, C.J. Drummond, F. Caruso and M. McCormick. Real time monitoring of the detergency process by using a quartz crystal microbalance. *Langmuir*, **14**, 575–577 (1998).

53. A. Weerawardena, C.J. Drummond, F. Caruso and M. McCormick. A quartz crystal microbalance study of the removal of solid organic soils from a hard surface in aqueous surfactant solution. *Colloids Surf. A*, **146**, 185–197 (1999).

54. A. Weerawardena, B.J. Boyd, C.J. Drummond and D.N. Furlong. Removal of a solid organic soil from a hard surface by glucose-derived surfactants: Effect of surfactant chain length, head group polymerization and anomeric configuration. *Colloids Surf. A*, **169**, 317–328 (2000).

55. B.J. Boyd, C.J. Drummond, I. Krodkiewska, A. Weerawardena, D.N. Furlong and F. Grieser. Alkyl chain positional isomers of dodecyl beta-D-glucoside: Thermotropic and lyotropic phase behavior and detergency. *Langmuir*, **17**, 6100–6107 (2001).

56. O. Favrat, J. Gavoille, L. Aleya and G. Monteil. Real time study of detergent concentration influence on solid fatty acid film removal processes. *J. Surfact. Deterg.*, **16**, 213–219 (2013).

57. X. Cui, X. Liu, A.S. Tatton, S.P. Brown, H. Ye and A. Marsh. Nanodiamond promotes surfactant-mediated triglyceride removal from a hydrophobic surface at or below room temperature. *ACS Appl. Mater. Interfaces*, **4**, 3225–3232 (2012).

58. K. Gotoh and M. Tagawa. Detachment behavior of Langmuir–Blodgett films of arachidic acid from a gold surface studied by the quartz crystal microbalance method. *Colloids Surf. A*, **196**, 145–152 (2002).

59. K. Gotoh. The role of liquid penetration in detergency of long-chain fatty acid. *J. Surfact. Deterg.*, **8**, 305–310 (2005).

60. S.P. Beaudoin, C.S. Grant and R.G. Carbonell. Removal of organic films from solid surfaces using aqueous solutions of nonionic surfactants. 1. Experiments. *Ind. Eng. Chem. Res.*, **34**, 3307–3317 (1995).

61. S.P. Beaudoin, C.S. Grant and R.G. Carbonell. Removal of organic films from solid surfaces using aqueous solutions of nonionic surfactants. 2. Theory. *Ind. Eng. Chem. Res.*, **34**, 3318–3325 (1995).

62. J.A. Kabin, A.E. Saez, C.S. Grant and R.G. Carbonell. Removal of organic films from rotating disks using aqueous solutions of nonionic surfactants: Film morphology and cleaning mechanisms. *Ind. Eng. Chem. Res.*, **35**, 4494–4506 (1996).

63. J.A. Kabin, S.L. Tolstedt, A.E. Saez, C.S. Grant and R.G. Carbonell. Removal of organic films from rotating disks using aqueous solutions of nonionic surfactants: Effect of surfactant molecular structure. *J. Colloid Interface Sci.*, **206**, 102–111 (1998).

64. J.A. Kabin, A.E. Saez, C.S. Grant and R.G. Carbonell. Removal rates of major and trace components of an organic film using aqueous nonionic surfactant solutions. *Ind. Eng. Chem. Res.*, **38**, 683–691 (1999).

65. J.A. Kabin, S.T. Withers, C.S. Grant, R.G. Carbonell and A.E. Saez. Removal of solid organic films from rotating disks using emulsion cleaners. *J. Colloid Interface Sci.*, **228**, 344–358 (2000).

66. J.Y.M. Chew, S.J. Torneijk, W.R. Paterson and D.I. Wilson. Solvent-based cleaning of emulsion polymerization reactors. *Chem. Eng. J.*, **117**, 61–69 (2006).

67. B. Hunek and E.L. Cussler. Mechanisms of photoresist dissolution. *AIChE J.*, **48**, 661–672 (2002).

Index

Page numbers followed by f and t indicate figures and tables, respectively.